AVIATION AND THE GLOBAL ATMOSPHERE

Although it is fewer than 100 years since the first powered flight, the aviation industry has undergone rapid growth and has become an integral and vital part of modern society. Projections into the future suggest that growth is likely to continue. It is therefore highly relevant to consider the current and possible future effects of aircraft on the atmosphere.

This Intergovernmental Panel on Climate Change (IPCC) Special Report is the most comprehensive assessment available of the effects of aviation on the global atmosphere. The report considers all the gases and particles emitted by aircraft into the atmosphere and the role they play in modifying the chemical properties of the atmosphere and initiating the formation of contrails. The report then considers how all this can modify the radiative properties of the atmosphere, leading to climate change, and how it can modify the ozone layer, leading to changes in ultraviolet radiation reaching the Earth. The report also considers how potential changes in aircraft technology, air transport operations, and the institutional, regulatory, and economic framework might affect emissions in the future.

This volume provides accurate, unbiased, policy-relevant information to serve the aviation industry, policymakers, environmental organizations, and researchers in global change, atmospheric chemistry, and economics.

Joyce E. Penner is a professor in the Department of Atmospheric, Oceanic, and Space Sciences at the University of Michigan. Prior to that she served as Division Leader of the Global Climate Research Division at the Lawrence Livermore National Laboratory. She is an Associate Editor for the *Journal of Geophysical Research* and the *Journal of Climate*. She has served on several scientific advisory committees, including the National Academy of Sciences Atmospheric Chemistry Committee and the National Academy of Sciences Panel on Aerosol Forcing and Climate Change. She has served as Secretary of the Atmospheric Sciences Section of the American Geophysical Union.

David H. Lister is a Technical Manager in the Propulsion Department of the UK Defence Research and Evaluation Agency (DERA). He has served as Project Manager in the European Community AERONOX program, as a member of the U.S. National Academy of Sciences/National Research Council Panel on Atmospheric Effects of Stratospheric Aircraft, as Chairman of ICAO/CAEP/WG3 (emissions) Technology and Certification Subgroup, and as lead author in the World Meteorological Organization's *Scientific Assessment of Ozone Depletion* (1994).

David J. Griggs is Head of the IPCC Working Group I Technical Support Unit at the Hadley Centre, UK Meteorological Office.

David J. Dokken is Project Administrator of the IPCC Working Group II Technical Support Unit, Washington, DC, USA.

Mack McFarland is Principal Scientist in Environmental Programs at DuPont Fluoroproducts, Wilmington, DE, USA.

Aviation and the Global Atmosphere

Edited by

Joyce E. Penner
University of Michigan

David H. Lister
*UK Defence Research
and Evaluation Agency*

David J. Griggs
UK Meteorological Office

David J. Dokken
*University Corporation
for Atmospheric Research*

Mack McFarland
DuPont Fluoroproducts

A Special Report of IPCC Working Groups I and III

in collaboration with the

Scientific Assessment Panel to the Montreal Protocol
on Substances that Deplete the Ozone Layer

Published for the Intergovernmental Panel on Climate Change

PUBLISHED BY THE PRESS SYNDICATE OF THE UNIVERSITY OF CAMBRIDGE
The Pitt Building, Trumpington Street, Cambridge, United Kingdom

CAMBRIDGE UNIVERSITY PRESS
The Edinburgh Building, Cambridge CB2 2RU, UK www.cup.cam.ac.uk
40 West 20th Street, New York, NY 10011-4211, USA www.cup.org
10 Stamford Road, Oakleigh, Melbourne 3166, Australia
Ruiz de Alarcón 13, 28014 Madrid, Spain

First published 1999

Printed in the United States of America

A catalog record for this book is available from the British Library

Library of Congress Cataloging-in-Publication Data available

ISBN 0 521 66300 8 hardback
ISBN 0 521 66404 7 paperback

Contents

Foreword

The Intergovernmental Panel on Climate Change (IPCC) was jointly established by the World Meteorological Organization (WMO) and the United Nations Environment Programme (UNEP) in 1988 to: (i) assess available information on the science, the impacts, and the economics of, and the options for mitigating and/or adapting to, climate change and (ii) provide, on request, scientific/technical/socio-economic advice to the Conference of the Parties (COP) to the United Nations Framework Convention on Climate Change (UNFCCC). Since then the IPCC has produced a series of Assessment Reports, Special Reports, Technical Papers, methodologies, and other products that have become standard works of reference, widely used by policymakers, scientists, and other experts.

This Special Report was prepared following a request from the International Civil Aviation Organization (ICAO) and the Parties to the Montreal Protocol on Substances that Deplete the Ozone Layer. The state of understanding of the relevant science of the atmosphere, aviation technology, and socio-economic issues associated with mitigation options is assessed and reported for both subsonic and supersonic fleets. The potential effects that aviation has had in the past and may have in the future on both stratospheric ozone depletion and global climate change are covered; environmental impacts of aviation at the local scale, however, are not addressed. The report synthesizes the findings to identify and characterize options for mitigating future impacts.

As is usual in the IPCC, success in producing this report has depended first and foremost on the enthusiasm and cooperation of experts worldwide in many related but different disciplines.

We would like to express our gratitude to all the Coordinating Lead Authors, Lead Authors, Contributing Authors, Review Editors, and Expert Reviewers. These individuals have devoted enormous time and effort to produce this report and we are extremely grateful for their commitment to the IPCC process.

We would also like to express our sincere thanks to:

- Robert Watson, the Chairman of the IPCC and Co-Chair of the Scientific Assessment Panel to the Montreal Protocol
- John Houghton, Ding Yihui, Bert Metz, and Ogunlade Davidson—the Co-Chairs of IPCC Working Groups I and III
- Daniel Albritton, Co-Chair of the Scientific Assessment Panel to the Montreal Protocol
- David Lister and Joyce Penner, the Coordinators of this Special Report
- Daniel Albritton, John Crayston, Ogunlade Davidson, David Griggs, Neil Harris, John Houghton, Mack McFarland, Bert Metz, Nelson Sabogal, N. Sundararaman, Robert Watson, and Howard Wesoky—the Science Steering Committee for this Special Report
- David Griggs, David Dokken, and all the staff of the Working Group I and II Technical Support Units, including Mack McFarland, Richard Moss, Anne Murrill, Sandy MacCracken, Maria Noguer, Laura Van Wie McGrory, Neil Leary, Paul van der Linden, and Flo Ormond, and Neil Harris who provided additional help
- N. Sundararaman, the Secretary of the IPCC, and his staff, Rudie Bourgeois, Cecilia Tanikie, and Chantal Ettori.

G.O.P. Obasi

Secretary-General
World Meteorological Organization

K. Töpfer

Executive Director
United Nations Environment Programme
and
Director-General
United Nations Office in Nairobi

Preface

Following a request from the International Civil Aviation Organization (ICAO) to assess the consequences of greenhouse gas emissions from aircraft engines, the IPCC at its Twelfth Session (Mexico City • 11-13 September 1996) decided to produce this Special Report, *Aviation and the Global Atmosphere*, in collaboration with the Scientific Assessment Panel to the Montreal Protocol. The task was initially a joint responsibility between IPCC Working Groups I and II but, following a change in the terms of reference of the Working Groups (Thirteenth Session of the IPCC • Maldives • 22 and 25-28 September 1997), the responsibility was transferred to IPCC Working Groups I and III, with administrative support remaining with the Technical Support Units of Working Groups I and II.

Although it is less than 100 years since the first powered flight, the aviation industry has undergone rapid growth and has become an integral and vital part of modern society. In the absence of policy intervention, the growth is likely to continue. It is therefore highly relevant to consider the current and possible future effects of aircraft engine emissions on the atmosphere. A unique aspect of this report is the integral involvement of technical experts from the aviation industry, including airlines, and airframe and engine manufacturers, alongside atmospheric scientists. This involvement has been critical in producing what we believe is the most comprehensive assessment available to date of the effects of aviation on the global atmosphere. Although this Special Report is the first IPCC report to consider a particular industrial subsector, other sectors equally deserve study.

The report considers all the gases and particles emitted by aircraft into the upper atmosphere and the role that they play in modifying the chemical properties of the atmosphere and initiating the formation of condensation trails (contrails) and cirrus clouds. The report then considers (a) how the radiative properties of the atmosphere can be modified as a result, possibly leading to climate change, and (b) how the ozone layer could be modified, leading to changes in ultraviolet radiation reaching the Earth's surface. The report also considers how potential changes in aircraft technology, air transport operations, and the institutional, regulatory, and economic framework might affect emissions in the future. The report does not deal with the effects of engine emissions on local air quality near the surface.

The objective of this Special Report is to provide accurate, unbiased, policy-relevant information to serve the aviation industry and the expert and policymaking communities. The report, in describing the current state of knowledge, also identifies areas where our understanding is inadequate and where further work is urgently required. It does not make policy recommendations or suggest policy preferences, thus is consistent with IPCC practice.

This report was compiled by 107 Lead Authors from 18 countries. Successive drafts of the report were circulated for review by experts, followed by review by governments and experts. Over 100 Contributing Authors submitted draft text and information to the Lead Authors and over 150 reviewers submitted valuable suggestions for improvement during the review process. All the comments received were carefully analyzed and assimilated into a revised document for consideration at the joint session of IPCC Working Groups I and III held in San José, Costa Rica, 12-14 April 1999. There, the Summary for Policymakers was approved in detail and the underlying report accepted.

We wish to express our sincere appreciation to the Report Coordinators, David Lister and Joyce Penner; to all the Coordinating Lead Authors, Lead Authors, and Review Editors whose expertise, diligence, and patience have underpinned the successful completion of this report; and to the many contributors and reviewers for their valuable and painstaking dedication and work. We thank the Steering Committee for their wise counsel and guidance throughout the preparation of the report. We are grateful to:

- ICAO for hosting the initial scoping meeting for the report and the final drafting meeting, and for translating the Summary for Policymakers into Arabic, Chinese, French, Russian, and Spanish (ICAO also provided technical inputs as requested)
- The government of Trinidad and Tobago for hosting the first drafting meeting
- The International Air Transport Association (IATA) for hosting the second drafting meeting
- The government of Costa Rica for hosting the Joint Session of IPCC Working Groups I and III (12-14 April 1999), where the Summary for Policymakers was approved line by line and the underlying assessment accepted.

In particular, we are grateful to John Crayston (ICAO), Steve Pollonais (government of Trinidad and Tobago), Leonie Dobbie (IATA), and Max Campos (government of Costa Rica) for their taking on the demanding burden of arranging for these meetings.

We also thank Anne Murrill of the Working Group I Technical Support Unit and Sandy MacCracken of the Working Group II Technical Support Unit for their tireless and good humored support throughout the preparation of the report. Other members of the Technical Support Units of Working Groups I and II also provided much assistance, including Richard Moss,

Mack McFarland, Maria Noguer, Laura Van Wie McGrory, Neil Leary, Paul van der Linden, and Flo Ormond. The staff of the IPCC Secretariat, Rudie Bourgeois, Cecilia Tanikie, and Chantal Ettori, provided logistical support for all government liaison and travel of experts from the developing and transitional economy countries.

Robert Watson, IPCC Chairman
John Houghton, Co-Chair of IPCC Working Group I
Ding Yihui, Co-Chair of IPCC Working Group I
Bert Metz, Co-Chair of IPCC Working Group III
Ogunlade Davidson, Co-Chair of IPCC Working Group III
N. Sundararaman, IPCC Secretary
David Griggs, IPCC Working Group I TSU
David Dokken, IPCC Working Group II TSU

SUMMARY FOR POLICYMAKERS

AVIATION AND THE GLOBAL ATMOSPHERE

A Special Report of Working Groups I and III of the Intergovernmental Panel on Climate Change

This summary, approved in detail at a joint session of IPCC Working Groups I and III (San José, Costa Rica • 12-14 April 1999), represents the formally agreed statement of the IPCC concerning current understanding of aviation and the global atmosphere.

Based on a draft prepared by:

David H. Lister, Joyce E. Penner, David J. Griggs, John T. Houghton, Daniel L. Albritton, John Begin, Gerard Bekebrede, John Crayston, Ogunlade Davidson, Richard G. Derwent, David J. Dokken, Julie Ellis, David W. Fahey, John E. Frederick, Randall Friedl, Neil Harris, Stephen C. Henderson, John F. Hennigan, Ivar Isaksen, Charles H. Jackman, Jerry Lewis, Mack McFarland, Bert Metz, John Montgomery, Richard W. Niedzwiecki, Michael Prather, Keith R. Ryan, Nelson Sabogal, Robert Sausen, Ulrich Schumann, Hugh J. Somerville, N. Sundararaman, Ding Yihui, Upali K. Wickrama, Howard L. Wesoky

CONTENTS

1. Introduction

This report assesses the effects of aircraft on climate and atmospheric ozone and is the first IPCC report for a specific industrial subsector. It was prepared by IPCC in collaboration with the Scientific Assessment Panel to the Montreal Protocol on Substances that Deplete the Ozone Layer, in response to a request by the International Civil Aviation Organization (ICAO)[1] because of the potential impact of aviation emissions. These are the predominant anthropogenic emissions deposited directly into the upper troposphere and lower stratosphere.

Aviation has experienced rapid expansion as the world economy has grown. Passenger traffic (expressed as revenue passenger-kilometers[2]) has grown since 1960 at nearly 9% per year, 2.4 times the average Gross Domestic Product (GDP) growth rate. Freight traffic, approximately 80% of which is carried by passenger airplanes, has also grown over the same time period. The rate of growth of passenger traffic has slowed to about 5% in 1997 as the industry is maturing. Total aviation emissions have increased, because increased demand for air transport has outpaced the reductions in specific emissions[3] from the continuing improvements in technology and operational procedures. Passenger traffic, assuming unconstrained demand, is projected to grow at rates in excess of GDP for the period assessed in this report.

The effects of current aviation and of a range of unconstrained growth projections for aviation (which include passenger, freight, and military) are examined in this report, including the possible effects of a fleet of second generation, commercial supersonic aircraft. The report also describes current aircraft technology, operating procedures, and options for mitigating aviation's future impact on the global atmosphere. The report does not consider the local environmental effects of aircraft engine emissions or any of the indirect environmental effects of aviation operations such as energy usage by ground transportation at airports.

2. How Do Aircraft Affect Climate and Ozone?

Aircraft emit gases and particles directly into the upper troposphere and lower stratosphere where they have an impact on atmospheric composition. These gases and particles alter the concentration of atmospheric greenhouse gases, including carbon dioxide (CO_2), ozone (O_3), and methane (CH_4); trigger formation of condensation trails (contrails); and may increase cirrus cloudiness—all of which contribute to climate change (see Box 1).

The principal emissions of aircraft include the greenhouse gases carbon dioxide and water vapor (H_2O). Other major emissions are nitric oxide (NO) and nitrogen dioxide (NO_2) (which together are termed NO_x), sulfur oxides (SO_x), and soot. The total amount of aviation fuel burned, as well as the total emissions of carbon dioxide, NO_x, and water vapor by aircraft, are well known relative to other parameters important to this assessment.

The climate impacts of the gases and particles emitted and formed as a result of aviation are more difficult to quantify than the emissions; however, they can be compared to each other and to climate effects from other sectors by using the concept of radiative forcing.[4] Because carbon dioxide has a long atmospheric residence time (≈100 years) and so becomes well mixed throughout the atmosphere, the effects of its emissions from aircraft are indistinguishable from the same quantity of carbon dioxide emitted by any other source. The other gases (e.g., NO_x, SO_x, water vapor) and particles have shorter atmospheric residence times and remain concentrated near flight routes, mainly in the northern mid-latitudes. These emissions can lead to radiative forcing that is regionally located near the flight routes for some components (e.g., ozone and contrails) in contrast to emissions that are globally mixed (e.g., carbon dioxide and methane).

The global mean climate change is reasonably well represented by the global average radiative forcing, for example, when evaluating the contributions of aviation to the rise in globally averaged temperature or sea level. However, because some of aviation's key contributions to radiative forcing are located mainly in the northern mid-latitudes, the regional climate response may differ from that derived from a global mean radiative forcing. The impact of aircraft on regional climate could be important, but has not been assessed in this report.

Ozone is a greenhouse gas. It also shields the surface of the earth from harmful ultraviolet (UV) radiation, and is a common air pollutant. Aircraft-emitted NO_x participates in ozone chemistry. Subsonic aircraft fly in the upper troposphere and lower stratosphere (at altitudes of about 9 to 13 km), whereas supersonic aircraft cruise several kilometers higher (at about 17 to 20 km) in the stratosphere. Ozone in the upper troposphere and lower stratosphere is expected to increase in response to NO_x increases and methane is expected to decrease. At higher altitudes, increases in NO_x lead to decreases in the stratospheric ozone layer. Ozone precursor (NO_x) residence times in these regions increase with altitude, and hence perturbations to ozone by aircraft depend on the altitude of NO_x injection and vary from regional in scale in the troposphere to global in scale in the stratosphere.

---

[1] ICAO is the United Nations specialized agency that has global responsibility for the establishment of standards, recommended practices, and guidance on various aspects of international civil aviation, including environmental protection.
[2] The revenue passenger-km is a measure of the traffic carried by commercial aviation: one revenue-paying passenger carried 1 km.
[3] Specific emissions are emissions per unit of traffic carried, for instance, per revenue passenger-km.
[4] Radiative forcing is a measure of the importance of a potential climate change mechanism. It expresses the perturbation or change to the energy balance of the Earth-atmosphere system in watts per square meter (Wm-2). Positive values of radiative forcing imply a net warming, while negative values imply cooling.

Box 1. The Science of Climate Change

Some of the main conclusions of the Summary for Policymakers of Working Group I of the IPCC Second Assessment Report, published in 1995, which concerns the effects of all anthropogenic emissions on climate change, follow:

- Increases in greenhouse gas concentrations since pre-industrial times (i.e., since about 1750) have led to a positive radiative forcing of climate, tending to warm the surface of the Earth and produce other changes of climate.
- The atmospheric concentrations of the greenhouse gases carbon dioxide, methane, and nitrous oxide (N_2O), among others, have grown significantly: by about 30, 145, and 15%, respectively (values for 1992). These trends can be attributed largely to human activities, mostly fossil fuel use, land-use change, and agriculture.
- Many greenhouse gases remain in the atmosphere for a long time (for carbon dioxide and nitrous oxide, many decades to centuries). As a result of this, if carbon dioxide emissions were maintained at near current (1994) levels, they would lead to a nearly constant rate of increase in atmospheric concentrations for at least two centuries, reaching about 500 ppmv (approximately twice the pre-industrial concentration of 280 ppmv) by the end of the 21st century.
- Tropospheric aerosols resulting from combustion of fossil fuels, biomass burning, and other sources have led to a negative radiative forcing, which, while focused in particular regions and subcontinental areas, can have continental to hemispheric effects on climate patterns. In contrast to the long-lived greenhouse gases, anthropogenic aerosols are very short-lived in the atmosphere; hence, their radiative forcing adjusts rapidly to increases or decreases in emissions.
- Our ability from the observed climate record to quantify the human influence on global climate is currently limited because the expected signal is still emerging from the noise of natural variability, and because there are uncertainties in key factors. These include the magnitude and patterns of long-term natural variability and the time-evolving pattern of forcing by, and response to, changes in concentrations of greenhouse gases and aerosols, and land-surface changes. Nevertheless, the balance of evidence suggests that there is a discernible human influence on global climate.
- The IPCC has developed a range of scenarios, IS92a-f, for future greenhouse gas and aerosol precursor emissions based on assumptions concerning population and economic growth, land use, technological changes, energy availability, and fuel mix during the period 1990 to 2100. Through understanding of the global carbon cycle and of atmospheric chemistry, these emissions can be used to project atmospheric concentrations of greenhouse gases and aerosols and the perturbation of natural radiative forcing. Climate models can then be used to develop projections of future climate.
- Estimates of the rise in global average surface air temperature by 2100 relative to 1990 for the IS92 scenarios range from 1 to 3.5°C. In all cases the average rate of warming would probably be greater than any seen in the last 10,000 years. Regional temperature changes could differ substantially from the global mean and the actual annual to decadal changes would include considerable natural variability. A general warming is expected to lead to an increase in the occurrence of extremely hot days and a decrease in the occurrence of extremely cold days.
- Average sea level is expected to rise as a result of thermal expansion of the oceans and melting of glaciers and ice-sheets. Estimates of the sea level rise by 2100 relative to 1990 for the IS92 scenarios range from 15 to 95 cm.
- Warmer temperatures will lead to a more vigorous hydrological cycle; this translates into prospects for more severe droughts and/or floods in some places and less severe droughts and/or floods in other places. Several models indicate an increase in precipitation intensity, suggesting a possibility for more extreme rainfall events.

Water vapor, SO_x (which forms sulfate particles), and soot[5] play both direct and indirect roles in climate change and ozone chemistry.

3.   How are Aviation Emissions Projected to Grow in the Future?

Global passenger air travel, as measured in revenue passenger-km, is projected to grow by about 5% per year between 1990 and 2015, whereas total aviation fuel use—including passenger, freight, and military[6]—is projected to increase by 3% per year, over the same period, the difference being due largely to improved aircraft efficiency. Projections beyond this time are more

uncertain so a range of future unconstrained emission scenarios is examined in this report (see Table 1 and Figure 1). All of these scenarios assume that technological improvements leading to reduced emissions per revenue passenger-km will continue in the future and that optimal use of airspace availability (i.e.,

---

[5] Airborne sulfate particles and soot particles are both examples of aerosols. Aerosols are microscopic particles suspended in air.

[6] The historical breakdown of aviation fuel burn for civil (passenger plus cargo) and military aviation was 64 and 36%, respectively, in 1976, and 82 and 18%, respectively, in 1992. These are projected to change to 93 and 7%, respectively, in 2015, and to 97 and 3%, respectively, in 2050.

Table 1: Summary of future global aircraft scenarios used in this report.

Scenario Name	Avg. Traffic Growth per Year (1990–2050)[1]	Avg. Annual Growth Rate of Fuel Burn (1990–2050)[2]	Avg. Annual Economic Growth Rate	Avg. Annual Population Growth Rate	Ratio of Traffic (2050/1990)	Ratio of Fuel Burn (2050/1990)	Notes
Fa1	3.1%	1.7%	2.9% *1990–2025* 2.3% *1990–2100*	1.4% *1990–2025* 0.7% *1990–2100*	6.4	2.7	Reference scenario developed by ICAO Forecasting and Economic Support Group (FESG); mid-range economic growth from IPCC (1992); technology for both improved fuel efficiency and NO_x reduction
Fa1H	3.1%	2.0%	2.9% *1990–2025* 2.3% *1990–2100*	1.4% *1990–2025* 0.7% *1990–2100*	6.4	3.3	Fa1 traffic and technology scenario with a fleet of supersonic aircraft replacing some of the subsonic fleet
Fa2	3.1%	1.7%	2.9% *1990–2025* 2.3% *1990–2100*	1.4% *1990–2025* 0.7% *1990–2100*	6.4	2.7	Fa1 traffic scenario; technology with greater emphasis on NO_x reduction, but slightly smaller fuel efficiency improvement
Fc1	2.2%	0.8%	2.0% *1990–2025* 1.2% *1990–2100*	1.1% *1990–2025* 0.2% *1990–2100*	3.6	1.6	FESG low-growth scenario; technology as for Fa1 scenario
Fe1	3.9%	2.5%	3.5% *1990–2025* 3.0% *1990–2100*	1.4% *1990–2025* 0.7% *1990–2100*	10.1	4.4	FESG high-growth scenario; technology as for Fa1 scenario
Eab	4.0%	3.2%			10.7	6.6	Traffic-growth scenario based on IS92a developed by Environmental Defense Fund (EDF); technology for very low NO_x assumed
Edh	4.7%	3.8%			15.5	9.4	High traffic-growth EDF scenario; technology for very low NO_x assumed

[1]Traffic measured in terms of revenue passenger-km.
[2]All aviation (passenger, freight, and military).

ideal air traffic management) is achieved by 2050. If these improvements do not materialize then fuel use and emissions will be higher. It is further assumed that the number of aircraft as well as the number of airports and associated infrastructure will continue to grow and not limit the growth in demand for air travel. If the infrastructure was not available, the growth of traffic reflected in these scenarios would not materialize.

IPCC (1992)[7] developed a range of scenarios, IS92a-f, of future greenhouse gas and aerosol precursor emissions based on assumptions concerning population and economic growth,

land use, technological changes, energy availability, and fuel mix during the period 1990 to 2100. Scenario IS92a is a mid-range emissions scenario. Scenarios of future emissions are not predictions of the future. They are inherently uncertain because they are based on different assumptions about the future, and

---

[7] **IPCC,** 1992: *Climate Change 1992: The Supplementary Report to the IPCC Scientific Assessment* [Houghton, J.T., B.A. Callander, and S.K.Varney (eds.)]. Cambridge University Press, Cambridge, UK, 200 pp.

the longer the time horizon the more uncertain these scenarios become. The aircraft emissions scenarios developed here used the economic growth and population assumptions found in the IS92 scenario range (see Table 1 and Figure 1). In the following sections, scenario Fa1 is utilized to illustrate the possible effects of aircraft and is called the reference scenario. Its assumptions are linked to those of IS92a. The other aircraft emissions scenarios were built from a range of economic and population projections from IS92a-e. These scenarios represent a range of plausible growth for aviation and provide a basis for sensitivity analysis for climate modeling. However, the high growth scenario Edh is believed to be less plausible and the low growth scenario Fc1 is likely to be exceeded given the present state of the industry and planned developments.

4.   What are the Current and Future Impacts of Subsonic Aviation on Radiative Forcing and UV Radiation?

The summary of radiative effects resulting from aircraft engine emissions is given in Figures 2 and 3. As shown in Figure 2, the uncertainty associated with several of these effects is large.

4.1.   Carbon Dioxide

Emissions of carbon dioxide by aircraft were 0.14 Gt C/year in 1992. This is about 2% of total anthropogenic carbon dioxide emissions in 1992 or about 13% of carbon dioxide emissions from all transportation sources. The range of scenarios considered here projects that aircraft emissions of carbon dioxide will continue to grow and by 2050 will be 0.23 to 1.45 Gt C/year. For the reference scenario (Fa1) this emission increases 3-fold

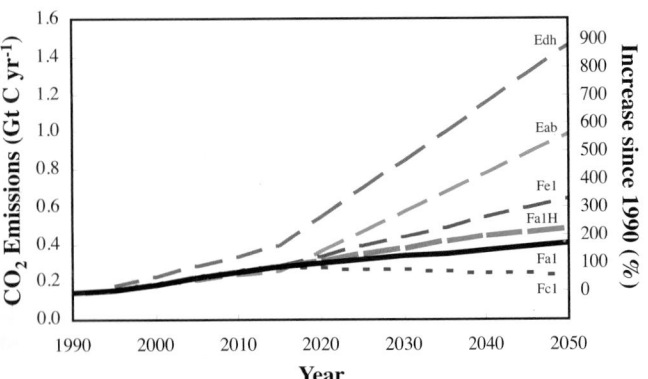

Figure 1: Total aviation carbon dioxide emissions resulting from six different scenarios for aircraft fuel use. Emissions are given in Gt C [or billion (10^9) tonnes of carbon] per year. To convert Gt C to Gt CO_2 multiply by 3.67. The scale on the righthand axis represents the percentage growth from 1990 to 2050. Aircraft emissions of carbon dioxide represent 2.4% of total fossil fuel emissions of carbon dioxide in 1992 or 2% of total anthropogenic carbon dioxide emissions. (Note: Fa2 has not been drawn because the difference from scenario Fa1 would not be discernible on the figure.)

by 2050 to 0.40 Gt C/year, or 3% of the projected total anthropogenic carbon dioxide emissions relative to the mid-range IPCC emission scenario (IS92a). For the range of scenarios, the range of increase in carbon dioxide emissions to 2050 would be 1.6 to 10 times the value in 1992.

Concentrations of and radiative forcing from carbon dioxide today are those resulting from emissions during the last 100 years or so. The carbon dioxide concentration attributable to aviation in the 1992 atmosphere is 1 ppmv, a little more than 1% of the total anthropogenic increase. This percentage is lower than the percentage for emissions (2%) because the emissions occurred only in the last 50 years. For the range of scenarios in Figure 1, the accumulation of atmospheric carbon dioxide due to aircraft over the next 50 years is projected to increase to 5 to 13 ppmv. For the reference scenario (Fa1) this is 4% of that from all human activities assuming the mid-range IPCC scenario (IS92a).

4.2.   Ozone

The NO_x emissions from subsonic aircraft in 1992 are estimated to have increased ozone concentrations at cruise altitudes in northern mid-latitudes by up to 6%, compared to an atmosphere without aircraft emissions. This ozone increase is projected to rise to about 13% by 2050 in the reference scenario (Fa1). The impact on ozone concentrations in other regions of the world is substantially less. These increases will, on average, tend to warm the surface of the Earth.

Aircraft emissions of NO_x are more effective at producing ozone in the upper troposphere than an equivalent amount of emission at the surface. Also increases in ozone in the upper troposphere are more effective at increasing radiative forcing than increases at lower altitudes. Due to these increases the calculated total ozone column in northern mid-latitudes is projected to grow by approximately 0.4 and 1.2% in 1992 and 2050, respectively. However, aircraft sulfur and water emissions in the stratosphere tend to deplete ozone, partially offsetting the NO_x-induced ozone increases. The degree to which this occurs is, as yet, not quantified. Therefore, the impact of subsonic aircraft emissions on stratospheric ozone requires further evaluation. The largest increases in ozone concentration due to aircraft emissions are calculated to occur near the tropopause where natural variability is high. Such changes are not apparent from observations at this time.

4.3.   Methane

In addition to increasing tropospheric ozone concentrations, aircraft NO_x emissions are expected to decrease the concentration of methane, which is also a greenhouse gas. These reductions in methane tend to cool the surface of the Earth. The methane concentration in 1992 is estimated here to be about 2% less than that in an atmosphere without aircraft. This aircraft-induced reduction of methane concentration is much smaller than the observed overall 2.5-fold increase since pre-industrial times. Uncertainties in the sources and sinks of methane preclude testing the impact of aviation on methane concentrations with atmospheric observations. *In the reference scenario (Fa1) methane would be about 5% less than that calculated for a 2050 atmosphere without aircraft.*

Changes in tropospheric ozone are mainly in the Northern Hemisphere, while those of methane are global in extent so that, even though the global average radiative forcings are of similar magnitude and opposite in sign, the latitudinal structure of the forcing is different so that the net regional radiative effects do not cancel.

4.4. Water Vapor

Most subsonic aircraft water vapor emissions are released in the troposphere where they are rapidly removed by precipitation

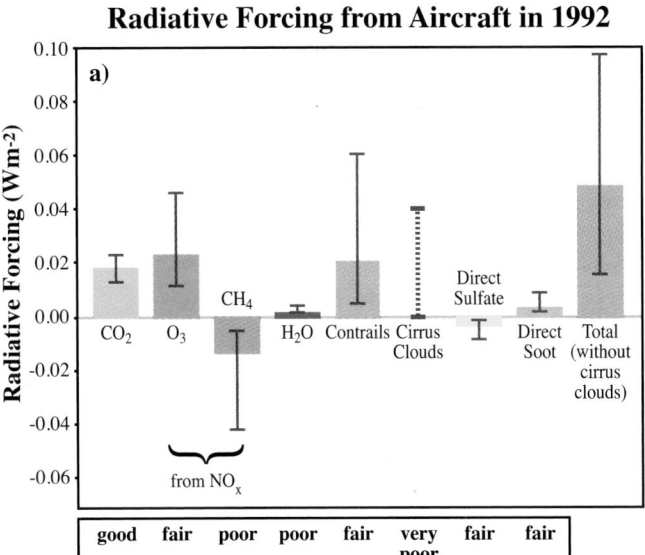

Radiative Forcing from Aircraft in 1992

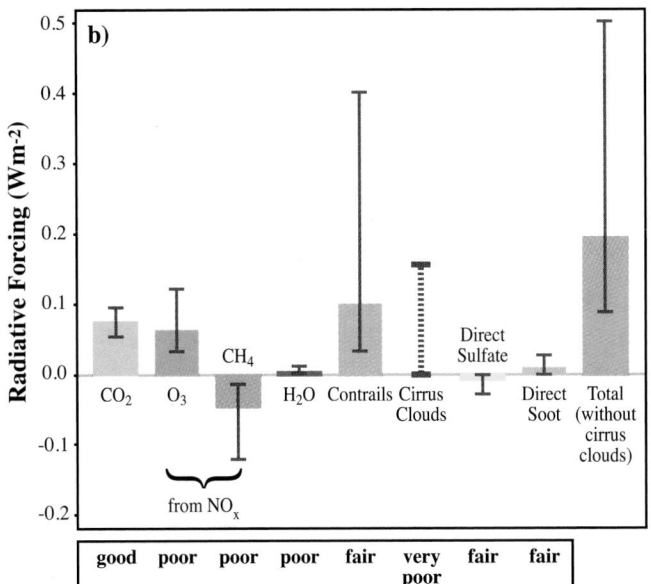

Radiative Forcing from Aircraft in 2050

within 1 to 2 weeks. A smaller fraction of water vapor emissions is released in the lower stratosphere where it can build up to larger concentrations. Because water vapor is a greenhouse gas, these increases tend to warm the Earth's surface, though for subsonic aircraft this effect is smaller than those of other aircraft emissions such as carbon dioxide and NO_x.

4.5. Contrails

In 1992, aircraft line-shaped contrails are estimated to cover about 0.1% of the Earth's surface on an annually averaged basis with larger regional values. Contrails tend to warm the Earth's surface, similar to thin high clouds. The contrail cover is projected to grow to 0.5% by 2050 in the reference scenario (Fa1), at a rate which is faster than the rate of growth in aviation fuel consumption. This faster growth in contrail cover is expected because air traffic will increase mainly in the upper troposphere where contrails form preferentially, and may also occur as a result of improvements in aircraft fuel efficiency. Contrails are triggered from the water vapor emitted by aircraft and their optical properties depend on the particles emitted or formed in the aircraft plume and on the ambient atmospheric conditions. The radiative effect of contrails depends on their optical properties and global cover, both of which are uncertain. Contrails have been observed as line-shaped clouds by satellites

Figure 2: Estimates of the globally and annually averaged radiative forcing (Wm^{-2}) (see Footnote 4) from subsonic aircraft emissions in 1992 (*2a*) and in 2050 for scenario Fa1 (*2b*). The scale in Figure 2b is greater than the scale in 2a by about a factor of 4. The bars indicate the best estimate of forcing while the line associated with each bar is a two-thirds uncertainty range developed using the best knowledge and tools available at the present time. (The two-thirds uncertainty range means that there is a 67% probability that the true value falls within this range.) The available information on cirrus clouds is insufficient to determine either a best estimate or an uncertainty range; the dashed line indicates a range of possible best estimates. The estimate for total forcing does not include the effect of changes in cirrus cloudiness. The uncertainty estimate for the total radiative forcing (without additional cirrus) is calculated as the square root of the sums of the squares of the upper and lower ranges for the individual components. The evaluations below the graph ("good," "fair," "poor," "very poor") are a relative appraisal associated with each component and indicates the level of scientific understanding. It is based on the amount of evidence available to support the best estimate and its uncertainty, the degree of consensus in the scientific literature, and the scope of the analysis. This evaluation is separate from the evaluation of uncertainty range represented by the lines associated with each bar. This method of presentation is different and more meaningful than the confidence level presented in similar graphs from *Climate Change 1995: The Science of Climate Change.*

over heavy air traffic areas and covered on average about 0.5% of the area over Central Europe in 1996 and 1997.

4.6.   Cirrus Clouds

Extensive cirrus clouds have been observed to develop after the formation of persistent contrails. Increases in cirrus cloud cover (beyond those identified as line-shaped contrails) are found to be positively correlated with aircraft emissions in a limited number of studies. About 30% of the Earth is covered with cirrus cloud. *On average an increase in cirrus cloud cover tends to warm the surface of the Earth. An estimate for aircraft-induced cirrus cover for the late 1990s ranges from 0 to 0.2% of the surface of the Earth. For the Fa1 scenario, this may possibly increase by a factor of 4 (0 to 0.8%) by 2050;* however, the mechanisms associated with increases in cirrus cover are not well understood and need further investigation.

4.7.   Sulfate and Soot Aerosols

The aerosol mass concentrations in 1992 resulting from aircraft are small relative to those caused by surface sources. Although aerosol accumulation will grow with aviation fuel use, aerosol mass concentrations from aircraft in 2050 are projected to remain small compared to surface sources. Increases in soot tend to warm while increases in sulfate tend to cool the Earth's surface. The direct radiative forcing of sulfate and soot aerosols from aircraft is small compared to those of other aircraft emissions. Because aerosols influence the formation of clouds, the accumulation of aerosols from aircraft may play a role in enhanced cloud formation and change the radiative properties of clouds.

4.8.   What are the Overall Climate Effects of Subsonic Aircraft?

The climate impacts of different anthropogenic emissions can be compared using the concept of radiative forcing. The best estimate of the radiative forcing in 1992 by aircraft is 0.05 Wm⁻² or about 3.5% of the total radiative forcing by all anthropogenic activities. For the reference scenario (Fa1), the radiative forcing by aircraft in 2050 is 0.19 Wm⁻² or 5% of the radiative forcing in the mid-range IS92a scenario (3.8 times the value in 1992). According to the range of scenarios considered here, the forcing is projected to grow to 0.13 to 0.56 Wm⁻² in 2050, which is a factor of 1.5 less to a factor of 3 greater than that for Fa1 and from 2.6 to 11 times the value in 1992. These estimates of forcing combine the effects from changes in concentrations of carbon dioxide, ozone, methane, water vapor, line-shaped contrails, and aerosols, but do not include possible changes in cirrus clouds.

Globally averaged values of the radiative forcing from different components in 1992 and in 2050 under the reference scenario (Fa1) are shown in Figure 2. Figure 2 indicates the best estimates of the forcing for each component and the two-thirds

uncertainty range.[8] The derivation of these uncertainty ranges involves expert scientific judgment and may also include objective statistical models. The uncertainty range in the radiative forcing stated here combines the uncertainty in calculating the atmospheric change to greenhouse gases and aerosols with that of calculating radiative forcing. For additional cirrus clouds, only a range for the best estimate is given; this is not included in the total radiative forcing.

The state of scientific understanding is evaluated for each component. This is not the same as the confidence level expressed in previous IPCC documents. This evaluation is separate from the uncertainty range and is a relative appraisal of the scientific understanding for each component. The evaluation is based on the amount of evidence available to support the best estimate and its uncertainty, the degree of consensus in the scientific literature, and the scope of the analysis. The total radiative forcing under each of the six scenarios for the growth of aviation is shown in Figure 3 for the period 1990 to 2050.

The total radiative forcing due to aviation (without forcing from additional cirrus) is likely to lie within the range from 0.01 to 0.1 Wm⁻² in 1992, with the largest uncertainties coming from contrails and methane. Hence the total radiative forcing may be about 2 times larger or 5 times smaller than the best estimate. For any scenario at 2050, the uncertainty range of radiative forcing is slightly larger than for 1992, but the largest variations of projected radiative forcing come from the range of scenarios.

Over the period from 1992 to 2050, the overall radiative forcing by aircraft (excluding that from changes in cirrus clouds) for all scenarios in this report is a factor of 2 to 4 larger

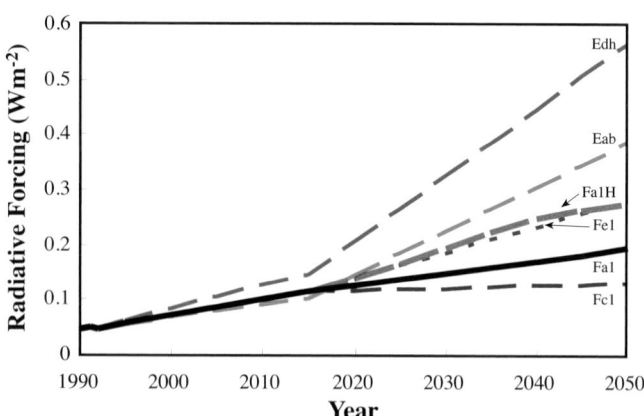

Figure 3: Estimates of the globally and annually averaged total radiative forcing (without cirrus clouds) associated with aviation emissions under each of six scenarios for the growth of aviation over the time period 1990 to 2050. (Fa2 has not been drawn because the difference from scenario Fa1 would not be discernible on the figure.)

---

[8] The two-thirds uncertainty range means there is a 67% probability that the true value falls within this range.

than the forcing by aircraft carbon dioxide alone. The overall radiative forcing for the sum of all human activities is estimated to be at most a factor of 1.5 larger than that of carbon dioxide alone.

The emissions of NO_x cause changes in methane and ozone, with influence on radiative forcing estimated to be of similar magnitude but of opposite sign. However, as noted above, the geographical distribution of the aircraft ozone forcing is far more regional than that of the aircraft methane forcing.

The effect of aircraft on climate is superimposed on that caused by other anthropogenic emissions of greenhouse gases and particles, and on the background natural variability. The radiative forcing from aviation is about 3.5% of the total radiative forcing in 1992. It has not been possible to separate the influence on global climate change of aviation (or any other sector with similar radiative forcing) from all other anthropogenic activities. Aircraft contribute to global change approximately in proportion to their contribution to radiative forcing.

4.9. *What are the Overall Effects of Subsonic Aircraft on UV-B?*

Ozone, most of which resides in the stratosphere, provides a shield against solar ultraviolet radiation. The erythemal dose rate, defined as UV irradiance weighted according to how effectively it causes sunburn, is estimated to be decreased by aircraft in 1992 by about 0.5% at 45°N in July. For comparison, the calculated increase in the erythemal dose rate due to observed ozone depletion is about 4% over the period 1970 to 1992 at 45°N in July.[9] The net effect of subsonic aircraft appears to be an increase in column ozone and a decrease in UV radiation, which is mainly due to aircraft NO_x emissions. Much smaller changes in UV radiation are associated with aircraft contrails, aerosols, and induced cloudiness. In the Southern Hemisphere, the calculated effects of aircraft emission on the erythemal dose rate are about a factor of 4 lower than for the Northern Hemisphere.

For the reference scenario (Fa1), the change in erythemal dose rate at 45°N in July in 2050 compared to a simulation with no aircraft is -1.3% (with a two-thirds uncertainty range from -0.7 to -2.6%). For comparison, the calculated change in the erythemal dose rate due to changes in the concentrations of trace species, other than those from aircraft, between 1970 to 2050 at 45°N is about -3%, a decrease that is the net result of two opposing effects: (1) the incomplete recovery of stratospheric ozone to 1970 levels because of the persistence of long-lived halogen-containing compounds, and (2) increases in projected surface emissions of shorter lived pollutants that produce ozone in the troposphere.

5. **What are the Current and Future Impacts of Supersonic Aviation on Radiative Forcing and UV Radiation?**

One possibility for the future is the development of a fleet of second generation supersonic, high speed civil transport (HSCT) aircraft, although there is considerable uncertainty whether any such fleet will be developed. These supersonic aircraft are projected to cruise at an altitude of about 19 km, about 8 km higher than subsonic aircraft, and to emit carbon dioxide, water vapor, NO_x, SO_x, and soot into the stratosphere. NO_x, water vapor, and SO_x from supersonic aircraft emissions all contribute to changes in stratospheric ozone. The radiative forcing of civil supersonic aircraft is estimated to be about a factor of 5 larger than that of the displaced subsonic aircraft in the Fa1H scenario. The calculated radiative forcing of supersonic aircraft depends on the treatment of water vapor and ozone in models. This effect is difficult to simulate in current models and so is highly uncertain.

Scenario Fa1H considers the addition of a fleet of civil supersonic aircraft that was assumed to begin operation in the year 2015 and grow to a maximum of 1,000 aircraft by the year 2040. For reference, the civil subsonic fleet at the end of the year 1997 contained approximately 12,000 aircraft. In this scenario, the aircraft are designed to cruise at Mach 2.4, and new technologies are assumed that maintain emissions of 5 g NO_2 per kg fuel (lower than today's civil supersonic aircraft which has emissions of about 22 g NO_2 per kg fuel). These supersonic aircraft are assumed to replace part of the subsonic fleet (11%, in terms of emissions in scenario Fa1). Supersonic aircraft consume more than twice the fuel per passenger-km compared to subsonic aircraft. *By the year 2050, the combined fleet (scenario Fa1H) is projected to add a further 0.08 Wm^{-2} (42%) to the 0.19 Wm^{-2} radiative forcing from scenario Fa1 (see Figure 4). Most of this additional forcing is due to accumulation of stratospheric water vapor.*

The effect of introducing a civil supersonic fleet to form the combined fleet (Fa1H) is also to reduce stratospheric ozone and increase erythemal dose rate. The maximum calculated effect is at 45°N where, in July, the ozone column change in 2050 from the combined subsonic and supersonic fleet relative to no aircraft is -0.4%. The effect on the ozone column of the supersonic component by itself is -1.3% while the subsonic component is +0.9%.

The combined fleet would change the erythemal dose rate at 45°N in July by +0.3% compared to the 2050 atmosphere without aircraft. The two-thirds uncertainty range for the combined fleet is -1.7% to +3.3%. This may be compared to the projected change of -1.3% for Fa1. Flying higher leads to larger ozone column decreases, while flying lower leads to smaller ozone column decreases and may even result in an ozone column increase for flight in the lowermost stratosphere. In addition, emissions from supersonic aircraft in the Northern Hemisphere stratosphere may be transported to the Southern Hemisphere where they cause ozone depletion.

---

[9] This value is based on satellite observations and model calculations. See **WMO**, 1999: *Scientific Assessment of Ozone Depletion: 1998.* Report No. 44, Global Ozone Research and Monitoring Project, World Meteorological Organization, Geneva, Switzerland, 732 pp.

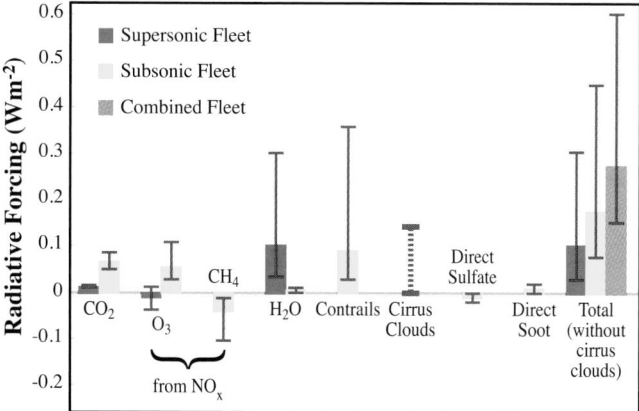

Radiative Forcing from Aircraft in 2050 with Supersonic Fleet

Figure 4: Estimates of the globally and annually averaged radiative forcing from a combined fleet of subsonic and supersonic aircraft (in Wm^{-2}) due to changes in greenhouse gases, aerosols, and contrails in 2050 under the scenario Fa1H. In this scenario, the supersonic aircraft are assumed to replace part of the subsonic fleet (11%, in terms of emissions in scenario Fa1). The bars indicate the best estimate of forcing while the line associated with each bar is a two-thirds uncertainty range developed using the best knowledge and tools available at the present time. (The two-thirds uncertainty range means that there is a 67% probability that the true value falls within this range.) The available information on cirrus clouds is insufficient to determine either a best estimate or an uncertainty range; the dashed line indicates a range of possible best estimates. The estimate for total forcing does not include the effect of changes in cirrus cloudiness. The uncertainty estimate for the total radiative forcing (without additional cirrus) is calculated as the square root of the sums of the squares of the upper and lower ranges. The level of scientific understanding for the supersonic components are carbon dioxide, "good;" ozone, "poor;" and water vapor, "poor."

6. What are the Options to Reduce Emissions and Impacts?

There is a range of options to reduce the impact of aviation emissions, including changes in aircraft and engine technology, fuel, operational practices, and regulatory and economic measures. These could be implemented either singly or in combination by the public and/or private sector. Substantial aircraft and engine technology advances and the air traffic management improvements described in this report are already incorporated in the aircraft emissions scenarios used for climate change calculations. Other operational measures, which have the potential to reduce emissions, and alternative fuels were not assumed in the scenarios. Further technology advances have the potential to provide additional fuel and emissions reductions. In practice, some of the improvements are expected to take place for commercial reasons. The timing and scope of regulatory, economic, and other options may

affect the introduction of improvements and may affect demand for air transport. Mitigation options for water vapor and cloudiness have not been fully addressed.

Safety of operation, operational and environmental performance, and costs are dominant considerations for the aviation industry when assessing any new aircraft purchase or potential engineering or operational changes. The typical life expectancy of an aircraft is 25 to 35 years. These factors have to be taken into account when assessing the rate at which technology advances and policy options related to technology can reduce aviation emissions.

6.1. Aircraft and Engine Technology Options

Technology advances have substantially reduced most emissions per passenger-km. However, there is potential for further improvements. Any technological change may involve a balance among a range of environmental impacts.

Subsonic aircraft being produced today are about 70% more fuel efficient per passenger-km than 40 years ago. The majority of this gain has been achieved through engine improvements and the remainder from airframe design improvement. A 20% improvement in fuel efficiency is projected by 2015 and a 40 to 50% improvement by 2050 relative to aircraft produced today. The 2050 scenarios developed for this report already incorporate these fuel efficiency gains when estimating fuel use and emissions. Engine efficiency improvements reduce the specific fuel consumption and most types of emissions; however, contrails may increase and, without advances in combuster technology, NO_x emissions may also increase.

Future engine and airframe design involves a complex decision-making process and a balance of considerations among many factors (e.g., carbon dioxide emissions, NO_x emissions at ground level, NO_x emissions at altitude, water vapor emissions, contrail/cirrus production, and noise). These aspects have not been adequately characterized or quantified in this report.

Internationally, substantial engine research programs are in progress, with goals to reduce Landing and Take-off cycle (LTO) emissions of NO_x by up to 70% from today's regulatory standards, while also improving engine fuel consumption by 8 to 10%, over the most recently produced engines, by about 2010. Reduction of NO_x emissions would also be achieved at cruise altitude, though not necessarily by the same proportion as for LTO. Assuming that the goals can be achieved, the transfer of this technology to significant numbers of newly produced aircraft will take longer—typically a decade. Research programs addressing NO_x emissions from supersonic aircraft are also in progress.

6.2. Fuel Options

There would not appear to be any practical alternatives to kerosene-based fuels for commercial jet aircraft for the next *several decades. Reducing sulfur content of kerosene will reduce SO_x emissions and sulfate particle formation.*

Jet aircraft require fuel with a high energy density, especially for long-haul flights. Other fuel options, such as hydrogen, may be viable in the long term, but would require new aircraft designs and new infrastructure for supply. Hydrogen fuel would eliminate emissions of carbon dioxide from aircraft, but would increase those of water vapor. The overall environmental impacts and the environmental sustainability of the production and use of hydrogen or any other alternative fuels have not been determined.

The formation of sulfate particles from aircraft emissions, which depends on engine and plume characteristics, is reduced as fuel sulfur content decreases. While technology exists to remove virtually all sulfur from fuel, its removal results in a reduction in lubricity.

6.3. Operational Options

Improvements in air traffic management (ATM) and other operational procedures could reduce aviation fuel burn by between 8 and 18%. The large majority (6 to 12%) of these reductions comes from ATM improvements which it is anticipated will be fully implemented in the next 20 years. All engine emissions will be reduced as a consequence. In all aviation emission scenarios considered in this report the reductions from ATM improvements have already been taken into account. The rate of introduction of improved ATM will depend on the implementation of the essential institutional arrangements at an international level.

Air traffic management systems are used for the guidance, separation, coordination, and control of aircraft movements. Existing national and international air traffic management systems have limitations which result, for example, in holding (aircraft flying in a fixed pattern waiting for permission to land), inefficient routings, and sub-optimal flight profiles. These limitations result in excess fuel burn and consequently excess emissions.

For the current aircraft fleet and operations, addressing the above-mentioned limitations in air traffic management systems could reduce fuel burned in the range of 6 to 12%. It is anticipated that the improvement needed for these fuel burn reductions will be fully implemented in the next 20 years, provided that the necessary institutional and regulatory arrangements have been put in place in time. The scenarios developed in this report assume the timely implementation of these ATM improvements, when estimating fuel use.

Other operational measures to reduce the amount of fuel burned per passenger-km include increasing load factors (carrying more passengers or freight on a given aircraft), eliminating non-essential weight, optimizing aircraft speed, limiting the use of auxiliary power (e.g., for heating, ventilation),

and reducing taxiing. The potential improvements in these operational measures could reduce fuel burned, and emissions, in the range 2 to 6%.

Improved operational efficiency may result in attracting additional air traffic, although no studies providing evidence on the existence of this effect have been identified.

6.4. Regulatory, Economic, and Other Options

Although improvements in aircraft and engine technology and in the efficiency of the air traffic system will bring environmental benefits, these will not fully offset the effects of the increased emissions resulting from the projected growth in aviation. Policy options to reduce emissions further include more stringent aircraft engine emissions regulations, removal of subsidies and incentives that have negative environmental consequences, market-based options such as environmental levies (charges and taxes) and emissions trading, voluntary agreements, research programs, and substitution of aviation by rail and coach. Most of these options would lead to increased airline costs and fares. Some of these approaches have not been fully investigated or tested in aviation and their outcomes are uncertain.

Engine emissions certification is a means for reducing specific emissions. The aviation authorities currently use this approach to regulate emissions for carbon monoxide, hydrocarbons, NO_x, and smoke. The International Civil Aviation Organization has begun work to assess the need for standards for aircraft emissions at cruise altitude to complement existing LTO standards for NO_x and other emissions.

Market-based options, such as environmental levies (charges and taxes) and emissions trading, have the potential to encourage technological innovation and to improve efficiency, and may reduce demand for air travel. Many of these approaches have not been fully investigated or tested in aviation and their outcomes are uncertain.

Environmental levies (charges and taxes) could be a means for reducing growth of aircraft emissions by further stimulating the development and use of more efficient aircraft and by reducing growth in demand for aviation transportation. Studies show that to be environmentally effective, levies would need to be addressed in an international framework.

Another approach that could be considered for mitigating aviation emissions is emissions trading, a market-based approach which enables participants to cooperatively minimize the costs of reducing emissions. Emissions trading has not been tested in aviation though it has been used for sulfur dioxide (SO_2) in the United States of America and is possible for ozone-depleting substances in the Montreal Protocol. This approach is one of the provisions of the Kyoto Protocol where it applies to Annex B Parties.

Voluntary agreements are also currently being explored as a means of achieving reductions in emissions from the aviation sector. Such agreements have been used in other sectors to reduce greenhouse gas emissions or to enhance sinks.

Measures that can also be considered are removal of subsidies or incentives which would have negative environmental consequences, and research programs.

Substitution by rail and coach could result in the reduction of carbon dioxide emissions per passenger-km. The scope for this reduction is limited to high density, short-haul routes, which could have coach or rail links. Estimates show that up to 10% of the travelers in Europe could be transferred from aircraft to high-speed trains. Further analysis, including trade-offs between a wide range of environmental effects (e.g., noise exposure, local air quality, and global atmospheric effects) is needed to explore the potential of substitution.

7.   Issues for the Future

This report has assessed the potential climate and ozone changes due to aircraft to the year 2050 under different scenarios. It recognizes that the effects of some types of aircraft emissions are well understood. It also reveals that the effects of others are not, because of the many scientific uncertainties. There has been a steady improvement in characterizing the potential impacts of human activities, including the effects of aviation on the global atmosphere. The report has also examined technological advances, infrastructure improvements, and regulatory or market-based measures to reduce aviation emissions. Further work is required to reduce scientific and other uncertainties, to understand better the options for reducing emissions, to better inform decisionmakers, and to improve the understanding of the social and economic issues associated with the demand for air transport.

There are a number of key areas of scientific uncertainty that limit our ability to project aviation impacts on climate and ozone:

- The influence of contrails and aerosols on cirrus clouds
- The role of NO_x in changing ozone and methane concentrations
- The ability of aerosols to alter chemical processes
- The transport of atmospheric gases and particles in the upper troposphere/lower stratosphere
- The climate response to regional forcings and stratospheric perturbations.

There are a number of key socio-economic and technological issues that need greater definition, including *inter alia* the following:

- Characterization of demand for commercial aviation services, including airport and airway infrastructure constraints and associated technological change
- Methods to assess external costs and the environmental benefits of regulatory and market-based options
- Assessment of the macroeconomic effects of emission reductions in the aviation industry that might result from mitigation measures
- Technological capabilities and operational practices to reduce emissions leading to the formation of contrails and increased cloudiness
- The understanding of the economic and environmental effects of meeting potential stabilization scenarios (for atmospheric concentrations of greenhouse gases), including measures to reduce emissions from aviation and also including such issues as the relative environmental impacts of different transportation modes.

AVIATION AND THE GLOBAL ATMOSPHERE

A Special Report of Working Groups I and III
of the Intergovernmental Panel on Climate Change

This Special Report was accepted by IPCC Working Groups I and III
(San José, Costa Rica • 12-14 April 1999), but not approved in detail.

1

Introduction

Lead Authors:
J.H. Ellis, N.R.P. Harris, D.H. Lister, J.E. Penner

Review Editor:
B.S. Nyenzi

CONTENTS

1.1. Background

Aviation is an integral part of the infrastructure of today's society. It plays an important role in the global economy; it supports both commerce (through business travel and air freight) and private travel. Aviation also plays an important role in military activity. As such, aviation affects the lives of citizens in every country in the world, regardless of whether they fly. The activities of the civil air transport industry have long been circumscribed by matters of public interest in addition to economic factors.

Of most importance historically are matters related to safety and environmental issues associated with local noise and air pollution.

Two global environmental issues have emerged for which aviation may have potentially important consequences: Climate change, including changes to weather patterns (i.e., rainfall, temperature, etc.), and, for supersonic aircraft, stratospheric ozone depletion and the resultant increase in UV-B radiation at the Earth's surface. Boxes 1-1 and 1-2 contain general

---

Box 1-1. Climate Change and the Framework Convention

Human activities release greenhouse gases into the atmosphere. The atmospheric concentrations of carbon dioxide, methane, nitrous oxide, chlorofluorocarbons, and tropospheric ozone have all increased over the past century. These rising levels of greenhouse gases are expected to cause climate change. By absorbing infrared radiation, these gases change the natural flow of energy through the climate system. The climate must somehow adjust to this "thickening blanket" of greenhouse gases to maintain the balance between energy arriving from the sun and energy escaping back into space. This relatively simple picture is complicated by increased amounts of sulfate aerosol from human activities that modulate incoming solar radiation and tend to cause a cooling effect on climate, at least on regional and hemispheric scales.

Global mean surface air temperatures have increased by 0.3-0.6°C since the late 19th century, and recent years have been among the warmest on record. Any human-induced effect on climate, however, is superimposed on natural climate variability resulting from climate fluctuations (e.g., El Niño) and external causes such as solar variability and volcanic eruptions. More sophisticated approaches are now being applied to the detection and attribution of the causes of change in climate by looking, for example, for spatial patterns expected from climate-forcing change by greenhouse gases and aerosols. To date, the balance of the evidence suggests that there is a discernible human influence on the global climate.

Model projections of future climate, based on the present understanding of climate processes and using emission scenarios (IS92) based on a range of economic and technological assumptions, estimate a rise in global mean temperature of 1–3.5°C (best estimate 2°C) between 1990 and 2100. In all cases, the average rate of warming would probably be greater than any in the past 10,000 years, though actual annual-to-decadal changes would include considerable natural variability. A general warming is expected to lead to an increase in the occurrence of extremely hot days and a decrease in the occurrence of extremely cold days. Regional temperature changes could differ substantially from the global mean value, and there are many uncertainties about the scale and impacts of climate change, particularly at the regional level. The mean sea level is expected to rise 15–95 cm (best estimate 50 cm) by 2100, with some flooding of low-lying areas. Forests, deserts, rangelands, and other unmanaged ecosystems would face new climatic stresses, partly as a result of changes in the hydrological cycle; many could decline or fragment, with some individual species of flora or fauna becoming extinct. Because of the delaying effect of the oceans, surface temperatures do not respond immediately to greenhouse gas emissions, so climate change would continue for many decades even if atmospheric concentrations were stabilized.

Achieving stabilized atmospheric concentrations of greenhouse gases would demand a major effort. For CO_2 alone, freezing global emissions at their current rates would result in a doubling of its atmospheric concentrations from pre-industrial levels soon after 2100. Eventually, emissions would have to decrease well below current levels for concentrations to stabilize at doubled CO_2 levels, and they would have to continue to fall thereafter to maintain a constant CO_2 concentration. The radiative forcing of greenhouse gas levels (including methane, nitrous oxide, and others, but not aerosols) could equal that caused by a doubling of pre-industrial CO_2 concentrations by 2030 and a trebling or more by 2100.

The international community is tackling this challenge through the United Nations Framework Convention on Climate Change (UNFCCC). Adopted in 1992, the Convention seeks to stabilize atmospheric concentrations of greenhouse gases at safe levels. More than 170 countries have become Parties to the Convention. Developed countries have agreed to take voluntary measures aimed at returning their emissions to 1990 levels by the year 2000, with further legally binding emissions cuts after the year 2000 proposed at Kyoto in late 1997. Developed countries have also agreed to promote financial and technological transfers to developing countries to help them address climate change.

Source: IPCC, 1996a.

Box 1-2. Stratospheric Ozone Depletion, UV-B Radiation, and the Montreal Protocol

Although ozone can be measured throughout much of the atmosphere, most of it is found in the stratosphere in a layer centered about 20 km above the Earth's surface. Stratospheric ozone is beneficial to life on Earth because it blocks much of the dangerous ultraviolet light (UV-B) radiated by the sun. If unnaturally high levels of UV-B radiation reach the Earth's surface, many forms of life can be harmed. For instance, UV-B can cause skin cancers in humans and may reduce crop yields.

Natural ozone amounts in the stratosphere result from a balance of production and loss processes involving chemistry, meteorology, and solar radiation. Since the 1960s, however, increases in atmospheric concentrations of human-generated chlorine- and bromine-containing compounds (principally chlorofluorocarbons and halons) have caused additional ozone loss. This trend has resulted in declines in stratospheric ozone amounts at middle and high latitudes in both hemispheres. The most dramatic manifestation is the Antarctic ozone hole, where more than half of the ozone is destroyed in a 6-week period each spring. In recent winters, similar features—but with half the ozone loss (20–30%)—have been observed over the Arctic, and at northern mid-latitudes a long-term decline of 5–10% has occurred over the past 20–30 years. Annual amounts of biologically active UV-B radiation have increased by about 10% over mid-latitudes since 1979. No significant loss of ozone or increase in UV-B radiation has been found in the tropics.

Concern that chlorofluorocarbons might destroy ozone was first raised in the 1970s. Following the general realization that these human-generated chemicals posed a real threat to the ozone layer, the Vienna Convention for the Protection of the Ozone Layer was adopted in 1985. Shortly afterwards, the Antarctic ozone hole was discovered, leading to renewed pressure to control ozone-depleting substances. In 1987, the Montreal Protocol on Substances that Deplete the Ozone Layer was agreed; it has since been ratified by more than 160 countries. Initially, the Montreal Protocol imposed clear limits on the future production of chlorofluorocarbons and halons only; as scientific evidence about ozone depletion has mounted, however, the Protocol has been modified to include other chemicals.

As a direct result of these controls, there have been marked reductions in emissions of these substances into the atmosphere; assuming full compliance with the Protocol, these reductions will result in reduced atmospheric amounts of chlorine and bromine. However, the magnitude of the ozone loss at any given time depends on a number of factors. Although the amount of chlorine and bromine clearly is important, other influences include the temperature of the ozone layer, the atmosphere's chemical composition, and long-term changes in atmospheric circulation related to climate change. How all of these factors evolve over coming decades will determine future ozone amounts as chlorine and bromine are reduced.

Source: WMO, 1999.

---

descriptions of the basic science and the political process related to these two issues (without addressing aviation in particular), respectively.

For both of these issues, the effects from aviation are part of a larger picture. Human-generated emissions at the Earth's surface can be carried aloft and affect the global atmosphere. The unique property of aircraft is that they fly several kilometers above the Earth's surface. The effects of most aircraft emissions depend strongly on the flight altitude and whether aircraft fly in the troposphere or stratosphere. The effects on the atmosphere can be markedly different from the effects of the same emissions at ground level.

A number of aircraft emissions can affect climate. Carbon dioxide (CO_2) and water (H_2O) do so directly; other effects (e.g., production of ozone in the troposphere, alteration of methane lifetime, formation of contrails and modified cirrus cloudiness) are indirect. The emissions that can affect stratospheric ozone (i.e., nitrogen oxides, particulates, and water vapor) do so indirectly by modifying the chemical balance in the stratosphere.

There has been sustained long-term growth in civil air transportation. For example, over the past 10 years, passenger traffic on scheduled airlines has increased by 60%. Over the next 10 to 15 years, demand for air travel is expected to grow by about 5% per year (Airbus, 1997; Boeing, 1997; Brasseur *et al.*, 1998), though there are likely to be regional variations in demand. In contrast, no such increase in the numbers of military aircraft is anticipated; they are expected to remain static or even decrease. As a consequence, fuel use and emissions produced by future military activities are expected to be a decreasing part of the total from aviation (see Chapter 9).

Aviation fuel currently corresponds to 2–3% of the total fossil fuels used worldwide. Of this total, the majority (> 80%) is used by civil aviation. By comparison, the whole transportation sector currently accounts for 20–25% of all fossil fuel consumption. Thus, the aviation sector consumes 13% of the fossil fuel used in transportation; it is the second biggest sector after road transportation, which consumes 80% (IPCC, 1996b).

Given the continued growth of aviation, a number of questions have been raised regarding the future effects of aviation emissions on the global environment. For example, if supersonic aircraft (which fly primarily in the stratosphere) were to be introduced in significant numbers, what special effects might there be, and what trade-offs might be possible?

Answering such questions involves consideration of a number of complex issues and assumptions about the future growth, technology trends, and operational practices of the aircraft industry. In the past, for example, fuel efficiency has improved dramatically over time, so total aviation fuel use did not increase as fast as passenger or freight traffic. Fuel use, CO_2 emissions, and NO_x emissions per passenger-kilometer have decreased, but the increase in total NO_x emissions has been larger than the increase in total fuel use. If it is not technologically possible to reduce all aircraft emissions simultaneously, then some kind of trade-off may be needed. The relative environmental benefits of further reductions in all emissions need to be carefully considered; this analysis is one of the underlying aims of this report.

This report covers issues relating both to the atmosphere and to the aviation industry that address these questions. It is the first such assessment of a single industry and its global environmental impact. (Issues such as local air quality and ground transportation around airports are outside its scope, however, and are not discussed.)

The chapter structure and the major relationships between the chapters are depicted schematically in Figure 1-1. This chapter provides a background for the main part of the report and indicates where detailed information on each subject can be found. In Section 1.2, the present-day aviation industry is described. In Section 1.3, aircraft emissions that are most relevant for climate change and stratospheric ozone depletion are summarized. In Section 1.4, the use and development of future scenarios are discussed, and in Section 1.5, possible options for mitigating the atmospheric effects of aviation are outlined.

1.2.   Aviation

Air transportation plays a substantial role in world economic activity, and society relies heavily on the benefits associated with aviation. The aviation industry includes suppliers and operators of aircraft, component manufacturers, fuel suppliers, airports, and air navigation service providers. Its customers represent every sector of the world's economy and every segment of the world's population.

The commercial sector of the industry is highly competitive, consisting in 1994 of about 15,000 aircraft operating over routes of approximately 15 million km in total length and serving nearly 10,000 airports. In 1994, more than 1.25 billion passengers used the world's airlines for business and vacation travel, and well in excess of a third of the value of the world's manufactured exports were transported by air. The aviation industry accounted for 24 million jobs for the world's workforce and provided US$1,140 billion in annual gross output. By the

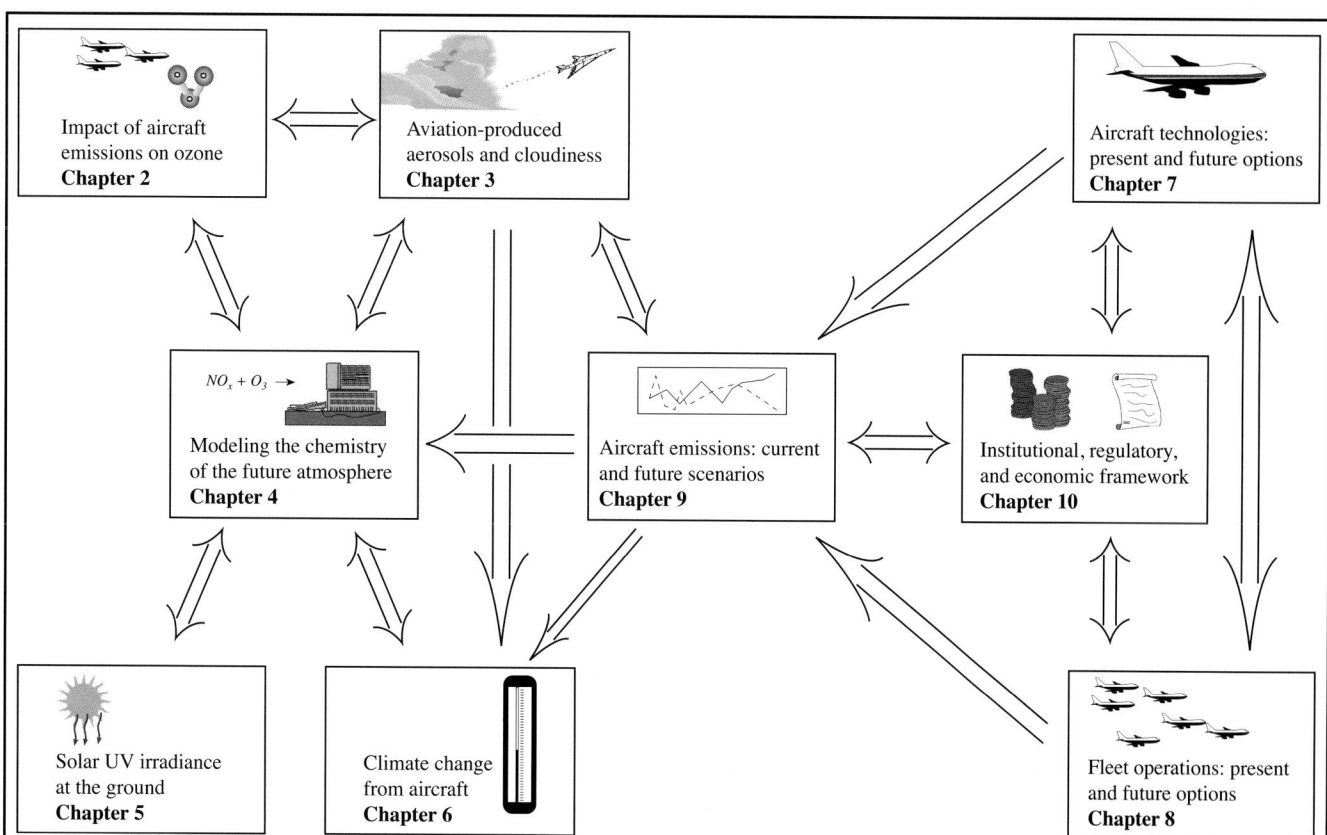

Figure 1-1: Major links between chapters in this report.

year 2010, aviation's global impact could exceed US$1,800 billion and more than 33 million jobs (IATA, 1994, 1996).

The 1944 Chicago Convention, to which 185 countries are now party, established the International Civil Aviation Organization (ICAO) as the United Nations' specialized agency with authority to develop standards and recommended practices regarding all aspects of international aviation—including certification standards for emissions and noise. These standards are published as Annexes to the Convention and are adopted by the ICAO Council, which is composed of 33 member nations elected by the entire ICAO membership. Individual countries either adopt ICAO standards or file differences with ICAO.

Since 1977, ICAO has promulgated international emissions and noise standards (ICAO, 1993a,b) for aircraft and aircraft engines that apply to all member states. ICAO, through its Committee on Aviation Environmental Protection (CAEP), has reviewed and revised these standards when warranted and has developed operational policies and procedures to mitigate further the environmental impacts of civil aviation. ICAO has also developed broader policy guidance on fuel taxation and charging principles that have relevance in the emissions context. In addition to the harmonization achieved through ICAO, international flights are subject to bilateral air service agreements between individual countries.

The commercial airline industry, though predominantly privately owned and managed, must rely on airport infrastructure and air navigation services that the industry neither owns nor controls. The overall growth of air traffic and the capacity limitations of airports and air navigation services have introduced congestion as a challenge for aviation. This congestion causes delays, introduces unreliability or inefficiencies for all system users, and produces considerable extra energy consumption and emissions. During 1996, for example, 15.4% of flights in Europe incurred an average delay of 16.7 minutes. In the United States of America, the average delay for domestic departures was 7.2 minutes.

Growth in demand for aviation averaged about 5% per year for the period 1980–95. The industry expects demand to continue to rise, though not monotonically. Aviation growth may be estimated reasonably well in the near term, but forecasts are subject to greater uncertainty beyond a 5–10 year period because of changes in factors such as the real cost of air travel, economic activity, new market opportunities, world disposable income trends, world political stability, tourism, and air transport liberalization.

An aircraft is a major investment, with a useful economic life of 25 years or more. Operation of an aircraft includes airframe and engine performance. The performance of an aircraft must

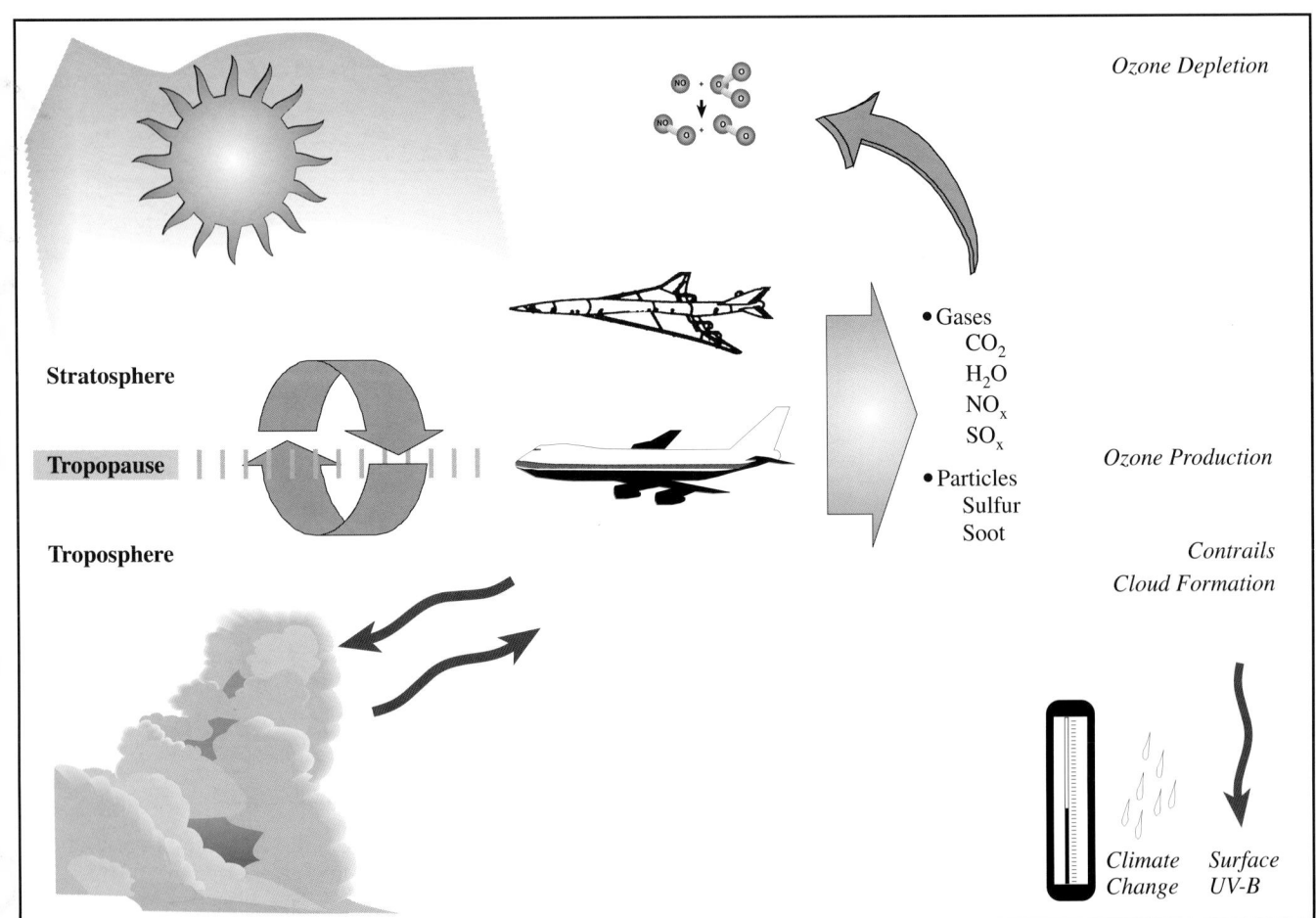

Figure 1-2: Impacts of aviation on the atmosphere.

address the overriding issue of safety, as well as mission or performance efficiencies, economics, and environmental objectives. ICAO is the responsible organization to ensure that such objectives are met on an internationally harmonized level as far as possible.

1.3. Emissions and the Environment

The global environmental issues addressed in this report are climate changes and depletion of stratospheric ozone (see Boxes 1-1 and 1-2). Major international assessments of these issues are made periodically, and no attempt is made here to revisit such reports (for more information, see IPCC, 1996a,b; WMO, 1999, and references therein). These reports are used for contextual information and as a background to the possible impact of aviation on the atmosphere. There have also been several recent assessments of the impacts of aviation in Europe and the United States of America (Stolarski *et al.*, 1995; Friedl *et al.*, 1997; Schumann *et al.*, 1997; Brasseur *et al.*, 1998).

Aviation represents only one of many perturbations associated with future scenarios. There are many sources of climate-active substances. For example, greenhouse gases are emitted from a wide range of industrial, domestic, and agricultural activities, and there are numerous sources of aerosols (e.g., sulfate from fossil fuel combustion and volcanoes and carbonaceous aerosols from fossil fuel and biomass burning). Increases in tropospheric ozone are expected to result from increasing methane, nitrogen oxides, carbon monoxide, and hydrocarbon emissions. Similarly, perturbations to stratospheric ozone result from chlorofluorocarbons (CFCs) and CFC-substitutes and from halon and methyl bromide emissions in addition to any potential perturbations resulting from aircraft emissions.

Some of the processes related to aviation emissions and their location in the atmosphere are shown in Figure 1-2. Subsonic aircraft fly in the troposphere and lower stratosphere, whereas supersonic aircraft fly in the stratosphere 80–85% of the time, with cruise altitudes several kilometers above those of subsonic aircraft. The differing chemical and physical processes in the two regions must be taken into account.

The troposphere is principally heated near the Earth's surface, and the temperature decreases with altitude. Warm, moist air tends to underlie cool, dry air, leading to frequent vertical turbulent motions that exchange air throughout this region. In contrast, the temperature in the stratosphere generally increases with altitude; the result, as reflected in its name, is that this region is stratified. Vertical motions are much slower here than in the troposphere. The stratosphere is also much drier than the troposphere, and clouds rarely form at this level.

Table 1-1 lists aircraft emissions that are important from an atmospheric perspective, with summaries of the roles that they play. These emissions can be usefully divided into two categories, depending on how they affect climate: Direct, as with CO_2 (where the emitted compound is the species that can modify climate), and indirect, where the climate species is not the same as the emitted species—as with modified cirrus cloud coverage resulting from particles and particle precursors.

1.3.1. Carbon Dioxide

The behavior of CO_2 within the atmosphere is simple and well understood. There are no important formation or destruction processes that take place in the atmosphere itself. Atmospheric sources and sinks occur principally at the Earth's surface and involve exchanges with the biosphere and the oceans (e.g., Schimel *et al.*, 1995, 1996). The effect of CO_2 on climate change is direct and depends simply on its atmospheric concentration. CO_2 molecules absorb outgoing infrared radiation emitted by the Earth's surface and lower atmosphere. The observed 25–30% increase in atmospheric CO_2 concentrations over the past 200 years has caused a warming of the troposphere and a cooling of the stratosphere.

There has been much discussion about how stabilization of CO_2 concentrations might be achieved in the future (e.g., IPCC, 1996a, 1997a,b). One of the most important factors is the accumulated emission between now and the time at which stabilization is reached. The way in which annual emissions vary over time is less important. Two findings from these IPCC reports are worth noting:

- If global anthropogenic CO_2 emissions were maintained near 1994 levels, the atmospheric concentration would continue to rise for more than 200 years; by the end of the 21st century, it would have reached about 500 ppmv (compared to its pre-industrial value of 280 ppmv).
- A range of carbon cycle models indicates that stabilization of atmospheric CO_2 concentrations at 450, 650, or 1000 ppmv could be achieved only if global anthropogenic CO_2 emissions drop to 1990 levels by approximately 40, 140, or 240 years from now, respectively, and substantially below 1990 levels subsequently.

The amount of CO_2 formed from the combustion of aircraft fuel is determined by the total amount of carbon in the fuel because CO_2 is an unavoidable end product of the combustion process (as is water). The subsequent transport and processing of this CO_2 in the atmosphere follows the same pathways as those of other CO_2 molecules emitted into the atmosphere from whatever source. Thus, CO_2 emitted from aircraft becomes well mixed and indistinguishable from CO_2 from other fossil fuel sources, and the effects on climate are the same. The rate of growth in aviation CO_2 emission is faster than the underlying global rate of economic growth, so aviation's contribution, along with those of other forms of transportation, to total emissions resulting from human activities is likely to grow in coming years. The radiative and climate implications are addressed in Chapter 6.

Table 1-1: *Species contributing to climate and ozone change.*

Emitted Species	Role and Major Effect at Earth's Surface
CO_2	*Troposphere and Stratosphere* Direct radiative forcing ⇨ warming
H_2O	*Troposphere* Direct radiative forcing ⇨ warming Increased contrail formation ⇨ radiative forcing ⇨ warming *Stratosphere* Direct radiative forcing ⇨ warming Enhanced PSC formation ⇨ O_3 depletion ⇨ enhanced UV-B Modifies O_3 chemistry ⇨ O_3 depletion ⇨ enhanced UV-B
NO_x	*Troposphere* O_3 formation in upper troposphere ⇨ radiative forcing ⇨ warming ⇨ reduced UV-B Decrease in CH_4 ⇨ less radiative forcing ⇨ cooling *Stratosphere* O_3 formation below 18–20 km ⇨ reduced UV-B O_3 formation above 18–20 km ⇨ enhanced UV-B Enhanced PSC formation ⇨ O_3 depletion ⇨ enhanced UV-B
SO_x and H_2SO_4	*Troposphere* Enhanced sulfate aerosol concentrations Direct radiative forcing ⇨ cooling Contrail formation ⇨ radiative forcing ⇨ warming Increased cirrus cloud cover ⇨ radiative forcing ⇨ warming Modifies O_3 chemistry *Stratosphere* Modifies O_3 chemistry
Soot	*Troposphere* Direct radiative forcing ⇨ warming Contrail formation ⇨ radiative forcing ⇨ warming Increased cirrus cloud cover ⇨ radiative forcing ⇨ warming Modifies O_3 chemistry *Stratosphere* Modifies O_3 chemistry

Notes:

1) Positive radiative forcing will tend to warm the Earth's surface; negative forcing will tend to cool the Earth's surface.

2) Any warming at the Earth's surface caused by increases in atmospheric CO_2 will be accompanied by a cooling of the stratosphere.

3) Any decrease in ozone, whether in the troposphere or stratosphere, will tend to enhance UV-B radiation and cool the Earth's surface. Conversely, any increase in ozone will tend to reduce UV-B and heat the Earth's surface. This dual effect is shown in this table only for tropospheric ozone.

4) Chemi-ions and metal particles are also emitted from aircraft. They may play an important role in contrail and enhanced cirrus formation (see Chapter 3), in which case their emission would lead to a tendency to warm the Earth's surface.

5) Hydrocarbons and CO are also emitted and contribute in a minor way to tropospheric ozone formation.

1.3.2. *Water*

The natural cycle of water in the atmosphere is also complex, involving a suite of closely coupled physical processes. This is particularly true in the troposphere, where there is continual cycling between water vapor, clouds, precipitation, and ground water. Water vapor and clouds have large radiative effects on climate and directly influence tropospheric chemistry. The stratosphere is much drier than the troposphere. Nevertheless, water vapor is important in determining radiative balance and chemical composition, most dramatically in polar ozone loss through the formation of polar stratospheric clouds (PSCs).

Emissions of water vapor by the global aircraft fleet into the troposphere are small compared with fluxes within the natural hydrological cycle; however, the effects of contrails and enhanced cirrus formation must be considered (Chapter 3). Water vapor resides in the troposphere for about 9 days. In the stratosphere, the time scale for removal of any aircraft water emissions is longer (months to years) than in the troposphere, and there is a greater chance for aircraft emissions to increase the ambient concentration. Any such increase could have two effects: A direct radiative effect with a consequent influence on climate, and a chemical perturbation of stratospheric ozone both directly and through the potentially increased occurrence of polar stratospheric clouds at high latitudes. The implication of releases in the stratosphere are discussed in Chapters 4 and 6.

1.3.3. Nitrogen Oxides

Nitrogen oxides (NO and NO_2 are jointly referred to as NO_x) are present throughout the atmosphere. They are very influential in the chemistry of the troposphere and the stratosphere, and they are important in ozone production and destruction processes. There are a number of sources (oxidation of N_2O, lightning, fossil fuel combustion) whose contribution to NO_x concentrations in the upper troposphere are not well quantified.

In all regions, the chemistry of the atmosphere is complex; aircraft NO_x emissions are best viewed as perturbing a web of chemical reactions with a resultant impact on ozone concentrations that differs with location, season, and so forth. In the upper troposphere and lower stratosphere, aircraft NO_x emissions tend to cause increased ozone amounts, so increased ozone and its greenhouse effects are the main issues for NO_x emissions from subsonic aircraft. The pathways of other atmospheric constituents are also affected. Principal among these effects for NO_x emissions is the reduction in the atmospheric lifetime and concentration of methane, another greenhouse gas. On the other hand, NO_x emissions at the higher altitudes (18 km or above) of supersonic aircraft tend to deplete ozone. These and other issues related to the atmospheric chemistry of aviation emissions are discussed in Chapters 2 (past and present) and 4 (future). The effects on ultraviolet fluxes at the Earth's surface are discussed in Chapter 5.

1.3.4. Particles and Particle Precursors

A similarly complex system of atmospheric processes and effects exists for particles. There are many types of particles, each with its own complex physics and chemistry. Natural types of particles include salt particles from sea spray, wind-blown soil, and sulfate aerosols produced from naturally emitted sulfur-containing gases. Aerosols resulting from human activities include sulfate aerosols and soot from fossil fuel burning. Carbonaceous aerosols are produced from biomass burning and fossil fuel burning.

Particles related to aviation (principally sulfate aerosols and soot particles) are discussed in Chapter 3 together with contrail and cloud formation. Aircraft engines actually emit a mixture of particles (including metal particles and chemi-ions) and gases (e.g., SO_2). These emissions evolve in the engine exhaust and the atmosphere to form a variety of particles mainly composed of soot from incomplete combustion and sulfuric acid (H_2SO_4) from sulfur in the aviation fuel. These particles are capable of seeding contrails and cirrus clouds, thus potentially changing the total cloud cover in the upper troposphere. The climate impact of clouds is a balance of their capabilities to reflect sunlight back to space and to trap outgoing infrared radiation from the Earth's surface. For high clouds, the latter effect is larger, and increased cirrus coverage would result in a warming tendency. (This effect is opposite in sign to that of surface emissions of SO_2, which mainly affect low-altitude clouds and produce a cooling effect.)

Particles are also involved in the chemical balance of the atmosphere. It is well established that the sulfate aerosol layer in the stratosphere is critically important in determining the NO_x budget there; any long-term changes in the surface area of particles would affect stratospheric NO_x, hence ozone. The chemical issues related to particles are discussed in Chapters 2 and 4.

1.3.5. Atmospheric Models

Atmospheric models attempt to describe the workings of the atmosphere; the detail of the description depends on factors such as the scientific understanding of the processes involved, the time scale of interest, and the available computer resources. Different models include different facets of the atmosphere system. For instance, state-of-the-art climate models are similar to weather prediction models, but additionally may include descriptions of the ocean and the biosphere so that the exchange of heat and carbon dioxide can be modeled. The next generation of models is likely to include chemical processes and be derived from the current generation of models that contain detailed descriptions of the chemistry.

A wide range of different types of atmospheric models are used within this report, depending on the problem of interest. Chemical transport models are used to calculate changes in chemical composition resulting from aviation emissions (Chapters 2 and 4); microphysical models are used to calculate changes in particle composition (Chapter 3), and radiative transfer and climate models are used to assess the possible impact on UV-B radiation (Chapter 5) and climate (Chapter 6). Model uncertainties arise from one of two main sources: Incorrect or poorly quantified descriptions of the processes involved, and missing processes. These uncertainties are typically reduced over time as the state of the underlying scientific knowledge evolves. Although it is difficult to quantify these uncertainties, one of the major aims of this report is to give a clear idea of the uncertainties associated with model calculations.

1.4. Emissions Scenarios

Calculations of projected future changes in atmospheric composition rely on a number of additional factors. For example, atmospheric models require forecasts of future emissions if realistic predictions of the future atmosphere are to be made. However, accurate forecasts of future demand are not possible. In this report, the future is explored on the basis of scenarios. Scenarios should not be interpreted as forecasts but as tools to explore a range of future outcomes.

To assess the possible future impact of aviation, plausible scenarios for other changes in the future composition of the atmosphere are required—in particular for background CO_2, NO_x, SO_2, CO, and hydrocarbon emissions. The IS92 scenarios used here were originally described in IPCC (1992). Six scenarios of future greenhouse gas and aerosol precursor emissions (IS92a-f) were developed, based on assumptions concerning population and economic growth, land use, technological changes, energy availability, and fuel mix during the period 1990 to 2100. Through an understanding of the global carbon cycle and atmospheric chemistry, these emissions can be used to project atmospheric concentrations of greenhouse gases and aerosols. In addition, scenarios for future atmospheric chlorine and bromine abundances have been calculated by assuming that the Montreal Protocol and its Amendments and Adjustments will be followed and effective. The IS92 scenarios are recognized to be imperfect (for instance, they assume that no regulatory interventions will be made, so they have been outdated by the UNFCCC process). The underlying assumptions, strengths, and weaknesses were assessed in IPCC (1995). The IS92a scenario is a mid-range emissions scenario and is used to describe future non-aviation emissions in the calculations presented in this report.

The future growth of aviation will depend heavily on factors such as economic growth (at global and regional levels), the demand for travel (in an age of rapid advances in information technology), the development of infrastructure to support air travel and available flight technology, and the availability and cost of fuel. Increases in demand will not translate directly into increases in emissions. Changes in engine efficiency, airplane design (size and shape), and operational practice are all expected to lead to more efficient use of fuel because there are strong commercial reasons for airlines and other operators to keep fuel costs down.

The scenarios used in this report have been developed using models of passenger demand on a regional and global basis that assume future economic growth rates as found in the IS92 scenarios (particularly IS92a). In all cases, it is assumed that infrastructure (e.g., airports) and technology will be developed so that this growth is not constrained. Different aircraft types and fuel use are included, so CO_2 and H_2O emissions—which depend solely on the amount of fuel burned—can be calculated directly. SO_2 emissions are estimated simply by assuming what the sulfur composition in the fuel will be. Emissions of NO_x, CO, and hydrocarbons depend strongly on combustor technology—particularly the mixing of fuel and air in the combustion chamber, as well as temperature and pressure. These emissions are estimated using semi-empirical relationships between in-flight fuel

flow and emissions of NO_x, CO, and hydrocarbons derived from ICAO engine certifications together with determined flight patterns. Emissions of all of these compounds for global air traffic are produced on a 3-D grid (latitude, longitude, altitude) that can be used in global atmospheric models. Little equivalent information regarding emissions exists for particulates other than total sulfur emissions.

In this report, future aviation emissions and their effects are assessed at two different times in the future—2015 and 2050. The technological assumptions in aviation demand models are relatively well determined for 2015, the first year in which future impacts are assessed. Time scales for the development of new aircraft types and technologies are too long for any radically different option to become available and enter service to any significant extent by 2015 (see Box 1-3). The most significant uncertainties relate to underlying economic growth and available aviation infrastructure, which will be critical in determining how many aircraft are flying in 2015 and what the relative numbers of each type will be (i.e., the fleet mix).

By 2050—the second year for which future impacts are assessed—many more technological options could be introduced, and uncertainties about what will happen are much larger. For example, a second generation of high-speed civil transport (HSCT) aircraft could be operational in significant numbers (many more than the current fleet of 13 Concordes) by the middle of the next century; these aircraft may well replace some of the subsonic market. This development would be important from an atmospheric perspective because HSCT emissions are released at significantly higher altitudes than those of subsonic aircraft. The scenarios used here thus cover a wider range of possibilities than for 2015. As for 2015, the calculated demand is based on the economic growth rates in the IS92 scenarios, and allowance is made for differential regional growth. Chapter 9 contains a description of the methodologies used to produce a range of scenarios of emissions from aviation. These scenarios assume idealized operational practices (i.e., direct routing, optimum flight profiles, and no delays for the assumed fleets). They therefore represent minimum fuel use and emissions.

1.5.   Mitigation

Environmental issues regarding emissions from aircraft were originally related to their contribution to local air quality in the vicinity of airports. These considerations led to the introduction of legislation in the United States resulting in domestic regulatory standards. Subsequently, ICAO developed international standards and recommended practices for the control of fuel venting and emissions of carbon monoxide, hydrocarbons, nitrogen oxides, and smoke from aircraft engines over a prescribed landing/take-off (LTO) cycle below 915 m (3,000 feet) (ICAO, 1981). Although the global environmental effects of aircraft emissions have been a matter of much debate at a scientific and technical level, there are no specific standards for the control of emissions from aircraft during cruise. However,

Box 1-3. Time Scales in Aviation and the Atmosphere

Understanding the time scales of the processes involved is important in assessing the impact aviation can have on the atmosphere now and in the future. It takes many years for a new aircraft design to progress from the drawing board into service. Once aircraft are operative, their emissions remain in the atmosphere for periods ranging from days to centuries, with some climatic effects felt on even longer time scales. Furthermore, although new technologies would have an immediate effect on emissions from new aircraft, any impact on the global abundance of short-lived atmospheric constituents would be limited by the rate of introduction of the new technology into the global fleet. A rough idea of the various time scales involved is provided.

The processes that remove trace species from the atmosphere can be chemical (e.g., the oxidation of methane), physical (e.g., in rain or by dry deposition onto land or sea), or biological (e.g., the uptake of CO_2 by plants). The rate of each process typically varies with season and location in the atmosphere. These rates can be combined to produce a rough estimate of how long each constituent remains in the atmosphere. A constituent with a short lifetime responds quickly to any change in emissions. A trace species with a long lifetime responds slowly to a change in emissions.

The atmospheric effects of H_2O, NO_x, SO_x, and soot are all relatively short-lived. Broadly speaking, the tropospheric lifetimes of these constituents are a couple of weeks or less; that of any ozone produced by NO_x is a month or so. The stratospheric time scales involved are longer but are all well under a decade. By contrast, emissions of carbon dioxide affect the atmosphere for a long time (about 100 years), with little difference for emissions into the stratosphere or troposphere.

The main factors affecting how quickly new aircraft are introduced are technological feasibility, certification, and commercial viability. Typically, new technology is likely to be a decade in its gestation, although this time scale may be reduced if there are significant market opportunities. The project launch of a new aircraft type by an airframe manufacturer is normally concurrent with the launch of new engines supplied by competing manufacturers. Development of the engine culminates in airworthiness and emissions certification, usually 3–5 years later—but the time scale for entry into service is dictated by the airframe manufacturer and its customer airlines. Once the engine has achieved airworthiness certification, it is installed on the airframe, and the aircraft typically then takes another year to complete the airworthiness and noise certification process before initial deliveries are made to customers.

With commercial airlines, individual aircraft will operate for 25 years or more in revenue service. A good product, including its derivatives, will have a substantial production period (possibly 25 years or longer); therefore, the overall time scale between introduction into service of an aircraft type and withdrawal from service may exceed 50 years.

Development of the infrastructure for air transportation (airports, air traffic control, etc.) can take years or even decades. This development is driven by overall increased demand for air transportation, both for passengers and freight. It is limited by the availability of financial resources and local environmental concerns about noise and increased ground traffic around new or expanded airports.

the LTO standards in place do indirectly limit emissions from an engine during climb and cruise.

The UNFCCC seeks to stabilize greenhouse gas concentrations in the atmosphere at a level that would prevent dangerous anthropogenic interference with the climate system. Its coverage includes emissions from all sources and all sectors, although it does not specifically refer to aviation. However, the Kyoto Protocol to the Convention, which was agreed in December 1997, includes two elements that are particularly relevant to aviation. First, the Kyoto Protocol requires developed countries to reduce their total national emissions from all sources by an average of about 5% for the years 2008–2012 compared with 1990 (with differences for individual countries). It contains a provision for so called "flexible mechanisms"—including

emissions trading, "joint implementation," and a "clean development mechanism." Secondly, the Kyoto Protocol contains a provision (Article 2) that calls on developed countries to pursue policies and measures for the limitation or reduction of greenhouse gases from "aviation bunker fuels," working through ICAO. These issues are discussed in Chapter 10.

Future mitigation options and strategies will need to consider the motivation for increasing operational efficiencies and reducing fuel use in light of other environmental effects of aviation, such as noise. The availability and cost of fuel in the overall budget of aircraft operators will continue to exert strong pressure for fuel efficiency. A number of technological improvements (e.g., to airframe aerodynamics, aircraft weight, and engine cycle performance) over the years have improved the fuel efficiency of subsonic aircraft and engines. These innovations have had a direct impact on the amount of CO_2 and H_2O emitted by aircraft (the less fuel consumed, the less CO_2 and H_2O emitted) and have reduced CO and hydrocarbon emissions. The effect on emissions of NO_x and particles is not as simple, however. The drive to improve fuel efficiency and reduce aircraft noise has resulted in a general trend to higher operating pressures and temperatures in engines and increased production of NO_x for a given type of combustor technology. Combustor design changes can offset this problem to some extent but may result in increased complexity and weight. Different considerations may apply to the potential second generation of civil supersonic aircraft. The current status and potential changes in the technology of engines and aircraft themselves, with consequences for emissions, are discussed in Chapter 7.

Alternative fuels to aviation kerosene are being investigated, and some of these fuels have some attractive environmental characteristics. For instance, hydrogen offers the potential for eliminating direct CO_2 emissions, though at the expense of increased H_2O production. None of the alternatives appear capable of eliminating both CO_2 and H_2O emissions. The use of such fuels also would require the development and implementation of new technology and infrastructure, and many factors would need to be considered, including overall energy use, energy density, availability, cost, indirect impacts through production, and environmental benefits. These issues are also discussed in Chapter 7.

Future emissions from aviation will also be influenced by the manner in which aircraft are operated. At present, there is non-optimum use of airspace and ground infrastructure. However, advances in digital communications technology and satellite systems should allow new flight management procedures involving greater use of computerized air traffic control systems. In principle, such systems could lead to reductions in the lengths of routes between certain cities and higher traffic volumes in heavily flown corridors. More efficient routing would directly reduce fuel use and emissions. Economic and environmental benefits also might be enhanced through greater use of meteorological information. Changes in flight altitudes and speeds could occur as a result of new aircraft designs and operating procedures; these changes would result in aircraft emissions occurring at different altitudes. These factors are described in Chapter 8.

The framework within which technical and operational changes occur is influenced by government and industry, with aircraft safety the most important objective. Operator fleet decisions are influenced primarily by aircraft mission, performance, and operating cost, though aircraft technology and regulatory acceptance are significant parameters. Economic instruments such as fuel taxes and emissions charges affect an aircraft's operating costs. Aircraft operating limitations such as emission caps could directly affect capital investment as well as operating costs. On the other hand, new and emerging market mechanisms such as aircraft emissions

trading are policy instruments that could introduce flexibility into regulatory compliance schemes. Such issues are discussed in Chapter 10.

When evaluating possible options for limiting certain emissions in the future, it is important to keep a proper perspective. This report is the first detailed assessment of the global environmental effects of a single industrial sector. Air transport is only one of a number of transport modes that use fossil fuel, either directly or indirectly. Each of these modes may have specific advantages, globally or nationally. Other sources of greenhouse gases and other emissions also contribute to the composition of the global atmosphere and are likely to change with time. The environmental consequences of all emissions (transport and non-transport) and the economic impacts associated with different policy options will need to be balanced.

In this report, the relative importance of various aircraft emissions are assessed according to the best available knowledge of atmospheric effects and in the light of current knowledge of future technological options. Economic analyses will be required to investigate the consequences of possible mitigation strategies. Ideally, these analyses would take into account the wide range of activities in the aeronautics and aviation industries and assign monetary value to emissions and their effects. The current state of such economic analysis is discussed in Chapter 10.

References

Airbus, 1997: Confirming very large demand. In: *Global Market Forecast 1997–2016*. Airbus Industrie, Toulouse, France, 27 pp.

Boeing, 1997: World air travel demand and aeroplane supply requirements. In: *1997 Current Market Outlook*. Boeing Commercial Airplane Group, Seattle, WA, USA, 51 pp.

Brasseur, G.P., R.A. Cox, D. Hauglustaine, I. Isaksen, J. Lelieveld, D.H. Lister, R. Sausen, U. Schumann, A. Wahner, and P. Wiesen, 1998: European scientific assessment of the atmospheric effects of aircraft emissions. *Atmospheric Environment*, **32**, 2327–2422.

Friedl, R.R., S. Baughcum, B. Anderson, J. Hallett, K-N Liou, P. Rasch, D. Rind, K. Sassen, H. Singh, L. Williams, and D. Wuebbles, 1997: *Atmospheric Effects of Subsonic Aircraft: Interim Assessment of the Advanced Subsonic Technology Program*. NASA Reference Publication 1400, National Aeronautics and Space Administration, Washington, DC, USA, 168 pp.

IATA, 1994: *The Economic Benefits of Air Transport*. Air Transport Action Group, International Air Transport Association, Geneva, Switzerland, 32 pp.

IATA, 1996: *Environmental Review 1997*. International Air Transport Association, Geneva, Switzerland, 103 pp.

ICAO, 1981:*International Standards and Recommended Practices— Environmental Protection. Annex 16 to the Convention on International Civil Aviation, Volume II: Aircraft Engine Emissions*. International Civil Aviation Organization, Montreal, Canada, 1st ed.

ICAO, 1993a: *International Standards and Recommended Practices— Environmental Protection. Annex 16 to the Convention on International Civil Aviation, Volume I: Noise*. International Civil Aviation Organization, Montreal, Canada, 3rd ed.

ICAO, 1993b: *International Standards and Recommended Practices, Environmental Protection. Annex 16 to the Convention on International Civil Aviation, Volume II: Aircraft Engine Emissions*. International Civil Aviation Organization, Montreal, Canada, 2nd ed.

IPCC, 1992: *Climate Change 1992: The Supplementary Report to the IPCC Scientific Assessment.* Prepared by IPCC Working Group I [Houghton, J.T., B.A. Callander, and S.K. Varney (eds.)]. Cambridge University Press, Cambridge, United Kingdom and New York, NY, USA, 200 pp.

IPCC, 1995: *Climate Change 1994: Radiative Forcing of Climate Change and an Evaluation of the IPCC IS92 Emission Scenarios.* [Houghton, J.T., L.G. Meira Filho, J. Bruce, H. Lee, B.A. Callander, E. Haites, N. Harris, and K. Maskell (eds.)]. Cambridge University Press, Cambridge, United Kingdom and New York, NY, USA, 339 pp.

IPCC, 1996a: *Climate Change 1995: The Science of Climate Change. Contribution of Working Group I to the Second Assessment Report of the Intergovernmental Panel on Climate Change* [Houghton, J.T., L.G. Meira Filho, B.A. Callander, N. Harris, A. Kattenberg, and K. Maskell (eds.)]. Cambridge University Press, Cambridge, United Kingdom and New York, NY, USA, 572 pp.

IPCC, 1996b: *Climate Change 1995: Impacts, Adaptations and Mitigation of Climate Change: Scientific-Technical Analyses. Contribution of Working Group II to the Second Assessment Report of the Intergovernmental Panel on Climate Change* [Watson, R.T., M.C. Zinyowera, and R.H. Moss (eds.)]. Cambridge University Press, Cambridge, United Kingdom and New York, NY, USA, 880 pp.

IPCC, 1997a: *Stabilization of Atmospheric Greenhouse Gases: Physical, Biological and Socio-economic Implications: Technical Paper III.* Intergovernmental Panel on Climate Change Working Group I [Houghton, J.T., L.G. Meira Filho, D. Griggs, and K. Maskell (eds.)]. World Meteorological Organization, Geneva, Switzerland, 52 pp.

IPCC, 1997b: *Implications of Proposed CO_2 Emissions Limitations: Technical Paper 4.* Intergovernmental Panel on Climate Change Working Group I [Houghton, J.T., L.G. Meira Filho, D.J. Griggs, and K. Maskell, (eds.)]. World Meteorological Organization, Geneva, Switzerland, 52 pp.

Schimel, D., I.G. Enting, M. Heimann, T.M.L. Wigley, D. Raynaud, D. Alves, and U. Siegenthaler, 1995: CO_2 and the carbon cycle. In: *Climate Change 1994: Radiative Forcing of Climate Change and an Evaluation of the IPCC IS92 Emission Scenarios.* [Houghton, J.T., L.G. Meira Filho, J. Bruce, H. Lee, B.A. Callander, E. Haites, N. Harris, and K. Maskell (eds.)]. Cambridge University Press, Cambridge, United Kingdom and New York, NY, USA, pp. 39–71.

Schimel, D., D. Alves, I. Enting, M. Heimann, F. Joos, D. Raynaud, T. Wigley, M. Prather, R. Derwent, D. Ehhalt, P. Fraser, E. Sanhueza, X. Zhou, P. Jonas, R. Charlson, H. Rodhe, S. Sadasivan, K.P. Shine, Y. Fouquart, V. Ramaswamy, S. Solomon, J. Srinivasan, D. Albritton, I. Isaksen, M. Lal, and D. Wuebbles, 1996a: Radiative forcing of climate change. In: *Climate Change 1995: The Science of Climate Change. Contribution of Working Group I to the Second Assessment Report of the Intergovernmental Panel on Climate Change* [Houghton, J.T., L.G. Meira Filho, B.A. Callander, N. Harris, A. Kattenberg, and K. Maskell (eds.)]. Cambridge University Press, Cambridge, United Kingdom and New York, NY, USA, pp. 65–131.

Schumann, U., A. Chlond, A. Ebel, B. Kärcher, H. Pak, H. Schlager, A. Schmitt, and P. Wendling (eds.), 1997: Pollutants from air traffic—results of atmospheric research 1992–1997. In: *Final Report on the BMBF Verbundprogramm, Schadstoffe in der Luftfahrt.* DLR-Mitteilung 97-04, Deutsches Zentrum für Luft- und Raumfahrt (German Aerospace Center), Oberpfaffenhofen and Cologne, Germany, 301 pp.

Stolarski, R.S., S. Baughcum, W. Brune, A. Douglass, D. Fahey, R. Friedl, S. Liu, A. Plumb, L. Poole, H. Wesoky and D. Worsnop, 1995: *Scientific Assessment of the Atmospheric Effects of Stratospheric Aircraft.* NASA Reference Publication 1381, National Aeronautics and Space Administration, Washington, DC, USA, 110 pp.

WMO, 1999: *Scientific Assessment of Ozone Depletion: 1998.* Global Ozone Research and Monitoring Project, Report No. 44, World Meteorological Organization, Geneva, Switzerland, 732 pp.

2

Impacts of Aircraft Emissions on Atmospheric Ozone

RICHARD DERWENT AND RANDALL FRIEDL

Lead Authors:
I.L. Karol, H. Kelder, V.W.J.H. Kirchhoff, T. Ogawa, M.J. Rossi, P. Wennberg

Contributors:
T. Berntsen, C. Brühl, D. Brunner, P. Crutzen, M. Danilin, F. Dentener, L. Emmons,
F. Flatoy, J.S. Fuglestvedt, T. Gerz, V. Grewe, D.A. Hauglustaine, G. Hayman,
Ø. Hov, D. Jacob, C. Johnson, M. Kanakidou, B. Kärcher, D. Kinnison,
A.A. Kiselev, I. Köhler, J. Lelieveld, J.A. Logan, J.-F. Müller, J.E. Penner, H. Petry,
G. Pitari, R. Ramaroson, F. Rohrer, E.Z. Rozanov, K. Ryan, R.J. Salawitch,
R. Sausen, U. Schumann, F. Slemr, D. Stevenson, F. Stordal, A. Strand,
A. Thompson, P. Valks, P. van Velthoven, G. Velders, Y. Wang, W. Wauben,
D. Weisenstein

Review Editor:
A. Wahner

CONTENTS

EXECUTIVE SUMMARY

- Aircraft emit a number of chemically active species that can alter the concentration of atmospheric ozone. The species with the greatest potential impact are nitric oxide (NO) and nitrogen dioxide (NO_2) (collectively termed NO_x), sulfur oxides, water, and soot.

- Ozone concentrations in the upper troposphere and lowermost stratosphere are expected to increase in response to NO_x increases and decrease in response to sulfur and water increases. At higher altitudes, increases in NO_x lead to decreases in ozone.

- Soot surfaces destroy ozone and possibly convert nitric acid to NO_x. However, because atmospheric soot reactions are highly unlikely to be catalytic and because ambient soot concentrations are low, the effect on ambient ozone is expected to be negligible.

- There is no direct observational evidence that aircraft emissions have altered ozone. Geographical and temporal variations in observed ozone trends for the upper troposphere and lowermost stratosphere are inconsistent with a major perturbation by aircraft. The largest increase in ozone from aircraft emissions is predicted to occur near the tropopause, where ozone variability is high. This variability and the presence of other factors inducing ozone change make it difficult to discern an aircraft effect at the predicted level in the existing ozone database.

- Aircraft emissions are calculated to have increased NO_x at cruise altitudes in northern mid-latitudes by approximately 20%. The uncertainty in this calculation is primarily related to uncertainties in the NO_x chemical lifetime and in the relative magnitude of the aircraft source compared to lightning, rapid vertical convection of surface NO_x, and other sources of upper tropospheric NO_x. The calculated increase is substantially smaller than the observed variability in NO_x.

- NO_x emissions from current aircraft are calculated to have increased ozone by about 6% in the region 30–60°N latitude and 9–13 km altitude. Calculated total ozone column changes in this latitude range are approximately 0.4%. Calculated effects are substantially smaller outside this region. Some of the uncertainty in these calculations is captured by the range of model results. However, the models are notably deficient in coupling representations of stratospheric and tropospheric chemistry and in describing exhaust plume processes, HO_x sources, and non-methane chemistry in the upper troposphere. In addition, there is high uncertainty associated with the model description of vertical and horizontal transport in the upper troposphere/ lower stratosphere.

- The effect of current aircraft particle and particle precursor emissions (i.e., soot, sulfur, and water) in the stratosphere on ozone is estimated to be smaller than, and of opposite sign to, the NO_x effect. Model representations of aerosol microphysics and chemistry are, however, largely incomplete.

- Aircraft-related increases in NO_x in the upper troposphere are calculated to increase the concentration of hydroxyl (OH) radicals by a few percent throughout the Northern Hemisphere. The OH change results in a corresponding decrease in the concentration of methane (CH_4). Uncertainties in the global budget of CH_4 and the factors that control OH preclude testing this calculation with atmospheric observations. Because the chemical processes that lead to the reduction of CH_4 are the same processes that increase ozone, calculated CH_4 and ozone effects are correlated.

2.1. Relating Aircraft Emissions to Atmospheric Ozone

2.1.1. Introduction

The chemical products of aircraft jet fuel combustion are emitted at the engine nozzle exit plane as part of a high-velocity plume. This gaseous and particulate stream is subject to chemical and dynamical processes that influence downstream composition. Eventually, plume constituents irreversibly mix with, and are diluted by, ambient air. Subsequently, some of the emitted species act in concert with other natural and anthropogenic chemicals to change ozone abundances in the Earth's atmosphere. The ultimate fates of these aircraft-derived species are determined by larger-scale chemical and transport processes.

Concerns about NO and NO_2 (i.e., NO_x) emissions from present-generation subsonic and supersonic aircraft operating in the upper troposphere (UT) and lower stratosphere (LS) were raised more than 20 years ago by Hidalgo and Crutzen (1977) because these emissions could change ozone levels locally by several percent or so. Despite extensive research and evaluation during the intervening years, WMO-UNEP (1995) concluded that assessments of ozone changes related to aviation remained uncertain and depended critically on NO_x chemistry and its representation in complex models. Because of large uncertainties in present knowledge of the tropospheric NO_x budget, little confidence has been placed in previous assessments of quantitative model results of subsonic aircraft effects on atmospheric ozone. Assessment tools and their input data continue to improve, however, and reconsideration is appropriate in the light of the extensive research results published since the WMO-UNEP (1995) assessment.

The research results published since the WMO-UNEP (1995) assessment have addressed a number of issues relevant to the assessment of ozone impacts of present aviation. These issues have included the development of improved aircraft NO_x emission inventories, updating of evaluated chemical kinetic and photochemical databases, studies of aircraft plume chemistry, and the development of three-dimensional (3-D) modeling tools. Reviews have also been published of U.S. (Friedl, 1997) and European (Schumann *et al.*, 1997; Brasseur *et al.*, 1998) research programs addressing ozone and other environmental impacts of present aviation.

In this chapter we evaluate, from a qualitative and quantitative standpoint, the impact on atmospheric ozone of aircraft exhaust species, emitted either directly from engines or produced as secondary products of processes occurring in aircraft plumes. Our evaluation is based primarily on global model calculations rather than ozone trends because expected changes are not easily discerned from observations, as discussed below. We use intermodel comparisons and atmospheric observations of ozone to test the physics and chemistry parameterized in these global models and identify areas of remaining uncertainty.

2.1.1.1. Aircraft Engine Emissions

Most present-day jet aircraft cruise in an altitude range (9–13 km) that contains portions of the UT and LS. Because these two atmospheric regions are characterized by different dynamics and photochemistry, the placement of aircraft exhaust into these regions must be considered when evaluating the impact of exhaust species on atmospheric ozone. Determination of the partitioning of exhaust into the two atmospheric regions is complicated by the highly variable and latitudinally dependent character of the tropopause (i.e., the transition between the stratosphere and troposphere). Comparisons of aircraft cruise altitudes with mean tropopause heights has led to estimates for stratospheric release of 20–40% of total emissions (Hoinka *et al.*, 1993; Baughcum, 1996; Schumann, 1997; Gettleman and Baughcum, 1999).

Carbon dioxide (CO_2) and water vapor (H_2O) are easily the most abundant products of jet fuel combustion (emission indices for CO_2 and H_2O are 3.15 kg/kg fuel burned and 1.26 kg/kg fuel, respectively). However, both species have significant natural background levels in the UT and the LS (Schumann, 1994; WMO-UNEP, 1995). and neither current aircraft emission rates nor likely future subsonic emission rates will affect the ambient levels by more than a few percent. Future supersonic aviation, on the other hand (which would emit at higher altitudes), could perturb ambient H_2O levels significantly at cruise altitudes. Regardless of the magnitude of the aircraft source, CO_2 does not participate directly in ozone photochemistry because of its thermodynamic and photochemical stability. It may participate indirectly by affecting stratospheric cooling, which can in turn lead to changes in atmospheric thermal stratification, increased polar stratospheric cloud (PSC) formation, and reduced ozone concentrations.

Aircraft water contributions, although relatively small in the troposphere, lead to the atmospheric phenomenon of contrail formation. Depending on the precise composition of contrail particles—which is largely determined by the specific processes occurring in the aircraft plume and by the ambient atmosphere composition and temperature—the particles may act as surfaces for a variety of heterogeneous reactions (Kärcher *et al.*, 1995; Louisnard *et al.*, 1995; WMO-UNEP, 1995; Schumann *et al.*, 1996; Danilin *et al.*, 1997; Kärcher, 1997; Karol *et al.*, 1997). The participation of contrails in atmospheric photochemistry is further addressed in Section 2.1.3.

NO_x constitutes the next most abundant engine emission (emission indices range from 5 to 25 g of NO_2 per kg of fuel burned) (Fahey *et al.*, 1995; WMO-UNEP, 1995; Schulte and Schlager, 1996; Schulte *et al.*, 1997). With respect to ozone photochemistry, NO_x is the most important and most studied component; its aircraft emission rates are sufficient to affect background levels in the UT and LS. Moreover, its active role in ozone photochemistry in the UT and LS has been well recognized (WMO-UNEP, 1985, 1995). A great deal of the recent scientific literature has focused on aircraft NO_x effects, and this chapter neccessarily reflects that focus.

Aircraft carbon monoxide (CO) emissions are of the same order of magnitude as NO_x emissions (i.e., 1–2 g kg^{-1} for the Concorde aircraft and 1–10 g kg^{-1} for subsonic aircraft) (Baughcum *et al.*, 1996). Like NO_x, CO is a key participant in tropospheric ozone production. However, natural and non-aircraft anthropogenic sources of CO are substantially larger than analogous NO_x sources, thereby reducing the role of aircraft CO emissions in ozone photochemistry to a level far below that of aircraft NO_x emissions (WMO-UNEP, 1995).

Emissions of sulfur dioxide (SO_2) and hydrocarbons from aircraft, at less than 1 g kg^{-1} fuel, are significantly less than the more prominent exhaust species discussed above (Spicer *et al.*, 1994; Slemr *et al.*, 1998). Their primary potential impacts are related to formation of sulfate and carbonaceous aerosols that may serve as sites for heterogeneous chemistry. This possibility is discussed in Section 2.1.3. Non-methane hydrocarbon (NMHC) emissions may also contribute to autocatalytic production of HO_x, provided that the reactivity of the NHMCs is sufficiently large relative to that of CH_4 to overcome their numerical inferiority. However, model studies have indicated that volatile organic emissions from aircraft have an insignificant impact on atmospheric ozone at cruise altitudes (Hayman and Markiewicz, 1996; Pleijel, 1998).

2.1.1.2.   *Plume and Wake Processing of Engine Emissions*

Although jet exhaust spends a relatively short time in the immediate vicinity behind the aircraft, a number of important processes occur during that time that influence exhaust gas and aerosol composition, hence the ozone-forming or ozone-depleting potential of the exhaust. The near-field evolution of jet aircraft exhaust wake can be divided into three distinct regimes—commonly termed jet, vortex, and plume dispersion. The time scales associated with these regimes are 0–10 s for the jet, 10–100 s for the vortex, and 100 s to tens of hours for plume dispersion—the latter time period defining the effective "lifetime" of the aircraft plume. The jet and vortex regimes are closely related; they are initiated at the exit plane of the engine nozzle and ended by atmospheric shear forces at distances of approximately 10–20 km behind the aircraft (Hoshizaki, 1975; Schumann, 1994).

Several fluid dynamic models are now available to study wake dynamics—namely, two-dimensional (2-D) jet mixing codes (Miake-Lye *et al.*, 1993; Beier and Schreier, 1994; Kärcher, 1994; Garnier *et al.*, 1996) and codes that capture the jet/vortex interaction and vortex break-up (Quackenbush *et al.*, 1993; Lewellen and Lewellen, 1996; Schilling *et al.*, 1996), some of them using vortex filament methods combined with large eddy simulations (LES) (Gerz and Ehret, 1997).

The small spatial and temporal scales of exhaust species distributions in near-field wakes hamper a robust comparison of model simulations with *in situ* observations of exhaust effluents. Nevertheless, the dynamic models have been successful in explaining the few observations of near-field tracer concentration,

temperature, and humidity (Anderson *et al.*, 1996; Garnier *et al.*, 1996; Gerz and Ehret, 1997; Gerz and Kärcher, 1997). The data and calculations reveal a strong suppression of plume mixing and dispersion during the vortex regime. Vortex systems are composed of cylindrical core regions, not well mixed radially and entraining only small amounts of ambient air. As a result, vortex plume temperatures and associated H_2O concentrations may be well defined from fluid dynamic simulations and known emission indices. Within the vortex, high concentrations of exhaust species interact with each other and with small amounts of ambient gases and particles over a range of temperatures that differ from those in the background atmosphere. It is likely that some of the chemical interactions occurring in the vortex regime will influence the eventual composition of aircraft-derived aerosol particles and gases.

The plume dispersion regime begins after disintegration of the wake vortex and extends to an area where the primary exhaust gas concentrations (i.e., NO_x, H_2O, CO, CO_2) are of the same order of magnitude as the corresponding ambient background levels. Results from modeling and observational studies of aged plumes (Karol *et al.*, 1997; Meijer *et al.*, 1997; Schlager *et al.*, 1997; Schumann *et al.*, 1998) show that most plumes mix with the background atmosphere according to a simple dilution law that can be approximated with a Gaussian plume model that includes estimated and measured atmospheric shear and diffusion parameters (Konopka, 1995; Schumann *et al.*, 1995; Durbeck and Gerz, 1996). The key observables for these models have been ice particles in visible contrails and measured CO_2 that serve as tracers of the plume mixing process. All of the studies have indicated that during the 10–20 hrs of plume dispersion, the plume cross-section may grow to 50–100 km in width and 0.3–1.0 km in height, with a corresponding exhaust species dilution ratio (R—the ratio of the plume mass to fuel mass) up to 10^8 as a result of ambient air entrainment. From analysis of more than 70 aircraft plume crossings by research aircraft in the North Atlantic flight corridor, Schumann *et al.* (1998) proposed that R can be approximated by R=7000 $(t/t_0)^{0.8}$, where $t_0 = 1$s for $0.006 < t < 10^4$ s. The relative rate, (dR/dt)/R, of ambient air entrainment into the plume is on the order of 10^{-3} s^{-1} in the first minutes of plume dispersion but decreases to on the order of 10^{-4} s^{-1} over a 1–2 hr period (Durbeck and Gerz, 1996).

In the plume dispersion stage, aircraft-derived gas and particle concentrations are still highly elevated over background levels, but they interact with large volumes of ambient species under temperature and pressure conditions of the background atmosphere. The composition and reactive characteristics of aircraft-derived particles fully evolve in the vortex and plume dispersion regions as a result of aerosol-precursor photochemistry and particle condensation, coagulation, and agglomeration processes. These particle-forming processes are described in further detail in Chapter 3. In addition, chemistry process model calculations indicate that a significant fraction of emitted NO_x is converted to other reactive nitrogen (NO_y) species in the plume dispersion region during the daylight (Karol *et al.*,

1997; Meijer *et al*., 1997; Petry *et al*., 1998). Observations of NO_y in aircraft plume compositions are consistent with these results (Schlager *et al*., 1997).

2.1.1.3. Atmospheric Ozone

Most interactions between ambient ozone and ozone-controlling gases and aircraft exhaust occur in the days and weeks following emission. Dispersion of exhaust on regional and global scales is dictated by the same large-scale atmospheric dynamic processes that control mixing of other natural and anthropogenic sources of gases and particles. During that time, aircraft-derived gases and particles participate in the natural chemical cycles that control ambient ozone levels. The following subsections provide an overview of ozone chemistry.

2.1.1.3.1. Stratospheric ozone

Approximately 80% of atmospheric ozone resides in the stratosphere, where it is produced via *in situ* photochemistry occurring predominantly in the tropical middle stratosphere, albeit with significant contributions from mid-latitudes. Stratospheric circulation patterns transport ozone from the tropical stratosphere poleward and then downward from the mid-stratosphere predominantly in the winter hemisphere. Stratospheric ozone is not only transported but also destroyed via photochemical reactions over the whole stratosphere. In addition, about 7–25% of the total ozone mass (WMO-UNEP, 1985; Wauben *et al*., 1998) is transported to the extratropical troposphere; this type of transport occurs most intensively in the winter and in the Northern Hemisphere. Different transport modes correspond to different time scales, ranging from days to years.

Ozone formation and destruction rates increase with height and change with latitude in the stratosphere. Consequently, ozone "lifetime" decreases with height from about a year in the LS to minutes in the upper stratosphere. At the uppermost altitudes, ozone lifetime is sufficiently short that its abundance is in local photochemical equilibrium (WMO-UNEP, 1985).

At lower altitudes, ozone is not in photochemical steady-state, and ozone transport by air motions of various scales becomes increasingly important. The primary mechanism for mean global stratospheric transport is referred to as the Brewer-Dobson circulation, with rising motion in the equatorial belt of the LS and air mass spreading to the poles in the middle and upper stratosphere, with more intensive transport into the winter hemisphere.

In summary, stratospheric ozone distributions are determined mainly by atmospheric motions in the nightime polar regions, by a mixture of transport and photochemistry in the lower and middle stratosphere, and by photochemistry in the upper stratosphere.

2.1.1.3.2. Tropospheric ozone

Sources of ozone in the troposphere are more numerous than in the stratosphere, as are the photochemical reactions participating in ozone production and loss. Although *in situ* photochemistry is the dominant source of tropospheric ozone, downward flux of stratospheric ozone represents a significant source, especially in the UT and in winter over high latitudes. Removal of tropospheric ozone occurs predominantly by photochemistry, with some contribution from surface deposition.

The lifetime of ozone in the troposphere varies with latitude and altitude; in general it is of the order of 1 month, a value that is smaller than the time scale for transport between the Northern and Southern Hemisphere troposphere, which is typically about 1 year (WMO-UNEP, 1985).

2.1.1.3.3. Stratosphere/troposphere exchange

As mentioned in Section 2.1.1.1, emissions from present aviation are injected near the tropopause. Dynamic, chemical, and radiative coupling between the stratosphere and troposphere are among the important processes that must be understood if we are to provide an adequate description and prediction of the impact of aviation on atmosphere and climate. Of special significance is the exchange of chemical species between the stratosphere and the troposphere. In the stratosphere, large-scale transport takes place via the Brewer-Dobson circulation, induced by momentum deposition by planetary gravity waves. This circulation is responsible for the observed difference between the stratospheric temperature and its radiative equilibrium value. However, this exchange involves a wide spectrum of scales ranging from large-scale ascent and descent via synoptic scales toward transport by waves, convection, and turbulence (Brewer, 1949; Holton *et al*., 1995; McIntyre, 1995). For the impact of subsonic aviation, the focus is on the exchange between the troposphere and the lowermost stratosphere (Hoskins *et al*., 1985). This part of the stratosphere is strongly coupled with the troposphere and is separated from the LS by a region enclosed between the 380 K and 400 K potential temperature surfaces (Holton *et al*., 1995).

For the transport, it is useful to distinguish between different regions of the globe:

- In the tropics, upward transport occurs mostly through deep convection, though small-scale vertical mixing by gravity waves might also play an important role. The tropics are the regions on Earth in which the largest net upward transport into the stratosphere occurs and which therefore directly influence the composition of the global middle stratosphere. Mixing between the tropical and mid-latitude lower stratosphere is influenced by the subtropical barrier. Between the tropical UT and subtropical LS, however, transport along isentropic surfaces is important (Minschwaner *et al*., 1996; Plumb, 1996; Volk *et al*., 1996).

- In mid-latitudes, the exchange between the troposphere and lowermost stratosphere flows in both directions, with a somewhat larger downward component (Siegmund *et al.*, 1996). Most of this transport is related to the occurrence of extra-tropical cyclones and blocking anticyclones. In cyclones, polar stratospheric air is drawn into the troposphere while subtropical tropospheric air is drawn into the stratosphere. The intermediate process of tropopause folding is followed by re-establishment of the tropopause. There is evidence that upward and lateral mixing of tropospheric air into the stratosphere remains limited to the lowest few kilometers of the mid-latitude LS (Dessler *et al.*, 1995; Boering *et al.*, 1996; Hintsa *et al.*, 1998).

- Around the polar vortices, the exchange in the stratosphere occurs along isentropic surfaces, from the polar vortex toward mid-latitudes, by filaments torn off from the vortex boundary. Vertical transport in the vortex itself mostly takes place in the form of large-scale descent caused by radiative cooling. Horizontal transport on the equatorward flank of the polar night jet is sharply coupled with vertical transport associated with diabatic descent caused by radiative cooling of warm air within the descending branch of the baroclinic circulation (Pierce *et al.*, 1993, 1994).

2.1.2. Effects of Aircraft Gaseous Emissions

As discussed above, the concentration of ozone is determined by transport from other locations and by local production and loss. Production and destruction rates of ozone are strongly influenced by the concentration of the free-radical catalysts NO and NO_2 (NO_x) and OH and HO_2 (HO_x). In the presence of NO, ozone is produced as a by-product when CO, CH_4, and other hydrocarbons are oxidized by OH. NO_x also influences the destruction rate of ozone directly as a catalyst (in the stratosphere) and indirectly as a result of reactions that couple NO_x to other reactive species such as the odd hydrogen radicals, OH and HO_2, and, in the stratosphere, the halogen free radicals chlorine monoxide (ClO) and bromine monoxide (BrO).

2.1.2.1. Production of Ozone

The production of ozone in the stratosphere is dominated by the photolysis of oxygen (O_2) by sunlight. Because radiation with short wavelengths (less than 242 nm) is screened out by O_2 and ozone in the upper atmosphere, this process is not very important in the troposphere. In the region where commercial aircraft fly, ozone (O_3) is produced mainly from the oxidation of CO:

$$OH + CO \rightarrow H + CO_2 \qquad (1)$$
$$H + O_2 + M \rightarrow HO_2 + M \qquad (2)$$
$$HO_2 + NO \rightarrow NO_2 + OH \qquad (3)$$
$$NO_2 + sunlight \rightarrow NO + O \qquad (4)$$
$$O + O_2 + M \rightarrow O_3 + M \qquad (5)$$

Net: $CO + 2 O_2 \rightarrow CO_2 + O_3$

(where M represents a gaseous third body such as N_2 or O_2). The oxidation of CH_4 also contributes to ozone formation but is less important in the UT than reaction 1.

The rates of these reactions depend directly on the concentrations of NO_x and HO_x. Increases in the concentration of NO_x from aircraft generally will increase the rate of ozone production by speeding the oxidation of CO and CH_4. Other aircraft emissions, such as H_2O, CO, and NMHCs, are not expected to significantly affect this background chemistry because natural sources of these compounds far exceed the perturbation from aviation (Friedl, 1997; Brasseur *et al.*, 1998). Within the plume, however, the production of particulate and subsequent contrail and cloud formation may influence NO_x and HO_x and therefore the production rate of ozone (see Section 2.1.3).

2.1.2.2. Sources and Sinks of NO_x

Aircraft emissions are one of many sources of NO_x in the troposphere and the stratosphere. In the stratosphere, NO_x is produced primarily from the oxidation of nitrous oxide (N_2O). N_2O is produced by numerous sources, and its concentration has been increasing at a rate of 0.5–0.8 ppbv yr^{-1} (IPCC, 1996). As a result of this source, NO_x concentrations in the LS are quite large, increasing from about 100 pptv at the tropopause to as much as 3000 pptv at an altitude of 20 km. Aircraft exhaust provides an additional source of NO_x, but there is no evidence that this source has appreciably altered the concentration of stratospheric NO_x.

The primary sources of NO_x in the troposphere are fossil fuel combustion, biomass burning, soil emissions, lightning, transport from the stratosphere, ammonia oxidation, and aircraft exhaust. The largest source is fossil fuel combustion; 95% of its emissions are in the Northern Hemisphere (Lee *et al.*, 1997). Biomass burning occurs primarily in the continental tropics. Soil emissions come from microbial denitrification and nitrification processes, the rate depends on soil type and temperature, ecosystem type, water content, and several other variables (Matthews, 1983; Müller, 1992; Williams *et al.*, 1992; Yienger and Levy, 1995). The contribution of NO_x produced by lightning is very uncertain because it is extremely difficult to measure directly. The distribution around thunderstorms is highly variable because NO production differs for cloud-to-cloud and cloud-to-ground strikes. Oxidation of ammonia (NH_3) of NO occurs at the surface, mainly in the tropics. Although the sources of NH_3 are fairly well known, the rates of reactions that result in NO are uncertain (Lee *et al.*, 1997), and, under some conditions, ammonia oxidation can be a sink for NO_x.

Various studies have estimated the emission of NO_x from these sources (Müller, 1992; Atherton *et al.*, 1996; Lee *et al.*, 1997). Additionally, estimates for individual sources have been produced: fossil fuel combustion (Dignon, 1992; Benkovitz *et al.*, 1996); biomass burning (Crutzen and Andreae, 1990; Müller, 1992; Atherton, 1995); soil emissions (Matthews, 1983; Williams *et al.*, 1992; Yienger and Levy, 1995); lightning (Price and Rind,

1992; Levy *et al.*, 1996; Ridley *et al.*, 1996); aircraft exhaust (Wuebbles *et al.*, 1993; Metwally, 1995; Baughcum *et al.*, 1996; Gardner *et al.*, 1997); and transport from the stratosphere (Kasibhatla *et al.*, 1991; Wang *et al.*, 1998a).

Compilations and evaluations of emission rates (Friedl, 1997; Lee *et al.*, 1997) have emphasized that the contribution from lightning remains highly uncertain. The relative impact of NO_x produced from aviation is critically dependent on the strength of this source. Despite this uncertainty, it is likely that averaged over the globe, the lightning source of NO_x in the UT is 2 to 8 times as large as the aircraft source. Although lightning, aircraft emissions, and transport of NO_x from the boundary layer during deep convection clearly provide significant sources of NO_x to the UT, the contribution of each in determining the total NO_x concentration is not well understood.

In situ observations have been made in the UT to help identify and quantify sources of NO_x. The concentration of NO measured in these field campaigns is extremely variable. Based on a number of *in situ* aircraft missions, however, typical NO concentrations in the UT are in the range 50–200 pptv (Emmons *et al.*, 1997; Schlager *et al.*, 1997; Tremmel *et al.*, 1998). Because of the large variability in the atmosphere, these relatively sparse observations cannot quantitatively define the strengths of the various NO_x sources in the UT. Recent aircraft campaigns—for example, SONEX, POLINAT-2, NOXAR—should, however, provide important constraints on the role of aircraft in perturbing NO_x. For example, a full year of observations of NO in the UT and lowermost stratosphere obtained onboard a commercial airplane have demonstrated the importance of NO_x sources associated with convection (Brunner *et al.*, 1998).

The NO_x species belong to a family of chemicals collectively called reactive nitrogen (NO_y). This family includes species that are coupled to NO_x on time scales of one day to several weeks, such as HNO_4, nitrogen pentoxide (N_2O_5), nitric acid (HNO_3), and, in the troposphere, peroxyacetylnitrate (PAN). The primary form of reactive nitrogen emitted into the atmosphere is NO. Once NO is in the atmosphere, it undergoes rapid interconversion with NO_2 and then slower conversion to other reactive nitrogen species such as HNO_3.

Reactions occurring on aerosol surfaces also play an important role in the chemical partitioning of NO_y. It is well-established that NO_x concentrations are reduced and HNO_3 concentrations are enhanced as a result of heterogeneous chemistry on aerosol surfaces. It is also possible that reactions occurring on soot may play the opposite role, converting HNO_3 back to NO_x (Section 2.1.3).

Observations of NO_y species have demonstrated that our understanding of the sources of NO_y and the partitioning of NO_y in the stratosphere is relatively good (within ± 30%) (Rinsland *et al.*, 1996; Gao *et al.*, 1997). Photochemical models of the partitioning and distribution of NO_y in the UT, however, have typically demonstrated poor agreement with observations. Because the reactions that convert NO_x, PAN, and HNO_3 occur

on time scales similar to transport, it is not clear whether the observed differences are caused by errors in the chemical description or by dynamic influences that are not captured in the models (Prather and Jacob, 1997).

2.1.2.3. Sources and Sinks of OH and HO₂

HO_x is produced in the UT and LS via a number of processes. One important primary source of OH is the reaction of water vapor with $O(^1D)$ produced in the photolysis of ozone:

$$O_3 + \text{sunlight} \rightarrow O(^1D) + O_2 \tag{6a}$$
$$O(^1D) + H_2O \rightarrow OH + OH \tag{7}$$

Additional sources of OH and HO_2 in the troposphere include the photolysis of acetone (Singh *et al.*, 1995), peroxides (Chatfield and Crutzen, 1984), and formaldehyde. In the stratosphere, the photolysis of HOBr and HNO_3 produced in reactions occurring on aerosols may also be important HO_x sources (Hanson *et al.*, 1996). The existence of important primary HO_x sources in addition to reactions 6a and 7 has been demonstrated by recent measurements of OH and HO_2 in the UT and LS (McKeen *et al.*, 1997; Jaeglé *et al.*, 1997, 1998; Brune *et al.*, 1998; Wennberg *et al.*, 1998). These primary sources of OH can be further amplified during the oxidation of CH_4 and other hydrocarbons.

OH is chemically coupled to HO_2 by reactions such as 1–3 that convert these species on time scales of seconds to minutes. Removal of HO_x (leading to the formation of H_2O) occurs on a longer time scale (5–30 min) and is dominated by processes such as:

$$OH + HO_2 \rightarrow H_2O + O_2 \tag{8}$$

$$OH + NO_2 + M \rightarrow HNO_3 + M \tag{9}$$
$$OH + HNO_3 \rightarrow H_2O + NO_3 \tag{10}$$
$$\text{net: } OH + OH + NO_2 \rightarrow H_2O + NO_3$$

$$HO_2 + NO_2 + M \rightarrow HNO_4 + M \tag{11}$$
$$OH + HNO_4 \rightarrow H_2O + NO_2 + O_2 \tag{12}$$
$$\text{net: } OH + HO_2 \rightarrow H_2O + O_2$$

$$HO_2 + HO_2 + M \rightarrow H_2O_2 + O_2 + M \tag{13}$$
$$OH + H_2O_2 \rightarrow H_2O + HO_2 \tag{14}$$
$$\text{net: } OH + HO_2 \rightarrow H_2O + O_2$$

NO_x and HO_x are therefore linked by a number of important reactions (3,9,10–12); the concentration of each depends on the concentration of the other.

2.1.2.4. The Response of Ozone to NOₓ in the Upper Troposphere

Figure 2-1 illustrates how the background photochemistry works to change ozone in the UT in response to variation in the

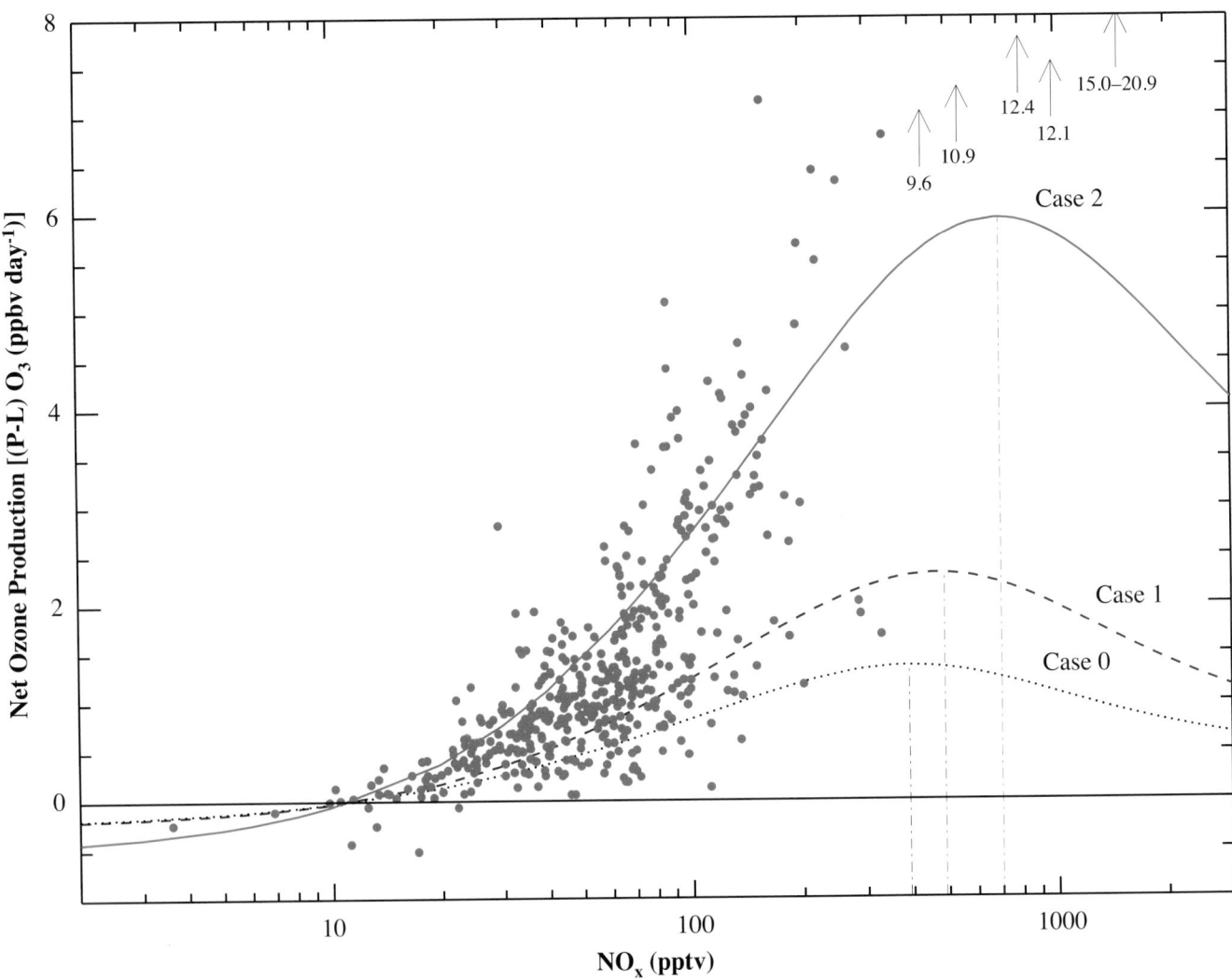

Figure 2-1: Net ozone production (24-hr average) as a function of NO_x in the upper troposphere (adapted from Jaeglé *et al.*, 1998). During the NASA-sponsored SUCCESS campaign (April–May 1996), simultaneous measurements of HO_2 and NO were obtained from the NASA DC-8 aircraft. These observations define the rate of ozone production via the chemistry outlined in Section 2.1.2.1. Also shown in this figure are three calculations for average tropospheric conditions experienced above 11 km during SUCCESS. Case 0 illustrates the production rate expected if the only primary source of HO_x is the reaction of O^1D with H_2O and CH_4. Case 1 is the rate calculated by assuming acetone is present at 510 ppbv, consistent with recent airborne measurements (Singh *et al.*, 1995; Arnold *et al.*, 1997). Case 2 assumes that a convective source of peroxides and formaldehyde provides additional HO_x production. These non-traditional HO_x sources dramatically increase the ozone production rate in the dry (<100 ppmv H_2O) upper troposphere.

concentration of NO_x (Jaeglé *et al.*, 1998). The points are calculated from *in situ* measurements of HO_x and NO obtained during the SUCCESS campaign. The lines show how the calculated response varies with assumptions about HO_x source strength. The response includes ozone production via reactions 4 and 5 and ozone destruction primarily via:

$$O_3 + HO_2 \rightarrow OH + 2\,O_2 \qquad (15)$$

In all models, the net ozone production rate (production rate - loss rate) increases rapidly with NO_x until a maximum is reached. At NO_x concentrations larger than 500 pptv, the net rate of ozone production is expected to decrease with increasing NO_x. Depending on the background concentration of NO_x,

additions of NO_x from aviation can increase or decrease the net ozone production rate. Thus, the background concentration of NO_x determines both the magnitude and the sign of the perturbation. Field measurements of NO_x in the middle and upper troposphere typically have found NO_x to be 50–200 pptv (Emmons *et al.*, 1997; Schlager *et al.*, 1997; Tremmel *et al.*, 1998). At these concentrations, the rate of net ozone production increases almost linearly with NO_x.

Figure 2-1 also illustrates that the rate of ozone production depends critically on HO_x source strength. In this figure, the calculation in Case 0 assumes that the primary source of HO_x is limited to the reaction of O^1D with H_2O and CH_4. This assumption is made in many (but not all) of the chemical transport models used to assess the effect of aviation in Section 2.2.1 and Chapter 4. Case 1 includes an additional source of HO_x from acetone photolysis (Singh *et al.*, 1995), assuming acetone is present at a concentration roughly consistent with recent measurements in the UT (Singh *et al.*, 1995; Arnold *et al.*, 1997). Case 2 assumes that in addition to acetone, peroxides and formaldehyde are transported to the UT by convection (Chatfield and Crutzen, 1984; Jaeglé *et al.*, 1997). It is clear from observations of HO_x obtained from the DC-8 (Brune *et al.*, 1998) and the ER-2 (Wennberg *et al.*, 1998) that HO_x sources in addition to $O^1D + H_2O$ are needed to explain measured concentrations of OH and HO_2. The effect of these HO_x sources is most pronounced in UT air when the water vapor mixing ratio is less than 100 ppmv. At median NO_x concentrations observed during these campaigns—50–100 pptv, typical of the UT (Brunner *et al.*, 1998)—the net ozone production rate is calculated to be 1–2 ppbv per day. This rate is significantly faster than would be calculated assuming only the simple O^1D HO_x chemistry.

The sensitivity of the net ozone production rate to assumptions about the sources of odd hydrogen is high and remains an area of significant uncertainty. The budget of acetone, for example, is poorly understood, and relatively few measurements of its concentration have been made in the UT. Observational constraints on the HO_x chemistry of the UT are just now becoming available; the number of measurements is expected to increase greatly over the next few years.

2.1.2.5. The Response of Stratospheric Ozone Destruction to NO_x

The concentration of NO_x also influences the rate at which ozone is destroyed in the atmosphere, particularly in the stratosphere and in the lower troposphere. In the stratosphere, because of the abundance of ozone and lower pressures, the concentration of atomic oxygen is sufficiently large that NO_x can catalytically destroy ozone:

$$O_3 + \text{sunlight} \rightarrow O + O_2 \qquad (6b)$$
$$O + NO_2 \rightarrow NO + O_2 \qquad (16)$$
$$NO + O_3 \rightarrow NO_2 + O_2 \qquad (17)$$
$$\text{net: } 2\,O_3 \rightarrow 3\,O_2$$

Reactions of the HO_x family also destroy ozone. In particular, reaction 15 leads to significant ozone loss in the LS and in the troposphere. Reaction 3 is in competition with reaction 15, so the rate of ozone loss by HO_x decreases with increasing NO.

Finally, ozone loss by halogen chemistry is important in the stratosphere. During winter, particularly in polar regions, it can dominate all other chemical destruction mechanisms. As with HO_x chemistry, NO_x interferes with this chemistry by binding to the reactive chlorine radical ClO:

$$ClO + NO_2 + M \rightarrow ClONO_2 + M. \qquad (18)$$

As a result of these coupling reactions, changes in the concentration of NO_x can lead to increased or decreased rates of stratospheric ozone destruction. When NO_x is low—as it is in most of the LS during winter, fall, and spring—most of the ozone loss occurs through HO_x and halogen chemistry. Under these conditions, enhancements of NO_x will decrease ozone destruction. On the other hand, at higher altitudes and during summer, NO_x-catalyzed ozone loss (reactions 16-17) can dominate the removal of lower stratospheric ozone, so enhancements in NO_x will speed ozone loss (Brühl *et al.*, 1998). These effects have been demonstrated by direct measurements of free radicals in the stratosphere (Wennberg *et al.*, 1994; Jucks *et al.*, 1997).

This chemistry is illustrated in Figure 2-2. A calculation is shown for typical mid-latitude springtime conditions. This entire profile is within the stratosphere, where catalytic ozone loss competes with and can exceed photochemical production. The left panel shows the fraction of ozone destroyed during the month of March as a result of catalysis by NO_x (squares), halogens (circles), and HO_x (crosses). For this latitude and season, the loss is dominated by halogen and hydrogen oxides below 20 km, whereas above 25 km, nitrogen oxides are most important. To illustrate how changes in NO_x perturb this chemistry, the right panel shows the effect of a uniform 20% increase in the concentration of NO_x. In regions where NO_x is high, ozone destruction increases. On the other hand, the opposite occurs in the LS because the increased NO_x decreases the loss of ozone by hydrogen and halogen radicals. Thus, as with the production rate of ozone in the troposphere, the response of ozone destruction with changes in NO_x is highly nonlinear. Because the photochemical lifetime of ozone in the LS is very long, the concentration of ozone in this region of the atmosphere is strongly influenced by transport. The change in ozone loss rates illustrated in Figure 2-2 does not translate directly into a change in ozone. For example, for a uniform 20% increase in NO_x, enhanced loss rates at high altitudes will reduce the transport of ozone to the LS. As a result, ozone concentrations in the LS can decrease even when the local ozone loss rate slows. Thus, the change in the ozone column with added NO_x is very sensitive to the altitude distribution of the perturbation. The subsonic aircraft fleet adds NO_x only to the lowermost stratosphere (< 13 km), where large-scale dynamics tend to prevent advection to higher altitude. As a result, injection of NO_x by the present fleet is thought to increase ozone in the LS.

2.1.2.6. Net Effects on Ozone

If the major direct impact of aircraft on the chemistry of the UT and lowermost stratosphere (below approximately 16 km) is an increase in the concentration of NO_x, we can say with high confidence that the ozone concentrations in this region will be higher than they would be in the absence of aviation. This increase occurs because NO speeds the catalytic oxidation rate of CO and reduces the destruction rate of ozone by HO_x and halogens (primarily in the stratosphere). In this context, it is important to note that the conventional troposphere-stratosphere

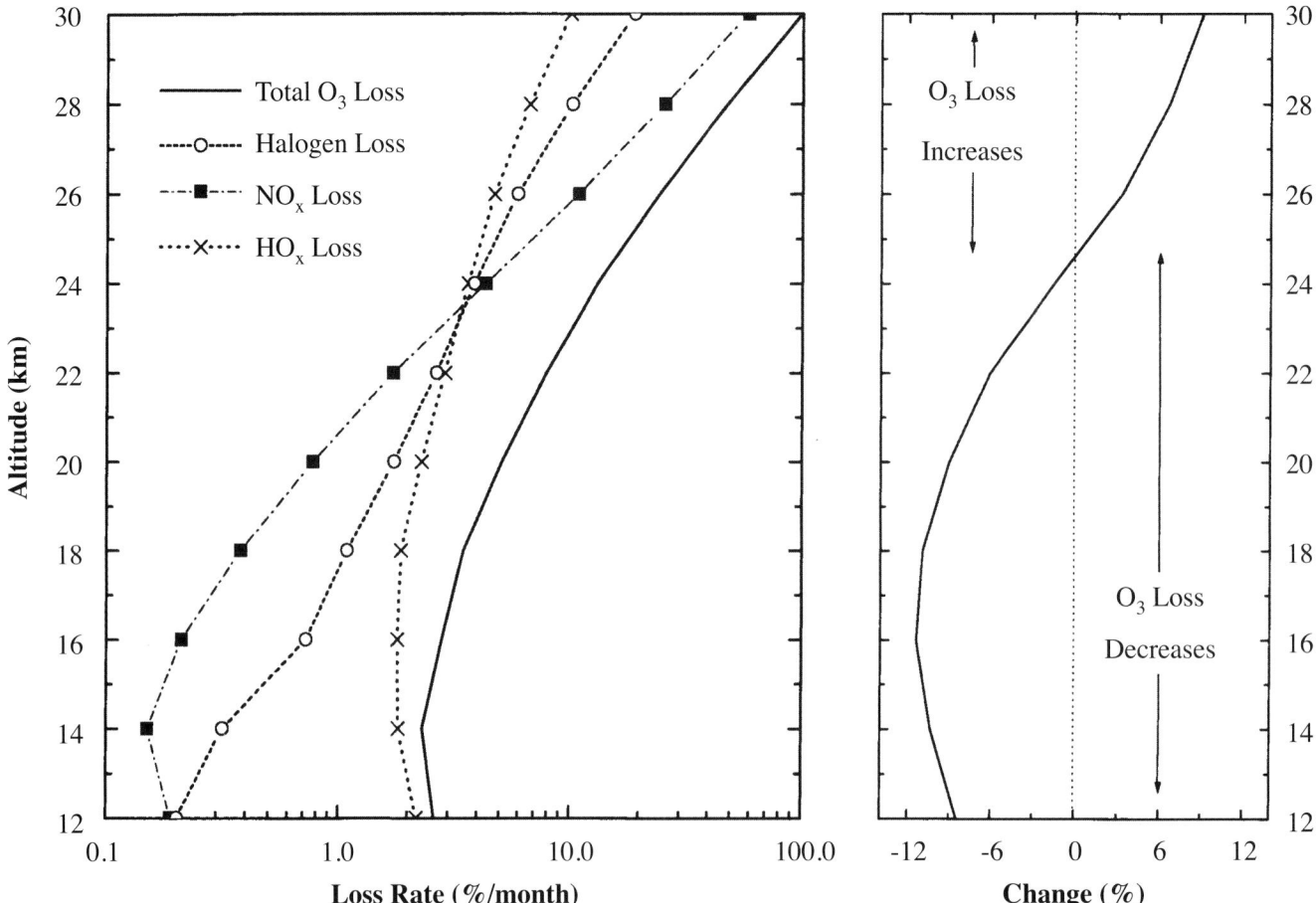

Figure 2-2: In the stratosphere, NO_x both catalytically destroys ozone and acts to interfere with the destruction of ozone by reactions of HO_x and halogens. Shown is a calculation of the rate of ozone loss in the lower stratosphere for springtime mid-latitude conditions during March. The loss from each family is illustrated. Ozone loss in the lowermost stratosphere is quite slow (a few percent per month). The right panel shows the effect on the removal rate of ozone when 20% more reactive nitrogen is added to the stratosphere uniformally (calculations provided by R. Salawitch using 1D-trajectory model).

boundary (i.e., the tropopause), reflecting important changes in atmospheric dynamics, does not coincide with the separation between net positive and negative NO_x-induced changes in ozone (see right panel of Figure 2-2). This distinction is important when considering future aviation scenarios that include a significant supersonic component.

The actual quantitative change in ozone from present aircraft operation is very sensitive to the meteorology of this region of the atmosphere; longer residence times will lead to larger NO_x increases and therefore higher ozone. As a result, only coupled chemical and dynamic models can estimate how large an increase is expected. To accurately predict the perturbation, these models must accurately describe the background NO_x concentration, the magnitude of the aircraft-induced NO_x perturbation, the sources of HO_x in this region of the atmosphere, and the meteorology. Finally, because ozone itself is relatively long-lived in the UT and LS, its concentration is strongly influenced by transport (as discussed in Section 2.1.1.3.3). The transport of ozone between different regions of the atmosphere significantly confounds attempts to assign causality to local ozone trends.

As described in Section 2.1.3, the production of particulates and aircraft-induced formation of clouds may be important for ozone. Chemical processes occurring in aerosols and on ice clouds may lead to increases in chlorine radical abundance in the stratosphere and suppress NO_x throughout the region.

2.1.3.   *Effects of Aircraft Aerosol Emissions*

There is strong laboratory and observational evidence for the importance of aerosols in atmospheric photochemical processes such as lower stratospheric ozone depletion (WMO-UNEP, 1995), tropospheric SO_2 oxidation (Calvert *et al.*, 1985), and tropospheric soluble trace gas removal. A large number of additional chemical processes involving aerosols has been postulated; however, a lack of information on the abundance, nature, and reactivity of the various aerosol types (e.g., dust, soot, sulfate, nitrate, organic) has hindered an accurate assessment. In this section, we summarize the connections between aerosol processes and ozone chemistry in the UT and LS. In addition, we assess the nature of aircraft-derived aerosol impacts on ozone abundances at aircraft cruising altitudes.

2.1.3.1. *Aerosols and Tropospheric Ozone Chemistry*

Reactions occurring on the surface of solid and liquid aerosols (i.e., heterogeneous reactions) or inside aqueous aerosols (i.e., homogeneous bulk reactions) can lead to a decrease in the production of ozone (reactions 1–5) by catalyzing the removal of NO_x and HO_x. Generally, heterogeneous reactions counteract reactions involving free radical species. From the ozone production perspective, removal of active species can occur by irreversible deposition of HO_x, NO_x, or NO_x and HO_x source species or by conversion of more reactive nitrogen- and hydrogen-containing species into less reactive ones. Conversion of N_2O_5 into HNO_3 on sulfate or ice particles is the best established example of the latter mechanism. Wet deposition of HNO_3, hydrogen peroxide (H_2O_2), and other soluble acids is an example of the former mechanism. Because removal of active species can also occur by gas-phase mechanisms, the importance of this aerosol chemistry depends on the relative rates of the gas and aerosol processes. The rates of heterogeneous processes depend on the available aerosol surface area as well as on their size distribution. As summarized in Chapter 3, typical mid-latitude upper tropospheric soot and sulfate aerosol surface areas are 10 μm^2 cm^{-3} (see also Pueschel *et al.*, 1997). Much larger water-ice surface areas, on the order of 10^4 μm^2 cm^{-3}, are found inside young contrails (Petzold *et al.*, 1997) and natural cirrus clouds (Dowling and Radke, 1990).

Aircraft impact studies have motivated laboratory investigations of heterogeneous soot reactions. Interest in reactions occurring on soot has also increased in response to suggestions, based on analysis of field observations (Hauglustaine *et al.*, 1996; Jacob *et al.*, 1996; Lary *et al.*, 1997), that aerosol-assisted conversion of HNO_3 to NO_x may occur under some unknown conditions. Processes like the HNO_3-to-NO_2 conversion, which result in a reduction of the oxidation state of an atmospheric species, are of particular interest because they proceed counter to the overall tendency of the troposphere to act as an oxidizing medium. For the specific HNO_3 example, the reduction mechanism operates at the expense of the oxidation of the soot substrate and would promote production of ozone by increasing levels of active NO_x.

Some of the most important heterogeneous reactions on soot involving NO_x and its reservoirs, which are of likely importance in the UT and have been studied in the laboratory, are presented below. In view of the fact that the number concentration of nonvolatile particles—the majority of which are presumed to be soot—in a young contrail is on the order of 10^4 particles cm^{-3} (Brasseur *et al.*, 1998), we start our discussion with heterogeneous oxidation-reduction reactions that may occur on soot. Heterogeneous kinetic studies involving "soot" have been performed on a variety of substrates, encompassing materials as diverse as commercially available amorphous carbon, carbonaceous material ("active carbon" or "carbon black"), and soot from hydrocarbon diffusion flames that use fuels such as hexane, toluene, ethylene, and acetylene. Therefore, comparisons of results obtained on different substrates should be made with caution.

The heterogeneous reaction of NO_2 with soot may be represented by the following reactions (Tabor *et al.*, 1993, 1994; Rogaski *et al.*, 1997; Gerecke *et al.*, 1998):

$$NO_2 + soot \rightarrow NO + [soot•O] \qquad (19a)$$

followed by the thermal decomposition of the oxygen adduct [soot•O]:

$$[soot•O] \rightarrow CO, CO_2 \qquad (19b)$$
$$NO_2 + soot + H_2O \rightarrow HONO + [soot•OH] \qquad (19c)$$
$$NO_2 + H(ads) \rightarrow HONO \qquad (19d)$$

where [soot•O] and [soot•OH] represent surface sites on soot that have been oxidized, hence deactivated by the heterogeneous reaction. It is not clear yet if soot participates as a reducing agent (reaction 19c) or simply as a reservoir of hydrogen (H(ads), reaction 19d) in its reaction with NO_2 to yield HONO.

Reactions 19a and 19c correspond to a reduction-oxidation reaction in which soot is the reducing agent. Soot is oxidized in the process and releases CO and CO_2 upon heating, although the primary oxidation product [soot•O] has not been characterized on a molecular level. It is thought that the surface of soot will be modified in the oxidation process, resulting in the accumulation of multiple functional groups bearing oxygen (Chughtai *et al.*, 1990).

The branching ratio between reactions 19a and 19c or 19d depends on the soot sampling location as well as the fuel used to produce the soot, albeit to a minor extent. In general, amorphous carbon does not generate HONO, whereas soot sampled early in its formation process may give rise to a HONO yield of up to 90% compared to the NO_2 taken up (Gerecke *et al.*, 1998). Measurements within exhaust plumes have established the presence of nitrous acid (HNO_2) and HNO_3 at concentrations well above background (Arnold *et al.*, 1992, 1994). The data indicate that about 0.6% of the NO_x is converted to HNO_2 and HNO_3. More recent data on exhaust plumes of five B-747s and one DC-10 increase this efficiency to 1–5% of the NO_x (Brasseur *et al.*, 1998). Therefore, care must be exercised when HONO concentrations measured in the plume are used to infer OH concentrations or emission indices for OH using known reaction parameters for OH + NO (Tremmel *et al.*, 1998).

Another pair of heterogeneous reactions occurring on soot have been shown to be of importance in the troposphere (Brouwer *et al.*, 1986):

$$N_2O_5 + H_2O + [soot] \rightarrow 2HNO_3 + [soot] \qquad (20a)$$
$$N_2O_5 + [soot] \rightarrow NO_2, NO + [soot•O] \qquad (20b)$$
$$[soot•O] \rightarrow CO, CO_2 \qquad (19b)$$

Reaction 20a is a hydrolysis reaction for which soot is the support material, whereas reaction 20b corresponds to a reduction-oxidation reaction in which soot is the reducing agent. The primary oxidation product [soot•O] of reaction 20b has not been identified, but it releases CO and CO_2 upon heating (reaction 19b), similar to the oxidation of soot by NO_2 in reaction 19a.

The heterogeneous reaction of the NO_x reservoir HNO_3 on soot also corresponds to a reduction-oxidation reaction analogous to the examples listed above (Rogaski *et al.*, 1997):

$$2HNO_3 + 2[soot] \rightarrow NO_2, NO + 2[soot\bullet O]$$
$$+ H_2O(ads) \text{ and/or } [soot\bullet OH] \qquad (21)$$

Finally, ozone reacts heterogeneously with soot in a redox reaction in which soot is the reducing agent (Stephens *et al.*, 1986; Fendel *et al.*, 1995; Rogaski *et al.*, 1997):

$$O_3 + [soot] \rightarrow O_2 + [soot\bullet O] \qquad (22)$$
$$[soot\bullet O] \rightarrow CO, CO_2 \qquad (19b)$$

It is important to note that reaction 22 does not correspond to a catalytic reaction because for every ozone molecule reacted, CO and CO_2 are formed according to reaction 19b. Therefore, soot is consumed as shown in the mass balance involving carbon (Stephens *et al.*, 1986). Thus, in all of the above reactions— with the possible exception of reaction 19d—soot appears to act as a reducing agent that is consumed in the course of the reduction-oxidation reaction.

Heterogeneous reactions of other potentially important reservoir and active compounds on soot are virtually unexplored. Examples may include reactions of HNO_4, H_2O_2, HOX, XNO_2, and $XONO_2$, with X=Cl, Br. HNO_2 does not, however, seem to undergo heterogeneous interaction on soot to any significant extent (Gerecke *et al.*, 1998).

Ice as a substrate in heterogeneous reactions has not received as much attention in laboratory studies as reactions occurring on other stratospherically relevant substrates such as supercooled and frozen sulfuric acid (H_2SO_4) hydrates, as well as type Ia and Ib polar stratospheric cloud (PSC) substrates. N_2O_5 undergoes a reactive uptake on tropospherically important substrates such as H_2SO_4/H_2O (Hanson and Ravishankara, 1991; Fried *et al.*, 1994) and water ice (Quinlan *et al.*, 1990; Hanson and Ravishankara, 1992), which results in HNO_3 formation. HNO_3 undergoes non-reactive uptake only on the aforementioned substrates (i.e., condensation) (Hanson, 1992; IUPAC, 1997a; JPL, 1997). N_2O_5 also undergoes a bimolecular reaction with hydrogen chloride (HCl) on ice, resulting in $ClNO_2$ in competition with the well-documented hydrolysis reaction leading to HNO_3 (IUPAC, 1997a; JPL, 1997; Seisel *et al.*, 1998):

$$N_2O_5 + HCl \rightarrow ClNO_2 + HNO_3 \qquad (23)$$

Reaction 23 occurs faster on ice than on H_2SO_4 because of limitations of HCl solubility in H_2SO_4. The increase in the rate of heterogeneous reactions on ice compared to acidic surfaces generally occurs for all reactions involving hydrohalic acids (IUPAC, 1997a; JPL, 1997). Reaction 23 competes with heterogeneous hydrolysis of N_2O_5, similar to reaction 20a. At temperatures near 200 K and at the limit of high HCl concentrations, 65% of the reaction proceeds via $ClNO_2$ formation; the rest proceeds through N_2O_5 hydrolysis.

NO, NO_2, and ozone do not interact with H_2SO_4 at concentrations encountered in the UT, or with water ice. Similarly, HONO does not interact with water ice down to temperatures of 180 K (Fenter and Rossi, 1996). However, HONO is taken up by H_2SO_4/H_2O binary mixtures (weight of $H_2SO_4 > 65\%$) from ambient temperatures down to 180 K and eventually yields nitrosylsulfuric acid (NSA) after protonation (Becker *et al.*, 1996; Fenter and Rossi, 1996; Zhang *et al.*, 1996; Kleffmann *et al.*, 1998; Longfellow *et al.*, 1998):

$$HONO + H_2SO_4 \rightarrow NO^+HSO_4^- + H_2O \qquad (24)$$

NSA has the ability to activate chlorine by reacting with HCl (reaction 25). The resulting NOCl product rapidly photolyzes in the UT into NO and Cl.

$$NO^+HSO_4^- + HCl \rightarrow NOCl + H_2SO_4 \qquad (25)$$

An important finding regarding halogen activation by reactions 24 and 25 is the fact that these reactions occur approximately 20 times faster on pure ice, bypassing NSA as an intermediate species altogether because of its instability on ice (Fenter and Rossi, 1996). The propensity of many bimolecular heterogeneous reactions to occur more efficiently on ice surfaces compared to sulfuric acid aerosols is considered particularly relevant with regard to chemical processes occurring on natural cirrus clouds in the UT and on those seeded by particles emitted from aircraft engines, owing to a change in the reaction mechanism. The relevant water-ice mechanism corresponds to a simple ionic displacement of the form

$$Cl^- + HONO \rightarrow NOCl + OH^- \qquad (26)$$

where NOCl may undergo photolysis to Cl and NO. Although reactions 24–26 may not be important under volcanically quiescent periods (Longfellow *et al.*, 1998), they may nevertheless play an important role in young contrails and perhaps in aged plumes, in view of the large number concentrations of volatile particles whose densities may be on the order of a few 10^4 particles cm^{-3} (Brasseur *et al.*, 1998).

The chemical transformation of NO to N_2O on acidic surfaces is too slow to be of significance in the present context (Wiesen *et al.*, 1994). However, an efficient heterogeneous reaction involving NO and SO_2 on solid and liquid aerosols could lead to the formation of N_2O within the plume on a time scale of tens of minutes (Pires *et al.*, 1996). In this chemistry, the interaction of HONO and SO_2 on humid surfaces, solid or liquid, may lead to N_2O, which also has been observed as a result of other laboratory experiments on heterogeneous chemistry of NO_2-soot interactions (Smith *et al.*, 1988).

Many of the other potentially important reservoir species— such as HNO_3, HNO_4, and H_2O_2—absorb onto sulfuric acid aerosols and ice particles but do not react with them (IUPAC, 1997a; JPL, 1997). These heterogeneous interactions are central to the removal of NO_x and HO_x species during denitrification and sedimentation events in the UT and LS.

The particle dimension of cirrus cloud crystals is on the order of 10 to 15 μm, which is approximately 10 times larger than PSC type IIs and 50 to 100 times larger than PSC type Is (Petzold and Schröder, 1998). This result, together with the fact that the pressure is higher in the UT than in the LS, necessitates a downward correction for heterogeneous rates in cases in which the uptake coefficient (γ) is larger than 5×10^2 because molecular diffusion of the gas toward the surface of the particle becomes rate limiting. This correction may be especially important for the case of a heterogeneous reaction on ice with an uptake coefficient exceeding 0.1 and may thus increase the uncertainty when extrapolating laboratory rate parameters to atmospheric concentrations. At present, most published models assume that the interfacial reaction is rate limiting and prescribe a rate constant of the form $k = \gamma <c> A/4$, where γ is the uptake coefficient, $<c>$ is the average molecular velocity, and A is the surface area of the atmospheric particle per unit volume.

2.1.3.2. Aerosols and Stratospheric Ozone Chemistry

It is now well-established that aerosols—consisting mainly of water, sulfate, and nitrate—play a crucial role in defining lower stratospheric levels of active NO_x and ClO_x (WMO-UNEP, 1995). Heterogeneous reactions in the stratosphere convert the inactive species HCl and chlorine nitrate ($ClONO_2$) into the photolytically labile Cl_2 species and the more active N_2O_5 species into the less active HNO_3 species. The net effect of these processes is to decrease NO_x levels and increase stratospheric levels of ClO_x and HO_x, hence the importance of ClO_x- and HO_x-catalyzed ozone loss relative to NO_x-catalyzed ozone loss (refer to Figure 2-2). Increases in the total surface area of lower stratospheric aerosol by additional aerosol sources such as aircraft further accentuate the importance of ClO_x- and HO_x-catalyzed chemistry and reduce the impact of aircraft NO_x emissions on ozone depletion. The net effect on ozone depletion will depend on the magnitude of the aerosol perturbation relative to the natural background and on given background levels of ClO_x and NO_x radicals, as well as on possible changes in the frequency of denitrification events induced by aircraft aerosols, water vapor, and HNO_3. Aircraft-induced aerosols will be imbedded in a highly variable natural stratospheric aerosol background. In particular, stratospheric sulfate surface areas—typically on the order of 1 μm^2 in volcanically quiescent periods—are nearly an order of magnitude larger than corresponding soot surface areas. During PSC type II events, surface areas of water-ice can approach 10 μm^2. Likewise, amplification of H_2SO_4 aerosol surface area can be up to a factor of 100 following volcanic eruptions.

A number of heterogeneous reactions are important to various degrees on the surfaces of ice particles (PSC type II), HNO_3/H_2O aerosols (PSC type Ia), and saturated ternary solutions of water, HNO_3, and H_2SO_4 (PSC type Ib). Rate parameters for heterogeneous reactions depend complexly on temperature, pressure, and humidity conditions. Consequently, modeling of stratospheric reactions must explicitly address differences in conditions between poles and mid-latitudes and between the tropopause and the LS.

The most important heterogeneous reaction in relation to ozone depletion on a global scale is the heterogeneous hydrolysis of N_2O_5 (reaction 20) (Hofmann and Solomon, 1989; Solomon *et al.*, 1996; Kotamarthi *et al.*, 1997). This and the following reactions are likely to be enhanced in the presence of increased atmospheric particulates such as contrails and aircraft-induced cirrus clouds, which correspond to PSC type II in the LS from a chemical point of view.

Chlorine and bromine activation on PSC type Ia and Ib as well as on PSC type II (ice) takes place according to the following reactions:

$$ClONO_2 + H_2O(s) \rightarrow HOCl + HNO_3 \qquad (27)$$
$$ClONO_2 + HCl(s) \rightarrow Cl_2 + HNO_3 \qquad (28)$$
$$HOCl + HCl(s) \rightarrow Cl_2 + H_2O \qquad (29)$$
$$BrONO_2 + H_2O(s) \rightarrow HOBr + HNO_3 \qquad (30)$$
$$BrONO_2 + HCl(s) \rightarrow BrCl + HNO_3 \qquad (31)$$
$$HOBr + HCl(s) \rightarrow BrCl + H_2O \qquad (32)$$

Reactions of N_2O_5 with HCl(s) and HONO with HCl(s) (reactions 23 and 26) are also possible (IUPAC, 1997a; JPL, 1997). Because of the strong decrease in the solubility of HCl with increasing acidity of H_2SO_4 solutions—thus with increasing stratospheric temperature—reactions 28, 29, 31, and 32 will be most important at high latitudes (Portmann *et al.*, 1996). Accordingly, these bimolecular reactions will proceed fastest on type II PSC surfaces in the polar vortex or on cirrus cloud particles in the tropopause region (Borrmann *et al.*, 1996; Solomon *et al.*, 1997). In addition, the hydrolysis reaction of $ClONO_2$ (reaction 27) is less efficient than the corresponding one involving $BrONO_2$ (reaction 30), on account of thermochemical differences between the reactions (Allanic *et al.*, 1997; Barone *et al.*, 1997; Oppliger *et al.*, 1997). One consequence of this difference is that reaction 27 is prone to surface saturation on solid substrates, whereas reaction 30 is a strong and continuous source of HOBr, thereby making $BrONO_2$ a relatively labile reservoir compound.

2.1.3.3. Aircraft Aerosols and Heterogeneous Reactivity

Heterogeneous reactions are important in defining ozone removal rates in the stratosphere and troposphere. Stratospheric and tropospheric aerosol particles differ markedly in composition, lifetime, and size distribution. In the stratosphere, the global background aerosol consists predominantly of sulfate; water-ice and carbonaceous particles are minor components on the global scale, although PSCs that contain water-ice can be important on smaller geographic scales. Tropospheric aerosols consist of a wide variety of constituents, including water, sulfuric acid, soot, mineral dust, sea salt, and organic particles.

Soot, sulfuric acid ("sulfate"), and water-ice particles are the main condensed-phase species found in the exhaust of jet aircraft. As mentioned above, all of these species are present in the background atmosphere, and all derive from natural and other non-aircraft anthropogenic sources. Consequently, the effect of aircraft aerosol emissions on atmospheric photochemistry, to a first approximation, is to increase the soot, sulfate, and water-ice surface areas available for heterogenous and multiphase processes discussed in the previous subsections. Large-scale aerosol loading from aircraft can be estimated roughly from knowledge of fleet emission rates and average residence time of particles deposited at given altitudes. A detailed discussion of aircraft particle loading appears in Chapter 3. For the purposes of this chapter, it suffices to note that aircraft soot and sulfur emissions are thought to be significant at cruise altitudes, whereas estimates of the perturbation from aircraft H_2O remain highly uncertain.

Beyond the simple aircraft aerosol loading approximation, chemistry occurring inside aircraft plumes carries the potential to produce volatile aerosols of significantly different heterogeneous reactivity relative to typical background aerosols. For instance, the water content of H_2SO_4 particles will vary over a wide range in an aircraft plume and wake, with concomitant effects on the rates of reactions 27–32. In addition, under cold lower stratospheric conditions, aircraft plume production of HNO_3-rich liquid aerosol may propogate to the larger scale and contribute to the formation of solid PSCs, hence enhance processing of inactive chlorine species (Kärcher, 1997). Finally, evidence is mounting that volatile aerosols formed under conditions of low fuel sulfur content may contain significant amounts of light fuel-bound organic constituents (Kärcher *et al.*, 1998b). The heterogeneous reactivity of such organic-containing aerosols is unknown and must await further *in situ* aerosol characterization.

Significant differences may also exist between aircraft-derived and ambient background soot particles. As discussed in Section 2.1.3.1, soot surfaces may act as reaction catalysts or consumables, depending on their properties. In addition, the surface properties likely determine the extent to which reduction-oxidation reactions occur on a particular soot particle. At present there is virtually no information on the properties of aircraft-derived or ambient background carbonaceous particles upon which to base an evaluation.

2.1.3.4.   *Net Effects on Ozone*

Based on our current understanding of heterogeneous chemistry, aircraft sulfate and water-ice particles will lower ozone concentrations in the UT and LS relative to what they would be if aircraft emissions contained only NO_x. This process occurs because the particles remove the ozone precursors HO_x and NO_x in the UT and liberate ozone-destroying ClO_x in the LS. The heterogeneous chemistry occurring on soot is much less well understood, so its role in atmospheric chemistry is much harder to define. However, because particle abundances of sulfate in the LS are much greater than those of soot, we can conclude with some confidence that present aircraft soot emissions are having little impact on stratospheric ozone, provided that their primary impact is on partitioning between NO_y and NO_x.

The extent to which aircraft aerosols offset the effects of aircraft NO_x emissions on atmospheric ozone depends on a variety of chemical and dynamical factors. To quantify this balance, we will need atmospheric models that combine representations of aerosol microphysics, gas and heterogeneous chemistry, and atmospheric dynamics.

2.1.4.   *Indirect Effects Involving Atmospheric CO and CH_4*

The tropospheric modeling tools detailed in Section 2.2.1 produce an increased global inventory of OH radicals in response to subsonic aircraft injection of NO_x. This characteristic model response to increased NO_x levels in air traffic corridors arises through the reaction sequence:

$$HO_2 + NO \rightarrow NO_2 + OH \tag{3}$$

so that increased ozone production is accompanied by an increase in the OH radical concentration and the concentration ratio of OH to HO_2. Increased OH levels in air traffic corridors then lead to decreased CO concentrations through the following reaction series:

$$OH + CO \rightarrow H + CO_2 \tag{1}$$
$$H + O_2 + M \rightarrow HO_2 + M \tag{2}$$

CO lifetimes may approach 2 months and are therefore much longer than NO_x lifetimes. As a result, a region of decreased CO concentrations spreads out from air traffic corridors toward lower altitudes and toward the Equator. Ultimately, as a result of subsonic aircraft NO_x emissions, decreased CO levels are found in tropical and sub-tropical regions where much of the CH_4 is oxidized. Because CO levels are lower, OH levels are higher; hence CH_4 concentrations decrease slightly, so the total flux through the OH + CH_4 reaction (35) remains in balance with CH_4 emissions:

$$OH + CH_4 \rightarrow CH_3 + H_2O \tag{35}$$

As a result, the global CH_4 distribution adjusts slowly to higher OH levels, and a new steady-state is established in which CH_4 concentrations are reduced slightly. This adjustment in CH_4 itself drives a further increase in OH levels, with the result that CH_4 concentrations build up more slowly over 10–15 years. These indirect or feedback processes have been described in some detail in previous IPCC (1995, 1996) reports.

As quantified in the following section, aircraft NO_x emissions produce an increase in the total global inventory of OH of about 2%, which should be reflected in a corresponding change in CH_4 loss rate (IPCC, 1995, 1996). The CH_4 chemical feedback will then amplify this change in loss rate, producing a decrease in CH_4 concentrations that is about 1.4 times the change in loss rate. CH_4 concentrations should then decrease by about 3%. These reduced CH_4 burdens will then decrease tropospheric ozone production, although the effect may be considered negligible (Fuglestvedt *et al.*, 1999). Because these readjustments take place over 10–15 years, they are exceedingly difficult to represent fully in global 3-D models, though they have been fully explored in 2-D models (Fuglestvedt *et al.*, 1996; Johnson and Derwent, 1996).

The stratospheric modeling tools described in Chapter 4 indicate that supersonic aircraft flying in the LS may lead to stratospheric ozone depletion. Under conditions of stratospheric ozone depletion, there is greater penetration of solar ultraviolet radiation through the stratosphere. N_2O transported up from the troposphere therefore has higher photolysis rates lower in the stratosphere, leading to a shorter photolysis lifetime. The global N_2O distribution will then slowly adjust over decades to the increase in stratospheric destruction, leading to decreased N_2O concentrations, assuming constant emissions. Because N_2O is an important greenhouse gas, there is an indirect radiative forcing impact of supersonic aircraft over 50–200 years. This indirect impact is relatively straightforward to take into account using modeling tools described in Chapter 4.

2.2.   Impacts of Present Aviation on Atmospheric Composition

2.2.1.   *Modeling Changes in Atmospheric Ozone from Present Aviation*

Because of the substantial chemical and dynamic differences between the stratosphere and the troposphere, individual atmospheric models have tended to focus on one or the other of the regions. This separation has worked well in addressing primarily stratospheric issues such as chlorofluorocarbons (CFCs) and tropospheric issues such as regional pollution. Subsonic aircraft emissions, insofar as they are injected into the interface between the two atmospheric regions, present a particularly complex modeling challenge. To date, almost all model studies of subsonic aircraft NO_x effects have been conducted with models that emphasize tropospheric chemistry; few contain an explicit representation of stratospheric chemistry. Most of the following discussion will center on those efforts. However, two models of the stratosphere—the Atmospheric Environmental Research (AER) and Commonwealth Scientific and Industrial Research Organisation (CSIRO) models used in Chapter 4—have calculated stratospheric ozone responses to 1992 subsonic fleet emissions of NO_x. We discuss these results in Section 2.2.1.2 and use them in conjunction with the tropospheric model results to arrive at a total atmospheric impact of NO_x emissions from present subsonic aviation.

Recognition of the importance of heterogeneous chemical processes has increased steadily over the past 10 years; all atmospheric models now contain at least some of the gas-surface interactions identified in laboratory studies. However, relatively few global atmospheric models represent aerosol microphysics in a way that allows for simulation of the impacts of aircraft aerosols on atmospheric particle densities and surface areas. As a consequence, there has not yet been much effort directed at modeling the effects of aircraft particle emissions on atmospheric ozone. We evaluate one such study in Section 2.2.1.3.

2.2.1.1.   *Effects of Aircraft NO_x Emissions on Tropospheric Ozone*

Model studies of the impact of NO_x emissions from aviation on ozone levels in the UT and LS have been performed with increasing sophistication since the 1970s. Most studies set up a base case without aircraft NO_x emissions and a perturbed case with the aircraft source added. Differences in atmospheric composition and, ultimately, radiative forcing between the two cases are attributed to the influence of aircraft emissions. There are detailed differences among the many model studies that arise because of differences in the aircraft emission inventories employed. There are, however, many other more crucial differences in the model studies: The extent and accuracy of the treatment of atmospheric chemistry processes; the treatment of other NO_x sources besides aircraft; the treatment of dispersion, transport, and advection of NO_x species; and the adequacy of the treatment of the underpinning description of the coupled chemistry and the global atmospheric circulation in the UT and LS.

Many of the model studies note that the bulk of aircraft NO_x emissions from subsonic aviation occurs in an atmospheric region that is sensitive to these emissions. Because of low background NO_x levels—typically 50–200 ppt—in the mid-latitude UT and LS, high NO/NO_2 ratios, and low HO_x concentrations, ozone production (in terms of ozone molecules) per NO_x molecule is more efficient here than anywhere else in the atmosphere. Because the greenhouse blanketing produced by a given atmospheric ozone increment is directly proportional to the temperature contrast between radiation absorbed and radiation emitted, radiative forcing efficiency is greatest for ozone changes near the tropopause (Lacis *et al.*, 1990), though its climate sensitivity may have a different response (Hansen *et al.*, 1997).

In completing our assessment of ozone impacts from present subsonic aircraft NO_x emissions, we have gathered results from 48 model studies, which are listed in the bibliography at the end of the chapter. In each case, the model results have been normalized by the assumed global aircraft NO_x emission source strength and put on the common basis of 0.5 Tg N yr^{-1}. There are, however, differences in the spatial distributions used in these inventories, and local NO_x and ozone concentration impacts will not necessarily scale accurately with global emission source strength. Despite this important reservation, global emission source strength is a useful index in a first examination of the model results.

Based on these results, we conclude that subsonic aviation has impacts on atmospheric composition in the principal traffic areas at 9–13 km altitude, 30–60°N latitude, averaged over chemically significant time scales of up to a few days. These impacts may be summarized as follows for normalized NO_x emissions of 0.5 Tg N yr^{-1}:

- Significant local increases in NO_x levels, by up to hundreds of ppt

- Significant local increases in ozone, by up to 12 ppb, over a region including air traffic corridors and polewards of them, averaged over available calculations for January and July
- Ozone increases of 2–14%, at their maximum in July, downstream and extending to the north of principal traffic areas, reflecting the long atmospheric lifetime of ozone in the UT and LS
- Ratios of ozone increments to NO_x increments on the order of 100, reflecting the high efficiency of the ozone production cycle in the UT and LS
- Globally averaged ozone inventory increases of 3–10 Tg, or 1–3%, up to an altitude of about 15 km
- Increases in tropospheric hydroxyl radical inventories of 0.3–3% globally up to an altitude of about 15 km.

Despite improvements in the realism with which processes have been represented in the models, the increase in dimensionality from 2-D to 3-D has not been accompanied by a change in the likely magnitude of local NO_x and ozone changes anticipated from aircraft NO_x emissions, when normalized by the assumed aircraft NO_x emission to 0.5 Tg N yr^{-1} (Figure 2-3).

The original 2-D (altitude-latitude) model studies of the tropospheric ozone impacts of subsonic NO_x emissions are close to the (mean + 1xsd) line from the entire ensemble of studies. 2-D model estimates have declined steadily so that by 1992 they were close to the (mean - 1xsd) line. The original 1-D studies showed a large amount of scatter and have not been

used extensively for assessment purposes for the past decade or so. 2-D channel (altitude-longitude) models have been applied to the aircraft assessment as an interim strategy prior to 3-D (altitude-latitude-longitude) model development. 2-D channel model results are toward the high end of the range of results.

3-D models have been employed increasingly from 1994 onward, with the results lying toward the low end of the overall spread of model results. 3-D model results still show a considerable amount of scatter, and there is no evidence that the move from 2-D to 3-D models or the selection of 3-D models in Chapter 4 has improved inter-model consistency or overall predictive confidence. Regardless of how 2-D model results compare with zonally averaged 3-D model results, a crucial advantage of 3-D models is that they can account for zonal asymmetries and, in principle, be evaluated more decisively against local observations. Consequently, more confidence can be gained in their predictive capabilities for aircraft assessments (see Sections 2.3.1.1 and 2.3.1.2).

Based on all model results compiled for this assessment, our estimate of the likely ozone increase at its maximum in the principal traffic areas—9–13 km altitude, 30–60°N latitude—from 1992 subsonic aviation and during summer is about 6%, or 8 ppb, for a global aircraft NO_x emission of 0.5 Tg N yr^{-1} (1.65 Tg NO_x as NO_2 yr^{-1}; see Chapter 9).

Chapter 4 focuses attention on the development of future scenarios for subsonic aircraft NO_x emissions; six 3-D global

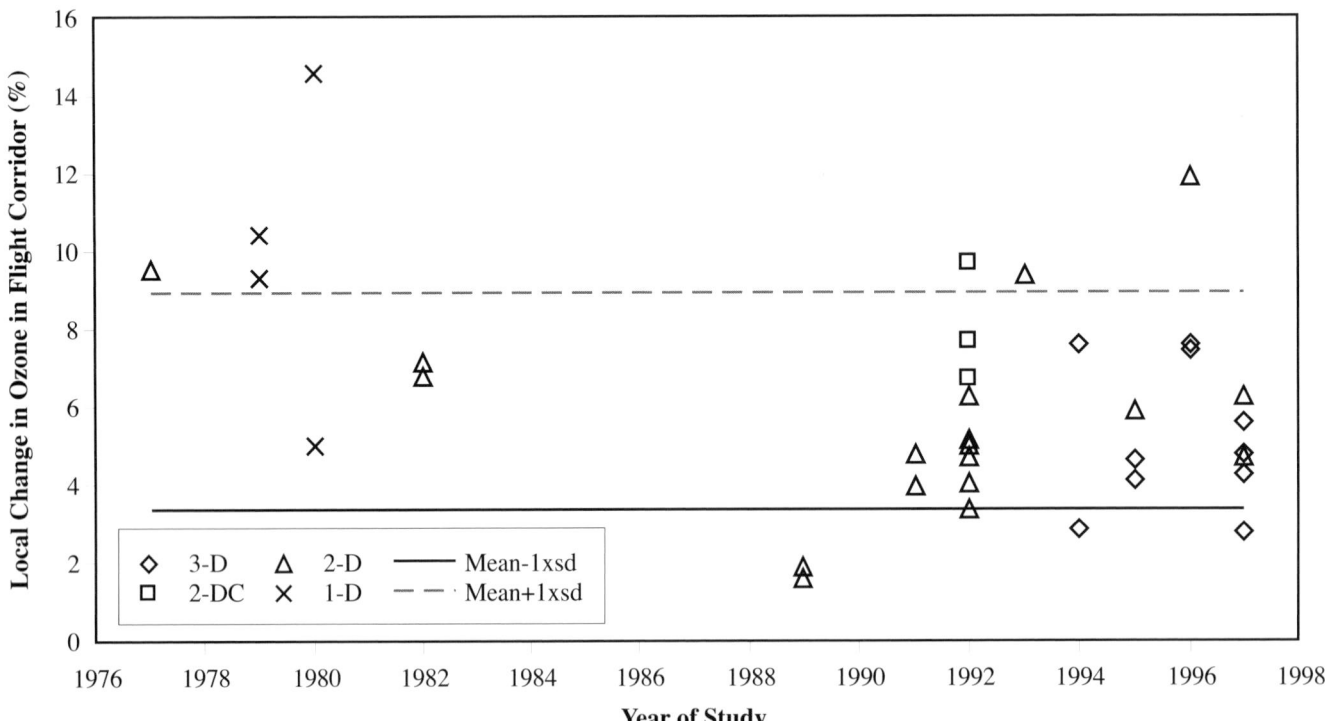

Figure 2-3: Maximum percentage increase in ozone concentrations, in principal traffic areas, in the 9- to 13-km altitude range, 30–60°N latitude range, during summer, normalized by assumed global aircraft NO_x emission to 0.5 Tg N yr^{-1} and plotted against the year the calculation was performed. Each point represents a particular model study, with the lines showing the one standard deviation (1xsd) range of results.

Table 2-1: *Calculated changes in ozone in principal traffic areas and 9- to 13-km region during summer for five selected models employed in Chapter 4 for the 1992 aircraft subsonic fleet NO_x emissions of 0.5 Tg N yr^{-1} (see Chapter 9), together with the range of previous studies shown in Figure 2-3.*

| Model | **Local Ozone Increases during July** | |
	ppb	%
BISA/IMAGES	4.4	4.3
HARVARD	6.0	4.4
UiO	5.5	3.5
UKMO	12.7	7.6
TM3	10.0	
Range of Previous 48 Studies	3–12	3–9

models have been employed to assess future impacts of subsonic aircraft. Table 2-1 shows ozone changes calculated in some of these models for 1992 subsonic operations compared with model results presented in previous studies. The latest model results in Table 2-1 appear to lie well within the central range of estimates from previous studies.

The estimates in Table 2-1 can be compared with previous assessments of the likely ozone impacts of NO_x emissions from present aviation. The WMO-UNEP (1995) assessment described model results available to the end of 1993 as preliminary and indicative of maximum ozone increases of 4–12% at around 10 km for 30–50°N. The above estimates encompass a narrower range than that given in the WMO-UNEP (1995) assessment and suggest that the true value is at the lower end of previous estimates. An assessment prepared for the European Commission (Brasseur *et al.*, 1998) concluded that the current fleet of commercial aircraft should have increased NO_x abundances by 50–100 ppt near 200 hPa, with a corresponding increase in ozone concentrations of 5–9 ppb (i.e., 4–8%), during summertime. This latter range is consistent with the range identified in the present study, based on the five selected 3-D models.

2.2.1.2. *Effects of Aircraft NOx Emissions on Stratospheric Ozone*

2-D models have been developed extensively over the past decade in response to the need for credible predictions of CFC impacts on stratospheric ozone and more recently for understanding the effects of future supersonic (high speed civil transport, or HSCT) aircraft fleets. Emphasis in these models has been placed on ClO_x, HO_x, and NO_x chemistry in the 20–40 km region of the atmosphere, where the CFC and HSCT effects are expected to be manifested. Model calculations for the 1992 subsonic aircraft fleet have been performed by the AER and CSIRO 2-D models, described in detail in Chapter 4. For these calculations, the models have used the approach described in Section 2.2.1.1; namely, they have set up a base case without

aircraft NO_x emissions and a perturbed case with the aircraft source added. They also have both used the subsonic fleet emission inventory described in Chapter 9. The results of the two studies are in close agreement and yield the following conclusions:

- Approximately 20% of subsonic emissions are released directly into the stratosphere; this result is on the low end of empirically based estimates, resulting in roughly a 1% increase in summertime NO_x levels existing just above the tropopause at 45°N latitude.
- The aircraft-induced increase in lowermost stratospheric NO_x causes an increase in ozone. Peak ozone increases are approximately 0.5% at northern mid-latitudes and at altitudes immediately above the tropopause. The total column ozone change resulting from stratospheric aircraft emissions is approximately 0.1%.
- A very small amount of aircraft NO_x is transported to altitudes above 22 km, where it acts to decrease ozone concentrations slightly.

2.2.1.3. *Effects of Aircraft Aerosol Emissions*

A number of previous model studies have examined the effects of HSCT sulfate aerosols on stratospheric chemistry; several new studies are discussed in Chapter 4. A key uncertainty in these calculations has been the degree to which emitted sulfur forms submicron sulfate particles in the aircraft plume (see Section 3.2.3). As part of the Chapter 4 studies, the AER and UNIVAQ stratospheric models have calculated the response of stratospheric ozone to the subsonic fleet (year 2015) for a range of assumed SO_2-to-particle conversion efficiencies. Details of those studies appear in Section 4.4.1.2. A rough estimate of the effect of sulfur emissions from the 1992 fleet can be made by scaling the 2015 results to 1992 emission levels. This calculation results in an estimate for the column ozone change at 45°N of approximately -0.1%.

The potential role of subsonic aircraft-generated soot in stratospheric ozone depletion has been investigated by Bekki (1997) using a 2-D model that includes heterogeneous reactions of HNO_3, NO_2, and ozone on soot. Using modeled aircraft soot aerosol distributions that were broadly consistent with observations, the model yields increased losses of ozone (approximately -0.2% column change at 45°N for the 1992 fleet) relative to the same model without soot chemistry. Moreover, inclusion of soot chemistry improves model agreement with the observed trend in LS ozone depletion. However, recent laboratory measurements indicating that soot is a reaction consumable do not support a major assumption of the Bekki model that soot acts as a catalyst.

The degree to which soot consumption affects the Bekki calculations can be estimated as follows: If the heterogeneous reaction between ozone and soot proceeds at the highest reported rate (the one used in the Bekki calculation), namely $\gamma(O_3) = 3.0 \times 10^{-3}$ (Stephens *et al.*, 1986; Fendel *et al.*, 1995; Rogaski *et al.*, 1997), the time required to consume the mass of soot present in the stratosphere would be approximately 10 minutes. It is important to note that this result is independent of the soot surface-to-volume ratio; it depends only upon the ozone concentration, which is taken to be 10^{11} cm^{-3} at 10-km altitude. At this reaction rate, only 5 hours would be required to consume the entire soot particulate population at 10 km (assuming 5×10^7 carbon atoms cm^{-3} at a soot loading of 1 ng m^{-3}) and cause the effective soot-ozone reaction rate to drop to zero.

Laboratory evidence suggests that the soot-ozone reaction rate decreases after consumption and alteration of the top surface carbon layer (Stephens *et al.*, 1986). Small reported laboratory γ values—on the order of 3.0×10^{-5}—likely reflect the reaction rate associated with a "deactivated" laboratory soot surface. Using this value for γ, the time required to consume individual particles of soot would be less than a day, and the time to consume all atmospheric soot would be lengthened to approximately 3 weeks. Although this change in reaction rate extends the lifetime of atmospheric soot, it comes at the expense of the overall ozone destruction rate. Accordingly, our analysis indicates that the mass of soot in the UT and LS does not seem to be sufficient to quantitatively affect ozone, regardless of the reaction rate chosen.

In support of this conclusion, we note that *in situ* measurements in the engine exhaust of a Concorde supersonic aircraft in the LS have been used recently to constrain heterogeneous reaction rates on soot particles in a plume model (Gao *et al.*, 1998). Gao *et al.* inferred low reactivity for the ozone-soot interaction; coupled with the measured abundance of the soot aerosol, Gao *et al.* reach the same conclusion that we derived from our qualititative arguments.

In summary, the few model calculations performed suggest that subsonic aircraft sulfate and soot emissions in the stratosphere act as substrates for ozone depletion. The magnitude of the depletions is likely to be small relative to ozone increases from NO$_x$ emissions, but the depletions are highly uncertain because of the poor state of understanding and model development relative to aerosol chemistry.

2.2.2.   Observing Changes in Atmospheric Ozone from Present Aviation

2.2.2.1.   Observed Ozone Trends in the Upper Troposphere and Lower Stratosphere

A decreasing trend in stratospheric ozone has been a pivotal diagnostic in the assessment of anthropogenic halocarbon release. Because the bulk of global ozone resides in the stratosphere, measurements of total column ozone—which can be made quite accurately—have served as a proxy for stratospheric ozone abundance. Downward trends in total ozone are now well-established throughout all seasons and all latitudes, except in the tropics (WMO-UNEP, 1999). Broad agreement on the magnitude of the total ozone trend exists between ground-based and satellite observational databases and model predictions based on chlorine-catalyzed ozone destruction. However, as discussed in Section 2.1, aircraft

engine emissions may induce changes of different magnitude and/or sign in tropospheric and stratospheric ozone densities. Therefore, to observe possible effects of aviation on the ozone layer, one is likely to have to focus on trends in the vertical ozone profile rather than overall column abundance.

Natural phenomena such as volcanic eruptions and seasonal and interannual climate variations may affect ozone density variations in the UT and LS. The time constants associated with these phenomena range from months (in the case of short-term climate variation) to years (for the occasional volcanic eruption) to possibly decades (for long-term climate change). Because extensive observational data on ozone are limited to the past several decades, it is not possible to completely deconvolute the impacts of various natural phenomena. The data record is sufficiently long, however, to allow characterization of periodic phenomena occuring on shorter time scales. Most of the anthropogenic forcings have been increasing secularly during the period of observation. Consequently, attempts to discriminate trend components can be carried out only with the aid of model predictions for each forcing.

Trend analyses of vertical ozone profiles have become possible only during the 1980s and 1990s as a result of data from the ground-based (Umkehr technique) and ozonesonde networks, and satellite-borne solar backscatter ultraviolet spectrometer (SBUV) and Stratospheric Aerosol and Gas Experiment (SAGE) I/II instruments (Logan, 1994; Miller *et al.*,1995; WMO-UNEP, 1995; Fortuin and Kelder, 1997; Harris *et al.*, 1997; WMO, 1998; see also Figure 2-4). The middle stratospheric trends derived from different data sets show broad agreement with each other. The negative trend peaking at ~40-km altitude and extending from 30 to 50 km in middle latitudes is ascribed to the simple Cl-ClO catalytic cycle of ozone destruction from enhanced atmospheric chlorine loading. A significant negative trend is also discerned in the lowermost stratosphere (i.e., between the troposphere and approximately 20-km altitude), where increased heterogeneous conversion of chlorine-containing reservoir species to reactive radical forms has been suggested as a factor in ozone destruction through catalytic cycles involving the ClO+ClO and ClO+BrO reactions.

Trends in UT ozone for the time period encompassing large growth in aircraft fuel consumption (1970 to the present) are available from a number of ozonesonde stations. All of the sonde stations at middle and high latitudes of the Northern Hemisphere show a stratospheric decrease at altitudes between the tropopause and ~24 km for the period 1970–96. Upper tropospheric trends vary substantially among stations, with increases of 10–20%/decade over Europe, decreases of 5–10%/decade over Canada and the eastern United States, and no trend over Japan (WMO-UNEP, 1999). Neither aircraft nor surface NO$_x$ emissions—both showing little geographical variation in their European, North American, and Asian trends (Logan, 1994)—are consistent with observed ozone trend variations.

Detailed time-series analysis of ozone trend data in an area of heavy aircraft traffic (i.e., west-central Europe) reveals that air traffic growth is unlikely to be a primary factor in the observed upper tropospheric trend. For example, the sonde observations at Hohenpeissenberg, Germany, clearly show a mean increase in ozone of ~10%/decade below ~9-km altitude between 1970–96 (Figure 2-5). However, most of the increase occurred before 1985 (WMO-UNEP, 1999), even though air traffic growth remained steady, and the ozone trend for the period 1980–96 is slightly negative in the UT. The lack of growth in ozone after 1985 mimics the lack of growth of surface emissions of NO_x (Logan, 1994). Decreases in the amount of ozone transported to the UT from the LS, because of reductions in stratospheric ozone abundance and/or weakening of dynamical transport, may also be a factor in the observed trend during the 1980s and 1990s.

In situ aircraft sampling of ozone in the 9–13 km region that has occurred sporadically over the past 20 years provides complementary data sets for use in understanding ozone climatology in the tropopause region. In 1994, a focused effort to collect climatological ozone data from aircraft platforms was initiated as the MOZAIC program (Marenco *et al.*, 1999; Thouret *et al.*, 1999). This database is now sufficiently long to address a number of important issues related to tropopause heights and seasonal variations, although it cannot yet address the issue of long-term trends.

2.2.2.2. Other Diagnostics of Large-Scale Aviation Impacts

As discussed in previous sections, ambient levels of NO_x and soot are likely to be affected by aircraft to a greater extent than ozone. Accordingly, a comprehensive set of NO_x and aerosol measurements taken over a wide range of locations and over the period of the past 20 years could provide a basis for evaluating aircraft impacts on these ozone-related species. Compared to the historical record for ozone, however, the

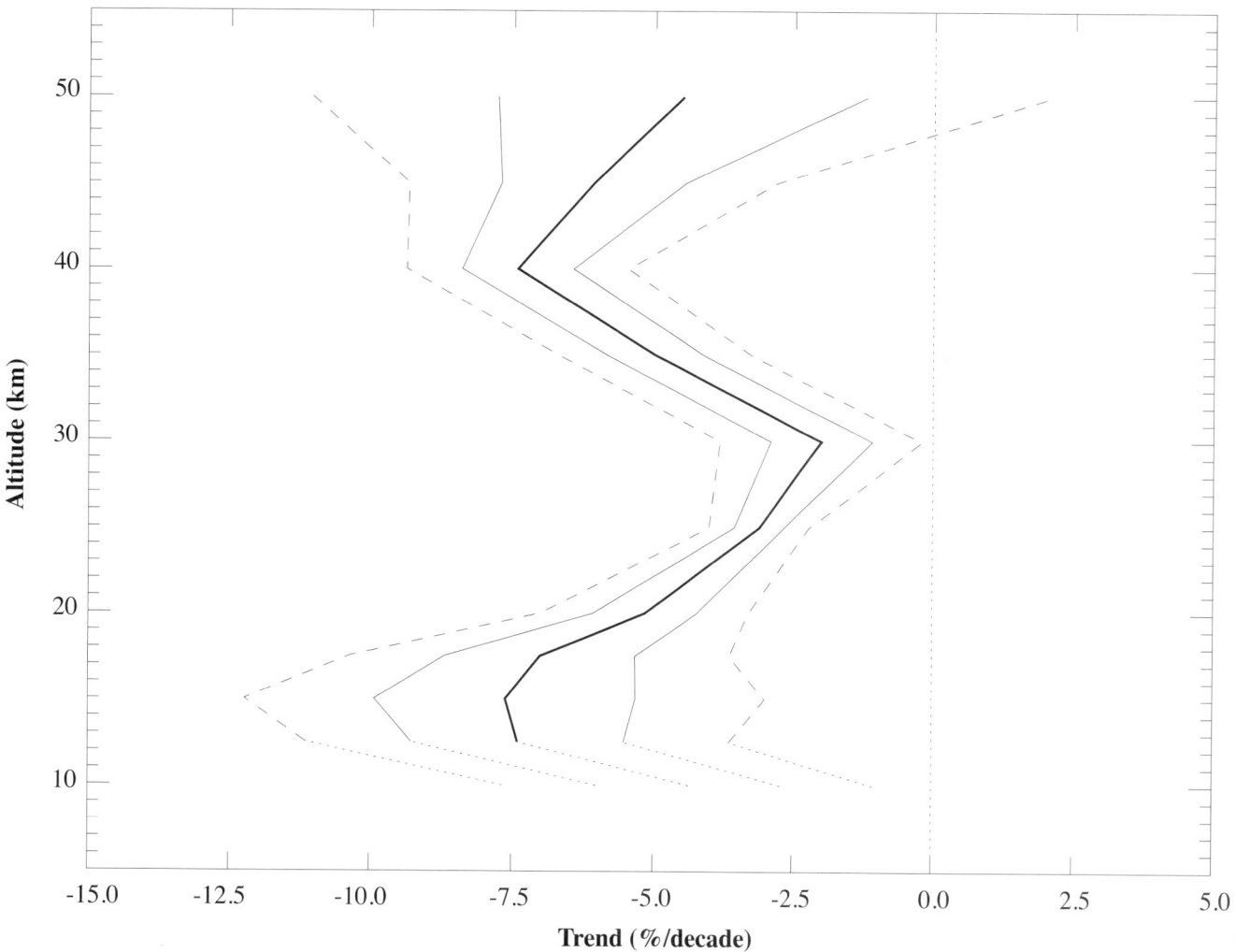

Figure 2-4: Estimate of mean trend using all four measurement systems (i.e., Umkehr, ozonesondes, SBUV, and SAGE I/II) at northern mid-latitudes (heavy solid line). Combined uncertainties are also shown as 1σ (light solid lines) and 2σ (dashed lines). Combined trends and uncertainties are extended down to 10 km as shown by the light dotted lines. The results below 15 km are a mixture of tropospheric and stratospheric trends, and the exact numbers should be viewed with caution. Combined trends have not been extended lower into the troposphere because the small sample of sonde stations have an additional unquantified uncertainty concerning their representativeness of mean trends (WMO, 1998).

available information on NO_x and aerosol is sparser and was obtained only by *in situ* sampling from aircraft and balloons (Hofmann, 1993; Blake and Kato, 1995; Emmons *et al.*, 1997). Satellite data are available for lower stratospheric aerosol, but the data record is relatively short and heavily influenced by recent volcanic eruptions. Analysis of NO_x and aerosol trends would be exceedingly difficult to interpret because the aircraft source would be convolved with many other increasing sources of anthropogenic NO_x and aerosol. In addition, the high degree of air mass variability in the troposphere places severe constraints on the atmospheric sampling strategy one would have to adopt to collect representative data.

In principle, aircraft signatures could be discerned from observation of the distribution of NO_x because the aircraft source is geographically distinct. *In situ* aircraft sampling efforts have begun to provide a global map of NO_x in the UT (Emmons *et al.*, 1997; see also Figure 2-6a). During the past few years, field campaigns have been performed specifically to investigate aircraft flight corridors. For example, observations in air traffic have been made by Schlager *et al.* (1996, 1997). The observation area was the major flight route in the eastern North Atlantic, and the parameters observed were NO_x, SO_2, and particles; observations were made perpendicular to flight tracks. Under special meteorological conditions associated with a stagnant anticyclone, measured

data indicated a large-scale accumulation of NO_x and particles from aircraft emissions.

Approximately 4,000 hours of NO_x measurements were collected from a B-747 platform during the Nitrogen Oxides and ozone measurements along Air Routes (NOXAR) project between spring 1995 and spring 1996, as shown in Figure 2-6b (Brunner, 1998). The NOXAR measurements demonstrated that, in addition to aircraft emissions, NO_x produced by lightning and NO_x emitted at the surface and transported upward by convection make large contributions to the NO_x abundance in the UT. These contributions were largest over and downstream of continents in summer. Finally, the recently completed SONEX and POLINAT II campaigns were designed specifically to quantify various NO_x sources in the UT. The findings of these latest studies are just now being reported.

2.3. Uncertainties in the Impact Assessment of Present Aviation: Implications for Use of Models in Predicting Future Change

2.3.1. Uncertainties in Modeling Aviation Impacts

2.3.1.1. Key Issues and Processes for Tropospheric Models

In performing assessments of the impacts of subsonic aircraft NO_x emissions on tropospheric composition, the modeling studies have pointed out several key issues and processes that have to be adequately addressed:

- Model spatial and time resolution
- Time resolution of meteorological data
- Subgrid-scale processes (e.g., plume processes)
- Tropospheric NO_x and NO_y sources
- Tropospheric gas-phase and heterogeneous chemistry
- Stratospheric gas-phase and heterogeneous chemistry
- Stratosphere-troposphere exchange
- Upper troposphere-lower stratosphere dynamics and convective transport
- Sources and sinks of water and NO_y.

These issues and processes are discussed in general terms in the paragraphs that follow. A detailed description of process representation in the assessment models is given in Chapter 4.

Over the years, there has been a steady increase in the spatial resolution of the models used in the assessment of subsonic aircraft impacts on tropospheric composition. Initially, the assessment models were one-dimensional (altitude), averaged around latitude circles and from north to south poles. Relatively quickly, researchers realized that interhemispheric gradients were crucial for ozone and anthropogenic trace gases, so much of the assessment work has been carried out with 2-D (altitude and latitude) models that average around latitude circles. With NO_x lifetimes of days or less and transport times around latitude circles of up to 2 weeks, researchers appreciated that 2-D models needed parameterizations to

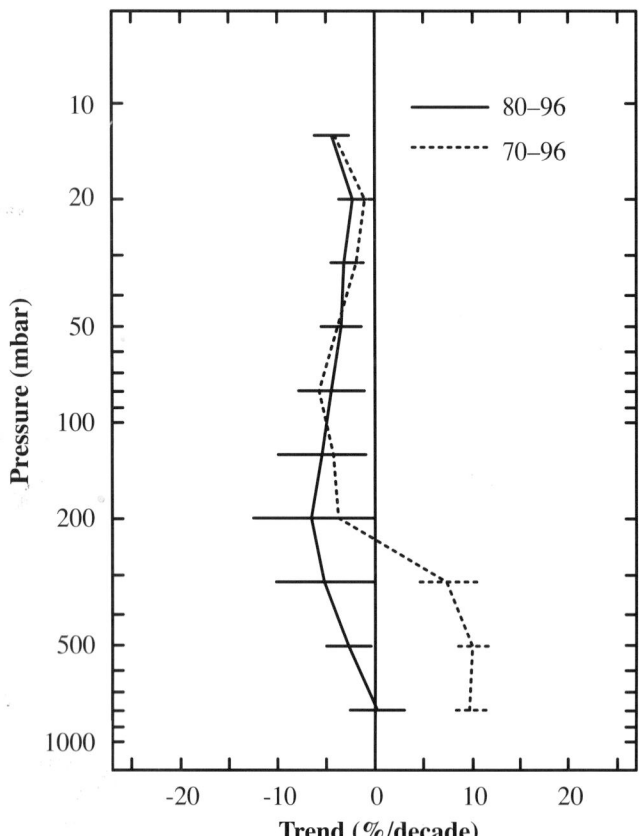

Figure 2-5: Annual trends for the periods 1970–96 and 1980–96 shown by ozonesonde measurements at Hohenpeissenberg, Germany. 95% confidence limits are shown [adapted from WMO (1998) by J. Logan].

represent the main features of the distributions of NO_x and other short-lifetime species. Some research teams investigated 2-D channel models (altitude and longitude) models, but the development of 3-D chemistry transport models (CTMs) has been the main thrust of tropospheric modeling efforts.

Global 3-D CTMs are now the main tools for the assessment of subsonic aircraft impacts in the troposphere. Typically, these models have horizontal resolutions of 2–6° by 2–6°, limited by the resolution of the general circulation model (GCM) from which they have been derived. Emissions databases are now available with higher spatial resolution, so model performance is limited by the meteorological databases used in CTMs and

the necessary computing time. Vertical resolution is a major limitation with current CTMs, in terms of the height taken as the top of the model and the number of layers into which the model domain is divided. Few models have enough vertical resolution to fully resolve the atmospheric boundary layer and tropopause domain and to describe the exchange of trace gases between the UT and the LS.

There is a major concern with 3-D CTMs regarding the adequacy of time resolution required in emission inventories and in meteorological data used to transport trace gases from their sources to their sinks. Initially, some CTMs used monthly averaged fields of horizontal and vertical winds, temperatures,

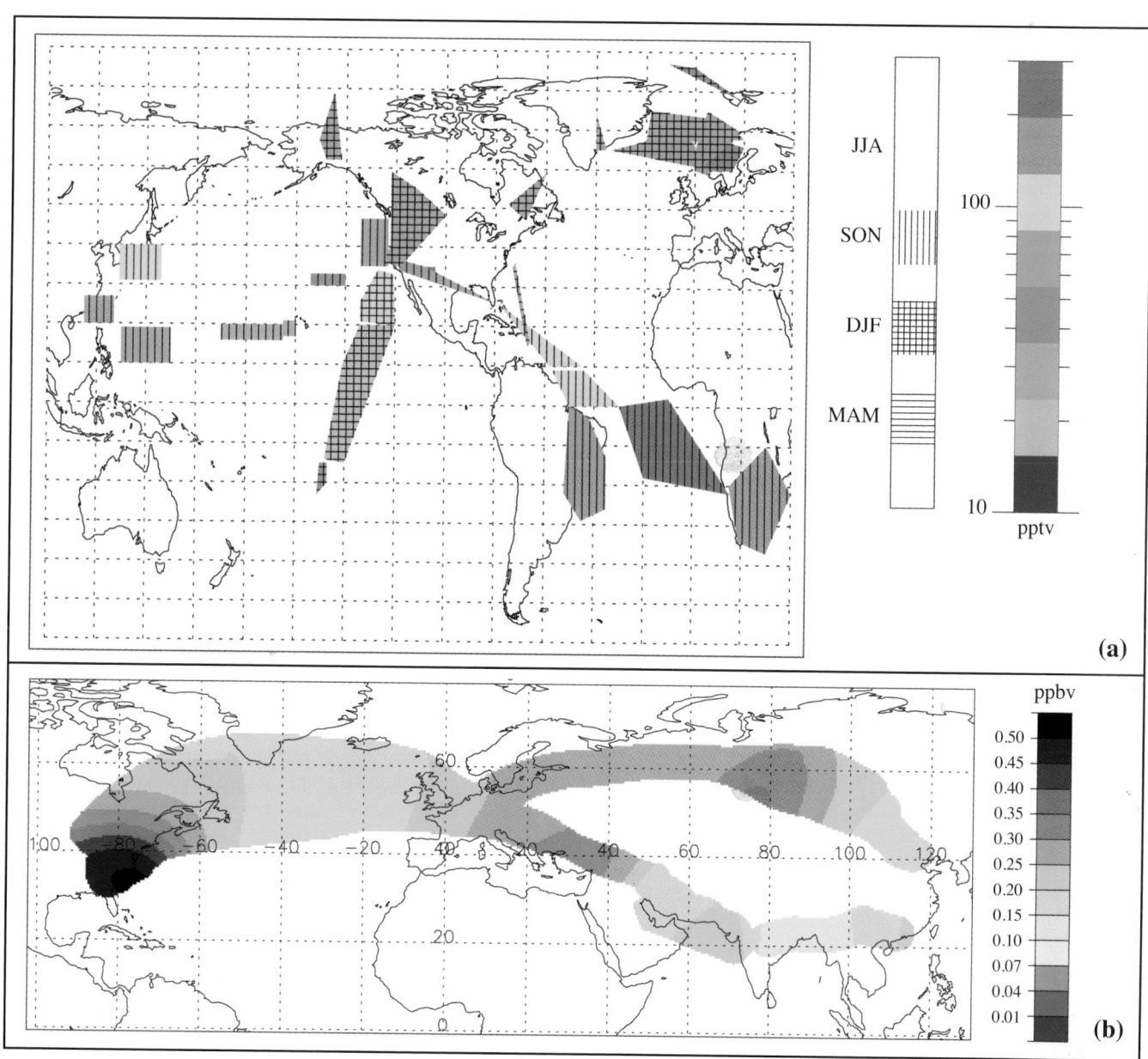

Figure 2-6: (a) Median NO_x mixing ratios measured between 9 and 12 km during a number of *in situ* aircraft campaigns (compiled in Emmons *et al.*, 1997); (b) NO_x concentration field in the altitude regions between 300 and 190 hPa obtained by the measurements of NOXAR. As an example, the results for summer are shown. The aircraft carrying the NOXAR instruments operated from Zürich to destinations in the USA and the far East (Beijing and Bombay-Hong Kong) from spring 1995 to spring 1996.

clouds, and humidities. To resolve major storm systems and convective events, the meteorological data have been updated in the CTMs on a steadily increasing frequency; fields are now usually updated every 6 hours. On this basis, it is possible to resolve the changing stability of the atmospheric boundary layer, the developing behavior of major weather systems, and large-scale convective events.

Time and spatial resolution are crucial issues in evaluating the impacts of subsonic aircraft. To evaluate whether the chosen time and spatial resolutions are adequate in each of the tropospheric assessment tools, a number of sensitivity studies should be carried out in the near future.

Concentration changes occurring in the aircraft plume and wake take place on a spatial scale (i.e., < 20 km) that is less than the smallest global atmospheric model scale (i.e., > 100 km). Consequently, global models do not typically treat aircraft near-field processes in detail. In fact, most current global model studies have input aircraft emissions inventories (i.e., emissions indexes, or EI) by simple dilution of the aircraft plume at the altitude of injection, with no chemical changes taking place in the near field. As a possible means of connecting near-field processes to the global model grid scale, Petry *et al.* (1998) and Karol *et al.* (1998) have proposed the concept of effective emissions index (EEI) to account for changes in species concentrations in the plume dispersion region resulting from photochemical reactions. As an example, $EEI(NO_x)$ will be less than the corresponding EIs because of plume processes that convert NO_x to NO_y. However, first estimates show that EEIs are very sensitive to temperature and light intensity, which results in a large variation of EEIs in latitude, altitude, and season (Karol *et al.*, 1997; Meijer *et al.*, 1997).

Recent ozone model studies have also pointed out the importance of background NO_x sources in understanding ozone tendencies with respect to increasing, or additional, NO_x sources. In a sense, the aircraft case is one example of a broader issue regarding nonlinearity between ozone impacts and NO_x levels. Model studies have shown that the magnitude of ozone changes from aircraft NO_x emissions may depend significantly on the amount of NO_x from non-aircraft sources. One difficulty with modeling background NO_x sources, however, is the short lifetime of NO_x (typically 1–5 days). In general, aircraft NO_x impacts on upper tropospheric and lower stratospheric ozone will be overstated if background NO_x concentrations are underestimated, and *vice versa*.

Over the past 2 decades, a significant amount of research has been committed to improving our understanding of background NO_x sources through model studies and observations of aircraft NO_x emissions. Much of the NO_x in the troposphere comes from surface NO_x sources, either via fast vertical transport as NO_x (Ehhalt *et al.*, 1992) or by conversion to temporary reservoir NO_y carriers such as PAN or HNO_3, followed by subsequent conversion back to NO_x. An accurate representation of the contribution made by surface NO_x sources to the UT and LS requires full treatment of boundary layer chemistry, exchanges

between the boundary layer and the free troposphere, deposition and wet scavenging, free tropospheric chemistry and transport to the upper troposphere by convection, atmospheric circulation, and synoptic-scale weather systems. An estimate of annual flux into the free troposphere from European surface NO_x sources, as a fraction of the total surface NO_x source, can be made from European Monitoring and Evaluation Program (EMEP) modeling studies (Tuovinen *et al.*, 1994). For NO_x, 52% of the emitted NO_x was deposited within the EMEP area of Europe and 48% was exported out of the model region during 1985–93. Approximately half of this material is vented into the free troposphere as NO_y (1.7 Tg N yr[-1]), with the remainder deposited elsewhere, without reaching the free troposphere.

In North America, a similar picture applies. Model calculations (Brost *et al.*, 1988) have estimated that about 1.8 Tg yr[-1] (of total North American emissions of 7.9 Tg N yr[-1]) is transported east to the Atlantic Ocean between the surface and 5.5-km altitude. That is, about 25% of the North American NO_x emissions remained airborne in the boundary layer or free troposphere as the air left North America. More recent studies (Jacob *et al.*, 1993; Horowitz *et al.*, 1998; Liang *et al.*, 1998) have derived a lower estimate of the transport out of North America (on the order of 6%).

A detailed model study of advective and convective venting of ozone, NO_x, and NO_y out of the boundary layer over northwest Europe during July and October–November 1991 showed that, of surface NO_x emissions, 7% were brought to the free troposphere as NO_x and 20% as NO_y during the summer (Flatoy and Hov, 1996), with slightly smaller percentages during the fall.

Transport from the surface is therefore undoubtedly an important contributor to background NO_x. However, it is difficult to evaluate how well this process is handled in each of the tropospheric assessment models used for aviation impact calculations.

Lightning is an important NO_x source in the UT (Chameides *et al.*, 1987; Lamarque *et al.*, 1996). Because of its sporadic nature and small spatial scale (tens of km), it is exceedingly difficult to represent quantitatively in even the most complex of tropospheric models. Most model studies include some representation of lightning NO_x, with global total emissions in the range 1–10 Tg N yr[-1]. However, there is no consensus on how to represent this source with time of day, season, altitude, or spatially, nor how to treat lightning in concert with convection, cloud processing, and wet scavenging.

Stratospheric NO_y is a further important source of NO_x in the UT and LS, through the photolysis of HNO_3 (Murphy *et al.*, 1993). There is a downward flux of NO_y from the stratosphere to the troposphere that globally balances the stratospheric NO_x source produced by the reaction of N_2O with $O(^1D)$. Few model studies of aircraft NO_x emissions extend high enough in altitude to include a full treatment of stratospheric NO_x and NO_y. Moreover, most do not include an explicit representation of stratospheric chemistry (i.e., halogen chemistry) from which to realistically calculate stratospheric NO_x. Instead, most models use a constant ratio of NO_y to ozone and describe the stratospheric NO_y source in the same way as the stratosphere-troposphere exchange of ozone. Typical ozone to NO_y ratios are assumed to be about 1000:1, giving a stratospheric NO_y source in the UT of about 0.5 Tg N yr^{-1}.

Although the issue of the magnitude of background NO_x levels has been clearly identified and much work has been performed to characterize surface, lightning, and stratospheric sources, there are still too few measurements of NO_x and NO_y in the UT and LS with which to assess quantitatively representations of background NO_x in the models summarized in Table 2-1. Recently published data (Emmons *et al.*, 1997) have begun to be used for evaluation of model performance (Wang *et al.*, 1998b).

As discussed in Section 2.1.2.4, the response of ozone to increasing NO_x depends on the strength of the HO_x source. Recent evidence (Brune *et al.*, 1998; Wennberg *et al.*, 1998) supports the presence of additional upper tropospheric HO_x sources from organic precursors that are not included in many current models. Improved model treatments of HO_x production from precursors such as acetone, peroxides, and aldehydes will require additional data on the mechanisms and kinetics of a number of NMHC reactions. The role of heterogeneous chemistry in influencing HO_x and NO_x levels in the UT has not been investigated fully yet and is expected to become an increasingly important issue for tropospheric models.

2.3.1.2. *Tropospheric Model Evaluation*

Previous IPCC reports (IPCC, 1996) identified tropospheric ozone modeling as one of the more difficult tasks in atmospheric chemistry. Difficulties arise, in part, from the large number of processes that control tropospheric ozone and its precursors and, in part, from the large range of spatial and temporal scales that must be resolved. Global 3-D CTMs attempt to simulate the life cycles of many trace gases and the impacts of subsonic aircraft NO_x emissions on them. We need to understand the level of confidence that is to be ascribed to these model studies.

There is a significant amount of scatter in current model assessments of the impacts of subsonic aircraft NO_x emissions on all aspects of tropospheric composition. With respect to reducing the range of uncertainty, it would be helpful if we could point to particular aspects of model performance and gauge models against specified benchmarks. Some model evaluation studies have begun the difficult task of identifying the current level of model performance and defining the level of confidence that should be placed in them. To this end, a number of model intercomparison exercises have been completed; some are in hand, and some are only at the planning stage. These exercises have involved the following elements:

- Transport of ^{222}Rn
- Fast photochemistry
- Transport of NO_x
- Comparison of model data and observations of tropospheric ozone.

2.3.1.2.1. *Transport of ^{222}Rn*

Twenty atmospheric models participated in the ^{222}Rn intercomparison for global CTMs (IPCC, 1996; Jacob *et al.*, 1997). Differences between model-calculated distributions of this short-lived (e-folding lifetime of 5.5 days) radioactive decay product emitted at the surface from soils were large, which enabled the drawing of conclusions about the general adequacy of transport schemes in CTMs. Owing to the lack of extensive observations, evaluation efforts to date have been restricted mainly to model-model intercomparisons.

The ^{222}Rn model intercomparison concluded that tropospheric CTMs based on 2-D models and monthly averaged 3-D models have a fundamental flaw in transporting tracers predominantly by diffusion; thus, these models cannot be viewed as reliable in simulating the global transport of tracers. Synoptic 3-D models need significantly improved representations of boundary layer processes, clouds, and convection. Large differences are found among established 3-D CTMs in the rates of global-scale meridional transport in the UT—particularly, interhemispheric transport. These latter differences are particularly relevant to the current issue of subsonic aircraft impacts.

2.3.1.2.2. *Fast photochemistry*

More than 20 model groups participated in the tropospheric photochemical model intercomparison exercise, PhotoComp—a tightly controlled experiment in which consistency was determined among models used to predict tropospheric ozone changes (IPCC, 1996; Olson *et al.*, 1997). A similar study, involving fewer models, was carried out as part of a U.S. National Aeronautics and Space Administration (NASA) assessment (Friedl, 1997). As with the radon case, there are no easy observational tests of model fast photochemistry, so model-model intercomparison exercises were carried out in both cases.

Over the intercomparison tests for fast photochemistry of the sunlit troposphere, modeled OH concentrations fell within a ±20% band, and ozone changes fell within a ±30% band. These obvious variations between model results did not correlate with other model differences, and no single model input parameter appeared to account for all of the spread in the results. Nevertheless, ozone photolysis rates used in the models accounted for about half of the root mean square (RMS) differences; further investigation of these parameters and their comparison with observations is called for. The results also became more uncertain in model experiments involving NMHCs. Further CTM development is required so that models have the required grid and time resolutions to simulate accurately the scales of chemistry required to describe the removal of NO_x and NMHCs while producing and destroying ozone, quantitatively.

2.3.1.2.3. *Transport of NO_x*

Passive transport of subsonic aircraft NO_x emissions has been studied with a hierachy of global CTMs (Friedl, 1997; van

Velthoven *et al.*, 1997). The 3-D CTMs showed that the monthly mean NO_x concentrations varied by a factor of three longitudinally and that the temporal variability of background NO_x in the air traffic corridor was about ±30% on synoptic time scales. Vertical redistribution by convection strongly affected the maximum NO_x concentrations at subsonic aircraft cruise altitudes.

A number of model deficiencies and biases were found, including the oscillatory nature of NO_x distributions obtained with a spectral advection scheme, the strong diffusion of GCMs into polar regions, and the too-intense interhemispheric exchange found in some 2-D CTMs. The intercomparisons concluded that assessment of the tropospheric impacts of subsonic aircraft NO_x emissions could be performed better with 3-D CTMs.

2.3.1.2.4.   *Comparison of model data and observations of tropospheric ozone*

An increasing number of activities are aimed at evaluating global model results in relation to ozone observations (Wang *et al.*, 1998b; Wauben *et al.*, 1998). However, there are too few ozone data, especially in the tropics, to allow for comprehensive evaluations. Comparisons are showing that model simulations are reproducing the broad features of monthly mean measured ozone concentrations. Some models do not produce the observed seasonality in the northern mid-latitude troposphere. Differences are most pronounced in the free troposphere, especially close to the tropopause (see Figure 2-7).

As part of the International Global Atmospheric Chemistry Project/Global Integration and Modeling Activity (IGAC/GIM) study, an intercomparison exercise is currently being attempted of ozone concentrations calculated by 12 global 3-D CTMs (Kanikidou *et al.*, 1998). Many of these CTMs have already performed assessments of the impacts of subsonic aircraft NO_x emissions on tropospheric ozone; their results have been included in Table 2-1. Furthermore, all of the tropospheric assessment models employed in Chapter 4 have submitted results to the IGAC/GIM intercomparison.

The GIM intercomparison extends the intercomparisons described above in that it employs some of the available observational database to evaluate intermodel differences. Figure 2-7 presents some of the model intercomparison results for seasonal cycles of ozone at 300 mb at three widely separated sites.

The GIM model intercomparison with monthly mean values of observations demonstrates that the models capture some of the considerable variability within the observations. The range in observations may approach 20 ppb at 500 mb and up to 40 ppb at 300 mb, with the ranges in the models significantly greater. These ranges are significantly greater than the tropospheric ozone impacts of about 8 ppb anticipated from subsonic aircraft NO_x emissions (Table 2-1).

2.3.1.3.   *Key Issues and Processes for Stratospheric Models*

CFC and HSCT assessment activities have engaged 2-D (height and latitude) and, to some extent, 3-D (height, latitude, and longitude) models focused on the stratosphere over the past 10 years. These efforts have served to highlight a number of critical stratospheric model issues:

- Aircraft plume processes
- Stratospheric transport
- Stratospheric gas-phase and heterogeneous chemistry
- Sulfate aerosol
- PSCs
- Soot.

The issue of soot has been raised only to a small extent by the HSCT studies, although it has assumed a more prominent role

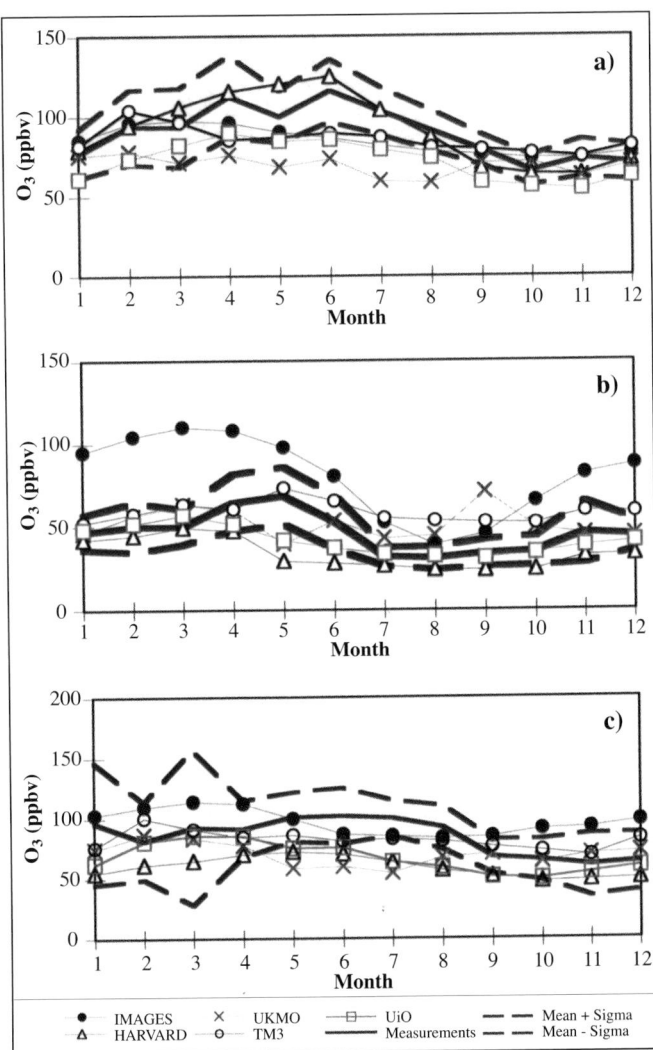

Figure 2-7: Comparison of modeled and observed ozone concentrations at 300 mb pressure-height for three locations: (a) Hohenpeissenberg, Germany (48°N, 11°E); (b) Hilo, Hawaii, USA (20°N, 155°W); and (c) Wallops Island, Virginia, USA (38°N, 76°W). Descriptions of the models are given in Chapter 4. Measurements are from ozonesondes.

in the subsonic aviation case (see Section 2.1.3). In the following paragraphs, we discuss these issues in the context of current model treatments of subsonic aviation impacts.

Aircraft emissions, whether supersonic or subsonic, are deposited primarily at northern mid-latitudes and over a limited vertical range. A key issue for models is how fast these emissions are dispersed to other regions of the atmosphere, such as the tropical stratosphere or the mid-latitude troposphere, where the response of ozone to the emissions will be substantially different. In addition, it is important to consider the chemistry occurring in the aircraft plume and wake before it has been expanded to the model grid scale. Initial attempts to combine near field, far field, and global models in series (Danilin *et al.*, 1997) are the first global impact studies to be based directly on detailed microphysics and chemical kinetics occurring in the aircraft plume and wake. An increasingly robust plume and wake observational database is being collected to validate this approach (Kärcher, 1998; Kärcher *et al.*, 1998b).

To date, most models used to assess the impact of aviation on the atmosphere have been 2-D, in which the time-consuming complexity of the real 3-D atmosphere is reduced to a manageable calculation by averaging around latitude circles. Because of this simplification, 2-D models do not adequately simulate all dynamic features of the atmosphere. Horizontal transport between mid-latitudes and the tropics (or polar vortex) is an inherently episodic, wave-driven process that is parameterized in 2-D models by eddy diffusion terms. Measurements of the NO_y-to-ozone ratio in the LS have provided evidence for distinctly different airmass characteristics that are not well represented in 2-D models (Murphy *et al.*, 1993; Minschwaner *et al.*, 1996; Volk et al., 1996; Schoeberl *et al.*, 1997). One method for improving the 2-D representation of tropical/extratropical air mass difference has been to reduce the horizontal eddy coefficient in the subtropical region. Efforts such as these have underscored the fact that an accurate model representation of tropical/mid-latitude air mass distinctions, including the extent of transport of tropical air into mid-latitudes, remains an important assessment uncertainty.

Model representation of bulk, global-scale vertical exchange between the stratosphere and the troposphere by diabatic circulation is likely adequate (Holton *et al.*, 1995). However, most models do not adequately resolve tropopause-folding events or stratosphere-troposphere exchange along isentropic surfaces. To the extent that these processes are important, calculated aviation impacts will be sensitive to model horizontal and vertical resolution.

Model representation of gas-phase photochemical links between ozone and atmospheric trace species such as HO_x and NO_x may be the most mature area of model construction, although rate parameter uncertainties increase with decreasing temperature. This representation is facilitated by the existence of evaluated compilations of photochemical rate parameters (IUPAC, 1997a,b; JPL, 1997). Because of the sensitivity of reaction rates to temperature and photolysis rates to solar zenith angle, model treatments must account for temperature and solar flux changes as air parcels move around the globe and encounter day and nighttime conditions. Diurnal variations in calculated radical concentrations can be reproduced either by invoking an explicit time marching kinetic scheme or by applying a correction factor to concentrations calculated from averaged solar zenith angles.

The dependence of reaction rate coefficients on temperature, especially for PSC processes, can present a particular problem for 2-D models, which are constrained to zonal-mean temperature fields. One strategy to address zonal variations has been to describe the zonal mean temperature by a probability distribution (Considine *et al.*, 1994). The applicability of this approach to PSC processes is an area of active investigation. Type II PSC particles, consisting of water-ice and uniformally formed at temperatures below 188 K, can be adequately captured in 2-D formulations. However, the temperature thresholds for PSC type I particle formation are highly variable because of the multitude of possible particle compositions, and they depend more heavily on the temperature histories of air parcels. Some of the type I PSCs considered in stratospheric models include solid nitric acid hydrates (e.g., trihydrate and dihydrate), mixed hydrates, and supercooled sulfate, nitrate, and water ternary solutions (Worsnop *et al.*, 1993; Carslaw *et al.*, 1994; Tabazadeh *et al.*, 1994; Fox *et al.*, 1995). Compositional details of modeled PSC type Is are important because they determine what the model will calculate for the size, density, and removal rates (by sedimentation) of the particles as well as the partitioning of NO_y between gaseous and condensed phases.

Finally, stratospheric models must describe background sulfate and carbonaceous aerosol formation and evolution adequately to gauge perturbations from aircraft SO_x and soot emissions. In past studies, models have merely prescribed aerosol suface area distributions based on satellite observations. Recognition that aircraft exhaust may contain a large number of small-diameter sulfate particles has motivated development of aerosol microphysical schemes (Weisenstein *et al.*, 1996).

2.3.1.4. *Stratospheric Model Evaluation*

The growing body of satellite, balloon, and aircraft chemical and meteorological data for the middle atmosphere has made it possible to devise tests of photochemistry and transport within stratospheric models. A number of 2-D and 3-D models have participated in two major intercomparison efforts, Models and Measurements (M&M) I and II (Prather and Remsberg, 1993; Park, 1999). These comparisons have focused on testing the ability of these models to estimate the atmospheric effects of a proposed fleet of supersonic aircraft that would operate near 20 km. As a direct result of the first M&M effort, a number of errors in the models were identified and corrected. Both M&M efforts have served to highlight important tests of model representations. Because of the supersonic aircraft focus, however, less analysis has been directed at model performance in the lowermost stratosphere, where subsonic aviation effects are expected. With the exception of ozone representation, rigorous tests of model representation of the dynamics and chemistry of the lowermost stratosphere and UT have not been performed to date. Poor agreement between model predictions and observations of ozone in this region of the atmosphere (typical errors greater than 50%) suggests that significant improvement will be required before stratospheric assessment models can be used to examine the impact of aviation (or, for that matter, any perturbation) on the lowermost stratosphere and UT. In the following paragraphs, we summarize comparison efforts for the following key issues (for altitudes above 15 km):

- Photochemistry
- Dynamics
- Comparison of model data and observations of stratospheric ozone.

2.3.1.4.1.   Photochemistry

The photochemical mechanisms employed by most of the models compare well with each other. Tests of the photochemical mechanisms were performed by comparing predicted concentrations of short-lived reactive chemicals from these models against a benchmark photo-stationary state model constrained by the distribution of precursors from each 2-D model. These comparisons provide a means of accounting for differences in the transport of long-lived species, such as NO_y, and O_3, within the models. The distribution of NO_y versus altitude and the mixing ratio of N_2O was markedly different among the various 2-D models. Most of the differences for calculated concentrations of hydrogen, nitrogen, and chlorine free radicals among the various 2-D models were shown to be caused by differences in NO_y and to a lesser degree ozone. The benchmark model has been tested extensively against atmospheric observations and has been shown to generally reproduce observed concentrations of OH, HO_2, NO, NO_2, and ClO in the stratosphere to within ±30%, provided precursor fields and aerosol surface areas are accurately known.

However, no significant tests of the model photochemistry of the lowermost stratosphere were performed during the recent M&M workshop. The chemistry of this region is considerably different. For example, the relatively high ratio of CO to ozone implies that ozone production from the oxidation of CO is much more important in this region than at higher altitudes. Furthermore, at the tropopause and below, saturated conditions often exist; therefore, chemical processes occurring on ice particles may be important. In addition, because this air is influenced by mixing of reactive trace gases from the lower troposphere, these models must consider transport of a larger number of reactive species than they typically do.

2.3.1.4.2.   Dynamics

Tests of the dynamics within the 2-D and 3-D models during both M&M I and II revealed a number of problems. In general,

the mean age of air within the stratosphere is much older than predicted by these models. Measurements of CO_2 (Boering *et al.*, 1996) and sulfur hexafluoride (SF_6) (Elkins *et al.*, 1996), both of which are increasing rapidly, provide a means of dating stratospheric air. The models had a high dispersion in predicted conversion rate of N_2O to NO_y. It is unclear whether this dispersion reflects errors in dynamics or chemistry related to the high-altitude sink of NO_y. This is a key point: If the assessment models are unable to accurately simulate observed concentrations of total NO_y, their ability to predict the influence of additional NO_y from aircraft on ozone will remain relatively uncertain. Transport in the lowermost stratosphere is considerably different, and in many ways even more difficult to represent in 2-D models, than transport at higher altitudes. This fact certainly does not bode well for the ability of current 2-D models to describe accurately the dynamic context within which the current subsonic fleet is operating.

2.3.1.4.3.   Comparison of model data and observations of stratospheric ozone

As part of the M&M II effort, results from a group of stratospheric models were compared with a recently developed ozone climatology (WMO, 1998). The data used for the climatology are from sonde stations and from SAGE II, the latter data set having been evaluated by comparison with other satellite, lidar, sonde, and Umkehr data. Although agreement between models and between models and observations is relatively good above 25 km, differences between modeled and observed ozone are found to increase rapidly below 25 km and are largest between 20 km and the tropopause. The modeled ozone tends to be larger than observed ozone by up to a factor of 2 at these altitudes.

Overestimation of LS ozone in some models may be ascribed partly to the fact that they have tropopauses at mid-latitudes that are either invariant or do not vary correctly with season. However, based on chemistry and dynamics tests described in the preceding subsections, it is likely that differences between models and observations are caused in large part by deficiencies in model transport representation.

2.3.1.5.   Implications for Modeling Aviation Impacts

Global tropospheric 3-D CTMs are now the main modeling tools for climate-chemistry studies, including the role of subsonic aircraft NO_x emissions. Although 3-D models with high temporal and spatial resolution have performed significantly better than 2-D or monthly averaged 3-D CTMs in the ^{222}Rn, PhotoComp, NO_x, and Ozone/GIM intercomparison exercises, key fundamental problems have been identified that are crucial to the representation of the impacts of subsonic NO_x emissions from aircraft.

3-D CTM studies have provided only preliminary estimates of subsonic impacts, which exhibit significant scatter, as Table 2-1 shows. At present, we are unable to rationalize these real differences in results between studies because there is no one aspect of input data or process parameterization that can account for the spread in model results. Furthermore, the extent of model evaluation is highly variable, and no models have been evaluated comprehensively against all of the key issues detailed in Section 2.3.1.1.

These same difficulties apply to the subset of models adopted in Chapter 4 to examine the future impact of subsonic aircraft. There is no suggestion that these models have any distinguishing features that identify them as being inherently more or less reliable for assessment of the tropospheric impacts of subsonic aircraft NO_x emissions. Furthermore, we have no concrete means of establishing a higher level of confidence in the models used in Chapter 4, compared with any of the similar 3-D models listed in Table 2-1.

Although the effects of present aviation on ozone are calculated to be much smaller in the stratosphere than in the troposphere —primarily because of the smaller fraction of exhaust released into the stratosphere—the performance of 2-D stratospheric models has not been extensively evaluated in the lowermost stratospheric region. Consequently, the results reported in Section 2.3.1.3 represent only preliminary estimates of subsonic aviation impacts on the stratosphere. The modeling situation is significantly better for evaluating the effects of future supersonic aircraft in that a number of intercomparisons have established the general quality of modeled middle stratosphere photochemistry. However, confident predictions of stratospheric effects of future aviation will require resolution of discrepancies between modeled and observed transport tracers.

2.3.2. *Uncertainties in Observing Aviation Impacts*

The data set resulting from ozonesondes is the only useful one for ozone trend analysis in the UT and LS. The error of an individual ozonesonde measurement has been evaluated to be ~5% in the LS, based on several intercomparison campaigns (WMO, 1998). The error is larger in the UT, where ozone densities, hence instrument signals, are substantially smaller. In addition, the background signals (i.e., dark current) of the sonde sensors have been checked relatively infrequently during the measurement period, giving rise to further measurement uncertainty. If the measurement error is random, one can improve the statistical significance of observed trends by increasing the observation frequency. Ozone densities vary greatly on time scales of days in the UT and LS, particularly in middle and high latitudes during the winter and spring. The cause of this variability is believed to be related to active dynamic transport associated with weather disturbances. Because the variability is largely random, it can be treated, to first order, as noise in the trend data. The variability is considered to be of the same order of magnitude (or larger) as noise from instrument measurement errors. The frequency of ozonesonde observation—once a week at

most stations—is not enough to document these variations properly.

Long-term trends of external forcings other than aircraft greatly complicate analysis of ozone trends. The long-term variation of atmospheric chlorine loading is relatively well-documented, allowing for the construction of credible models to predict stratospheric ozone depletion. However, changes in gases important in UT photochemistry—such as NO_x, oxygenated hydrocarbons, and water vapor—are much less well characterized. Feedbacks on tropospheric gases from climatic changes (e.g., greenhouse warming) may also have an impact on ozone in the UT and LS, but even the sign of this effect on ozone levels is uncertain.

In summary, because the database for ozone observations in the UT and LS is still relatively limited and because uncertainties in observational data, as well as model representations of non-aircraft ozone forcing phenomena, are quite large, it is presently impossible to associate a trend in ozone to aircraft operation with meaningful statistical significance.

2.4. Conclusions and Overall Assessment of Present Aviation Impacts on Ozone

Currently, there is no experimental evidence for a large geographical effect of aircraft emissions on ozone anywhere in the troposphere. Furthermore, the only evidence for an effect on NO_x—the major ozone precursor in aircraft emissions anywhere outside the immediate vicinity (i.e., a few miles) of a jet engine's exhaust—has been obtained during a stagnant meterological condition when exhaust products built up over several days. Nevertheless, our understanding of UT/LS chemical and dynamical processes continues to improve and has progressed to a point where one can predict with some confidence the cruise-level effects of aviation.

Based on our current overall understanding of UT and LS processes, we are confident that NO_x emissions from present subsonic aircraft lead to increased NO_x and ozone concentrations at cruise altitudes, especially in air traffic corridors between and over Northern Hemisphere continents and at altitudes of 9–13 km. Based on the relatively large number of tropospheric model calculations, we are reasonably confident that tropospheric ozone increases from aircraft NO_x have been on the order of 8 ppb, equivalent to 6% of the ozone density in the principal traffic areas.

Model studies, which have internal uncertainties associated with process parameterization and external uncertainties associated with the strengths of other very large NO_x sources, have returned effects as low as 2% of the ozone density in high-traffic areas and as high as 14% in those areas. One of the major current limitations to the models' credibility in assessing aircraft emissions is the identification and quantification of background NO_x levels and sources. Recent HO_x measurements allow for a much better understanding of ozone production in the UT, and these measurements have shown that additional HO_x sources are necessary to explain the observations. Moreover, these additional HO_x sources cause the sensitivity of ozone production from NO_x emissions to be higher than previously thought.

Much less confidence is attached to our understanding of the effects of NO_x emissions in the lowermost stratosphere and aerosol emissions in the troposphere and stratosphere. The available data suggest that these effects are smaller than (and, in the case of aerosols, of opposite sign) those of NO_x emissions in the UT (see Figure 2-8).

Based on our model predictions, the impact of present subsonic NO_x emissions on ozone is well within the range of interannual variability of ozone concentrations in the UT as measured with ozonesondes. Furthermore, expressed as a decadal trend, the impact of subsonic NO_x emissions on upper tropospheric and lower stratospheric ozone is smaller than or comparable to the span of confidence limits in the ozone trend analysis for mid-latitude stations. Finally, we note that aircraft NO_x emissions should lead to decreased CH_4 concentrations; however, any impact should be undetectable in the CH_4 record.

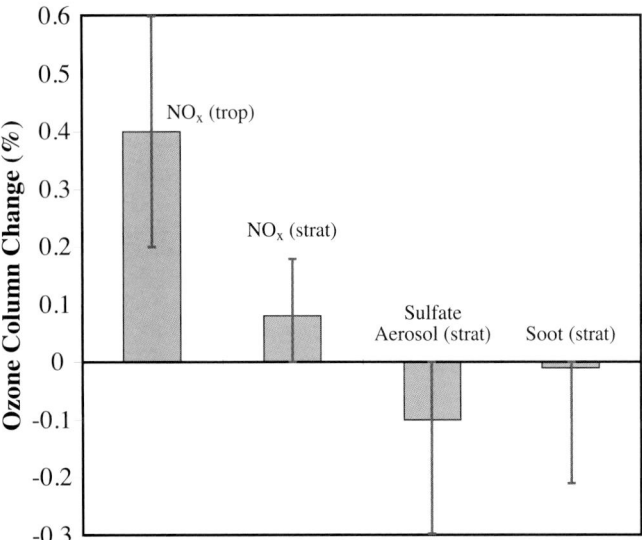

Figure 2-8: Estimates of northern mid-latitude total ozone column changes (%) from NO_x emission in the troposphere and stratosphere and aerosol emissions in the stratosphere from present subsonic aviation. The height of the rectangular bars indicates the mid-range estimate; the error bars represent an estimate of the two-thirds probability uncertainty range (i.e., a 67% chance that the true value falls within this range) as determined from either the dispersion among global model results or from consideration of process understanding. Independent of the specified uncertainty range, we judge the state of scientific understanding regarding aviation's NO_x and particle effects to be fair and poor, respectively. This evaluation is based on the level of consensus in the scientific literature and the amount of evidence available to support the estimates and their uncertainties.

References

Allanic, A., R. Oppliger, and M.J. Rossi, 1997: Real-time kinetics of the uptake of HOBr and $BrONO_2$ on ice and in the presence of HCl in the temperature range 190–200K. *Journal of Geophysical Research*, **D102**, 23529.

Anderson, M.R., R.C. Miake-Lye, R.C. Brown, and C.E. Kolb, 1996: Calculation of exhaust plume structure and emissions of the ER-2 aircraft in the stratosphere. *Journal of Geophysical Research*, **101**, 4025–4032.

Arnold, F., J. Schneider, K. Gollinger, H. Schlager, P. Schulte, D.E. Hage, P.D. Whitefield, and P. van Velthoven, 1997: Observations of upper tropospheric sulfur dioxide- and acetone-pollution: potential implications for hydroxyl radical and aerosol formation. *Geophysical Research Letters*, **24**, 57–60.

Arnold, F., J. Schneider, M. Klemm, J. Scheid, T. Stilp, H. Schlager, P. Schulte, and M.E. Reinhardt, 1994: Mass spectrometric measurement of SO_2 and reactive nitrogen gases in exhaust plumes of commercial jet airliners at cruise altitude. In: *Impact of Emissions from Aircraft and Spacecraft upon the Atmosphere* [Schumann, U. and D. Wurzel (eds.)]. Proceedings of an international scientific colloquium, 18–20 April 1994, Cologne, Germany. DLR-Mitteilung 94-06, Deutsches Zentrum für Luft- und Raumfahrt (German Aerospace Center), Oberpfaffenhoffen and Cologne, Germany, pp. 323–328.

Arnold, F., J. Scheid, T. Stilp, H. Schlager, and M.E. Reinhardt, 1992: Measurements of jet aircraft emissions at cruise altitude. I: The odd-nitrogen gases NO, NO_2, HNO_2 and HNO_3. *Geophysical Research Letters*, **19**, 2421–2424.

Atherton, C.S., 1995: *Biomass Burning Sources of Nitrogen Oxides, Carbon Monoxide, and Non-Methane Hydrocarbons*. UCRL-ID-122583, Lawrence Livermore National Laboratory, Livermore, CA, USA, 15 pp.

Atherton, C.S., S. Grotch, D.D. Parrish, J.E. Penner, and J.J. Walton, 1996: The role of anthropogenic emissions of NO_x on tropospheric ozone over the North Atlantic ocean—a 3-dimensional global model study. *Atmospheric Environment*, **30**, 1739–1749.

Barone, S.B., M.A. Zondlo, and M.A. Tolbert, 1997: A kinetic and product study of the hydrolysis of $ClONO_2$ on type Ia polar stratospheric cloud materials at 185K. *Journal of Physical Chemistry*, **A101**, 8643–8652.

Baughcum, S.L., 1996: *Aircraft Emissions Deposited in the Stratosphere and Within the Arctic Polar Vortex*. NASA-CR-4714, National Aeronautics and Space Administration, Langley Research Center, Hampton, VA, USA, 131 pp.

Baughcum, S.L., T.G. Tritz, S.C. Henderson, and D.C. Pickett, 1996: *Scheduled Civil Aircraft Emission Inventories for 1992: Database Development and Analysis*. NASA-CR-4700, National Aeronautics and Space Administration, Langley Research Center, Hampton, VA, USA, 205 pp.

Becker, K.H., J. Kleffmann, R. Kurtenbach, and P. Wiesen, 1996: Solubility of nitrous acid (HONO) in sulfuric acid solutions. *Journal of Physical Chemistry*, **100**, 14984–14990.

Beier, K. and F. Schreier, 1994: Modelling of aircraft exhaust emissions and infrared spectra for measurements of nitrogen oxide. *Annales Geophysicae*, **12**, 920–943.

Bekki, S., 1997: On the possible role of aircraft-generated soot in the middle latitude ozone depletion. *Journal of Geophysical Research*, **102**, 10751–10758.

Benkovitz, C.M., J. Dignon, J. Pacyna, T. Scholtz, L. Tarrasson, E. Voldner, and T.E. Graedel, 1996: Global inventories of anthropogenic emissions of SO_2 and NO_x. *Journal of Geophysical Research*, **101**, 29239–29254.

Blake, D.F. and K. Kato, 1995: Latitudinal distribution of black carbon soot in the upper troposphere and lower stratosphere. *Journal of Geophysical Research*, **100**, 7195–7202.

Boering, K.A., S.C. Wofsy, B.C. Daube, H.R. Schneider, M. Loewenstein, and J.R. Podolske, 1996: Stratospheric mean ages and transport rates from observations of carbon dioxide and nitrous oxide. *Science*, **274**, 1340–1343.

Borrmann, S., S. Solomon, J.E. Dye, and B. Luo, 1996: The potential of cirrus clouds for heterogeneous chlorine activation. *Geophysical Research Letters*, **23**, 2133–2136.

Brasseur, G.P., R.A. Cox, D. Hauglustaine, I. Isaksen, J. Lelieveld, D.H. Lister, R. Sausen, U. Schumann, A. Wahner, and P. Wiesen, 1998: European scientific assessment of the atmospheric effects of aircraft emissions. *Atmospheric Environment*, **32**, 2327–2422.

Brewer, A.W., 1949: Evidence for a world circulation provided by the measurements of helium and water vapour distribution in the stratosphere. *Quarterly Journal of the Royal Meteorological Society*, **75**, 351–363.

Brost, R.A., R.B. Chatfield, J.P. Greenberg, P.L. Haagenson, S. Madronich, B.A. Ridley, and R.R. Zimmerman, 1988: Three-dimensional modeling of transport of chemical species from continents to the Atlantic Ocean. *Tellus,* **40B,** 358–379.

Brouwer, J., M.J. Rossi, and D.M. Golden, 1986: Reaction of N_2O_5 on carbonaceous surfaces. *Journal of Physical Chemistry,* **90,** 4599.

Brühl, C., P.J. Crutzen, and J.U. Grooß, 1998: High-latitude, summertime NO_x activation and seasonal ozone decline in the lower stratosphere: Model calculation based on observations by HALOE on UARS. *Journal of Geophysical Research,* **103,** 3587–3597.

Brune, W.H., I.C. Fallona, D. Tan, A.J. Weinheimer, T. Campos, B.A. Ridley, S.A. Vay, J.E. Collins, G.W. Sachse, L. Jaeglé, and D.J. Jacob, 1998: Airborne in-situ OH and HO_2 observations in the cloud-free troposphere and lower stratosphere during SUCCESS. *Geophysical Research Letters,* **25,** 1701–1704.

Brunner, D., J. Staehelin, and D. Jeker, 1998. Large-scale nitrogen oxide plumes in the tropopause region and implications for ozone. *Science,* **282,** 1305–1309.

Calvert, J.G., A. Lazrus, G.L. Kok, B.G. Heikes, J.G. Walega, J.G. Lind, and C.A. Cantrell, 1985: Chemical mechanisms of acid generation in the troposphere. *Nature,* **317,** 27–35.

Carslaw, K.S., B.P. Luo, S.L. Clegg, T. Peter, P. Brimblecombe, and P.J. Crutzen, 1994: Stratospheric aerosol growth and HNO_3 gas-phase depletion from coupled HNO_3 and water-uptake by liquid particles. *Geophysical Research Letters,* **21,** 2479–2482.

Chameides, W.L., D.D. Davis, J. Bradshaw, M. Rodgers, S. Sandholm, and D.B. Bai, 1987: An estimate of the NO_x production rate in electrified clouds based on NO observations from the GTE CITE-1 fall 1983 operation. *Journal of Geophysical Research,* **92,** 2153–2156.

Chatfield, R.B. and P.J. Crutzen, 1984: Sulfur dioxide in remote oceanic air: Cloud transport of reactive precursors. *Journal of Geophysical Research,* **89,** 7111–7132.

Chughtai, A.R., W.F. Welch, Jr., M.S. Akhter, and D.M. Smith, 1990: A spectroscopic study of gaseous products of soot-oxides of nitrogen/water reactions. *Applied Spectroscopy,* **44,** 294–298.

Considine, D.B., A.R. Douglass, and C.H. Jackman, 1994: Effects of a polar stratospheric cloud parameterization on ozone depletion due to stratospheric aircraft in a 2-dimensional model. *Journal of Geophysical Research,* **99,** 18879–18894.

Crutzen, P.J. and M.O. Andreae, 1990: Biomass burning in the tropics—impact on atmospheric chemistry and biogeochemical cycles. *Science,* **250,** 1669–1678.

Danilin, M.Y., J.M. Rodriguez, M.K.W. Ko, D.K. Weisenstein, R.C. Brown, R.C. Miake-Lye, and M.R. Anderson, 1997: Aerosol particle evolution in an aircraft wake: implications for the high-speed civil transport fleet impact on ozone. *Journal of Geophysical Research,* **102,** 21453–21463.

Dessler, A.E., E.J. Hintsa, E.M. Weinstock, J.G. Anderson, and K.R. Chan, 1995: Mechanisms controlling water vapor in the lower stratosphere: a tale of two stratospheres. *Journal of Geophysical Research,* **100,** 23167–23172.

Dignon, J., 1992: NO_x and SO_x emissions from fossil fuels: a global distribution. *Atmospheric Environment,* **A26,** 1157–1163.

Dowling, D.R. and L.F. Radke, 1990: A summary of physical properties of cirrus clouds. *Journal of Applied Meteorology,* **29,** 970–978.

Durbeck, T. and T. Gerz, 1996: Dispersion of aircraft exhausts in the free atmosphere. *Journal of Geophysical Research,* **101,** 26007–26015.

Ehhalt, D.H., F. Rohrer, and A. Wahner, 1992: Sources and distribution of NO_x in the upper troposphere at northern mid-latitudes. *Journal of Geophysical Research,* **97,** 3725–3738.

Elkins, J.W., D.W. Fahey, J.M. Gilligan, G.S. Dutton, T.J. Baring, C.M. Volk, R.E. Dunn, R.C. Myers, S.A. Montzka, P.R. Wamsley, A.H. Hayden, J.H. Butler, T.M. Thompson, T.H. Swanson, E.J. Dlugokencky, P.C. Novelli, D.F. Hurst, J.M. Lobert, S.J. Ciciora, R.J. McLaughlin, T.L. Thompson, R.H. Winkler, P.J. Fraser, L.P. Steele, and M.P. Lucarelli, 1996: Airborne gas chromatograph for in situ measurements of long-lived species in the upper troposphere and lower stratosphere. *Geophysical Research Letters,* **23,** 347–350.

Emmons, L.K., M.A. Carroll, D.A. Hauglustaine, G.P. Brasseur, C. Atherton, J. Penner, S. Sillman, H. Levy II, F. Rohrer, W.M.F. Wauben, P.F.J. van Velthoven, Y. Wang, D.J. Jacob, P. Bakwin, R. Dickerson, B. Doddridge, C. Gerbig, R. Honrath, G. Hubler, D. Jaffe, Y. Kondo, J.W. Munger, A. Torres, and A. Volz-Thomas, 1997: Climatologies of NO_x and NO_y: a comparison of data and models. *Atmospheric Environment,* **31,** 1851–1903.

Fahey, D.W., E.R. Keim, K.A. Boering, C.A. Brock, J.C. Wilson, H.H. Jonsson, S. Anthony, T.F. Hanisco, P.O. Wennberg, R.C. Miake-Lye, R.J. Salawitch, N. Louisnard, E.L. Woodbridge, R.-S. Gao, S.G. Donelly, R. Wamsley, L.A. Del Negro, B.C. Daube, S.C. Wofsy, C.R. Webster, R.D. May, K.K. Kelly, M. Loewenstein, J.R. Podolske, and K.R. Chan, 1995: Emission measurements of the Concorde supersonic aircraft in the lower stratosphere. *Science,* **270,** 70–74.

Fendel, W., D. Matter, H. Burtscher, and A. Schmidt-Olt, 1995: Interaction between carbon or iron aerosol particles and ozone. *Atmospheric Environment,* **29,** 967–973.

Fenter, F.F. and M.J. Rossi, 1996: Heterogeneous kinetics of HONO on H_2SO_4 solutions and on ice: activation of HCl. *Journal of Physical Chemistry,* **100,** 13765–13775.

Flatoy, F. and Ø. Hov, 1996: Three-dimensional model studies of the effect of NO_x emissions from aircraft on ozone in the upper troposphere over Europe and the North Atlantic. *Journal of Geophysical Research,* **101,** 1401–1422.

Fortuin, J.P.F. and H. Kelder, 1997: Possible links between ozone and temperature profiles. *Geophysical Research Letters,* **23,** 1517–1520.

Fox, L.E., D.R. Worsnop, M.E. Zahniser, and S.C. Wofsy, 1995: Metastable phases in polar stratospheric aerosols. *Science,* **267,** 351–355.

Fried, A., B.E. Henry, J.G. Calvert, and M. Mozurkewich, 1994: The reaction probability of N_2O_5 with sulfuric acid aerosols at stratospheric temperatures and compositions. *Journal of Geophysical Research,* **99,** 3517–3532.

Friedl, R.R. (ed.), 1997: *Atmospheric Effects of Subsonic Aircraft: Interim Assessment Report of the Advanced Subsonic Technology Program.* NASA Reference Publication 1400, National Aeronautics and Space Administration, Goddard Space Flight Center, Greenbelt, MD, USA, 168 pp.

Fuglestvedt, J.S., T.K. Berntsen, I.S.A. Isaksen, H. Mao, X.-Z. Liang, and W.-C. Wang, 1999: Climatic forcing of nitrogen oxides through changes in tropospheric ozone and methane; global 3D model studies. *Atmospheric Environment,* **33,** 961–977.

Fuglestvedt, J.S., I.S.A. Isaksen, and W.C. Wang, 1996: Estimates of indirect global warming potentials for CH_4, CO, and NO_x. *Climatic Change,* **34,** 405–437.

Gao, R.-S., B. Kärcher, E.R. Kein, and D.W. Fahey, 1998: Constraining the heterogeneous loss of O_3 on soot particles with observations in jet engine exhaust plumes. *Geophysical Research Letters,* **25,** 3323–3326.

Gao, R.-S., D.W. Fahey, R.J. Salawitch, S.A. Lloyd, D.E. Anderson, R. DeMajistre, C.T. McElroy, E.L. Woodbridge, R.C. Wamsley, S.G. Donnelly, L.A. Del Negro, M.H. Proffitt, R.M. Stimpfle, D.W. Kohn, S.R. Kawa, L.R. Lait, M. Loewenstein, J.R. Podolske, E.R. Keim, J.E. Dye, J.C. Wilson, and K.R. Chan, 1997: Partitioning of the reactive nitrogen reservoir in the lower stratosphere of the southern hemisphere: observations and modeling. *Journal of Geophysical Research,* **102,** 3935–3949.

Gardner, R.M., K. Adams, T. Cook, F. Deidewig, S. Ernedal, R. Falk, E. Fleuti, E. Herms, C.E. Johnson, M. Lecht, D.S. Lee, M. Leech, D. Lister, B. Masse, M. Metcalfe, P. Newton, A. Schmitt, C. Vandenbergh, and R. Van Drimmelen, 1997: The ANCAT/EC global inventory of NO_x emissions from aircraft. *Atmospheric Environment,* **31,** 1751–1766.

Garnier, F., L. Jacquin, and A. Laverdant, 1996: Engine jet entrainment in the near field of an aircraft: impact of aircraft emissions upon the atmosphere. In: *Impact of Aircraft Emissions upon the Atmosphere.* Proceedings of an international scientific colloquium, 15–18 October 1996, Paris, France. Office Nationale d'Etudes et de Recherches Aerospatiales, Chatillon, France, Vol. I, pp. 483–489.

Gerecke, A., A. Thielmann, L. Gutzwiller, and M.J. Rossi, 1998: The chemical kinetics of HONO formation resulting from heterogeneous interaction of NO_2 with flame soot. *Geophysical Research Letters,* **25,** 2453–2456.

Gerz, T. and T. Ehret, 1997: Wingtip vortices and exhaust jets during the jet regime of aircraft wakes. *Aerospace Science Technology,* **1,** 463–474.

Gerz, T. and B. Kärcher, 1997. Dilution of aircraft exhaust and entrainment rates for trajectory box models. In: *Impact of Aircraft Emissions upon the Atmosphere*. Proceedings of an international scientific colloquium, 15–18 October 1996, Paris, France. Office Nationale d'Etudes et de Recherches Aerospatiales, Chatillon, France, Vol. I, pp. 271–276.

Gettelman, A. and S.L. Baughcum, 1999. Direct deposition of subsonic aircraft emissions into the stratosphere. *Journal of Geophysical Research*, (in press).

Hanson, D.R., 1992: The uptake of HNO_3 onto ice, NAT, and frozen sulfuric-acid. *Geophysical Research Letters*, **19**, 2063–2066.

Hanson, D.R. and A.R. Ravishankara, 1991: The reaction probabilities of $ClONO_2$ and N_2O_5 on 40 to 75% sulfuric acid solutions. *Journal of Geophysical Research*, **96**, 17307–17314.

Hanson, D.R. and A.R. Ravishankara, 1992: Investigation of the reactive and nonreactive processes involving $ClONO_2$ and HCl on water and nitric acid doped ice. *Journal of Physical Chemistry*, **96**, 2682–2691.

Hanson, D.R., A.R. Ravishankara, and E.R. Lovejoy, 1996: Reaction of $BrONO_2$ with H_2O on submicron acid aerosol and the implications for the lower stratosphere, *Journal of Geophysical Research*, **101**, 9063–9069.

Harris, N.R.P., G. Ancellet, L. Bishop, D.J. Hofmann, J.B. Kerr, R.D. McPeters, M. Prendez, W.J. Randel, J. Shaehelin, A. Volz-Thomas, J. Zawodny, and C.S. Zerofros, 1997: Trends in stratospheric and free tropospheric ozone. *Journal of Geophysical Research*, **102**, 1571–1590.

Hauglustaine, D.A., B.A. Ridley, S. Solomon, P.G. Hess, and S. Madronich, 1996: HNO_3/NO_x ratio in the remote troposphere during MLOPEX-2: evidence for nitric acid reduction on carbonaceous aerosols. *Geophysical Research Letters*, **23**, 2609–2612.

Hayman, G.D. and M. Markiewicz, 1996: Chemical modelling of the aircraft exhaust plume. In: *Pollution from Aircraft Emissions in the North Atlantic Flight Corridor* [Schumann, U. (ed.)]. EUR-16978-EN, Office for Publications of the European Communities, Brussels, Belgium, pp. 280–297.

Hidalgo, H. and P.J. Crutzen, 1977: The tropospheric and stratospheric composition perturbed by NO_x emissions of high-altitude aircraft. *Journal of Geophysical Research*, **82**, 5833–5866.

Hintsa, E.J., P.A. Newman, H.H. Jonsson, C.R. Webster, R.D. May, R.L. Herman, L.R. Lait, M.R. Schoeberl, J.W. Elkins, P.R. Wamsley, G.S. Dutton, T.P. Bui, D.W. Kohn, and I.G. Anderson, 1998: Dehydration and denitrification in the Arctic polar vortex during the 1995–1996 winter. *Geophysical Research Letters*, **25**, 501–504.

Hoinka, K.P., M.E. Reinhardt, and W. Metz, 1993: North Atlantic air traffic within the lower stratosphere: cruising times and corresponding emissions. *Journal of Geophysical Research*, **98**, 23, 113–123, 131.

Hofmann, D.J. and S. Solomon, 1989: Ozone destruction through heterogeneous chemistry following the eruption of El Chichon. *Journal of Geophysical Research*, **94**, 5029–5041.

Hofmann, D.J., 1993: 20 years of balloon-borne tropospheric aerosol measurements at Laramie, Wyoming. *Journal of Geophysical Research*, **98**, 12753–12766.

Holton, J.R., P.H. Haynes, M.E. McIntyre, A.R. Douglass, R.B. Rood, and L. Pfister, 1995: Stratosphere-troposphere exchange. *Review of Geophysics*, **33**, 403–409.

Horowitz, L.W., J.Y. Liang, G.M. Gardner, and D.J. Jacob, 1998: Export of reactive nitrogen from North-America during summertime—sensitivity to hydrocarbon chemistry. *Journal of Geophysical Research*, **103**, 13451–13476.

Hoshizaki, H., L.B. Anderson, R.J. Conti, N. Farlow, J.W. Meyer, T. Overcamp, K.O. Redler, and V. Watson, 1975: Aircraft wake microscale phenomena. In: *CIAP Monograph 3, The Stratosphere Perturbed by Propulsion Effluent*. DOT-TST-75-53, U.S. Department of Transportation, Washington, DC, USA, 77 pp.

Hoskins, B.J., M.E. McIntyre, and A.W. Robertson, 1985: On the use and significance of isentropic potential vorticity maps. *Quarterly Journal of the Royal Meteorological Society*, **111**, 877–946.

IPCC, 1996: *Climate Change 1995: The Science of Climate Change. Contribution of Working Group I to the Second Assessment Report of the Intergovernmental Panel on Climate Change* [Houghton, J.T., L. Meira Filho, B. Callander, N. Harris, A. Kattenberg, and K. Maskell (eds.)] Cambridge University Press, Cambridge, United Kingdom and New York, NY, USA, pp. 285–357.

IPCC, 1995: *Climate Change 1994: Radiative Forcing of Climate Change and an Evaluation of the IPCC IS92 Emissions Scenarios* [Houghton, J.T., L.G. Meira Filho, J. Bruce, H. Lee, B.A. Callander, E. Haites, N. Harris, and K. Maskell (eds.)]. Cambridge University Press, Cambridge and New York, pp. 72–126.

IUPAC, 1997a: Evaluated kinetic and photochemical data for atmospheric chemistry. Supplement V. *Journal of Physical Chemistry Reference Data*, **26**, 509–1011.

IUPAC, 1997b: Evaluated kinetic and photochemical data for atmospheric chemistry. Supplement VI. *Journal of Physical Chemistry Reference Data*, **26**, 1329–1499.

Jacob, D.J., J.A. Logan, G.M. Gardner, R.M. Yevich, C.M. Spivakovsky, S.C. Wofsy, S. Sillman, and M.J. Prather, 1993: Factors regulating ozone over the United States and its export to the global atmosphere. *Journal of Geophysical Research*, **98**, 14817–14826.

Jacob, D.J., M.J. Prather, P.J. Rasch, R.-L. Shia, Y.J. Balkanski, S.R. Beagley, D.J. Bergmann, W.T. Blackshear, M. Brown, M. Chiba, M.P. Chipperfield, J de Grandpre, J.E. Dignon, J. Feichter, C. Genthon, W.L. Grose, P. Kasibhatla, I. Kohler, M.A. Kritz, K. Law, J.E. Penner, M. Ramonet, C.E. Reeves, D.A. Rotman, D.Z. Stockwell, P.F.J. van Velthoven, G. Verver, O. Wild, H. Yang, and P. Zimmermann, 1997: Evaluation and intercomparison of global atmospheric transport models using ^{222}Rn and other short-lived tracers. *Journal of Geophysical Research*, **102**, 5953–5970.

Jacob, D.J., B.G. Heikes, S.-M. Fan, J.A. Logan, D.L. Mauzerall, J.D. Bradshaw, H.B. Singh, G.L. Gregory, R.W. Talbot, D.R. Blake, G.W. Sachse, 1996: The origin of ozone and NO_x in the tropical troposphere: a photochemical analysis of aircraft observations over the south Atlantic basin. *Journal of Geophysical Research*, **101**, 24235–24250.

Jaeglé, L., D.J. Jacob, P.O. Wennberg, C.M. Spivakovsky, T.F. Hanisco, E.J. Lanzendorf, E.J. Hintsa, D.W. Fahey, E.R. Keim, M.H. Proffitt, E. Atlas, F. Flock, S. Schauffler, C.T. McElroy, C. Midwinter, L. Pfister, and J.C. Wilson, 1997: Observed OH and HO_2 in the upper troposphere suggest a major source from convective injection of peroxides. *Geophysical Research Letters*, **24**, 3181–3184.

Jaeglé, L., D.J. Jacob, W.H. Brune, D. Tan, I.C. Faloona, A.J. Weinheimer, B.A. Ridley, T.L. Campos, and G.W. Sachse, 1998: Sources of HO_x and production of ozone in the upper troposphere over the United States. *Geophysical Research Letters*, **25**, 1709–1712.

Johnson, C.E. and R.G. Derwent, 1996: Relative radiative forcing consequences of global emissions of hydrocarbons, carbon monoxide and NO_x from human activities estimated with a zonally-averaged two-dimensional model. *Climatic Change*, **34**, 439–462.

JPL, 1997: *Chemical Kinetics and Photochemical Data for Use in Stratospheric Modeling, Evaluation Number 12*. JPL-97-4, NASA Panel for Data Evaluation, Jet Propulsion Laboratory, National Aeronautics and Space Administration, Pasadena, CA, USA, 266 pp.

Jucks, K.W., D.G. Johnson, K.V. Chance, W.A. Traub, R.J. Salawitch, and R.A. Stachnik, 1997: Ozone production and loss rate measurements in the middle stratosphere. *Journal of Geophysical Research*, **101**, 28785–28792.

Kanakidou, M., F.J. Dentener, T.K. Berntsen, W.J. Collins, D.A. Hauglustaine, S. Houweling, I. Isaksen, M. Krol, M.G. Lawrence, J.F. Müller, N. Poisson, G.J. Roelofs, Y. Wang, and W.M.F. Wauben, 1998: 3-D global simulations of tropospheric CO distribution - results of the GIM/IGAC intercomparison 1997 exercise. In: *Proceedings of the International Conference on Atmospheric Carbon Monoxide and Its Environmental Effect, Portland State University, Portland, OR, USA, December 3–6, 1997.*

Kärcher, B., 1998: Physicochemistry of aircraft-generated liquid aerosols, soot and ice particles, 1. model description. *Journal of Geophysical Research*, **103**, 17111–17128.

Kärcher, B., F. Yu, F.P. Schröder, and R.P. Turco, 1998a: Ultrafine aerosol particles in aircraft plumes: analysis of growth mechanisms. *Geophysical Research Letters*, **25**, 2793–2796.

Kärcher, B., R. Busen, A. Petzold, F.P. Schröder, U. Schumann, and E.J. Jensen, 1998b: Physicochemistry of aircraft-generated liquid aerosols, soot and ice particles, 2. comparison with observations and sensitivity studies. *Journal of Geophysical Research*, **103**, 17129–17147.

Kärcher, B., 1997: Heterogeneous chemistry in aircraft wakes: constraints for uptake coefficients. *Journal of Geophysical Research,* **102,** 19119–19135.

Kärcher, B., T. Peter, and R. Ottmann, 1995: Contrail Formation: Homogeneous nucleation of H_2SO_4/H_2O droplets. *Geophysical Research Letters,* **22,** 1501–1504.

Kärcher, B., 1994: Transport of exhaust products in the near trail of a jet engine under atmospheric conditions. *Journal of Geophysical Research,* **99,** 14509–14517.

Karol, I.L., Y.E. Ozolin, and E.V. Rozanov, 1997: Box and Gaussian plume models of the exhaust composition evolution of the subsonic transport aircraft in and out of the flight corridor. *Annales Geophysicae,* **15,** 88–96.

Karol, I.L., Y.E. Ozolin, and E.V. Rozanov, 1998: Effective emission indices (EEI) of the subsonic aircraft exhausts and their dependence on the external conditions. In: *Proceedings of the 1998 Conference on the Atmospheric Effects of Aviation, Virginia Beach, VA, USA, April 27–May 1 1998.* NASA, Goddard Space Flight Center, Greenbelt, MD, USA.

Kasibhatla, P.S., H. Levy II, W.J. Moxim, and W.L. Chameides, 1991: The relative impact of stratospheric photochemical production on tropospheric NO_y levels: a model study. *Journal of Geophysical Research,* **96,** 18631–18646.

Kleffmann, J., K.-H. Becker, and P. Wiesen, 1998: Heterogeneous NO_2 conversion processes on acid surfaces: possible atmospheric implications. *Atmospheric Environment,* **32,** 2721–2729.

Konopka, P., 1995: Analytical Gaussian solutions for anisotropic diffusion in a linear shear-flow. *Journal of Non-Equilibrium Thermophysics,* **20,** 78–91.

Kotamarthi, V.R., J.M. Rodriguez, N.D. Sze, Y. Kondo, R. Pueschel, G. Ferry, J. Bradshaw, S. Sandholm, G. Gregory, D. Davis, and S. Liu, 1997: Evidence of heterogeneous chemistry on sulfate aerosols in stratospherically influenced air masses sampled during PEM-West B. *Journal of Geophysical Research,* **102,** 28425–28436.

Lacis, A.A., D.J. Wuebbles, and J.A. Logan, 1990: Radiative forcing of climate by changes in the vertical distribution of ozone. *Journal of Geophysical Research,* **95,** 9971–9981.

Lamarque, J.-F., G.P. Brasseur, and P.G. Hess, 1996: Three-dimensional study of the relative contributions of the different nitrogen sources in the troposphere. *Journal of Geophysical Research,* **101,** 22955–22968.

Lary, D.J., A.M. Lee, R. Toumi, M.J. Newchurch, M. Pirre, and J.B. Renard, 1997: Carbon Aerosols and atmospheric photochemistry. *Journal of Geophysical Research,* **102,** 3671–3682.

Lee, D.S., I. Köhler, E. Grobler, F. Rohrer, R. Sausen, L. Gallardo-Klenner, J.G.J. Olivier, F.J. Dentener, and A.F. Bouwman, 1997: Estimations of global NO_x emissions and their uncertainties. *Atmospheric Environment,* **31,** 1735–1749.

Levy II, H., W.J. Moxim, and P.S. Kasibhatla, 1996: A global 3-dimensional time-dependent lightning source of tropospheric NO_x. *Journal of Geophysical Research,* **101,** 22911–22922.

Lewellen, D.C. and W.S. Lewellen, 1996: Large-eddy simulations of the vortex-pair breakup in aircraft wakes. *AIAA Journal,* **34,** 2337–2345.

Liang, J.Y., L.W. Horowitz, D.J. Jacob, Y.H. Wang, A.M. Fiore, J.A. Logan, G.M. Gardner, and J.W. Munger, 1998: Seasonal budgets of reactive nitrogen species and ozone over the United States, and export fluxes to the global atmosphere. *Journal of Geophysical Research,* **103,** 13435–13450.

Logan, J.A., 1994: Trends in the vertical distribution of ozone: an analysis of ozonesonde data. *Journal of Geophysical Research,* **99,** 25553–25585.

Longfellow, C.A., T. Imamura, A.R. Ravishankara, and D.R. Hanson, 1998: HONO solubility and heterogeneous reactivity on sulfuric acid surfaces. *Journal of Physical Chemistry,* **A102,** 3323–3332.

Louisnard, N., C. Baundoin, G. Billet, F. Garnier, D. Hills, T. Mentel, P. Mirabel, J. Petit, J. Schultz, D. Taleb, J. Thlibi, A. Wahner, and P. Woods, 1995: Physics and chemistry in the aircraft wake. In: *AERONOX. The Impact of NO_X Emissions from Aircraft Upon the Atmosphere at Flight Altitudes 8–15 km* [Schumann, U., (ed.)]. EUR-16209-EN, Office for Publications of the European Communities, Brussels, Belgium, pp. 195–309.

Marenco, A., V. Thouret, P. Nedelec, H.G. Smit, M. Helten, D. Kley, F. Kärcher, P. Simon, K. Law, J. Pyle, G. Poschmann, R. Von Wrede, C. Hume, and T. Cook, 1999: Measurement of ozone and water vapor by Airbus in-service aircraft: the MOZAIC program, an overview. *Journal of Geophysical Research,* **103,** 25631–25642.

Matthews, E., 1983: Global vegetation and land use: New high-resolution data bases for climate studies. *Journal of Climatic Applied Meteorology,* **22,** 474–487.

McIntyre, M.E., 1995: The stratospheric polar vortex and sub-vortex—fluid dynamics and mid-latitude ozone loss. *Philosophical Transactions of the Royal Society,* **A352,** 227–240.

McKeen, S.A., T. Gierczak, J.B. Burkholder, P.O. Wennberg, T.F. Hanisco, E.R. Keim, R.-S. Gao, S.C. Liu, A.R. Ravishankara, and D.W. Fahey, 1997: The photochemistry of acetone in the upper troposphere: a source of odd-hydrogen radicals. *Geophysical Research Letters,* **24,** 3177–3180.

Meijer, E.W., P. van Velthoven, W. Wauben, H. Kelder, J. Beck, and G. Velders, 1997: The effects of the conversion of nitrogen oxides in aircraft exhaust plumes in global models. *Geophysical Research Letters,* **24,** 3013–3016.

Metwally, M. 1995: *Jet Aircraft Engine Emissions Database Development— 1992 Military, Charter, and Nonscheduled Traffic.* NASA-CR-4684, National Aeronautics and Space Administration, Hampton, VA, USA, 61 pp.

Miake-Lye, R.C., M. Martinez-Sanchez, R.C. Brown, and C.E. Kolb, 1993: Plume and wake dynamics, mixing and chemistry behind a high speed civil transport aircraft. *Journal of Aircraft,* **30,** 467–479.

Miller, A.J., G.C. Tiao, G.C. Reinsel, D. Wuebbles, L. Bishop, J. Kerr, R.M. Nagatani, J.J. Deluisi, and C.L. Mateer, 1995: Comparisons of observed ozone trends in the stratosphere through examination of Umkehr and balloon ozonesonde data. *Journal of Geophysical Research,* **100,** 11209–11217.

Minschwaner, K., A.E. Dessler, J.W. Elkins, C.M. Volk, D.W. Fahey, M. Loewenstein, J.R. Podolske, A.E. Roche, and K.R. Chan, 1996: Bulk properties of isentropic mixing into the tropics and the lower stratosphere. *Journal of Geophysical Research,* **101,** 9433–9439.

Müller, J.-F., 1992: Geographical distribution and seasonal variation of surface emissions and deposition velocities of atmospheric trace gases. *Journal of Geophysical Research,* **97,** 3787–3804.

Murphy, D.M., D.W. Fahey, M.H. Proffitt, S.C. Liu, K.R. Chan, C.S. Eubank, S.R. Kawa, and K.K. Kelly, 1993: Reactive nitrogen and its correlation with ozone in the lower stratosphere and upper troposphere. *Journal of Geophysical Research,* **98,** 8751–8773.

Olson, J., M. Prather, T. Berntsen, G. Carmichael, R. Chatfield, P. Connell, R. Derwent, L. Horowitz, S. Jin, M. Kanakidou, P. Kasibhatla, R. Kotamarthi, M. Kuhn, K. Law, J.E. Penner, L. Perliski, S. Sillman, F. Stordal, A. Thompson, and O. Wild, 1997: Results from the Intergovernmental Panel on Climatic Change photochemical model intercomparison (PhotoComp). *Journal of Geophysical Research,* **102,** 5979–5991.

Oppliger, R., A. Allanic, and M.J. Rossi, 1997: Real-time kinetics of the uptake of $ClONO_2$ on ice and in the presence of HCl in the temperature range $160 \leq T/K \leq 200$. *Journal of Physical Chemistry,* **A101,** 1903–1911.

Park, J.H., M.K.W Ko, C.H. Jackman, and K.H. Sage (eds.), 1999: *Models and Measurements II.* National Aeronautics and Space Administration, Langley Research Center, Hampton, VA, USA, 470 pp. (in press).

Petry, H., J. Hendricks, M. Möllhoff, E. Liepert, A. Meier, A. Ebel, and R. Sausen, 1998: Chemical conversion in the dispersing plume: calculation of effective emission indices. *Journal of Geophysical Research,* **103,** 5759–5772.

Petzold, A. and F. Schröder, 1998: Jet Engine exhaust aerosol characterization. *Aerosol Science and Technology,* **28,** 62–76.

Petzold, A., R. Busen, F.P. Schröder, R. Baumann, M. Kuhn, J. Ström, D.E. Hagen, P.D. Whitefield, D. Baumgardner, F. Arnold, S. Borrmann, and U. Schumann, 1997: Near-field measurements on contrail properties from fuels with different sulfur content. *Journal of Geophysical Research,* **102,** 29867–29881.

Pierce, R.B., T.D. Fairlie, W.L. Grose, R. Swinbank, and A. O'Neill, 1994: Mixing processes within the polar night jet. *Journal of Atmospheric Chemistry,* **51,** 2957–2972.

Pierce, R.B., W.T. Blackshear, T.D. Fairlie, W.L. Grose, and R.E. Turner, 1993: The interaction of radiative and dynamical processes during a simulated sudden stratospheric warming. *Journal of Atmospheric Science,* **50,** 3829–3851.

Pires, M.R., H. van den Bergh, and M.J. Rossi, 1996: The heterogeneous formation of N₂O over bulk condensed phases in the presence of SO₂ at high humidities. *Journal of Atmospheric Chemistry,* **25,** 229–250.

Pleijel, K., 1998: Impact from emitted NOₓ and VOC in an aircraft plume: model results for the free troposphere. IVL Report B1245, Gothenburg, Sweden.

Plumb, R.A., 1996: A "tropical pipe" model of stratospheric transport, *Journal of Geophysical Research,* **101,** 3957–3972.

Portmann, R.W., S. Solomon, R.R. Garcia, L.W. Thomason, L.R. Poole, and M.P. McCormick, 1996: Role of aerosol variations in anthropogenic ozone depletion in the polar regions. *Journal of Geophysical Research,* **101,** 22991–23006.

Prather, M.J. and D.J. Jacob, 1997: A persistent imbalance in HOₓ and NOₓ photochemistry of the upper troposphere driven by deep tropical convection. *Geophysical Research Letters,* **24,** 3189–3192.

Prather, M.J. and E.E. Remsberg (eds.), 1993: *The Atmospheric Effects of Stratospheric Aircraft: Report of the 1992 Models and Measurements Workshop.* NASA Reference Publication 1292, National Aeronautics and Space Administration, USA.

Price, C. and D. Rind, 1992: A simple lightning parameterization for calculating global lightning distributions. *Journal of Geophysical Research,* **97,** 9919–9933.

Pueschel, R.F., K.A. Boering, S. Verma, S.D. Howard, G.V. Ferry, J. Goodman, D.A. Allen, and P. Hamill, 1997: Soot aerosol in the lower stratosphere: pole-to-pole variability and contributions by aircraft. *Journal of Geophysical Research,* **102,** 13113–13118.

Quackenbush, T.R., M.E. Teske, and A.J. Bilanin, 1993: *Computation of Wake/Exhaust Mixing Downstream of Advanced Transport Aircraft.* Paper presented at American Institute of Aeronautical Engineers 24th Fluid Dynamics Conference, Orlando, FL, USA, 6–9 July 1993.

Quinlan, M.A., C.M. Reihs, D.M. Golden, and M.A. Tolbert, 1990: Heterogeneous reactions on model polar stratospheric cloud surfaces: Reaction of N₂O₅ on ice and nitric acid trihydrate. *Journal of Physical Chemistry,* **94,** 3255–3260.

Ridley, B.A., J.E. Dye, J.G. Walega, J. Zheng, F.E. Grahek, and W. Rison, 1996: On the production of active nitrogen by thunderstorms over New Mexico. *Journal of Geophysical Research,* **101,** 20985–21005.

Rinsland, C.P., M.R. Gunson, R.J. Salawitch, H.A. Michelsen, R. Zander, M.J. Newchurch, M.M. Abbas, M.C. Abrams, G.L. Manney, A.Y. Chang, F.W. Irion, A. Goldman, and E. Mahieu, 1996: ATMOS/ATLAS-3 measurements of stratospheric chlorine and reactive nitrogen partitioning inside and outside the November 1994 Antarctic vortex. *Geophysical Research Letters,* **23,** 2365–2368.

Rogaski, C.A., D.M. Golden, and L.R. Williams, 1997: Laboratory measurements of H₂O uptake on amorphous carbon: Effects of chemical treatment with NO₂, SO₂, H₂SO₄, and HNO₃. *Geophysical Research Letters,* **24,** 381–384.

Schilling, V., S. Siano, and D. Etling, 1996: Dispersion of aircraft emissions due to wake vortices in stratified shear flows: A two-dimensional numerical study. *Journal of Geophysical Research,* **101,** 20965–20974.

Schlager, H, P. Schulte, H. Ziereis, F. Arnold, J. Ovarlez, P. van Velthoven, and U. Schumann, 1996: Airborne observations of large scale accumulations of air traffic emissions in the North Atlantic flight corridor with a stagnant anticyclone. In: *Impact of Aircraft Emissions upon the Atmosphere.* Proceedings of an international scientific colloquium, 15–18 October 1996, Paris, France, Office Nationale d'Etudes et de Recherches Spatiales, Chatillon, France, Vol. I, pp. 247–252.

Schlager, H., P. Konopka, P. Schulte, U. Schumann, H. Ziereis, F. Arnold, M. Klemm, D.E. Hagen, P.D. Whitefield, and J. Ovarlez, 1997: *In situ* observations of air traffic emission signatures in the North Atlantic flight corridor. *Journal of Geophysical Research,* **102,** 10739–10750.

Schoeberl, M.R., A.E. Roche, J.M. Russell, J.M. Ortland, P.B. Hays, and J.W. Waters, 1997: An estimation of the dynamical isolation of the tropical lower stratosphere using UARS wind and trace gas observations of the quasi-biennial oscillation. *Geophysical Research Letters,* **24,** 53–56.

Schulte, P. and H. Schlager, 1996. In-flight measurements of cruise altitude nitric oxide emission indices of commercial jet aircraft. *Geophysical Research Letters,* **23,** 165–168.

Schulte, P., H. Schlager, H. Ziereis, U. Schumann, S. Baughcum, and F. Deidewig, 1997: NOₓ emission indices of subsonic long-range jet aircraft at cruise altitude: *in situ* measurements and predictions. *Journal of Geophysical Research,* **102,** 21431–21442.

Schumann, U., 1994: On the effect of emissions from aircraft engines on the state of the atmosphere. *Annales Geophysicae,* **12,** 365–384.

Schumann, U., 1997: The impact of nitrogen oxides emissions from the aircraft upon the atmosphere at flight altitudes—results from the AERONOX Project. *Atmospheric Environment,* **31,** 1723–1733.

Schumann, U., P. Konopka, R. Baumann, R. Busen, T. Gerz, H. Schlager, P. Schulte, and H. Volkert, 1995: Estimate of diffusion parameters of aircraft exhaust plumes near the tropopause from nitric oxide and turbulence measurements. *Journal of Geophysical Research,* **100,** 14147–14162.

Schumann, U., J. Ström, R. Busen, R. Baumann, K. Gierens, M. Krautstrunk, F.P. Schröder, and J. Stingl, 1996: *In situ* observations of particles in jet aircraft exhausts and contrails for different sulphur containing fuels. *Journal of Geophysical Research,* **101,** 6853–6869.

Schumann, U., A. Chlond, A. Ebel, B. Kärcher, H. Pak, H. Schlager, A. Schmitt, and P. Wendling, 1997: Pollutants from air traffic—results of atmospheric research 1992–1997. In: *Final Report on the BMBF Verbundprogramm, Schadstoffe in der Luftfahrt* [Schumann, U. and A. Chlond, A. Ebel, B. Kärcher, H. Pak, H. Schlager, A. Schmitt, and P. Wendling (eds.)]. DLR-Mitteilung 97-04, Deutsches Zentrum für Luft- und Raumfahrt, Oberpfaffenhofen and Cologne, Germany. p. 291.

Schumann, U., H. Schlager, F. Arnold, R. Baumann, P. Haschberger, and O. Klemm, 1998: Dilution of aircraft exhaust plumes at cruise altitudes. *Atmospheric Environment,* **32,** 3097–3103.

Seisel, S., B. Flückiger, and M.J. Rossi, 1998: The heterogeneous reaction of N₂O₅ and HBr on ice. Comparison with N₂O₅ + HCl. *Berliner Berichte der Bunsen - Gesellschaft - Physical Chemistry,* **102,** 811–820.

Siegmund, P.C., P.F.J. van Velthoven, and H. Kelder, 1996: Cross-tropopause transport in the extratropical northern winter hemisphere. diagnosed from high-resolution ECMWF data. *Quarterly Journal of the Royal Meteorological Society,* **122,** 1921–1941.

Singh, H.B., M. Kanakidou, P.J. Crutzen, and D.J. Jacob, 1995: High concentrations and photochemical fate of oxygenated hydrocarbons in the global troposphere. *Nature,* **378,** 50–54.

Slemr, F., H. Giehl, J. Slemr, R. Busen, P. Schulte, and P. Haschberger, 1998: In-flight measurement of aircraft non-methane hydrocarbon emission indices. *Geophysical Research Letters,* **25,** 321–324.

Smith, D.M., W.F. Welch, S.M. Graham, A.R. Chughtai, B.G. Wicke, and K.A. Grady, 1988: Reaction of nitrogen oxides with black carbon: an FT-IR study. *Applied Spectroscopy,* **42,** 674–680.

Solomon, S., R.W. Portmann, R.R. Garcia, L.W. Thomason, L.R. Poole, and M.P. McCormick, 1996: The role of aerosol variations in anthropogenic ozone depletion at northern midlatitudes. *Journal of Geophysical Research,* **101,** 6713–6727.

Solomon, S., S. Boormann, R.R. Garcia, R. Portmann, L. Thomason, L.R. Poole, D. Winkler, and M.P. McCormick, 1997: Heterogeneous chlorine chemistry in the tropopause region. *Journal of Geophysical Research,* **102,** 21411–21429.

Spicer, C.W., M.W. Holdren, R.M. Higgin, and T.F. Lyon, 1994: Chemical composition and photochemical reactivity of exhaust from aircraft turbine engines. *Annales Geophysicae,* **12,** 944–955.

Stephens, S., M.J. Rossi, and D.M. Golden, 1986: The heterogeneous reaction of ozone on carbonaceous surfaces. *International Journal of Chemical Kinetics,* **18,** 1133–1149.

Tabazadeh, A., R.P. Turco, K. Drdla, M.Z. Jacobson, and O.B. Toon, 1994: A study of type-I polar stratospheric cloud formation. *Geophysical Research Letters,* **21,** 1619–1622.

Tabor, K., L. Gutzwiller, and M.J. Rossi, 1993: The chemical kinetics of the heterogeneous interaction of NO₂ with amorphous carbon. *Geophysical Research Letters,* **20,** 1431.

Tabor, K., L. Gutzwiller, and M.J. Rossi, 1994: The chemical kinetics of the heterogeneous reaction of NO₂ on amorphous carbon. *Journal of Physical Chemistry,* **98,** 6172.

Thouret, V., A. Marenco, P. Nedelec, and C. Grouhel, 1999: Ozone climatologies at 9–12 km altitude as seen by the MOZAIC airborne program between September 1994 and August 1996. *Journal of Geophysical Research,* **103,** 25653–25679.

Tremmel, H.G., H. Schlager, P. Konopka, P. Schulte, F. Arnold, M. Klemm, and B. Droste-Franke, 1998: Observations and model calculations of jet aircraft exhaust products at cruise altitude and inferred initial OH emissions. *Journal of Geophysical Research,* **103,** 10803–10816.

Tuovinen, J.-P., K. Barrett, and H. Styve, 1994: *Transboundary Acidifying Pollution in Europe: Calculated Fields and Budgets, 1985–93.* Report EMEP/MSC-W 1/94, Norwegian Meteorological Institute, Oslo, Norway, 70 pp.

van Velthoven, P.F.J., R. Sausen, C. Johnson, H. Kelder, I. Kohler, A. Kraus, R. Ramaroson, F. Rohrer, D. Stevenson, A. Straud, and W.M.F. Wauben, 1997: The passive transport of NO$_x$ emissions from aircraft studied with a hierarchy of models. *Atmospheric Environment,* **31,** 1783–1799.

Volk, C.M., J.W. Elkins, D.W. Fahey, R.J. Salawitch, G.S. Dutton, J.M. Gilligan, M.H. Proffitt, M. Loewenstein, J.R. Podolske, K. Minschwaner, J.J. Margitan, and K.R. Chan, 1996: Quantifying transport between the tropical and mid-latitude lower stratosphere. *Science,* 272, 1763–1768.

Wang, Y.H., D.J. Jacob, and J.A. Logan, 1998a: Global simulation of tropospheric O$_3$-NO$_x$-hydrocarbon chemistry 3—origin of tropospheric ozone and effects of nonmethane hydrocarbons. *Journal of Geophysical Research,* **103,** 10757–10767.

Wang, Y.H., J.A. Logan, and D.J. Jacob, 1998b: Global simulation of tropospheric O$_3$-NO$_x$-hydrocarbon chemistry 2—model evaluation and global ozone budget. *Journal of Geophysical Research,* **103,** 10727–10755.

Wauben, W.M.F., J.P.F. Fortuin, P.F.J. van Velthoven, and H. Kelder, 1998: Comparison of modelled ozone distributions with sonde and satellite observations. *Journal of Geophysical Research,* **103,** 3511–3530.

Weisenstein, D.K., M.W.K. Ko, N.-D. Sze, and J.M. Rodriguez, 1996: Potential impact of SO$_2$ emissions from stratospheric aircraft on ozone. *Geophysical Research Letters,* **23,** 161–164.

Wennberg, P.O., R.C. Cohen, R.M. Stimpfle, J.P. Koplow, J.G. Anderson, R.J. Salawitch, D.W. Fahey, E.L. Woodbridge, E.R. Keim, R.-S. Gao, C.R. Webster, R.D. May, D.W. Toohey, L.M. Avallone, M.H. Proffitt, M. Loewenstein, J.R. Podolske, K.R. Chan, and S.C. Wofsy, 1994: Removal of stratospheric O$_3$ by radicals: *in situ* measurements of OH, HO$_2$, NO, NO$_2$, ClO, and BrO. *Science,* **266,** 398–404.

Wennberg, P.O., T.F. Hanisco, L. Jaeglé, D.J. Jacob, E.J. Hintsa, E.J. Lanzendorf, J.G. Anderson, R.-S. Gao, E.R. Keim, S. Donnelly, L. Del Negro, D.W. Fahey, S.A. McKeen, R.J. Salawitch, C.R. Webster, R.D. May, R. Herman, M.H. Proffitt, J.J. Margitan, E. Atlas, C.T. McElroy, J.C. Wilson, C. Brock, and P. Bui, 1998: Hydrogen radicals, nitrogen radicals and the production of ozone in the upper troposphere. *Science,* **279,** 49–53.

Wiesen, P., J. Kleffmann, R. Kurtenbach, and K.-H. Becker, 1994: Emission of nitrous oxide and methane from aero engines. *Geophysical Research Letters,* **21,** 2027–2030.

Williams, E.J., G.L. Hutchinson, and F.C. Fehsenfeld, 1992: NO$_x$ and N$_2$O emissions from soil. *Global Biogeochemical Cycles,* **6,** 351–388.

WMO, 1998: *Ozone Changes as a Function of Altitude: Report of the Ozone Trends Panel.* Report no. 43, Global Ozone Research and Monitoring Project, World Meteorological Organization, Geneva, Switzerland.

WMO-UNEP, 1999: *Scientific Assessment of Ozone Depletion: 1998.* Global Ozone Research and Monitoring Project, World Meteorological Organization, (in press).

WMO-UNEP, 1995: *Scientific Assessment of Ozone Depletion: 1994.* Report no. 37, Global Ozone Research and Monitoring Project, World Meteorological Organization, Geneva, Switzerland, 500 pp.

WMO-UNEP, 1985: *Atmospheric Ozone 1985.* Report no. 16, Global Ozone Research and Monitoring Project, World Meteorological Organization, Geneva, Switzerland, 1095 pp.

Worsnop, D.R., L.E. Fox, M.S. Zahniser, and S.C. Wofsy, 1993: Vapor pressures of solid hydrates of nitric acid—implications for polar stratospheric clouds. *Science,* 259, 71–74.

Wuebbles, D.J., D. Maiden, R.K. Seals, Jr., S.L. Baughcum, M. Metwally, and A. Mortlock, 1993: Emission scenarios development: Report of the Emissions Scenarios Committee. In: *The Atmospheric Effects of Stratospheric Aircraft: A Third Program Report* [Stolarski, R.S. and H.L. Wesoky (eds.)]. NASA Reference Publication 1313, National Aeronautics and Space Administration, Langley Research Center, Hampton, VA, USA, 428 pp.

Yienger, J.J. and H. Levy II, 1995: Empirical model of global soil-biogenic NO$_x$ emissions. *Journal of Geophysical Research,* **100,** 11447–11464.

Zhang, R., M.-T. Leu, and L.F. Keyser, 1996: Heterogeneous chemistry of HONO on liquid sulfuric acid: A new mechanism of chlorine activation on stratospheric sulfate aerosols. *Journal of Physical Chemistry,* **100,** 339–345.

Bibliography of Model Studies

Beck, J.P., C.E. Reeves, F.A.A.M. de Leeuw, and S.A. Penkett, 1992: The effect of aircraft emissions on tropospheric ozone in the Northern Hemisphere. *Atmospheric Environment,* **A26,** 17–29.

Brasseur, G.P., J.-F. Müller, and C. Granier, 1996: Atmospheric impact of NO$_x$ emissions by subsonic aircraft: a three dimensional model study. *Journal of Geophysical Research,* **101,** 1423–1428.

Collins, W.J., D.S. Stevenson, C.E. Johnson, and R.G. Derwent, 1997: Tropospheric ozone in a global-scale three-dimensional Lagrangian model and its response to NO$_x$ emission controls. *Journal of Atmospheric Chemistry,* **26,** 223–274.

Dameris, M., V. Grewe, I. Köhler, R. Sausen, C. Brühl, J.-U. Grooß, and B. Steil, 1997: Impact of aircraft NO$_x$-emissions on tropospheric and stratospheric ozone. Part II: 3-D model results. *Atmospheric Environment,* **32,** 3185–3199.

Derwent, R.G., 1982: Two-dimensional model studies of the impact of aircraft exhaust emissions on tropospheric ozone. *Atmospheric Environment,* **16,** 1997–2007.

Ehhalt, D.H. and F. Rohrer, 1994: The impact of commercial aircraft on tropospheric ozone. In: *Chemistry of the Atmosphere.* Royal Society of Chemistry Special Publication No. 170 [Bandy, A.R. (ed.)]. Cambridge, UK.

Ehhalt, D.H., F. Rohrer, and A. Wahner, 1992: Sources and distribution of NO$_x$ in the upper troposphere at northern mid-latitudes. *Journal of Geophysical Research,* **97,** 3725–3738.

Fuglestvedt, J.S., I.S.A. Isaksen, and W.-C. Wang, 1996: Estimates of indirect global warming potentials for CH$_4$, CO and NO$_x$. *Climatic Change,* **34,** 405–437.

Grassi, B., G. Pitari, L. Ricciardulli, and G. Visconti, 1996: HSCT impact on ozone: comparison of 2-D and 3-D model experiment results. In: *Proceedings of the XVIII Quadrennial Ozone Symposium, 1996, L'Aquila, Italy* [Bojkov, R.D. and F. Visconti (eds.)].

Grooß, J.-U., C. Brühl, and T. Peter, 1998: Impact of aircraft emissions on tropospheric and stratospheric ozone. Part I: Chemistry and 2D-model results. *Atmospheric Environment,* **32,** 3173–3184.

Hauglustaine, D.A., C. Granier, G.P. Brasseur, and G. Mégie, 1994: Impact of present aircraft emissions of nitrogen oxides on tropospheric ozone and climate forcing. *Geophysical Research Letters,* **21,** 2031–2034.

Hidalgo, H., and P.J. Crutzen, 1977: The tropospheric and stratospheric composition perturbed by NO$_x$ emissions of high-altitude aircraft. *Journal of Geophysical Research,* **82,** 5833–5866.

Isaksen, I.S.A., 1980: The tropospheric ozone budget and possible man-made effects. In: *Proceedings of the Quadrennial Ozone Symposium.* World Meteorological Organization, Geneva, Switzerland.

Isaksen, I.S.A., F. Stordal, and T. Berntsen, 1989: *Model Studies of Effects of Highflying Supersonic Commercial Transport on Stratospheric and Tropospheric Ozone.* Report no. 76, Institute of Geophysics, University of Oslo, Norway.

Johnson, C.E. and J. Henshaw, 1991: The impact of NO$_x$ emissions from tropospheric aircraft. AEA-EE-0127, Atomic Energy Authority Energy and Environment, Harwell Laboratory, Oxfordshire, UK, 64 pp.

Johnson, C. and R. Kingdon, 1995: The 2-D chemistry model TROPOS. In: *AERONOX. The Impact of NO$_x$ Emissions from Aircraft upon the Atmosphere at Flight Altitudes 8–15 km* [Schumann, U. (ed.)]. EUR-16209-EN, Office for Publications of the European Communities, Brussels, Belgium, pp. 348–350.

Johnson, C.E. and D.S. Stevenson, 1996: Effect on tropospheric oxidant concentrations resulting from aircraft NO$_x$ emissions. In: *Pollution from Aircraft Emissions in the North Atlantic Flight Corridor (POLINAT)* [Schumann, U. (ed.)]. Air Pollution Research Report no. 58, EUR-6978-EN, Office for Publications of the European Communities, Brussels, Belgium, pp. 175–202.

Johnson, C.E., J. Henshaw, and G. McInnes, 1992: Impact of aircraft and surface emissions of nitrogen oxides on tropospheric ozone and global warming. *Nature,* **355,** 69–72.

Karol, I.L. and A.A. Kiselev, 1996a: Non-stationary 2-D photochemical model. In: *Research Activities in Atmospheric and Ocean Modelling* [Staniforth, A. (ed.)]. WMO/TD publication no. 792, World Meteorological Organization, Geneva, Switzerland, pp. 7.25–7.26.

Karol, I.L. and A.A. Kiselev, 1996b: Photochemical transformation of aircraft exhausts at their transition from the plume to the large scale dispersion in the northern temperate belt. In: *Impact of Aircraft Emissions upon the Atmosphere.* Proceedings of an international scientific colloquium, 15–18 October 1996, Paris, France. Office Nationale d'Etudes et de Recherches Aerospatiales, Chatillon, France, Vol. II, 235–240.

Köhler, I. and R. Sausen, 1997: Revised contributions of NO_x emissions from aircraft to the atmospheric NO_x content in the upper troposphere of the northern mid-latitudes and POLINAT region. In: *POLINAT-2. Pollution from Aircraft Emissions in the North Atlantic Flight Corridor* [Schumann, U. (ed.)]. Progress report April 1996–March 1997. Deutsches Zentrum für Luft- und Raumfahrt, Oberpfaffenhofen, Germany, pp. 23–36.

Köhler, I., R. Sausen, and R. Reinberger, 1997. Contribution of aircraft emissions to the atmospheric NO_x content. *Atmospheric Environment,* **31,** 1801–1818.

Kraus, A.B., F. Rohrer, E.S. Grobler, and D.H. Ehhalt, 1996: The global tropospheric distribution of NO_x estimated by a three-dimensional chemical tracer model. *Journal of Geophysical Research,* **101,** 18587–18604.

Lee, D.S., I. Köhler, E. Grobler, F. Rohrer, R. Sausen, L. Gallardo-Klenner, J.G.J. Olivier, F.J. Dentener, and A.F. Bouwman, 1997: NO_x emission database for AERONOX, including surface and lightning emissions. *Atmospheric Environment,* **31,** 1735–1750.

Liu, S.C., D. Kley, M. McFarland, J.D. Mahlman, and H. Levy, 1980: On the origin of tropospheric ozone. *Journal of Geophysical Research,* **85,** 7546–7552.

Luther, F.M., J.S. Chang, W.M. Duewer, J.E. Penner, R.L. Tarp, and D.J. Wuebbles, 1979: Potential environmental effects of aircraft emissions. LLNL-UCRL-52861, Lawrence Livermore National Laboratory, Livermore, CA, USA.

Müller, J.-F. and G. Brasseur, 1995: IMAGES: a three-dimensional chemical transport model of the global troposphere. *Journal of Geophysical Research,* **100,** 16445–16490.

Penner, J.E., C.S. Atherton, J. Dignon, S.J. Ghan, J.J. Walton, and S. Hameed, 1991: Tropospheric nitrogen: a three-dimensional study of sources, distribution and deposition. *Journal of Geophysical Research,* **96,** 959–990.

Penner, J.E., D.J. Bergmann, J.J. Walton, D. Kinnison, M.J. Prather, D. Rotman, C. Price, K.E. Pickering, and S.L. Baughcum, 1998: An evaluation of upper troposphere NO_x with two models. *Journal of Geophysical Research,* **103,** 22097–22113.

Petry, H., J. Hendricks, M. Möllhoff, E. Lippert, A. Meier, A. Ebel, and R. Sausen, 1997: Impact of aircraft exhaust on the atmosphere: box model studies and 3-D mesoscale numerical case studies of seasonal differences. In: *Impact of Aircraft Emissions upon the Atmosphere.* Proceedings of an international scientific colloquium, 15–18 October 1996, Paris, France. Office Nationale d'Etudes et de Recherches Aerospatiales, Chatillon, France, Vol. II, pp. 241–246.

Ramaroson, R. 1995a: The 3-D chemical transport model MEDIANTE. In: *AERONOX. The Impact of NO_x Emissions from Aircraft upon the Atmosphere at Flight Altitudes 8–15 km* [Schumann, U. (ed.)]. EUR-16209-EN, Office for Publications of the European Communities, Brussels, Belgium, pp. 352–353.

Ramaroson, R. 1995b: The 3-D chemical transport model MEDIANTE. In: *AERONOX. The Impact of NO_x Emissions from Aircraft upon the Atmosphere at Flight Altitudes 8–15 km* [Schumann, U. (ed.)]. EUR-16209-EN, Office for Publications of the European Communities, Brussels, Belgium, pp. 356–357.

Rozanov, E.V., T.A. Egorova, V.A. Zubov, and Y.E. Ozolin, 1996: The model evaluation of subsonic aircraft effect on the ozone and radiative forcing. In: *Impact of Aircraft Emissions upon the Atmosphere.* Proceedings of an international scientific colloquium, 15–18 October 1996, Paris, France. Office Nationale d'Etudes et de Recherches Aerospatiales, Chatillon, France, Vol. II, pp. 641–646.

Rozanov, E.V., T.A. Egorova, V.A. Zubov, Y.E. Ozolin, and S. Jagovkina, 1997: The model evaluation of subsonic aircraft effect on ozone and radiative forcing. In: *Research Activities in Atmospheric and Ocean Modelling* [Staniforth, A. (ed.)]. WMO/TD Publication no. 792, World Meteorological Organization, Geneva, Switzerland, pp. 7.58–7.59.

Steil, B., M. Dameris, C. Brühl, P.J. Crutzen, V. Grewe, M. Ponater, and R. Sausen, 1998: Development of a chemistry module for GCMS—first results of a multiannual integration. *Annales Geophysicae,* **16,** 205–228.

Stevenson, D.S., W.J. Collins, C.E. Johnson, and R.G. Derwent, 1997: The impact of aircraft nitrogen oxide emissions on tropospheric ozone studied with a 3D Lagrangian model including fully diurnal chemistry. *Atmospheric Environment,* **31,** 1837–1850.

Stordal, F. and U. Pedersen, 1992: *Regional and Global Air Pollution from Aircraft.* NILU Report OR 32/92, Norsk Institut for Luftforskning, Lillestrom, Norway, 37 pp.

Strand, A. and Ø. Hov, 1993: A two-dimensional zonally averaged transport model including convective motions and a new strategy for the numerical solution. *Journal of Geophysical Research,* **98,** 9023–9037.

Strand, A. and Ø. Hov, 1994: A two-dimensional global study of tropospheric ozone production. *Journal of Geophysical Research,* **99,** 22877–22895.

Strand, A. and Ø. Hov, 1995: The impact of man-made and natural NO_x emissions on upper troposphere ozone: a two-dimensional model study. *Atmospheric Environment,* **30,** 1291–1303.

The, T.H.P., 1997: *Description of a Chemistry Toolbox: A New Implementation of MOGUNTIA.* RIVM report 722201 008, Rijksinstitut voor Volksgezondheid en Milieu, Bilthoven, The Netherlands.

Veenstra, D.L., J.P Beck, T.H.P. The, and J.G.J. Olivier, 1995: *The Impact of Aircraft Exhaust Emissions on the Atmosphere: Scenario Studies with a Three-Dimensional Global Model.* RIVM Report no. 722201 003, Rijksinstitut voor Volksgezondheid en Milieu, Bilthoven, The Netherlands.

Wang, Y., D.J. Jacob, and J.A. Logan, 1997a: Global simulation of tropospheric O_3-NO_x-hydrocarbon chemistry. 1. Model formulation. *Journal of Geophysical Research,* **103,** 10713–10725.

Wang, Y., J.A. Logan, and D.J. Jacob, 1997b: Global simulation of tropospheric O_3-NO_x-hydrocarbon chemistry. 2. Model evaluation and global ozone budget. *Journal of Geophysical Research,* **103,** 10727–10755.

Wauben, W.M.F., P.F.J. van Velthoven, and H. Kelder, 1997: A 3D chemistry transport model study of changes in atmospheric ozone due to aircraft NO_x emissions. *Atmospheric Environment,* **31,** 1819–1836.

3

Aviation-Produced Aerosols and Cloudiness

DAVID W. FAHEY AND ULRICH SCHUMANN

Lead Authors:
S. Ackerman, P. Artaxo, O. Boucher, M.Y. Danilin, B. Kärcher, P. Minnis, T. Nakajima, O.B. Toon

Contributors:
J.K. Ayers, T.K. Berntsen, P.S. Connell, F.J. Dentener, D.R. Doelling, A. Döpelheuer, E.L. Fleming, K. Gierens, C.H. Jackman, H. Jäger, E.J. Jensen, G.S. Kent, I. Köhler, R. Meerkötter, J.E. Penner, G. Pitari, M.J. Prather, J. Ström, Y. Tsushima, C.J. Weaver, D.K. Weisenstein

Review Editor:
K.-N. Liou

CONTENTS

EXECUTIVE SUMMARY

- Current aircraft engines emit aerosol particles and gaseous aerosol precursors into the upper troposphere and lower stratosphere that may affect air chemistry and climate. Aircraft engines also directly emit soot and metal particles. Liquid aerosol precursors include water vapor, oxidized sulfur in various forms, chemi-ions (charged molecules), nitrogen oxides, and unburned hydrocarbons.

- Large numbers (about 10^{17}/kg fuel) of small (radius 1 to 10 nm) volatile particles are formed in the exhaust plumes of cruising aircraft, as shown by *in situ* observations and model calculations. These new particles initially form from sulfuric acid, chemi-ions, and water vapor; they grow in size by coagulation and uptake of water vapor and other condensable gases. The conversion fraction of fuel sulfur to sulfuric acid in the young plume is inferred to be likely in the range of 0.4 to about 20%.

- Subsonic aircraft emissions near the tropopause at northern mid-latitudes are a significant source of soot mass and sulfate aerosol surface area density and number concentration, according to existing measurements and model results. Aircraft generate far less aerosol than that emitted and produced at the Earth's surface or by strong volcanic eruptions. Aircraft emissions injected directly at 9- to 12-km altitudes are more important than similar surface emissions because of longer atmospheric residence times in the upper troposphere. The impact of present aircraft emissions on the formation of polar stratospheric clouds is much smaller than what is expected for a projected fleet of supersonic aircraft.

- Regional enhancements in concentrations of aircraft-produced aerosol have been observed near air traffic corridors. Global changes in sulfate aerosol properties at subsonic air traffic altitudes were small over the past few decades. The contribution of aircraft emissions to changes or possible trends in these regions is difficult to determine because of the large variability of natural sources.

- Contrails are visible line clouds that form behind aircraft flying in sufficiently cold air as a result of water vapor emissions. Contrail formation can be accurately predicted for given atmospheric temperature and humidity conditions. In the exhaust, water droplets form on soot and sulfuric acid particles, then freeze to form contrail particles. Models suggest that contrails would also form without soot and sulfur emissions by activation and freezing of background particles. Increasing fuel sulfur content results

in more and smaller ice particles. In the future, aircraft with more fuel-efficient engines will produce lower exhaust temperatures for the same concentration of emitted water vapor, hence will tend to cause contrails at higher ambient temperatures and over a larger altitude range.

- Persistent contrails often develop into more extensive contrail cirrus in ice-supersaturated air masses. Ice particles in such persistent contrails grow by uptake of water vapor from the surrounding air. The area of the Earth covered by persistent contrails is controlled by the global extent of ice-supersaturated air masses and the number of aircraft flights in those air masses. Present contrail cover will increase further as air traffic increases. The properties of persistent contrails depend on the aerosol formed in exhaust plumes. Regions of ice-supersaturation vary with time and location and are estimated to cover an average of 10 to 20% of the Earth's surface at mid-latitudes. Ice-supersaturation in these regions is often too small to allow cirrus to form naturally, so aircraft act as a trigger to form cirrus clouds.

- The mean coverage of line-shaped contrails is currently greatest over the United States of America, Europe, and the North Atlantic; it amounts to 0.5% on average over central Europe during the daytime. The mean global linear contrail coverage represents the minimum change in cirrus cloud coverage from air traffic; its present value is estimated to be 0.1% (possibly 0.02 to 0.2%).

- Aviation-induced aerosol present in exhaust plumes and accumulated in the background atmosphere may indirectly affect cirrus cloud cover or other cloud properties throughout the atmosphere. Observations and models are not yet sufficient to quantify the aerosol impact on cirrus cloud properties.

- Satellite and surface-based observations of seasonal and decadal changes in cirrus cover and frequency in main air traffic regions suggest a possible relationship between air traffic and cirrus formation. Cirrus changes in main air traffic regions suggest global cirrus cover increases of up to 0.2% of the Earth's surface since the beginning of jet aviation, in addition to the 0.1% cover by line-shaped contrails. Observed cirrus cover changes have not been conclusively attributed to aircraft emissions or to other causes.

- Contrails cause a positive mean radiative forcing at the top of the atmosphere. They reduce both the solar radiation reaching the surface and the amount of longwave radiation leaving the Earth to space. Contrails reduce the daily temperature range at the surface and cause a heating of the troposphere, especially over warm and bright surfaces. The radiative effects of contrails depend mainly on their coverage and optical depth.

- For an estimated mean global linear-contrail cover of 0.1% of the Earth's surface and contrail optical depth of 0.3, radiative forcing is computed to be 0.02 W m^{-2}, with a maximum value of 0.7 W m^{-2} over parts of Europe and the United States of America. Radiative forcing by linear contrails is uncertain by a factor of about 3 to 4 (range from 0.005 to 0.06 W m^{-2}), reflecting uncertainties in contrail cover (x 2) and contrail mean optical depth (x 3).

- In the current atmosphere, the direct radiative forcing of accumulated aircraft-induced aerosol is smaller than that of contrails. The optical depth of aircraft-induced aerosol is less than 0.0004 in the zonal mean and is much smaller than that of stratospheric aerosol from large volcanic eruptions or mean tropospheric aerosol abundances.

- Indirect radiative forcing is caused by aviation-induced cirrus that is produced in addition to line-shaped contrail cirrus. This forcing is likely positive and may be larger than that from line-shaped contrails. Radiative forcing from additional cirrus may be as large as 0.04 W m^{-2} in 1992. Indirect forcing from other cloud effects has not yet been determined and may be either positive or negative.

- In the future, contrail cloudiness and radiative forcing are expected to increase more strongly than global aviation fuel consumption because air traffic is expected to increase mainly in the upper troposphere, where contrails form preferentially, and because aircraft will be equipped with more fuel-efficient engines. More efficient engines will cause contrails to occur more frequently and over a larger altitude range for the same amount of air traffic. For the threefold increase in fuel consumption calculated for a 2050 scenario (Fa1), a fivefold increase in contrail cover and a sixfold increase in radiative forcing are expected. The contrail cover would increase even more strongly if the number of cruising aircraft increases more than their fuel consumption. For other 2050 scenarios (Fc1 and Fe1), the expected cirrus cover increases by factors of 3 and 9, respectively, compared to 1992. Higher cruise altitudes will increase contrail cover in the subtropics; lower cruise altitudes will increase contrail cover in polar regions. Future climate changes may cause further changes in expected aircraft-induced cirrus cover.

- The future aerosol impact of aviation will increase with fuel consumption. The trends depend on future fuel-sulfur content, engine soot emissions, and the efficiency with which fuel sulfur is transformed into aerosol behind the aircraft.

- Aerosol microphysical and chemical processes are similar in subsonic and supersonic aircraft plumes. Aerosol properties will differ because soot emission levels, aerosol formation potential, and plume dilution properties vary with engine type and atmospheric conditions at cruise altitudes. Significant increases in stratospheric aerosol are expected for the operation of a large fleet of supersonic aircraft, at least for non-volcanic periods.

3.1. Introduction

Recent advances in our understanding of heterogeneous chemistry in the lower stratosphere and the role of aerosols and clouds in climate forcing have increased the need to understand the influence of these aircraft emissions on atmospheric composition. Aerosol particles from aviation—comprising soot, metals, sulfuric acid, water vapor, and possibly nitric acid and unburned hydrocarbons—may influence the state of the atmosphere in many ways. These particles may provide surfaces for heterogeneous chemical reactions, both in the exhaust plume and on regional and global scales; represent a sink for condensable atmospheric gases; absorb or scatter radiation directly; and change cloud properties that may affect radiation indirectly. Persistent contrails can directly cause additional cirrus clouds to form. In addition, aerosol particles may enhance sedimentation and precipitation of atmospheric water vapor, hence affecting the hydrological cycle and the budget of other gases and particles. Changes in cloud formation properties and cloud cover may also affect actinic fluxes in the atmosphere and ultraviolet-B (UV-B) radiation at the surface.

This chapter addresses the following questions related to aviation-induced aerosol particles:

- What are the processes that produce aerosols and contrails in the plume of a jet aircraft engine?
- What is the relationship of the aircraft aerosol source to background aerosol abundances and trends in the atmosphere?
- Why do persistent contrails form, and what are their properties?
- What is the relationship of aircraft-induced aerosol and contrails to cirrus cloudiness and trends?
- What are the radiative properties of aviation-induced aerosol, contrails, and cirrus clouds, and how do they affect the Earth-atmosphere system?
- What changes in the effects of aviation-induced aerosol might occur in response to future changes in climatological conditions or aircraft operating parameters?

Section 3.2 describes the theoretical and experimental basis of the emission and formation of aerosol in aircraft plumes. Section 3.3 describes findings and calculations of regional and global aerosol distributions and their trends, quantifies the change in aerosol mass and surface density from present aviation emissions at global scales, and compares aircraft sources with volcanic and other natural or anthropogenic sources. Section 3.4 reviews recent results on contrails and cirrus clouds, provides estimates for regional and global contrail coverage, and describes measured contrail particle properties. Section 3.5 presents recent evidence that may relate changes in cirrus cloudiness and related climate parameters to aircraft emissions. Section 3.6 describes changes in radiative fluxes from contrails as a function of various cloud parameters using three one-dimensional radiation transport models, and presents a computation of global radiative forcing from contrails. Finally, Section 3.7 identifies climatological and aircraft operational parameters that may influence the

future importance of aviation-induced aerosol and cloudiness and estimates future contrail cover for a fixed climate.

The material in this chapter relies on Chapters 7 and 9 to describe emissions present at the engine exit plane and total emissions from global air traffic; Chapters 2 and 4 to discuss the chemical implications of changes in aerosol properties; and Chapters 5 and 6 to describe changes in UV-B radiation and climate from aerosols.

3.2. Aerosol Emission and Formation in Aircraft Plumes

Aircraft jet engines directly emit aerosol particles and condensable gases such as water vapor (H_2O), sulfuric acid (H_2SO_4), and organic compounds, which lead to the formation of new, liquid (volatile) particles in the early plume by gas-to-particle conversion (nucleation) processes. Other gas-phase species and charged molecular clusters (chemi-ions, or CIs) are also generated at emission, including nitric acid (HNO_3) and nitrous acid (HNO_2). Emission and formation of H_2SO_4 depend on fuel sulfur content, or sulfur emission index [EI(S)], and the conversion fraction of fuel sulfur to H_2SO_4. Formation of HNO_3 and HNO_2 depends on reactions of nitrogen oxides ($NO_x = NO + NO_2$) with hydroxyl radicals (OH). Particle formation depends on mixing of exhaust gases with ambient air, plume cooling rate, plume chemistry, and ambient aerosol properties. Soot particles formed during fuel combustion and emitted metallic particles constitute the solid (nonvolatile) particle fraction present in exhaust plumes. Under certain thermodynamic conditions, emitted water vapor condenses and freezes to form water-ice particles, thereby producing a condensation trail (contrail). These line clouds evaporate rapidly if the ambient humidity is low but may change the size and chemical composition of the remaining liquid aerosol particles. If the humidity is above ice saturation, contrails persist and grow through further deposition of ambient water.

An invisible aerosol trail is always left behind cruising aircraft. Aerosol and contrail formation processes in an aging plume determine the number, surface area, and mass of particles that are formed per mass of fuel consumed. Exhaust particle properties change in the presence of a contrail. Exhaust particle morphology and surface properties and aircraft-induced perturbations of background aerosol surface areas (Section 3.3) are of central importance for ozone changes caused by heterogeneous chemical reactions (Chapters 2 and 4). Particle number and freezing probability are key for the formation of ice (cirrus) clouds after passage of an aircraft in a region where otherwise no clouds would form (Section 3.4). Finally, aviation-produced aerosol can directly or indirectly influence the radiation budget of the atmosphere (Section 3.6 and Chapter 6). For recent reviews see Schumann (1996a), Fabian and Kärcher (1997), Friedl (1997), and Brasseur *et al.* (1998).

The following subsections provide a description of volatile aerosol precursors and the formation of volatile aerosol particles, a characterization of emitted soot and metal particles, a review of contrail and ice formation, and a discussion of the mutual interactions between these particle types. Comments on reducing the impact of aerosols are given in Section 3.7.4.

3.2.1.   *Volatile Aerosol Precursors*

3.2.1.1.   *Water Vapor*

Water vapor is present in aircraft exhaust in known amounts because the emission index is specified by the stoichiometry of near-complete fuel combustion (Chapter 7). Water vapor concentrations of a few percent at the engine exhaust nozzle far exceed the concentrations of other aerosol precursor gases. Ambient water vapor also participates in aerosol processes, with concentrations that vary widely depending on flight altitude and meteorological processes. Because of its abundance and thermodynamic properties, water vapor participates in nearly all aerosol formation and nucleation processes (e.g., Pruppacher and Klett, 1997).

3.2.1.2.   *Sulfur Species*

Aviation fuels (kerosene) contain sulfur in trace amounts. In the current world market, the sulfur content—hence the EI(S)—of aviation fuels is near 0.4 g S/kg fuel or 400 parts per million by mass (ppm; 1 ppm = 0.0001%), with an upper limit specification of 3 g S/kg (Chapter 7). Of importance for the formation of plume aerosol is the partitioning of sulfur at the engine exhaust nozzle into sulfur dioxide (SO_2) and fully oxidized sulfur S(VI) compounds, sulfur trioxide and sulfuric acid (S(VI) = SO_3 + H_2SO_4). Most fuel sulfur is expected to be emitted as SO_2 based on combustion kinetics and some observations (Miake-Lye *et al.*, 1993, 1998; Arnold *et al.*, 1994; Schumann *et al.*, 1998). However, a fraction of the SO_2 can be converted into S(VI) by gas phase chemical reactions with OH, oxygen atoms (O), and H_2O inside the engine. The fractional conversion depends on details of combustion conditions, turbine flow properties, blade cooling effects (Chapter 7), and mixing (Chapter 2). Further oxidation can occur in the plume, where the rate-limiting step is thought to be oxidation of SO_2 by OH to form SO_3 (Stockwell and Calvert, 1983) or liquid-phase reactions of SO_2 with H_2O_2, O_3, metals (Jacob and Hofmann, 1983), or HNO_3 (Fairbrother *et al.*, 1997). Once SO_3 is formed, the gas-phase reaction with emitted H_2O to form H_2SO_4 is fast (< 0.1 s) under plume conditions (Reiner and Arnold, 1993; Kolb *et al.*, 1994; Lovejoy *et al.*, 1996). Gaseous H_2SO_4 and $HSO_4^-(H_2SO_4)_n$ (mostly with n = 1,2) ion clusters have been observed in jet exhaust (Frenzel and Arnold, 1994; Arnold *et al.*, 1998a,b).

The chemical lifetime of exhaust OH in the early jet regime is determined by reactions with emitted NO_x and by OH self-reactions, the latter leading to the formation of hydrogen peroxide (H_2O_2) (Kärcher *et al.*, 1996a; Hanisco *et al.*, 1997). Measurements indicate OH exit concentrations below 1 ppmv

(Tremmel *et al.*, 1998). For an OH concentration of 0.5 to 1 ppmv at the engine's nozzle exit plane and without SO_3 emissions, the OH-induced pathway alone yields about 0.3 to 1% S-to-H_2SO_4 conversion in the plume (Miake-Lye *et al.*, 1993; Danilin *et al.*, 1994; Kärcher *et al.*, 1996a). Model calculations indicate overall S(VI) conversion fractions in the range of 2 to 10% for various supersonic and subsonic jet engines (Brown *et al.*, 1996a; Lukachko *et al.*, 1998; Chapter 7), consistent with some earlier SO_3 measurements behind gas turbines (e.g., CIAP, 1975; Hunter, 1982).

3.2.1.3.   *Chemi-ions*

A large number of chemi-ions (CIs) are expected to be present in aircraft exhaust because ion production via high-temperature chemical reactions is known to occur in the combustion of carbon-containing (not necessarily sulfur-containing) fuels (e.g., Burtscher, 1992). In the jet regime, some recent models indicate that CIs effectively promote formation and growth of electrically charged droplets containing H_2SO_4 and H_2O (Yu and Turco, 1997). In addition, CIs may contribute to the activation of exhaust soot. Positive ions include H_3O^+ and organic molecules like CHO^+, $C_3H_3^+$, and larger molecules (Calcote, 1983), whereas the free electrons rapidly attach to other molecules to form negative ions with sulfate and nitrate cores. Measurements of positive CIs in exhaust plumes are not available, and only very few *in situ* measurements of negative CIs are available to date.

Arnold *et al.* (1998a) measured a total negative CI concentration of 3 x 10^7 cm^{-3} (about 3 x 10^{15}/kg fuel) at plume ages of around 10 ms in the exhaust of a jet engine on the ground, consistent with approximately 10^9 cm^{-3} at a plume age of 1 ms (Yu *et al.*, 1998).

That concentration represents a lower bound from diffusion losses of these particles within sampling devices prior to detection and the limited detection range of the employed mass spectrometer. The fact that these measurements yielded only a fraction of CIs in the plume has been partially confirmed by in-flight measurements (Arnold *et al.*, 1998b) showing smaller total CI count rates for high-sulfur fuel compared with low-sulfur fuel. Therefore, current CI data are consistent with a CI emission index of about 2–4 x 10^{17}/kg, corresponding to a concentration of about 2 x 10^9 cm^{-3} at the engine exit. This value has been estimated numerically based on coupled ion-ion recombination kinetics and plume mixing (Yu and Turco, 1997; Kärcher *et al.*, 1998b; Yu *et al.*, 1998). Although not directly comparable, CIs have been observed in hydrocarbon flames at concentrations of about 10^8 to 10^{11} cm^{-3} (Keil *et al.*, 1984), supporting the estimated concentration of negative CIs.

3.2.1.4   *Nitrogen Species*

The primary nitrogen emission from aircraft is in the form of NO_x (Chapter 2). In reactions of NO_x with OH in the plume, gaseous HNO_2 and HNO_3 are formed. Despite larger reaction rates, less HNO_3 is formed than HNO_2 because the ratio of NO_2 to NO at the engine exit is small (< 0.2) (e.g., Schulte *et al.*, 1997) (see Chapter 7). HNO_3 can also form in the plume even in the absence of NO_2 emissions (Kärcher *et al.*, 1996a). *In situ* measurements in young plumes revealed both HNO_2 and HNO_3 concentrations above background levels (Arnold *et al.*, 1992, 1994; Tremmel *et al.*, 1998). HNO_3 can be more abundant in plumes than H_2SO_4, especially for low EI(S) values. These acids (especially HNO_3) are important because they can be taken up by water-soluble exhaust particles and form stable condensed phases such as nitric acid trihydrate (NAT = $HNO_3 \bullet 3H_2O$) and liquid ternary ($H_2O/H_2SO_4/HNO_3$) solutions under cold and humid plume conditions (Arnold *et al.*, 1992; Kärcher, 1996).

3.2.1.5. Hydrocarbons

Aircraft engines emit non-methane hydrocarbons (NMHCs) as a result of incomplete fuel combustion. These species include alkenes (mostly ethene), aldehydes (mostly formaldehyde), alkines (mostly ethine), and a few aromates. A few (8 to 10) species were found to account for up to 80% of NMHC emissions (Spicer *et al.*, 1992, 1994). High levels of carbonyl compound

emissions (on the order of 0.2 ppmv) also have been observed in a combustor (Wahl *et al.*, 1997). Some in-flight data indicate that NMHCs with up to 8 carbon atoms have EIs in the range 0.05 to 0.2 g C/kg fuel and represent approximately 70% of total NMHC emissions (Slemr *et al.*, 1998). However, the current database on NMHC emissions and on partitioning between individual compounds is small and perhaps not representative for all engine types. Some emitted NMHCs might act as aerosol-forming agents in nascent plumes and may be adsorbed or dissolved in plume particles, thereby possibly contributing to the total amount of volatile aerosol found in plumes (Kärcher *et al.*, 1998b). In addition, the presence of trace NMHCs amounts may facilitate nucleation (e.g., Katz *et al.*, 1977) and alter the hygroscopic behavior and growth rates of particles (Saxena *et al.*, 1995; Cruz and Pandis, 1997). Engines also may emit volatile particles containing engine oils or other lubricants, but this effect has not been quantified.

Aircraft also occasionally introduce hydrocarbons by jettisoning fuel at low altitudes in the troposphere. Most of the fuel evaporates while it falls to the ground (Quackenbush *et al.*, 1994), which leads to a small increase of hydrocarbons in this region. Because of the small amounts of fuel released in this way, no essential impacts on atmospheric aerosols are expected.

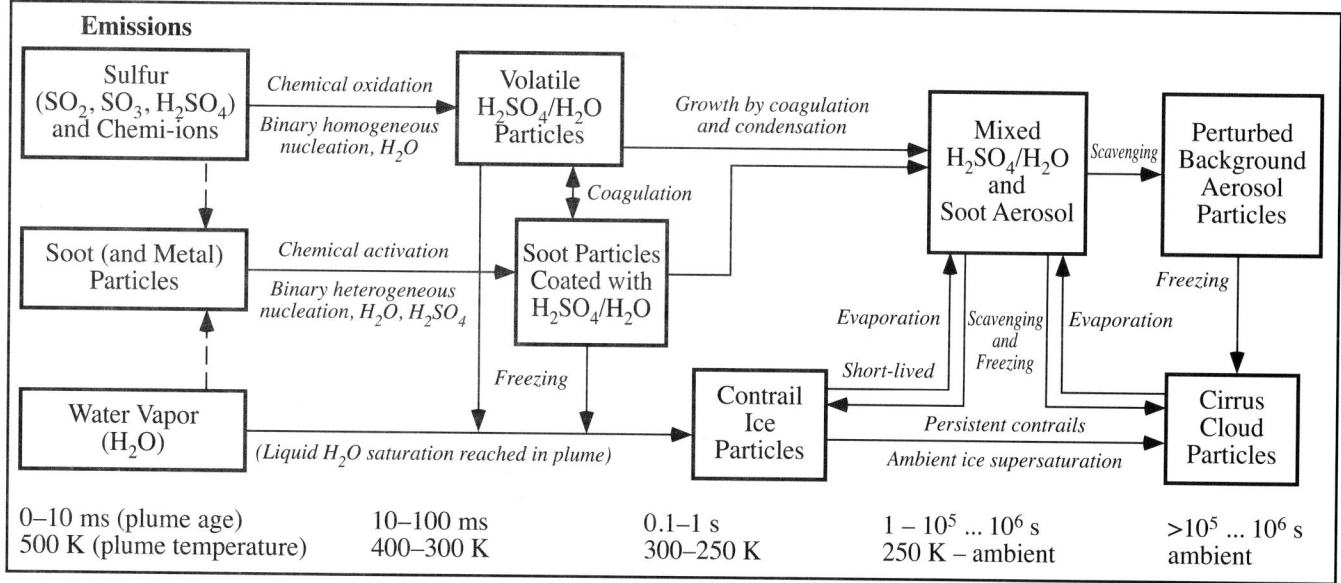

Figure 3-1: Aerosol and contrail formation processes in an aircraft plume and wake as a function of plume age and temperature. Reactive sulfur gases, water vapor, chemi-ions (charged molecules), soot aerosol, and metal particles are emitted from the nozzle exit planes at high temperatures. H_2SO_4 increases as a result of gas-phase oxidation processes. Soot particles become chemically activated by adsorption and binary heterogeneous nucleation of SO_3 and H_2SO_4 in the presence of H_2O, leading to the formation of a partial liquid H_2SO_4/H_2O coating. Upon further cooling, volatile liquid H_2SO_4/H_2O droplets are formed by binary homogeneous nucleation, whereby chemi-ions act as preferred nucleation centers. These particles grow in size by condensation and coagulation processes. Coagulation between volatile particles and soot enhances the coating and forms a mixed H_2SO_4/H_2O-soot aerosol, which is eventually scavenged by background aerosol particles at longer times. If liquid H_2O saturation is reached in the plume, a contrail forms. Ice particles are created in the contrail mainly by freezing of exhaust particles. Scavenging of exhaust particles and further deposition of H_2O leads to an increase of the ice mass. The contrail persists in ice-supersaturated air and may develop into a cirrus cloud. Short-lived and persistent contrails return residual particles into the atmosphere upon evaporation. Scavenging time scales are highly variable and depend on exhaust and background aerosol size distributions and abundances, as well as on wake mixing rates (see Section 3.3).

3.2.2.   *Volatile Particles*

3.2.2.1.   *Basic Processes*

Volatile particles form in the exhaust plume of an aircraft as a result of nucleation processes associated with the emission of aerosol precursors (Hofmann and Rosen, 1978). Typical aerosol parameters in a young plume are included in Table 3-1 for reference. The newly formed particles grow by condensation (uptake of gaseous species) and coagulation (particles collide and attach) in the expanding plume. Coagulation processes involving charged particles originating from CI emissions are more effective because charge forces enhance collision rates. These processes are schematically presented in Figure 3-1. Key processes at young plume ages are determined mainly

from the results of simulation models because of the lack of suitable plume measurements.

Previous models (Miake-Lye et al., 1994; Kärcher et al., 1995; Zhao and Turco, 1995; Brown et al., 1996b; Taleb et al., 1997; Gleitsmann and Zellner, 1998a,b) show that particles form when the condensing species, primarily H_2SO_4/H_2O, reach concentrations critical for binary homogeneous nucleation in the expanding and cooling exhaust gas. Because concentrations fall below nucleation thresholds as the plume further dilutes, the amount of H_2SO_4 in the early stages of the plume (< 1 s) controls the formation of new volatile particles. More recent models emphasize the role of CI emissions in volatile particle formation and growth (Yu and Turco, 1997, 1998a,b; Kärcher, 1998a; Kärcher et al., 1998a).

Table 3-1: *Summary of number mean radius, number density, and surface area density for sulfate and soot particles in aircraft plumes and in the background atmosphere, and for ice particles in contrails and cirrus. Flight levels of subsonic (supersonic) aircraft are in the 10–12 km (16–20 km) range. Also included are estimates of zonal mean perturbations to sulfate and soot properties caused by the 1992 aircraft fleet.*

	Radius (μm)	**Number Density** (cm^{-3})	**Surface Area Density** (μm^2 cm^{-3})
Sulfate			
Plume (1 s)[a]	0.002	$(1–2) \times 10^7$	500–1000
Background (10–12 km)[b]	0.01–0.1	50–1000	1–6 (10–40)
Background (20 km, non-volcanic)[c]	0.07	5–10	0.5–1
Background (20 km, volcanic)[d]	0.2–0.5	10–100	10–40
Soot			
Plume (1 s)[e]	0.01–0.03	$5 \times 10^4 – 5 \times 10^5$	50–5000
Background (10–20 km)[f]	0.05–0.1	0.01–0.1	$3 \times 10^{-5} – 3 \times 10^{-2}$
Ice			
Young Contrail (0.1–0.5 s)[g]	0.3–1	$10^4–10^5$	$10^4–10^5$
Persistent Contrail (10 min to 1 h)[h]	1–15	10–500	$10^3–10^4$
Young Cirrus[i]	5–10	1	$10^2–10^4$
1992 Aircraft Perturbation[j]			
Sulfate Aerosol (10–12 km, 50–60°N)	0.01	90–900	0.1–1.1
Soot (10–12 km, 50–60°N)	0.02	3–30	0.02–0.2

[a] Detectable only by ultrafine particle counters (particles smaller than 2-3 nm radius are not detected). Calculations by Yu and Turco (1997) for average FSC consistent with observed data.

[b] Properties highly variable; size distributions often bimodal. Ranges include small (> 10 nm) particles. Large particle mode (~100 nm) often similar to stratospheric aerosol particles (Hofmann, 1993; Yue et al., 1994; Schröder and Ström, 1997; Solomon et al., 1997; Hofmann et al., 1998). High range of values inferred from satellite extinction data and represents mixtures of aerosols and subvisible clouds.

[c,d] Yue et al., 1994; Kent et al., 1995; Borrmann et al., 1997; Thomason et al., 1997b.

[e] Hagen et al., 1992; Rickey, 1995; Petzold et al., 1999.

[f] Only largest soot particles with longest atmospheric lifetimes are measured by wire impactors (Sheridan et al., 1994; Blake and Kato, 1995; Pueschel et al., 1997). Uncertainties in total surface area introduced by fractal geometry of particles.

[g] Kärcher et al., 1996b, 1998a; Petzold et al., 1997.

[h] Values representative of contrail core for low ice-supersaturation (Heymsfield et al., 1998a; Schröder et al., 1998b) (see also Sections 3.4.4 and 3.6.3). Far larger particles are observed for large ice-supersaturation (Knollenberg, 1972; Gayet et al., 1996).

[i] Ström et al., 1997; Schröder et al., 1998b. Larger values are observed in warm cirrus clouds (Heymsfield, 1993; see also Sections 3.4.4 and 3.6.3).

[j] Results of fuel tracer simulations discussed in Section 3.3.4. Values shown represent upper bounds to zonal mean perturbations caused by emissions of the 1992 aircraft fleet. Results are representative of flight levels at northern mid-latitudes and are calculated using the range of values of computed tracer concentrations from all models and assuming a fuel sulfur content of 0.4 g/kg fuel, a 5% conversion of sulfur to sulfate aerosol, an EI(soot) of 0.04 g/kg fuel, and a mean particle size of 10(20) nm for sulfate (soot) particles.

Figure 3-2 shows the size distributions of exhaust particles at a plume age of 1 s, as inferred from models and a few measurements. In the radius range below 10 nm, volatile particles containing H_2SO_4 and H_2O dominate the overall distribution. Model analyses of near-field particle measurements strongly suggest that the volatile particle size distribution exhibits a bimodal structure (Yu and Turco, 1997), with smaller particles formed by the aggregation of homogeneously nucleated clusters of hydrated H_2SO_4 molecules (neutral mode) and larger particles formed by rapid scavenging of charged molecular clusters by CIs (ion-mode). Only particles from the ion mode are expected to grow beyond the smallest detectable sizes (radius ~2 to 3 nm) of particle counters. Soot and ice contrail particles are significantly larger than the volatiles. An approximate stratospheric size distribution is shown for comparison. In contrast to soot and ice particles (see below), volatile particle spectra are mainly derived from numerical simulation models because of the lack of size-resolved, *in situ* particle measurements in the nanometer size range. However, the use (in field measurements) of multiple particle counters with different lower size-detection limits allows derivation of the mean sizes of observable particles as a function of fuel sulfur content and plume age (Kärcher *et al.,*

1998b; Yu *et al.,* 1998), thereby providing strong criteria to test the validity of model results.

Once formed, the new volatile particles interact with non-volatile and contrail ice particles through the processes of coagulation, freezing, condensation, and evaporation (Figure 3-1). Calculations show that the new liquid particles grow and shrink as a function of relative humidity, whereas H_2SO_4 molecules that enter the droplets stay in the liquid phase because of their very low saturation vapor pressure (Mirabel and Katz, 1974). They also suggest that volatile particles may take up HNO_3 and H_2O in the near field (Kärcher, 1996) to form particles with compositions similar to those found in cold regions of the stratosphere. These particles may persist in cold (< 200 K), HNO_3-rich stratospheric air but will be short-lived (< 1 min) otherwise. As the plume continues to dilute with ambient air, abundant newly formed volatile particles remain at nanometer sizes and therefore add substantially to the overall aerosol surface area and abundance (Danilin *et al.,* 1997). Their efficiency for heterogeneous chemistry and cloud formation, however, is size- and composition-dependent (Kärcher, 1997). They may be too small to act as efficient cloud- or ice-forming nuclei in the background atmosphere unless the air mass containing the aerosol is lifted or cooled or the relative humidity increases. Although studies exist on heterogeneous plume processing along selected trajectories (Danilin *et al.,* 1994), systematic investigations of heterogeneous chemistry coupled to plume aerosol dynamics remain to be performed (Chapter 2).

The evolution of volatile particles is significantly altered if a contrail forms. In contrails, volatile particles have to grow to sizes greater than about 100 nm via uptake of ambient H_2O before most of them freeze (Section 3.2.4.2). As ice particles grow in size by deposition of H_2O, they may also scavenge other volatile and soot particles (Anderson *et al.,* 1998a,b; Schröder *et al.,* 1998a). Thus, contrails are expected to contain fewer small particles than non-contrail plumes because of enhanced scavenging losses. After evaporation of contrail ice crystals, the residual volatile and soot cores remain as particles in the atmosphere (Figure 3-1). This contrail processing is expected to modify the particle size distribution and composition and may lead to efficient cloud condensation nuclei production (Yu and Turco, 1998b).

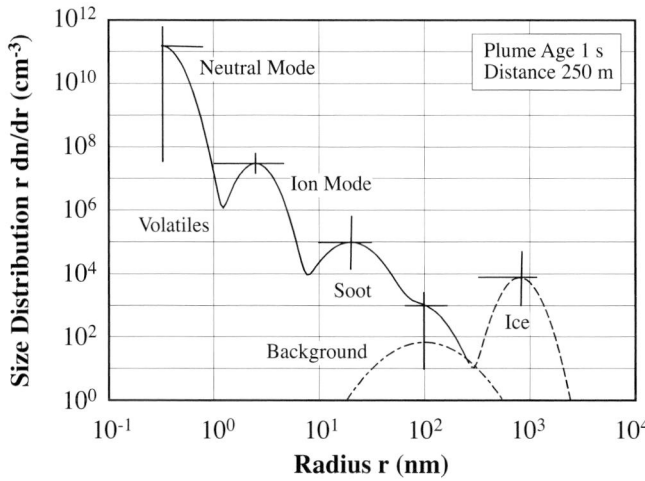

Figure 3-2: Size distribution of various aerosol types present in young jet aircraft exhaust plumes (adapted from Kärcher, 1998a). Shown are approximate size distributions (solid line) versus radius for volatile particles (neutral and chemi-ion modes) and soot (primary and secondary modes), when no contrail is produced by the aircraft. If a contrail is formed, these size distributions change (not shown), and ice particles are created (dashed line). Size distributions of soot and ice particles have been measured *in situ*. Mean sizes and numbers of ion mode particles have been deduced from particle counter measurements, and corresponding size distributions of volatile particles have been inferred from simulation models. Bars indicate the approximate range of variability resulting from variations of fuel sulfur content, engine emission parameters, and ambient conditions, as suggested by evaluation of current observations and modeling studies. An approximate background aerosol size distribution is included for comparison (dot-dashed line) (see Table 3-1).

3.2.2.2. Observations and Modeling of Volatile Particles and Sulfur Conversion

Volatile particle abundances observed *in situ* in the plumes (mostly young, < 100 s) of subsonic and supersonic aircraft are summarized in Figure 3-3a. The data have been compiled from various field studies (Fahey *et al.,* 1995a,b; Schumann *et al.,* 1997; Anderson *et al.,* 1998a,b; Schröder *et al.,* 1998a). The results show EIs for ultrafine volatile aerosol particles (nominal radii > 2 to 3 nm) in the range of 10^{15} to 10^{16}/kg fuel for low to average fuel sulfur content values and exceeding 10^{17}/kg fuel for high-sulfur fuel. Besides the obvious dependence on fuel sulfur content, the spread in EI values may be explained by differences in the emission characteristics of the engines, variations in the lower size detection limits of the particle counters, and differences in plume ages at the time of the observations. The increase in ultrafine particle abundance with increasing fuel sulfur content for the Advanced Technology Testing Aircraft System (ATTAS), T-38, and B757 aircraft strongly suggests an important role for fuel sulfur in the growth of volatiles from molecular clusters to detectable particles.

Only a few observations have been analyzed using detailed microphysical simulation models (Brown *et al.*, 1996a; Danilin

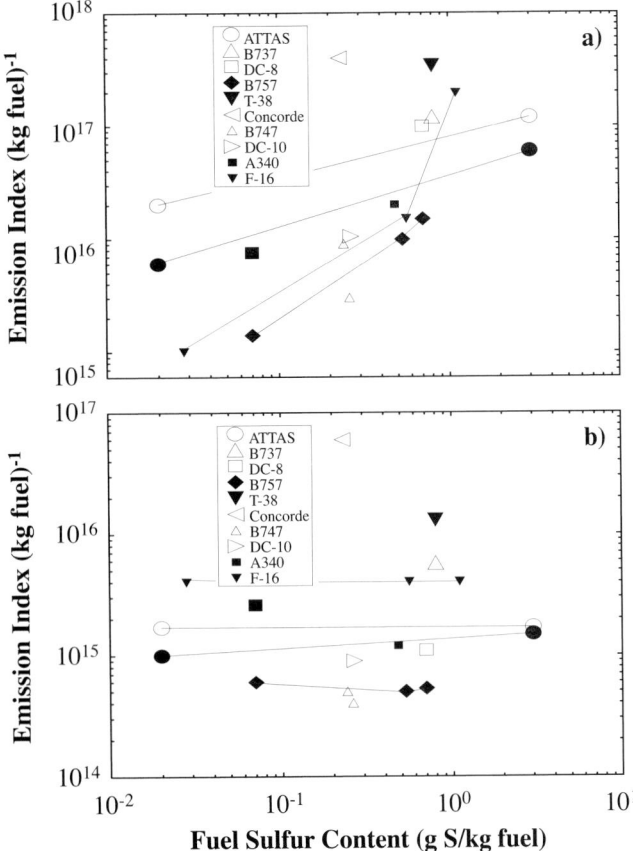

Figure 3-3: (a) Emission indices of detectable volatile particles in number per kg fuel measured *in situ* in plumes of various subsonic aircraft and the supersonic Concorde. The aircraft were examined as part of several missions in Europe and the United States of America; aircraft types are indicated in the legends. Ultrafine volatile particles with radii greater than 2 to 3 nm were measured at plume ages less than 20 s (ATTAS, B737, DC-8, B757, T-38, F-16), with radii greater than 4 to 5 nm at ages of 0.25 to 1 h (Concorde), and with radii greater than 6 nm at ages of 1 to 2 min (B747, DC-10, A340). Observations involving the same aircraft are connected by lines. Fuel sulfur content values were determined by chemical analysis of fuel or inferred from *in situ* SO_2 measurements (for B747 and DC-10). During the observations, a contrail either did (filled symbols) or did not (open symbols) form. (b) Same as (a), but for the emission indices of soot particles. Values from Howard *et al.* (1996) for a modern jet engine (about 10^{13}/kg) are below the range of values shown.

et al., 1997; Kärcher and Fahey, 1997; Yu and Turco, 1997, 1998a; Andronache and Chameides, 1998; Kärcher *et al.*, 1998a,b). Simulations show better agreement between calculated and observed particle concentrations when ion effects are taken into account. More important, the description of plume microphysics using binary homogeneous nucleation failed to explain a field measurement (Yu *et al.*, 1998). In two cases, condensation nucleus observations in the exhaust of the ATTAS and the Concorde (Figure 3-3a) have been explained in detail with a model that includes CI emissions on the order of 10^{17}/kg fuel. The observable (ion mode) particles have mean radii of about 2 to 4 nm in the young plume, for EI(S) ranging from average to high values. For decreasing levels of available H_2SO_4, the ion mode particles decrease in size. The number of detected particles falls below 10^{17}/kg fuel when the mean radius of their size distribution becomes smaller than the detection limit of the particle counters.

The extent of conversion of fuel sulfur to S(VI) necessary to explain the observed mass of volatile aerosol in young plumes seems to be variable. Direct measurements of H_2SO_4 have provided a lower bound of ~0.4% (for high-sulfur fuel, 2.7 g/kg fuel) and an upper bound of ~2.5% (for low-sulfur fuel, 0.02 g/kg fuel) for the conversion fraction in one case (Curtius *et al.*, 1998), consistent with calculated SO_3 emission levels (Brown *et al.*, 1996a,c). For the low fuel-sulfur case, it has been demonstrated that the observed volatile particles cannot be mainly composed of H_2SO_4 (Kärcher *et al.*, 1998b). In other cases, conversion fractions have been indirectly inferred from mass balance arguments involving observed or inferred particle size and number distributions, available sulfur as measured in fuel samples, and assumptions of aerosol composition. The dependence of the conversion fraction on EI(S) differs in the few studies performed to date (Fahey *et al.*, 1995a; Schumann *et al.*, 1996; Hagen *et al.*, 1998; Miake-Lye *et al.*, 1998; Pueschel *et al.*, 1998). The values deduced from these measurements range from the minimum value of 0.4% to more than 20%. Some of the indirect analyses may be affected by uncertainties (possibly about 20%) regarding the sulfur content in the fuels. Other experimental uncertainties are associated with these determinations, and the range of aircraft engines and operating parameters adds to the observed variability.

Generalization of these results cannot be done with confidence because of limited empirical knowledge of the S(VI) conversion fraction and emission levels of CIs representative of the exhaust as it enters the atmosphere. Part of the observed volatile aerosol may be composed of HNO_2 (Zhang *et al.*, 1996; Kärcher, 1997) or NMHCs (Kärcher *et al.*, 1998b), or it may result from unrecognized sulfur oxidation reactions (Danilin *et al.*, 1997; Miake-Lye *et al.*, 1998). Uptake of gaseous SO_2 and subsequent heterogeneous oxidation to sulfate in the new volatile particles is likely small (Kärcher, 1997), but the possibility of other condensational or aqueous-phase growth mechanisms has not yet been fully explored. On the other hand, CI emissions of about 10^{17}/kg fuel are consistent with observations (see Section 3.2.1.3), although they need to be confirmed by further measurements at the engine exit.

The best estimate for the S-to-H_2SO_4 conversion fraction in young plumes is 5%, with an estimated variability, or uncertainty range, of 1 to 20%. Combustion models predict a range of sulfur conversion up to about 10% and a potential for slightly higher values because of turbine blade cooling effects (Sections 3.2.1.2 and 7.6). Low conversion fractions of 2% are sufficient to explain observed volatile particle concentrations in the young plume behind the ATTAS when the effects of CIs are taken into account in simulation models of plume chemistry and microphysics (Kärcher *et al.*, 1998a,b; Yu and Turco, 1998a,b; Yu *et al.*, 1998). In contrast, the Concorde observations can be explained by assuming that about 20% of the fuel sulfur is converted to SO_3 before leaving the engine exit in such simulations (Yu and Turco, 1997).

In situ measurements detailing particle volatility and size distributions such as those included in Figure 3-3 have involved relatively young plumes. Further observations in aging plumes (> 1 h) as they dilute with the background atmosphere are currently lacking. Without detailed observations of the microphysical evolution and chemical composition of volatile exhaust particles from the engine exhaust plume to the global scale, important uncertainties remain in assessing the potential global impact of exhaust products on chemistry and cloudiness.

3.2.3. Soot and Metal Particles

3.2.3.1. Soot

Aircraft jet engines directly emit solid soot particles. Soot encompasses all primary, carbon-containing products from incomplete combustion processes in the engine. Besides the pure (optically black) carbon fraction, these products may also contain nonvolatile (gray) organic compounds (e.g., Burtscher, 1992; Bockhorn, 1994). Soot parameters of importance for understanding plume processes are concentration and size distribution at the engine exit, nucleating and chemical properties, and freezing ability.

Soot emissions for current aircraft engines are specified under the International Civil Aviation Organization (ICAO) using smoke number measurements (Chapter 7). The smoke number is dominated by the largest soot particles collected onto a filter. Sampling soot particles smaller than about 300 nm on such filters becomes inefficient. Correlations between smoke number and soot mass concentrations (e.g., Champagne, 1988) are used to estimate the soot mass EI from ICAO certification data. A mean value has been estimated to be approximately 0.04 g/kg fuel for the present fleet (Döpelheuer, 1997). Because soot emissions depend strongly on engine types, power settings, and flight levels, additional information is generally needed to relate smoke number to emissions under flight conditions. However, details of the size distribution and physicochemical properties of soot under flight conditions are generally not known and cannot be inferred from smoke number data.

Soot particle measurements for a variety of contemporary engines show values that scatter around 10^{15}/kg fuel (Figure 3-3b). Thus, soot is about 100 times less abundant in the plume than volatile aerosol particles; no significant, if any, dependence exists between soot and fuel sulfur content (Petzold *et al.*, 1997, 1999; Anderson *et al.*, 1998a; Paladino *et al.*, 1998). Values of 10^{14} to 10^{15}/kg fuel are consistent with a mass range of 0.01 to 0.2 g/kg fuel for individual engines using estimated size distributions. The older Concorde and T-38 engines show exceptionally high number EIs, whereas one modern subsonic engine emits much fewer soot particles (about 10^{13}/kg fuel) (Howard *et al.*, 1996). Soot particles are composed of individual, nearly spherical particles (spherules), which have a number mean radius between 10 and 30 nm and exceed the size of volatile aerosol particles in a young plume (Hagen *et al.*, 1992; Rickey, 1995) (Figure 3-2). Several spherical soot particles may aggregate and form a complex chain structure that may change with time (Goldberg, 1985). The smallest soot particles will be most rapidly immersed in background aerosol droplets by coagulation, consistent with the fact that only larger individual soot particles or agglomerates with radii larger than about 50 to 100 nm are observable at cruising levels (Sheridan *et al.*, 1994). Reported estimates of soot surface area at the engine exit are in the range of 5,000 to 10^5 μm^2 cm^{-3} (Rickey, 1995; Petzold *et al.*, 1999). These values continually decrease as the plume dilutes.

Much less information is available concerning the hydration properties of exhaust soot. In the initial stages of formation, graphite-like soot particles are hydrophobic. However, laboratory observations have shown that n-hexane soot particles and other black carbons are partially hydrated (e.g., Chughtai *et al.*, 1996). Soot particles fresh from jet engines very likely become hydrophilic by oxidation processes or deposition of water-soluble species present in the exhaust. Irregular surface features and chemically active sites can also increase chemical reactivity and amplify heterogeneous nucleation processes.

A clear correlation between fuel sulfur content and soluble mass fractions found on fresh exhaust soot suggests that soot hydrates more effectively with increasing EI(S) and that H_2SO_4 is the primary soluble constituent (Whitefield *et al.*, 1993). Hydration of carbon particles was observed under water-subsaturated conditions after treatment with gaseous H_2SO_4 (Wyslouzil *et al.*, 1994). This increase in H_2O adsorption is in qualitative agreement with an analysis of the wetting behavior of graphitic carbon under plume conditions (Kärcher *et al.*, 1996b). Heterogeneous nucleation of H_2SO_4 hydrates on soot was found to be unlikely under plume conditions. Soot hydration properties may also change after treatment with OH and ozone (Kärcher *et al.*, 1996b; Kotzick *et al.*, 1997).

Production of water-soluble material by the interaction of soot with SO_2 is unlikely because the sticking probabilities of gaseous SO_2 on amorphous carbon are too small (Andronache and Chameides, 1997; Rogaski *et al.*, 1997). However, SO_3 and H_2SO_4 might easily adsorb on soot prior to volatile particle formation and may explain measured soluble mass fractions on soot (Kärcher, 1998b). Sulfur may also become incorporated into soot already within the engines, possibly via S-containing hydrocarbons involved in soot formation (Petzold and Schröder, 1998). Scavenging of small volatile droplets constitutes another soot activation pathway (Zhao and Turco, 1995; Brown *et al.*, 1996b; Schumann *et al.*, 1996). The resulting liquid H_2SO_4/H_2O coating increases with plume age and may enhance the ice-forming ability of soot, which is only poorly known (Section 3.2.4), or it may suppress reactions identified in the laboratory using dry soot surfaces (Gao *et al.*, 1998).

3.2.3.2   Metal Particles

Aircraft jet engines also directly emit metal particles. Their sources include engine erosion and the combustion of fuel containing trace metal impurities or metal particles that enter the exhaust with the fuel (Chapter 7). Metal particles—comprising elements such as Al, Ti, Cr, Fe, Ni, and Ba—are estimated to be present at the parts per billion by volume (ppbv) level at nozzle exit planes (CIAP, 1975; Fordyce and Sheibley, 1975). The corresponding concentrations of 10^7 to 10^8 particles/kg fuel (assuming 1-μm radius; see below) are much smaller than for soot. Although metals have been found as residuals in cirrus and contrail ice particles (Chen *et al.*, 1998; Petzold *et al.*, 1998; Twohy and Gandrud, 1998), their number and associated mass are considered too small to affect the formation or properties of more abundant volatile and soot plume aerosol particles.

3.2.4.   *Contrail and Ice Particle Formation*

3.2.4.1.   *Formation Conditions and Observations*

Contrails consist of ice particles that mainly nucleate on exhaust soot and volatile plume aerosol particles. Contrail formation is caused by the increase in relative humidity (RH) that occurs in the engine plume as a result of mixing of warm and moist exhaust gases with colder and less humid ambient air (Schmidt, 1941; Appleman, 1953). The RH with respect to liquid water must reach 100% in the young plume behind the aircraft for contrail formation to occur (Höhndorf, 1941; Appleman, 1953; Busen and Schumann, 1995; Jensen *et al.*, 1998a). The thermodynamic relation for formation depends on pressure, temperature, and RH at a given flight level; fuel combustion properties in terms of the emission index of H_2O and combustion heat; and overall efficiency η (Cumpsty, 1997). η, defined as the fraction of fuel combustion heat that is used to propel the aircraft, can be computed from engine and aircraft properties (Schumann, 1996a; see also Section 3.7). Only the fraction (1-η) of the combustion heat leaves the engine with the exhaust gases. As the value of η increases, exhaust plume temperatures decrease for a given concentration of emitted water vapor, hence contrails form at higher ambient temperatures and over a larger range of altitudes in the atmosphere (Schmidt, 1941).

Several recent studies reported formation and visibility of contrails at temperatures and humidities as predicted by thermodynamic theory for a variety of aircraft and ambient conditions (Busen and Schumann, 1995; Schumann, 1996b; Schumann *et al.*, 1996; Jensen *et al.*, 1998a; Petzold *et al.*, 1998). These data are compiled in Figure 3-4. The mixing process in the expanding exhaust plume is close to isobaric, so the specific excess enthalpy and water content of the plume decrease with a fixed ratio as plume species dilute from engine exit to ambient values. Hence, plume conditions follow straight "mixing lines" in a plot of H_2O partial pressures versus temperature (Schmidt, 1941) (Figure 3-4). The thermodynamic properties of H_2O are such that the saturation pressures over liquid water and water-ice (solid and dashed lines) increase exponentially with temperature. Therefore, within the first second in the plume, the exhaust RH increases to a maximum, then decreases to ambient values. The ambient temperature reaches threshold values for contrail formation when the mixing lines touch the liquid saturation curve in Figure 3-4b. Contrails persist when mixing-line endpoints fall between the liquid and ice saturation pressures—that is, when the ambient atmosphere is ice-supersaturated. Without ambient ice supersaturation, contrail ice crystals evaporate on time scales of seconds to minutes. Short-lived contrails may also form without ambient water vapor if ambient temperatures are sufficiently low.

Contrails become visible within roughly a wingspan distance behind the aircraft, implying that the ice particles form and grow large enough to become visible within the first tenths of a second of plume age. Ice size distributions peak typically at 0.5 to 1 μm number mean radius (Figure 3-2). A lower limit concentration of about 10^4 cm^{-3} of ice-forming particles in the plume (at plume ages between 0.1 and 0.3 s) is necessary for a contrail to have an optical depth above the visibility threshold (Kärcher *et al.*, 1996b). These values and the corresponding mean radii of 1 μm of contrail ice particles are in agreement with *in situ* measurements in young plumes (Petzold *et al.*, 1997). Initial ice particle number densities increase from 10^4 to 10^5 cm^{-3} and mean radii decrease from 1 to 0.3 μm when the ambient temperature is lowered by 10 K from a typical threshold value of 222 K (Kärcher *et al.*, 1998a). Although aerosol and ice particle formation in a contrail are influenced by the fuel sulfur content (Andronache and Chameides, 1997, 1998), it has only a small (< 0.4 K) impact on the threshold temperature for contrail formation (Busen and Schumann, 1995; Schumann *et al.*, 1996).

Simulations of contrail formation further suggest that contrails would also form without soot and sulfur emissions by activation and freezing of background particles (Jensen *et al.*, 1998b; Kärcher *et al.*, 1998a). However, the resulting contrails would have fewer and larger particles.

Ice particle size spectra within and at the edge of young contrails systematically differ from each other (Petzold *et al.*, 1997). Ambient aerosol may play a larger role in contrail regions that nucleate at the plume edges, where the ratio of ambient to soot particles is largest and when ambient temperatures are low

(212 K) (Jensen *et al.*, 1998b). Ice particles may also nucleate on ambient droplets in the upwelling limbs of vortices and could contribute to contrail ice mass (Gierens and Ström, 1998). Metal (and soot) particles have been found as inclusions in contrail ice particles larger than 2 to 3 μm in radius (Twohy and Gandrud, 1998), but these particles are numerically unimportant compared with other plume particles.

Contrail ice crystals evaporate quickly when the ambient air is subsaturated with respect to ice, unless the particles are coated with other species such as HNO_3 (Diehl and Mitra, 1998). Simulations suggest that a few monolayers of HNO_3 may condense onto ice particle surfaces and form NAT particles in stratospheric contrails (Kärcher, 1996). These particles would be thermodynamically stable and longer lived and would cause a different chemical perturbation than would short-lived stratospheric contrails composed of water ice.

However, the relevance of this effect on larger scales has not yet been studied because no parameterization of NAT particle nucleation in aircraft plumes exists for use in atmospheric models (Chapter 4).

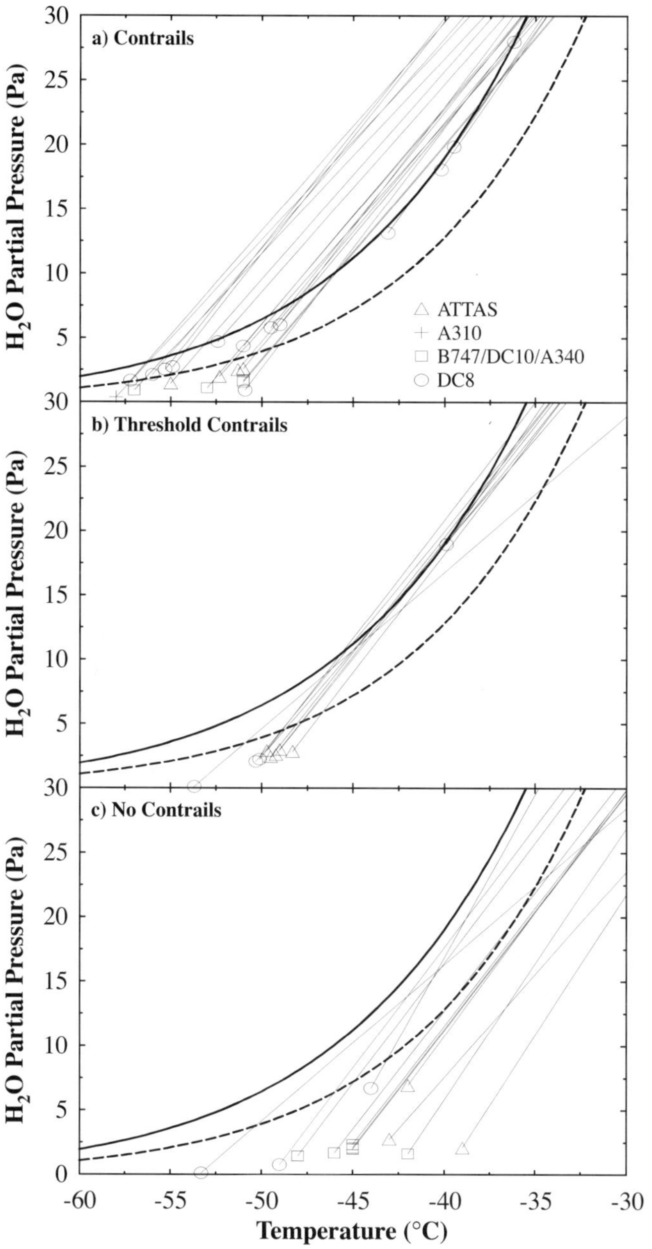

3.2.4.2. Freezing of Contrail Particles

In a young contrail, activated particles first grow to sizes $> 0.1\,\mu$m by water uptake before many of them freeze homogeneously to form water-ice particles (Kärcher *et al.*, 1995; Brown *et al.*, 1997). The fraction of H_2SO_4/H_2O droplets that freezes depends on the actual droplet composition, which affects the homogeneous freezing rate, the time evolution of H_2O supersaturation and temperature in the plume, and possible competition with heterogeneous freezing processes involving soot (see Figure 3-1).

Pure water droplets freeze homogeneously (without the presence of a foreign substrate) at a rate that grows in proportion to droplet volume and becomes very large when the droplet is cooled to the homogeneous freezing limit near about -45°C (Pruppacher, 1995). Acidic solutions freeze at lower temperatures than pure water. Freezing is often induced heterogeneously by solid material immersed inside a droplet (immersion freezing) or in contact with its surface (contact freezing). Prediction of heterogeneous freezing rates requires detailed knowledge about the ice-forming properties of droplet inclusions (Pruppacher and Klett, 1997). If homogeneous and heterogeneous freezing processes are possible, the most efficient freezing

Figure 3-4: Water vapor partial pressure and temperature measurements and calculations from various contrail studies. Symbols indicate measured ambient conditions of temperature and H_2O abundance behind various identified aircraft. Measurements are grouped using visual observations of contrails into three categories: (a) The existence of contrails is confirmed, (b) contrails were just at the limit of formation or disappearance, and (c) contrails were not observed. Liquid and ice saturation pressure are given by full and dashed curves, respectively. The thin line connected to each symbol represents the mixing line of plume states between ambient and engine exit conditions. The mixing lines have altitude- and engine-dependent slopes given by the expression $EI(H_2O)\ c_p\ p\ [0.622\ Q(1 - \eta)]^{-1}$, which includes the emission index of water vapor, the specific heat capacity of air c_p, the ambient pressure p, the ratio of molar masses of water and air (0.622), and the effective specific combustion heat $Q(1 - \eta)$, where η is the overall efficiency of propulsion for the aircraft and Q is the fuel-specific heat of combustion. Because contrails are observed only when the mixing line crosses [in (a)] or at least touches [in (b)] the liquid water saturation curve, these observations are consistent with the modified Schmidt-Appleman criterion used to predict contrail occurrence (from Kärcher *et al.*, 1998a).

mode takes up available H_2O and may prevent the growth of other particle modes.

When the ambient temperature is near the threshold value for contrail formation, models suggest that volatile aerosol particles take up only a little water and stay below the critical size (radii > 2 to 5 nm) required for growth and subsequent freezing (Kärcher et al., 1995). This critical size—hence freezing probability—depends on maximum supercooling reached in the expanding plume. Particle growth rates—hence freezing rates—are larger in cooler and more humid ambient air, so volatile particles may contribute considerably to ice crystal nucleation at temperatures below the contrail threshold value. This result is supported by observations of contrails and their microphysical properties for different fuel sulfur levels (Petzold et al., 1997) and environmental temperatures

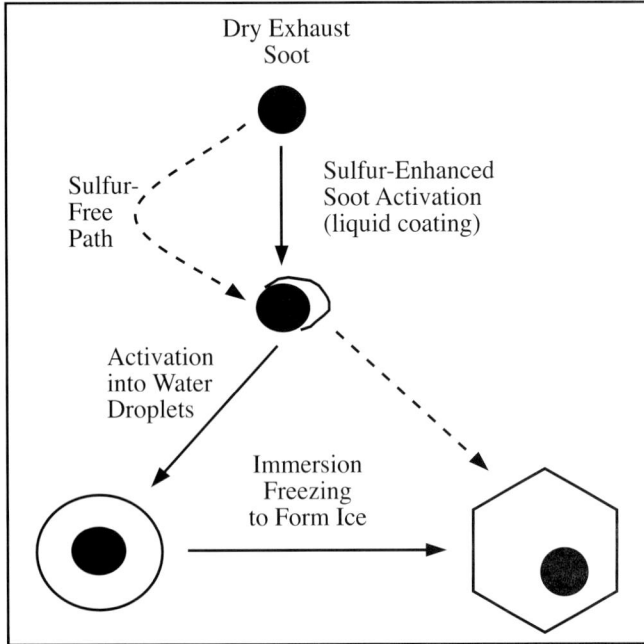

Figure 3-5: Soot activation and heterogeneous freezing in young aircraft exhaust plumes (Kärcher, 1998a).
Observational evidence suggests the existence of a sulfur-free pathway (starting with the dashed arrow) for freezing of ice at threshold formation conditions. This pathway is probably dominant for low and very low fuel sulfur levels. The sulfur-enhanced path (solid arrows) is controlled by adsorption of oxidized sulfur molecules, water vapor, and scavenging of H_2SO_4/H_2O droplets. Activation into water droplets occurs when the relative humidity in the plume exceeds 100%. A few ice particles may nucleate below liquid water saturation (dashed arrow). Well below threshold formation temperatures, homogeneous freezing of volatile droplets from the nucleation mode (in which case ice crystals initially contain no soot inclusions) is thought to dominate over soot-induced immersion freezing in the formation of contrail ice particles. The hexagon is a schematic representation of ice particle shapes that are close to spherical in young contrails but may vary in aging contrails. Soot cores may reside inside the ice particles or be attached at their surfaces.

(Freudenthaler et al., 1996). Volatile particles grown on charged droplets are more easily activated than neutral mode droplets (compare in Figure 3-2), therefore may play an enhanced role in contrail formation (Yu and Turco, 1998b). Ambient particles also contribute to ice crystal nucleation in the contrail (Twohy and Gandrud, 1998).

In contrail particle formation, heterogeneous freezing processes involving soot (see Figures 3-1 and 3-5) compete with homogeneous freezing of volatile plume particles. Volatile droplets will be prevented from freezing if rapid freezing of soot-containing particles occurs. Although this analysis is supported by model simulations and some observations (Gierens and Schumann, 1996; Kärcher et al., 1996b, 1998a; Schumann et al., 1996; Brown et al., 1997; Konopka and Vogelsberger, 1997; Schröder et al., 1998a; Twohy and Gandrud, 1998), the freezing probability of soot is poorly known because unique evidence that soot is directly involved in ice formation is difficult to obtain from in situ measurements. On the other hand, fresh soot particles do not act as efficient ice (deposition) nuclei in the exhaust (Rogers et al., 1998), consistent with the absence of contrails at temperatures above the liquid water saturation threshold.

Contrails observed near threshold formation conditions are thought to result from freezing of water on soot particles (Kärcher et al., 1996b; Schumann et al., 1996; Brown et al., 1997) (Figure 3-5). This result is supported by laboratory experiments (DeMott, 1990; Diehl and Mitra, 1998) that provide evidence that soot may induce ice formation by heterogeneous immersion freezing at temperatures colder than about 250 K. Water activation of soot may result from the formation of at least a partial surface coating of H_2SO_4/H_2O droplets, which likely develops for average to high fuel sulfur levels (Figure 3-5). Hence, more fuel sulfur leads to a greater number of ice particles. However, observations demonstrate that the number of ice particles (diameter > 300 nm) in young contrails increases by only about 30% when the fuel content increases from 6 to 2700 ppm (Petzold et al., 1997), as model simulations of contrail formation also show (Kärcher et al., 1998a).

Contrails at threshold conditions appear to be formed for very low (2 ppm) fuel sulfur content in the same manner as for average fuel sulfur content (260 ppm) (Busen and Schumann, 1995), but their properties differ measurably for larger fuel sulfur content (Schumann et al., 1996). This result suggests that soot may take up water even at zero fuel sulfur content, though this uptake may be enhanced in the presence of sulfur emissions (Kärcher et al., 1998a; see Figure 3-5).

The presence of liquid coatings may alter the chemical reactivity of dry exhaust soot, which is poorly known (Chapter 2). Soot particles acting as freezing nuclei have the potential to alter cirrus cloud properties (see Section 3.4). Present observations do not rule out the possibility that aircraft soot particles can act as freezing nuclei in cirrus formation, perhaps even without a H_2SO_4/H_2O coating. Information is lacking on how the chemical reactivity and freezing properties of soot might change in aging plumes from interactions with background gases and particles or as a result of aerosol processing in contrails.

3.3. Regional and Global-Scale Impact of Aviation on Aerosols

3.3.1. Global Aircraft Emissions and Aerosol Sources

Aircraft emissions may cause changes in the background distribution of soot and sulfuric acid aerosols at regional and global scales. In this section, observations and model results are used to evaluate these potential aircraft-induced changes.

Soot and sulfur mass emissions from aircraft are small compared with other global emissions from anthropogenic and natural sources (Table 3-2). However, aircraft emissions occur in the upper troposphere and lower stratosphere, where background values are lower and removal processes are much less effective than near the Earth's surface. Moreover, aircraft aerosol particles tend to be smaller than background particles, so small emission masses may still cause large changes in aerosol number and surface area densities. In addition, aerosol particles from aircraft can participate in the formation of contrails and clouds in the

upper troposphere, hence potentially alter the radiative balance of the atmosphere (Section 3.6).

About 93% of all aviation fuel is consumed in the Northern Hemisphere and 7% in the Southern Hemisphere (Baughcum *et al.*, 1996; see Chapter 9). Within the Northern Hemisphere, 76% of aviation fuel is consumed at mid- and polar latitudes (> 30°N). The geographical and altitude distribution of current aviation fuel consumption implies that the largest changes in aerosol and gas composition from aviation will be at northern mid-latitudes at altitudes of 10 to 12 km.

3.3.2. Sulfate Aerosol

3.3.2.1. Stratosphere

The background stratospheric sulfate layer is believed to be formed largely via the transport of carbonyl sulfide (OCS) into the stratosphere, its subsequent conversion to H_2SO_4 (Crutzen, 1976), and condensation of H_2SO_4 onto small particles nucleated primarily near the equatorial tropopause (Brock *et al.*, 1995; Hamill *et al.*, 1997). Current global photochemical models estimate that the natural source from OCS contributes 0.03 to

Table 3-2: Emission indices and estimated global emission rates of exhaust products of the present (1992) aircraft fleet using representative emission indices. Emission sources other than aircraft and estimated magnitudes of these emissions are listed in the last two columns. Values in parentheses indicate estimated range (adapted from Fabian and Kärcher, 1997; Schumann, 1994).

Fuel and Emissions	Emission Index (g pollutant/kg fuel)	Emission Rate (1992 fleet) (Tg yr⁻¹)	Comparable Emissions (Tg yr⁻¹)	Comparable Emission Source
Fuel	–	140 (139–170)[a]	3140	Total consumption of petrol
H_2O	1260	176	45 525000	CH_4 oxidation in the stratosphere Evaporation from Earth's surface
NO_x (as NO_2)	14 (12–16)[a]	2	2.9 ± 1.4 90 ± 35	Flux from the stratosphere All anthropogenic sources
Soot	0.04 (0.01–0.1)[b]	0.006	12[c]	Fossil fuel combustion and biomass burning
Sulfur	0.4 (0.3–0.5)	0.06	65[d] 10–50 2.7[f] 4.0[g]	Total from fossil fuel combustion Natural source, mostly as DMS[e] Non-eruptive volcanoes Eruptive volcanoes
C_xH_y	0.6 (0.2–3.0)	0.1	90	Anthropogenic emissions at Earth's surface

[a] From Chapter 9.
[b] Döpelheuer, 1997.
[c] Liousse *et al.*, 1996.
[d] Benkovitz *et al.*, 1996.
[e] Watson *et al.*, 1992.
[f] Spiro *et al.*, 1992.
[g] Chin *et al.*, 1996.

0.06 Tg S yr[-1] into the stratosphere (Chin and Davis, 1995; Weisenstein *et al.,* 1997). Additional sources of stratospheric sulfur may be required to balance the background sulfur budget (Chin and Davis, 1995), such as a strong convective transport of SO_2 precursors (Weisenstein *et al.,* 1997). Large increases in H_2SO_4 mass in the stratosphere often occur in periods following volcanic eruptions (Trepte *et al.,* 1993). Increased H_2SO_4 increases the number and size of stratospheric aerosol particles (Wilson *et al.,* 1993). The relaxation to background values requires several years, as Figure 3-6 illustrates with aerosol extinction measurements derived from satellite observations. The relative effect of aircraft emissions will be reduced in periods of strong volcanic activity, particularly in the stratosphere, because the aircraft source of aerosol becomes small compared with the volcanic source (see Table 3-1).

The current subsonic fleet injects ~0.02 Tg S yr[-1] into the stratosphere under the assumption that one-third of aviation fuel is consumed in the stratosphere (Hoinka *et al.,* 1993; Berger *et al.,* 1994) (see Table 3-2). This amount is 1.5 to 3

times less than natural sources of stratospheric sulfur in nonvolcanic periods. The enhanced sulfate aerosol surface area in the stratosphere affects ozone photochemistry through surface reactions that reduce nitrogen oxides and release active chlorine species (Weisenstein *et al.,* 1991, 1996; Bekki and Pyle, 1993; Fahey *et al.,* 1993; Borrmann *et al.,* 1996; Solomon *et al.,* 1997). The chemical impact of sulfate aerosol in the stratosphere is discussed in Chapters 2 and 4.

3.3.2.2. *Troposphere*

Anthropogenic and natural sources of sulfur are much larger in the troposphere than in the stratosphere (Table 3-2). Anthropogenic emissions of sulfur exceed natural sources by factors of 2 to 3 on a global scale, and emission in the Northern Hemisphere exceeds that in the Southern Hemisphere by a factor of 10 (Langner and Rodhe, 1991). Accordingly, aerosol abundance is larger in the upper troposphere than in the stratosphere and larger in the Northern Hemisphere troposphere than in the

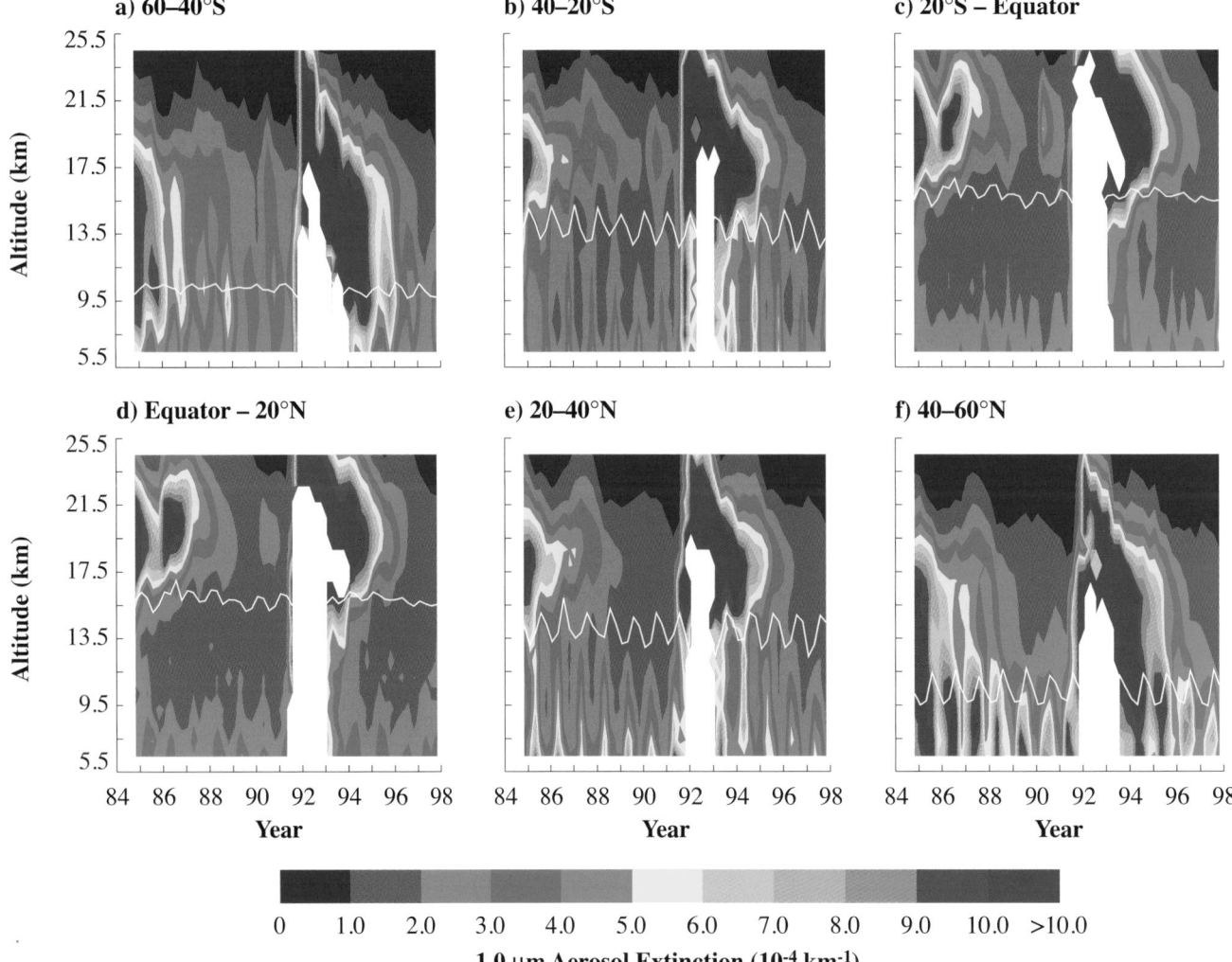

Figure 3-6: Aerosol extinction at 1.02 μm from SAGE II satellite observations at altitudes of 6.5 to 24.5 km. The time period covered is October 1984 to late 1997, with a resolution of 3 months. Each panel indicates mean values observed in a 20° latitude band. The white line indicates the location of the tropopause (extension of Plate 1 in Kent *et al.,* 1995).

Southern Hemisphere counterpart (Hofmann, 1993; Benkovitz *et al.*, 1996; Rosen *et al.*, 1997; Thomason *et al.*, 1997a,b) (Figure 3-6). Tropospheric aerosol concentrations are much larger than lower stratospheric concentrations under nonvolcanic conditions. Condensation nucleus number densities exceeding 1,000 cm^{-3} are not uncommon in the troposphere (Schröder and Ström, 1997; Hofmann *et al.*, 1998), whereas values in the lower stratosphere are less than 50 cm^{-3} in nonvolcanic periods (Wilson *et al.*, 1993).

The effect of aircraft sulfur emissions on aerosol in the upper troposphere and lower stratosphere is far larger than comparison of their amount with global sulfur sources suggests. The major surface sources of tropospheric sulfate aerosol include SO_2 and dimethyl sulfide (DMS), both of which have tropospheric lifetimes of less than 1 week (Langner and Rodhe, 1991; Weisenstein *et al.*, 1997). There is large variability in upper tropospheric aerosol particle number and size (Hofmann, 1993; Thomason *et al.*, 1997b) because of variability in tropospheric meteorology and the short lifetime of sulfur source gases. Surface emissions are known to reach the upper troposphere under certain conditions, such as during deep mid-latitude and tropical convection (Arnold *et al.*, 1997; Prather and Jacob, 1997; Dibb *et al.*, 1998; Talbot *et al.*, 1998). Only a small fraction of the surface sulfur emissions reaches the upper troposphere, however, because of the large removal rates of sulfur species near the surface.

3.3.2.3.   *Differences between the Upper Troposphere and Lower Stratosphere*

The effects of aircraft emissions on aerosol particles and aerosol precursors depend on the amounts emitted into the troposphere and stratosphere. In addition to aerosol abundance and sources, stratospheric and tropospheric aerosols also differ in composition and residence time. Typical parameters for sulfate at 12 and 20 km at northern mid-latitudes are summarized in Table 3-1.

Sulfate is considered to be the dominant component of stratospheric aerosol; soot and metals are considered to be minor components (Pueschel, 1996). The composition of tropospheric aerosol, particularly near the surface, also includes ammonium, minerals, dust, sea salt, and organic particles (Warneck, 1988). The role of minor components of aerosol composition in affecting heterogeneous reaction rates is not fully understood. In general, the lower stratosphere contains highly concentrated H_2SO_4/H_2O particles (65–80% H_2SO_4 mass fraction) as a result of low relative humidity in the stratosphere and low temperatures (Steele and Hamill, 1981; Carslaw *et al.*, 1997). Higher H_2O abundances (by a factor of 10 or more) and similar temperatures cause particles in the upper troposphere to be more dilute (40–60% H_2SO_4). Surface reactions that activate chlorine are particularly effective on dilute H_2SO_4 particles and cirrus cloud particles at low temperatures in the tropopause region (Chapter 2).

Aircraft emissions injected into the stratosphere have greater potential to perturb the aerosol layer than those emitted into the

troposphere, because in the stratosphere the background concentrations are lower and the residence times are longer. The initial residence time (1/e-folding time) of most of the stratospheric sulfate aerosol mass from volcanic eruptions is about 1 year as a result of aerosol sedimentation rates (Hofmann and Solomon, 1989; Thomason *et al.*, 1997b; Barnes and Hofmann, 1997) (see also Figure 3-6). The residence time of the remaining aerosol mass contained in smaller particles is several years. The residence time of upper tropospheric aerosol particles is much smaller, ranging from several days (Charlson *et al.*, 1992) to between 10 and 15 days (Balkanski *et al.*, 1993; Schwartz, 1996). Tropospheric particles are larger than those in the stratosphere (Hofmann, 1990), therefore sediment faster. They are also removed by cloud scavenging and rainout.

3.3.3.   *Observations of Aircraft-Produced Aerosol and Sulfate Aerosol Changes*

Observations of aircraft-induced aerosols have increased substantially in recent years (see Section 3.2). Concentrations of aerosol particles and aerosol precursor gases well above background values have been observed in the exhaust plumes of aircraft operating in the upper troposphere and lower stratosphere. Although aircraft emissions are quickly diluted by mixing with ambient air to near background values, the accumulation of emissions in flight corridors used in the routing of commercial air traffic has the potential to cause notable atmospheric changes.

Estimated changes from aircraft emissions (Schumann, 1994; WMO, 1995) are small compared with natural variability, hence are not always apparent in observational data sets. However, regional enhancements in concentrations of aircraft-produced aerosol have been observed near air traffic corridors. During measurement flights across the North Atlantic flight corridor over the eastern Atlantic, signatures of NO_x, SO_2, and condensation nuclei (CN) were clearly evident in the exhaust plumes of 22 aircraft that passed the corridor at this altitude in the preceding 3 h, with values exceeding background ambient levels by 30, 5, and 3 times, respectively (Schlager *et al.*, 1997). A mean CN/NO_x abundance ratio of 300 cm^{-3} ppbv^{-1} was measured. This ratio corresponds to a mean particle emission index of about 10^{16} kg^{-1} and implies CN increases of 30 cm^{-3} in corridor regions where aircraft increase NO_x by 0.1 ppbv (cf. Chapter 2). The regional perturbation was found to be detectable at scales of more than 1,000 km under special meteorological conditions within a long-lasting stagnant anticyclone (Schlager *et al.*, 1996). In an analysis of 25 years of balloon measurements in Wyoming in the western United States of America, subsonic aircraft are estimated to contribute about 5–13% of the CN concentration at 8–13 km, depending on the season (Hofmann *et al.*, 1998). This estimate provides only a lower bound of the aircraft contribution because smaller aircraft-produced particles (radius < 10 nm) are not detected. Additionally, regular lidar measurements have been made of aerosol optical depth at aircraft altitudes (10–13 km) in an area of heavy air traffic in southern Germany (Jäger *et al.*, 1998).

Table 3-3: *Non-volcanic upper tropospheric annual mean optical depth and % change per year, along with standard deviation, from SAGE satellite observations during 1979–97. Values in parentheses are for the Southern Hemisphere (adapted from Kent et al., 1998).*

Latitude Band	Annual Mean Optical Depth (10^{-4})	Change per Year (%)
80–60° N(S)	25.1 ± 4.7 (2.9 ± 0.9)	-0.4 ± 0.2 (-0.7 ± 0.6)
60–40° N(S)	18.1 ± 4.7 (7.0 ± 1.0)	0.4 ± 0.2 (1.4 ± 0.3)
40–20° N(S)	19.3 ± 2.9 (15.2 ± 2.5)	0.2 ± 0.1 (1.2 ± 0.2)
20–0° N(S)	19.6 ± 0.9 (18.2 ± 1.6)	0.2 ± 0.1 (0.8 ± 0.1)
Hemisphere N(S)		0.1 ± 0.1 (0.9 ± 0.3)
Globe		0.5 ± 0.2

Large optical depths on the order of 0.1 that could be attributed to the accumulation of aircraft aerosol were observed very rarely at this location.

Global changes in sulfate aerosol properties at subsonic air traffic altitudes were small over the last few decades. The examination of long-term changes in aerosol parameters suggests that aircraft operations up to the present time have not substantially changed the background aerosol mass. Multiyear observations for the upper troposphere and lower stratosphere are available from satellite and balloon platforms and ground-based lidar systems. Long-term variations of the optical depth in the upper troposphere from the SAGE satellite (Figure 3-6 and Table 3-3) have been analyzed with periods of volcanic influence excluded (Kent *et al.*, 1998). Data indicate that changes are less than about 1% yr[-1] between 1979 and 1998, when observations are averaged over either hemisphere. A significant change in aerosol amounts is also not found in the 15- to 30-km region examined with lidar over Mauna Loa (20°N) for the period 1979 to 1996 (Barnes and Hofmann, 1997). Similarly, aerosol mass data above the tropopause derived from lidar soundings show no trend in a region of heavy air traffic in Germany (48°N) over the last 22 years (Figure 3-7) (Jäger and Hofmann, 1991; Jäger *et al.*, 1998).

Long-term changes in aerosol parameters measured *in situ* are also small. *In situ* measurements are important because the number of particles in the upper troposphere and lowermost stratosphere is dominated by sizes that are too small (< 0.15-μm radius) to be remotely detected. Long-term changes in CN are small in the 10- to 12-km region of the mid-latitude troposphere, where most of the current aircraft fleet operates (Hofmann, 1993). The 5% yr[-1] increase in larger particle (radius > 0.15 μm) abundances found in lower stratospheric balloon measurements made between 1979 and 1990 was considered consistent with the accumulation of aircraft sulfur emissions (Hofmann, 1990, 1991). However, the absence of a change in observed stratospheric CN number suggests that the trend in the larger particles is the result of the growth of existing particles rather than nucleation of new particles. Model results show that the contribution of the current subsonic fleet to aerosol mass amounts between 15 and 20 km is about 100 times smaller than the observed aerosol amounts (Section 3.3.4; Bekki and Pyle, 1992).

Previous balloon-borne CN counters did not measure particles below 10 nm in radius, which are now detected with more modern CN counters and dominate aerosol number in aircraft plumes. Moreover, the attribution of aerosol changes to aircraft is complicated by changes in surface sources of sulfur and episodic strong injections of sulfur from volcanic eruptions (Hitchman *et al.*, 1994; Barnes and Hofmann, 1997; Thomason *et al.*, 1997a,b). Hence, the contribution of aircraft emissions to changes or possible trends in these regions is difficult to determine at present.

3.3.4. Modeling Sulfate Aerosol Perturbations Caused by Aircraft

3.3.4.1. Subsonic Aircraft

Global models are required to evaluate the atmospheric impact of aerosol generated by subsonic aircraft (Friedl, 1997; Brasseur *et al.*, 1998). The global distribution of tropospheric

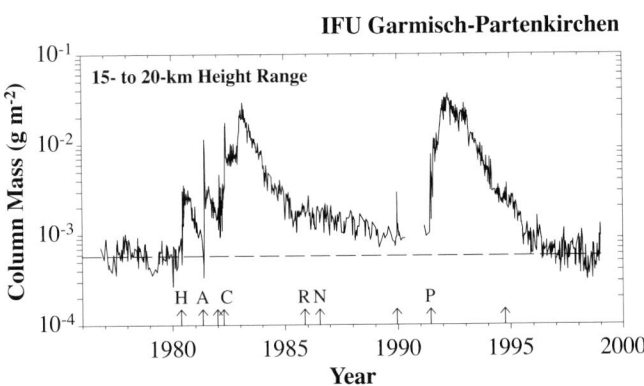

Figure 3-7: Aerosol-mass column between 15- and 20-km altitude, derived from backscatter measurements made by lidar at Garmisch-Partenkirchen, Germany, between 1976 and end of 1998. Dates of major volcanic eruptions are marked on the horizontal axis (H, St. Helens; A, Alaid; C, El Chichón; R, Ruiz; N, Nyamuragira; P, Pinatubo). The dashed line indicates the average value of 1979, which can be considered a background value free from volcanic influence (adapted from Jäger and Hofmann, 1991; Jäger *et al.*, 1998).

sulfur species has been investigated using various three-dimensional (3-D) models (Langner and Rodhe, 1991; Penner *et al.*, 1994; Chin *et al.*, 1996; Feichter *et al.*, 1996; Pham *et al.*, 1996; Schwartz, 1996). Because most models have been developed to investigate regional or global effects of surface emissions in the lower or middle troposphere, few have addressed the potential impact of aviation sources on aerosol parameters in the upper troposphere and lower stratosphere.

A systematic model study has been carried out with a suite of two-dimensional (2-D) and 3-D atmospheric models to determine upper bounds for the accumulation of aviation aerosol in the atmosphere (Danilin *et al.*, 1998). Each model computed the steady-state global distribution of a passive tracer emitted into the model atmosphere with the same rate and distribution as aviation fuel use, based on the NASA 1992 database (see Chapter 9). The only sink for the passive tracer is below 400 hPa (approximately 7 km), where it is removed with a 1/e-folding time of 5 days. The resultant global tracer distribution can be used to provide estimates of steady-state concentration change from a specific emission by multiplying the tracer value by the associated aircraft engine emission index (EI). Figure 3-8 shows steady-state tracer distributions in tracer-to-air mass mixing ratio units and annually and zonally averaged fuel source used in the simulation. Table 3-4 summarizes the main results of these simulations.

All models predict the largest perturbation at mid-latitudes in the Northern Hemisphere in the altitude range of 10–12 km.

However, the magnitude of the perturbation varies by a factor of 10, ranging from 12.6 (ECHAM3) to 122 ng g^{-1} (GSFC-2D), reflecting differences in model resolution and current uncertainties in modeling of global atmospheric dynamics and turbulent diffusion. To mitigate the effects of model resolution, the tracer amount was summed in the 8- to 16-km altitude region between 30 and 90°N (shown by the thick dashed line in Figure 3-8). This region contains 34 (AER, ECHAM3) to 61% (GSFC-2D) of the total accumulated tracer. The absolute amount of tracer mass in this volume ranges from 2.9 (ECHAM3) to 14.5 Tg (GSFC-2D). The amount of the tracer above 12 km, which serves to diagnose the fraction of aircraft emissions transported toward the stratospheric ozone maximum, ranges from 14 (UCI/GISS, GSFC-2D) to 45% (UMICH, AER) of each model's global tracer amount. The global residence time of the fuel tracer, defined as the ratio of the steady-state tracer mass to the tracer source, varies from 21 days (TM3, ECHAM3) to 65 days (GSFC-2D, LLNL). The lower values are similar to the global residence times (approximately 18 days) found for air parcels uniformly released at 11 km between 20 and 60°N and followed with a trajectory model using assimilated wind fields (Schoeberl *et al.*, 1998). The 1/e-folding aircraft emissions lifetime of 50 days computed by Gettelman (1998) is consistent with the results of the fuel tracer experiment described here (Danilin *et al.*, 1998).

The model simulation results indicate that aircraft contribute little to the sulfate mass near the tropopause. For example, the sulfate aerosol mass density from the GSFC-2D model (see

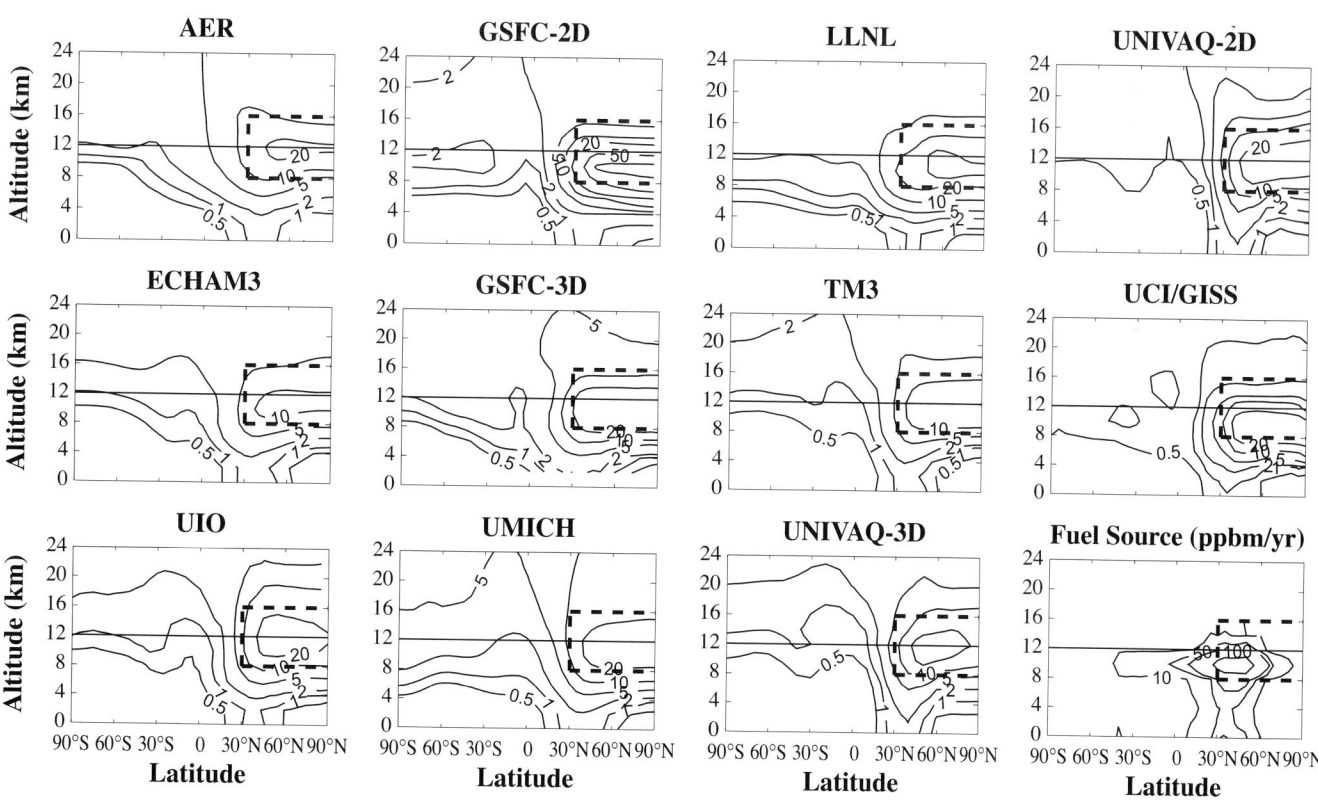

Figure 3-8: Zonally and annually averaged distribution of fuel tracer in ng(tracer)/g(air), according to indicated models. The fuel source is shown in the bottom right panel. The thick dashed line shows the region between 8–16 km and 30–90°N. The thin horizontal line indicates 12-km altitude (from Danilin *et al.*, 1998).

Table 3-4: *Results from the 1992 fuel tracer simulations (other results included in Table 3-1).*

Model[a]	Max. Tracer Value (ng g[-1])	Latitude of Max. (°N)	Global Residence Time (days)	Tracer 8–16 km 30–90°N (%)	Tracer >12km (%)	Max. Tracer Column[b] (µg cm[-2])	Global Soot Column (ng cm[-2])	Global SO$_4$ Column[c] (ng cm[-2])
2-D Models								
AER	26.7	55	38	34	45	6.6	0.11	3.5
GSFC-2D	122	55	62	61	16	22.9	0.20	5.9
LLNL	72.5	65	65	42	38	14.5	0.20	6.0
UNIVAQ-2D	36.4	60	23	58	33	7.7	0.08	2.2
3-D Models								
ECHAM3	12.6	50	22	34	31	4.1	0.07	2.0
GSFC-3D	46.7	50	52	44	29	11.7	0.16	4.9
TM3	20.1	80	21	45	40	4.9	0.07	2.0
UCI/GISS	34.4	55	27	49	14	8.2	0.09	2.6
UiO	28.2	55	29	50	40	7.8	0.09	2.7
UMICH	30.4	65	45	37	44	9.6	0.14	4.2
UNIVAQ-3D	38.4	50	25	50	41	7.7	0.08	2.3

[a] Models are denoted as follows: **2-D**—Atmospheric and Environmental Research (AER) (Weisenstein *et al.*, 1998), Goddard Space Flight Center (GSFC-2D) (Jackman *et al.*, 1996), Lawrence Livermore National Laboratory (LLNL) (Kinnison *et al.*, 1994), University of L'Aquila (UNIVAQ-2D) (Pitari *et al.*, 1993); **3-D**—German Aerospace Center (DLR) (ECHAM3) (Sausen and Köhler, 1994), Goddard Space Flight Center (GSFC-3D) (Weaver *et al.*, 1996), Royal Netherlands Meteorological Institute (KNMI) (TM3) (Wauben *et al.*, 1997), University of California at Irvine (UCI/GISS) (Hannegan *et al.*, 1998), University of Oslo (UiO) (Berntsen and Isaksen, 1997), University of Michigan (UMICH) (Penner *et al.*, 1991), UNIVAQ-3D (Pitari, 1993).

[b] Column amounts calculated between 0–60 km for all models except ECHAM3 and TM3 (0–32 km) and UiO (0–26 km).

[c] These values are calculated from model results with assumptions of EI(soot) of 0.04 g/kg fuel, EI(sulfur) of 0.4 g/kg fuel, and 100% conversion of sulfur to H$_2$SO$_4$.

Figure 3-8) is 0.055 µg m[-3] at 10 km at 55°N, in contrast to background concentrations of 1 to 2 µg m[-3] (Yue *et al.*, 1994). Other recent model studies (Danilin *et al.*, 1997; Kjellström *et al.*, 1998) show similar results for aerosol mass; these studies further conclude that aircraft emissions may noticeably enhance the background number and surface area densities (SAD) of sulfate aerosol (see Tables 3-1 and 3-4) because of the smaller radii of aircraft-produced particles.

Fuel tracer simulation also provides estimates of soot and sulfate column amounts that can be used to calculate the direct radiative forcing of current subsonic fleet emissions (see Chapter 6). Maximum tracer column values are located near 50 to 60°N and range from 4.1 (ECHAM3) to 22.9 µg cm[-2] (GSFC-2D). To calculate instantaneous direct radiative forcing at the top of the atmosphere from aircraft soot emissions, globally averaged tracer column values (which are smaller than their maximum values by a factor of 3 to 4) are first multiplied by EI(soot) to obtain soot column values. For aircraft sulfur emissions, the tracer column is scaled by EI(S), the ratio of molar mass of SO$_4$ and S, and a 100% conversion fraction of sulfur to sulfate (see Table 3-4). For the suite of models in Table 3-4, the upper bound for the average soot column is 0.1 ng cm[-2], with a range from 0.07 to 0.20 ng cm[-2]; for the average sulfate column the upper bound is 2.9 ng cm[-2], with a range from 2 to 6 ng cm[-2]. If photochemical oxidation lifetime and tropospheric washout rate are taken into account, a 50%

conversion fraction of sulfur to sulfate is a more suitable value than 100%. In this case, the average sulfate column is 1.4 ng cm[-2], with a range of 1 to 3 ng cm[-2].

In addition to the passive tracer simulation, the AER 2-D model also calculated the evolution of aerosol using a sulfur photochemistry and aerosol microphysics model designed for stratospheric conditions (Weisenstein *et al.*, 1997). This model calculates about 3.4 ng cm[-2] for the perturbation in sulfate aerosol at 55°N, consistent with the AER model value in Table 3-4 but almost 30 times smaller than the background sulfate column amounts (~100 ng cm[-2]). Figure 3-9 depicts the annually averaged increase of sulfate aerosol SAD calculated with the AER 2-D model, assuming 5% conversion of sulfur emissions into new particles (as recommended in Section 3.2) with a radius of 5 nm and fuel with 0.4 g S/kg. The maximum SAD perturbation, located at about 10–12 km in northern mid-latitudes, is about 0.3 µm[2] cm[-3], which is comparable to ambient values in nonvolcanic periods (Hofmann and Solomon, 1989; Thomason *et al.*, 1997b).

Table 3-1 presents upper-bound estimates of soot and sulfate aerosol number and surface area densities from 1992 fuel simulations. Present aircraft emissions noticeably increase the number and surface area densities of aerosol particles in the tropopause region despite the large CN background concentration in the upper troposphere. The estimates use the range of values

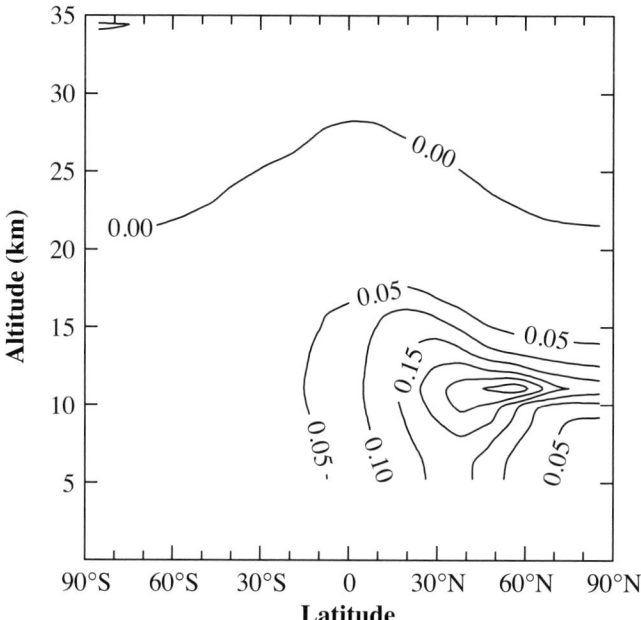

Figure 3-9: Latitude and altitude distribution of annually averaged increase of surface area density of sulfate aerosol (in μm^2 cm^{-3}), calculated with AER 2-D model assuming a 1992 aircraft fuel-use scenario, 0.4 g S/kg fuel, and 5% conversion of sulfur emissions into new particles with a radius of 5 nm (adapted from Weisenstein *et al.*, 1997).

of computed tracer concentrations from all models and the effective EIs of soot and sulfate mass and assume a mean particle size of 10(20) nm for sulfate (soot) particles and a 5% conversion of sulfur to sulfate aerosol. The results represent order-of-magnitude estimates of zonal mean maximum values at 12 km and can be compared with background aerosol properties in the lowermost stratosphere as given in Table 3-1. For these estimates, 5% of the emitted sulfur dioxide is assumed to be converted to sulfuric acid before dispersal out of the zonal region of maximum air traffic.

Tracer simulations strongly suggest that aircraft emissions are not the source of observed decadal H_2O changes at 40°N. The simulation results can be scaled by $EI(H_2O)$ to provide an upper bound (neglecting precipitation from the upper troposphere) for the accumulation of water vapor above the tropopause as a result of aircraft emissions. The LLNL model shows the largest tracer accumulation at 40°N, with equivalent H_2O values smoothly decreasing from 55 ppbv at 10 km to 12 ppbv at 24 km. These values are small in comparison to current ambient values of 59 ppmv at 10–12 km and 4.2 ppmv at 22–24 km (Oltmans and Hofmann, 1995). Assuming 5% yr^{-1} growth in fuel consumption and $EI(H_2O)$ of 1.23 kg/kg, the change in aircraft-produced H_2O ranges from 3.4 ppbv yr^{-1} at 10 km to 0.8 ppbv yr^{-1} at 24 km. These values represent a change of +0.006% yr^{-1} at 10 km and +0.018% yr^{-1} at 24 km and are more than of 20 times smaller than those found in long-term balloon observations (Oltmans and Hofmann, 1995).

3.3.4.2. *Supersonic Aircraft*

The impact of a future fleet of supersonic aircraft on sulfate aerosol abundance in the stratosphere has been discussed using measurements and model results (Bekki and Pyle, 1993; Fahey *et al.*, 1995a; Stolarski *et al.*, 1995; Weisenstein *et al.*, 1996, 1998). The results suggest that aerosol surface area density will be substantially greater than nonvolcanic background values in proposed fleet scenarios (Section 3.7 and Chapter 4). The consequences for stratospheric ozone changes depend on the simultaneous emissions of nitrogen oxides and chlorine and aerosol loadings of the atmosphere (Solomon *et al.*, 1997; Weisenstein *et al.*, 1998; see also Chapter 2).

3.3.5. **Soot**

The primary atmospheric source of soot or black carbon particles is combustion of fossil fuels and biomass burning at the Earth's surface, with total emission values near 12 Tg C yr^{-1} (Liousse *et al.*, 1996). This value exceeds reasonable estimates of the aircraft source of black carbon by several orders of magnitude (Bekki, 1997). For example, aircraft are estimated to have emitted 0.0015 to 0.015 Tg C as soot into the atmosphere in 1992 [with EI(soot) of 0.01 to 0.1 g C/kg fuel] (Friedl, 1997; Rahmes *et al.*, 1998). As in the case of sulfate aerosol, deposition and scavenging of black carbon near surface sources creates large vertical gradients in the lower atmosphere, with soot concentrations falling by 1 to 2 orders of magnitude between the surface and the lower stratosphere (Penner *et al.*, 1992; Cooke and Wilson, 1996; Liousse *et al.*, 1996). A possible meteoritic source of soot in the lower stratosphere has been considered but is not well quantified at present (Chuan and Woods, 1984).

Few direct measurements of soot abundance are available in the upper troposphere and lower stratosphere. The most extensive measurements in these regions are from aircraft impactor measurements (Pueschel *et al.*, 1992, 1997; Blake and Kato, 1995). The accuracy of such measurements depends on knowledge of the impactor for small soot particles. The results (Figure 3-10) are considered to represent a lower limit for soot number and mass because of size-selective sampling and because of scavenging of soot by background aerosol particles. Features of the measurements include a large gradient between the Northern and Southern Hemispheres and large variability with altitude at northern mid-latitudes. The large vertical variability of soot at northern mid-latitudes cannot be explained by the accumulation of aircraft soot emissions. With typical soot mass densities observed to be approximately 1 ng m^{-3}, soot is estimated to represent approximately 0.01% of the stratospheric aerosol mass (Pueschel *et al.*, 1992). In other sampling flights over southern Germany, measurements of absorbing mass (probably soot) at 8–12 km altitude and partly within cirrus clouds showed concentrations above 10 ng m^{-3}, with higher values correlated with local aviation fuel consumption (Ström and Ohlsson, 1998).

Modeling studies of soot distribution in the upper troposphere and lower stratosphere differ on the importance of aircraft

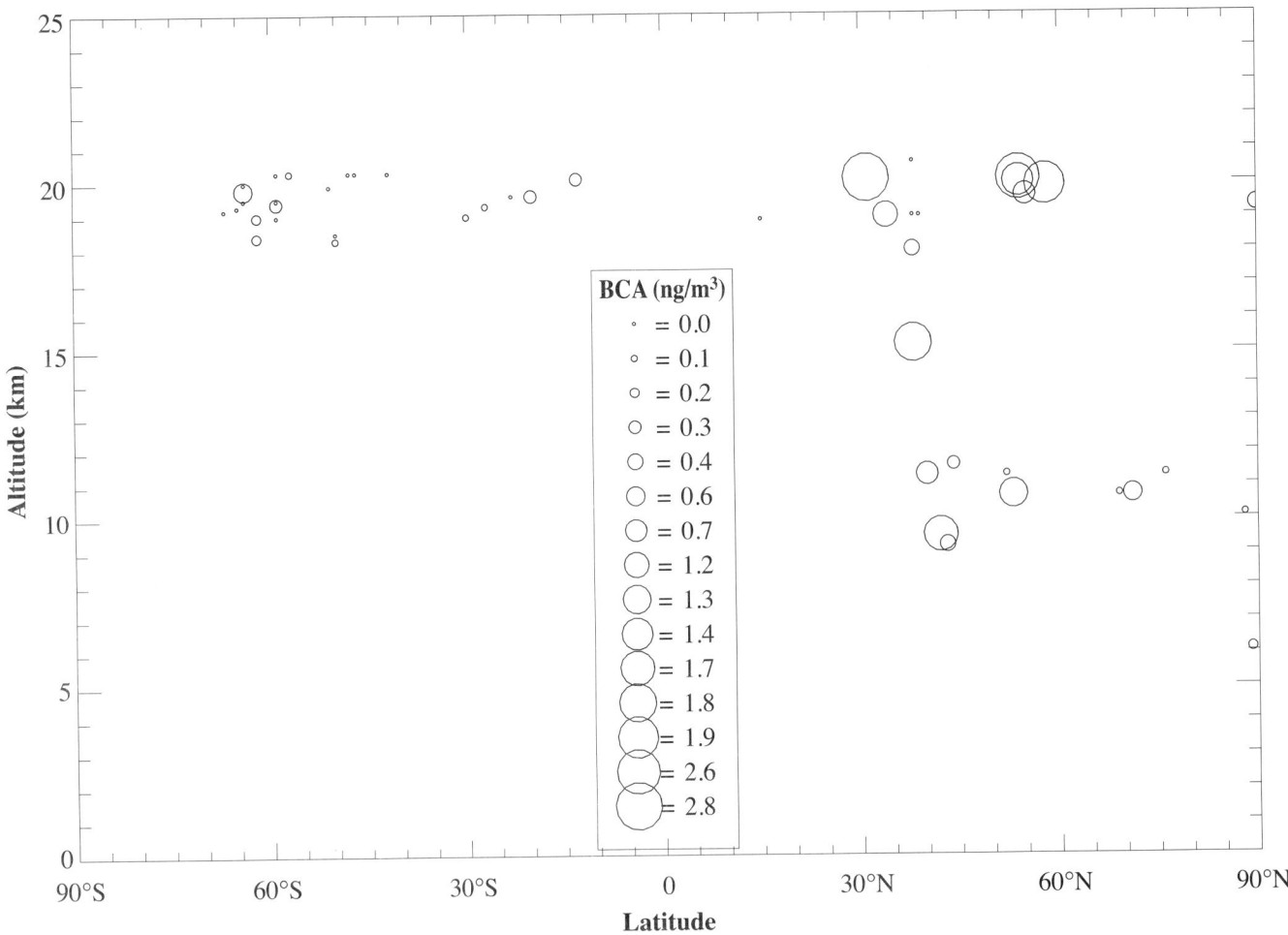

Figure 3-10: Latitude and altitude distribution of measured values of soot concentrations (BCA = black carbon aerosol) in upper troposphere and lower stratosphere (in ng m^{-3}). Measurements were obtained between November 1991 and April 1992 (Northern Hemisphere) and between April and October 1994 (Southern Hemisphere) (adapted from Pueschel *et al.*, 1997).

sources of soot. Some model simulations suggest that ground-level sources of soot (12 Tg C yr^{-1}) are as important as aircraft sources in the upper troposphere and predict maximum soot values there [2 to 5 ng m^{-3} in Liousse *et al.* (1996) and 10 to 50 ng m^{-3} in Cooke and Wilson (1996)] that exceed observed values near 200 hPa at northern mid-latitudes (Pueschel *et al.*, 1997). Other model results show that current aircraft could be a noticeable source of the soot near the tropopause at northern mid-latitudes (Bekki, 1997; Danilin *et al.*, 1998; Rahmes *et al.*, 1998). Tracer simulation results from the AER 2-D model (Section 3.5.1) were multiplied by EI(soot) of 0.04 g/kg fuel (Döpelheuer, 1997) to estimate the global distribution of soot. The results (Figure 3-11) show maximum values of 0.6 ng m^{-3} near 12 km at northern mid-latitudes. These values are in the middle of the range of other model results and are in the range of observed values (Figure 3-10). However, because the fleet-mean EI(soot) is uncertain and may range from 0.01 to 0.1 g/kg fuel, the effective range of the maximum in Figure 3-11 is 0.15 to 1.5 ng m^{-3}. At 20 km, fuel tracer simulations show aircraft-induced zonal mean soot perturbations to be approximately 100 times smaller than maximum observed values (Figure 3-10).

Observations of soot in the upper troposphere and lower stratosphere are too limited to provide an estimate of any

long-term changes in soot concentrations in those regions. The possible consequences of heterogeneous reactions on soot (Bekki, 1997; Lary *et al.*, 1997) are discussed in Chapter 2.

3.3.6. Polar Stratospheric Clouds and Aircraft Emissions

During winter in the polar regions, low temperatures lead to the formation of polar stratospheric cloud (PSC) particles, which contain H_2SO_4, HNO_3, and H_2O (e.g., WMO, 1995; Carslaw *et al.*, 1997; Peter, 1997). PSCs activate chlorine, leading to significant seasonal ozone losses in the lower stratosphere, particularly in the Southern Hemisphere (WMO, 1995). PSC formation may be enhanced by the atmospheric accumulation of aircraft emissions of NO_x, H_2O, and sulfate, as well as through direct formation in aircraft plumes in polar regions (Section 3.2 and Chapter 4). If aircraft emissions change the frequency, abundance, or composition of PSCs, the associated ozone loss may also be modified (Peter *et al.*, 1991; Arnold *et al.*, 1992; Considine *et al.*, 1994; Tie *et al.*, 1996; Del Negro *et al.*, 1997). The effects of subsonic aircraft emissions on PSCs and stratospheric ozone are expected to be smaller than those of similar emissions from supersonic aircraft because subsonic emissions occur in the 10- to 12-km region, whereas supersonic emissions will most likely occur in the 15- to 20-km region. Ambient temperatures in the 10- to 12-km region are usually too high (> 200 K) for PSCs to form with available H_2O and HNO_3, and ozone and total inorganic chlorine concentrations are much lower than near 20 km.

The impact of the subsonic fleet on PSC formation has not been well studied. The results of the fuel tracer simulation discussed in Section 3.3.4 can be used to estimate the increase of PSC SAD as a result of aircraft emissions of H_2O and NO_x. Assuming an $EI(H_2O)$ of 1,230 g/kg, $EI(NO_x)$ of 15 g/kg, complete conversion of NO_x to HNO_3, and formation of NAT particles at threshold temperatures, AER model results for a 1992 subsonic fleet show an additional condensation of HNO_3 on NAT particles ranging from 0.02 ppbv at 60°N to 0.12 ppbv at 85°N at 20 km in January. These values provide an increase of 0.08 μm^2 cm^{-3} in PSC SAD at 20 km and 85°N, assuming a unimodal distribution of PSC particles with diameter of 1 μm. The increase in spatial extent of PSCs both vertically and latitudinally is small in the model. PSC increases are very sensitive to background temperature, H_2O, and HNO_3 values and will differ considerably among models. The increases are not likely to significantly alter ozone changes in polar winter because the SAD increases are much less than typical values of 1–10 μm^2 cm^{-3} calculated for PSC events, and satellite data observations show that the probability of PSC formation below 14 km in the Arctic is generally very low (< 1%) (Poole and Pitts, 1994).

An important caveat related to the assessment of additional PSC formation as a result of aircraft emissions is that plume processes are not included. Global models generally assume that aircraft emissions are homogeneously distributed in a model grid box that is much larger than an aircraft plume. The consequences of this assumption have not yet been fully evaluated. In one model study, reactions on PSCs did not affect ozone chemistry in a subsonic plume at northern mid-latitudes in April (Danilin *et al.*, 1994). A further caveat is that estimated PSC changes from aircraft emissions have not accounted for projected cooling of the stratosphere, which may enhance PSC formation.

The chemical implications of increased PSC formation for ozone chemistry and atmospheric composition are further discussed in Chapter 2. The effects of future aircraft fleets on additional PSC formation and subsequent ozone response are presented in Chapter 4.

3.4. Contrail Occurrence and Persistence and Impact of Aircraft Exhaust on Cirrus

Aircraft cause visible changes in the atmosphere by forming contrails that represent artificially induced cirrus clouds. The conditions under which contrails form are discussed in Section 3.2.4. This section describes the formation, occurrence, and properties of persistent contrails and how they compare with natural cirrus.

3.4.1. Cirrus and Contrails

Cirrus clouds (Liou, 1986; Pruppacher and Klett, 1997) contain mainly ice crystals (Weickmann, 1945). The distinctive properties of cirrus and contrails derive from the physics of ice formation. Ice particle nucleation occurs either through homogeneous nucleation (when pure water droplets or liquid aerosol particles freeze) or through heterogeneous nucleation, when freezing of the liquid is triggered by a solid particle or surface that is in contact with the liquid or suspended within the liquid. Both processes depend strongly on temperature and relative humidity (Heymsfield and Miloshevich, 1995; see Section 3.2.4.2).

Comparisons of model simulations with observations suggest that we may understand homogeneous nucleation (Ström *et al.*, 1997; Jensen *et al.*, 1998b) and are making headway in understanding the potential contribution of heterogeneous nucleation (DeMott *et al.*, 1998; Rogers *et al.*, 1998). Each of these freezing mechanisms requires that the atmosphere be highly supersaturated with respect to the vapor pressure of ice before crystals can form. For instance, supersaturations in excess of 40–50% with respect to ice are needed for sulfuric acid particles to freeze homogeneously (Tabazadeh *et al.*, 1997) at temperatures above 200 K. Observations of relative humidity with respect to ice at the leading edges of wave clouds are consistent with the requirement of large ice supersaturations for nucleation of ice on the bulk of the atmospheric aerosol (Heymsfield *et al.*, 1998a; Jensen *et al.*, 1998c). Hence, there is a large supersaturation range in which

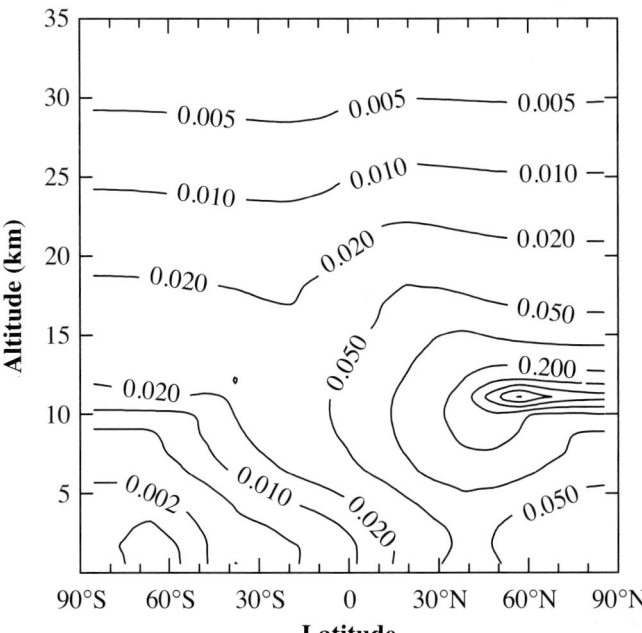

Figure 3-11: Latitude and altitude distribution of annually averaged increase in soot mass density calculated with the AER 2-D model, assuming a 1992 aircraft fuel-use scenario and a soot emission index of 0.04 g C/kg fuel. Contour values are in ng m^{-3} (adapted from Weisenstein *et al.*, 1997).

heterogeneous nuclei could lead to cirrus formation before the bulk of the atmospheric particles freeze (DeMott *et al.*, 1997). This potential for heterogeneous nuclei to cause ice formation at ice supersaturations that are relatively low compared to those needed to freeze sulfate particles leads to concern about the role of aircraft exhaust in modifying ambient clouds (Jensen and Toon, 1997).

Ice crystal number densities are limited by competition between increasing saturation as a result of cooling in vertical updrafts and decreasing saturation as a result of growth of ice crystals. The depletion of vapor as a result of growth of the first few ice crystals nucleated prevents further ice nucleation on the remaining particles (Jensen and Toon, 1994). Once ice crystals form and take up available water vapor, supersaturation declines and further nucleation of ice ceases. This selectivity causes ice crystals in cirrus to be larger relative to droplets in liquid water-containing clouds—apart from differences in saturation vapor pressures over ice compared to water, which causes more water vapor to be available for deposition on ice particles than on water droplets. Large ice particles may precipitate rapidly. Many cirrus clouds have a "fuzzy" appearance because rapid precipitation causes optically thin edges of clouds to be diffuse, and precipitation allows particles to spread in the wind, forming long tails of cloud. Ice crystal nucleation also depends on available aerosol in the upper troposphere, the properties of which are only poorly known (Ström and Heintzenberg, 1994; Podzimek *et al.*, 1995; Sassen *et al.*, 1995; Schröder and Ström, 1997). In some locations, upper tropospheric particles are dominated by sulfates (Yamato and Ono, 1989; Sheridan *et al.*, 1994). However, more recent data show that minerals, organic compounds, metals, and other substances may often be present in significant quantities (Chen *et al.*, 1998; Talbot *et al.*, 1998).

Cirrus clouds occur mainly in the upper troposphere. The mean tropopause altitude is about 16 km in the tropics and 10 km (250 hPa) north of 45°N latitude (Hoinka, 1998). The tropopause temperature at northern mid-latitudes varies typically between -40 and -65°C; it may reach below -80°C in the tropics. At mid-latitudes, the upper troposphere often is humid enough for cirrus and persistent contrails to form. The stratosphere is commonly so dry that cirrus (PSCs) and persistent contrails form only in polar winter (for T > -60°C and p < 250 hPa, a 10-ppmv H_2O mixing ratio corresponds to less than 15% relative humidity). Ground-based observers report a mean cirrus cover of 13% over oceans and 23% over land (Warren *et al.*, 1986, 1988). Satellite data (Wang *et al.*, 1996; Wylie and Menzel, 1999) identify larger cloud cover (~40%) by subvisible (optical depth τ at 0.55 μm below ~0.03) and semi-transparent (0.1 < τ < 0.6) cirrus clouds. Cirrus clouds are typically 1.5 (0.1 to 4) km thick; are centered at 9 (4 to 18) km altitude; have a crystal concentration of 30 (10^{-4} to 10^4) L^{-1}, with crystal lengths of 250 (1 to 8000) μm (see Dowling and Radke, 1990); and have an optical depth at 0.55 μm of about 0.3 (0.01 to 30) (Wylie and Menzel, 1999). These typical values of cirrus crystal concentration and mean particle size may be biased by early studies that failed to make adequate measurements of ice

crystals smaller than about 50 μm. More recent studies suggest that crystal concentrations in cirrus are often on the order of 1 cm^{-3} (Ström *et al.*, 1997; Schröder *et al.*, 1998b), although crystal concentrations in wave clouds can exceed 10 cm^{-3}. PSCs (between the tropopause and ~25-km altitude) and noctilucent clouds (~80-km altitude) contain ice and might therefore be included as very high altitude forms of cirrus clouds. Systems of cirrus clouds have a lifetime that may reach several hours or even days (Ludlam, 1980), but the lifetimes of particles within clouds is much shorter.

Persistent contrail formation requires air that is ice-supersaturated (Brewer, 1946). Ice-supersaturated air is often free of visible clouds (Sassen, 1997) because the supersaturation is too small for ice particle nucleation to occur (Heymsfield *et al.*, 1998b). Supersaturated regions are expected to be quite common in the upper troposphere (Ludlam, 1980). The presence of persistent contrails demonstrates that the upper troposphere contains air that is ice-supersaturated but will not form clouds unless initiated by aircraft exhaust (Jensen *et al.*, 1998a). Aircraft initiate contrail formation by increasing the humidity within their exhaust trails, whereas local atmospheric conditions govern the subsequent evolution of contrail cirrus clouds. Indeed, the ice mass in long-lasting contrails originates almost completely from ambient water vapor (Knollenberg, 1972).

Ice-supersaturated air masses are often formed when ice-saturated air masses are lifted by ambient air motions. While the air lifted, it may remain cloud-free until it is cooled adiabatically to near-liquid saturation (Ludlam, 1980). Other evidence for large supersaturation occurring in the upper troposphere is provided by cirrus fallstreaks that grow while falling through supersaturated air layers (Ludlam, 1980) and by a few localized humidity measurements (Brewer, 1946; Murphy *et al.*, 1990; Ovarlez *et al.*, 1997; Heymsfield *et al.*, 1998b). Recent humidity measurements by commercial aircraft show that—in flights between Europe, North and South America, Africa, and Asia—14% of flight time was in air masses that were ice-supersaturated with a mean value of 15% (Helten *et al.*, 1998; Gierens *et al.*, 1999).

3.4.2 Cirrus and Contrail Models

Small-scale, regional, and global models have been used to study cirrus clouds and contrails. Small-scale models simulate the details of cloud formation in a single parcel of air along some defined trajectory (e.g., Jensen *et al.*, 1994b). High-resolution 2-D models have simulated the dynamics and microphysics of cirrus (Starr and Cox, 1985) and contrails (Gierens, 1996; Chlond, 1998). Regional and global models describe the large-scale dynamics of clouds with simpler representations of microphysical cloud processes (Sundqvist, 1993; Lohmann and Roeckner, 1995; Fowler *et al.*, 1996; Westphal *et al.*, 1996). Regional-scale (typically 10-km horizontal grid-scale) and global-scale models are not able to resolve the vertical motions and small-scale temperature and humidity fluctuations (Gierens *et al.*, 1997) that drive supersaturations involved in cirrus nucleation. However, global models with prescribed or parameterized contrail cover have been useful for understanding the radiative impact of a change in cloud cover as a result of aircraft emissions (Ponater *et al.*, 1996).

3.4.3. Contrail Occurrence

At plume ages between 1 min and 1 h, contrails grow much faster horizontally (to several km width) than vertically (200 to

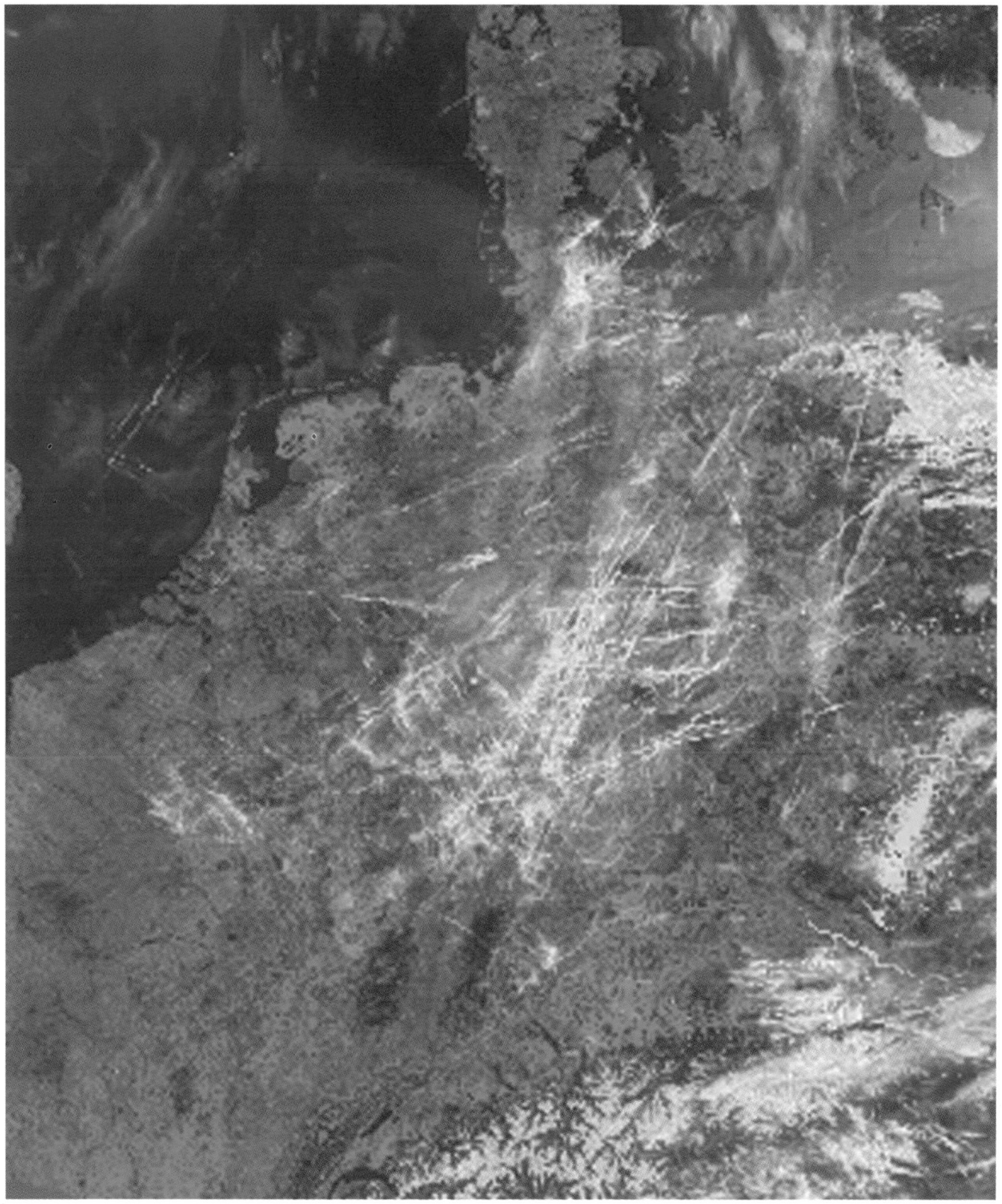

Figure 3-12: Contrails over central Europe on 0943 UTC 4 May 1995, based on NOAA-12 AVHRR satellite data (from Mannstein, 1997).

400 m), especially in highly sheared environments (Freudenthaler *et al.*, 1995, 1996; Sassen, 1997). Young contrails spread as a result of turbulence created by aircraft vortices (Lewellen and Lewellen, 1996; Gerz *et al.*, 1998; Jensen *et al.*, 1998a,b,c), shear in the ambient wind field (Freudenthaler *et al.*, 1995; Schumann *et al.*, 1995; Dürbeck and Gerz, 1996; Gierens, 1996), and possibly radiatively driven mixing (Jensen *et al.*, 1998d).

Contrails often become wide and thick enough to induce radiative disturbances that are sufficient to be detectable in multispectral satellite observations. They have been observed at 1-km spatial resolution with instruments such as the Advanced Very High Resolution Radiometer (AVHRR) on board National Oceanic and Atmospheric Administration (NOAA) polar-orbiting satellites (e.g., Lee, 1989) and at 4-km resolution in the infrared with the Geostationary Operational Environmental Satellite (GOES) (Minnis *et al.*, 1998a). The AVHRR channels in the 11- to 12-μm range (4 and 5) are particularly suited to detect thin ice clouds because of the different emissivity of ice particles in this spectral range (King *et al.*, 1992; Minnis *et al.*, 1998c). Figure 3-12 shows, for example, a mid-European scene derived from AVHRR data in these channels, combined with the visible channel to represent the surface, for a day when many line-shaped contrails were formed by heavy air traffic (Mannstein, 1997). The figure shows that aircraft trigger contrail cirrus that evolves into cirrus clouds that are much more extensive in scale than the initial contrails. Such spread and deformed contrail cirrus can no longer be distinguished from naturally occurring cirrus. In Figure 3-12, contrails that still have a line-shaped appearance cover about 5% of the scene.

Aged contrails often cannot be distinguished from cirrus, which poses an observational problem in determining the frequency and area of coverage by contrails. An important example of the persistence of contrails and their evolution into more extensive cirrus is shown in Figure 3-13. An initial oval contrail observed in GOES-8 satellite images diffused as it was advected over California until it no longer resembled its initial shape 3 h later (Minnis *et al.*, 1998a). The exhaust from this single aircraft flying for less

than 1 h in a moist atmosphere caused a cirrus cloud that eventually covered up to 4,000 km² and lasted for more than 6 h. Other contrails and contrail clusters were observed to develop over periods of 7 to 17 h, spreading to cover areas of 12,000 to 35,000 km². Such dispersed contrails are usually indistinguishable from natural cirrus; hence, satellite detection algorithms based on the linear structures of young contrails will not detect these dispersed contrails.

At present, observations of actual frequency or areal coverage of Earth by contrails are limited to a few selected regions.

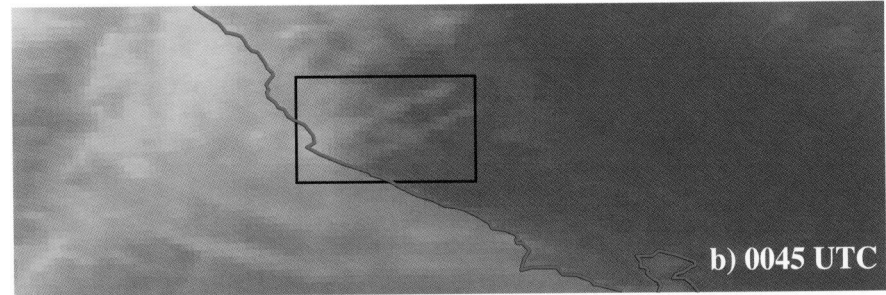

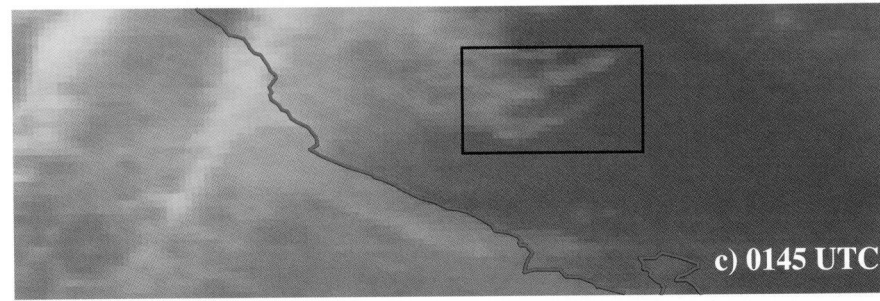

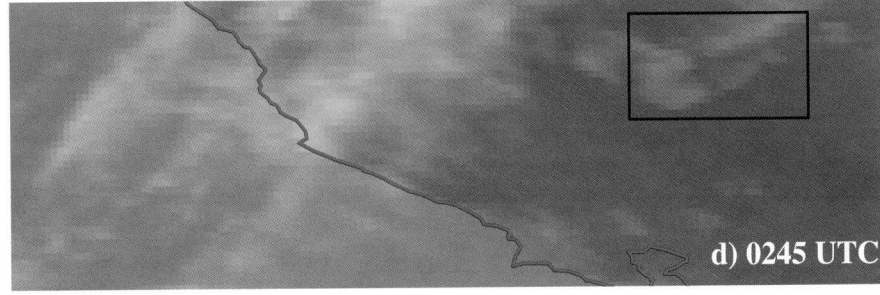

Figure 3-13: Time series of GOES-8 satellite images showing the evolution of a contrail from an initial oval shape to extensive cirrus clouds (from Minnis *et al.*, 1998a). The NASA DC-8 flew an oval flight pattern several times off the coast of California on 12 May 1996 (a), resulting in a visible contrail 15 minutes later (b). This contrail spread as it was advected over California (c), until it no longer resembled its initial shape 3 hours later (d). Satellite photographs courtesy of L. Nguyen of AS&M, Inc., Hampton, VA, USA.

Contrail frequency refers to the probability that a contrail will be observed somewhere within the scene being viewed; area of coverage refers to the fraction of the area of the scene in which contrails are observed. Estimates of contrail coverage or occurrence have been made directly or indirectly from surface (Detwiler and Pratt, 1984; Minnis *et al.*, 1997) and satellite observations (Joseph *et al.*, 1975; Carleton and Lamb, 1986; Lee, 1989; DeGrand *et al.*, 1990; Schumann and Wendling, 1990; Betancor-Gothe and Grassl, 1993; Bakan *et al.*, 1994; Mannstein *et al.*, 1999).

Data for the frequency of contrails are available from surface-based observations at 19 locations across the continental United States of America for every hour during 1993–94 (Minnis *et al.*, 1997). The data indicate that contrail frequency peaks around February/March and is at a minimum during July. Annual mean persistent contrail frequency (not the cover) for the 19 sites was 12%. When related to fuel use and extended to the remainder of the country, mean annual contrail frequency for the United States of America is estimated at 9% (Minnis *et al.*, 1997). The relationship between fuel consumption and contrail frequency from this data set is shown in Figure 3-14. The correlation implies that contrail coverage is limited mainly by the number of aircraft flights, not by atmospheric conditions at cruise altitudes. Pilots flying over the former Soviet Union have reported that contrails occur most frequently in winter and spring and less often during summer (Mazin, 1996). Sky photographs taken from 1986 to 1996 over Salt Lake City, Utah, reveal a seasonal cycle in contrail frequency, with a maximum in fall and winter and a minimum in July (Sassen, 1997). These data are similar to observations from a site 64 km north (Minnis *et al.*, 1997) of Salt Lake City. Sassen (1997) and Minnis *et al.* (1997) found that contrails occur with cirrus in approximately 80% of observations. The coincidence of cirrus and contrails suggests that cloud-free supersaturated regions are usually interspersed with areas in which natural clouds have actually formed.

From an analysis of AVHRR infrared images taken over the northeast Atlantic and Europe, a mean contrail cover of 0.5% was derived for 1979–81 and 1989–92 (Bakan *et al.*, 1994). A distinct seasonal cycle was found, with a southward displacement of the maximum cover during winter. Maximum coverage (~2%) occurred during summer, centered along the North Atlantic air routes. Using AVHRR channel 4 and 5 differences and a pattern-recognition algorithm to differentiate line-shaped clouds from fuzzy cirrus clouds, the contrail cover—defined as line-shaped clouds over mid-Europe—was systematically evaluated for all of 1996 using nearly all noon passages of the NOAA-14 satellite (see Figure 3-15) (Mannstein *et al.*, 1999). Contrails are not uniformly distributed; instead, they lie along air traffic corridors and accumulate near upper air route crossings. As with observations over the United States and the former Soviet Union, maximum coverage occurred during winter and spring. At noon on average over the year, line-shaped contrails cover about 0.5% of the area of the mid-European region shown in this figure. This coverage represents the lower bound for the actual contrail cover because the algorithm cannot identify non-line-shaped contrails. Contrail coverage at night over Europe is one-third of the noon-time contrail cover (Mannstein *et al.*, 1999).

Contrails often occur in clusters within regions that are cold and humid enough to allow persistent contrails to form. Contrail clusters observed in satellite data indicate that these air masses cover 10 to 20% of the area over mid-Europe (Mannstein *et al.*, 1999) and parts of the United States of America (Carleton and Lamb, 1986; Travis and Changnon, 1997), consistent with the fraction of air masses expected to be ice-supersaturated at cruise altitudes. Hence, as air traffic increases in these regions, persistent contrail coverage is also expected to increase, possibly up to a limit of 10–20%.

Estimates of global coverage by air masses that are sufficiently cold and humid for persistent contrails can be obtained using meteorological analysis data of temperature and humidity, a model to estimate the frequency of ice-saturation in each grid cell as a function of the analyzed relative humidity, and the Schmidt-Appleman criterion (depending on η; see Section 3.2.4.1) for contrail formation conditions. Using 11 years of meteorological data from the European Center for Medium Range Weather Forecast (ECMWF), the cover of suitable air masses is found to be largest in the upper troposphere, especially in the tropics (global mean value of 16%) (Sausen *et al.*, 1998). This coverage is similar in size to the area covered by clusters of contrails in satellite data and the frequency of ice-supersaturated air masses observed along commercial aircraft routes (Gierens *et al.*, 1999). The expected contrail cover is then computed from the product of the air mass coverage and the fuel consumption

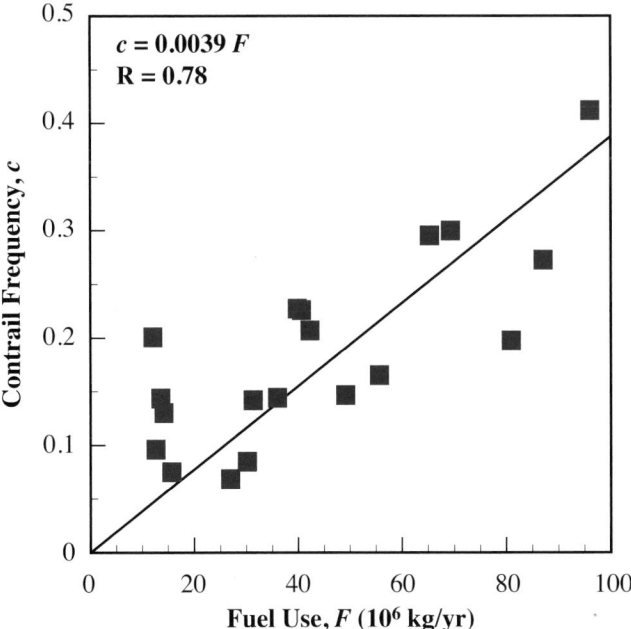

Figure 3-14: Correlation of mean annual contrail frequency and estimated May 1990 aircraft fuel usage above 7-km altitude. Contrail frequencies are based on 1993 to 1994 hourly surface observations from 19 stations in the United States of America (Minnis *et al.*, 1997). The fit to the data (solid line) is shown. R is the correlation coefficient.

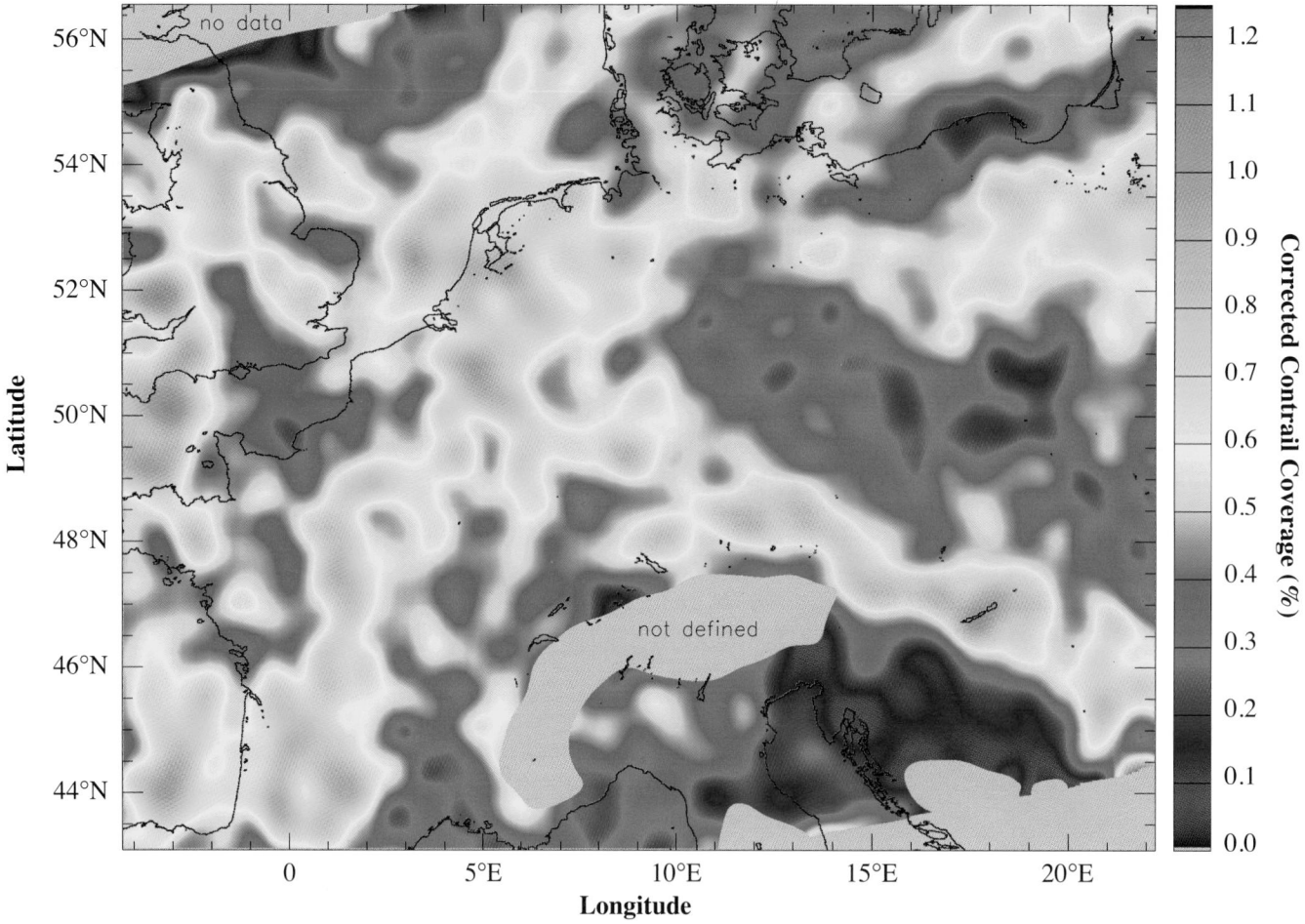

Figure 3-15: Annual mean corrected contrail coverage at noon over mid-Europe in 1996, as derived from AVHRR data from the NOAA-14 satellite (from Mannstein *et al.*, 1999).

rate in the same region, with the latter chosen as one of several possible measures of air traffic. The product is scaled to give a 0.5% mean contrail cover in the European/Atlantic region (30°W to 30°E, 35 to 75°N) as considered by Bakan *et al.* (1994). The resulting contrail cover (see Figure 3-16) for 1992 fuel emissions and $\eta = 0.3$ is 0.087% (about 0.1%) in the global mean, with a local maximum of 5% over the eastern United States of America. The computed contrail coverage over southeastern Asia is only slightly smaller than that over Europe and North America. Although air traffic is much less extensive over southeastern Asia, this high contrail coverage may result because this region more often has ice-supersaturated air. However, the predicted contrail coverage over North America and southeastern Asia has not yet been verified by observations. Predicted regional differences are sensitive to the representation of the number of aircraft in operation. For example, in some regions, a large fuel consumption density is caused by a few large aircraft (Gierens *et al.*, 1998). Studies have not been performed to evaluate the accuracy of humidity values provided by meteorological data. The results depend linearly on scaling by the cover observed in the reference region, as provided here by Bakan *et al.* (1994). Other data imply global cover values between 0.02 and 0.1% (Gierens *et al.*, 1998). A value larger than 0.1% (possibly 0.2%) cannot be excluded because the analysis uses only thermodynamic contrail formation conditions

and the scaling is based solely on observed line-shaped contrail cover.

3.4.4. Contrail Properties

The relatively small particles present in newly formed contrails serve to distinguish contrail radiative properties from those of most natural cirrus (Grassl, 1970; Ackerman *et al.*, 1998; see also Section 3.6). Measurements of contrail particles with impactors and optical probes (see also Section 3.2.4 and Table 3-1) reveal a wide variety of size, shape, and spectral size distributions. The results depend on plume age, ambient humidity, ambient aerosol, and other parameters. At a plume age of 30 to 70 s, ice particles have been found to form a single-mode log-normal size distribution with a volume-equivalent radius in the range of 0.02 to 10 μm, a mean radius of about 2 μm, and maximum dimension of 22 μm. Particle shapes are mainly hexagonal plates, along with columns and triangles. The axial ratios of the columns were found to be less than 2, and the shapes of the crystals were already established for particles of about 1-μm radius (Goodman *et al.*, 1998). Contrail particle sizes increase with time in humid air. Ice particles observed in 2-min-old contrails typically have radii of 2 to 5 μm with shapes that are almost spherical, indicating frozen solution droplets (Schröder *et al.*, 1998b). On the other hand, some contrail particles strongly polarize light, indicating non-spherical shapes (Freudenthaler *et al.*, 1996; Sassen, 1997). After 10 min to 1 h, contrail particle size distributions may range from 32 to 100 μm or even 75 μm to 2 mm (Knollenberg, 1972; Strauss *et al.*, 1997).

Particle properties within a contrail core differ from those at the edges of the contrail. In the center of contrails, insufficient water vapor is available to allow the large number of ice particles to grow to large crystals. At the edges, ice supersaturation is greater, thereby sometimes allowing ice crystals to form up to 300 μm in diameter (Heymsfield *et al.*, 1998a). The larger particles have various shapes, with the largest being bullet rosettes (Lawson *et al.*, 1998). Such large particles are within the natural variability of cirrus particle sizes. As a consequence, old dispersed contrails appear to have particle sizes similar to those in surrounding cirrus (Duda *et al.*, 1998; Minnis *et al.*, 1998a).

Aircraft create a large number of new ice crystals that grow from ambient water vapor. As a contrail spreads, some evidence suggests that ice crystals are neither lost nor created in significant numbers. In one remote-sensing study, the total number of particles (about 2.6×10^9 ice crystals per cm of contrail length) in a contrail was found to remain constant as the contrail dispersed and the particle concentration (particles per unit volume) dropped (Spinhirne *et al.*, 1998). *In situ* measurements (Schröder *et al.*, 1998b) reveal 10^8 to 10^9 particles larger than

4.5-μm diameter formed per cm of contrail length. In contrast, changes in particle number will occur when some particles eventually sediment out of the contrail and new particles are nucleated as a result of air motions induced by the contrail. The ice water content of new contrails increases with time to exceed the amount of water emitted by the aircraft by more than two orders of magnitude. However, the water content per unit volume remained approximately constant in the case observed by Spinhirne *et al.* (1998). The optical depth of contrails remains nearly constant during the first hour despite horizontal spreading (Jäger *et al.*, 1998).

3.4.5. *Impact of Aircraft Exhaust on Cirrus Clouds and Related Properties*

Aircraft may perturb natural cirrus through the addition of water vapor, soot, and sulfate particles and by inducing vertical motions and turbulent mixing (Gierens and Ström, 1998). Observations of cirrus coverage in certain regions have found perturbations from anthropogenic aerosol (Ström *et al.*, 1997). Persistent contrails are often associated with or embedded in natural cirrus (Minnis *et al.*, 1997; Sassen, 1997). Such in-cloud contrails may be formed slightly above the Schmidt-Appleman temperature threshold because ambient ice particles that enter the engine inlet increase the humidity in the exhaust plume (Jensen *et al.*, 1998a; Kärcher *et al.*, 1998a). Some evidence associates in-cloud contrails with regions of enhanced absorbing

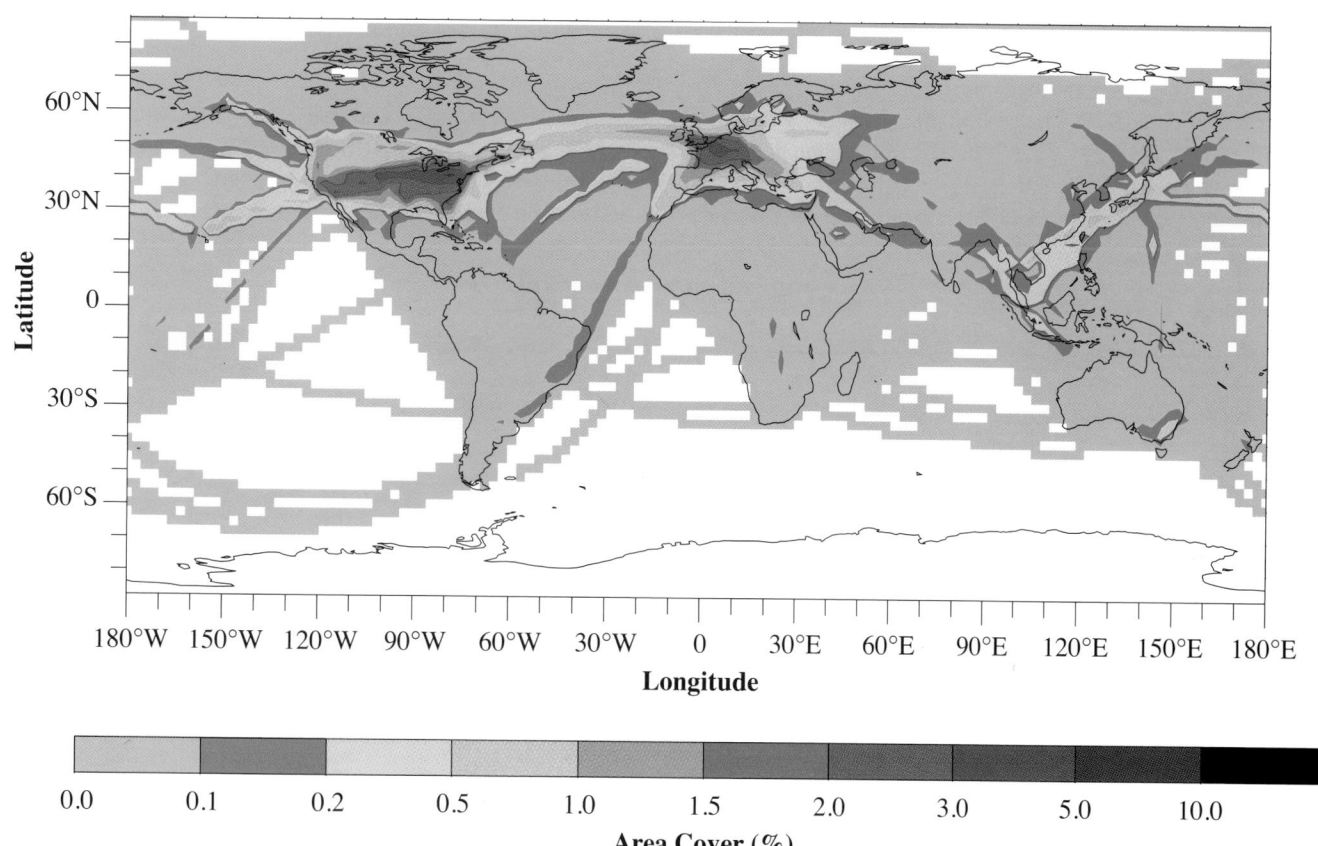

Figure 3-16: Persistent contrail coverage (in % area cover) for the 1992 aviation fleet, assuming linear dependence on fuel consumption and overall efficiency of propulsion η of 0.3. The global mean cover is 0.1% (from Sausen *et al.*, 1998).

material and enhanced ice crystal number densities (Ström and Ohlsson, 1998). The presence of HNO_3 may increase the hygroscopic growth of supercooled cloud droplets (Laaksonen *et al.*, 1997), and HNO_3 dissolved in sulfate solution droplets may change their freezing behavior. The importance of such perturbations has not been quantified.

Soot particles originating from aircraft exhaust may act as freezing nuclei. In an atmosphere with few freezing nuclei, this perturbation could lead to an expansion of cirrus cover, a change in average particle size, and related changes in cloud surface area and optical depth (Jensen and Toon, 1997)—hence have consequences for radiative forcing (see Section 3.6.5). For aircraft to alter cirrus properties, exhaust particles would need to be more efficient or more abundant than ice nuclei currently existing in the atmosphere. If the predominant particle that freezes to form ice contains sulfate, aircraft soot could serve as important ice nuclei because sulfate particles require a relatively high supersaturation before freezing occurs. Over the central United States of America, however, as many as 0.1 to 0.2 cm^{-3} freezing nuclei (effective at -35°C) have been found in the upper troposphere (DeMott *et al.*, 1998), indicating that the number of freezing nuclei is not always small. Some observations show that aircraft exhaust does not contain large numbers of ice nuclei active at temperatures above -35°C (Rogers *et al.*, 1998). Soot and metals were found to be significant, but not dominant, components of ice nuclei in contrails (Chen *et al.*, 1998; Petzold *et al.*, 1998; Twohy and Gandrud, 1998).

Aircraft measurements in and near clouds have indicated the presence of light-absorbing material contained inside ice crystals. The distribution pattern and the amount of measured absorbers suggest that the material is related to aircraft soot (Ström and Ohlsson, 1998) (Figure 3-17). For the same abundance of aerosol particles, clouds perturbed by absorbing material contained 1.6 to 2.8 times more ice crystals than unperturbed

portions of clouds. These observations suggest that aircraft-produced particles enhance cloud ice particle concentrations, although they have not revealed the physical mechanism involved. Specifically, exhaust soot particles may have been involved in ice crystal formation within the cirrus or formed contrail ice particles within the exhaust plume before being incorporated into the cloud. These observations are roughly consistent with calculations of a cirrus cloud forming in a region of recent exhaust trails (Jensen and Toon, 1997).

Cirrus clouds may also be perturbed by enhanced sulfate aerosol. Small sulfate particles (e.g., 10-nm radius) are unlikely to be ice nuclei or cloud condensation nuclei. Larger particles would be more efficient. Such large particles may originate as ambient sulfate particles enlarged by growth from sulfur gases emitted by aircraft or by processing of liquid exhaust particles in short-lived contrails (see Section 3.2). Because ice nucleation processes tend to be self-limiting (Jensen and Toon, 1994), cirrus cloud properties would change only slightly if the number of sulfate particles increased while the size distribution and composition of ambient sulfate particles remained constant. On the other hand, large sulfate particles—such as those induced by the Mt. Pinatubo volcanic cloud—could lead to a significant increase in the number of ice crystals, optical depth, and radiative forcing of cirrus clouds (Jensen and Toon, 1992). Cloud modifications would require a large increase in the number of larger sulfate particles; such modifications would occur primarily in cirrus that were very cold (T < -50°C) and weakly forced by slow updrafts. Lidar observations of a cirrus cloud embedded in aerosol from the Mt. Pinatubo eruption show significantly higher than normal ice crystal concentrations (Sassen, 1992; Sassen *et al.*, 1995). Satellite data analyzed after the Mt. Pinatubo eruption suggest that the volcanic aerosol reduced the occurrence of cirrus clouds with high extinction coefficients, increased the occurrence of clouds with low extinction coefficients, and increased extinction in optically

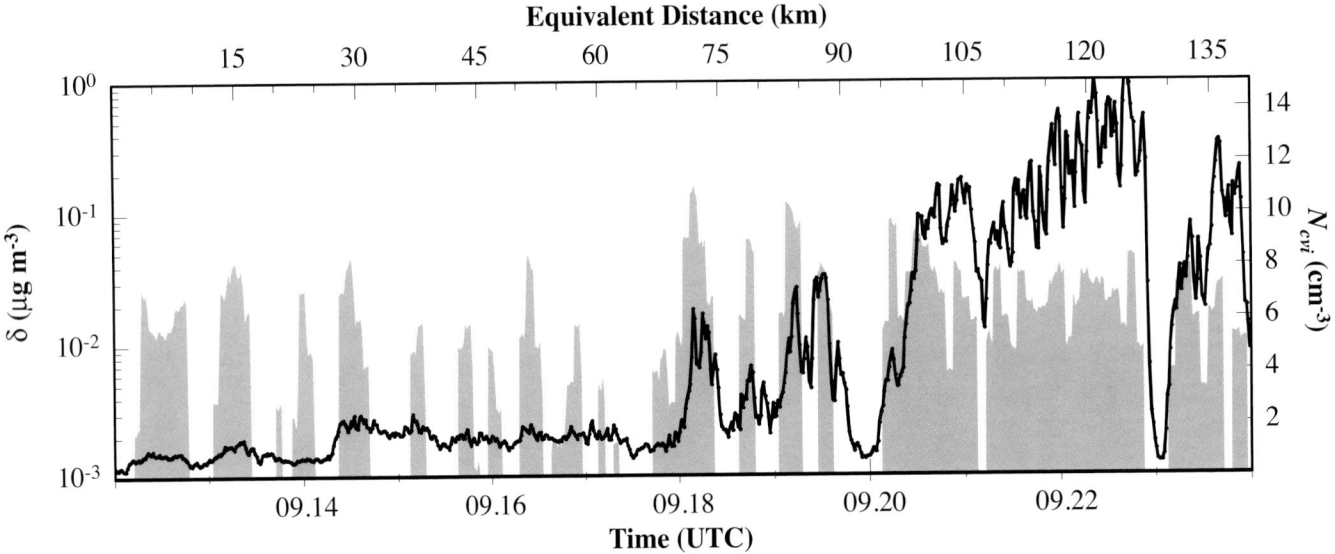

Figure 3-17: Time series of ice crystal concentration (N_{cvi}, thick line) and concentration of light-absorbing material contained in ice crystals (δ, shaded area) for a flight through a cirrus layer in a region with heavy air traffic (adapted from Ström and Ohlsson, 1998).

thick clouds (Minnis *et al.*, 1993; Wang *et al.*, 1995). Hence, some available models and data suggest that aircraft exhaust could play a significant role in modifying the properties of clouds. However, the magnitude and nature of these modifications are not well understood.

Sedimentation of large particles in persistent contrails may remove water vapor from the upper troposphere, possibly reducing radiative heating by water vapor, and cause seeding of lower level clouds (Murcray, 1970; Knollenberg, 1972). Sedimentation of ice crystals, which has been observed occasionally (Konrad and Howard, 1974; Schumann, 1994; Heymsfield *et al.*, 1998a), becomes important only in strongly supersaturated air (Hauf and Alheit, 1997) when large ice crystals form and have the potential to fall through lower-lying saturated air in which they will not evaporate readily. Because no attempts have been made to quantify the precipitation rate from contrails, the significance of this precipitation has not been assessed.

3.5. Long-Term Changes in Observed Cloudiness and Cloud-Related Parameters

Contrails have long been considered possible modifiers of regional climate (Murcray, 1970; Changnon, 1981). Contrails may increase total cloud and cirrus cloud amounts, and consequently change the Earth's radiation balance. As a result, surface and upper tropospheric temperatures may change (Detwiler, 1983; Frankel *et al.*, 1997). In the following sections, we examine changes in cloudiness and other climate parameters for their possible relationship to aircraft operations. Some data indicate changes in observed cirrus cloudiness. These observations are used to provide a first estimate of an upper bound on the increase in contrail-cirrus coverage since the beginning of the jet air traffic era. Limitations in attributing observed trends to aircraft are discussed in Section 3.5.3.

3.5.1. *Changes in the Occurrence and Cover of Cirrus Clouds*

3.5.1.1. *Surface Observations*

Although most studies reporting trends in cloud cover have considered total cloud cover (Henderson-Sellers, 1989, 1992; Angell, 1990; Plantico *et al.*, 1990; Karl *et al.*, 1993), we focus here on studies reporting trends in cirrus cloud cover, which is more relevant to the issue of aviation effects on cloudiness. Observations at Hohenpeissenberg, Germany, indicate that the frequency of high clouds during sunny hours increased from 45% in 1954 to 70% in 1995 (Vandersee, 1997; Winkler *et al.*, 1997). Such large changes are not atypical of regional cloud climatologies (e.g., Rebetez and Beniston, 1998). Over the same period, global radiation during sunshine hours decreased by about 10%, indicating that the observed cloud trend is not an artifact. Similarly, cirrus frequency increased between 1964 and 1990 under cloudy-with-sun conditions by 27% over Hamburg and 15% over Hohenpeissenberg (in northern and

southern Germany, respectively) (Liepert *et al.*, 1994; Liepert, 1997). These changes are not directly attributable to aircraft, however; instead, they might be caused by an increase in tropopause altitude or higher relative humidity in the upper troposphere (Winkler *et al.*, 1997). Cirrus cover over Salt Lake City and Denver has increased from about 12% annual mean cover to 20% in the period from the early 1960s (i.e., since the beginning of the jet aircraft era) to the 1980s, possibly because of increased air traffic over those cities. These trends are significantly lower in the 1990s, and similar observations over Chicago and San Francisco show no obvious trends (Liou *et al.*, 1990; Frankel *et al.*, 1997).

Global trends of observed total and cirrus cover can be deduced from ground-based cloud observations over land and ocean (Warren *et al.*, 1986, 1988; Hahn *et al.*, 1994, 1996) for 1971–91. For the period 1982–91, mean global trends in cirrus occurrence frequency were 1.7 and 6.2% per decade over land and ocean, respectively (Boucher, 1999). The decadal trend was 5.6% over North America and 13.3% over the heavy air traffic region located in the northeastern United States of America. The computed average change in cirrus occurrence (1987–91 relative to 1982–86) as a function of aviation fuel use at the observation locations is shown in Figure 3-18. The results indicate a statistically significant (97% confidence level) increase in cirrus occurrence in the North Atlantic flight corridor compared with the rest of the North Atlantic Ocean (Boucher, 1998, 1999).

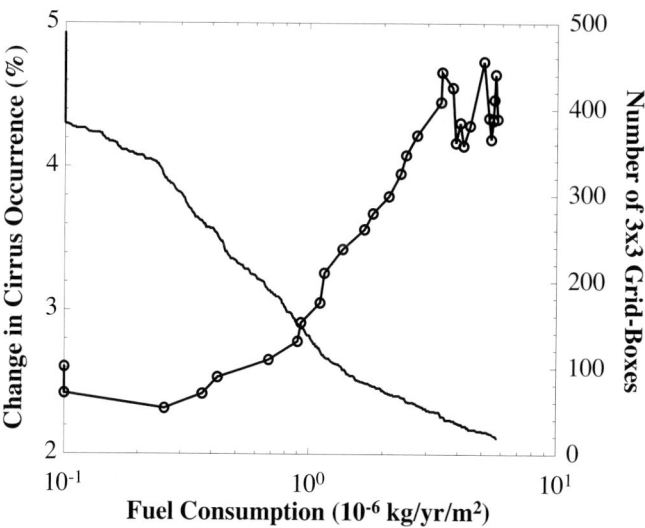

Figure 3-18: Change in cirrus occurrence frequency between 1987–1991 and 1982–1986 (in %) computed from surface observations over the North Atlantic Ocean averaged over grid boxes with fuel consumption above 8 km greater than indicated on the x-axis (open circles). Data are from surface observations of Hahn *et al.* (1996) and are averaged over 3°x3° grid boxes in the region 99°W to 21°E, 0°N to 72°N of the Atlantic Ocean. Only cloud reports satisfying the illuminance criterion defined by Hahn *et al.* (1995) are included in the statistics. The bold line indicates the number of grid boxes used for the average (adapted from Boucher, 1999).

From the same source of observations, cirrus cover was analyzed for the periods 1971–81 and 1982–91 and for a combined data set extending from 1971–92 (Minnis *et al.*, 1998b). Data were averaged for gross air traffic regions (ATRs) and the rest of the globe. ATRs are rectangular areas on the globe that contain most air traffic routes and constitute 26% of the available regions with data. The ATR trends are larger for cirrus clouds than for total cloudiness and are most significant in the first period (1971–81), with the largest annual values found in the United States of America (3%), western Asia (1.6%), and the North Pacific (1.7%). No significant ATR trends were found over Europe. Changes for the rest of the globe were significant only over land but with values less than those over the United States of America and western Asia. Averaged over all ATRs, the cirrus cover increase between 1971 and 1981 amounts to 1.5% per decade, compared with 0.1% for the rest of the globe. From 1982–91, cirrus cover increased over all areas except over non-ATR land and North Atlantic regions, where it changed by -0.4 and 0%, respectively.

The combined 1971–92 cirrus cover data set shows somewhat different results from the two separate data sets. The mean trend for land was 0% per decade for ATRs, compared to -1.1% per decade for other land regions. Over the United States of America, however, the ATR trend is significant at 1.2% per decade. Over oceans, the mean ATR trend is 1.2% per decade and 0.6% per decade over the rest of the oceans. The Pacific ATR trend is strongest, at 1.5% per decade. Averaging of apparent trends from the two separate data sets gives results that differ from the mean trends in the combined data set. However, the relative difference in the mean trends between the ATR regions

and the rest of the globe is roughly 1.1% in both cases, indicating that cirrus coverage in areas with significant air traffic is increasing relative to that over the remainder of the globe.

Using ground-based data sets, Figure 3-19 shows the seasonal trends in cirrus coverage over the United States of America and Europe. The average trend of the two separate data sets is compared with the seasonal cycle of contrail occurrence frequency over the United States of America and cover over Europe. Contrail occurrence over the United States of America was derived from 1 year of surface station observations (Minnis *et al.*, 1997); contrail coverage over mid-Europe was computed from satellite data (Mannstein *et al.*, 1999). Over the United States of America, the seasonal trends in cirrus coverage (statistically significant at least at the 75% confidence level) are roughly in phase with the seasonal cycle of contrail occurrence (see Figure 3-19a), suggesting that cirrus changes are related to contrail occurrence. In contrast, the seasonal cycle of cirrus cover trends for Europe does not resemble the seasonal cycle of contrail cover (see Figure 3-19b). The European data, which are not statistically significant, show that the trends observed for the United States of America are not so obvious for other air traffic regions.

3.5.1.2. ISCCP Observations

Data from the International Satellite Cloud Climatology Project (ISCCP, C2) (Rossow and Schiffer, 1991) between 50°S and 70°N from 1984 to 1990 have also been inspected for trends in total and high cloud cover over land and ocean

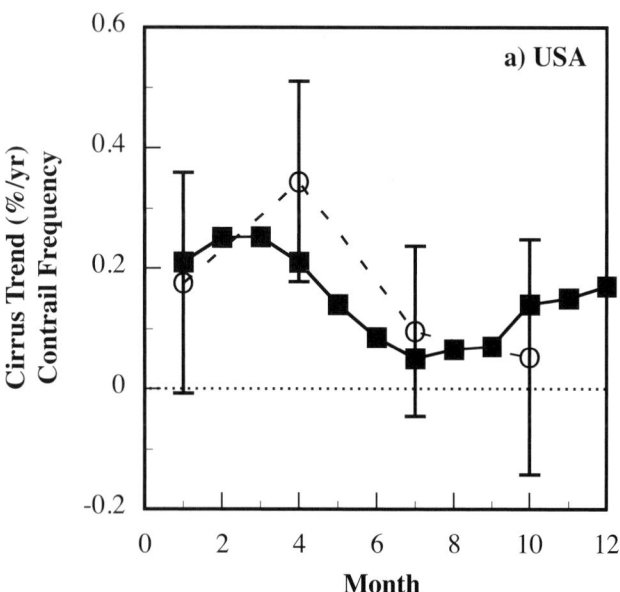
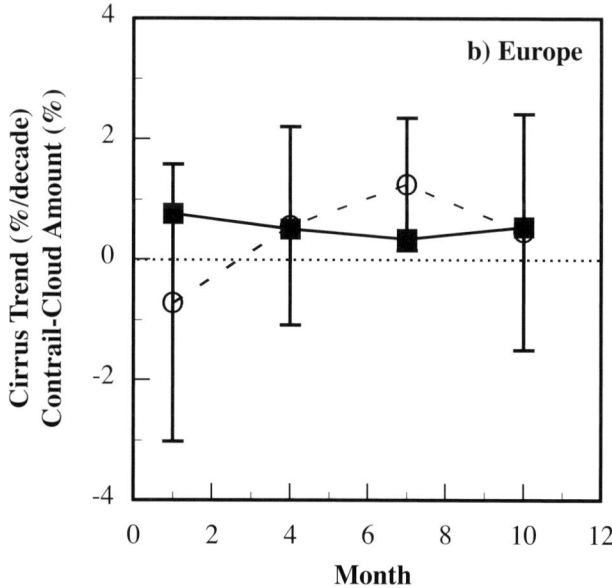

Figure 3-19: Observed monthly variation of contrail occurrence frequency (solid squares) and seasonal trends of cirrus cover (open circles) for (a) the United States of America (%/yr) and (b) Europe (%/decade). Contrail data from the United States of America are based on surface observations of contrail frequencies in 1993–1994 (Minnis *et al.*, 1997). The mid-Europe contrail data are derived from satellite analysis of line-shaped contrail cover for 1996 (Mannstein *et al.*, 1999). Seasonal mean trends of observed fractional cirrus cover in the time period 1971 to 1991 are mean values derived from separate data sets for 1971 to 1982 (Warren *et al.*, 1986) and 1982 to 1991 (Hahn *et al.*, 1996). Error bars indicate the standard error of the computed trends (from Minnis *et al.*, 1998b).

Table 3-5: *Computed trend values and differences of trend values in cirrus/high cloud amounts (% of Earth surface cover per decade) from surface and ISCCP satellite data sets taken between 1971 and 1991 for regions categorized as having a mean value of contrail coverage less than and greater than 0.5% as computed by Sausen et al. (1998) for 1992 aircraft operations (Minnis et al., 1998b). Differences in trends between contrail cover regions are shown as "trend difference" values. The last two rows (Global ISCCP, Global Surface) indicate global (ocean + land) trends in respective data sets. The time periods of the data sets are also indicated. Statistical significance refers to level of confidence in difference between trends for the two regions. An asterisk indicates insufficient resolution for analysis. Dashes indicate a confidence level less than 95%.*

| | Period | Trend (% per decade) in Regions with Computed Contrail Cover | | Trend Difference (% per decade) | Statistical Significance of Difference (%) |
		< 0.5%	> 0.5%		
Ocean from ISCCP	1984–90	4.3	5.9	1.6	—
Land from ISCCP	1984–90	1.2	4.7	3.5	95
Ocean from surface	1971–81	*	*		*
Land from surface	1971–81	-0.21	2.1	2.3	99
Ocean from surface	1982–91	0.22	1.0	0.78	—
Land from surface	1982–91	-0.01	0.95	0.96	—
Ocean from surface	1971–92	0.75	0.24	-0.51	—
Land from surface	1971–92	-1.5	0.10	1.6	99
Global ISCCP	1984–90	3.2	5.2	1.9	99
Global surface	1984–90	0.0	1.0	1.0	—

regions (Minnis *et al.*, 1998b). The trends are similar to those found in the 1982–91 surface observations, though details differ. Over all land areas, total cloudiness decreased by 0.5% per decade; high cloud cover grew at a rate of 1.2% per decade. Over the United States of America, total cloudiness decreased by 2.3% per decade, but high cloud cover increased by 5.5% per decade. The trends over Europe were of the same sign but half the magnitude. Over Asia, total cloud cover increased by 4.8% per decade, and high cloudiness grew by 2.7% per decade. Thus, the satellite data suggest a relatively stronger increase in cirrus over the United States of America than surface observations suggest. Over ocean, total cloudiness decreased by 1.2% per decade, whereas high cloud cover was enhanced by 3.7% per decade overall. Some of the discrepancies between the satellite and surface data may originate from the different spatial sampling patterns, calibration, and orbit drift issues peculiar to the satellite data (Klein and Hartmann, 1993; Brest *et al.*, 1997). The satellite covers all regions almost equally, whereas surface observations are from fixed inhabited locations or from well-traveled ship routes.

Further analysis of the ISCCP satellite and the surface data sets was undertaken to isolate and quantify the effects of contrails (Minnis *et al.*, 1998b). The apparent trends were calculated separately for regions having a mean value of contrail coverage less than and greater than 0.5% as computed by Sausen *et al.* (1998) for 1992 aircraft operations (see Section 3.4.3). The mean values of the computed contrail coverage in the two regions are 0.04 and 1.4%, respectively (Figure 3-16). Satellite data and surface observations show larger increases in cirrus

cloudiness where contrails are expected to occur most frequently than in all other areas (Table 3-5). Satellite data show that, over land, the increase in high cloudiness in contrail regions was almost four times that in other areas. Over ocean, the difference, though less significant statistically, amounts to a differential increase of 1.6% cover per decade in contrail areas with respect to non-contrail areas. Cirrus trends over contrail regions as derived from surface observations are 0.8 to 2.3% per decade greater than those in the remainder of the globe for the two periods from 1971–92. The single 1971–92 surface data set shows a significant increase (1.6%) in cirrus over land contrail regions compared to remaining land areas. Over oceans, the relative difference is negative but insignificant because of the small number of samples and large variance. Although the differences in these trends are significant at confidence levels of 95% only for the land and global ISCCP data and the 1971–81 land surface data, they show consistent tendencies.

3.5.1.3. HIRS Observations

High-Resolution Infrared Radiation Sounder (HIRS) data from NOAA satellites have been analyzed for the period June 1989 to February 1997 to determine total and high cloud cover (Wylie and Menzel, 1999). During this period, high clouds observed by the NOAA-10 and -12 satellites increased by 4–5% over land and ocean in the Northern Hemisphere (from 23 to 65°N) but only by about 2% in the tropics. High clouds increased by about 3% over southern mid-latitude oceans. The trend values inferred from the NOAA-11 and -14 satellites are different and somewhat more uncertain because of orbit drift. Over oceans, they also indicate a larger high cloud increase in northern mid-latitudes (3.3%) than over the tropics (-0.4%). Further analysis of HIRS data is required to determine the extent of any contrail impact.

3.5.1.4.   SAGE Observations

Data from the Stratospheric Aerosol and Gas Experiment (SAGE) II satellite instrument indicate that the frequency of subvisible cirrus clouds near 45°N is twice that at 45°S (Wang *et al.*, 1996). Aviation, as well as hemispheric differences in atmospheric conditions and background aerosol (Chiou *et al.*, 1997; Rosen *et al.*, 1997), may contribute to such differences (Sausen *et al.*, 1998).

3.5.1.5.   Upper Bound for Aviation-Induced Changes in Cirrus Clouds

The line-shaped contrail cover and global extrapolation described in Section 3.4.3 (see Figure 3-16; Sausen *et al.*, 1998) provide only a lower bound for aviation-induced changes in cirrus cloud cover because they are based on satellite observations that identify only contrails and additional cirrus clouds that are line-shaped. Although estimates of an upper bound of aviation-induced cirrus cover have not yet been established, evidence for a correlation of long-term increases in cirrus cloudiness with air traffic has been published (Liou *et al.*, 1990; Frankel *et al.*, 1997; Boucher, 1998, 1999). Here, observations of cloudiness changes described above are used to provide a preliminary estimate of this upper bound.

Differences in trends derived from observations indicate a stronger mean increase of cirrus amounts in regions with large computed contrail cover than in regions with low computed contrail cover, at least over land (see Table 3-5). The trend difference values vary and are of different statistical significances. The values are considered more meaningful over land because there is less air traffic over the oceans. Of regions with large computed contrail cover, only 14% occurs over oceans, and most of these regions occur near the coasts. In addition, less correlation is expected between computed and observable contrail cover over oceans because actual flight tracks often deviate significantly from idealized great-circle routes. Over land, surface-based observations for 1971–81 suggest a differential increase in cirrus cover of 2.3% per decade. The 1982–91 trend difference is smaller but still positive, and the combined 22-year trend difference is 1.6% per decade. The 7 years of ISCCP satellite data suggest even larger trend differences, both globally and over land. If a trend difference of 1.6% per decade is adopted as most representative of available data, and if that trend is assumed to have persisted for the 3 decades since the beginning of the jet aircraft era (the end of the 1960s), then the current increase in cirrus coverage from aircraft is 4.8% in areas with contrail cover greater than 0.5%. This value is about three times the currently

computed linear-contrail cover of 1.4% in those areas and is equivalent to about 0.3% coverage of the Earth's surface, 0.2% more than for line-shaped contrails. If no counteracting process took place that reduced cloudiness over this same period, this value gives an upper bound for total aviation-induced cloudiness.

3.5.2.   *Changes in Other Climate Parameters*

3.5.2.1.   *Sunshine Duration and Surface Radiation*

Observations from 100 stations over the United States of America showed that the mean value for total cloud cover over the years 1970–88 increased by $2.0 \pm 1.3\%$ relative to the years 1950–68 (Angell, 1990). Sunshine duration decreased by only $0.8 \pm 1.2\%$, presumably because cirrus clouds are often too thin to reduce measured sunshine duration. Sunshine duration decreased by an average of 3.7% per decade over Germany between 1953 and 1989 (Liepert *et al.*, 1994). Contrails were estimated to be insufficient to account for the diminished sunshine. In an analysis of radiation trends in Germany from 1964 to 1990, solar radiation during cloudy periods with sun decreased strongly (by 20 to 80 W m^{-2}) (Liepert, 1997). The change might be attributable to aircraft, but the strong reduction in diffuse radiation revealed a more turbid atmosphere, which cannot be explained by increasing cirrus coverage alone.

3.5.2.2.   *Temperature*

A substantial decrease in diurnal surface temperature range (DTR) has been observed on all continents (Karl *et al.*, 1984; Rebetez and Beniston, 1998). The reasons for this decrease are not clear; it could be caused by a change in aerosol burden, cloud amount, cloud ceiling height (Hansen *et al.*, 1995), soil moisture, or absorption of solar radiation by increased water vapor in the atmosphere (Roeckner *et al.*, 1998). A weak correlation was found between contrail occurrence data and DTR over the United States of America (DeGrand *et al.*, 1990; Travis and Changnon, 1997). The relationship was strongest during the summer and autumn and most significant over the southwestern United States of America. Rebetez and Beniston (1998) find large DTR trends in the Swiss Alps (a region with heavy air traffic) in correlation with low-level clouds but no reduction of DTR at higher elevation sites, contradicting a potential major impact on DTR from aviation-induced cirrus.

From data taken between 1901 and 1977 over the midwestern United States of America, increases of moderated temperatures (below average maximum and above average minimum), decreases in sunshine and clear days, and shifts in cloud cover, occurred after 1960; these changes were concentrated in the main east-west air corridor (Changnon, 1981). There are, however, many reasons for changes in global mean surface temperatures since 1880 (Halpert and Bell, 1997). The maximum increase in tropospheric temperatures occurred at northern mid-latitudes in the lower troposphere (Tett *et al.*, 1996; Parker *et al.*, 1997). However, radiative transfer models predict maximum heating rates right below contrails and cirrus in the upper troposphere (Liou, 1986; Section 3.6). Studies using a full-feedback global circulation model (GCM) indicate a nearly constant increase in temperature with altitude for an increase in thin cirrus cloudiness (Hansen *et al.*, 1997). Thus, existing studies do not support the conclusion that aircraft-induced contrails and cloudiness caused changes in global tropospheric temperatures.

3.5.3. *Limitations in Observed Climate Changes*

The observations cited above suggest that some of the apparent trends in observed cirrus cloudiness and sunshine duration may be caused by air traffic. However, any observed change in cirrus cloudiness or climatic conditions may have other natural or anthropogenic causes (IPCC, 1996). Natural variability includes the El Niño Southern Oscillation (ENSO), the North Atlantic Oscillation (NAO), and the Pacific North America pattern (PNA). For example, the NAO index increased overall from the mid-1960s until 1995 (Hurrell, 1995; Halpert and Bell, 1997). However, there is no evident trend in the NAO index (Hurrell, 1995) and mean position of the Iceland low (Kapala *et al.*, 1998; Mächel *et al.*, 1998) over the period 1982–91 when significant increases in cirrus occurrence and cirrus amount were observed over the North Atlantic Ocean (Boucher, 1998, 1999). Changes in cirrus and subsequent changes in sunshine duration may reflect changes in cyclonic activity (Weber, 1990), ocean surface temperature and related atmospheric dynamics (Wallace *et al.*, 1995), upper troposphere temperature and relative humidity, tropopause altitude (Hoinka, 1998; Steinbrecht *et al.*, 1998), or volcanic effects (Minnis *et al.*, 1993)—or they may be a result of global climate change caused by increased greenhouse gases. Changes in anthropogenic or natural emissions of aerosol or aerosol precursors at the ground might also influence cirrus cloud formation (Plantico *et al.*, 1990; Hansen *et al.*, 1996; Hasselmann, 1997). Cirrus trend values deduced from surface observations may also be influenced by a variety of issues. For example, reported cirrus frequencies depend in an unknown manner on how observers classify contrails as cirrus. Reliable cloud observations are obtained mainly during daytime periods (Hahn *et al.*, 1995), which do not always correlate with the main air traffic periods. The significance of the statistical relationship between cloud changes and air traffic is limited because of the limited duration of satellite and surface cloud observations and the brief period of aircraft emissions in comparison to long-term climate variations. Thus, existing studies are not complete enough to conclude that aircraft-induced contrails and cloudiness have caused observable changes in surface and tropospheric temperatures or other climate parameters.

3.6. Radiative Properties of Aerosols, Contrails, and Cirrus Clouds

Aircraft emissions have an impact on the Earth's radiation budget and climate through direct and indirect changes in aerosols and cloudiness. Recent climate assessments have stressed the importance of natural and anthropogenic changes in aerosols on direct radiative forcing (Charlson *et al.*, 1990; Schwartz, 1996). Aerosols and contrails have direct effects (scattering and absorbing solar and longwave radiation) and indirect effects (modifying the formation of cloud particles and radiative properties of clouds). Several studies have addressed the direct impact of contrails (e.g., Detwiler and Pratt, 1984; Grassl, 1990; Liou *et al.*, 1990; Sassen, 1997). The indirect effect of contrails has not yet been investigated in detail. The direct radiative impacts of aircraft soot emissions (Pueschel *et al.*, 1992, 1997) and sulfate aerosol have been evaluated as being small (Friedl, 1997; Brasseur *et al.*, 1998). The indirect radiative effect of aircraft-induced aerosols on clouds is essentially unknown. In fact, the indirect radiative effect of non-aviation aerosol has been studied for liquid water clouds (IPCC, 1996), but the indirect effect of changing cirrus is not yet known either. Here, the discussion focuses on the impact of aircraft-generated aerosol and that of contrails and changed cirrus clouds.

Radiative forcing is defined as the net radiative flux change at some level in the atmosphere calculated in response to a perturbation, such as a change in cloud cover. The definition of radiative forcing used here is the "instantaneous" or "static" flux change at the top of the atmosphere (TOA) (IPCC, 1996) (see Chapter 6). A positive net flux change represents an energy gain, hence a net heating of the Earth system.

3.6.1. *Direct Radiative Impact of Aerosols*

The direct radiative impact of aircraft-induced aerosol is much smaller than that of volcanic stratospheric or regional tropospheric aerosol and smaller than other aircraft-induced radiative forcing values (see Chapter 6). Radiative forcing by aerosol depends on many parameters (Haywood and Ramaswamy, 1998). For aerosol in the size range 0.1 to 1 μm, the increase in solar reflectance (albedo effect) exceeds the trapping of terrestrial radiation (greenhouse effect), causing a surface cooling (Lacis *et al.*, 1992, Minnis *et al.*, 1993; Minnis *et al.*, 1998c). Aircraft-induced aerosols could have an importance similar to that of natural stratospheric aerosol variations if their optical depth in the solar range were of comparable magnitude to that of natural stratospheric aerosol. The optical depth of aerosol is proportional to its column load. The maximum zonally averaged column load from soot emissions and aircraft-induced sulfuric acid is estimated (see Section 3.3) to be less than 1 and 6 ng cm^{-2}, respectively. Small sulfuric acid particles are nonabsorbent but scatter strongly in the shortwave spectrum, with mass scattering efficiencies of 7 (4–10) m^2 g^{-1} in the solar range depending on relative humidity (Boucher and Anderson, 1995; Lacis and Mishchenko, 1995). Soot is a strong absorber of solar energy, with mass extinction coefficients of about 10 m^2 g^{-1} (Pueschel *et al.*, 1992; Penner, 1995; Petzold and Schröder, 1998). As a consequence of the product of given column load and extinction efficiency, the maximum change in zonal-mean optical depth from aircraft soot and sulfate aerosol is less than 4 x 10^{-4}. For comparison, the solar optical depth of stratospheric aerosol varies typically between 0.005 and 0.15, depending on volcanic aerosol loading (Sato *et al.*, 1993). Regionally, within the main flight corridors, a particle concentration change of approximately 30 cm^{-3} at 0.1-μm mean diameter over a vertical layer of 2 km cannot be ruled out (Schumann *et al.*, 1996; Friedl, 1997; Schlager *et al.*, 1997), implying a mass load on the order of 10 ng cm^{-2} and a solar optical depth of about 0.001. This regional change is small compared with other regional variations. Tropospheric aerosol layers with optical depth of 0.1 to 0.5 occur frequently off the coasts of North America and Europe (Russell *et al.*, 1997).

3.6.2. Radiative Properties of Cirrus Clouds

Contrails are ice clouds with radiative effects similar to thin cirrus cloud layers (Liou, 1986; Raschke *et al.*, 1998). At the TOA, thin layers of cirrus clouds or contrails tend to enhance radiative forcing and hence the greenhouse effect because they cause only a small reduction of the downward solar flux but have relatively larger impact on the upward terrestrial radiative flux. An increase in cloud cover by thin cirrus clouds may therefore cause an increase in the net energy gain of the planet (Stephens and Webster, 1981; Fu and Liou, 1993). Contrails and cirrus clouds also have an impact on the radiative energy budget at the Earth's surface. Although radiative forcing at the TOA is most important for long-term and global climate changes, forcing at the surface may have short-term regional consequences.

Longwave (LW) radiative forcing by cirrus or contrails is greatest when clear-sky radiative flux to space is large (i.e., larger over warm than over cool surfaces, larger in a dry than in a humid atmosphere) and cloud emissivity is large (Ebert and Curry, 1992; Fu and Liou, 1993). For thin clouds, emissivity increases with ice water path, which is the product of the ice water content (IWC) of the cloud and its geometrical depth. The emissivity of ice particles in a cirrus layer is much larger, in particular at 8 to 12 μm (King *et al.*, 1992; Minnis *et al.*, 1998c), than that of the same amount of water in gaseous form. Hence, absorption and emission from an atmospheric layer increase when ice particles form at the expense of ambient water vapor (Detwiler, 1983; Meerkötter *et al.*,1999).

Shortwave (SW) radiative forcing of cirrus clouds is determined mainly by solar zenith angle, surface albedo, and cloud optical depth (which increases with ice water path) (Ebert and Curry, 1992; Platt, 1997). For fixed ice water path, clouds containing smaller particles have larger optical depth and exhibit larger solar albedo (Twomey, 1977; Betancor-Gothe and Grassl, 1993). Aspherical particles cause a larger albedo than spherical ones (Kinne and Liou, 1989; Gayet *et al.*, 1998). SW forcing is negative when the cloud causes an increase of system albedo. In general, SW forcing has a greater magnitude over dark surfaces than over bright surfaces.

The net radiative forcing of clouds is the sum of SW and LW flux changes and may be positive or negative. Thin cirrus clouds cause a small but positive radiative forcing at the TOA; thick cirrus clouds may cause cooling (Stephens and Webster, 1981; Fu and Liou, 1993). In the global mean, an increase in cirrus cloud cover warms the Earth's surface (Hansen *et al.*, 1997). Maximum heating is reached at intermediate ice water path, corresponding to an optical depth of about 2 to 3, and the effect shifts to small cooling for optical depths greater than about 10 to 20 (Platt, 1981; Jensen *et al.*, 1994a). Net forcing also varies with particle size but less than its two spectral components. For thin cirrus, smaller particles (but > 3-μm radius) tend to cause stronger heating by increasing cloud albedo less strongly than emissivity (Fu and Liou, 1993).

3.6.3. Radiative Properties of Contrail Clouds

Contrails are radiatively important only if formed in ice-supersaturated air, where they may persist and spread to several-kilometer lateral widths and a few hundred meters vertical depth (Detwiler and Pratt, 1984; Jäger *et al.*, 1998) (see also Section 3.4.4). In contrast, shorter lived contrails have much smaller spatial and temporal scales, hence contribute much less to the radiation budget (Ponater *et al.*, 1996). The impact of contrails on transmission of radiation depends on their optical depth. In the solar range (near a wavelength of 0.55 μm), the optical depth of observed persistent contrails varies typically between 0.1 and 0.5 (Kästner *et al.*, 1993; Sassen, 1997; Jäger *et al.*, 1998; Minnis *et al.*, 1998a). Occasionally, very thick contrails (on the order of 700 m) with optical depths greater than 1.0 are found at higher temperatures (up to -30°C) (Schumann and Wendling, 1990; Gayet *et al.*, 1996). The optical depth in the 10-μm range is about half that near 0.55 μm (Duda and Spinhirne, 1996). Particles in young persistent contrails are typically smaller (mean diameter 10 to 30 μm) than in other cirrus cloud types (greater than 30 μm) (Brogniez *et al.*, 1995; Gayet *et al.*, 1996) but grow as the contrail ages and may approach the size of natural cirrus particles within a time scale on the order of 1 h (see Section 3.4) (Minnis *et al.*, 1998a). The number density of ice crystals in contrails (on the order of 10 to 200 cm^{-3}) is much larger than in cirrus clouds (Sassen, 1997; Schröder *et al.*, 1998b). Reported IWC values in aged contrails vary between 0.7 and 18 mg m^{-3} (Gayet *et al.*, 1996; Schröder *et al.*, 1998b), consistent with results from numerical studies (Gierens, 1996). As for cirrus clouds (Heymsfield, 1993; Heymsfield *et al.*, 1998b), the IWC of contrails is expected to depend on ambient temperature (Meerkötter *et al.*, 1999) because the amount of water mass available between liquid and ice saturation (for temperatures < -12°C) increases with temperature (Ludlam, 1980). Hence, contrails may be optically thicker at lower altitudes and higher temperatures. Contrail particles have been found to contain soot (see Section 3.2.3). Soot in or on ice particles may increase absorption of solar radiation by the ice particles, hence reduce the albedo of the contrails. The importance of soot depends on the type of internal or external mixing and the volume fraction of soot enclosures. Because soot particles are typically less than 1 μm in diameter, their impact on the optical properties of ice particles in aged contrails is likely to be small. Contrail particles often deviate from a spherical shape (see Section 3.4.4).

The magnitude and possibly even the sign of the mean net radiative forcing of contrails depends on the diurnal cycle of contrail cover. For the same contrail cover, the net radiative forcing is larger at night. Satellite data reveal a day/night contrail cover ratio of about 2 to 3 (Bakan *et al.*, 1994; Mannstein, 1997). Aviation fuel consumption inventories suggest a longitude-dependent noon/midnight contrail cover ratio of 2.8 as a global mean value (Schmitt and Brunner, 1997).

3.6.4. Radiative Forcing of Line-Shaped Contrail Cirrus

Model studies indicate the importance of contrails in changing the Earth's radiation budget. One-dimensional (1-D) models represent contrails as plane-parallel clouds in a homogeneously layered atmosphere and use area-weighted sums for fractional contrail cover (Fortuin *et al.*, 1995; Strauss *et al.*, 1997; Meerkötter *et al.*, 1999). As a consequence, contrail forcing grows linearly with contrail cover in these models. Inhomogeneity effects may be large in natural cirrus (Kinne *et al.*, 1997) and small for vertically thin contrail clouds (Schulz, 1998) but may be important for thick and narrow contrails. Computations of radiative forcing by contrails have been done for fixed atmospheric temperatures in the North Atlantic flight corridor (Fortuin *et al.*, 1995). Normalized to 100% contrail cover (as provided by 1-D models), these computations found a net forcing in the range of -30 to +60 W m^{-2} in summer and 10 to 60 W m^{-2} in winter. The negative forcing values apply to clouds that are much thicker than typical contrails. A radiative convective model used to simulate the climatic conditions of a mid-European region (Strauss *et al.*, 1997) found a radiative flux change of almost 30 W m^{-2} at the tropopause for 100% contrail cover with 0.55-μm optical depth of 0.28, and a surface temperature increase on the order of 0.05 K for a 0.5% increase in current contrail cloud cover. With a 2-D radiative convective model, a 1 K increase was found in surface temperature over most of the Northern Hemisphere for an additional cirrus cover of 5% (Liou *et al.*, 1990). The potential effects of contrails on global climate were simulated with a GCM that introduced additional cirrus cover with the same optical properties as natural cirrus in air traffic regions with large fuel consumption (Ponater *et al.*, 1996). The induced temperature change was more than 1 K at the Earth's surface in Northern mid-latitudes for 5% additional cirrus cloud cover in the main traffic regions.

Table 3-6: *Instantaneous TOA radiative flux changes averaged over a day for shortwave (SW), longwave (LW), and net (= SW + LW) radiation for 100% contrail cover in various regions and seasons, with prescribed surface albedo, contrail ice water content (IWC), and computed optical depth τ of contrail at 0.55 μm. Results are for spherical particles (model M, upper values) and hexagons (four-stream version of model FL, lower values).[a]*

Region	Surface Albedo	IWC (mg m^{-3})	τ	SW (Wm^{-2})	LW (Wm^{-2})	Net (Wm^{-2})
Mid-latitude summer continent, 45°N	0.2	21	0.52	-13.4	51.6	38.2
				-22.0	51.5	29.5
Mid-latitude winter continent, 45°N	0.2	7.2	0.18	-4.2	18.4	14.2
				-4.6	18.3	13.7
Mid-latitude winter continent with snow, 45°N	0.7	7.2	0.18	-2.3	18.4	16.1
				-2.0	18.3	16.3
North Atlantic summer ocean, 55°N	0.05	21	0.52	-21.5	53.3	31.8
				-32.7	50.9	18.2
Tropical ocean (Equator, June)	0.05	23	0.57	-16.0	63.0	47.0
				-25.9	57.4	31.5
Subarctic summer ocean, 62°N	0.05	28.2	0.70	-30.8	55.7	24.9
				-45.3	49.1	3.7
Subarctic winter ocean ice, 62°N	0.7	7.2	0.18	-0.6	14.6	14.0
				-0.7	13.2	12.5

[a]The contrail is embedded as a homogeneous cirrus cloud of 200-m vertical depth with a top at 11-km altitude (9 km in the subarctic); temperature-dependent IWC as listed; spherical ice particles with measured size spectrum (Strauss *et al.*, 1997) (volume-mean particle diameter of 16 μm); an otherwise clear atmosphere with continental or maritime aerosol (WMO, 1986) of 0.28 or 0.08 total 0.55-μm optical depth; a Lambertian surface with a spectrally constant SW albedo as listed; and a LW emissivity of 1. Reference atmospheres are prescribed according to McClatchey *et al.* (1972). Results are normalized for 100% contrail cover. See Meerkötter *et al.* (1999) for further details.

Table 3-7: *Sensitivity of daily mean of instantaneous net forcing at top of the atmosphere by contrails to a range of values for various parameters[a] for 100% contrail cover. The first two rows contain results from models FL, M, and N; the others are from model N.*

Case		Optical Depth τ at 0.55 μm	Net Forcing (Wm-2)
Reference (models FL, M, and N)		0.52	37.1–37.2
Different aspherical particles (models FL, M, and N)		0.4	22–36
Parameters	**Range**		
Solar zenith angle	60°–21°	0.52	37–49
Ice water content (IWC)	7.2–42 mg m-3	0.2–1.0	19–51
Particle diameter	10–40 μm	0.85–0.21	41–20
Surface temperature	289–299 K	0.52	35–39
Cloud cover/optical depth of underlying clouds	0–1 / 0–23	0.52	37–40
Surface albedo	0.05–0.3	0.52	31–40
Relative humidity	reference – 80%	0.52	37–31
Contrail vertical depth (for fixed ice water path)	200 m – 1 km	0.52	37.1–36.7
Lower contrail top (for fixed IWC)	11–10 km	0.52	37–31
Lower contrail top (for temperature-dependent IWC)	11–10 km	0.52–1.32	37–45

[a] Details in Meerkötter *et al.* (1999). Reference case for mid-latitude summer (McClatchey *et al.*, 1972) as in Table 3-6; spherical particles; solar zenith angle of 60°; diurnal sunshine fraction of 50%; IWC of 21 mg m-3; volume mean particle diameter of 32 μm; surface temperature of 294 K; no low-level clouds; relative humidity of reference profile (McClatchey *et al.*, 1972) varying from 77% at the surface to 11% at 12-km altitude; contrail depth 200 m; and contrail top at 11-km altitude.

Meerkötter *et al.* (1998) applied three established radiation transfer models to compute the static radiative forcing due to a prescribed additional cloud cover by contrails. These models, which assume plane parallel cloud cover, are the two- and four-stream models of Fu and Liou (1993) (FL); the matrix operator method of Plass *et al.* (1973), also used by Strauss *et al.* (1997) (M); and the four-stream model of Nakajima and Tanaka (1986, 1988) (N). The FL and M models have participated in model comparison exercises (Ellingson and Fouquart, 1990). Table 3-6 on the previous page shows the results of a parameter study that was carried out for various regions and seasons using models M and FL for spherical and hexagonal particles. The contrails cause a net forcing that is positive in all cases after summing over negative SW and positive LW flux changes. The magnitude of SW forcing is larger over dark ocean than over bright snow surfaces and larger for hexagonal ice particles than for spheres. LW forcing is larger in the tropics than in polar regions. Despite the small ocean albedo, net forcing is largest in the tropics because it has the warmest lower atmosphere. These results are consistent with the expectation that net forcing is smallest over cool and dark surfaces. Net forcing over the mid-latitude continent is larger in summer than in winter. Hexagonal particles cause a larger albedo than spherical ones, therefore less net forcing. These results show that contrail heating generally prevails over cooling in the atmosphere-surface system. However, this finding does not preclude situations in which contrails cause a net cooling—for example, very cold surface, high atmospheric humidity, low surface albedo, very small particles (< 10 μm; the limit depends on the ice water path in the cloud), or large optical contrail depth (> 10).

TOA radiative forcing depends mainly on the cover and on parameters that determine the solar optical depth of contrails and to a minor degree on other parameters. Forcing is very weakly sensitive to the methods used for radiative transfer calculations. This sensitivity can be seen from Table 3-7, which lists the results of a parameter sensitivity study with models FL, M, and N (Meerkötter *et al.*, 1999). Except for the first two rows of Table 3-7, results are from model N only because the results of the three models agree with each other to within 3%. The largest model differences are found for aspherical particles, which are represented by different shapes in various spectral regions in the models, but all models show smaller net forcing for aspherical particles than for spherical ones, as expected (Kinne and Liou, 1989). The results also depend rather strongly on solar zenith angle, ice water content, particle diameter, surface albedo, and relative humidity. Low-level clouds with large cover and optical depth reduce SW and LW forcing of the cloud-free case below a contrail, causing a small net increase of forcing. Variations in surface temperature (here ± 5 K) cause small LW flux changes. Lowering the altitude of a contrail for fixed IWC reduces the LW effect slightly. Lowering the altitude of the contrail and using the IWC that is expected for the higher temperature at lower levels makes the lower contrail optically thicker and radiatively more effective than the higher contrail. Hence, TOA radiative forcing by contrails grows with increasing surface temperature, surface albedo, and IWC. For the same IWC, contrails with small ice particles are more effective in radiative forcing than contrails with larger particles. Representative forcing values are approximately 25 to 40 W m-2 for 100% contrail cover and 0.55-μm optical depth of 0.5.

Radiative forcing by contrails depends strongly on the optical depth of the contrails and is different at the surface than at the TOA. Figure 3-20 shows computed SW, LW, and net change in radiative fluxes at the TOA (actually 50 km) and at the surface for 100% contrail cloud cover for the mid-latitude summer continental reference case with spherical ice particles. The ice water content was varied to yield different values of the 0.55-μm optical depth τ. The trends are consistent with those found in cirrus studies (Fu and Liou, 1993). At the TOA, LW forcing is larger than SW forcing, giving a net heating of the atmosphere that is maximum near $\tau = 3$. The flux increases slightly less than linearly with optical depth for small values of τ. The net forcing changes sign and becomes negative for $\tau > 10$ (not shown), but contrails are probably never that optically thick. This analysis indicates that contrails heat the atmosphere below them.

In all cases, TOA and top of the troposphere radiative flux changes from contrails differ by only about 10%; therefore, the instantaneous or static flux change gives a reasonable approximation for the adjusted radiative forcing as considered in IPCC (1996).

At the Earth's surface, LW flux changes are much smaller because water vapor closes much of the infrared radiation window in the lower atmosphere. In contrast, SW flux changes are only a little smaller than at TOA. Hence, the daytime SW contribution dominates and cools the surface in the daily mean.

A reduction of solar radiation by 40 W m^{-2} has been measured locally in the shadow of contrails, although the simultaneous change in infrared flux in the shadow of contrails was very small (Sassen, 1997). Hence, the Earth's surface is locally cooled in the shadow of contrails. This analysis does not exclude warming of the entire atmosphere-surface system driven by the net flux change at TOA. As radiation-convection models show, for example, vertical heat exchange in the atmosphere may cause a warming of the surface even when it receives less energy by radiation (Strauss *et al.*, 1997). Cooling of the atmosphere below contrails is also suggested by measurements of solar and infrared upward and downward fluxes above and below a few persistent contrails (Kuhn, 1970). These measurements show a strong (10–20%) reduction of net downward radiation just below approximately 500-m-thick contrails with little change in LW fluxes. Further investigation is required to demonstrate how these results depend on the geometry and age of contrails.

Contrail cirrus induce a heat source by the change in divergence of solar and infrared radiation fluxes mainly within but also below the contrail in the upper troposphere (Liou *et al.*, 1990; Strauss *et al.*, 1997; Meerkötter *et al.*, 1999). In the atmosphere below a contrail, the change in heat source is on the order of 0.3 K/day for 100% cover. When the contrail is located above a thick lower level cloud, the atmosphere is heated only above the lower cloud; the heat source is essentially zero below the lower cloud.

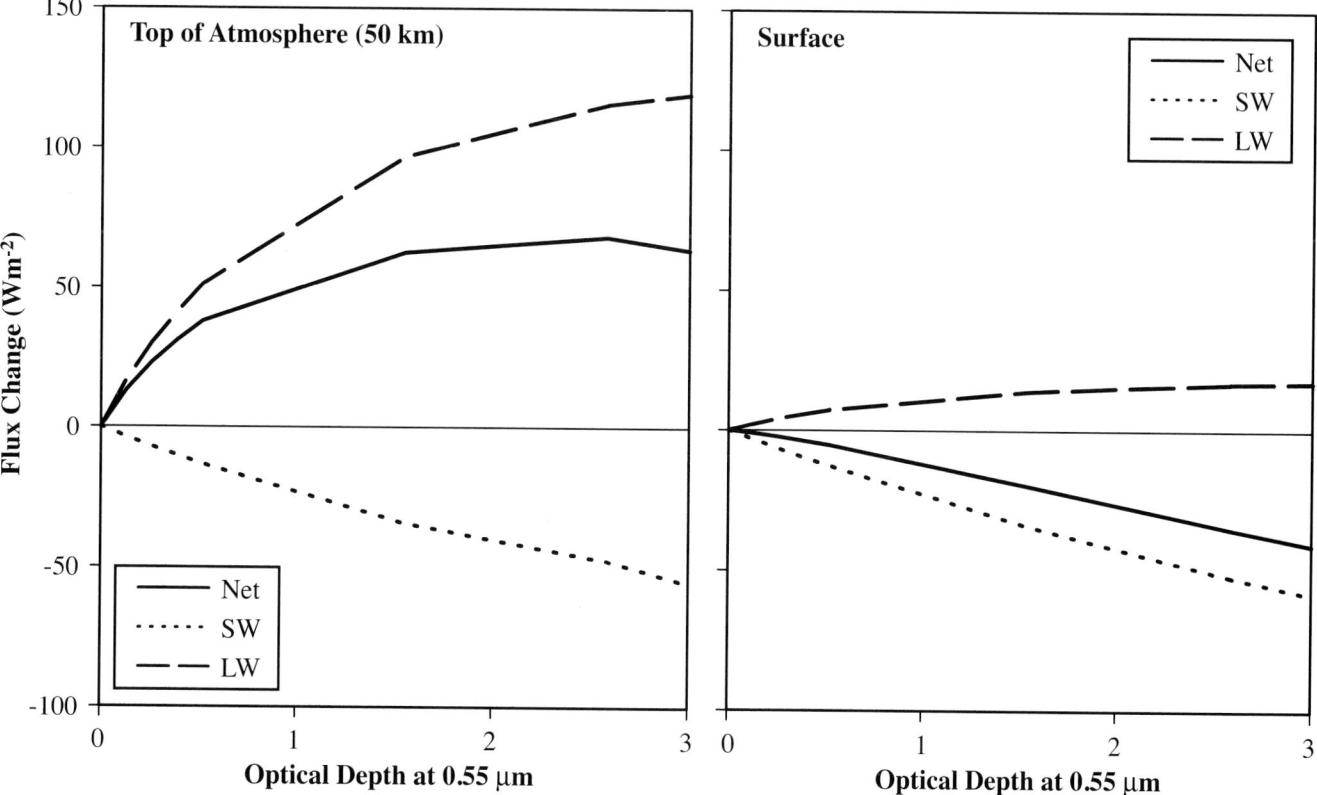

Figure 3-20: Shortwave (SW), longwave (LW), and net instantaneous radiative flux change from contrails with 100% cover under mid-latitude summer conditions averaged over a day as a function of optical depth (0.55 μm): Near the top of the atmosphere (50 km) and surface (0 km) (from Meerkötter *et al.*, 1999).

In the diurnal cycle, radiative forcing by contrails is positive and strongest during the night because of the absence of negative SW forcing (see Table 3-6). For small optical depth, net forcing at TOA is also positive during the day, hence always positive regardless of the diurnal cycle of contrail cover. Negative SW forcing is maximum not at noon but during morning or afternoon hours, when the solar zenith angle is near 70°. Other than at TOA, the net flux change is negative at the surface even for thin contrails during the day, in particular at intermediate zenith angles. The maximum day-night difference in net radiative forcing at the surface for cloudless summer mid-latitude conditions is about 20 W m^{-2} for constant 100% contrail cover with optical depth of 1 and constant surface temperature, and slightly less when accounting for the daily temperature cycle.

The global distribution of radiative forcing from contrails can be estimated using a radiation transfer model, the expected contrail cover, and a realistic representation of the cloudy atmosphere and surface (Minnis *et al.*, 1999). Global contrail cover for 1992, for example, is shown in Figure 3-16 (see Section 3.4.3). Global mean values for TOA radiative flux are summarized in Table 3-8, along with details of the calculations. As expected, SW forcing is negative and LW and net forcings are positive. All three values increase less than linearly with solar optical depth τ. The resulting 1992 global distribution of TOA net forcing for τ of 0.3 is shown in Figure 3-21. Forcing is largest in regions of heavy air traffic, with maximum values over northeast France (0.71 W m^{-2}) and near New York (0.58 W m^{-2}). Although the contrail amount is higher over the northeast United States of America, net contrail forcing over Europe is greater because of the greater LW forcing term. The global mean net forcing for τ between 0.3 and 0.5 is about 0.02 W m^{-2}; the zonal mean value is largest near 40°N, with a value five times larger than the global mean.

Computed results for global radiative forcing by contrails depend on assumed values for contrail cover and mean optical depth of contrails. Neither is well known. Here, the computed global contrail cover was normalized to yield 0.5% observed cover by line-shaped contrails over Europe, guided by satellite data. Because satellite data mainly reveal thicker contrails, a larger cover resulting from the presence of optically thin contrail cirrus might not be accurately detected. Therefore, the results for larger τ values (0.3 to 0.5) are considered in combination with the given contrail cover (0.1% globally) as being most representative for real forcing conditions. Hence, in the diurnal and annual mean, a global 0.1% increase in thin contrail cloud cover causes a net heating of the Earth-atmosphere system of approximately 0.02 W m^{-2}. The difference between the largest and smallest net forcing values in Table 3-8 suggests an error bound on the order of 0.01 W m^{-2}. Based on Section 3.5, the actual global cover may (with 2/3 probability) be 2 to 3 times smaller or larger than the derived line-shaped contrail cover. The optical depth value is likely known to a factor of 2 to 3. Thus, for assumed Gaussian behavior of individual uncertainties, the radiative forcing value may differ from the best estimate by a factor of about 3 to 4.

Based on computations using estimates of line-shaped contrail occurrence for the 1992 fuel scenario (0.1% cover), the best estimate of global-mean radiative forcing is positive, has a value of about 0.02 W m^{-2} with an uncertainty factor of 3 to 4, and may range from 0.005 to 0.06 W m^{-2} for present climate conditions. Certainly, the state of our understanding is only fair. Future investigations may result in considerable changes to the best estimates.

Global radiative forcing by contrails obviously is much smaller than that attributed to other anthropogenic changes in the past century — 1.5 W m^{-2}, which represents about the median of the range of values given in IPCC (1996) but comparable to the forcing by past CO_2 emissions by aircraft (Brasseur *et al.*, 1998; see Chapter 6). A mean radiative forcing of 0.02 W m^{-2} induces a vertically averaged heat source in the troposphere equivalent to approximately 0.0002 K day^{-1}. The atmosphere reacts to this

Table 3-8: Shortwave (SW), longwave (LW), and net radiative flux changes (Wm^{-2}) at top of atmosphere in global mean as caused by contrails in 1992 and 2050 scenario Fa1 (see Section 3.7.2 and Chapter 9), for contrail cover as shown in Figure 3-16 and for solar optical depths of 0.1, 0.3, and 0.5 for contrails (Minnis et al., 1999). The last line gives the results for temperature-dependent ice water content (IWC) with variable optical depth.[a]

Optical Depth at 0.55 μm	1992			2050 (Scenario Fa1)		
	SW	LW	Net	SW	LW	Net
0.1	-0.0030	0.0111	0.0081	-0.018	0.067	0.049
0.3	-0.0081	0.0246	0.0165	-0.049	0.148	0.099
0.5	-0.0124	0.0327	0.0203	-0.075	0.197	0.122
Variable (from temperature-dependent IWC)	-0.0038	0.0135	0.0097	-0.024	0.084	0.060

[a]Cloud amount, cloud distribution, and cloud optical depth are 1986 ISCCP 3-h data interpolated to 1 h, for 4 months (January, April, June, October), with mean cloud cover of 68% of the Earth surface. Ice particles are modeled as hexagons with mean diameters of 20 μm. Water clouds consist of droplets with 60-μm mean diameter. Surface skin temperature and surface albedo are taken from ISCCP and Staylor and Wilber (1990). Winter and standard temperature and moisture profiles are assumed for pressures < 50 hPa, and numerical weather analysis monthly mean profiles of National Meteorological Center (NMC; now National Center for Environmental Protection, NCEP) are used for pressures > 50 hPa. Continental and marine aerosols were also included in the model. Contrails are assigned at a pressure of 200 hPa and assumed to be 220-m thick with aspherical ice particles of 24-μm volume mean diameter. IWC in contrails was adjusted to result in an optical depth of 0.1, 0.3, or 0.5. A case with IWC set to half the amount of water available for ice formation from vapor at 100% humidity relative to liquid saturation (temperature-dependent IWC) is also considered (variable optical depth). Net TOA forcing is computed with model FL as difference of results with and without contrails.

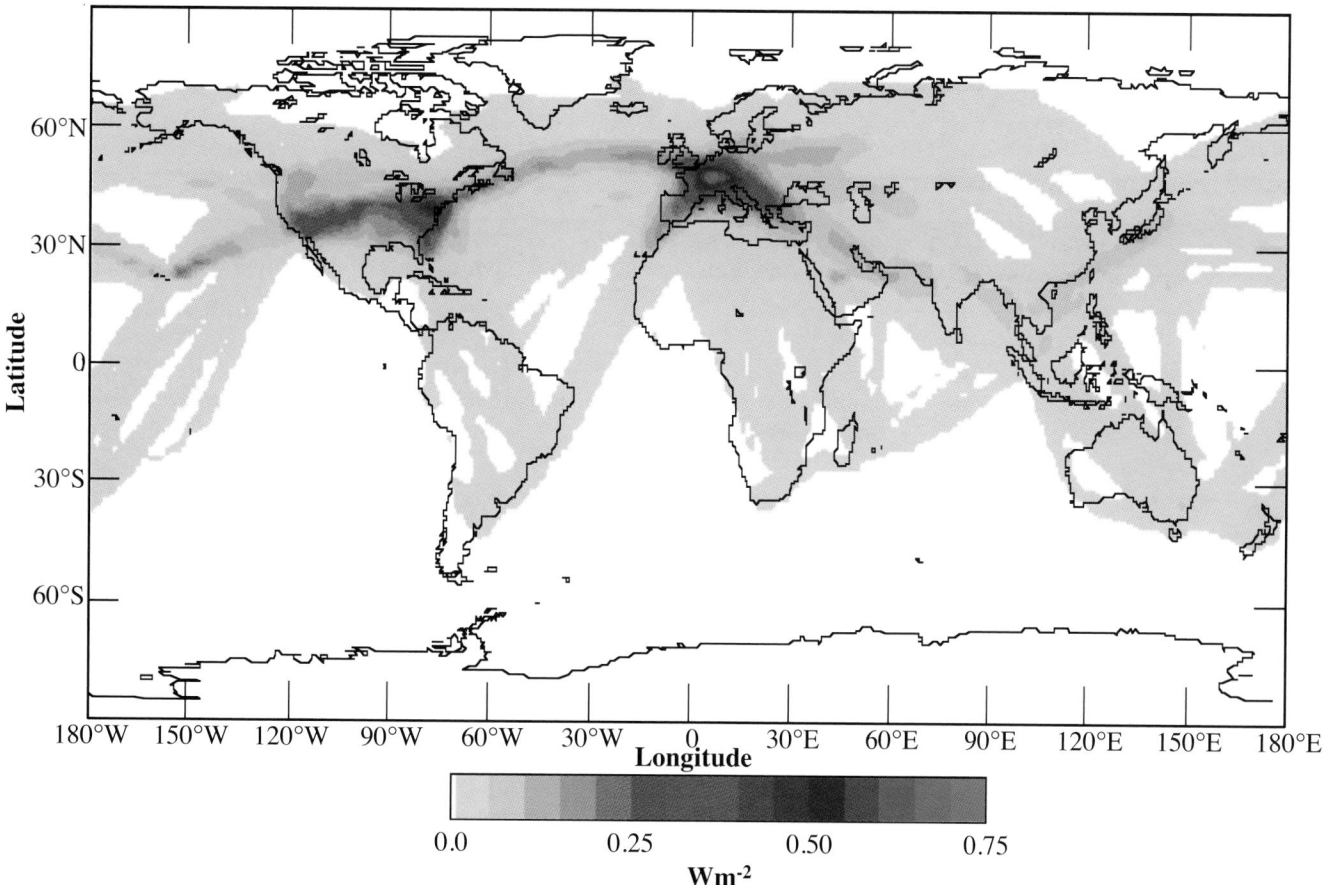

Figure 3-21: Global distribution of net instantaneous radiative forcing at the top of atmosphere in daily and annual average for present (1992) climatic conditions and analyzed contrail cover (see Figure 3-16) and 0.55-μm optical depth of 0.3 (Minnis *et al.*, 1999).

heat source in a complex manner (Ponater *et al.*, 1996) and may take decades to reach a steady-state temperature response. In steady state, the mean surface temperature may increase by about 0.01 to 0.02 K globally if climate sensitivity from contrail forcing is comparable to that of well-mixed greenhouse gases (IPCC, 1996; see Chapter 6.2.1). In comparison to the global mean value, annually averaged zonal mean values are larger by a factor of about 5, and regional values are larger by a factor of up to 40 (see Figure 3-21). The pattern of the climate response differs from the pattern of radiative forcing, in particular at small regional scales. Zonal mean steady-state temperature changes of between 0.01 and 0.1 K appear to be possible for present contrail cover when previous studies as scaled to the cover and radiative forcing found here (Liou *et al.*, 1990; Ponater *et al.*, 1996; Strauss *et al.*, 1997). Regional contrail forcing may have short-term consequences for the daily temperature range in such a region if contrail forcing persists for at least a day. These temperature changes appear to be too small to be detectable in comparison to atmospheric temperature variations.

3.6.5. Radiative Impact of Additional or Changed Cirrus and Other Indirect Effects

Besides forcing by line-shaped contrail cirrus, the global radiation balance may also be perturbed if there is a significant indirect effect of aircraft-induced aerosol, water vapor, and contrails on the coverage and properties (particle size distribution, number density, and composition) of "natural" cirrus clouds. In addition, there might be other indirect cloud-related effects, such as changes in humidity and precipitation. The principal effects might be similar to those caused by volcanic aerosol mixed down into the troposphere (see Section 3.4.5).

Radiative forcing will be enhanced by any increase in the cover of thin cirrus caused by aircraft beyond that of line-shaped contrail cirrus. If the additional global cirrus cover is as large as 0.2% (the estimated upper bound for 1992; see Section 3.5.1.5), and if the optical properties of this additional cirrus are the same as for line-shaped persistent contrails (optical depth of about 0.3, as also found often for natural cirrus; see Section 3.4.1), then the radiative forcing from the additional cirrus may be as large as 0.04 W m^{-2}—which is twice the value for line-shaped contrail cirrus.

Aircraft emissions may also change the properties of natural cirrus clouds (see Section 3.4.5). In a high-traffic region, cirrus was found to be affected by soot emissions from aircraft, causing an approximate doubling of the ice particle concentration (Ström and Ohlsson, 1998). Smaller particles cause larger optical depth for constant ice water content. Radiative forcing is strongly sensitive to particle size (see Table 3-7). As Figure 3-20

Table 3-9: Global radiative forcing by contrails and indirect cloud effects in 1992 and 2050 (scenario Fa1). No entry indicates insufficient information for best-estimate value.

Radiative Forcing	Best Estimate or Range	Uncertainty Range with 2/3 Probability	Status of Understanding
1992			
Line-shaped contrail cirrus	0.02 Wm^{-2}	0.005–0.06 Wm^{-2}	fair
Additional aviation-induced cirrus clouds	0–0.04 Wm^{-2}	—	very poor
Other indirect cloud effects	—	either sign, unknown magnitude	very poor
2050			
Line-shaped contrail cirrus	0.10 Wm^{-2}	0.03–0.4 Wm^{-2}	fair
Additional aviation-induced cirrus clouds	0–0.16 Wm^{-2}	—	very poor
Other indirect cloud effects	—	either sign, unknown range	very poor

indicates, an increase in optical depth causes additional heating if the cirrus cloud was optically thin but cooling if it was optically thick (Wyser and Ström, 1998). One recent study suggests that the indirect heating effect of aviation-induced changes in cirrus ice particle number density for fixed cloud cover may be positive and comparable to or even larger than that from increases in cloud cover (Meerkötter *et al.*, 1999).

While the impact of particle changes on radiative forcing by cirrus clouds may be studied parametrically, our understanding is very poor with respect to other indirect effects. Although comprehensive investigations are missing, there is no evidence that any of the indirect effects are important. Table 3-9 summarizes the assessment of global radiative forcing by aviation-induced cloudiness for 1992. The results for 2050 are explained in Section 3.7.

3.7. Parameters of Future Changes in Aircraft-Produced Aerosol and Cloudiness

The future effects of aircraft depend on trends in climate and air traffic amount and changes in the technical properties of aircraft. Our current understanding of the formation of aviation-induced aerosol and cloudiness can be used to estimate how future changes may affect the impacts of aviation and to identify mitigation options that would be effective in reducing these impacts.

3.7.1 Changes in Climate Parameters

If climate change occurs in the future, atmospheric parameters related to aerosols and contrails will also have changed. Of particular importance to aviation-induced aerosol and cloudiness are changes in temperature and humidity in the upper troposphere and lower stratosphere; changes in the height, temperature, and humidity of the tropopause region; changes in the abundance of particles; and changes in cloudiness. Table 3-10 summarizes how changes in these parameters may be reflected in aviation-related impacts.

General circulation models of the atmosphere predict that the climate of 2050 will reflect global warming from the accumulation of greenhouse gases. In this new climate, models predict increases in the amounts of cirrus clouds, the height of the tropopause, and upper tropospheric temperature (IPCC, 1996; Timbal *et al.*, 1997). A higher tropopause would cause more contrails, at least at high latitudes. Observed temperature changes (e.g., Parker *et al.*, 1997) do not reveal the expected temperature increase in the upper troposphere. Some models predict a higher tropopause if the surface temperature increases (about 200-m altitude increase for 1 K surface temperature increase) (Thuburn and Craig, 1997). Increases on the order of 100 m were analyzed in polar regions and at mid-latitudes (Hoinka, 1998; Steinbrecht *et al.*, 1998). Such changes may be forced by cooling of the lower stratosphere as a result of changes in ozone concentration (Hansen *et al.*, 1997) and increases in moisture as a result of increasing methane concentrations. Stratospheric temperatures between 50 and 100 hPa have decreased by about 1 to 2 K since 1980 (Ramaswamy *et al.*, 1996; Halpert and Bell, 1997). An increase in water vapor concentration has been observed in the lower stratosphere, with the largest trend (0.8%/yr) in the 18- to 20-km region (Oltmans and Hofmann, 1995). Because few contrails currently form in the lower stratosphere, small changes in stratospheric conditions will not create significant changes in contrail abundance. Aerosol loading in the troposphere and lower stratosphere may increase because of changed climate conditions and increased surface emissions. Surface emissions from fossil fuel burning were projected to grow by a factor of 1.5 to 2.1 from 1990 to 2040 (Wolf and Hidy, 1997).

3.7.2 Changes in Subsonic Aircraft

By the year 2050, the number of aircraft flying in the upper troposphere is expected to have increased significantly (see Chapter 9). In scenario Fa1, global annual aviation fuel consumption in 2050 will have increased by a factor of 3.2 compared with 1992, with a larger increase (factor of 4.3) above 500 hPa. Scenarios Fc1, Fe1, and Eah (see Chapter 9) assume increases by factors of 1.8, 5, and 14, respectively, in

Table 3-10: *Parameters affecting future changes in aircraft-produced aerosols and cloudiness and their impacts on contrails and aerosol abundance. Symbols indicate sign of change in the parameter and in the impact. Question marks indicate uncertainty in sign or importance of an impact. The symbol x indicates lack of a known or important impact.*

Parameter	Sign of Change	Global Contrail and Induced-Cirrus Coverage	Global Contrail Radiative Forcing	Aerosol Abundance
Upper troposphere temperature	+?	–	–	x
Lower stratosphere temperature	–	x	x	+
Humidity of lower stratosphere	+	x	x	+
Humidity of upper troposphere	+?	+	+	+
Tropopause altitude	+	+	+	x
Number of aircraft	+	+	+	+
Global aviation fuel consumption	+	+	+	+
Overall efficiency of propulsion	+	+	+	x
Cruise altitude at mid-latitudes	+	-	-	+
Cruise altitude in the tropics	+	+	+	x
Traffic in tropical regions	+	+	+	x
Soot emissions	–?	–?	–?	x
Fuel sulfur content	–?	–?	–?	–
Fuels with higher hydrogen content	+?	+	?	–

total fuel consumption compared to 1992. The frequency of contrail formation is expected to increase with traffic because large regions of the atmosphere are humid and cold enough to allow persistent contrails to form and because such regions are not at present fully covered with optically thick cirrus or contrail clouds (see Sections 3.4.1 and 3.4.3). The number of aircraft may grow slightly less rapidly than fuel consumption when smaller aircraft are replaced by larger ones. This factor is important because the amount of persistent contrail cover may depend mainly on the number of aircraft triggering contrails and less on fuel consumption.

As aircraft engines become more fuel efficient, contrails will form more frequently at lower flight levels because exhaust plumes of more efficient engines are cooler for the same water content (see Section 3.2.4.1). The overall efficiency η (Cumpsty, 1997) with which engines convert fuel combustion heat into propulsion of cruising subsonic aircraft was close to 0.22 in the 1950s, near 0.37 for modern engines in the early 1990s, and may reach 0.5 for new engines to be built by 2010 (see Figure 3-22). An increase of η from 0.3 to 0.5 in a standard atmosphere increases the threshold formation temperature of contrails by about 2.8 K (equivalent to 700-m lower altitude) (Schumann, 1996a).

The change in persistent contrail coverage because of changed traffic has been determined using thermodynamic analysis of meteorological data from 1983 to 1992 and fuel consumption data (Sausen *et al.*, 1998) (see Section 3.4). This method has been used to estimate future changes in contrail cover resulting from changes in air traffic, assuming a fixed climate, fuel consumption scenarios, and expected engine performance specifications (Gierens *et al.*, 1998). The computed contrail cover (Figure 3-23) for the 2050 Fa1 fuel scenario using present

analysis data and η of 0.5 shows a global and annual mean contrail cover of 0.47%, with values of 0.26% and 0.75% for

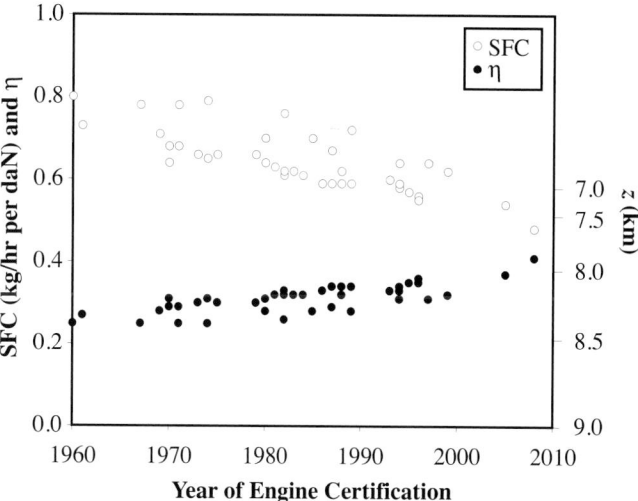

Figure 3-22: Trend in overall efficiency of propulsion η (solid circles), computed from aircraft specific fuel consumption (SFC) data (open circles; data as in Figure 7-9), according to $\eta = V(Q\,SFC)^{-1}$, with V as the aircraft speed ($\sim$240 m s^{-1}) and Q as the specific heat of combustion of aviation fuels (43 MJ kg^{-1}). Solid circles also denote the critical altitude z (right axis) above which contrails form (for 100% relative humidity in the mid-latitude standard atmosphere) for the years 1960 to 2010. SFC data were taken from a figure *Subsonic Engine-Specific Fuel Consumption at Cruise Versus Certification Date* originating from a NASA report and presented by H.G. Aylesworth at the Working Group Meeting CAEP-4, WG3 Emissions, International Civil Aviation Organization, 20-23 May 1997, Savannah, GA, USA.

scenarios Fc1 and Fe1. Values may be as high as 1 to 2% for scenario Eah, which does not specify the spatial distribution of future traffic and in which contrail cover may become limited by the amount of cloud-free ice-supersaturated air masses. In comparison, values are 0.087% for the 1992 DLR fuel inventory with η of 0.3 and 0.38% for the 2050 scenario with η of 0.3. Hence, contrail cover is expected to increase by a factor of about 5 over present cover for a 3.2-fold increase in annual aviation fuel consumption from 1992 to 2050, even under constant climate conditions. Increased efficiency of propulsion by future engines causes about 20% of the computed increase in contrail cover. In the year 2050, the maximum contrail coverage is expected to occur over Europe (4.6%, 4 times more than 1992), the United States of America (3.7%, 2.6 times more), and southeast Asia (1.2%, 10 times more). Contrail-induced increases in cirrus cloud cover may depend also on wind shear, vertical motions, and existing cirrus cover, which this thermodynamic analysis does not take into account. In addition, changes in climate conditions may influence future contrail formation conditions.

Radiative forcing from contrails was calculated for 2050 using the Fa1 fuel scenario and the same method as described in Section 3.6.3 (Minnis *et al.*, 1999). For the contrail cover shown in Figure 3-23, values of SW, LW, and net forcing were found to be about 6 times larger than in 1992 (see Table 3-8). The increase in radiative forcing from 1992 to 2050 is larger than the increase in contrail cover (factor of 5) during the same period because additional contrails in the subtropics and over Asia over relatively warm and cloud-free surfaces are more effective in increasing radiative forcing. The global distribution of radiative forcing calculated with this procedure is shown in Figure 3-24 for an assumed optical depth of 0.3. Radiative forcing grows more strongly globally than in regions of present peak traffic. Global mean forcing is 0.1 W m^{-2} in this computation, with maximum values of 3.0 and 1.4 W m^{-2} (3.3 and 2.4 times more than 1992) over northeast France and the eastern United States of America, respectively.

For an optical depth of 0.3, the best-estimate value of the global radiative forcing in 2050 (scenario Fa1) is 0.10 W m^{-2}. The uncertainty range is a little larger than in 1992, and estimated to amount to a factor of 4. Hence, the likely range of forcing extends from 0.03 to 0.4 W m^{-2} (see Table 3-9). The forcing for other scenarios has not been computed in detail, but rough estimates scale with the fuel consumption. The climatic consequences of this forcing are discussed in Chapter 6.

An estimate of the range of aviation-induced cirrus cloudiness in 2050, as required for this assessment, is not available in the scientific literature. For 1992, a range for the best estimate of the additional aviation-induced cirrus clouds was derived from decadal trends in high fuel-use regions (0–0.2% global cover; Section 3.5.1.5). For the 2050 time period, a different approach

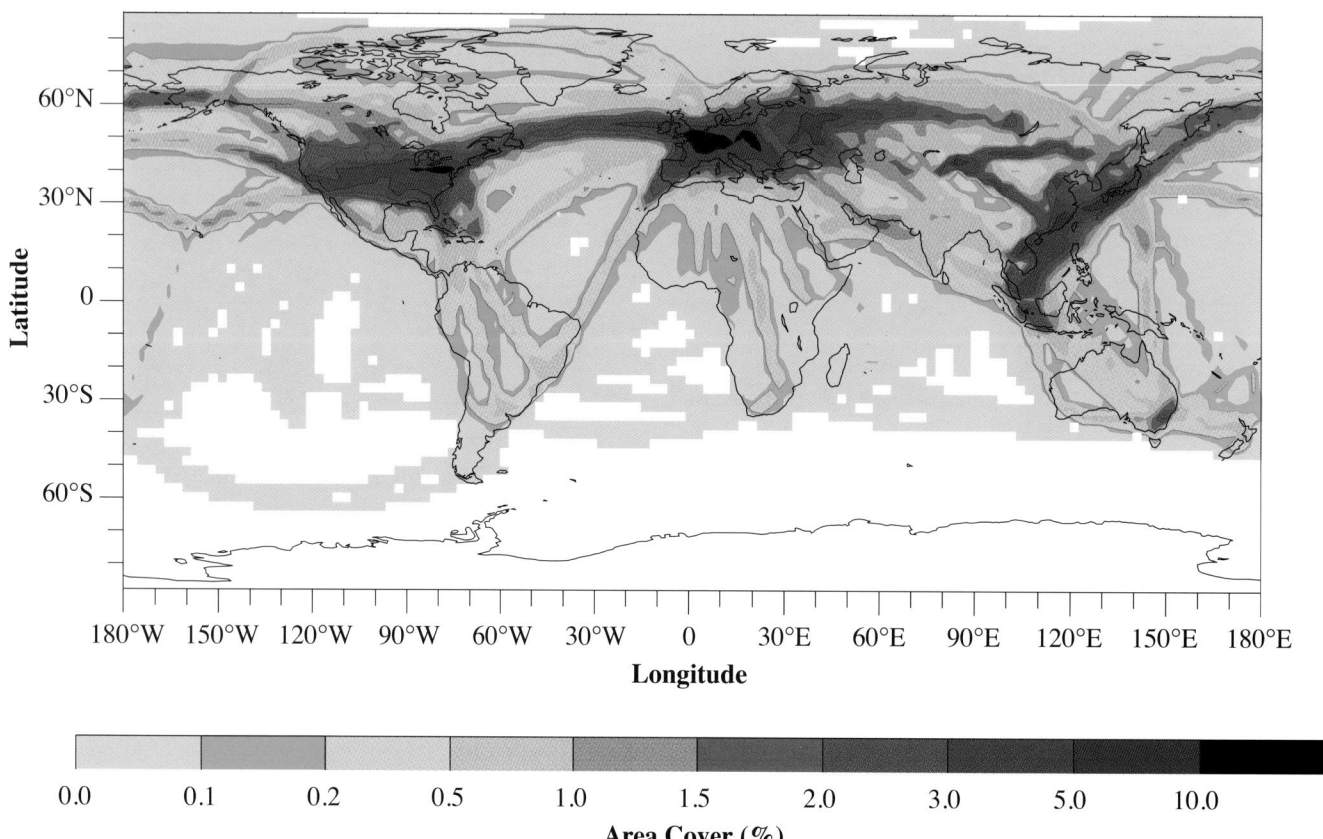

Figure 3-23: Persistent contrail coverage (in % area cover) based on meteorological analysis data and on fuel emission database for 2050 (Fa1 fuel consumption scenario for 2050), assuming linear dependence on fuel consumption and overall efficiency of propulsion η of 0.5; global mean cover is 0.5%. Compare with Figure 3-16 (from Gierens *et al.*, 1998).

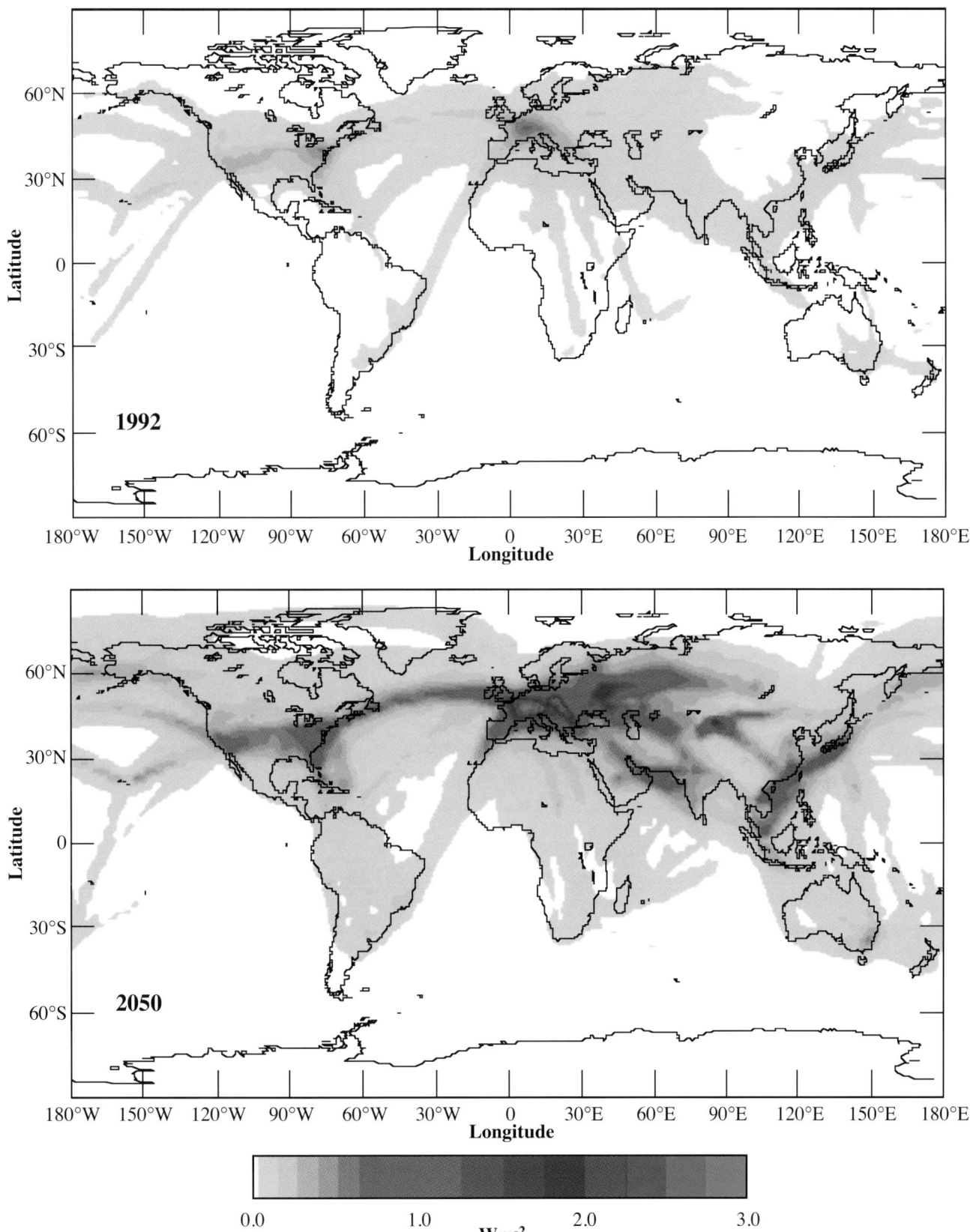

Figure 3-24: Global distribution of net radiative forcing at top of atmosphere in daily and annual average for contrails with 0.55-μm optical depth of 0.3: For 1992, as in Figure 3-21; and for the 2050 scenario with contrail cover as shown in Figure 3-23 (Minnis *et al.*, 1999).

is required. Observed contrail frequencies and trends in cirrus occurrence have been found to correlate with aviation fuel consumption (see Figures 3-14 and 3-18). Therefore, the aviation-induced cirrus cloudiness between 1992 and 2050 is projected to grow in proportion to the total aviation fuel consumption in the upper troposphere. This fuel consumption grows by a factor of 4 between 1992 and 2050 in scenario Fa1. Hence, the best-estimate of additional global cirrus cover in 2050 would range from 0 to 0.8%. For the same radiative sensitivity as in 1992, the associated radiative forcing could be between 0 and 0.16 W m^{-2} or up to 1.6 times the value given for line-shaped contrail cirrus in 2050 (see Table 3-9). The forcing could be outside this range if future aviation causes strong changes in the optical properties of the cirrus clouds. Saturation effects (Sausen *et al.*, 1999) will likely limit any increase in cirrus cover in heavy air traffic regions. Because of these uncertainties the status of understanding of radiative forcing from additional aviation-induced cirrus clouds in 2050 is very poor. The assessment of the other indirect effects (Section 3.6.5) is beyond the scope of present understanding.

Modern subsonic aircraft cruise most efficiently at flight altitudes of 9 to 13 km. Trends in aircraft cruise altitudes are discussed in Chapter 7. If mean flight levels of global air traffic were to increase, the frequency of persistent contrails in the troposphere at mid-latitudes would be reduced and the frequency in the upper troposphere in the tropics would be increased. In addition, the formation of polar stratospheric clouds in the lower polar stratosphere (Peter *et al.*, 1991) may be enhanced as a result of increased emissions in the stratosphere. At mid-latitudes, a 1-km flight-level increase causes a moderate reduction of contrail cover because of increased flights in the dry stratosphere (e.g., 12% less contrail cover over the North Atlantic compared with the nominal-altitude cover). Despite these changes, the global change in contrail cover from an altitude increase is small because of compensating changes in the tropics. The stronger increase of contrail cover in the tropics may cause a stronger positive radiative forcing because of the warmer surface in the tropics compared with mid-latitudes (see Table 3-7). A reduction in flight levels generally has the opposite effect (more contrails at high latitudes and fewer contrails in the tropics). Results for Europe and parts of the United States of America are different in that computed contrail coverage decreases slightly for both an increase and a decrease in mean flight levels because air traffic currently occurs in the cold and humid upper troposphere in those regions, and a shift toward the drier stratosphere or the warmer mid-troposphere reduces contrail coverage (Sausen *et al.*, 1998). A change in mean altitude of contrails may change their radiative impact even for constant areal coverage. A contrail at higher altitude in the troposphere will likely contain less ice mass and produce less radiative forcing, therefore, despite lower ambient temperatures (see Table 3-7).

Trends in soot emissions would be important if soot influences ice particle formation (see Section 3.4) or the chemistry of ozone (see Chapter 2). A soot mass emission index of 0.5 to 1 g kg^{-1} (and larger) is not uncommon for older aircraft engines. The soot mass emission index of the present aircraft

fleet is estimated as 0.04 g kg^{-1} (see Chapter 7). The soot emission index decreased with new engine technology until about 1980 but has showed no significant trend thereafter (Döpelheuer, 1997). The mass of soot emitted may decrease despite increases in fuel consumption. No data exist on trends for the number and surface area of soot aerosol emissions.

Atmospheric models and fuel consumption scenarios suggest that aircraft emissions contribute little to the tropospheric mass of sulfate and soot aerosol in today's atmosphere and in 2050 (see Section 3.3). However, aircraft-induced particles will increase with growing emission rates of condensable sulfur compounds and soot particle mass. A reduction in fuel sulfur content is not to be expected for the near future (see Chapter 7). The fraction of condensable sulfur compounds formed from fuel-sulfur depends on the details of the chemistry between the combustor and the engine exit (Brown *et al.*, 1996a; Lukachko *et al.*, 1998). The dependence of this fraction on expected changes in engine technology is not known (see Chapter 7).

Engines burning liquid hydrogen (liquid methane) instead of kerosene (Wulff and Hourmouziadis, 1997) emit 2.6 (1.5) times more water vapor for the same amount of combustion heat. Therefore, such engines trigger contrails at about 1 to 2 km lower altitude (4 to 10 K higher ambient temperature) than comparable kerosene engines. Therefore, an increase in contrail coverage is expected with such fuels. Because of larger water emissions, such contrails will grow to larger diameters before evaporating in ice-subsaturated ambient air. On the other hand, aircraft using hydrogen (methane) fuels will emit no (little) soot and sulfur compounds, hence may cause contrails that have fewer and larger ice particles, smaller optical thickness, and a lesser impact on radiative fluxes (Schumann, 1996a) (compare Table 3-7).

3.7.3. *Expected Changes for Supersonic Aircraft*

The expected emissions of future high speed civil transports (HSCTs) flying above 16-km altitude would substantially add to aerosol amounts in the stratosphere. A fleet of 500 HSCTs is expected to consume about 72 Tg fuel yr^{-1} in 2015 (Baughcum and Henderson, 1998). This level of consumption will cause emissions of sulfur and soot of 14.4 and 2.9 Gg yr^{-1}, for emission indices of 0.2 g S kg^{-1} and 0.04 g soot kg^{-1}, respectively. Microphysical calculations by the AER 2-D model (Weisenstein *et al.*, 1997) show that 28 Gg of sulfate will accumulate in the global atmosphere, assuming that 10% of sulfur emissions are converted in the plume to new particles with a radius of 10 nm. The globally averaged aircraft-produced sulfate column is equal to 5.4 ng SO$_4$ cm^{-2}, with a maximum of 13.6 ng SO$_4$ cm^{-2} near 50°N. This value is about twice that computed for present subsonic aviation (Table 3-4). The annually and zonally averaged perturbation of sulfate aerosol SAD as shown in Figure 3-25 is used for scenario SA5 in Chapter 4 and in calculations in Chapters 5 and 6. The chemical consequences of these SAD changes are discussed in detail in Section 4.3.3.

Though supersonic aircraft may have better engine efficiency than subsonic aircraft (0.38 for the Concorde), supersonic aircraft are expected to form few persistent contrails because the probability of ice-supersaturated air at cruise altitudes is small, except in the polar regions and near the tropical tropopause (Miake-Lye *et al.*, 1993). However, the accumulation of supersonic aircraft emissions in the polar atmospheres and local H_2O, HNO_3, and aerosol concentration increases in aircraft plumes may enhance the occurrence of polar stratospheric clouds. The impact on tropospheric cloud formation of supersonic aircraft cruising in the stratosphere is very likely much smaller than the impact of major volcanic events.

3.7.4. Mitigation Options

In the following discussion, options related to aircraft and aircraft operations are briefly considered for the reduction of volatile and nonvolatile particle emission and formation and for the reduction of contrail formation and contrail impact.

Volatile particle growth is controlled mainly by oxidized sulfur, chemi-ions, and water vapor present in aircraft exhaust. With current engines and fuels, no practical options exist to reduce water vapor emission indices. The oxidation of sulfur depends on the emission of SO_3 or the formation of H_2SO_4 in the engine and plume. The emission of SO_3 depends on the details of the reactive flow in and beyond the engine combustion chambers (Chapter 7). The processes controlling condensable sulfur oxides and chemi-ion production are not yet sufficiently

understood for a meaningful assessment of mitigation options. A reduction of sulfur content in fuel reduces plume levels of SO_3 and H_2SO_4, but not necessarily by the same factor (Brown *et al.*, 1996a). In addition, for low fuel sulfur content, volatile particles may remain that result from the emissions of other condensable material (Yu *et al.*, 1998; see Section 3.2) and thus require separate mitigation strategies.

Options to reduce soot emissions require changes in the combustion process (Chapter 7). Soot may be activated by H_2SO_4 and possibly other exhaust species. If soot activation by H_2SO_4 is to be avoided, fuel sulfur contents of less than 10 ppm would be required.

Simulations suggest that contrails would form even without any soot and sulfur emissions by activation and freezing of background particles (Jensen *et al.*, 1998b; Kärcher *et al.*, 1998a). Hence, the formation of contrails cannot be avoided completely by reducing exhaust aerosol emissions. Contrails formed in plumes with fewer exhaust particles are likely to be composed of fewer and larger particles, have smaller optical depths (Schumann, 1996a), hence cause less radiative forcing. Reduced soot and sulfate particle emissions may also lead to the formation of cirrus clouds with fewer but larger particles and less radiative forcing.

An increase in engine efficiency may change the global effects of contrails. Improvements in engine efficiency measured as specific fuel consumption (SFC) per unit thrust or overall efficiency, η, would reduce fuel consumption at cruise altitudes for a given amount of air traffic. Because more efficient engines increase the altitude range over which persistent contrails form (see Section 3.2.4.1 and 3.7.2), contrail frequency and cover would likely increase for a given air traffic amount. On the other hand, the number of ice crystals forming per aircraft-km would likely be reduced for lower SFC because aerosol and aerosol precursor emissions would be reduced. Fewer ice crystals could result in less radiative impact for a given amount of air traffic in altitude regions where contrails form at present. Hence, the balance of changes in contrail occurrence and the radiative impact that would result from changes in engine efficiency depend on a variety of factors, not all of which are well known enough at present.

Reducing the frequency of contrails for a given amount of air traffic could otherwise be effected by reducing the number of flights in the humid and cold regions of the upper and middle troposphere. Numerical weather prediction schemes may be used to predict and circumvent such regions on long-distance flights. Contrail-forming regions could also be avoided by flying at generally higher altitudes, but the climatic impact of contrails may not be reduced because of counteracting effects. For example, higher flight altitudes at low latitudes could increase contrails, possibly causing a net increase instead of a decrease in global radiative forcing by contrails. In addition, more flights in the lower stratosphere could result in enhanced aerosol and chemical impacts not related to contrails.

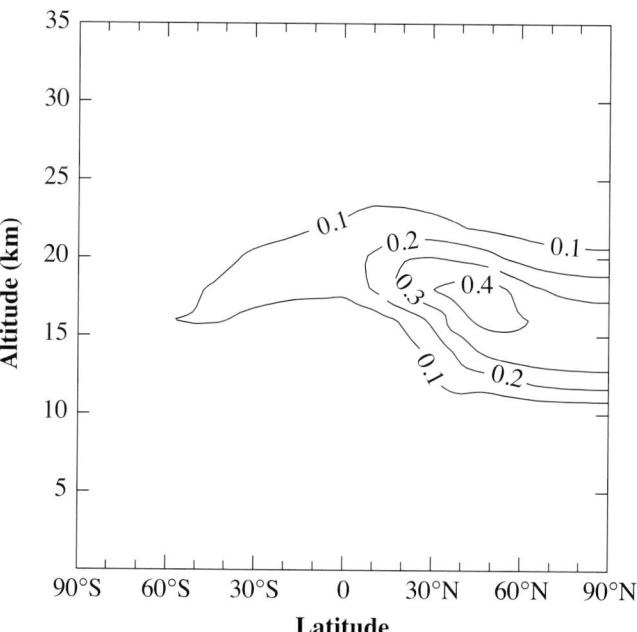

Figure 3-25: Annually and zonally averaged perturbation of sulfate aerosol surface area density (in μm^2 cm^{-3}) caused by an HSCT fleet of 500 aircraft flying at Mach 2.4 according to AER 2-D model (Weisenstein *et al.*, 1997). A sulfur emission index of 0.2 g kg^{-1} and a 10% conversion to sulfate particles with 10-nm radius in the plume are assumed in these calculations.

References

Ackerman, S.A., C. Moeller, K.I. Strabala, H.E. Gerber, L.E. Gumley, W.P. Menzel, and S-C. Tsay, 1998: Retrieval of effective microphysical properties of clouds: a wave cloud case study. *Geophysical Research Letters, 25,* 1121–1124.

Anderson, B.E., W.R. Cofer, J.W. Barrick, D.R. Bagwell, C.H. Hudgins, and G.D. Nowicki, 1998a: Airborne observations of aircraft aerosol emissions. 1. Total and nonvolatile particle emission indices. *Geophysical Research Letters, 25,* 1689–1692.

Anderson, B.E., W.R. Cofer, D.R. Bagwell, J.W. Barrick, C.H. Hudgins, and K.E. Brunke, 1998b: Airborne observations of aircraft aerosol emissions. 2. Factors controlling volatile particle production. *Geophysical Research Letters, 25,* 1693–1696.

Andronache, C. and W.L. Chameides, 1997: Interactions between sulfur and soot emissions from aircraft and their role in contrail formation. 1. Nucleation. *Journal of Geophysical Research, 102,* 21443–21451.

Andronache, C. and W.L. Chameides, 1998: Interactions between sulfur and soot emissions from aircraft and their role in contrail formation. 2. Development. *Journal of Geophysical Research, 103,* 10787–10802.

Angell, J.K., 1990: Variation in United States cloudiness and sunshine duration between 1950 and the drought year of 1988. *Journal of Climate, 3,* 296–308.

Appleman, H., 1953: The formation of exhaust contrails by jet aircraft. *Bulletin of the American Meteorological Society, 34,* 14–20.

Arnold, F., J. Scheid, T. Stilp, H. Schlager, and M.E. Reinhardt, 1992: Measurements of jet aircraft emissions at cruise altitude. I. The odd-nitrogen gases. *Geophysical Research Letters, 19,* 2421–2424.

Arnold, F., J. Schneider, M. Klemm, J. Scheid, T. Stilp, H. Schlager, P. Schulte, and M.E. Reinhardt, 1994: Mass spectrometric measurements of SO₂ and reactive nitrogen gases in exhaust plumes of commercial jet airliners at cruise altitude. In: *Impact of Emissions from Aircraft and Spacecraft upon the Atmosphere* [Schumann, U. and D. Wurzel (eds.)]. Proceedings of an international scientific colloquium, 18–20 April 1994, Cologne, Germany. DLR-Mitteilung 94-06, Deutsches Zentrum für Luft- und Raumfahrt (German Aerospace Center), Oberpfaffenhofen and Cologne, Germany, pp. 323–328.

Arnold, F., J. Schneider, K. Gollinger, H. Schlager, P. Schulte, P.D. Whitefield, D.E. Hagen, and P. van Velthoven, 1997: Observation of upper tropospheric sulfur dioxide- and acetone-pollution: potential implications for hydroxyl radical and aerosol formation. *Geophysical Research Letters, 24,* 57–60.

Arnold, F., T. Stilp, R. Busen, and U. Schumann, 1998a: Jet engine exhaust chemi-ion measurements: implications for gaseous SO₃ and H₂SO₄. *Atmospheric Environment, 32,* 3073–3077.

Arnold, F., K.-H. Wohlfrom, M.W. Klemm, J. Schneider, K. Gollinger. U. Schumann, and R. Busen, 1998b: First gaseous ion composition measurements in the exhaust plume of a jet aircraft in flight: implications for gaseous sulfuric acid, aerosols, and chemi-ions. *Geophysical Research Letters, 25,* 2137–2140.

Bakan, S., M. Betancor, V. Gayler, and H. Grassl, 1994: Contrail frequency over Europe from NOAA-satellite images. *Annales Geophysicae, 12,* 962–968.

Balkanski, Y.J., D.J. Jacob, G.M. Gardner, W.C. Graustein, and K.K. Turekian, 1993: Transport and residence times of tropospheric aerosols inferred from a global three-dimensional simulation of ²¹⁰Pb. *Journal of Geophysical Research, 98,* 20573–20586.

Barnes, J.E. and D.J. Hofmann, 1997: Lidar measurements of stratospheric aerosol over Mauna Loa Observatory. *Geophysical Research Letters, 24,* 1923–1926.

Baughcum, S.L. and S.C. Henderson, 1998: *Aircraft Emission Scenarios Projected in Year 2015 for the NASA Technology Concept Aircraft (TCA) High Speed Civil Transport.* NASA/CR-1998-207635, National Aeronautics and Space Administration, Hampton, VA, USA, 21 pp.

Baughcum, S.L. T.G. Tritz, S.C. Henderson, and D.C. Pickett, 1996: *Scheduled Civil Aircraft Emission Inventories for 1992: Database Development and Analysis.* NASA-CR-4700, National Aeronautics and Space Administration, Hampton, VA, USA, 205 pp.

Bekki, S., 1997: On the possible role of aircraft generated soot in the middle latitude ozone depletion. *Journal of Geophysical Research, 102,* 10751–10758.

Bekki, S. and J.A. Pyle, 1992: Two-dimensional assessment of the impact of aircraft sulphur emission on the stratospheric sulphate aerosol layer. *Journal of Geophysical Research, 97,* 15839–15847.

Bekki, S. and J.A. Pyle, 1993: Potential impact of combined NOₓ and SOₓ emissions from future high speed civil transport aircraft on stratospheric aerosols and ozone. *Geophysical Research Letters, 20,* 723–726.

Benkovitz, C.M., M.T. Scholtz, J. Pacyna, L. Tarrason, J. Dignon, E.V. Voldner, P.A. Spiro, J.A. Logan, and T.E. Graedel, 1996: Global gridded inventories of anthropogenic emissions of sulfur and nitrogen. *Journal of Geophysical Research, 101,* 29239–29253.

Berger, B., U. Schumann, and D. Wurzel, 1994: Fuel consumption by airliners above and below the tropopause analyzed from operational flight plan data. In: *Impact of Emissions from Aircraft and Spacecraft upon the Atmosphere* [Schumann, U. and D. Wurzel (eds.)]. Proceedings of an international scientific colloquium, 18–20 April 1994, Cologne, Germany. DLR Mitteilung 94-06, Deutsches Zentrum für Luft- und Raumfahrt, Oberpfaffenhofen and Cologne, Germany, pp. 71–75.

Berntsen, T. and I.S.A. Isaksen, 1997: A global three-dimensional CTM for the troposphere. 1. Model description and CO and ozone results. *Journal of Geophysical Research, 102,* 21239–21280.

Betancor-Gothe, M. and H. Grassl, 1993: Satellite remote sensing of the optical depth and mean crystal size of thin cirrus and contrails. *Theoretical and Applied Climatology, 48,* 101–113.

Blake, D.F. and K. Kato, 1995: Latitudinal distribution of black carbon soot in the upper troposphere and lower stratosphere. *Journal of Geophysical Research, 100,* 7195–7202.

Bockhorn, H. (ed.), 1994: *Soot Formation in Combustion. Mechanisms and Models.* Springer-Verlag, Berlin, Germany, 596 pp.

Borrmann, S., S. Solomon, J.E. Dye, and B.P. Luo, 1996: The potential of cirrus clouds for heterogeneous chlorine activation. *Geophysical Research Letters, 23,* 2133–2136.

Borrmann, S., S. Solomon, J.E. Dye, D. Baumgardner, K.K. Kelly, and K.R. Chan, 1997: Heterogeneous reactions on stratospheric background aerosol, volcanic sulfuric acid droplets, and type I polar stratospheric clouds: effects of temperature fluctuations and differences in particle phase. *Journal of Geophysical Research, 102,* 3639–3648.

Boucher, O. and T.L. Anderson, 1995: GCM assessment of the sensitivity of direct climate forcing by anthropogenic sulfate aerosols to aerosol size and chemistry. *Journal of Geophysical Research, 100,* 26117–26134.

Boucher, O., 1998: *Is the Observed Trend in Cirrus Occurrence Due to Aviation?* Note Interne du LOA No. 1, Laboratoire d'Optique Atmosphérique, UFR de Physique, Université de Lille-I, Villeneuve d'Ascq, France, 17 pp.

Boucher, O., 1999: Influence of air traffic on cirrus occurrence. *Nature, 397,* 30–31.

Brasseur, G.P., R.A. Cox, D. Hauglustaine, I. Isaksen, J. Lelieveld, D.H. Lister, R. Sausen, U. Schumann, A. Wahner, and P. Wiesen, 1998: European scientific assessment of the atmospheric effects of aircraft emissions. *Atmospheric Environment, 32,* 2329–2418.

Brest, C.L., W.B. Rossow, and M.D. Roiter, 1997: Update of radiance calibrations for ISCCP. *Journal of Atmospheric and Oceanic Technology, 14,* 1091–1109.

Brewer, A.W., 1946: Condensation trails. *Weather, 1,* 34–40.

Brock, C.A., P. Hamill, J.C. Wilson, H.H. Jonsson, and K.R Chan, 1995: Particle formation in the upper tropical troposphere: a source of nuclei for the stratospheric aerosol. *Science, 270,* 1650–1653.

Brogniez, G., J.-C. Buriez, V. Giraud, F. Parol, and C. Vanbauce, 1995: Determination of effective emittance and a radiatively equivalent microphysical model of cirrus from ground-based and satellite observations during the International Cirrus Experiment: the 18 October 1989 case study. *Monthly Weather Review, 123,* 1025–1036.

Brown, R.C., M.R. Anderson, R.C. Miake-Lye, C.E. Kolb, A.A. Sorokin, and Y.I. Buriko, 1996a: Aircraft exhaust sulfur emissions. *Geophysical Research Letters, 23,* 3603–3606.

Brown, R.C., R.C. Miake-Lye, M.R. Anderson, C.E. Kolb, and T.J. Resch, 1996b: Aerosol dynamics in near-field aircraft plumes. *Journal of Geophysical Research, 101,* 22939–22953.

Brown, R.C., R.C. Miake-Lye, M.R. Anderson, and C.E. Kolb, 1996c: Effect of aircraft exhaust sulfur emissions on near field plume aerosols. *Geophysical Research Letters, 23,* 3607–3610.

Brown, R.C., R.C. Miake-Lye, M.R. Anderson, and C.E. Kolb, 1997: Aircraft sulfur emissions and the formation of visible contrails. *Geophysical Research Letters, 24,* 385–388.

Burtscher, H. 1992: Measurements and characteristics of combustion aerosols with special consideration of photoelectric charging and charging by flame ions. *Journal of Aerosol Science,* **23,** 549–595.

Busen, R. and U. Schumann, 1995: Visible contrail formation from fuels with different sulfur contents. *Geophysical Research Letters,* **22,** 1357–1360.

Calcote, H.F., 1983: Ionic mechanisms of soot formation. In: *Soot in Combustion Systems and its Toxic Properties* [Lahaye, J. and G. Prado (eds.)]. Plenum Press, London, United Kingdom, pp. 197–215.

Carleton, A.M. and P.J. Lamb, 1986: Jet contrails and cirrus clouds: a feasibility study employing high-resolution satellite imagery. *Bulletin of the American Meteorological Society,* **67,** 301–309.

Carslaw, K.S., T. Peter, and S.L. Clegg, 1997: Modeling the composition of liquid stratospheric aerosols. *Review of Geophysics,* **35,** 125–154.

Champagne, D.L., 1988: *Standard Measurement of Aircraft Gas Turbine Engine Exhaust Smoke.* ASME 71-GT-88, American Society of Mechanical Engineers, New York, NY, USA, 11 pp.

Changnon, S.A., 1981: Midwestern sunshine and temperature trends since 1901: possible evidence of jet contrail effects. *Journal of Applied Meteorology,* **20,** 496–508.

Charlson, R.J., J. Langner, and H. Rodhe, 1990: Sulphate aerosol and climate. *Nature,* **348,** 22.

Charlson, R.J., S.E. Schwartz, J.M. Hales, R.D. Cess, J.A. Coakley, Jr., J.E. Hansen, and D.J. Hofmann, 1992: Climate forcing by anthropogenic aerosols. *Science,* **255,** 423–430.

Chen, Y., S.M. Kreidenweis, L.M. McInnes, D.C. Rogers, and P.J. DeMott, 1998: Single particle analyses of ice nucleating aerosols in the upper troposphere and lower stratosphere. *Geophysical Research Letters,* **25,** 1391–1394.

Chin, M. and D.D. Davis, 1995: A reanalysis of carbonyl sulfide as a source of stratospheric background sulfur aerosol. *Journal of Geophysical Research,* **100,** 8993–9005.

Chin, M., D.J. Jacob, G.M. Gardner, M.S. Foreman-Fowler, P.A. Spiro, and D.L. Savoie, 1996: A global three-dimensional model of tropospheric sulfate. *Journal of Geophysical Research,* **101,** 18667–18690.

Chiou, E.W., M.P. McCormick, and W.P. Chu, 1997: Global water vapor distributions in the stratosphere and upper troposphere derived from 5.5 years of SAGE II observations (1986–1991). *Journal of Geophysical Research,* **102,** 19105–19118.

Chlond, A., 1998: Large-eddy simulation of contrails. *Journal of Atmospheric Sciences,* **55,** 796–819.

Chuan, R.L. and D.C. Woods, 1984: The appearance of carbon aerosol particles in the lower stratosphere. *Geophysical Research Letters,* **11,** 553–556.

Chughtai, A.R., M.E. Brooks, and D.M. Smith, 1996: Hydration of black carbon. *Journal of Geophysical Research,* **101,** 19505–19514.

CIAP, 1975: *Monograph 2: Propulsion Effluents in the Stratosphere.* Final Report of the Climatic Impact Assessment Program. DOT-TST-75-52, Department of Transportation, Washington, DC, USA, 746 pp.

Considine, D.B., A.R. Douglass, and C.H. Jackman, 1994: Effects of polar stratospheric cloud parameterization on ozone depletion due to stratospheric aircraft in a two-dimensional model. *Journal of Geophysical Research,* **99,** 18879–18894.

Cooke, W.F. and J.J.N. Wilson, 1996: A global black carbon aerosol model. *Journal of Geophysical Research,* **101,** 19395–19409.

Crutzen, P.J., 1976: The possible importance of OCS for the sulfate layer of the stratosphere. *Geophysical Research Letters,* **3,** 73–76.

Cruz, C.N. and S.N. Pandis, 1997: A study of the ability of pure secondary organic aerosol to act as cloud condensation nuclei. *Atmospheric Environment,* **31,** 2205–2214.

Cumpsty, N., 1997: *Jet Propulsion.* Cambridge University Press, Cambridge, United Kingdom, and New York, NY, USA, 281 pp.

Curtius, J., B. Sierau, F. Arnold, R. Baumann, R. Busen, P. Schulte, and U. Schumann, 1998: First direct sulfuric acid detection in the exhaust plume of a jet aircraft in flight. *Geophysical Research Letters,* **25,** 923–926.

Danilin, M.Y., A. Ebel, H. Elbern, and H. Petry, 1994: Evolution of the concentrations of trace species in an aircraft plume: trajectory study. *Journal of Geophysical Research,* **99,** 18951–18972.

Danilin, M.Y., J.M. Rodriguez, M.K.W. Ko, D.K. Weisenstein, R.C. Brown, R.C. Miake-Lye, and M.R. Anderson, 1997: Aerosol particle evolution in an aircraft wake: implications for the high speed civil transport fleet impact on ozone. *Journal of Geophysical Research,* **102,** 21453–21463.

Danilin, M.Y., D.W. Fahey, U. Schumann, M.J. Prather, J.E. Penner, M.K.W. Ko, D.K. Weisenstein, C.H. Jackman, G. Pitari, I. Köhler, R. Sausen, C.J. Weaver, A.R. Douglass, P.S. Connell, D.E. Kinnison, F.J. Dentener, E.L. Fleming, T.K. Berntsen, I.S.A. Isaksen, J.M. Haywood, and B. Kärcher, 1998: Aviation fuel tracer simulation: model intercomparison and implications. *Geophysical Research Letters,* **25,** 3947–3950.

DeGrand, J.Q., A.M. Carleton, and P.J. Lamb, 1990: A mid-season climatology of jet condensation trails from high-resolution satellite data. In: *Proceedings of the Seventh Conference on Atmospheric Radiation, July 23–27, 1990, San Francisco, CA.* American Meteorological Society, Boston, MA, USA, 309–311.

Del Negro, L.A., D.W. Fahey, S.G. Donnelly, R.S. Gao, E.R. Keim, R.C. Wamsley, E.L. Woodbridge, J.E. Dye, D. Baumgardner, B.W. Gandrud, J.C. Wilson, H.H. Jonsson, M. Loewenstein, J.R. Podolske, C.R. Webster, R.D. May, D.R. Worsnop, A. Tabazadeh, M.A. Tolbert, K.K. Kelly, and K.R. Chan, 1997: Evaluating the role of NAT, NAD, and liquid $H_2SO_4/H_2O/HNO_3$ solutions in Antarctic polar stratospheric cloud aerosol: observations and implications. *Journal of Geophysical Research,* **102,** 13255–13282.

DeMott, P.J., 1990: An explanatory study of ice nucleation by soot aerosols. *Journal of Applied Meteorology,* **19,** 1072–1079.

DeMott, P.J., D.C. Rogers, and S.M. Kreidenweis, 1997: The susceptibility of ice formation in upper tropospheric clouds to insoluble aerosol components. *Journal of Geophysical Research,* **102,** 19575–19584.

DeMott, P.J., D.C. Rogers, S.M. Kreidenweis, Y. Chen, C.H. Twohy, D. Baumgardner, A.J. Heymsfield, and K.R. Chan, 1998: The role of heterogeneous freezing nucleation in upper tropospheric clouds: inferences from SUCCESS. *Geophysical Research Letters,* **25,** 1387–1390.

Detwiler, A., 1983: Effects of artificial and natural cirrus clouds on temperatures near the ground. *Journal of Weather Modification,* **15,** 45–55.

Detwiler, A. and R. Pratt, 1984: Clear-air seeding: opportunities and strategies. *Journal of Weather Modification,* **16,** 46–60.

Dibb, J.E., R.W. Talbot, and M.B. Loomis, 1998: Tropospheric sulfate distribution during SUCCESS: contributions from jet exhaust and surface sources. *Geophysical Research Letters,* **25,** 1375–1378.

Diehl, J. and S.K. Mitra, 1998: A laboratory study of the effects of a kerosene burner exhaust on ice nucleation and the evaporation rate of ice crystals. *Atmospheric Environment,* **32,** 3145–3151.

Döpelheuer, A., 1997: *Berechnung der Produkte unvollständiger Verbrennung aus Luftfahrttriebwerken.* IB-325-09-97, Deutsche Zentrum für Luft- und Raumfahrt, Cologne, Germany, 38 pp.

Dowling, D.R. and L.F. Radke, 1990: A summary of physical properties of cirrus clouds. *Journal of Applied Meteorology,* **29,** 970–978.

Duda, D.P. and J.D. Spinhirne, 1996: Split-window retrieval of particle size and optical depth in contrails located above horizontally inhomogeneous ice clouds. *Geophysical Research Letters,* **23,** 3711–3714.

Duda, D.P., J.D. Spinhirne, and W.D. Hart, 1998: Retrieval of contrail microphysical properties during SUCCESS by the split-window method. *Geophysical Research Letters,* **25,** 1149–1152.

Dürbeck, T. and T. Gerz, 1996: Dispersion of aircraft exhausts in the free atmosphere. *Journal of Geophysical Research,* **101,** 26007–26015.

Ebert, E.E. and J.A. Curry, 1992: A parameterization of ice cloud optical properties for climate models. *Journal of Geophysical Research,* **97,** 3831–3836.

Ellingson, R.G. and Y. Fouquart, 1990: *The Intercomparison of Radiation Codes in Climate Models.* WCRP-39, WMO/TD Report no. 371, World Meteorological Organization, Geneva, Switzerland, 49 pp.

Fabian, P. and B. Kärcher, 1997: The impact of aviation upon the atmosphere: an assessment of present knowledge, uncertainties, and research needs. *Physics and Chemistry of the Earth,* **22,** 503–598.

Fahey, D.W., S.R. Kawa, E.L. Woodbridge, P. Tin, J.C. Wilson, H.H. Jonsson, J.E. Dye, D. Baumgardner, S. Borrmann, D.W. Toohey, L.M. Avallone, M.H. Proffitt, J. Margitan, M. Loewenstein, J.R. Podolske, R.J. Salawitch, S.C. Wofsy, M.K.W. Ko, D.E. Anderson, M.R. Schoeberl, and K.R. Chan, 1993: *In situ* measurements constraining the role of sulphate aerosols in mid-latitude ozone depletion. *Nature,* **363,** 509–514.

Fahey, D.W., E.R. Keim, K.A. Boering, C.A. Brock, J.C. Wilson, S. Anthony, T.F. Hanisco, P.O. Wennberg, R.C. Miake-Lye, R.J. Salawitch, N. Louisnard, E.L. Woodbridge, R.S. Gao, S.G. Donnelly, R. Wamsley, L.A. DelNegro, B.C. Daube, S.C. Wofsy, C.R. Webster, R.D. May, K.K. Kelly, M. Lowenstein, J.R. Podolske, and K.R. Chan, 1995a: Emission measurements of the Concorde supersonic aircraft in the lower stratosphere. *Science,* **270,** 70–74.

Fahey, D.W., E.R. Keim, E.L. Woodbridge, R.S. Gao, K.A. Boering, B.C. Daube, S.C. Wofsy, R.P. Lohmann, E.J. Hintsa, A.E. Dessler, C.R. Webster, R.D. May, C.A. Brock, J.C. Wilson, R.C. Miake-Lye, R.C. Brown, J.M. Rodriguez, M. Lowenstein, M.H. Proffitt, R.M. Stimpfle, S.W. Bowen, and K.R. Chan, 1995b: *In situ* observations in aircraft exhaust plumes in the lower stratosphere at mid-latitudes. *Journal of Geophysical Research,* **100,** 3065–3074.

Fairbrother, D.H., D.J.D. Sullivan, and H.S. Johnston, 1997: Global thermodynamical atmospheric modeling: search for new heterogeneous reactions. *Journal of Physical Chemistry,* **101,** 7350–7358.

Feichter, J., E. Kjellström, H. Rodhe, F. Dentener, J. Lelieveld, and G.-J. Roelofs. 1996: Simulation of the tropospheric sulfur cycle in a global climate model. *Atmospheric Environment,* **30,** 1693–1708.

Fordyce, J.S. and D.W. Sheibley, 1975: Estimate of contribution of jet aircraft operation to trace element concentration at or near airports. *Journal of the Air Pollution Control Association,* **25,** 721–724.

Fortuin, J.P.F., R. van Dorland, W.M.F. Wauben, and H. Kelder, 1995: Greenhouse effects of aircraft emissions as calculated by a radiative transfer model. *Annales Geophysicae,* **13,** 413–418.

Fowler, L.D., D.A. Randall, and S.A. Rutledge, 1996: Liquid and ice cloud microphysics in the CSU general circulation model. Part I: Model description and microphysical processes. *Journal of Climate,* **9,** 489–529.

Frankel, D., K.N. Liou, S.C. Ou, D.P. Wylie, and P. Menzel, 1997: Observations of cirrus cloud extent and their impacts to climate. In: *Proceedings, Ninth Conference on Atmospheric Radiation, February 2–7, 1997, Long Beach, CA.* American Meteorological Society, Boston, MA, USA, Vol. 13.1, pp. 414–417.

Frenzel, A. and F. Arnold, 1994: Sulfuric acid cluster ion formation by jet engines: implications for sulfuric acid formation and nucleation. In: *Impact of Aircraft Emissions upon the Atmosphere* [Schumann, U. and D. Wurzel, (eds.)]. Proceedings of an international scientific colloquium, 18–20 April 1994, Cologne, Germany. DLR-Mitteilung 94-06, Deutsches Zentrum für Luft- und Raumfahrt, Oberpfaffenhoffen and Cologne, Germany, pp. 106–112.

Freudenthaler, V., F. Homburg, and H. Jäger, 1995: Contrail observations by ground-based scanning lidar: cross-sectional growth. *Geophysical Research Letters,* **22,** 3501–3504.

Freudenthaler, V., F. Homburg, and H. Jäger, 1996: Optical parameters of contrails from lidar measurements: linear depolarization. *Geophysical Research Letters,* **23,** 3715–3718.

Friedl, R.R. (ed.), 1997: *Atmospheric Effects of Subsonic Aircraft: Interim Assessment Report of the Advanced Subsonic Technology Program.* NASA Reference Publication 1400, National Aeronautics and Space Administration, Goddard Space Flight Center, Greenbelt, MD, USA, 168 pp.

Fu, Q. and K.N. Liou, 1993: Parameterization of the radiative properties of cirrus clouds. *Journal of Atmospheric Sciences,* **50,** 2008–2025.

Gao, R.S., B. Kärcher, E.R. Keim, and D.W. Fahey, 1998: Constraining the heterogeneous loss of O_3 on soot particles with observations in jet engine exhaust plumes. *Geophysical Research Letters,* **25,** 3323–3326.

Gayet, J.-F., G. Febvre, G. Brogniez, H. Chepfer, W. Renger, and P. Wendling, 1996: Microphysical and optical properties of cirrus and contrails: cloud field study on 13 October 1989. *Journal of Atmospheric Sciences,* **53,** 126–138.

Gayet, J.-F., F. Auriol, S. Oshchepkov, F. Schröder, C. Duroure, G. Febvre, J.-F. Fournol, O. Crépel, P. Personne, and D. Daugereon, 1998: *In situ* measurements of the scattering phase function of stratocumulus, contrails and cirrus. *Geophysical Research Letters,* **25,** 971–974.

Gerz, T., T. Dürbeck, and P. Konopka, 1998: Transport and effective diffusion of aircraft emissions. *Journal of Geophysical Research,* **103,** 25905–25913.

Gettelman, A., 1998: The evolution of aircraft emissions in the stratosphere. *Geophysical Research Letters,* **25,** 2129–2132.

Gierens, K., 1996: Numerical simulations of persistent contrails. *Journal of Atmospheric Sciences,* **53,** 3333–3348.

Gierens, K. and U. Schumann, 1996: Colors of contrails from fuels with different sulfur contents. *Journal of Geophysical Research,* **101,** 16731–16736.

Gierens, K. and J. Ström, 1998: A numerical study of aircraft wake induced ice cloud formation. *Journal of Atmospheric Sciences,* **55,** 3253–3263.

Gierens, K., U. Schumann, H.G.J. Smit, M. Helten, and G. Zangl, 1997: Determination of humidity and temperature fluctuations based on MOZAIC data and parameterisation of persistent contrail coverage for general circulation models. *Annales Geophysicae,* **15,** 1057–1066.

Gierens, K., R. Sausen, and U. Schumann, 1998: A diagnostic study of the global distribution of contrails. Part II: Future air traffic scenarios. *Theoretical and Applied Climatology,* (in press).

Gierens, K., U. Schumann, M. Helten, H. Smit, and A. Marenco, 1999: A distribution law for relative humidity in the upper troposphere and lower stratosphere derived from three years of MOZAIC measurements. *Annales Geophysicae,* (in press).

Gleitsmann, G. and R. Zellner, 1998a: The effects of ambient temperature and relative humidity on particle formation in the jet regime of commercial aircrafts: a modelling study. *Atmospheric Environment,* **32,** 3079–3087.

Gleitsmann, G. and R. Zellner, 1998b: A modelling study of the formation of cloud condensation nuclei in the jet regime of aircraft plumes. *Journal of Geophysical Research,* **103,** 19543–19556.

Goldberg, E.D., 1985: *Black Carbon in the Environment.* Wiley-Interscience, New York, NY, USA, 198 pp.

Goodman, J., R.F. Pueschel, E.J. Jensen, S. Verma, G.V. Ferry, S.D. Howard, S.A. Kinne, and D. Baumgardner, 1998: Shape and size of contrail ice particles. *Geophysical Research Letters,* **25,** 1327–1330.

Grassl, H., 1970: Determination of cloud drop size distributions from spectral transmission measurements. *Beiträge zur Physik der Atmosphäre,* **43,** 255–284.

Grassl, H., 1990: Possible climatic effects of contrails and additional water vapour. In: *Air Traffic and the Environment—Background, Tendencies, and Potential Global Atmospheric Effects* [Schumann, U. (ed.)]. Springer-Verlag, Heidelberg, Germany, pp. 124–137.

Hagen, D.E., M.B. Trueblood, and P.D. Whitefield, 1992: A field sampling of jet exhaust aerosols. *Particle Science Technology,* **10,** 53–63.

Hagen, D.E., P. Whitefield, J. Paladino, M. Trueblood, and H. Lilenfeld, 1998: Particulate sizing and emission indices for a jet engine exhaust sampled at cruise. *Geophysical Research Letters,* **25,** 1681–1684.

Hahn, C.J., S.G. Warren, and J. London, 1994: *Climatological Data for Clouds Over the Globe from Surface Observations, 1982–1991: The Total Cloud Edition.* Report no. NDP026A, Carbon Dioxide Information Analysis Center, Oak Ridge National Laboratory, Oak Ridge, TN, USA, 42 pp. [Also available from Data Support Section, National Center for Atmospheric Research, Boulder, CO, USA.]

Hahn, C. J., S.G. Warren, and J. London, 1995: The effect of moonlight on observation of cloud cover at night, and application to cloud climatology. *Journal of Climate,* **8,** 1429–1446.

Hahn, C. J., S.G. Warren, and J. London, 1996: *Edited Synoptic Cloud Reports from Ships and Land Stations over the Globe, 1982–1991.* Report no. NDP026B, Carbon Dioxide Information Analysis Center, Oak Ridge National Laboratory, Oak Ridge, TN, USA, 45 pp.

Halpert, M.S. and G.D. Bell, 1997: Climate assessment for 1996. *Bulletin of the American Meteorological Society,* **78,** S1–S49.

Hamill, P., E.J. Jensen, P.B. Russell, and J.J. Bauman, 1997: The life cycle of stratospheric aerosol particles. *Bulletin of the American Meteorological Society,* **78,** 1395–1410.

Hanisco, T.F., P.O. Wennberg, R.C. Cohen, J.G. Anderson, D.W. Fahey, E.R. Keim, R.S. Gao, R.C. Wamsley, S.G. Donnelly, L.A. DelNegro, R.S. Salawitch, K.K. Kelly, and M.H. Proffitt, 1997: The role of HO_x in super- and subsonic aircraft exhaust plumes. *Geophysical Research Letters,* **24,** 65–68.

Hannegan, B., S. Olsen, M. Prather, X. Zhu, D. Rind, and J. Lerner, 1998: The dry stratosphere: a limit on cometary influx. *Geophysical Research Letters,* **25,** 1649–1652.

Hansen, J.E., M. Sato, and R. Ruedy, 1995: Long-term changes in diurnal temperature range: implications about mechanisms of global climate change. *Atmospheric Research,* **37,** 175–209.

Hansen, J., R. Ruedy, M. Sato, and R. Reynolds, 1996: Global surface air temperature in 1995: return to pre-Pinatubo level. *Geophysical Research Letters,* **23,** 1665–1668.

Hansen, J., M. Sato, and R. Ruedy, 1997: Radiative forcing and climate response. *Journal of Geophysical Research,* **102,** 6831–6864.

Hasselmann, K., 1997: Are we seeing global warming? *Science,* **276,** 914–915.

Hauf, T. and R. Alheit, 1997: Transport of pollutants by gravitational settling of ice crystals. In: *Pollutants from Air Traffic – Results of Atmospheric Research 1992–1997* [Schumann, U., A. Chlond, A. Ebel, B. Kärcher, H. Pak, H. Schlager, A. Schmitt, and P. Wendling (eds.)]. DLR-Mitteilung 97-04, Deutsches Zentrum für Luft- und Raumfahrt, Oberpfaffenhofen and Cologne, Germany, pp. 197–206.

Haywood, J.M. and V. Ramaswamy, 1998: Global sensitivity studies of the direct radiative forcing due to anthropogenic sulfate and black carbon aerosols. *Journal of Geophysical Research*, **103**, 6043–6058.

Helten, M., H. G. J. Smit, W. Sträter, D. Kley, P. Nedelec, M. Zöger, and R. Busen, 1998: Calibration and performance of automatic compact instrumentation for the measurement of relative humidity from passenger aircraft. *Journal of Geophysical Research*, **103**, 25643–25652.

Henderson-Sellers, A., 1989: North American total cloud amount variations this century. *Paleogeography, Paleoclimatology, and Paleoecology*, **75**, 175–194.

Henderson-Sellers, A., 1992: Continental cloudiness changes this century. *Geo Journal*, **27**, 255–262.

Heymsfield, A.J. 1993: Microphysical structures of stratiform and cirrus clouds. In: *Aerosol-Cloud-Climate Interactions* [Hobbs, P.V. (ed.)]. Academic Press, San Diego, CA, USA, pp. 97–121.

Heymsfield, A.J. and L.M. Miloshevich, 1995: Relative humidity and temperature influences on cirrus formation and evolution: observations from wave clouds and FIRE II. *Journal of Atmospheric Sciences*, **52**, 4302–4326.

Heymsfield, A.J., R.P. Lawson, and G.W. Sachse, 1998a: Growth of ice crystals in a precipitating contrail. *Geophysical Research Letters*, **25**, 1335–1338.

Heymsfield, A.J., L.M. Miloshevich, C. Twohy, G. Sachse, and S. Oltmans, 1998b: Upper tropospheric relative humidity observations and implications for cirrus ice nucleation. *Geophysical Research Letters*, **25**, 1343–1346.

Hitchman, M., M. McKay, and C.R. Trepte, 1994: A climatology of stratospheric aerosol. *Journal of Geophysical Research*, **99**, 20689–20700.

Hofmann, D.J., 1990: Increase of the stratospheric background sulfuric acid aerosol mass in the past 10 years. *Science*, **248**, 996–1000.

Hofmann, D.J., 1991: Aircraft sulfur emissions. *Nature*, **349**, 659.

Hofmann, D.J., 1993: Twenty years of balloon-borne tropospheric aerosol measurements at Laramie, Wyoming. *Journal of Geophysical Research*, **98**, 12753–12766.

Hofmann, D.J. and J.M. Rosen, 1978: Balloon observations of a particle layer injected by stratospheric aircraft at 23 km. *Geophysical Research Letters*, **5**, 511–514.

Hofmann, D.J. and S. Solomon, 1989: Ozone destruction through heterogeneous chemistry following the eruption of El Chichon. *Journal of Geophysical Research*, **94**, 5029–5041.

Hofmann, D.J., R.S. Stone, M.E. Wood, T. Deshler, and J.M. Harris, 1998: An analysis of 25 years of balloon-borne aerosol data in search of a signature of the subsonic commercial aircraft fleet. *Geophysical Research Letters*, **25**, 2433–2436.

Höhndorf, F., 1941: *Beitrag zum Problem der Vermeidung von Auspuffwolken hinter Motorflugzeugen.* Forschungsbericht no. 1371, Deutsche Luftforschung, Aerologisches Institut, Deutsche Forschungsanstalt für Segelflug e.V., 15 pp. [Also available from Air Documents Division, T-2, AMC, Wright Field, Ohio, KA, USA, Microfilm no. R 2317 F 834.].

Hoinka, K.P., 1998: Statistics of the global tropopause pressure. *Monthly Weather Review*, **126**, 3303–3325.

Hoinka, K.-P., M.-E. Reinhardt, and W. Metz, 1993: North Atlantic air traffic within the lower stratosphere: cruising times and corresponding emissions. *Journal of Geophysical Research*, **98**, 23113–23131.

Howard, R.P., R.S. Hiers, P.D. Whitefield, D.E. Hagen, J.C. Wormhoudt, R.C. Miake-Lye, and R. Strange, 1996: *Experimental Characterization of Gas Turbine Emissions at Simulated Flight Altitude Conditions.* AEDC-TR-96-3, Arnold Engineering Development Center, National Technical Information Service, Arnold Air Force Base, TN, USA, 159 pp.

Hunter, S.C., 1982: Formation of SO_3 in gas turbines. *Transactions of the ASME Journal of Engineering Power*, **104**, 44–51.

Hurrell, J., 1995: Decadal trends in the North Atlantic oscillation: regional temperatures and precipitation. *Science*, **269**, 676–679.

IPCC, 1996: *Climate Change 1995: The Science of Climate Change. Contribution of Working Group I to the Second Assessment Report of the Intergovernmental Panel on Climate Change* [Houghton, J.T., L.G. Meira Filho, B.A. Callander, N. Harris, A. Kattenberg, and K. Maskell (eds.)]. Cambridge University Press, Cambridge, United Kingdom and New York, NY, USA, 572 pp.

Jackman, C.H., E.L. Fleming, S. Chandra, D.B. Considine, and J.E. Rosenfield, 1996: Past, present, and future modeled ozone trends with comparisons to observed trends. *Journal of Geophysical Research*, **101**, 28753–28767.

Jacob, D.J. and M.R. Hoffmann, 1983: A dynamic model for the production of H^+, NO_3^-, and SO_4^{2-} in urban smog. *Journal of Geophysical Research*, **88**, 6611–6621.

Jäger, H. and D. Hofmann, 1991: Mid-latitude lidar backscatter to mass, area, and extinction conversion model based on *in situ* aerosol measurements from 1980 to 1997. *Applied Optics*, **30**, 127–138.

Jäger, H., V. Freudenthaler, and F. Homburg, 1998: Remote sensing of optical depth of aerosols and clouds cover related to air traffic. *Atmospheric Environment*, **32**, 3123–3127.

Jensen, E.J. and O.B. Toon, 1992: The potential effects of volcanic aerosols on cirrus cloud microphysics. *Geophysical Research Letters*, **19**, 1759–1762.

Jensen, E.J. and O.B. Toon, 1994: Ice nucleation in the upper troposphere: sensitivity to aerosol number density, temperature, and cooling rate. *Geophysical Research Letters*, **21**, 2019–2022.

Jensen, E.J. and O.B. Toon, 1997: The potential impact of soot particles from aircraft exhaust on cirrus clouds. *Geophysical Research Letters*, **24**, 249–252.

Jensen, E.J., S. Kinne, and O.B. Toon, 1994a: Tropical cirrus cloud radiative forcing: sensitivity studies. *Geophysical Research Letters*, **21**, 2023–2026.

Jensen, E.J., O.B. Toon, D.L. Westphal, S. Kinne, and A.J. Heymsfield, 1994b: Microphysical modeling of cirrus. 1. Comparison with 1986 FIRE IFO measurements. *Journal of Geophysical Research*, **99**, 10421–10442.

Jensen, E.J., O.B. Toon, S. Kinne, G.W. Sachse, B.E. Anderson, K.R. Chan, C.H. Twohy, B. Gandrud, A. Heymsfield, and R.C. Miake-Lye, 1998a: Environmental conditions required for contrail formation and persistence. *Journal of Geophysical Research*, **103**, 3929–3936.

Jensen, E.J., O.B. Toon, R.F. Pueschel, J. Goodman, G.W. Sachse, B.E. Anderson, K.R. Chan, D. Baumgardner, and R.C. Miake-Lye, 1998b: Ice crystal nucleation and growth in contrails forming at low ambient temperatures. *Geophysical Research Letters*, **25**, 1371–1374.

Jensen, E.J., O.B. Toon, A. Tabazadeh, G.W. Sachse, B.E. Anderson, K.R. Chan, C.W. Twohy, B. Gandrud, S.M. Aulenbach, A.J. Heymsfield, J. Hallett, and B. Gary, 1998c: Ice nucleation processes in upper tropospheric wave-clouds observed during SUCCESS. *Geophysical Research Letters*, **25**, 1363–1366.

Jensen, E.J., A.S. Ackerman, D.E. Stevens, O.B. Toon, and P. Minnis, 1998d: Spreading and growth of contrails in a sheared environment. *Journal of Geophysical Research*, **103**, 31557–31567.

Joseph, J.H., Z. Levin, Y. Mekler, G. Ohring, and J. Otterman, 1975: Study of contrails observed from the ERST 1 satellite imagery. *Journal of Geophysical Research*, **80**, 366–372.

Kapala, A., H. Mächel, and H. Flohn, 1998: Behaviour of the centres of action above the Atlantic since 1881. Part II: Associations with regional climate anomalies. *International Journal of Climatology*, **18**, 23–36.

Kärcher, B., 1996: Aircraft-generated aerosols and visible contrails. *Geophysical Research Letters*, **23**, 1933–1936.

Kärcher, B., 1997: Heterogeneous chemistry in aircraft wakes: constraints for uptake coefficients. *Journal of Geophysical Research*, **102**, 19119–19135.

Kärcher, B., 1998a: Physicochemistry of aircraft-generated liquid aerosols, soot, and ice particles. 1. Model description. *Journal of Geophysical Research*, **103**, 17111–17128.

Kärcher, B., 1998b: On the potential importance of sulfur-induced activation of soot particles in nascent jet aircraft exhaust plumes. *Atmospheric Research*, **46**, 293–305.

Kärcher, B. and D.W. Fahey, 1997: The role of sulfur emissions in volatile particle formation in jet aircraft exhaust plumes. *Geophysical Research Letters*, **24**, 389–392.

Kärcher, B., T. Peter, and R. Ottmann, 1995: Contrail formation: homogeneous nucleation of H_2SO_4/H_2O droplets. *Geophysical Research Letters*, **22**, 1501–1504.

Kärcher, B., M.M. Hirschberg, and P. Fabian, 1996a: Small-scale chemical evolution of aircraft exhaust species at cruising altitude. *Journal of Geophysical Research*, **101**, 15169–15190.

Kärcher, B., T. Peter, U.M. Biermann, and U. Schumann, 1996b: The initial composition of jet condensation trails. *Journal of Atmospheric Sciences*, **53**, 3066–3083.

Kärcher, B., R. Busen, A. Petzold, F.P. Schröder, U. Schumann, and E.J. Jensen, 1998a: Physicochemistry of aircraft-generated liquid aerosols, soot, and ice particles. 2. Comparison with observations and sensitivity studies. *Journal of Geophysical Research*, **103**, 17129–17148.

Kärcher, B., F. Yu, F.P. Schröder, and R.P. Turco, 1998b: Ultrafine aerosol particles in aircraft plumes: analysis of growth mechanisms. *Geophysical Research Letters,* **25,** 2793–2796.

Karl, T.R., G. Kukla, and J. Gavin, 1984: Decreasing diurnal temperature range in the United States and Canada from 1941 through 1980. *Journal of Climate and Applied Meteorology,* **23,** 1489–1504.

Karl, T.R., P.D. Jones, R.W. Knight, G. Kukla, N. Plummer, V. Razuvayev, K.P. Gallo, J. Lindseay, R.J. Charlson, and T.C. Peterson, 1993: A new perspective on recent global warming: asymmetric trends of daily maximum and minimum temperature. *Bulletin of the American Meteorological Society,* **74,** 1007–1023.

Kästner, M., K.T. Kriebel, R. Meerkötter, W. Renger, G.H. Ruppersberg, and P. Wendling, 1993: Comparison of cirrus height and optical depth derived from satellite and aircraft measurements. *Monthly Weather Review,* **121,** 2708–2717.

Katz, J.L., F.C. Wen, T. McLaughlin, R.J. Reusch, and R. Partch, 1977: Nucleation on photoexcited molecules. *Science,* **196,** 1203–1205.

Keil, D.G., R.J. Gill, D.B. Olson, and H.F. Calcote, 1984: Ion concentration in premixed acetylene-oxygen flames near soot threshold. In: *The Chemistry of Combustion Processes* [Sloane, T.M. (ed.)]. American Chemical Society, Washington, DC, USA, pp. 33–43.

Kent, G.S., P.H. Wang, M.P. McCormick, and K.M. Skeens, 1995: Multiyear Stratospheric Aerosol and Gas Experiment II measurements of upper tropospheric aerosol characteristics. *Journal of Geophysical Research,* **100,** 13875–13899.

Kent, G.S., C.R. Trepte, and P.L. Lucker, 1998: Long-term Stratospheric Aerosol and Gas Experiment I and II measurements of upper tropospheric aerosol extinction. *Journal of Geophysical Research,* **103,** 28863–28874.

King, M.D., Y.J. Kaufman, W.P. Menzel, and D. Tanre, 1992: Remote sensing of cloud, aerosol, and water vapor properties from the moderate resolution imaging spectrometer (MODIS). *IEEE Transactions on Geoscience and Remote Sensing,* **30,** 2–27.

Kinne, S., and K.-N. Liou, 1989: The effects of the nonsphericity and size distribution of ice crystals on the radiative properties of cirrus clouds. *Atmospheric Research,* **24,** 273–284.

Kinne, S., T.P. Ackerman, M. Shiobara, A. Uchiyama, A.J. Heymsfield, L. Miloshevich, J. Wendell, E.W. Eloranta, C. Purgold, and R.W. Bergstrom, 1997: Cirrus cloud radiative and microphysical properties from ground observations and *in situ* measurements during FIRE 1991 and their application to exhibit problems in cirrus cloud radiative transfer modeling. *Journal of Atmospheric Science,* **54,** 2320–2344.

Kinnison, D.E., K.E. Grant, P.S. Connell, D.A. Rotman, and D.J. Wuebbles, 1994: The chemical and radiative effects of the Mt. Pinatubo eruption. *Journal of Geophysical Research,* **99,** 25705–25731.

Kjellström, E., J. Feichter, R. Sausen, and R. Hein, 1998: The contribution of aircraft emissions to the atmospheric sulfur budget. *Atmospheric Environment,* (in press). [Also available as Report no. CM-93, Department of Meteorology, University of Stockholm, Sweden.]

Klein, S.A. and D.L. Hartmann, 1993: Spurious changes in the ISCCP data set. *Geophysical Research Letters,* **20,** 455–458.

Knollenberg, R.G., 1972: Measurements of the growth of ice budget in a persisting contrail. *Journal of Atmospheric Sciences,* **29,** 1367–1374.

Kolb, C.E., J.T. Jayne, D.R. Worsnop, M.J. Molina, R.F. Meads, and A.A. Viggiano, 1994: Gas phase reaction of sulfur trioxide with water vapor. *Journal of the American Chemical Society,* **116,** 10314–10315.

Konopka, P. and W. Vogelsberger, 1997: Köhler equation for finite systems: a simple estimation of possible condensation mechanisms in aircraft contrails. *Journal of Geophysical Research,* **102,** 16057–16064.

Konrad, T.G. and J.C. Howard, 1974: Multiple contrail streamers observed by radar. *Journal of Applied Meteorology,* **13,** 563–572.

Kotzick, R., U. Panne, and R. Niessner, 1997: Changes in condensation properties of ultrafine carbon particles subjected to oxidation by ozone. *Journal of Aerosol Science,* **28,** 725–735.

Kuhn, P.M., 1970: Airborne observations of contrail effects on the thermal radiation budget. *Journal of Atmospheric Sciences,* **27,** 937–942.

Laaksonen, A., J. Hienola, M. Kulmala, and F. Arnold, 1997: Supercooled cirrus cloud formation modified by nitric acid perturbation of the upper troposphere. *Geophysical Research Letters,* **24,** 3009–3012.

Lacis, A.A. and M.I. Mishchenko, 1995: Climate forcing, climate sensitivity, and climate response: A radiative modeling perspective of atmospheric aerosols. In: *Aerosol Forcing of Climate* [Charlson, R.J. and J. Heintzenberg (eds.)]. J. Wiley, Chichester, United Kingdom, pp. 11–42.

Lacis, A., J. Hansen, and M. Sato, 1992: Climate forcing by stratospheric aerosols. *Geophysical Research Letters,* **19,** 1607–1610.

Langner, J. and H. Rodhe, 1991: A global three-dimensional model of the tropospheric sulfur cycle. *Journal of Atmospheric Chemistry,* **13,** 225–263.

Lary, D.J., A.M. Lee, R. Toumi, M. Newchurch, M. Pierre, and J.B. Renard, 1997: Carbon aerosols and atmospheric photochemistry. *Journal of Geophysical Research,* **102,** 3671–3682.

Lawson, R.P., A.J. Heymsfield, S.M. Aulenbach, and T.L. Jensen, 1998: Shapes, sizes and light scattering properties of ice crystals in cirrus and a persistent contrail during SUCCESS. *Geophysical Research Letters,* **25,** 1331–1334.

Lee, T.F., 1989: Jet contrail identification using the AVHRR infrared split window. *Journal of Applied Meteorology,* **28,** 993–995.

Lewellen, D.C. and W.S. Lewellen, 1996: Large eddy simulations of the vortex-pair breakup in aircraft wakes. *AIAA Journal,* **34,** 2337–2345.

Liepert, B.G., 1997: Recent changes in solar radiation under cloudy conditions. *International Journal of Climatology,* **17,** 1581–1593.

Liepert, B., P. Fabian, and H. Grassl, 1994: Solar radiation in Germany—observed trends and an assessment of their causes. Part I: Regional approach. *Beiträge zur Physik der Atmosphäre,* **67,** 15–29.

Liou, K., 1986: Influence of cirrus clouds on weather and climate processes: a global perspective. *Monthly Weather Review,* **114,** 1167–1199.

Liou, K.-N., S.C. Ou, and G. Koenig, 1990: An investigation of the climatic effect of contrail cirrus. In: *Air Traffic and the Environment: Background, Tendencies, and Potential Global Atmospheric Effects* [Schumann, U. (ed.)]. Springer-Verlag, Berlin, Germany, pp. 154–169.

Liousse, C., J.E. Penner, C. Chuang, J.J. Walton, H. Eddleman, and H. Cachier, 1996: A global three-dimensional model study of carbonaceous aerosols. *Journal of Geophysical Research,* **101,** 19411–19432.

Lohmann, U. and E. Roeckner, 1995: Influence of cirrus cloud radiative forcing on climate and climate sensitivity in a general circulation model. *Journal of Geophysical Research,* **100,** 16305–16323.

Lovejoy, E.R., D.R. Hanson, and L.G. Huey, 1996: Kinetics and products of the gas phase reaction of SO_3 with water. *Journal of Physical Chemistry,* **100,** 19911–19916.

Ludlam, F.H, 1980: *Clouds and Storms.* The Pennsylvania State University Press, University Park, PA, USA.

Lukachko, S.P., I.A. Waitz, R.C. Miake-Lye, R.C. Brown, and M.R. Anderson, 1998: Production of sulfate aerosol precursors in the turbine and exhaust nozzle of an aircraft engine. *Journal of Geophysical Research,* **103,** 16159–16174.

Mächel, H., A. Kapala, and H. Flohn, 1998: Behaviour of the centres of action above the Atlantic since 1881. Part I: Characteristics of seasonal and interannual variability. *International Journal of Climatology,* **18,** 1–22.

Mannstein, H., 1997: Contrail observations from space using NOAA-AVHRR data. In: *Impact of Aircraft upon the Atmosphere.* Proceedings of an international scientific colloquium, 15–18 October 1996, Paris, France, Office National d'Etudes et de Recherches Aerospatiales, Chatillon, France, Vol. II, pp. 427–431.

Mannstein, H., R. Meyer, and P. Wendling, 1999: Operational detection of contrails from NOAA-AVHRR-data. *International Journal of Remote Sensing.* (in press).

Mazin, I.P., 1996: Aircraft condensation trails. *Izvestiya, Atmospheric, and Oceanic Physics,* **32,** 1–13.

McClatchey, R.A.M., R.W. Fenn, J.E.A. Selby, F.E. Volz, and J.S. Garing, 1972: *Optical Properties of the Atmosphere.* AFCRL-72-0497, Air Force Cambridge Research Laboratory, Bedford, MA, USA, 3rd ed., 113 pp.

Meerkötter, R., U. Schumann, D.R. Doelling, P. Minnis, T. Nakajima, and Y. Tsushima, 1999: Radiative forcing by contrails. *Annales Geophysicae,* (in press).

Miake-Lye, R.C., M. Martinez-Sanchez, R.C. Brown, and C.E. Kolb, 1993: Plume and wake dynamics, mixing and chemistry behind an HSCT aircraft. *Journal of Aircraft,* **30,** 467–479.

Miake-Lye, R.C., R.C. Brown, M.R. Anderson, and C.E. Kolb, 1994: Calculation of condensation and chemistry in an aircraft contrail. In: *Impact of Emissions from Aircraft and Spacecraft upon the Atmosphere* [Schumann, U. and D. Wurzel (eds.)]. Proceedings of an international scientific colloquium, 18–20 April 1994, Cologne, Germany. DLR-Mitteilung 94-06, Deutsches Zentrum für Luft- und Raumfahrt, Oberpfaffenhofen and Cologne, Germany, pp. 274–279.

Miake-Lye, R.C., B.E. Anderson, W.R. Cofer, H.A. Wallio, G.D. Nowicki, J.O. Ballenthin, D.E. Hunton, W.B. Knighton, T.M. Miller, J.V. Seeley, and A.A. Viggiano, 1998: SO_x oxidation and volatile aerosol in aircraft exhaust plumes depend on fuel sulfur content. *Geophysical Research Letters*, **25**, 1677–1680.

Minnis, P., E.F. Harrison, L.L. Stowe, G.G. Gibson, F.M. Denn, D.R. Doelling, and W.L. Smith, Jr., 1993: Radiative climate forcing by the Mount Pinatubo eruption. *Science*, **259**, 1411–1415.

Minnis, P., J.K. Ayers, and S.P. Weaver, 1997: *Surface-Based Observations of Contrail Occurrence Frequency Over the U.S., April 1993–April 1994*. NASA Reference Publication 1404, National Aeronautics and Space Administration, Hampton, VA, USA, 79 pp.

Minnis, P., D.F. Young, D.P. Garber, L. Nguyen, W.L. Smith, Jr., and R. Palikonda, 1998a: Transformation of contrails into cirrus during SUCCESS. *Geophysical Research Letters*, **25**, 1157–1160.

Minnis, P., U. Schumann, D.R. Doelhing, J.K. Ayers, R. Palikonda, L. Nguyen, D.F. Young, and K.M. Gierens, 1998b: Contrails: another factor in climate change? *Science*, (under revision).

Minnis, P., D.P. Garber, D.F. Young, R.F. Arduini, and Y. Takano, 1998c: Parameterizations of reflectance and effective emittance for satellite remote sensing of cloud properties. *Journal of Atmospheric Science*, **55**, 3313–3339.

Minnis, P., U. Schumann, D.R. Doelling, K.M. Gierens, and D.W. Fahey, 1999: Global distribution of contrail radiative forcing. *Geophysical Research Letters*, (in press).

Mirabel, P. and J.L. Katz, 1974: Binary homogeneous nucleation as a mechanism for the formation of aerosols. *Journal of Chemical Physics*, **60**, 1138–1144.

Murcray, W.B., 1970: On the possibility of weather modification by aircraft contrails. *Monthly Weather Review*, **98**, 745–748.

Murphy, D.M., K.K. Kelly, A.F. Tuck, and M.H. Proffitt, 1990: Ice saturation at the tropopause observed from the ER-2 aircraft. *Geophysical Research Letters*, **17**, 353–356.

Nakajima, T. and M. Tanaka, 1986: Matrix formulations for the transfer of solar radiation in a plane-parallel scattering atmosphere. *Journal of Quantum Spectroscopic Radiation Transfer*, **35**, 13–21.

Nakajima, T., and M. Tanaka, 1988: Algorithms for radiative intensity calculations in moderately thick atmospheres using a truncation approximation. *Journal of Quantum Spectroscopic Radiation Transfer*, **40**, 51–69.

Oltmans, S.J. and D.J. Hofmann, 1995: Increase in lower-stratospheric water vapour at mid-latitude Northern Hemisphere site from 1981 to 1994. *Nature*, **374**, 146–149.

Ovarlez, J., H. Ovarlez, R.M. Philippe, and E. Landais, 1997: Water vapor measurements during the POLINAT campaigns. In: *Pollution from Aircraft Emissions in the North Atlantic Flight Corridor (POLINAT)* [Schumann, U. (ed.)]. EUR-16978-EN, Office for Publications of European Communities, Brussels, Belgium, pp. 70–81.

Paladino, J., P. Whitefield, D. Hagen, A.R. Hopkins, and M. Trueblood, 1998: Particle concentrations characterization for jet engine emissions under cruise conditions. *Geophysical Research Letters*, **25**, 1697–1700.

Parker, D.E., M. Gordon, D.P.N. Cullum, D.M.H. Sexton, C.K. Folland, and N. Rayner, 1997: A new gridded radiosonde temperature data base and recent temperature trends. *Geophysical Research Letters*, **24**, 1499–1502.

Penner, J.E., C.S. Atherton, J. Dignon, S.J. Ghan, J.J. Walton, and S. Hameed, 1991: Tropospheric nitrogen: a three-dimensional study of sources, distribution, and deposition. *Journal of Geophysical Research*, **96**, 959–990.

Penner, J.E., R.E. Dickinson, and C.A. O'Neil, 1992: Effects of aerosol from biomass burning on the global radiation budget. *Science*, **256**, 1432–1434.

Penner, J.E., C.E. Atherton, and T. Graedel, 1994: Global emissions and models of photochemically active compounds. In: *Global Atmospheric Biospheric Chemistry* [Prinn, R.G. (ed.)]. Plenum, New York, NY, USA, pp. 223–248.

Penner, J.E., 1995: Carbonaceous aerosols influencing atmospheric radiation: black and organic carbon. In: *Aerosol Forcing of Climate* [Charlson, R.J. and J. Heintzenberg (eds.)]. J. Wiley, Chichester, United Kingdom, pp. 91–108.

Peter, T., 1997: Microphysics and heterogeneous chemistry of polar stratospheric clouds. *Annual Review of Physical Chemistry*, **48**, 785–822.

Peter, T., C. Brühl, and P.J. Crutzen, 1991: Increase in the PSC-formation probability caused by high-flying aircraft. *Geophysical Research Letters*, **18**, 1465–1468.

Petzold, A. and F.P. Schröder, 1998: Jet engine exhaust aerosol characterization. *Aerosol Science Technology*, **28**, 62–76. Correction in *Aerosol Science Technology*, **29**, 355–356.

Petzold, A., R. Busen, F.P. Schröder, R. Baumann, M. Kuhn, J. Ström, D.E. Hagen, P.D. Whitefield, D. Baumgardner, F. Arnold, S. Borrmann, and U. Schumann, 1997: Near field measurements on contrail properties from fuels with different sulfur content. *Journal of Geophysical Research*, **102**, 29867–29881.

Petzold, A., J. Ström, S. Ohlsson, and F.P. Schröder, 1998: Elemental composition and morphology of ice-crystal residual particles in cirrus clouds and contrails. *Atmospheric Research*, **49**, 21–34.

Petzold, A., J. Ström, F.P. Schröder, and B. Kärcher, 1999: Carbonaceous aerosol in jet engine exhaust: emission characteristics and implications for heterogeneous chemistry. *Atmospheric Environment*, (in press).

Pham, M., J.-F. Müller, G.P. Brasseur, C. Granier, and G. Mégie, 1996: A 3-D model study of the global sulfur cycle: contributions of anthropogenic and biogenic sources. *Atmospheric Environment*, **30**, 1815–1822.

Pitari, G., 1993: A numerical study of the possible perturbation of stratospheric dynamics due to Pinatubo aerosols: implications for tracer transport. *Journal of Atmospheric Sciences*, **50**, 2443–2461.

Pitari, G., V. Rizzi, L. Ricciardulli, and G. Visconti, 1993: HSCT impact: the role of sulfate, NAT, and ice aerosols studied with a 2-D model including aerosol physics. *Journal of Geophysical Research*, **98**, 23141–23164.

Plantico, M.S., T.R. Karl, G. Kukla, and J. Gavin, 1990: Is recent climate change across the United States related to rising levels of anthropogenic greenhouse gases? *Journal of Geophysical Research*, **95**, 16617–16637.

Plass, G.N., G.W. Kattawar, and F.E. Catchings, 1973: Matrix-operator theory of radiative transfer. *Applied Optics*, **12**, 314–329.

Platt, C.M.R., 1981: The effect of cirrus of varying optical depth on the extraterrestrial net radiative flux. *Quarterly Journal of the Royal Meteorological Society*, **107**, 671–678.

Platt, C.M.R., 1997: A parameterization of the visible extinction coefficient of ice clouds in terms of ice water content. *Journal of Atmospheric Sciences*, **54**, 2083–2098.

Podzimek, J., D.E. Hagen, and E. Robb, 1995: Large aerosol particles in cirrus type clouds. *Atmospheric Research*, **38**, 263–282.

Ponater, M., S. Brinkop, R. Sausen, and U. Schumann, 1996: Simulating the global atmospheric response to aircraft water vapour emissions and contrails: a first approach using a GCM. *Annales Geophysicae*, **14**, 941–960.

Poole, L.R. and M.C. Pitts, 1994: Polar stratospheric cloud climatology based on Stratospheric Aerosol Measurement II observations from 1978 to 1989. *Journal of Geophysical Research*, **99**, 13083–13089.

Prather, M. and D.J. Jacob, 1997: A persistent imbalance in HO_x and NO_x photochemistry of the upper troposphere driven by deep tropical convection. *Geophysical Research Letters*, **24**, 3189–3192.

Pruppacher, H.R., 1995: A new look at homogeneous ice nucleation in super-cooled water drops. *Journal of Atmospheric Sciences*, **52**, 1924–1933.

Pruppacher, H.R. and J.D. Klett, 1997: *Microphysics of Clouds and Precipitation*. Kluwer, Dordrecht, The Netherlands, 954 pp.

Pueschel, R.F., 1996: Stratospheric aerosols: formation, properties, effects. *Journal of Aerosol Science*, **27**, 383–402.

Pueschel, R.F., D.F. Blake, K.G. Snetsinger, A.D.A. Hansen, S. Verma, and K. Kato, 1992: Black carbon (soot) aerosol in the lower stratosphere and upper troposphere. *Geophysical Research Letters*, **19**, 1659–1662.

Pueschel, R.F., K.A. Boering, S. Verma, S.D. Howard, G.V. Ferry, J. Goodman, D.A. Allen, and P. Hamill, 1997: Soot aerosol in the lower stratosphere: pole-to-pole variability and contribution by aircraft. *Journal of Geophysical Research*, **102**, 13113–13118.

Pueschel, R.F., S. Verma, G.V. Ferry, S.D. Howard, S. Vay, S.A. Kinne, J. Goodman, and A.W. Strawa, 1998: Sulfuric acid and soot particle formation in aircraft exhaust. *Geophysical Research Letters*, **25**, 1685–1588.

Quackenbush, T.R., M.E. Teske, and C.E. Polymeropoulos, 1994: A model for assessing fuel jettisoning effects. *Atmospheric Environment*, **28**, 2751–2759.

Rahmes, T.F., A.H. Omar, and D.J. Wuebbles, 1998: Atmospheric distributions of soot particles by current and future aircraft fleets and resulting radiative forcing on climate. *Journal of Geophysical Research*, **103**, 31657–31667.

Ramaswamy, V., M.D. Schwartzkopf, and W.J. Randel, 1996: Fingerprint of ozone depletion in the spatial and temporal pattern of recent lower-stratospheric cooling. *Nature,* **382,** 616–618.

Raschke, E., P. Flamant, Y. Fouquart, P. Hignett, H. Isaka, P.R. Jonas, H. Sundqvist, and P. Wendling, 1998: Cloud-radiation studies during the European Cloud and Radiation Experiment (EUCREX). *Surveys in Geophysics,* **19,** 89–138.

Rebetez, M. and M. Beniston, 1998: Changes in sunshine duration are correlated with changes in daily temperature range this century: an analysis of Swiss climatology data. *Geophysical Research Letters,* **25,** 3611–3613.

Reiner, T. and F. Arnold, 1993: Laboratory flow reactor measurements of the reaction $SO_3 + H_2O + M \rightarrow H_2SO_4 + M$: implications for gaseous H_2SO_4 and aerosol formation. *Geophysical Research Letters,* **20,** 2659–2662.

Rickey, J.E., 1995: *The Effect of Altitude Conditions on the Particle Emissions of a J85-GE-5L Turbojet Engine.* NASA-TM-106669, National Aeronautics and Space Administration, Lewis Research Center, Washington, DC, USA, 52 pp.

Roeckner, E., L. Bengtsson, J. Feichter, J. Lelieveld, and H. Rodhe, 1998: Transient climate change simulations with a coupled atmosphere-ocean GCM including the tropospheric sulfur cycle. Report no. 266, Max-Planck-Institut für Meteorologie, Hamburg, Germany, July 1998, 48 pp.

Rogaski, C.A., D.M. Golden, and L.R. Williams, 1997: Reactive uptake and hydration experiments on amorphous carbon treated with NO_2, SO_2, HNO_2 and H_2SO_4. *Geophysical Research Letters,* **24,** 381–384.

Rogers, D.C., P. DeMott, S.M. Kreidenweis, and Y. Chen, 1998: Measurements of ice-nucleating aerosols during SUCCESS. *Geophysical Research Letters,* **25,** 1383–1386.

Rosen, J.M., N.T. Kjome, and J.B. Liley, 1997: Tropospheric aerosol backscatter at a mid-latitude site in the Northern and Southern Hemisphere. *Journal of Geophysical Research,* **102,** 21329–21339.

Rossow, W.B. and R.A. Schiffer, 1991: ISCCP cloud data products. *Bulletin of the American Meteorological Society,* **72,** 2–20.

Russell, P.B., S.A. Kinne, and R.W. Bergstrom, 1997: Aerosol climate effects: local radiative forcing and column closure experiments. *Journal of Geophysical Research,* **102,** 9397–9407.

Sassen, K., 1992: Evidence for liquid-phase cirrus cloud formation from volcanic aerosols: climatic implications. *Science,* **257,** 516–519.

Sassen, K., 1997: Contrail-cirrus and their potential for regional climate change. *Bulletin of the American Meteorological Society,* **78,** 1885–1903.

Sassen, K., D.O. Starr, G.G. Mace, M.R. Poellot, S.H. Melfi, W.L. Eberhard, J.D. Spinhirne, E.W. Eloranta, D.E. Hagen, and J. Hallett, 1995: The 5–6 December 1991 FIRE IFO II jet stream cirrus case study: possible influences of volcanic aerosols. *Journal of Atmospheric Sciences,* **52,** 97–123.

Sato, M., J.E. Hansen, M.P. McCormick, and J.B. Pollack, 1993: Stratospheric aerosol optical depths, 1850–1990. *Journal of Geophysical Research,* **98,** 22987–22994.

Sausen, R., and I. Koehler, 1994: Simulating the global transport of nitrogen oxides emissions from aircraft. *Annales Geophysicae,* **12,** 394–402.

Sausen, R., K. Gierens, M. Ponater, and U. Schumann, 1998: A diagnostic study of the global distribution of contrails. Part I: Present day climate. *Theoretical and Applied Climatology,* **61,** 127–141.

Saxena, P., L.M. Hildemann, P.H. McMurry, and J.H. Seinfeld, 1995: Organics alter hygroscopic behavior of atmospheric particles. *Journal of Geophysical Research,* **100,** 18755–18770.

Schlager, H., P. Schulte, H. Ziereis, F. Arnold, J. Ovarlez, P. van Velthoven, and U. Schumann, 1996: Airborne observations of large scale accumulations of air traffic emissions in the North Atlantic flight corridor within a stagnant anticyclone. In: *Impact of Aircraft Emissions upon the Atmosphere.* Proceedings of a scientific colloquium, 15–18 October 1996, Paris, France. Office National d'Etudes et de Recherches Aerospatiales, Chatillon, France, Vol. 1, pp. 247–252.

Schlager, H., P. Konopka, P. Schulte, U. Schumann, H. Ziereis, F. Arnold, M. Klemm, D.E. Hagen, and P.D. Whitefield, 1997: *In situ* observations of air-traffic emission signatures in the North Atlantic flight corridor. *Journal of Geophysical Research,* **102,** 10739–10750.

Schmidt, E., 1941: Die Entstehung von Eisnebel aus den Auspuffgasen von Flugmotoren. In: *Schriften der Deutschen Akademie der Luftfahrtforschung.* Verlag R. Oldenbourg, Munich and Berlin, Germany, Vol. 44, pp. 1–15.

Schmitt, A. and B. Brunner, 1997: Emissions from aviation and their development over time. In: *Pollutants from Air Traffic – Results of Atmospheric Research 1992–1997* [Schumann, U., A. Chlond, A. Ebel, B. Kärcher, H. Pak, H. Schlager, A. Schmitt, and P. Wendling (eds.)]. DLR-Mitteilung 97-04, Deutsches Zentrum für Luft- und Raumfahrt, Cologne, Germany, pp. 37–52.

Schröder, F. and J. Ström, 1997: Aircraft measurements of submicrometer aerosol particles (> 7 nm) in the mid-latitude free troposphere and tropopause region. *Atmospheric Research,* **44,** 333–356.

Schröder, F.P., B. Kärcher, A. Petzold, R. Baumann, R. Busen, C. Hoell, and U. Schumann, 1998a: Ultrafine aerosol particles in aircraft plumes: *in situ* observations. *Geophysical Research Letters,* **25,** 2789–2792.

Schröder, F., B. Kärcher, C. Duroure, J. Ström, A. Petzold, J.-F. Gayet, B. Strauss, and P. Wendling, 1998b: On the transition of contrails into cirrus clouds. *Journal of Atmospheric Sciences,* (submitted).

Schoeberl, M.R., C.H. Jackman, and J.E. Rosenfield, 1998: A Lagrangian estimate of aircraft effluent lifetime. *Journal of Geophysical Research,* **103,** 10817–10825.

Schulte, P., H. Schlager, H. Ziereis, U. Schumann, S.L. Baughcum, and F. Deidewig, 1997: NO_x emission indices of subsonic long-range jet aircraft at cruise altitude: *in situ* measurements and predictions. *Journal of Geophysical Research,* **102,** 21431–21442.

Schulz, J., 1998: On the effect of cloud inhomogeneity on area-averaged radiative properties of contrails. *Geophysical Research Letters,* **25,** 1427–1430.

Schumann, U., 1994: On the effect of emissions from aircraft engines on the state of the atmosphere. *Annales Geophysicae,* **12,** 365–384.

Schumann, U., 1996a: On conditions for contrail formation from aircraft exhausts. *Meteorologische Zeitschrift,* **5,** 4–23.

Schumann, U. (ed.), 1996b: *Pollution from Aircraft Emissions in the North Atlantic Flight Corridor (POLINAT).* Air Pollution Research Report 58, Report EUR-16978-EN, Office for Official Publications of the European Communities, Luxembourg, 303 pp.

Schumann, U. and P. Wendling, 1990: Determination of contrails from satellite data and observational results. In: *Air Traffic and the Environment—Background, Tendencies, and Potential Global Atmospheric Effects* [Schumann, U. (ed.)]. Springer-Verlag, Heidelberg, Germany, pp. 138–153.

Schumann, U., P. Konopka, R. Baumann, R. Busen, T. Gerz, H. Schlager, P. Schulte, and H. Volkert, 1995: Estimate of diffusion parameters of aircraft exhaust plumes near the tropopause from nitric oxide and turbulence measurements. *Journal of Geophysical Research,* **100,** 14147–14162.

Schumann, U., J. Ström, R. Busen, R. Baumann, K. Gierens, M. Krautstrunk, F.P. Schröder, and J. Stingl, 1996: *In situ* observations of particles in jet aircraft exhausts and contrails for different sulfur containing fuels. *Journal of Geophysical Research,* **101,** 6853–6869.

Schumann, U., F. Arnold, B. Droste-Franke, T. Dürbeck, C. Feigl, F. Flatoy, I.J. Ford, D.E. Hagen, G.D. Hayman, A.R. Hopkins, O. Hov, H. Huntrieser, I.S.A. Isaksen, C.E. Johnson, J.E. Jonson, H. Kelder, G. Kirchner, M. Klemm, I. Köhler, P. Konopka, A. Kraabol, H. Ovarlez, J. Overlez, J. Paladino, R. Sausen, H. Schlager, J. Schneider, P. Schulte, D.S. Stevenson, F. Stordal, H. Teitelbaum, P. van Velthoven, W. Wauben, P.D. Whitefield, and H. Ziereis, 1997: Pollution from aircraft emission in the North Atlantic flight corridor—overview on the results of the POLINAT project. In: *Impact of Aircraft upon the Atmosphere.* Proceedings of an international scientific colloquium, 15–18 October 1996, Paris, France. Office National d'Etudes et de Recherches Aerospatiales, Chatillon, France, Vol. I, pp. 63–68.

Schumann, U., H. Schlager, F. Arnold, R. Baumann, P. Haschberger, and O. Klemm, 1998: Dilution of aircraft exhaust plumes at cruise altitudes. *Atmospheric Environment,* **32,** 3097–3103.

Schwartz, S.E., 1996: The whitehouse effect—shortwave radiative forcing of climate by anthropogenic aerosols: an overview. *Journal of Aerosol Science,* **27,** 359–382.

Sheridan, P.J., C.A. Brock, and J.C. Wilson, 1994: Aerosol particles in the upper troposphere and lower stratosphere: elemental composition and morphology of individual particles in northern midlatitudes. *Geophysical Research Letters,* **21,** 2587–2590.

Slemr, F., H. Giehl, J. Slemr, R. Busen, P. Haschberger, and P. Schulte, 1998: In-flight measurements of aircraft non-methane hydrocarbon emission indices. *Geophysical Research Letters,* **25,** 321–324.

Solomon, S., S. Borrmann, R.R. Garcia, R. Portmann, L. Thomason, L.R. Poole, D. Winker, and M.P. McCormick, 1997: Heterogeneous chlorine chemistry in the tropopause region. *Journal of Geophysical Research,* **102,** 21411–21429.

Spicer, C.W., M.W. Holdren, D.L. Smith, D.P. Hughes, and M.D. Smith, 1992: Chemical composition of exhaust from aircraft turbine engines. *Journal of Engineering and Gas Turbine Power,* **114,** 111–115.

Spicer, C.W., M.W. Holdren, R.M. Riggin, and T.F. Lyon, 1994: Chemical composition and photochemical reactivity of exhaust from aircraft turbine engines. *Annales Geophysicae,* **12,** 944–955.

Spinhirne, J.D., W.D. Hart, and D.P. Duda, 1998: Evolution of the morphology and microphysics of contrail cirrus from airborne remote sensing. *Geophysical Research Letters,* **25,** 1153–1156.

Spiro, P.A., D.J. Jacob, and J.A. Logan, 1992: Global inventory of sulfate emissions with 1° x 1° resolution. *Journal of Geophysical Research,* **97,** 6023–6036.

Starr, D.O. and S.K. Cox, 1985: Cirrus clouds. Part I: A cirrus cloud model. *Journal of Atmospheric Sciences,* **42,** 2663–2681.

Staylor, W.F. and A.C. Wilber, 1990: Global surface albedo from ERBE data. In: *Proceedings of the Seventh Conference on Atmospheric Radiation, San Francisco, CA, July 23–27, 1990.* American Meteorological Society, Boston, MA, USA, pp. 231–236.

Steele, H.M. and P. Hamill, 1981: Effects of temperature and humidity on the growth and optical properties of sulfuric acid-water droplets in the stratosphere. *Journal of Atmospheric Sciences,* **12,** 517–523.

Steinbrecht, W., H. Claude, U. Köhler, and K.P. Hoinka, 1998: Correlations between tropopause height and total ozone: implications for long-term changes. *Journal of Geophysical Research,* **103,** 19183–19192.

Stephens, G.L. and P.J. Webster, 1981: Clouds and climate: sensitivity of simple systems. *Journal of Atmospheric Sciences,* **38,** 235–247.

Stockwell, W.R. and J.G. Calvert, 1983: The mechanism of the HO-SO$_2$ reaction. *Atmospheric Environment,* **17,** 2231–2235.

Stolarski, R.S., S.L. Baughcum, W.H. Brune, A.R. Douglass, D.W. Fahey, R.R. Friedl, S.C. Liu, R.A. Plumb, L.R. Poole, H.L. Wesoky, and D.R. Worsnop, 1995: *1995 Scientific Assessment of the Atmospheric Effects of Stratospheric Aircraft.* NASA Reference Publication 1381, National Aeronautics and Space Administration, Washington, DC, USA, 110 pp.

Strauss, B., R. Meerkötter, B. Wissinger, P. Wendling, and M. Hess, 1997: On the regional climatic impact of contrails: microphysical and radiative properties of contrails and natural cirrus clouds. *Annales Geophysicae,* **15,** 1457–1467.

Ström, J. and J. Heintzenberg, 1994: Water vapor, condensed water and crystal concentration in orographically influenced cirrus clouds. *Journal of Atmospheric Sciences,* **51,** 2368–2383.

Ström, J., B. Strauss, T. Anderson, F. Schröder, J. Heintzenberg, and P. Wendling, 1997: *In situ* observations of the microphysical properties of young cirrus clouds. *Journal of Atmospheric Sciences,* **54,** 2542–2553.

Ström, J. and S. Ohlsson, 1998: *In situ* measurements of enhanced crystal number densities in cirrus clouds caused by aircraft exhaust. *Journal of Geophysical Research,* **103,** 11355–11361.

Sundqvist, H., 1993: Inclusion of ice phase of hydrometeors in cloud parameterizations for mesoscale and large-scale models. *Beiträge zur Physik der Atmosphäre,* **66,** 137–147.

Tabazadeh, A., E.J. Jensen, and O.B. Toon, 1997: A model description for cirrus nucleation from homogeneous freezing of sulfate aerosols. *Journal of Geophysical Research,* **102,** 23845–23850.

Talbot, R.W., J.E. Dibb, and M.B. Loomis, 1998: Influence of vertical transport on free tropospheric aerosols over the central USA in springtime. *Geophysical Research Letters,* **25,** 1367–1370.

Taleb, D.E., R. McGraw, and P. Mirabel, 1997: Time lag effects on the binary homogeneous nucleation of aerosols in the wake of an aircraft. *Journal of Geophysical Research,* **102,** 12885–12890.

Tett, S.F.B., J.F.B. Mitchell, D.E. Parker, and M.R. Allen, 1996: Human influence on the atmospheric vertical temperature structure: detection and observations. *Science,* **274,** 1170–1173.

Thomason, L.W., G.S. Kent, C.R. Trepte, and L.R. Poole, 1997a: A comparison of the stratospheric aerosol background periods of 1979 and 1989–1991. *Journal of Geophysical Research,* **102,** 3611–3616.

Thomason, L.W., L.R. Poole, and T. Deshler, 1997b: A global climatology of stratospheric aerosol surface area density deduced from Stratospheric Aerosol and Gas Experiment II measurements: 1984–1994. *Journal of Geophysical Research,* **102,** 8967–8976.

Thuburn, J. and G.C. Craig, 1997: GCM tests of theories for the height of the tropopause. *Journal of Atmospheric Sciences,* **54,** 869–882.

Timbal, B., J.-F. Mahfouf, J.-F. Royer, U. Cubasch, and J.M. Murphy, 1997: Comparison between doubled CO$_2$ time-slice and coupled experiments. *Journal of Climate,* **10,** 1463–1469.

Tie, X., G.P. Brasseur, C. Granier, A. deRudder, and N. Larsen, 1996: Model study of polar stratospheric clouds and their effect on stratospheric ozone. 2: Model results. *Journal of Geophysical Research,* **101,** 12575–12584.

Travis, D.J. and S.A. Chagnon, 1997: Evidence of jet contrail influences on regional-scale diurnal temperature range. *Journal of Weather Modification,* **29,** 7–83.

Tremmel, H.G., H. Schlager, P. Konopka, P. Schulte, F. Arnold, M. Klemm, and B. Droste-Franke, 1998: Observations and model calculations of jet aircraft exhaust products at cruise altitude and inferred initial OH emissions. *Journal of Geophysical Research,* **103,** 10803–10816.

Trepte, C.R., R.E. Viega, and M.P. McCormick, 1993: The poleward dispersal of Mount Pinatubo volcanic aerosol. *Journal of Geophysical Research,* **98,** 18563–18573.

Twohy, C.H. and B.W. Gandrud, 1998: Electron microscope analysis of residual particles from aircraft contrails. *Geophysical Research Letters,* **25,** 1359–1362.

Twomey, S., 1977: *Atmospheric Aerosols.* Elsevier Press, Amsterdam, The Netherlands, 293 pp.

Vandersee, W., 1997: Strahlungsmeßreihe am MOHp. *Promet,* **26(1/2),** 8–16.

Wahl, C., M. Kapernaum, P. Wiesen, J. Kleffmann, and R. Kurtenbach, 1997: Measurements of trace species in the exhaust of a reverse flow combustor. In: *Impact of Aircraft upon the Atmosphere.* Proceedings of an international scientific colloquium, 15–18 October 1996, Paris, France. Office National d'Etudes et de Recherches Aerospatiales, Chatillon, France, Vol. I, pp. 107–112.

Wallace, J.M., Y. Zhang, and J.A. Renwick, 1995: Dynamic contribution to hemispheric mean temperature trends. *Science,* **270,** 780–783.

Wang, P.-H., P. Minnis, and G.K. Yue; 1995: Extinction coefficient (1 μm) properties of high-altitude clouds from solar occultation measurements (1985–1990): evidence of volcanic aerosol effect. *Journal of Geophysical Research,* **100,** 3181–3199.

Wang, P.-H., P. Minnis, M.P. McCormick, G.S. Kent, and K.M. Skeens, 1996: A 6-year climatology of cloud occurrence frequency from SAGE II observations (1985–1990). *Journal of Geophysical Research,* **101,** 29407–29429.

Warneck, P., 1988: *Chemistry of the Natural Atmosphere.* Academic Press, San Diego, CA, USA, 757 pp.

Warren, S.G., C.J. Hahn, J. London, R.M. Chervin, and R.L. Jenne, 1986: *Global Distribution of Total Cloud Cover and Cloud Type Amounts over Land.* NCAR Technical Note TN-273+STR, National Center for Atmospheric Research, Boulder, CO, USA, 29 pp.

Warren, S.G., C.J. Hahn, J. London, R.M. Chervin, and R.L. Jenne, 1988: *Global Distribution of Total Cloud Cover and Cloud Type Amounts over Ocean.* NCAR Technical Note NCAR/TN-317+STR, National Center for Atmospheric Research, Boulder, CO, USA, 42 pp.

Watson, R.T., L.G. Meira Filho, E. Sanhuenza, and A. Janetos, 1992: Greenhouse gases: sources and sinks. In: *Climate Change 1992: The Supplemental Report to the IPCC Scientific Assessment.* Prepared by IPCC Working Group I [Houghton, J.T., B.T. Callander, and S.K. Varney (eds.)] and WMO/UNEP. Cambridge University Press, Cambridge, United Kingdom and New York, NY, USA, 23–46.

Wauben, W.M.F., P.F.J. van Velthoven, and H. Kelder, 1997: A 3-D CTM study of changes in atmospheric ozone due to aircraft NO$_x$ emissions. *Atmospheric Environment,* **31,** 1819–1836.

Weaver, C.J., A.R. Douglass, and D.B. Considine, 1996: A 5-year simulation of supersonic aircraft emission transport using a 3-D model. *Journal of Geophysical Research,* **101,** 20975–20984.

Weber, G.-R., 1990: Spatial and temporal variation of sunshine in the Federal Republic of Germany. *Theoretical and Applied Climatology,* **41,** 1–9.

Weickmann, H., 1945: Formen und Bildung atmosphärischer Eiskristalle. *Beiträge zur Physik der Atmosphäre,* **28,** 12–52.

Weisenstein, D.K., M.K.W. Ko, J.M. Rodriguez, and N.D. Sze, 1991: Impact of heterogeneous chemistry on model-calculated ozone change due to HSCT aircraft. *Geophysical Research Letters,* **18,** 1991–1994.

Weisenstein, D.K., M.K.W. Ko, N.D. Sze, and J.M. Rodriguez, 1996: Potential impact of SO$_2$ emissions from stratospheric aircraft on ozone. *Geophysical Research Letters,* **23,** 161–164.

Weisenstein, D.K., G.K. Yue, M.K.W. Ko, N.D. Sze, J.M. Rodriguez, and C.J. Scott, 1997: A two-dimensional model of sulfur species and aerosol. *Journal of Geophysical Research,* **102,** 13019–13035.

Weisenstein, D.K., M.K.W. Ko, I.G. Dyominov, G. Pitari, L. Ricciardulli, G. Visconti, and S. Bekki, 1998: The effects of sulfur emissions from HSCT aircraft: a 2-D model intercomparison. *Journal of Geophysical Research,* **103,** 1527–1547.

Westphal, D.L., S. Kinne, P. Pilewskie, J.M. Alvarez, P. Minnis, D.F. Young, S.G. Benjamin, W.L. Eberhard, R.A. Kropfli, S.Y. Matrosov, J.B. Snider, T.A. Uttal, A.J. Heymsfield, G.G. Mace, S.H. Melfi, D.O. Starr, and J.J. Soden, 1996: Initialization and validation of a simulation of cirrus using FIRE-II data. *Journal of Atmospheric Sciences,* **53**, 3397–3429.

Whitefield, P.D., M.B. Trueblood, and D.E. Hagen, 1993: Size and hydration characteristics of laboratory simulated jet engine combustion aerosols. *Particle Science Technology,* **11**, 25–36.

Wilson, J.C., H.H. Jonsson, C.A. Brock, D.W. Toohey, L.M. Avallone, D. Baumgardner, J.E. Dye, L.R. Poole, D.C. Woods, R.J. DeCoursey, M. Osborn, M.C. Pitts, K.K. Kelly, K.R. Chan, G.V. Ferry, M. Loewenstein, J.R. Podolske, and A. Weaver, 1993: *In situ* observations of aerosol and chlorine monoxide after the 1991 eruption of Mount Pinatubo: effect of reactions on sulfate aerosol. *Science,* **261**, 1140–1143.

Winkler, P., L. Gantner, and U. Köhler, 1997: Hat sich wegen der langfristigen Ozonabnahme die UV-Strahlung erhöht? In: *Deutscher Wetterdienst,* Offenbach am Main, Germany, Vol. 49, 37 pp.

WMO, 1986: *A Preliminary Cloudless Standard Atmosphere for Radiation Computation.* WMO/TD Report no. 24, World Climate Research Programme, World Meteorological Organization, Geneva, Switzerland, 45 pp.

WMO, 1995: *Scientific Assessment of Ozone Depletion: 1994.* World Meteorological Organization, Geneva, Switzerland, 536 pp.

Wolf, M.E. and G.M. Hidy, 1997: Aerosols and climate: anthropogenic emissions and trends for 50 years. *Journal of Geophysical Research,* **102**, 11113–11121.

Wulff, A. and J. Hourmouziadis, 1997: Technology review of aeroengine pollutant emissions. *Aerospace Science Technology,* **1**, 557–572.

Wylie, D.P. and W.P. Menzel, 1999: Eight years of high cloud statistics using HIRS. *Journal of Climate,* **12**, 170–184.

Wyser, K. and J. Ström, 1998: A possible change in cloud radiative forcing due to aircraft exhaust. *Geophysical Research Letters,* **25**, 1673–1676.

Wyslouzil, B.E., K.L. Carleton, D.M. Sonnenfroh, and W.T. Rawlins, 1994: Observation of hydration of single, modified carbon aerosols. *Geophysical Research Letters,* **21**, 2107–2110.

Yamato, M. and A. Ono, 1989: Chemical and physical properties of stratospheric aerosol particles in the vicinity of tropopause folding. *Journal of the Meteorological Society of Japan,* **67**, 147–165.

Yu, F. and R.P. Turco, 1997: The role of ions in the formation and evolution of particles in aircraft plumes. *Geophysical Research Letters,* **24**, 1927–1930.

Yu, F. and R.P. Turco, 1998a: The formation and evolution of aerosols in stratospheric aircraft plumes: numerical simulations and comparisons with observations. *Journal of Geophysical Research,* **103**, 25915–25934.

Yu, F. and R.P. Turco, 1998b: Contrail formation and impacts on aerosol properties in aircraft plumes: effects of fuel sulfur content. *Geophysical Research Letters,* **25**, 313–316.

Yu, F., R.P. Turco, B. Kärcher, and F.P. Schröder, 1998: On the mechanisms controlling the formation and properties of volatile particles in aircraft wakes. *Geophysical Research Letters,* **25**, 3839–3842.

Yue, G.K., L.R. Poole, P.-H. Wang, and E.W. Chiou, 1994: Stratospheric aerosol acidity, density, and refractive index deduced from SAGE II and NMC temperature data. *Journal of Geophysical Research,* **99**, 3727–3738.

Zhang, R., M.-T. Leu, and L.F. Keyser, 1996: Heterogeneous chemistry of HONO on liquid sulfuric acid: a new mechanism of chlorine activation on stratospheric sulfate aerosols. *Journal of Physical Chemistry,* **100**, 339–345.

Zhao, J. and R.P. Turco, 1995: Nucleation simulations in the wake of a jet aircraft in stratospheric flight. *Journal of Aerosol Science,* **26**, 779–795.

4

Modeling the Chemical Composition of the Future Atmosphere

IVAR ISAKSEN AND CHARLES JACKMAN

Lead Authors:
S. Baughcum, F. Dentener, W. Grose, P. Kasibhatla, D. Kinnison, M.K.W. Ko,
J.C. McConnell, G. Pitari, D.J. Wuebbles

Contributors:
T. Berntsen, M. Danilin, R. Eckman, E. Fleming, M. Gauss, V. Grewe, R. Harwood,
D. Jacob, H. Kelder, J.-F. Müller, M. Prather, H. Rogers, R. Sausen, D. Stevenson,
P. van Velthoven, M. van Weele, P. Vohralik, Y. Wang, D. Weisenstein

Review Editor:
J. Austin

Support:
E. Abilova, D. Brown, K. Sage

CONTENTS

EXECUTIVE SUMMARY

Although aircraft generally fly in corridors, their atmospheric effects are expected to propagate far beyond those regions.

For subsonic aircraft flying in the troposphere and the lowermost part of the stratosphere, emissions of oxides of nitrogen (NO_x) are the primary cause for the model-calculated increase in ozone (O_3) in the upper troposphere (UT). Emissions of NO_x, water vapor (H_2O), and sulfate from supersonic aircraft cruising in the stratosphere are expected to decrease the column abundance of O_3.

The effects of aircraft emissions are strongly dependent on flight altitudes: Emissions in the troposphere and the stratosphere have distinctly different effects on O_3. Although the global chemical transport models used in this assessment attempt to simulate both the troposphere and the stratosphere, these models have been developed for simulating either the troposphere or the stratosphere, but not both. Thus, the assessment of O_3 impact focuses on subsonic aircraft effects using tropospheric models and supersonic effects using stratospheric models. This approach is probably valid for supersonic aircraft, which cruise exclusively in the stratosphere (19 km), but it may be problematic for subsonic aircraft that have a sizable amount of emissions in the lowest levels of the stratosphere (around 12 km at mid-latitudes). The impact of subsonic emissions on the stratosphere has not been fully evaluated using models and requires further investigation.

Calculations for Subsonic Aircraft Scenarios

Model calculations were designed to assess the effects of added NO_x. They concluded that there would be an increase in ozone concentration and a decrease in methane (CH_4) concentration.

NO_x—All models compute increases of NO_x in the upper troposphere (UT)/lower stratosphere (LS) of 50–150 pptv at 12 km at mid-latitudes for 2015; these increases are significant compared to the average background levels of 50–200 pptv.

O_3—The O_3 increase is restricted to northern mid- and high-latitudes, with maximum increases in the UT and LS. Subsonic aircraft are predicted to cause a maximum annual average O_3 increase of 7–11 ppbv in 2015 in the latitude band 30–60°N at 10–13 km altitude. This result corresponds to an increase of approximately 5–10%.

The calculated increase in global average O_3 with increasing NO_x emission from aircraft is in broad agreement among different models. Although the models show general linearity for O_3 increase from NO_x emission, O_3 production is less efficient at high NO_x emission. The global average O_3 increase from aircraft NO_x emissions is not very sensitive to projected changes in atmospheric composition for the model scenarios investigated here.

The O_3 increase may be mitigated somewhat by emitted sulfate, leading to production of chlorine monoxide (ClO), bromine monoxide (BrO), and hydrogen dioxide (HO_2) in the LS. This process, however, has not been included in the model calculations presented in this report.

CH_4—As a result of increases in tropospheric hydroxyl (OH) caused by aircraft NO_x emissions, CH_4 removal rates are computed to increase by 1.6–2.9% in 2015 and 2.3–4.3% in 2050. The possible influence of changes in CH_4 on climate are discussed in Chapter 6.

Calculations for Supersonic Aircraft Scenarios

Because the supersonic fleet is still in its design stage, the range of emissions, fleet size, and cruise altitude covered by supersonic scenarios is larger than for subsonic aircraft. For a nominal cruise altitude of 19 km, the largest impacts of proposed supersonic aircraft occur in the stratosphere.

The predicted decrease of O_3 in the stratosphere is most sensitive to emissions of H_2O, oxides of sulfur (SO_x), and NO_x.

- For low emission indices of NO_x [5–10 g nitrogen dioxide (NO_2) kg^{-1} fuel], the predicted decrease in O_3 is dominated by emitted H_2O.
- The amount of emitted sulfur dioxide (SO_2) is important in determining the magnitude of the calculated ozone decrease. The response to sulfur depends on how much SO_2 is converted to sulfate particles in the aircraft exhaust plume, with larger depletions accompanying larger conversions.
- The calculated change in O_3 also depends on background sulfate surface area density (SAD), which is variable because of volcanic input. The computed change in O_3 is not very sensitive to projected changes in chlorine loading and trace gases for an assumed fleet size of 500 aircraft, using a background sulfate SAD condition in the model simulation.
- Changes in cruise altitude produce different results: Higher (lower) cruise altitudes result in larger (smaller) O_3 depletions.

Results reported here correspond to changes in O_3 when a supersonic fleet of aircraft replaces a portion of subsonic flights. The baseline computations assume that supersonic aircraft have a cruise altitude of 19 km, an emission index for water $EI(H_2O)=1230$ (1230 g H_2O kg^{-1} fuel), $EI(NO_x)=5$ (5 g NO_2 kg^{-1} fuel), and a range of values for $EI(SO_2)$.

H_2O—The increase in H_2O is calculated to be a maximum of 0.4–0.7 ppmv in the Northern Hemisphere mid-to high-latitude LS for a fleet of 500 aircraft, compared to a background of 3–4 ppmv.

Total Active Nitrogen (NO_y)—The increase in NO_y is calculated to be a maximum of 0.6–1.0 ppbv in the Northern Hemisphere mid-to high-latitude LS for a fleet of 500 aircraft, compared to a background of 3–10 ppbv.

SO_2—Emission and conversion of SO_2 to sulfate particles in the plume are still very uncertain for supersonic aircraft; thus, we summarize a range of scenarios. The SAD would increase 20–100% between 15 and 20 km in the 30–90°N latitude band for $EI(SO_2)=0.4$ and a range of assumptions about gas to particle conversion in the plume.

O_3—Each model shows different distributions of O_3 depletion and enhancement that probably reflect different methodologies of transport and chemistry incorporated in individual models. The results summarized below are from different models over a range of scenarios.

- The calculated range for Northern Hemisphere annual average total O_3 change is -1.3 to 0.0% for a fleet of 500 aircraft with an $EI(NO_x)=5$ in 2015 flying in a low background sulfate SAD stratosphere. The O_3 change is computed to be more positive in a higher background sulfate SAD stratosphere.
- The calculated range for Northern Hemisphere annual average total O_3 change is -1.4 to -0.1% for a fleet of 1,000 aircraft with an $EI(NO_x)=5$ in 2050.
- Model simulations show that O_3 depletion occurs throughout most of the stratosphere, except in the tropical LS. Some models calculate an O_3 increase in the lowest part of the mid-latitude stratosphere.

Carry-Through Computations to Chapters 5 and 6

Most Likely Values

For each scenario, a single model was used to propagate changes in constituents in Chapters 5 and 6.

Uncertainties

The various model computations exhibit a range of results that to some degree reflects uncertainties. However, this range does not define the uncertainties and, indeed, it is very difficult at this time to quantify them. Factors that contribute include the following:

- Deficiencies in model representation of transport processes
- Deficiencies in model representation of chemical processes
- Unknown or missing chemical and physical processes
- Limited knowledge about the chemical composition of the future atmosphere and subsequent changes in atmospheric temperatures and winds resulting from climate change effects
- Limitations in model resolution and dimensionality.

Uncertainties in the computations were estimated as follows:

- For the subsonic case, the uncertainty was taken to be a factor of 2, and was based on the differences among the various tropospheric models. The factor of 2 uncertainty was chosen for both the 2015 and 2050 atmospheres. This factor does not reflect additional uncertainties in future emissions, chemistry, and climate. Therefore, the confidence of the calculations is "fair" for 2015 and "poor" for 2050.
- For the supersonic case, the range of all model calculations was passed on as a qualitative indication of the uncertainty. The estimate of the likely uncertainty range is significantly larger than this model range. For example, the central value for the Northern Hemisphere annual average total O_3 change is -0.8% for a fleet of 1,000 aircraft in 2050, with a model range of -1.4 to -0.1%. A more likely uncertainty range for the possible Northern Hemisphere annual average total O_3 change from this supersonic fleet of aircraft is +1 to -3.5%, and a "fair" confidence is associated with this range.

4.1. Introduction

To assess the impact on the chemical composition of the atmosphere from a current and future fleet of aircraft, it is necessary to take the following steps:

- Identify emissions associated with aircraft operation
- Evaluate how each emission would change concentration of corresponding species in the atmosphere
- Determine how those changes could alter concentrations of other species in the atmosphere.

For aircraft with conventional engines that use hydrocarbon fuels, emissions include H_2O, carbon dioxide (CO_2), NO_x, oxidation products from sulfur impurities, hydrocarbons, carbon monoxide (CO), and soot. These products have the potential to perturb the atmosphere if their emissions are large enough to change background concentrations substantially. Other emissions, such as metallic particulates from engine wear and paints, have been considered and are thought to have minimal atmospheric effects (Stolarski and Wesoky, 1993). One environmental concern is how emissions could affect O_3 in the atmosphere—both in their potential to deplete O_3 in the lower stratosphere, leading to increases in UV-B radiation at the ground, and in their potential to increase O_3 in the upper troposphere, leading to greenhouse warming. In addition, increases in water vapor in the LS could have a direct effect on radiative balance as well as chemistry.

To evaluate how emission of a species could change its background concentration, one could estimate the expected change in concentration from emission rate and residence time and compare that with the background concentration, or one could compare the emission rate directly with sources that sustain the background concentration. Consideration of either criterion points to different impacts from material emitted either to the UT or the LS. The residence time for material emitted in the UT is typically on the order of weeks, whereas residence time in the LS is on the order of months to a few years (Johnston, 1989). For NO_x and H_2O, background sources also differ in the UT and LS. In the UT, NO_x is affected by surface sources as well as lightning sources in the whole troposphere. In contrast, NO_x in the LS is sustained by downward transport of NO_x and nitric acid (HNO_3) from the middle stratosphere and transport of NO_x produced by lightning in the tropical UT. In the case of H_2O, the background concentration in the UT is orders of magnitude larger than that in the LS.

The character of the O_3 budget is also very different in the UT and LS. In the UT, the transport and chemical time scales are on the order of weeks. The chemical transformation is dominated by reactions among oxides of hydrogen (HO_x) and NO_x radicals, which affect local production and removal rates of O_3. Local concentrations of HO_x species in the UT are controlled by concentrations of water, hydrocarbons, NO_x, and CO, each of which is affected by how contributions from surface sources are redistributed in the UT by convection. In the LS, transport and chemical time scales are on the order of months. The O_3

budget in the LS is maintained by a balance between transport and chemistry (chemical production balanced by transport out of the region in the tropical LS, and chemical removal balanced by transport into the region in the mid-latitude LS). Addition of NO_x and H_2O to the LS would modify the chemical production and destruction rates of O_3. However, the efficiency of the added radicals in removing O_3 is dependent on the amount of chlorine radicals in the background atmosphere and the extent of surface reactions that occur on sulfate particles and polar stratospheric clouds (PSCs). Previous modeling studies (Danilin *et al.*, 1997; Weisenstein *et al.*, 1998) have shown that sulfur emissions from supersonic aircraft can increase the surface area of the sulfate layer by about 50% in the Northern Hemisphere LS. Effects from the current subsonic fleet are less clear. Subsonic aircraft cruise in the troposphere or the very lowest part of the stratosphere (just above the tropopause); thus, the stratospheric impact from subsonic aircraft sulfur emissions would probably be less than that computed for the projected supersonic fleet. Whether any observed trend in the sulfate layer in the past decade can be ascribed to subsonic aircraft is currently under debate (see Hofmann, 1991, and Section 3.3.4.1). The amount of PSCs also would be increased as a result of H_2O and NO_x emissions from aircraft (see Section 3.3.6).

Numerical models of the atmosphere are used to calculate these changes. By solving a system of equations, these models simulate the transport and chemical interactions of trace gases to obtain their spatial and seasonal distributions. A typical model keeps track of the distributions of 50 species that interact via more than 100 reactions. Transport in the models is driven by winds and parameterization of mixing, which change with seasons. There are several ways to classify the models into different classes. One is by dimensionality: Two-dimensional (2-D) versus three-dimensional (3-D). 3-D models simulate the distributions of trace gases as functions of altitude, latitude, longitude, as well as season. 2-D models of the stratosphere simulate the zonal mean (averaged over longitudes) concentrations of species, taking advantage of the fact that many of the trace gases have uniform concentrations along latitude circles in the stratosphere. Another way to distinguish different types of models is to note whether the transport circulation is fixed or calculated in a consistent way with the model-generated trace gas concentrations.

General circulation models (GCMs) calculate temperature and transport circulation along with chemical composition. Alternatively, chemistry-transport models (CTMs) simulate the distribution of trace gases using temperature and transport circulation either from pre-calculated GCM results or derived from observations.

Because of intrinsic differences in chemistry and dynamics that control O_3 and precursor species in the UT and LS, different models have been developed to examine the different regions. Models for the troposphere require better resolution immediately above the planetary boundary layer and a proper description of convection that carries material from the boundary layer to the free troposphere. The chemical scheme in these models places more emphasis on the role of non-methane hydrocarbons (NMHCs), acetone, and peroxyacetalnitrate (PAN). Models with emphasis on the stratosphere concentrate on large-scale transport from the equatorial LS to the mid-latitudes and the exchange of material between the stratosphere and the UT. The chemical scheme pays more attention to the coupling between the nitrogen, hydrogen, and halogen species and their sources in the stratosphere. The computation requirements are such that it has not been possible to develop a model that will treat both the UT and the LS in a satisfactory manner. Historically, two sets of models have been used to evaluate aircraft impact in the UT and LS. This approach is clearly unsatisfactory because a portion of the subsonic fleet operates in the LS. In this report, we essentially continue to use this approach.

A large number of model studies of the impact of NO_x emissions from subsonic aircraft have been performed over the past 20 years (see Chapter 2 for an overview). During the past few years, these studies have been based on 3-D CTMs. Recent assessments of the atmospheric effects of aircraft emissions were completed by the National Aeronautics and Space Administration (NASA) (Friedl, 1997) and the European Community (Brasseur *et al.*, 1998). For these reports, 3-D CTM studies of the ozone perturbation from the present-day aircraft fleet were performed with several models. The model studies used the NASA database in Friedl (1997) and the Deutsches Zentrum für Luft- und Raumfahrt (German Aerospace Center) (DLR)-2 (Schmitt and Brunner, 1997) database in Brasseur *et al.* (1998). Although there are clear differences in the calculated perturbations caused by aircraft emissions, all model calculations show significant increases in NO_x concentration in the UT (up to 100% above those calculated without aircraft) in the latitude band where traffic is most frequent (30–60°N). Corresponding increases in O_3 concentration in the UT are up to 10% above those calculated without aircraft. Comparisons revealed significant differences in calculated O_3 perturbations among models, both in magnitude and in seasonal variation.

A projected fleet of high speed civil transport (HSCT) aircraft would fly at supersonic speeds in the LS at altitudes where stratospheric O_3 concentrations are large and particularly vulnerable to emissions from these aircraft. The effects of NO_x and H_2O emissions from this projected fleet on stratospheric ozone were thought to be most important when NASA's Atmospheric Effects of Stratospheric Aircraft (AESA) program started in 1989 (Prather *et al.*, 1992). Projected changes in the O_3 column as a result of aircraft emission of NO_x and H_2O from six 2-D models were presented by Stolarski *et al.* (1995) for various scenarios of fleet size and EI(NO_x) in g NO_2 kg^{-1} fuel. The model predictions in Stolarski *et al.* (1995) showed generally that supersonic fleet sizes of 500–1,000 aircraft would result in some depletion of Northern Hemisphere averaged total column O_3. More recently, the role of SO_2 aircraft emissions has been studied carefully and found to have a potentially major influence on resultant O_3 perturbation computed in models (Weisenstein *et al.*, 1996, 1998).

The new model studies in this chapter focus on the impact on atmospheric chemical composition from a current and future

fleet of subsonic aircraft flying in the UT and LS, and include a fleet of supersonic aircraft flying in the LS in one of the technology options. In the case of subsonic traffic, the estimated impact on atmospheric composition is based on 3-D CTMs; in the case of supersonic transports, the estimated impact is based on a combination of 2-D and 3-D CTMs.

This chapter has many ties to other chapters in this document. The aircraft fleet emissions used in model computations for current (circa 1992) and future (roughly 2015 and 2050) aircraft are described in Chapter 9. A discussion of the validity of models used here to accurately represent the present atmosphere is contained in Chapter 2. Information about the interaction between aerosols and aircraft emissions is presented in Chapter 3. Model computations of the distribution of a passive tracer emitted according to aircraft fuel burn are also presented in Chapter 3, and are used here to generate soot distributions by scaling to EI(soot). This tracer experiment is also used to evaluate the different transport characteristics of the models with respect to stratosphere-troposphere exchange and upper tropospheric mixing. Results presented in this chapter are incorporated in Chapters 5 and 6.

An evaluation of the effects of subsonic and supersonic engine effluent on atmospheric trace constituents is presented in Sections 4.2 and 4.3, respectively. A discussion of uncertainties in model results is given in Section 4.4, and a discussion concerning the selection of model simulations used for some of the computations in Chapters 5 and 6 is presented in Section 4.5.

4.2.   Model Studies of Subsonic Aircraft

In this section, we discuss the results of global 3-D CTMs used to assess the effects of subsonic aircraft on atmospheric concentrations of O_3, NO_x, and OH. The models differ in their formulations of vertical and horizontal resolution, transport, boundary conditions, and chemistry. Therefore, a wide range of results is to be expected. A short presentation of models and assumptions follows.

4.2.1.   *Models Used in Subsonic Aircraft Assessment*

Table 4-1 lists the six CTMs used and the names of the associated investigators.[1] Readers are referred to a Technical Report on

---

[1] The model results presented in this chapter are a summary of much model output. Supplemental material regarding the effects of subsonic aircraft are retrievable over the Internet. Subsonic model simulation information includes figures, tables, and text and is available at a NASA Langley Research Center computer until 31 December 2000:

   Host: uadp1.larc.nasa.gov
   Username: anonymous
   Password: your e-mail address
   Directory: IPCC_TECH_REPORTS/subsonic

Information concerning subsonic model simulations may also be viewed (retrieved) over the Web by going to the following URL: <ftp://uadp1.larc.nasa.gov/IPCC_TECH_REPORTS/subsonic/>.

Table 4-1: *Description of models used in the subsonic assessment.*

Model	Institution	References
ECHAM3/CHEM	German Aerospace Research Establishment, Germany	Roeckner *et al.* (1992); Steil *et al.* (1998)
HARVARD	Harvard University, USA	Wang *et al.* (1997a,b)
IMAGES/BISA	Belgian Institute for Space Aeronomy, Belgium	Müller and Brasseur (1995); Brasseur *et al.* (1996)
TM3/KNMI	Royal Netherlands Meteorological Institute, The Netherlands	Wauben *et al.* (1997a,b)
UKMO	United Kingdom Meteorological Office, UK	Collins *et al.* (1997); Stevenson *et al.* (1997)
UiO	University of Oslo, Norway	Berntsen and Isaksen (1997); Jaffe *et al.* (1997)

Subsonic Aircraft Effects, which is presently available over the Internet for additional details.

4.2.1.1. Off-Line vs. On-Line Models

All of the models in Table 4-1, except the ECHAM3/CHEM model, are off-line models; that is, they are driven using meteorological fields derived either from GCMs or from analysis of observations. The temporal resolution of the various meteorological fields used to drive the models ranges from 40 minutes to a day. One exception is the IMAGES/BISA model, which uses monthly-average meteorological fields and includes a parameterization to account for shorter-term variability in transport. In all of the off-line models, 1 year of wind fields is recycled in multiyear simulations to get the steady-state atmosphere. On-line calculations provide the potential capability of examining chemistry-climate interactions when model-calculated fields are used in radiation calculations. In ECHAM3/CHEM, the evolution of chemical fields is calculated on-line in a GCM, but the calculated chemical fields do not feed back into the dynamic calculations in this application. The model, therefore, operates in a similar way to the off-line models.

4.2.1.2. Model Resolution

Typically, these models have horizontal resolutions of 3–6°, with the exception of the UiO model, which has a horizontal resolution of 8°x10°. In the vertical dimension, the IMAGES/BISA model has 25 levels; the TM3/KNMI and ECHAM3/CHEM models have 19 levels; and the HARVARD, UKMO, and UiO models have nine levels. Four of the models (ECHAM3/CHEM, HARVARD, TM3/KNMI, UiO) have a top layer located at 10 mb; the IMAGES/BISA and UKMO models have top layers located at 50 mb and 100 mb, respectively. Because vertical model levels are defined on sigma coordinates and not on pressure coordinates, the number of model levels between fixed pressure surfaces can vary in time. Between the surface and 850 mb, the HARVARD, UiO, and UKMO models have about two vertical levels; the ECHAM3/CHEM and TM3/KNMI models have about five vertical levels, and the IMAGES/BISA model has eight vertical levels. In the UT/LS region between 100 and 300 mb, the HARVARD model has one vertical level, the UKMO model has about two levels, and the other four models have about four vertical levels.

4.2.1.3. Coupling to the Stratosphere

With the exception of the ECHAM3/CHEM model, the models have little or no representation of explicit stratospheric chemistry. Instead, either the cross-tropopause fluxes of O_3 and NO_y are specified or the mixing ratios of these species are specified in the LS based on observations. In the TM3/KNMI, UiO, and IMAGES/BISA models, however, the upper boundaries are higher in an attempt to minimize their influence on regions of maximum perturbations by aircraft. It should be noted that this condition may not be satisfied for the IMAGES/BISA model because the model top is at 50 mb.

4.2.1.4. Tropospheric NO_x Sources

All of the models include anthropogenic and biogenic tropospheric NO_x sources. For present-day conditions, the magnitudes of surface NO_x sources in the various models are ~21 Tg nitrogen (N) yr[-1] from surface-based fossil-fuel combustion, 5–12 Tg N yr[-1] from biomass burning, and 4–6 Tg N yr[-1] from soils. The present-day magnitude of the lightning source is 5 Tg N yr[-1] in the IMAGES/BISA, TM3/KNMI, UKMO, and UiO models; 4 Tg N yr[-1] in the ECHAM3/CHEM model (increased to 5 Tg N yr[-1] in the 2015 and 2050 simulations); and 3 Tg N yr[-1] in the HARVARD model. It should be noted, however, that the simulated impact of lightning on NO_y species in the troposphere can differ from model to model even if the magnitude of the lightning source of NO_x is the same, as a result of differences in factors such as duration and intensity of convective events, land/ocean differences in convection, height of NO_x emissions, and so forth. Sensitivity tests were run to evaluate the effect of this lightning assumption on calculated aircraft perturbation.

4.2.1.5. Tropospheric Chemistry

Most of the models include a comprehensive description of the CH_4-CO-NO_x-HO_x-O_3 chemical system. With the exception of ECHAM3/CHEM and TM3/KNMI, the models include representations of NMHC chemistry. However, the details of NMHC chemistry differ significantly from model to model. The ECHAM3/CHEM model includes a stratosphere with a chemistry scheme more suited to the stratosphere, however, it does not include some of the species that are important for tropospheric chemistry.

4.2.1.6. Tropospheric Transport

In addition to transport by resolved-scale winds, all models considered here include parameterizations of vertical transport by sub-grid-scale processes such as convection and turbulent mixing in the boundary layer. Again, the manner in which these processes are parameterized differs from model to model. In this context, it is worth noting that four of the models (or their close counterparts) used in this exercise (ECHAM3/CHEM, HARVARD, UKMO, and TM3/KNMI) were also involved in a model intercomparison exercise sponsored by the World Climate Research Program (WCRP) in 1993 (Jacob *et al.*, 1997). As part of this exercise, each model simulated a scenario in which a fictitious tracer with a 5.5-day e-folding lifetime was emitted in the Northern Hemisphere mid-latitude UT. The vertical gradient in the simulated fields was similar in several of the participating models. However, there were significant inter-model differences in the simulated rates of meridional tracer transport in the UT.

4.2.2. Definition of Scenarios

This section describes the scenarios for aircraft emissions evaluated for this assessment. The premises for current (circa 1992) and future (roughly 2015 and 2050) aircraft fleet emissions, along with descriptions of actual emissions databases, are given in Chapter 9. The assumptions used for the background atmosphere in model calculations of the effects of aircraft emissions on O_3 are important and influence the results. In the following sections, we discuss the basis for background atmospheres used in model calculations and the aircraft scenarios evaluated.

4.2.2.1. Background Atmospheres

Boundary conditions for CH_4 in the background atmosphere are 1714, 2052, and 2793 ppbv for the years 1992, 2015, and 2050, respectively. These amounts are based on the IPCC

IS92a scenario (IPCC, 1992, 1995). Updated projections for future CH_4 concentrations (WMO, 1999) are smaller than those assumed here. Recent observations by Dlugokencky *et al.* (1998) show that CH_4 levels currently are leveling off. If this trend continues during the next century, with little or no increase in the CH_4 concentration, the increase in background O_3 will also be substantially less than that calculated in these studies.

For shorter lived gases—such as CO, NO_x, and volatile organic compounds (VOCs)—the participating models use their standard boundary conditions for the 1992 cases. For 2015 and 2050, most model calculations assume that emissions are increased by the same factors at all locations relative to 1992 emissions, as shown in Table 4-2. Such constant increases were necessitated by difficulties in 3-D models to readily change emission inputs for assessment studies.

A special sensitivity study was also conducted for 2050 with the UiO model using a geographically varying emission increase (IPCC, 1995). A summary of these factors is presented in Table 4-3. Such regional differential factors are applied only to energy-related sources; biomass burning factors are applied as in the standard case (using Table 4-2).

4.2.2.2. Aircraft Emission Scenarios

The 3-D aircraft scenarios described in Chapter 9 form the basis for the assessment. The scenarios evaluated by the participating models are summarized in Table 4-4. Summaries of global emissions for these scenarios are given in Tables 9-4 and 9-5. Only a few scenarios are considered for subsonic assessment calculations because computational requirements for the 3-D models are high. The subsonic scenarios in Table 4-4 are generally analyzed relative to corresponding background atmospheres for 1992, 2015, or 2050.

In model calculations, aircraft effluents are put into the models as follows: Gridded fuel burn data (kg fuel/day) are first mapped into the model grid. The amount of material emitted into each grid box is given by the product of the fuel burn and the emission index. The emitted material is put into the grid box at each time step at the equivalent rate. In this approach, we ignore the effect of plume processing and assume that emitted

Table 4-2: Increase from 1992 to 2015 and 2050 for emissions of CO, NO_x, and VOCs (based on IPCC scenario IS92a).

	Source	2015	2050
CO	Energy	+15%	+66%
	Biomass burning	+9%	+21%
NO_x	Energy	+45%	+107%
	Biomass burning	+7%	+22%
VOCs	Energy-related sources (not isoprene)	+23%	+66%

Table 4-3: *Factors of increase from 1992 to 2050 for energy sources of CO, NO$_x$, and VOCs as applied to different regions in sensitivity studies.*

	NO$_x$	CO	VOCs
OECD countries	0.83	0.25	1.06
Eastern Europe and Soviet Union	1.00	1.00	1.50
Centrally planned Asia (excluding Korea)	3.33	4.67	6.00
North Korea	2.19	1.40	2.11
Middle East	5.53	3.54	5.31
Southeast Asia	2.77	1.77	2.66
South Asia	9.25	5.93	8.89
Africa (without South Africa)	7.19	4.60	6.91
South Africa	3.17	2.03	3.04
Latin America	4.85	3.10	4.66

material is instantaneously mixed into the grid box. For the subsonic assessment, NO$_x$ is the only aircraft emission considered. Because most models do not calculate the hydrological cycle in the troposphere, emitted water is not calculated. Sulfur, CO, and unburned hydrocarbons are also ignored.

The basic scenarios examine some of the important aspects in understanding the calculated environmental impact of aircraft. However, a number of uncertainties remain in the treatment of chemical and physical processes that may influence the effects from aircraft emissions. Therefore, a series of special sensitivity calculations was designed to investigate the most important of the recognized uncertainties. The subsonic aircraft sensitivity scenarios, as described later, examine uncertainties in the background atmosphere, the treatment of chemical and dynamical processes in the UT and LS, and different analyses of aircraft emissions.

It has not been possible (for practical reasons) for each of the modeling groups to run all of the scenarios set up for these 3-D model studies of aircraft perturbations. Each modeling group has completed a limited number of model simulations.

4.2.3. Model Results for Subsonic Aircraft Emissions

4.2.3.1. Ozone Perturbation

Figure 4-1 presents annual zonal average increases of O$_3$ volume mixing ratios caused by aircraft NO$_x$ emissions predicted by the six models for the year 2015. As this figure shows, the models treat the tropopause significantly differently, which leads to qualitatively different O$_3$ distributions and calculated O$_3$ perturbations near the tropopause. The UiO model calculates a maximum increase of O$_3$ of about 9 ppbv around 40–80°N at an elevation of 10–13 km. Throughout most of the Northern Hemisphere, increases larger than 1 ppbv are calculated. The IMAGES/BISA and HARVARD models calculate somewhat smaller peak perturbations of about 7 ppbv. In contrast, the TM3/KNMI and the UKMO models calculate maximum changes of about 11 ppbv. The UKMO model computes large increases up to the 16-km level, probably as a result of relatively large vertical exchange rates in the vicinity of the tropopause. In contrast to the other models, the ECHAM3/CHEM model predicts the highest O$_3$ perturbations in the Northern Hemisphere and Southern Hemisphere LS. Tropospheric changes are smaller than in the other models, however. The difference may be partly a result of the length of the ECHAM3/CHEM model run. This model has a representation of the stratosphere and reported results from the average of the last 10 years of a 15-year simulation—long enough to propagate aircraft emissions and O$_3$ perturbation in the Northern Hemisphere to the Southern Hemisphere (via the stratosphere). The other 3-D models do not account for such stratospheric transport processes because they constrain O$_3$ concentrations in their upper model levels: They fix their upper model layer concentrations using observations, or they prescribe O$_3$ fluxes from the stratosphere into the troposphere. The use of these boundary conditions could lead to a calculated impact on stratospheric O$_3$ that is too small. It may also be that the ECHAM3/CHEM model has too efficient transport in the LS.

The effect of constraining concentrations and fluxes at the upper boundary of the 3-D models was checked by running the stratospheric 2-D Atmospheric and Environmental Research, Inc.

Table 4-4: *Base background scenarios and subsonic aircraft scenarios.**

Model Scenarios

A 1992 Base (background atmosphere, no aircraft)

B 1992 Base + Subsonic Aircraft (Chapter 9, NASA 1992)

C 2015 Base (background atmosphere, no aircraft)

D 2015 Base + Subsonic Aircraft (Chapter 9, NASA 2015)

E 2050 Base (background atmosphere, no aircraft)

F 2050 Base + Subsonic Aircraft (Chapter 9, Fa1)

G 2050 Base + Subsonic Aircraft (Chapter 9, Fe1)

*When these scenarios are used in assessing supersonic aircraft influences, the sulfate distribution in the stratosphere is set to the stratospheric background SA0 (see section 4.3).

(AER) model for the same subsonic scenario. Consistent with the 3-D models, the AER model calculates a maximum O_3 increase of 8–10 ppbv in the Northern Hemisphere at an altitude of 8–12 km.

In the stratosphere at 16 km, small increases of 2 ppbv in the Southern Hemisphere and 6 ppbv in the Northern Hemisphere are calculated—somewhat higher, but consistent with most 3-D models.

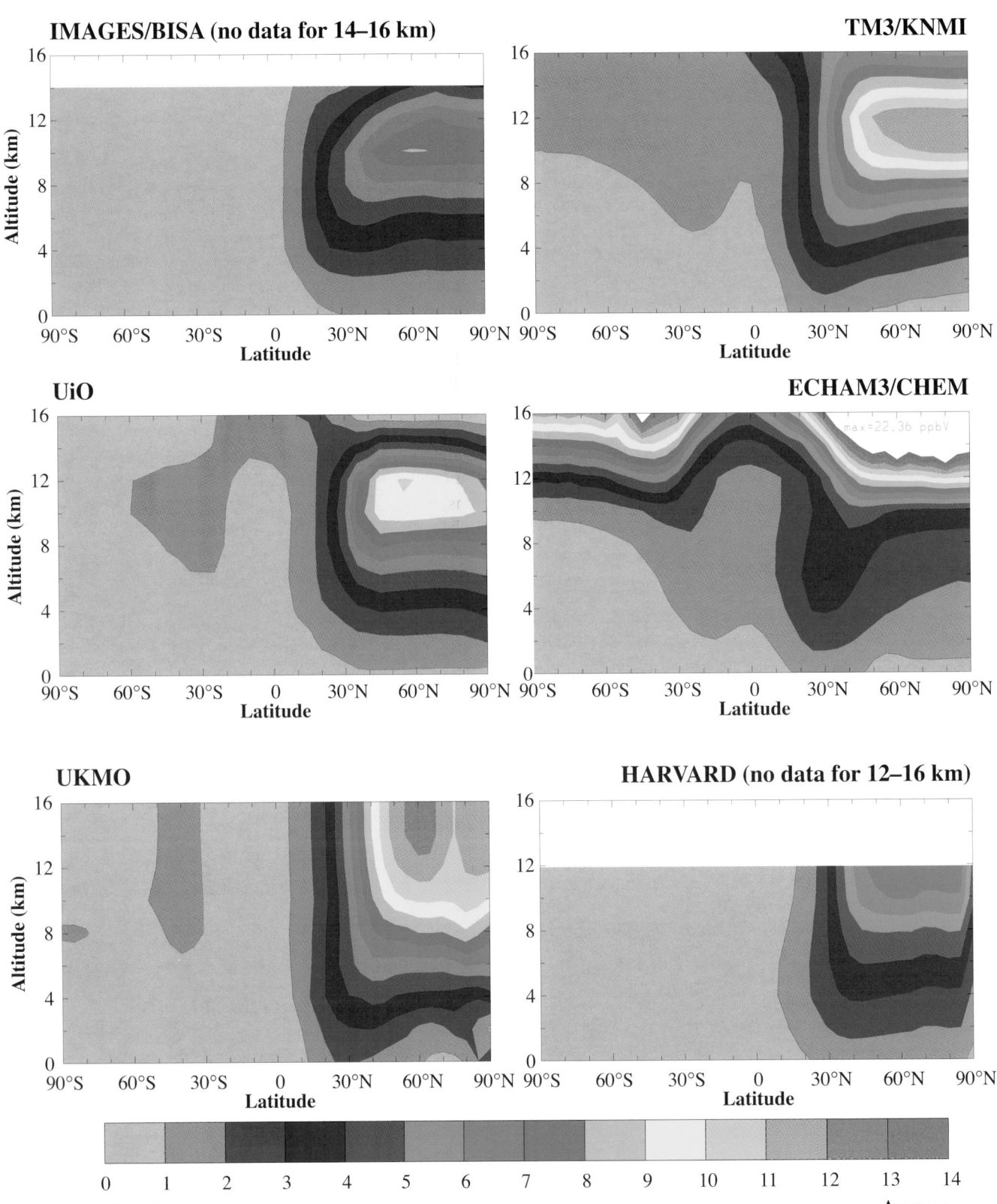

Figure 4-1: Annual (2015) and zonal average increases of ozone volume mixing ratios [ppbv] from aircraft emissions calculated by six 3-D models. The IMAGES/BISA model does not give results above 14 km, and the HARVARD model does not give results above 12 km.

Calculated O_3 increases are strongest in the UT and the LS. In the lower troposphere (< 6 km), the increase is reduced by a factor of about 5 in mixing ratio compared to the UT. All models calculate that about 85% of the O_3 increase for 1992 is in the Northern Hemisphere; for 2015 and 2050, the portions are about 80 and 75%, respectively.

Although emissions of precursor NO_x are spatially distributed heterogeneously, the resulting O_3 increases are distributed more uniformly as a result of the combined effects of strong longitudinal mixing and the relatively long residence time of O_3 in the free troposphere and LS. All models show efficient transport of excess O_3 from source regions at mid-latitudes to high latitudes, where the residence time of O_3 is particularly long as a result of decreased deposition (Stevenson *et al.*, 1997; Wauben *et al.*, 1997; Berntsen and Isaksen, 1999).

There may be a strong seasonal cycle in the calculated impact of aircraft emissions on O_3. For example, using the same emission scenarios, the UiO and the UKMO models calculate a 40% larger increase of O_3 in the Northern Hemisphere in April compared to July (Stevenson *et al.*, 1997; Berntsen and Isaksen, 1999). Other models find much weaker seasonal cycles (e.g., IMAGES/BISA and ECHAM3/CHEM), or find maximum increases in summer (e.g., TM3/KNMI and HARVARD). These seasonal differences are probably associated with different background NO_x conditions in the different models (see Section 4.2.3.2).

4.2.3.2. NO_x Perturbation

Figure 4-2 shows calculated zonal average increases of NO_x from aircraft emissions in July 2015. In the Northern Hemisphere, all but one model calculate increases of up to 150 pptv. These increases can be compared with background levels of 50–200 pptv at northern mid-latitudes in the 12-km region. In the stratosphere, the ECHAM3/CHEM model calculates larger increases, probably as a result of more efficient transport to the stratosphere. The height distribution of NO_x increases is very similar among models. All models also predict noticeable increases in upper tropospheric NO_x at low latitudes in the Southern Hemisphere. Only small increases are estimated in the lower troposphere. Background NO_x conditions, however, are rather different in the 3-D CTMs. For instance, at 12 km at 50°N, calculated background NO_x mixing ratios may vary, depending on the season, by a factor of 2 to 4. Such large differences could be important for O_3 production because of the nonlinear dependency of net O_3 production on NO_x concentrations, although, for the model scenarios explored, the global O_3 increase appears to be almost linear for most of the anticipated NO_x injections in the models (see Figure 4-3).

4.2.3.3.   *Future Total Ozone Increases from Aircraft Emissions and Comparison with Increases from Other Sources*

Figure 4-3 presents the increase of global O_3 abundance up to 16 km from aircraft emissions. The annual emissions of 0.5,

1.27, and 2.17 Tg N correspond to (projected) emissions for 1992, 2015, and 2050, respectively. Calculated O_3 increases range from 4 to 7 Tg in 1992, 9 to 17 Tg in 2015, and 19 to 24 Tg in 2050. For each CTM, the global O_3 increase scales almost linearly with aircraft NO_x emissions—even for the 2050 high-demand sensitivity study G (3.46 Tg N yr-1). However, O_3 production is less efficient at high NO_x emissions. The nearly linear response of global O_3 increase to aircraft NO_x emissions was not anticipated, given the well-known nonlinear O_3 production as a function of NO_x (discussed in Chapter 2). The main explanation seems to be that aircraft NO_x and associated reservoir species (e.g., HNO_3) are transported out of aircraft corridors, where net O_3 production depends more linearly on NO_x. We should recognize, however, that all of the global models used in this study have a coarse resolution that may systematically overestimate the O_3 production. Secondary effects are background increases in CH_4 and CO (as a result of enhanced surface emissions in 2015 and 2050), leading to somewhat more efficient O_3 production per NO_x molecule emitted, and the shift of emissions from Northern Hemisphere mid-latitudes toward the tropics, where background NO_x concentrations are smaller.

To further test linearity in O_3 increases to NO_x emission beyond the upper limit selected in these model studies, an extremely high NO_x emission of 1.5 times the high-demand, low-technology case (Scenario G) was run with the UiO model. This scenario showed only slight nonlinearity at lower NO_x emissions. This extreme simulation indicated that a level of nonlinearity was reached at northern latitudes where O_3 increases of only 10% were obtained, whereas at southern latitudes, where emissions are smaller, O_3 increases (approximately 50%) were nearly linear with increases in NO_x emissions.

Figure 4-4 shows the global increases of total O_3 from aircraft emissions in 2015 and 2050 relative to those in 1992—that is, the difference of O_3 budgets for scenarios listed in Table 4-4 (D with respect to B and F with respect to B). The same figure also shows the increases of O_3 in 2015 and 2050 from the effects of changes in surface emissions (Section 4.2.2.1)—that is, the differences for scenarios C with respect to A and E with respect to A. Aircraft emissions account for approximately 15–30% of the total O_3 increase in 2015 and 15–20% in 2050. It should be noted, however, that the projections of aircraft emissions and the IPCC IS92a scenario underlying the increase of surface emissions are extremely uncertain. Changes in aircraft or surface emissions scenarios could change the relative contribution from aircraft emissions to O_3 perturbations significantly. Using scenario G (high demand), an approximately 45% higher increase of O_3 from aircraft is calculated in 2050 by the UiO model (see sensitivity studies).

4.2.3.4.   *Influence of Changing OH on CH_4 Lifetime*

As discussed in Section 2.1.4, aircraft NO_x emissions lead to higher OH concentrations. In the troposphere, CH_4 is removed mainly by reaction with the OH radical. Therefore, a higher OH concentration will lead to more rapid removal of CH_4 from the atmosphere. Table 4-5 presents the chemical lifetime of CH_4 and changes from aircraft emissions for scenarios A–F. The lifetime in Table 4-5 is defined as the CH_4 amount up to 300 hPa divided by the amount annually destroyed by chemical processes. There are large differences in CH_4 lifetimes calculated by the models for base cases A, C, and E. It is beyond the scope of this report to assess what causes these differences, but it can be generally said that global OH is very sensitive to photolysis

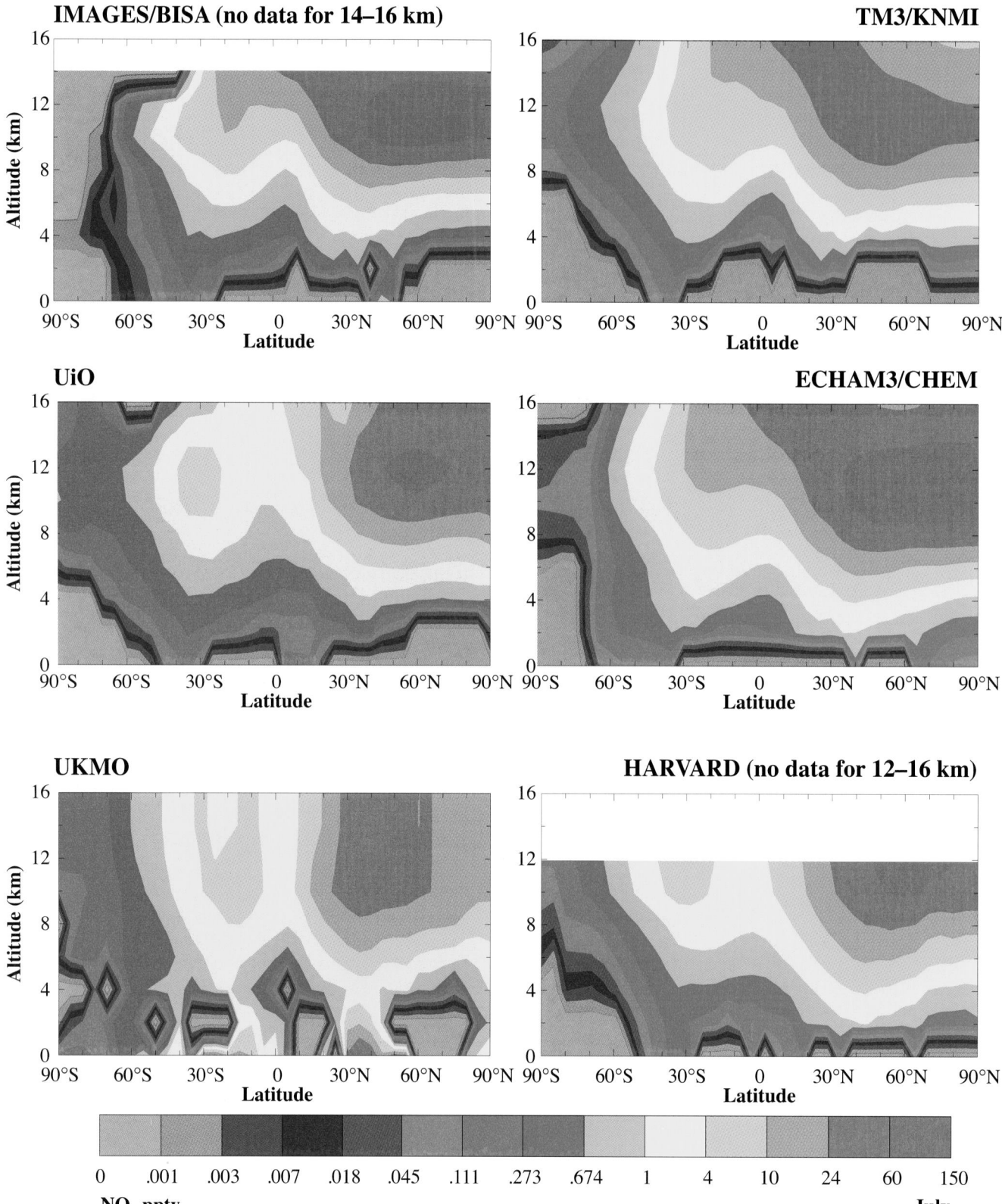

Figure 4-2: July zonal average increase in NO_x [pptv] from aircraft.

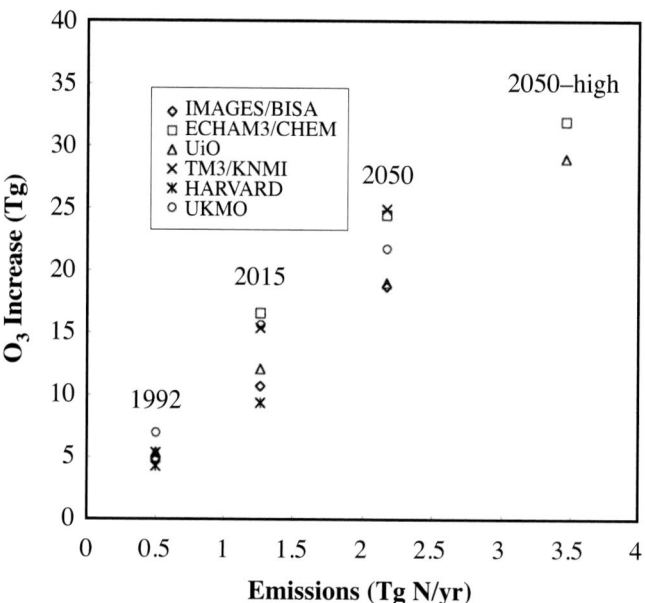

Figure 4-3: Increase in annual average global O_3 abundance (Tg O_3) up to 16 km from present and future aircraft emissions.

rates, parameterization of lightning NO_x emissions, and the amount and distribution of surface NO_x and other emissions. Comparing simulations A, C, and E, which show enhancements from changes in surface emissions, CH_4 lifetimes increase by 0.5–3.2% from 1992 to 2015 and by 7–12% from 1992 to 2050. The decrease of OH concentrations is a result of the strong effect of anthropogenic CO emissions and higher background CH_4 concentrations, which dominate the effect of surface emissions of NO_x. The models are rather consistent in their estimates of changes of CH_4 lifetimes from aircraft emissions. Comparing simulations with and without aircraft emissions, CH_4 lifetimes are calculated to decrease globally by 1.2–1.5% in 1992, 1.6–2.9% in 2015, and 2.3–4.3% in 2050.

Changes in calculated CH_4 lifetime from aircraft emissions for the three time periods considered are surprisingly similar in the model studies. With the exception of the ECHAM3/CHEM model, which gives smaller perturbations than the other models because it uses a fixed mixing ratio boundary condition for CO, the differences among models for aircraft impacts are within 20%. This perturbation of CH_4 residence time from aircraft emissions is significantly larger than that obtained in previous

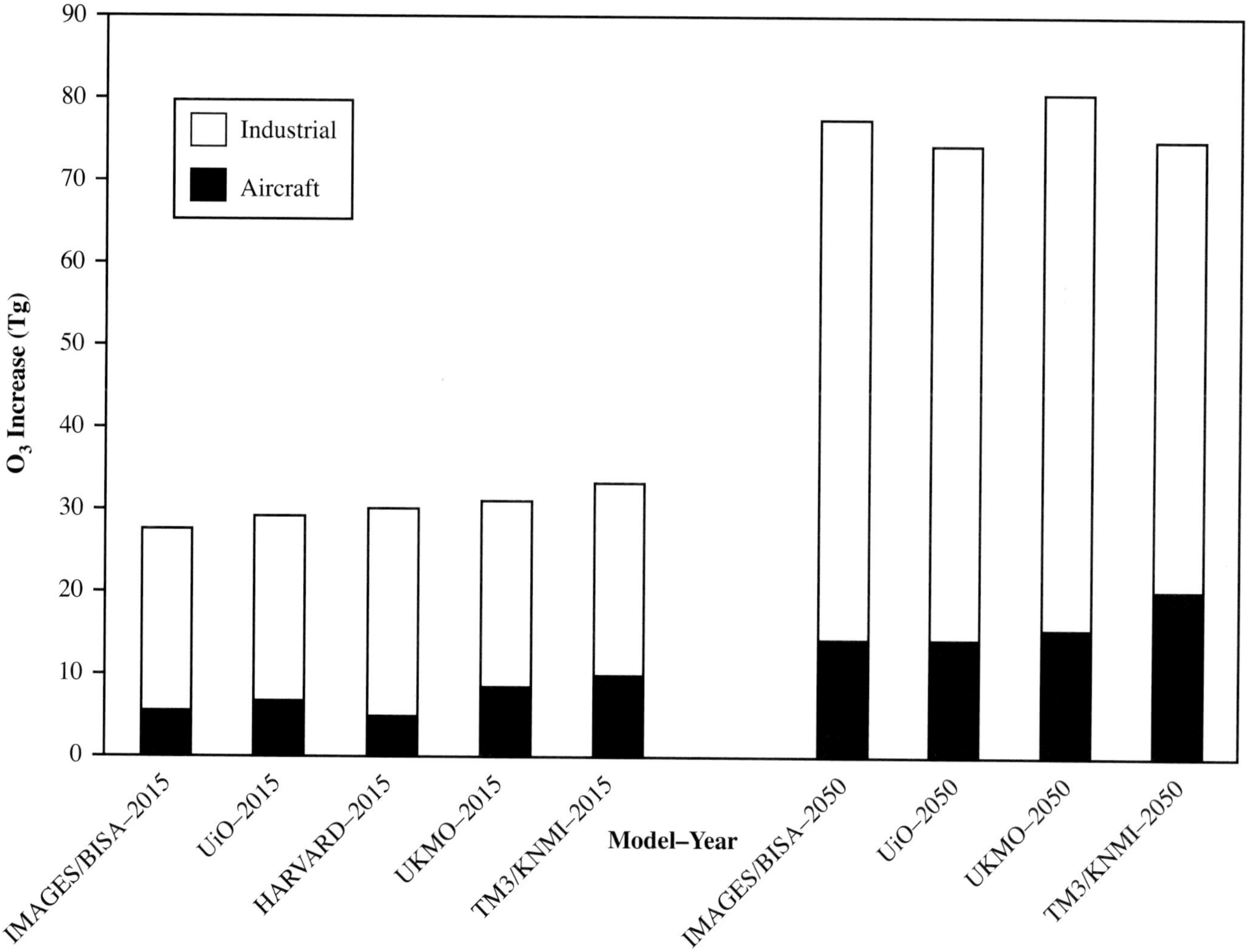

Figure 4-4: Increases in global total tropospheric ozone abundances (Tg O_3) in 2015 and 2050 from aircraft and other anthropogenic (industrial) emissions relative to 1992.

Table 4-5: *Chemical lifetime (in years) of methane [columns A, C, and E] up to 300 hPa (~10 km) and changes of this lifetime (%) (columns B, D, and F) by including aircraft emissions (nc = not calculated).*

| Scenario/ | 1992 | | 2015 | | 2050 | |
Model	A	B	C	D	E	F
IMAGES/BISA[a]	6.60	-1.2%	6.81	-2.6%	7.36	-3.7%
ECHAM3/CHEM	nc	nc	6.46	-1.6%	6.51	-2.3%
HARVARD	9.33	-1.2%	9.43	-2.6%	nc	nc
UiO	8.52	-1.3%	8.59	-2.6%	9.48	-3.9%
UKMO[b]	10.52	-1.5%	10.69	-2.9%	11.26	-4.3%
TM3/KNMI	8.97	-1.4%		-2.6%		-3.5%

[a]Uses fixed lower boundary conditions for CO.

[b]Lifetime up to 100 hPa; a lower lifetime is expected for integration up to 300 hPa.

studies (IPCC, 1995; Fuglestvedt *et al.*, 1996) using 2-D models. CH_4 loss is dominated by OH changes in the tropical and subtropical regions of the lower troposphere. These previous studies showed OH changes that were largely restricted to the UT, where OH perturbations have little impact on CH_4 residence time. Figure 4-5a shows the perturbation in the zonally averaged OH field (July) for 2015 aircraft emissions (given in 10^6 molecules/cm^{-3}) from the UiO model. The figure shows that the perturbations extend well into the lower troposphere at most latitudes in the Northern Hemisphere. One explanation for this result could be that CO, which accounts for most of the OH loss, has a sufficiently long lifetime to be transported over large distances. The impact on CO in one region could influence CO (Figure 4-5b) and OH in other regions (e.g., low-latitude lower troposphere), leading to the estimated impact on CH_4. Similar CO changes were found, for example, in the TM3/KNMI model. In addition to changes in CO, relatively small O_3 changes are predicted in the warm humid tropics. These changes also lead to somewhat increased OH production, hence a decrease in CH_4 lifetime. Thus, changes in CH_4 lifetime are related in direct and indirect ways to changes in O_3 concentrations and probably should be assessed together. The difference in estimated residence time compared with previous 2-D studies could therefore be a result of very different transport parameterizations in 2-D and 3-D models.

The reduction in CH_4 lifetime would lead to a nearly uniform CH_4 reduction globally because of the relatively long residence time computed for CH_4. This result would be in contrast to O_3, for which changes would occur on large regional scales. Finally, it should be noted that for computational reasons, the experiments in this assessment were performed using fixed CH_4 concentrations at the Earth's surface (see Section 4.2.2.1), and CH_4 concentrations at the surface were not allowed to adapt to higher OH abundances (positive feedback). Hence, even larger CH_4 decreases are to be expected if CH_4 ground flux boundary conditions are used. However, such calculations are much more computationally intensive. IPCC (1995) and Fuglestvedt *et al.* (1996) showed that the feedback factor is uncertain and model-dependent but is estimated to be in the range 1.2–1.5. Adopting a factor of 1.4 increases the percentage changes, in the CH_4 lifetimes shown in Table 4-5, to -2.2 to

-4.1% in 2015 and -3.2 to -6.0% in 2050. Changes in CH_4 lifetime of this order will lead to global average radiative forcing (see Chapter 6) similar to global average radiative forcing perturbations from aircraft-induced O_3 changes, but with an opposite sign (CH_4 will be reduced).

4.2.3.5. Sensitivity Studies

In this section, we focus on a limited set of sensitivity studies that help define the uncertainty range of the model calculations. Ideally, a large number of simulations should have been performed by all the participating models. However, only a limited number of model simulations was possible because of time constraints and the demand of computer resources for 3-D CTM studies. Therefore, only one or two models have performed each of the sensitivity studies. Also, some uncertainties cannot be addressed properly. For example, modeled upper tropospheric and lower stratospheric NO_x and NO_y concentrations are extremely uncertain and are difficult to compare to measurements because of large temporal and spatial variations and a limited number of observations (see Chapter 2). There may be additional uncertainties from unknown processes that feed back on increases in NO_x and O_3 in a future modified atmosphere.

The sensitivity studies focus on the impact on O_3. The following studies were performed:

- Sensitivity of aircraft-induced O_3 perturbations to background NO_x levels from lightning. This finding is obtained by increasing global average NO_x production from 5 Tg N yr^{-1}, which is used in the reference case, to 12 Tg N yr^{-1}. The same spatial distribution is used in both cases.
- Sensitivity of O_3 perturbation to different regional growth in emissions. In the base case (IS92a), a uniform growth rate was used for the surface emission of pollutants. The sensitivity run was performed with the same global growth rate as in the base case but with the different regional growth rates given in Table 4-3.
- Sensitivity to different projections of aircraft emissions. Instead of NASA 2015 emissions, the ANCAT-2015

Delta OH (10⁶ molecules/cm³), July

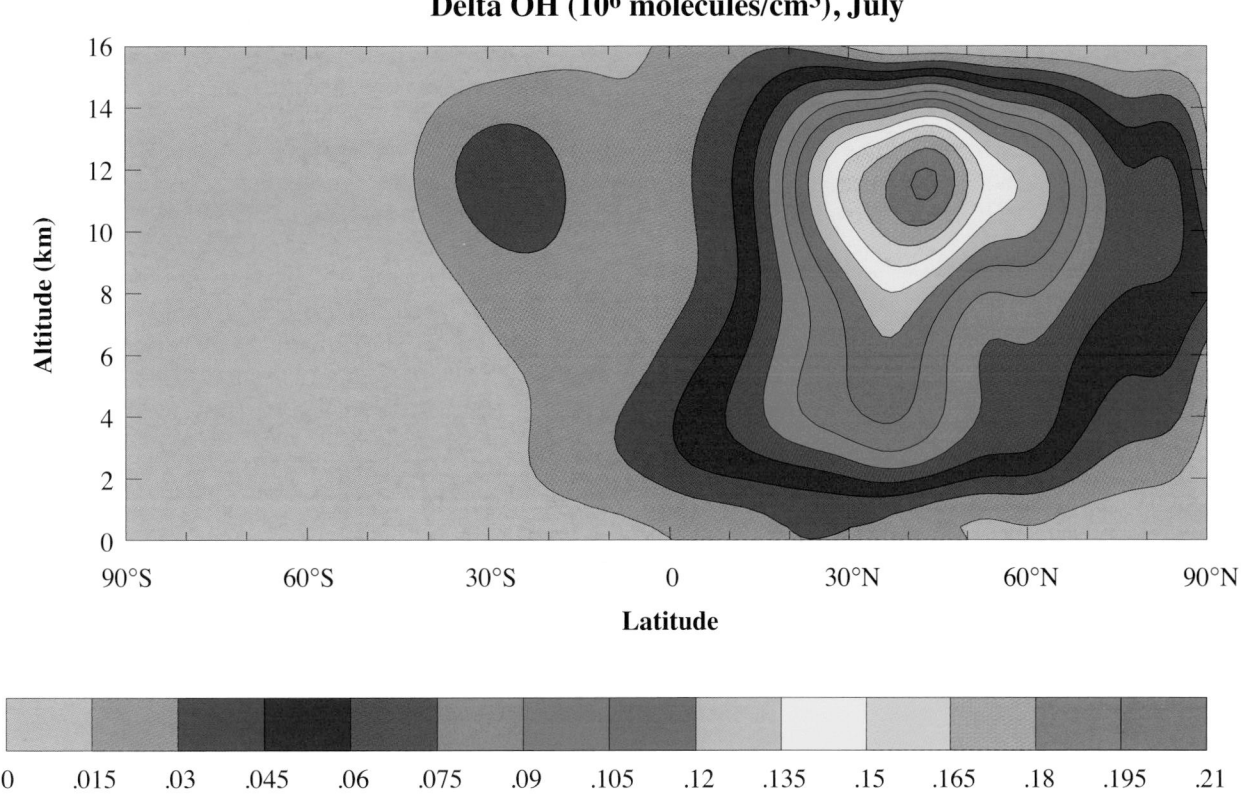

Delta CO (%), July

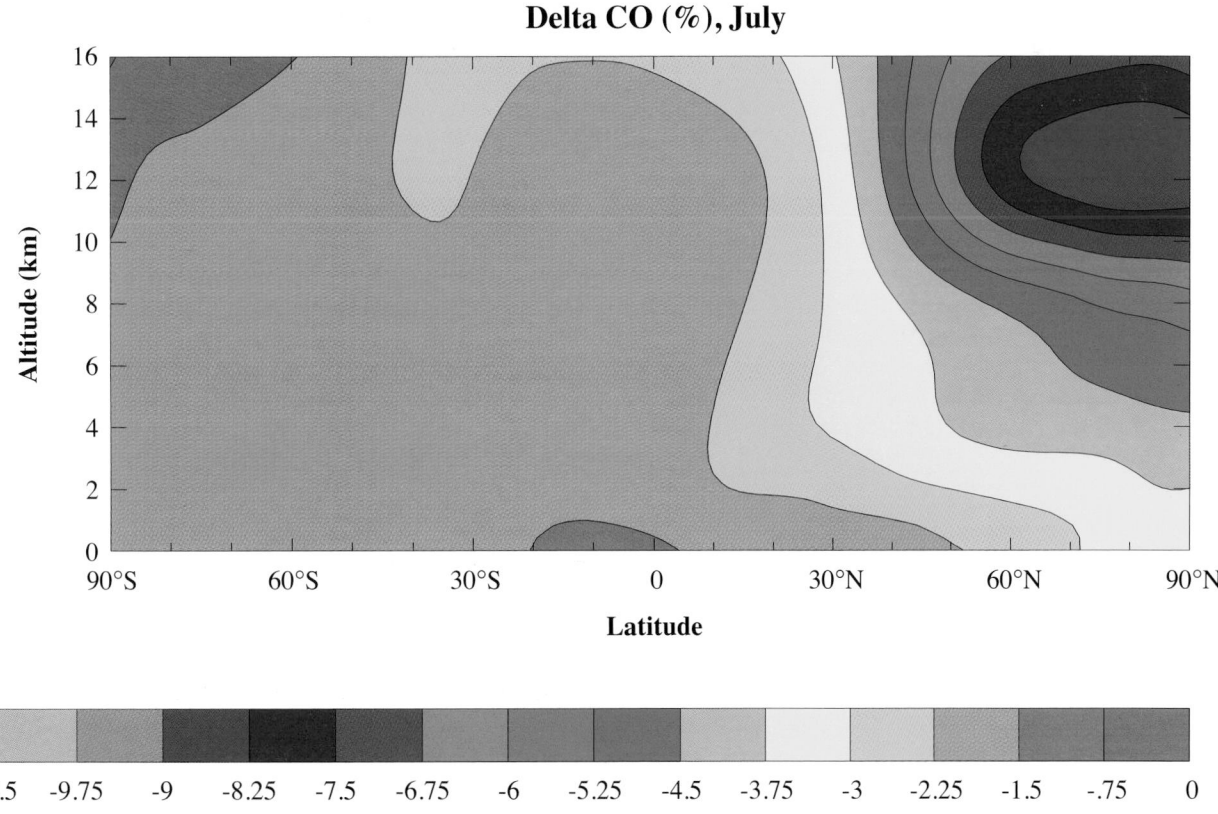

Figure 4-5: Zonally and monthly averaged change in concentration of OH (10⁶ molecules/cm³) and CO (%) in July as a result of emissions from aircraft in 2015, calculated by the UiO model.

aircraft emission data set was used. Total ANCAT emissions for the year 2015 are about 15% larger than the NASA emissions (see Section 9.3.4). Further differences relate to the location and seasonal variations of emissions.

- Sensitivity to inclusion of NMHC chemistry. This study was conducted by making runs in which NMHC chemistry was excluded and comparing the results with those in which NMHC chemistry was included.
- Sensitivity to neglecting heterogeneous chemistry on background sulfate aerosols. This sensitivity was estimated by comparing results with and without the heterogeneous nitrogen pentoxide (N_2O_5) + H_2O reaction on aerosol. The aerosol surface area was derived from model calculations of the sulfur cycle. Hydrolysis of N_2O_5 on wet aerosol converts active NO_x into the reservoir species HNO_3, which is effectively removed by rain out. Hence, in the base case, less NO_y is present in the free troposphere and LS, and emissions by aircraft are more effective in producing O_3.
- Sensitivity to interannual variability in meteorology.
- Sensitivity to uncertainty in emissions in 2050. This study is carried out by using scenario G instead of scenario F from Table 4-4. Total NO_x emission changes from 2 to 3.5 Tg N yr^{-1}.

The results of the sensitivity studies are presented in Table 4-6. The table presents the relative sensitivity r (%) of each process by comparing aircraft-induced increases in global O_3 for the base case and the sensitivity case:

$$r = [(O_3,2-O_3,1)/O_3,1]*100\% \qquad (1)$$

where $O_3,1$ is the global O_3 increase (kg) up to 16 km for the base case in 2015 (i.e., scenario D–C) and $O_3,2$ is the same increase calculated for the sensitivity study (i.e., scenario D'–C').

Table 4-6: *Relative sensitivity (%) of global ozone perturbations from aircraft emissions.*

Sensitivity Case	IMAGES/ BISA	ECHAM3/ CHEM	TM3/ KNMI	UiO
Lightning				-16
Surface emissions IS92a				-11
NASA-ANCAT		-20		
NMHC chemistry	-35		-10	
Exclusion of N_2O_5 removal on aerosol	-10		0	
Interannual variability		±6.3		
Scenario G - F				+44

The sensitivity studies show that increases in background NO_x from lightning (sensitivity study 1) and different growth rates in surface emissions in different regions (sensitivity study 2) have only a slight impact on O_3 perturbation from aircraft emissions. In both cases, O_3 perturbations are reduced. This finding shows that O_3 production in the UT and LS is limited by NO_x, rather than by hydrocarbons. Interannual variability in meteorology (sensitivity study 6) and exclusion of heterogeneous removal of N_2O_5 in the models also led to a small change in global average O_3 perturbation (sensitivity study 5). Excluding hydrocarbon chemistry (sensitivity study 4) would have a significant impact on O_3, resulting in a smaller perturbation. Furthermore, there were significant differences in the results between the two models used to perform the study. Comparison of results with two different emission scenarios (sensitivity study 3) showed a noticeable impact on global O_3 perturbation. Sensitivity study 7, which was set up to test the response of O_3 perturbation to greatly increased NO_x emission from aircraft (estimated upper limit in 2050), showed that the response is nearly linear and similar to what is computed for smaller NO_x perturbations.

It should be noted that the comparisons in Table 4-6 are made for global average O_3 perturbations; sensitivities are larger on regional and seasonal scales.

4.3. Model Studies of Supersonic Aircraft

This section contains an evaluation of the effects of supersonic engine effluent on stratospheric O_3, NO_y, H_2O, and other trace constituents. The assessment of the potential future supersonic aircraft fleet was conducted primarily with 2-D and 3-D CTMs. These global models are designed to represent the dynamic, chemical, and physical processes of the stratosphere. Chapter 2 discusses the validity of these models to accurately represent the present atmosphere.

Because these future aircraft have not yet been designed, we treat our study of supersonic aircraft in a parametric way. This assessment performed calculations for more than 50 supersonic scenarios and examined variables such as $EI(NO_x)$ and fleet size, ambient SAD and chlorine radical abundances, SO_2 gas-to-particle conversion in the aircraft plume, cruise altitude sensitivity, and model cold aerosol chemistry representation. Results from this section are incorporated in Chapters 5 and 6.

4.3.1. Models Used in Supersonic Aircraft Assessment

This section presents calculated responses of O_3 to HSCT aircraft from nine numerical models of the stratosphere. These nine modeling groups have been responsible for most recent publications on the subject. We restrict our analysis to this set of results because it is difficult to compare model results unless the models are produced in the same way. For example, it is difficult to compare the results here with recent publications of Dameris *et al.* (1998) because they isolated

Table 4-7: *Models that contributed results to this report.*

Model Name	Institution	Model Team
2-D Models		
AER	Atmospheric and Environmental Research, Inc., USA	Malcolm Ko, Debra Weisenstein, Courtney Scott, Jose Rodriguez, Run-Lie Shia, N.D. Sze
CSIRO	Commonwealth Scientific and Industrial Research Organization Telecommunications and Industrial Physics, Australia	Keith Ryan, Ian Plumb, Peter Vohralik, Lakshman Randeniya
GSFC	NASA Goddard Space Flight Center, USA	Charles Jackman, David Considine, Eric Fleming
LLNL	Lawrence Livermore National Laboratory, USA	Douglas Kinnison, Peter Connell, Keith Grant, Douglas Rotman
THINAIR	University of Edinburgh, UK	Robert Harwood, Vicky West
UNIVAQ	University of L'Aquila, Italy	Giovanni Pitari, Barbara Grassi, Lucrezia Ricciardulli, Guido Visconti
3-D Models		
LARC	NASA Langley Research Center, USA	William Grose, Richard Eckman
SCTM1	University of Oslo, Norway	Michael Gauss, Ivar Isaksen
SLIMCAT	University of Cambridge, UK	Helen Rogers, Martyn Chipperfield

the effect of NO_x emission by excluding other emissions from their simulation.

The six 2-D models and three 3-D models are listed in Table 4-7. The 3-D models obtain 3-D distributions of the species by solving 3-D mass continuity and chemistry equations. The 2-D models calculate the zonal-mean concentrations of the species by solving zonal-mean mass continuity equations. Transport of the zonal-mean concentration is affected by zonal-mean (vertical and horizontal) winds and eddy diffusion fluxes that simulate the effects of zonally asymmetric motions. Chemical packages are used to compute zonal-mean production and loss rates for each species. In this section, we highlight some of the differences in model formulations that may have contributed to differences in model predictions. The reader is referred to the Technical Report on Supersonic Aircraft Effects for additional details.[2]

4.3.1.1. Domain and Resolution

The lower boundary of the SLIMCAT model is at 335 K potential temperature (~ 10 km). All other models have lower boundary at the ground. The top boundary varies from 60–90 km. Vertical resolutions in the lower stratosphere range are 1.2 (AER), 1.5 (LLNL, SCTM1), 2 (GSFC, CSIRO), and 3 km (UNIVAQ, LARC, THINAIR, and SLIMCAT). Horizontal resolutions for the 2-D models are 5° latitude for LLNL and CSIRO and 10°

latitude for the rest. For the 3-D models, horizontal resolutions are 5.5°x5.5° for LARC, 7.5°x7.5° for SLIMCAT, and 10° longitude x 7.8° latitude for SCTM1.

All of the models use temperature lapse rates to define the location of the tropopause, which changes with season and latitude. Most 2-D models use temperature from climatology, whereas the 3-D models use temperature from GCM output or objectively analyzed temperatures that are based on measurements—for example from National Centers for Environmental Prediction (NCEP) or European Centre for Medium-range Weather Forecasts (ECMWF). The troposphere

---

[2] The model results presented in this chapter are a summary of much model output. Supplemental material regarding the effects of supersonic aircraft are retrievable over the Internet. Supersonic model simulation information includes figures, tables, and text and is available at a NASA Langley Research Center computer until 31 December 2000:

> Host: uadp1.larc.nasa.gov
> Username: anonymous
> Password: your e-mail address
> Directory: IPCC_TECH_REPORTS/supersonic

Information concerning supersonic model simulations may also be viewed (retrieved) over the Web by going to the following URL: <ftp://uadp1.larc.nasa.gov/IPCC_TECH_REPORTS/supersonic/>.

is distinguished from the stratosphere in the 2-D models by assigning large values of horizontal diffusion coefficient (K_{yy} typically 1–1.5×10^6 m²/sec) and vertical diffusion coefficient ($K_{zz} \sim 4$–10 m²/sec) to simulate rapid mixing. Studies indicate that stratosphere-troposphere exchange may be dominated by transport from the mid- and high-latitude lower stratosphere to the troposphere (Eluszkiewicz, 1996). In a 2-D model, this transport will manifest itself as eddy flux along isentropic surfaces across the tropopause boundary (Shia *et al.*, 1993). If this premise is true, calculated residence time of aircraft emissions in the LS would be sensitive to the horizontal and vertical resolutions of the models because resolution constrains the location and seasonal variation of the tropopause.

4.3.1.2. *Temperature, Transport Parameters, and Solvers*

The THINAIR model is the only model that calculates temperature and transport circulation consistent with calculated O_3. The rest of the models generated results in CTM mode; in other words, results are obtained using pre-calculated temperature and transport fields. Different methods are used to compute residual mean circulation and eddy diffusion coefficients. Transport circulation for the UNIVAQ 2-D model is taken from a low-resolution spectral GCM (Pitari *et al.*, 1992). Winds and temperature for the LARC and SCTM1 models are from off-line GCM simulations, and those for SLIMCAT are from the UKMO analysis. There is no accepted method to validate computed transport parameters. Temperature is used to compute temperature-dependent reaction rate constants and, in some models, to predict the surface areas of PSCs. Different numerical schemes are used to solve the mass-continuity equations. Given the different methods used in deriving the transport parameters in the models, it is not surprising that there are large differences in calculated distributions of trace gases in the models. Large differences in model-simulated distributions of chemically inert tracers such as sulfur hexafluoride point to transport differences as a major contributor. The research community is trying to identify a climatological database for zonal-mean distributions of trace gases through an ongoing exercise (Models and Measurements II, Park *et al.*, 1999) that can be used to diagnose transport parameters.

4.3.1.3. *Chemistry*

Previous model intercomparison exercises have shown that chemistry solvers in most models calculate the same partitioning of radicals under the same constraints (solar zenith angle, overhead O_3, local temperature, local sulfate surface area, and local concentrations of the long-lived species) when they are used as box models in photochemical steady-state. Because the 2-D models transport zonal-mean concentrations, zonal-mean production and loss rates are needed in the mass-continuity equations for long-lived species. Different techniques are used to obtain zonal mean production and loss rates, including integrating diurnally varying concentrations of radicals obtained by explicit time marching to compute the zonal-mean

rates or using diurnally averaged radical concentrations calculated from average solar zenith angles corrected by pre-calculated correction factors.

Reaction rate constants are from DeMore *et al.* (1997). There is no recommendation in DeMore *et al.* (1997) for the yield of hydrogen chloride (HCl) from the OH+ClO reaction. Lipson *et al.* (1997) recently measured a temperature-dependent yield of 5–6%. Most models assume that the HCl yield from the OH+ClO reaction to be 0%. However, the CSIRO, LARC, and SLIMCAT models assume a 5% yield of HCl from OH+ClO. Because of the nonlinear dependence of reaction rate constants on temperature, the zonal average reaction rate constant calculated using local temperature as the air parcel moves around the globe zonally is different from the rate constant calculated using zonal-mean temperature. Some 2-D models chose to account for this effect by using a zonal temperature distribution based on observations; other models just used zonal mean temperature.

The following heterogeneous reactions were identified in DeMore *et al.* (1997) as possible reactions on sulfate or PSC. Most models assume that the rate constant is in the form $\gamma v A/4$, where γ is the reaction probability, v is the thermal velocity of the reactant, and A is the surface area of the sulfate or PSC.

$$N_2O_5 + H_2O \rightarrow 2HNO_3$$
$$ClONO_2 + H_2O \rightarrow HNO_3 + HOCl$$
$$BrONO_2 + H_2O \rightarrow HNO_3 + HOBr$$
$$ClONO_2 + HCl \rightarrow HNO_3 + Cl_2$$
$$HOCl + HCl \rightarrow Cl_2 + H_2O$$
$$HOBr + HCl \rightarrow BrCl + H_2O$$

These reactions occur on the surface of particles (sulfate or PSC). In addition, all but two of the nine models (LLNL and UNIVAQ) include the reaction $N_2O_5 + HCl \rightarrow HNO_3 + ClNO_2$ on PSC. In most cases, H_2O and HCl molecules are adsorbed or dissolved in the particles, and the reaction proceeds as the

Table 4-8: *Background surface concentrations in 1992, 2015, and 2050 for long-lived gases.*

	1992	2015	2050
CFC-11 (pptv)	268	220	120
CFC-12 (pptv)	503	470	350
CFC-13 (pptv)	82	80	60
CCl_4 (pptv)	132	70	35
HCFC-22 (pptv)	100	250	15
CH_3CCl_3 (pptv)	135	3	0
HCFC-141b (pptv)	2	12	0
Halon-1301 (pptv)	3	1.4	0.9
Halon-1211 (pptv)	2	1.1	0.2
CH_3Cl (pptv)	600	600	600
CH_3Br (pptv)	10	10	10
CH_4 (ppbv)	1714	2052	2793
N_2O (ppbv)	311	333	371
CO_2 (ppmv)	356	405	509

other reactant collides with the particle. There are slight variations in the temperature dependence of γ for reactions on sulfate particles used in the various models. Previous analyses (Murphy and Ravishankara, 1994; Borrmann *et al.*, 1997; Michelsen *et al.*, 1999) have shown that γ is very sensitive to temperature for some reactions. As the air parcel goes around the globe, it experiences temperatures that are both lower and higher than the zonal mean temperature. Ignoring zonal asymmetry in temperature by using zonal mean values in calculations may underestimate the O_3 impact from HSCT (Pitari *et al.*, 1993; Considine *et al.*, 1994; Grooß *et al.* 1994; Weisenstein *et al.*, 1998). Because the sulfate surface area is specified in the calculations, the temperature-dependence comes from γ and v. Some models compute an effective γv based on a temperature-weighted γ and v from zonal-mean temperature, whereas other models compute a temperature-weighted product of γv. Some of the listed reactions also occur on type I PSCs (assumed to be nitric acid trihydrate (NAT)) or type II PSCs (assumed to be "ice"). Exact treatments in each model vary; see the Technical Report on Supersonic Aircraft Effects for details.

4.3.2. Definition of Scenarios

This section describes the scenarios for supersonic aircraft in 2015 and 2050 that were evaluated for this assessment. The premise for the future supersonic (proposed hypothetical HSCT) aircraft fleet is given in Chapter 9, along with descriptions of actual emissions databases. A subsonic fleet is included in all scenarios but is modified when an HSCT fleet is also present. The background atmosphere for the model scenarios was appropriate to either 2015 or 2050, though parametric studies of reactive chlorine (Cl_y) concentration and background SAD were also performed.

4.3.2.1. Background Atmospheres

Boundary conditions for the background atmosphere for long-lived gases for 1992, 2015, and 2050 (Table 4-8) are based on the IPCC IS92a scenario (IPCC, 1992, 1995). The boundary conditions used for halocarbons correspond to (a) 3.7, 3.0, and 2.0 ppbv of Cl_y in the upper stratosphere at 50 km at the Equator for 1992, 2015, and 2050, respectively; and (b) 15, 12.5, and 11.1 pptv of reactive bromine (Br_y) in the upper stratosphere at 50 km at the Equator for 1992, 2015, and 2050, respectively.

Background sulfate SAD was supplied for the unperturbed stratospheric aerosol case as described in Table 8-8 in WMO (1992). This case is designated SA0 aerosols. To approximate the effects of an active period of volcanic eruptions (e.g., the past 2 decades), some scenarios also evaluate the effects of 4xSA0. Treatments of PSCs and associated chemical and dynamical processes differ among models.

4.3.2.2. Aircraft Emission Scenarios

The supersonic aircraft scenarios for 500 and 1,000 HSCTs are based on recent work (Baughcum and Henderson, 1998) carried out for the NASA technology concept aircraft HSCT, which would cruise supersonically in the 17–20 km altitude range. The NASA subsonic scenarios described in Chapter 9 account for displacement of subsonic air traffic by supersonic aircraft. Baseline computations assume supersonic aircraft emissions with $EI(H_2O)=1230$ (1230 g H_2O kg[-1] fuel), $EI(NO_x)=5$ (5 g NO_2 kg[-1] fuel), and a range of sulfate emission levels. Parametric studies were conducted around these baseline HSCT scenarios by investigating NO_x emission index, fleet size, flight altitude, and background atmospheric conditions. These parametric studies are appropriate because the technology of a commercially viable supersonic airplane is not yet well-defined and results are needed to determine the sensitivity of the O_3 impact to the technology level. A description of each of the scenarios evaluated by the participating models is given in Tables 4-4, 4-9, 4-10, 4-11, and 4-12.

Table 4-9: Sulfate surface area density (SAD) fields used in this assessment. SAD distributions derived in the coupled AER 2-D/ Sulfate Microphysical models (i.e., SA1) are obtained by calculating the difference between perturbed and reference SADs in the model. For these calculations, an EI(S)=0.2 was assumed. This difference or aircraft-produced SAD is then added to the volcanically clean reference distribution (SA0). The cruise altitude for most HSCT scenarios is at a standard (Std) height.

SAD Name	HSCT Fleet Size	HSCT Cruise Altitude	SO₂ Gas to Particle Conversion	Reference
SA0	n/a	n/a	n/a	SAGEII, WMO (1992)
SA1	500	Std	50%	AER 2-D Model
SA2	1000	Std	50%	AER 2-D Model
SA2-2km	1000	-2 km	50%	AER 2-D Model
SA3	500	Std	100%	AER 2-D Model
SA4	1000	Std	100%	AER 2-D Model
SA5	500	Std	10%	AER 2-D Model
SA6	1000	Std	10%	AER 2-D Model
SA6-2km	1000	-2 km	10%	AER 2-D Model
SA7	500	Std	0%	AER 2-D Model

Table 4-10*: Extra baseline scenarios that include the fleet of "subsonic-only" aircraft necessary for comparison with certain supersonic scenarios. Bold italicized text highlights difference from scenario D.*

Scenario	Year	Cl_y (ppbv)	PSC Representation	SAD Representation
D	2015	3.0	Included	SA0
D3	2015	3.0	Included	***4 x SA0***
D4	2015	3.0	***Removed***	SA0
D5	2015	***1.0***	Included	SA0
D6	2015	***4.0***	Included	SA0
D7	2015	3.0	Included	***Sulfate microphysics with SO_2 gas-to-particle conversion of 50%***
D8	2015	3.0	Included	***Sulfate microphysics with SO_2 gas-to-particle conversion of 100%***
D9	***2050***	***2.0***	Included	SA0

The HSCT scenarios use a number and letter designation preceded by the letter S (e.g., the first scenario is S1a). HSCT scenarios contain subsonic aircraft as well as HSCT commercial aircraft, with the combination accounting for the same passenger demand as in subsonic-only scenarios for 2015 and 2050. The HSCT scenarios in Tables 4-11 and 4-12 are generally analyzed relative to the corresponding 2015 and 2050 base plus subsonic scenarios.

The effects of sulfur, NO_x, H_2O, CO, and NMHC (as CH_4) are simulated in the scenarios. The treatment of sulfur emission is discussed later in this section. For the other species, the aircraft effluents are put into the model as follows: Gridded fuel burn data (kg fuel/day) are first mapped into the model grid; the amount of material emitted into each grid box is given by the product of the fuel burn and the EI; and the emitted material is put into the grid box at each time step with the equivalent rate.

In this approach, we ignore the effect of plume processing and assume that these emitted materials are instantaneously mixed into the grid box. This assumption is probably valid in most of the stratosphere, though it may not be valid in the cold polar lower stratosphere during winter. In these regions, chemical processes are strongly nonlinear, thus raising concerns about the assumption. To date there have been no detailed wake model calculations supporting or rejecting this assumption. This important caveat should be remembered when considering the model results presented in this section.

The effect of sulfur emission on SAD depends on how much SO_2 is converted to small particles in the aircraft plume. Conversion of SO_2 to aerosol particles within the aircraft plume was found to perturb stratospheric SAD to a much

greater extent than equivalent SO_2 emissions (Weisenstein *et al.*, 1996, 1998). Atmospheric measurements in the plume of the Concorde (Fahey *et al.*, 1995) and other aircraft, along with modeling studies (Danilin *et al.*, 1997; Kärcher and Fahey, 1997; Yu and Turco, 1998), suggest a minimum 10% conversion, with much larger conversion rates possible. However, near-field particle observations (as discussed in Chapters 3 and 7) suggest a typical conversion of sulfur to sulfuric acid (H_2SO_4) of less than 10% with a large, yet not fully understood variability under the assumption that the aerosol composition is H_2SO_4/H_2O. Model sensitivity studies were designed to examine the full range of this uncertainty with conversion fractions of 0, 10, 50, and 100%.

In most scenarios, the reference atmosphere is consistent with a SAD using SA0. If the HSCT scenario assumes no sulfur emissions, SA0 is used for the HSCT case as well. In some HSCT calculations, the effect of increased SAD from SO_2 emissions by aircraft is also considered, using a range of different sulfate conversion fractions (Tables 4-9 through 4-12). The additional SAD fields were constructed by calculating sulfate surface area for background and HSCT conditions using the AER microphysical model coupled to their 2-D CTM (Weisenstein *et al.*, 1997). The SAD perturbation (HSCT-background) was derived in absolute units and was added to the SA0 background SAD distribution. The SA5 case assumes a 10% conversion rate for 500 HSCTs with an EI(SO_2) of 0.4. This case also roughly approximates a 5% conversion rate, for an EI(SO_2)=0.8, with the sulfur content of the fuel maintaining its current value. Figure 3-25 shows the annually zonally averaged perturbation of SAD (in μm^2 cm^{-3}) used in the SA5 scenario. SA6 is similar to SA5 but is based on 1,000 HSCTs. A 0% conversion with EI(SO_2)=0.4 is considered in SA7 for an aircraft

Table 4-11: *Percentage changes in total column ozone for each assessment model. The top value is for the Northern Hemisphere average; the bottom value is for the Southern Hemisphere average. The cruise altitude for most HSCT scenarios is at a standard (Std) height. Source gas boundary conditions are for the year 2015. Model results have been rounded off to one significant figure for clarity.*

IPCC Scenario	Cl_y (ppbv)	Fleet Size	$EI(NO_x)$	Altitude (km)	SAD Desc.	Ref. Atm.	AER-2D	GSFC-2D	UNIVAQ-2D	LLNL-2D	CSIRO-2D	THINAIR-2D	SLIMCAT-3D	SCTM1-3D	LARC-3D
S1a No H2O	3.0	500	5	Std	SA0	D	-0.07	-0.1	+0.2	-0.01	-0.09	-	-0.2	-0.2	-
							-0.03	+0.03	+0.1	+0.1	-0.07	-	-0.09	-0.007	-
S1b H2O only	3.0	500	0	Std	SA0	D	-0.6	-0.4	-0.4	-0.3	-0.3	-	-0.6	-	-
							-0.3	-0.8	-0.2	-0.3	-0.07	-	-0.7	-	-
S1c	3.0	500	5	Std	SA0	D	-0.3	-0.4	-0.002	-0.2	-0.2	-0.2	-0.4	-0.3	-0.05
							-0.1	-0.8	+0.02	-0.2	-0.1	-0.2	-0.6	-0.1	-0.1
S1d	3.0	500	10	Std	SA0	D	-0.3	-0.6	+0.2	-0.3	-0.3	-0.5	-0.5	-	+0.07
							-0.1	-0.7	+0.1	-0.1	-0.2	-0.3	-0.7	-	-0.03
S1e	3.0	500	15	Std	SA0	D	-0.3	-0.8	+0.4	-0.4	-0.5	-0.9	-	-	-
							-0.05	-0.7	+0.2	-0.01	-0.3	-0.5	-	-	-
S1f	3.0	500	5	Std	SA1	D	-1.0	-1.1	-0.4	-0.7	-0.5	-	-	-	-0.4
							-0.4	-1.1	-0.1	-0.5	-0.1	-	-	-	-0.3
S1g	3.0	500	10	Std	SA1	D	-0.8	-1.1	-0.2	-0.7	-0.5	-	-	-	-
							-0.3	-1.0	-0.01	-0.4	-0.2	-	-	-	-
S1h	3.0	500	5	Std	SA3	D	-1.1	-1.3	-0.5	-0.8	-0.5	-	-	-	-
							-0.4	-1.2	-0.2	-0.6	-0.1	-	-	-	-
S1i	3.0	1000	5	Std	SA0	D	-0.7	-0.9	-0.06	-0.5	-0.5	-0.4	-	-	-
							-0.3	-1.4	+0.005	-0.3	-0.2	-0.3	-	-	-
S1j	3.0	1000	10	Std	SA0	D	-0.7	-1.4	+0.1	-0.7	-0.7	-1.1	-	-	-
							-0.2	-1.4	+0.2	-0.2	-0.3	-0.6	-	-	-
S1k	3.0	500	5	Std	SA5	D	-0.8	-0.7	-0.2	-0.5	-0.4	-	-	-	-
							-0.3	-0.9	-0.05	-0.3	-0.1	-	-	-	-
S1l	3.0	500	5	Std	SA7	D	-0.6	-	-	-	-	-	-	-	-
							-0.3	-	-	-	-	-	-	-	-
S2a	3.0	500	5	+2 km	SA0	D	-0.7	-1.0	-0.1	-0.6	-0.6	-0.3	-	-	-
							-0.2	-1.2	-0.06	-0.3	-0.2	-0.2	-	-	-
S2b	3.0	500	5	-2 km	SA0	D	-0.1	-0.09	+0.01	-0.05	-0.05	-0.09	-	-	-
							-0.04	-0.2	+0.03	-0.01	-0.03	-0.09	-	-	-
S2c	3.0	500	10	-2 km	SA0	D	-	+0.1	-	-0.03	-0.003	-	-	-	-
							-	-0.06	-	+0.07	-0.03	-	-	-	-
S2d	3.0	500	15	-2 km	SA0	D	-	+0.3	-	-	-	-	-	-	-
							-	+0.07	-	-	-	-	-	-	-
S2e	3.0	1000	5	-2 km	SA0	D	-	-0.2	-	-0.1	-0.09	-0.2	-	-	-
							-	-0.4	-	-0.02	-0.05	-0.2	-	-	-
S3a H2O only	3.0	500	0	Std	4xSA0	D3	-0.5	-0.5	-0.4	-0.4	-0.3	-	-	-	-
							-0.2	-0.7	-0.1	-0.3	-0.1	-	-	-	-
S3b	3.0	500	5	Std	4xSA0	D3	-0.2	-0.04	+0.07	-0.2	-0.1	-	-	-	-
							-0.09	-0.5	+0.06	-0.2	-0.09	-	-	-	-
S3c	3.0	500	10	Std	4xSA0	D3	+0.04	+0.2	+0.4	+0.001	+0.03	-	-	-	-
							+0.02	-0.4	+0.2	-0.04	-0.08	-	-	-	-
S3d	3.0	500	15	Std	4xSA0	D3	+0.2	+0.4	+0.6	+0.09	+0.1	-	-	-	-
							+0.1	-0.3	+0.3	+0.07	-0.07	-	-	-	-

Table 4-11 (continued)

IPCC Scenario	Cl_y (ppbv)	Fleet Size	$EI(NO_x)$	Altitude (km)	SAD Desc.	Ref. Atm.	AER-2D	GSFC-2D	UNIVAQ-2D	LLNL-2D	CSIRO-2D	THINAIR-2D	SLIMCAT-3D	SCTM1-3D	LARC-3D
S3e	3.0	500	5		Std	4xSA0+ SA1-SA0	-0.5 -0.2	-0.4 -0.6	-0.09 -0.01	-0.3 -0.3	-0.2 -0.1	- -	- -	- -	- -
S3f	3.0	500	5		Std	4xSA0+ SA3-SA0	-0.6 -0.2	-0.5 -0.7	-0.2 -0.02	-0.4 -0.3	-0.3 -0.1	- -	- -	- -	- -
S4a No PSCs	3.0	500	0		Std	SA0	-0.5 -0.3	-0.4 -0.3	-0.3 -0.1	- -	-0.2 -0.07	- -	- -	- -	- -
S4b No PSCs	3.0	500	5		Std	SA0	-0.2 -0.1	-0.4 -0.3	+0.02 +0.001	- -	-0.2 -0.1	- -	- -	- -	- -
S4c No H_2O, PSCs	3.0	500	5		Std	SA0	-0.1 -0.1	-0.2 -0.02	+0.3 +0.1	- -	-0.09 -0.07	- -	- -	- -	- -
S5a	1.0	500	5		Std	SA0	-0.3 -0.1	-0.5 -0.5	-0.006 -0.02	- -	- -	- -	- -	- -	- -
S5b	1.0	500	5		Std	SA1	-0.5 -0.2	-0.7 -0.7	-0.3 -0.1	- -	- -	- -	- -	- -	- -
S6a	4.0	500	5		Std	SA0	-0.4 -0.1	-0.4 -0.8	+0.03 -0.01	- -	- -	- -	- -	- -	- -
S6b	4.0	500	5		Std	SA1	-1.0 -0.4	-1.1 -1.0	-0.4 -0.2	- -	- -	- -	- -	- -	- -
S7a Microphysics	3.0	500	5		Std	50% Particle	-1.0 -0.4	- -	-0.5 -0.3	- -	- -	- -	- -	- -	- -
S7b Microphysics	3.0	500	5	+2 km	Std	50% Particle	-1.4 -0.5	- -	-0.7 -0.4	- -	- -	- -	- -	- -	- -
S7c Microphysics	3.0	500	5	-2 km	Std	50% Particle	-0.6 -0.3	- -	-0.2 -0.1	- -	- -	- -	- -	- -	- -
S7d Microphysics	3.0	500	10		Std	50% Particle	-0.9 -0.4	- -	-0.2 -0.2	- -	- -	- -	- -	- -	- -
S8a Microphysics	3.0	500	5		Std	100% Particle	-1.1 -0.5	- -	-0.5 -0.3	- -	- -	- -	- -	- -	- -

fleet of 500 planes. The assumption of 50% conversion with $EI(SO_2)=0.4$ is considered here in SA1 and SA2 for aircraft fleets of 500 and 1,000 planes, respectively. The assumption of 100% conversion with $EI(SO_2)=0.4$ is considered in SA3 and SA4 as an upper limit. SA3 assumes 500 aircraft, whereas SA4 assumes 1,000 aircraft. SAD was also computed for fleets flying 2 km lower. A summary of all SAD distributions used in these scenarios is given in Table 4-9.

4.3.3. Model Results for Supersonic Aircraft Emissions

4.3.3.1. NO_y and H_2O Enhancement

Supersonic aircraft emissions of NO_x and H_2O were incorporated within six 2-D models and three 3-D models. These supersonic aircraft emissions are primarily deposited within the Northern Hemisphere LS. Figure 4-6 shows calculated perturbations in NO_y and H_2O from HSCT emissions in June 2015 with

$EI(NO_x)=5$ (scenario S1c-D). Maximum perturbations occur in the Northern Hemisphere LS and range from 0.6 to 1.0 ppbv for NO_y and from 0.4 to 0.7 ppmv for H_2O. Based on the calculated perturbation in NO_y and H_2O, one can conclude that the transport fields of these multi-dimensional assessment models are significantly different. For example, the GSFC model shows a significantly larger calculated perturbation in Northern Hemisphere lower stratospheric NO_y than the other models (20–40% higher peak NO_y abundance). This model also transports more NO_y and H_2O higher into the middle stratosphere (MS) and into the Southern Hemisphere than the other models. By contrast, the AER model tends to isolate supersonic emission increases in NO_y and H_2O within the Northern Hemisphere LS. The THINAIR and LLNL models calculate a NO_y perturbation from supersonic aircraft in the Northern Hemisphere LS that approaches that of the GSFC model. However, these models transport much less aircraft-enhanced NO_y and H_2O to the Southern Hemisphere. CSIRO model NO_y and H_2O perturbations are similar to those of

Table 4-12: *Percentage changes in total column ozone for each assessment model. The top value is for the Northern Hemisphere average; the bottom value is for the Southern Hemisphere average. Source gas boundary conditions are for the year 2050. Model results have been rounded off to one significant figure for clarity.*

IPCC Scenario	Cl_y (ppbv)	Fleet Size	$EI(NO_x)$	Altitude (km)	SAD Desc.	Ref. Atm.	AER-2D	GSFC-2D	UNIVAQ-2D	LLNL-2D	CSIRO-2D	THINAIR-2D	SLIMCAT-3D	SCTM1-3D	LARC-3D
S9a	2.0	500	5	Std	SA0	D9	-0.3	-0.5	+0.03	-0.2	-0.3	-	-	-	-
							-0.1	-0.8	+0.03	-0.1	-0.1	-	-	-	-
S9b	2.0	500	10	Std	SA0	D9	-0.4	-0.8	+0.1	-	-0.4	-	-	-	-
							-0.09	-0.8	+0.08	-	-0.2	-	-	-	-
S9c	2.0	500	5	Std	SA1	D9	-0.5	-0.8	-0.3	-	-0.4	-	-	-	-
							-0.2	-1.0	-0.08	-	-0.04	-	-	-	-
S9d	2.0	1000	5	Std	SA0	D9	-0.6	-1.0	-0.08	-0.5	-0.5	-	-	-	-
							-0.2	-1.4	-0.03	-0.3	-0.2	-	-	-	-
S9e	2.0	1000	5	Std	SA2	D9	-0.8	-1.3	-0.5	-0.8	-0.6	-	-	-	-
							-0.3	-1.7	-0.2	-0.6	-0.09	-	-	-	-
S9f	2.0	1000	5	Std	SA4	D9	-0.9	-1.4	-0.6	-0.9	-0.6	-	-	-	-
							-0.3	-1.8	-0.2	-0.6	-0.07	-	-	-	-
S9g	2.0	1000	10	Std	SA0	D9	-0.7	-1.6	+0.02	-0.8	-0.9	-	-	-	-
							-0.2	-1.5	+0.08	-0.2	-0.4	-	-	-	-
S9h	2.0	1000	5	Std	SA6	D9	-0.8	-1.1	-0.3	-0.6	-0.6	-	-	-	-
							-0.3	-1.5	-0.09	-0.4	-0.2	-	-	-	-
S10a	2.0	1000	5	-2 km	SA0	D9	-0.2	-0.3	+0.05	-0.1	-0.1	-	-	-	-
							-0.05	-0.5	+0.05	-0.02	-0.05	-	-	-	-
S10b	2.0	1000	10	-2 km	SA0	D9	-0.2	-0.6	+0.2	-0.2	-0.09	-	-	-	-
							+0.01	-0.5	+0.1	+0.04	-0.05	-	-	-	-
S10c	2.0	1000	5	-2 km	SA6-2km	D9	-0.4	-0.5	-0.1	-0.3	-0.2	-	-	-	-
							-0.1	-0.6	+0.001	-0.2	-0.04	-	-	-	-
S10d	2.0	1000	5	-2 km	SA2-2km	D9	-0.5	-0.6	-0.1	-0.4	-0.2	-	-	-	-
							-0.1	-0.6	-0.02	-0.2	-0.03	-	-	-	-
S11a	2.0	1000 Fa1	5	Std	SA0	F	-0.5	-0.9	-0.1	-	-0.5	-	-	-	-
							-0.2	-1.4	-0.03	-	-0.2	-	-	-	-
S12a	2.0	1000 Fe1	5	Std	SA0	G	-0.5	-0.9	-0.1	-	-0.5	-	-	-	-
							-0.2	-1.4	-0.03	-	-0.2	-	-	-	-

LLNL, except that the maximum NO_y abundance increase in the Northern Hemisphere LS is significantly less. The SLIMCAT model shows greater NO_y and H_2O enhancements in the Southern Hemisphere than the GSFC model but smaller Northern Hemisphere perturbations. Aircraft-induced NO_y enhancements in the middle stratosphere are greatest in the THINAIR, GSFC, and SLIMCAT models for Northern Hemisphere and tropical regions. Overall, the AER, CSIRO, LLNL, and THINAIR models show the greatest Northern Hemisphere isolation of supersonic aircraft effluent among the participating assessment models. The UNIVAQ, LARC, and SCTM1 models derive a maximum Northern Hemisphere LS NO_y abundance increase similar to CSIRO. However, these models derive relatively small horizontal gradients in the LS horizontal spread in NO_y and H_2O, extending from Northern Hemisphere mid-latitudes to Southern Hemisphere high latitudes, suggesting a weak tropical/mid-latitude barrier to mixing. The

UNIVAQ, SLIMCAT, SCTM1, and LARC models generate Southern Hemisphere LS NO_y and H_2O magnitudes that are similar to those of the GSFC model. To first order, the spread in model-derived changes in total O_3 for the ensemble of scenarios shown in Tables 4-11 and 4-12 is a direct result of large differences between supersonic aircraft-induced perturbations in the NO_y and H_2O fields.

4.3.3.2. Profile and Column Ozone Change

The percentage change in profile O_3 for June is shown in Figure 4-6c for scenario S1c-D. Above 25 km, in all models, local O_3 abundance is reduced as a result of including supersonic aircraft emissions of NO_x and H_2O. This result is not surprising because all assessment models represent the NO_x (middle stratosphere) and HO_x (upper stratosphere) chemical families

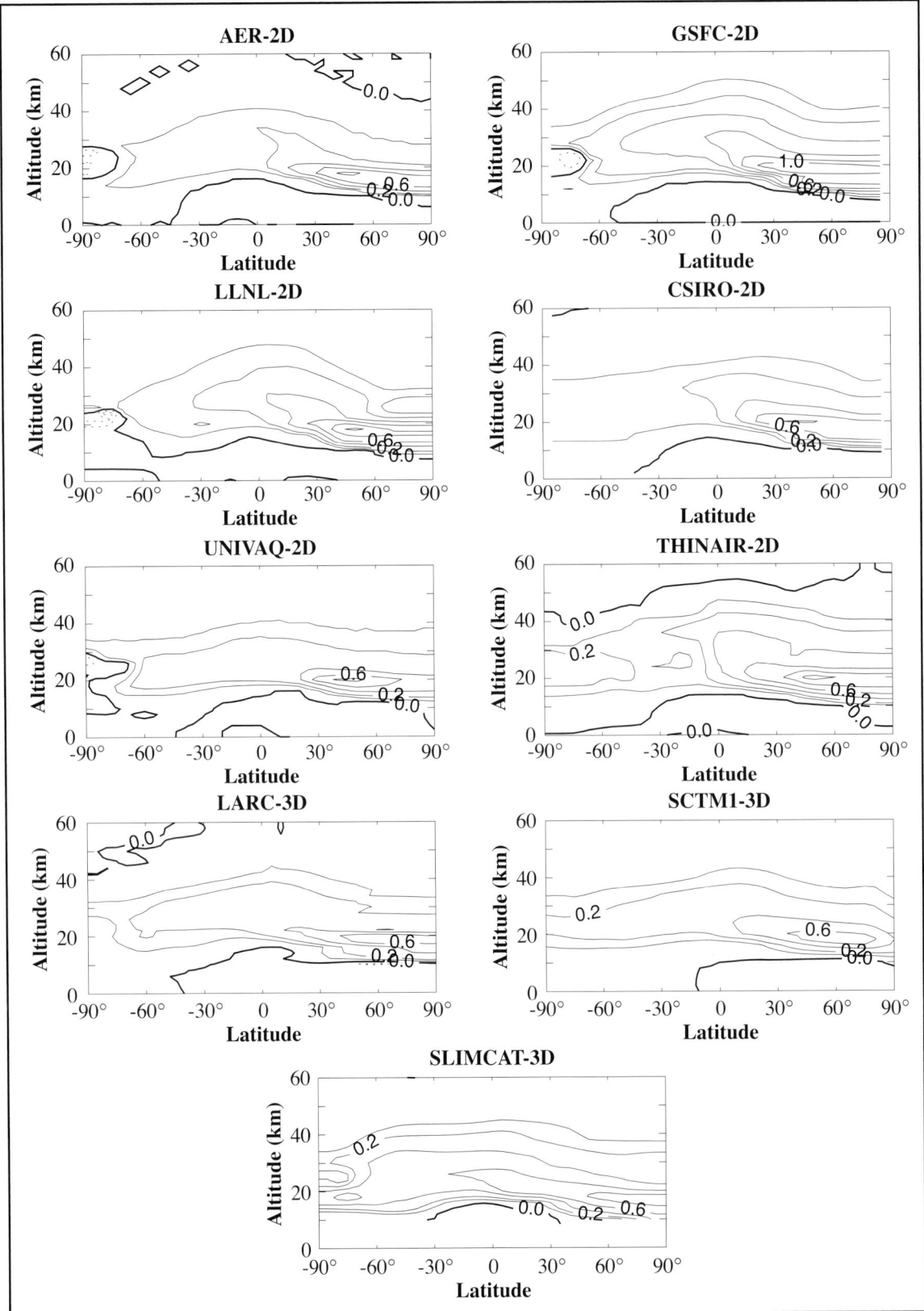

Figure 4-6a: NO_y change (ppbv) for S1c [$EI(NO_x)=5$, SA0, 500 supersonic aircraft + subsonic aircraft] relative to D [subsonic aircraft only] in June 2015.

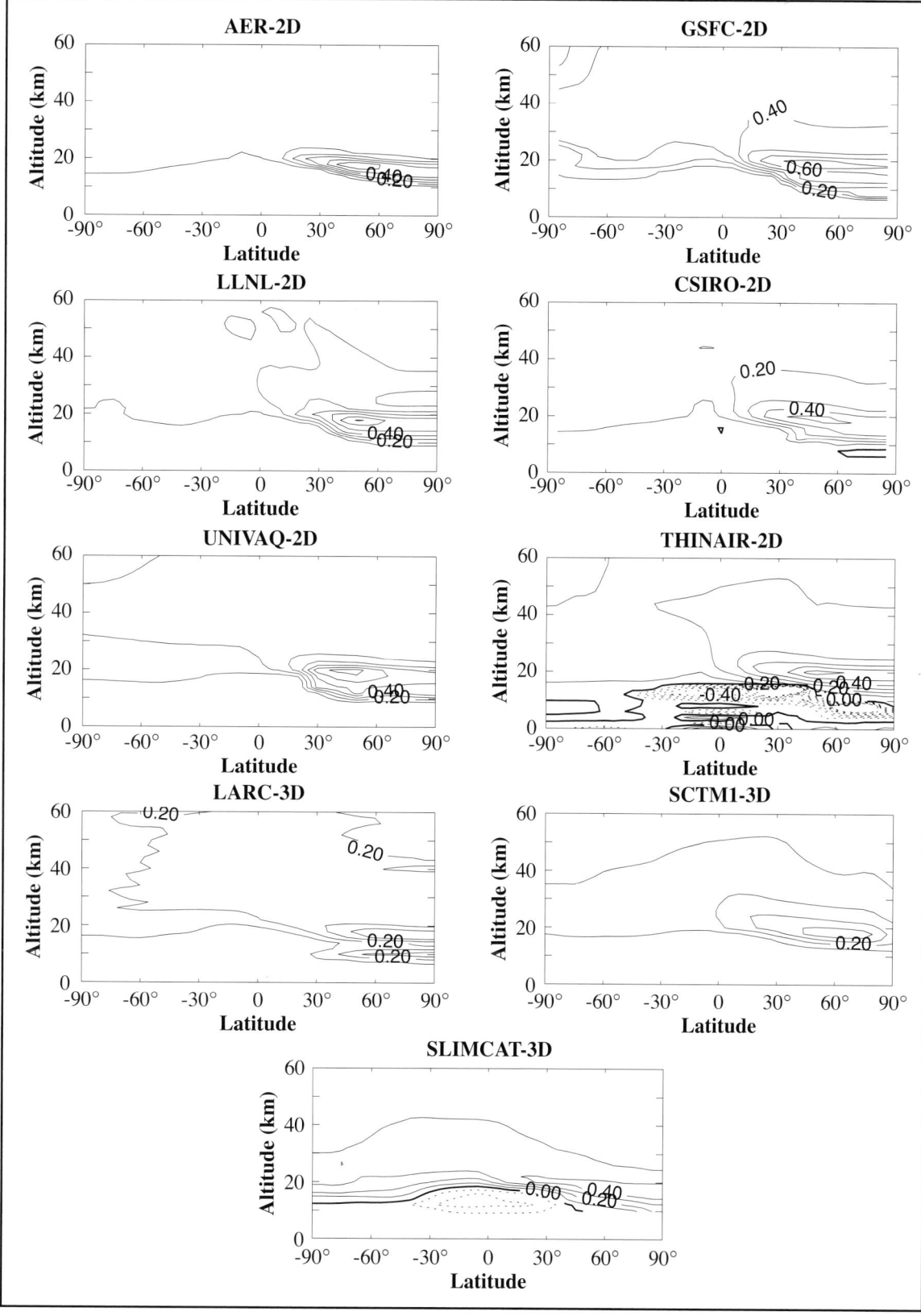

Figure 4-6b: H₂O change (ppmv) for S1c [EI(NO$_x$)=5, SA0, 500 supersonic aircraft + subsonic aircraft] relative to D [subsonic aircraft only] in June 2015.

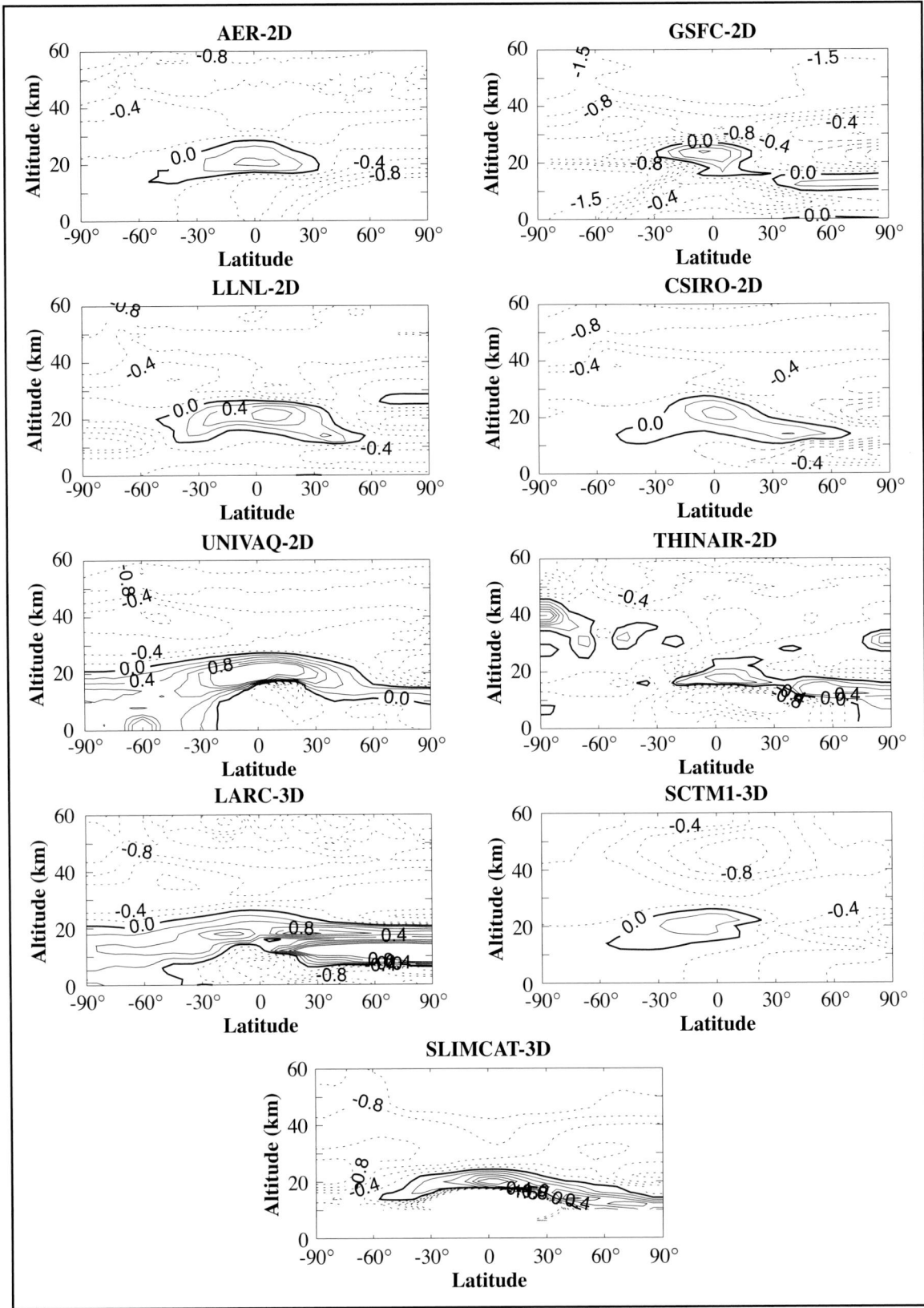

Figure 4-6c: O_3 profile change (%) for S1c [EI(NO_x)=5, SA0, 500 supersonic aircraft + subsonic aircraft] relative to D [subsonic aircraft only] in June 2015.

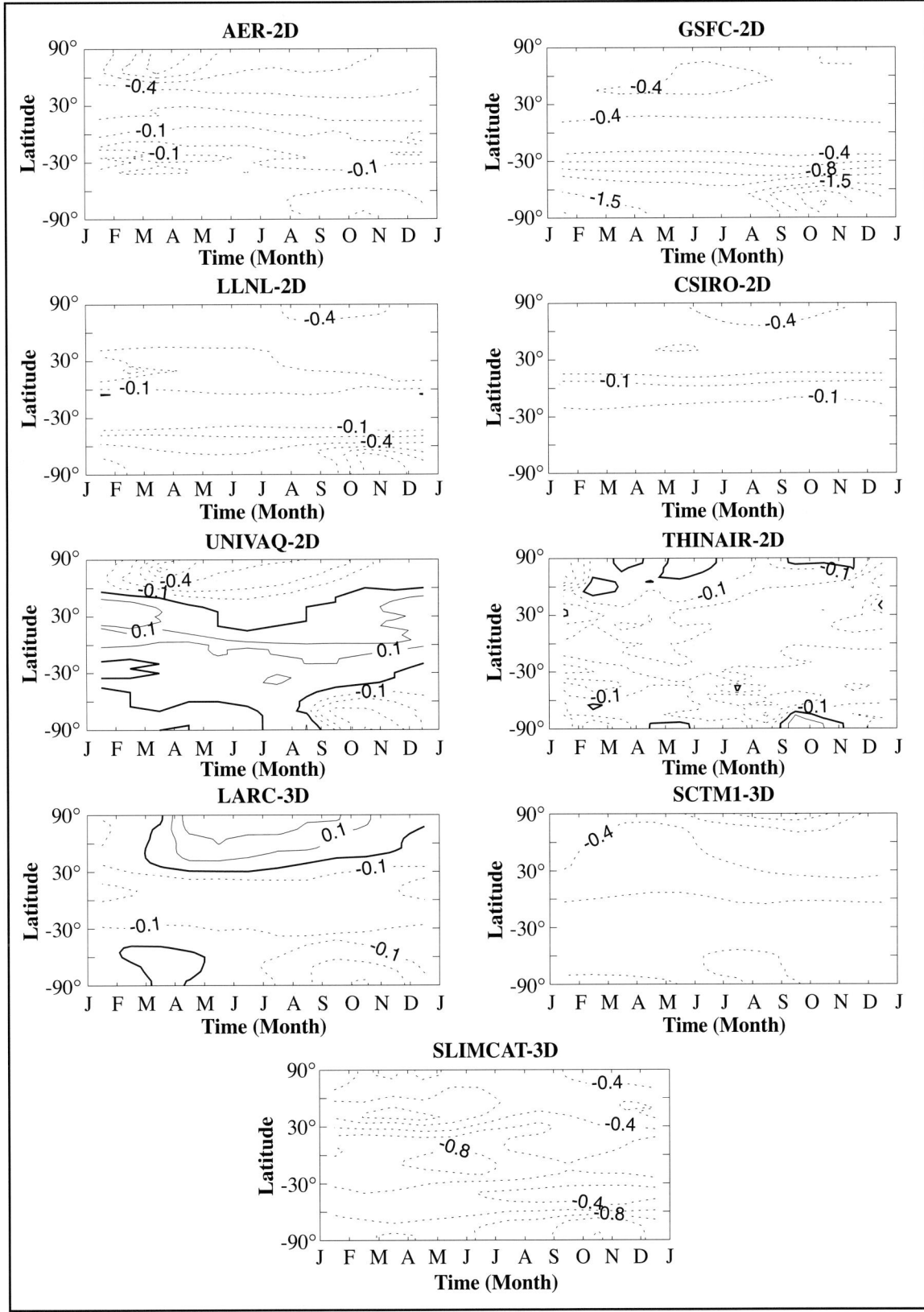

Figure 4-6d: Total column O$_3$ change (%) for S1c [EI(NO$_x$)=5, SA0, 500 supersonic aircraft + subsonic aircraft] relative to D [subsonic aircraft only].

as prime contributors to odd-oxygen loss above 25 km, in approximately the same relative importance. Therefore, sensitivity to NO_x and H_2O emissions from supersonic aircraft exhaust is similar in these regions and roughly proportional to calculated perturbations in NO_y and H_2O. Below 25 km, the relative odd-oxygen loss partitioning between NO_x, HO_x, and ClO_x-BrO_x chemical families is highly model-dependent. This situation is evident in the calculated local O_3 change. Models that are more HO_x dominant in the LS tend to show a more positive response to added aircraft NO_x. The opposite is true for models that are more NO_x dominant. These variations are a result of different methodologies of transport and chemistry incorporated in the individual models.

The percentage change in total column O_3 is shown in Figure 4-6d for scenario S1c-D. Six of the models (AER, CSIRO, GSFC, LLNL, SCTM1, and UNIVAQ) indicate smaller depletions in the tropics and larger depletions at mid- and high latitudes. The other three models (LARC, SLIMCAT, and THINAIR) show more complicated features. The LARC model predicts increases in Northern Hemisphere mid- and high latitudes and decreases in the tropics. The SLIMCAT model predicts decreases of more than 0.6% in the tropics, with less depletion predicted at Northern Hemisphere mid- and high latitudes and Southern Hemisphere mid-latitudes. The THINAIR model generally predicts larger depletions at lower latitudes. Decreases in Northern Hemisphere polar latitudes of greater than 1% are shown in the AER model results. The GSFC and SLIMCAT models predict decreases in the Southern Hemisphere polar spring of greater than 1%.

4.3.3.3. Column Ozone Sensitivity to Supersonic Emission of NO_x and H_2O

Sensitivity studies of supersonic engine emissions of H_2O and NO_x were carried out. These studies investigated individual and combined impacts on O_3 from H_2O and NO_x emissions within an atmosphere that was volcanically clean (SA0). In Figure 4-7a, the change in annual average Northern Hemisphere total column O_3 as a function of $EI(NO_x)$ is shown for a fleet of 500 supersonic aircraft in the year 2015 (Scenarios S1b-e). The reference atmosphere included emissions from a representative 2015 fleet of subsonic aircraft. Under the conditions of these scenarios, the models predict that water emissions [corresponding to $EI(NO_x)=0$] have the most significant impact. For $EI(NO_x)=0$, the participating models all derive a total O_3 depletion in the 0.2–0.6% range.

For low emission indices of NO_x [$EI(NO_x)=5$ to 10], the predicted decrease in O_3 is dominated by emitted H_2O; in most models, the addition of NO_x emissions leads to less predicted O_3 depletion. Because O_3 loss in the lower stratosphere tends to be dominated by HO_x constituents and supersonic aircraft emissions are most important in the lower stratosphere (see Figure 4-6), the predicted importance of H_2O emissions is understandable.

Several models show relatively little sensitivity to NO_x emissions over the $EI(NO_x)=5$ to 15 range. Including NO_x emission at an

$EI(NO_x)=5$ level, S1c had little effect on total column O_3 change for the LLNL, GSFC, and CSIRO models [relative to $EI(NO_x)=0$]. For the AER, SLIMCAT, and UNIVAQ models, added NO_x buffers the larger total O_3 depletion derived when H_2O is the only emitted constituent. When NO_x emission is included, six of the models (CSIRO, GSFC, LLNL, SCTM1, SLIMCAT, and THINAIR) calculate a larger O_3 depletion as $EI(NO_x)$ increases from 5 to 15 (S1c-e). The AER model showed little sensitivity to $EI(NO_x)$ in this range. Within the same range, the LARC and UNIVAQ models showed a positive O_3 change slope. In general, the model-derived O_3 change is in better agreement among participating models at $EI(NO_x)=5$, with a spread among models from approximately 0% total

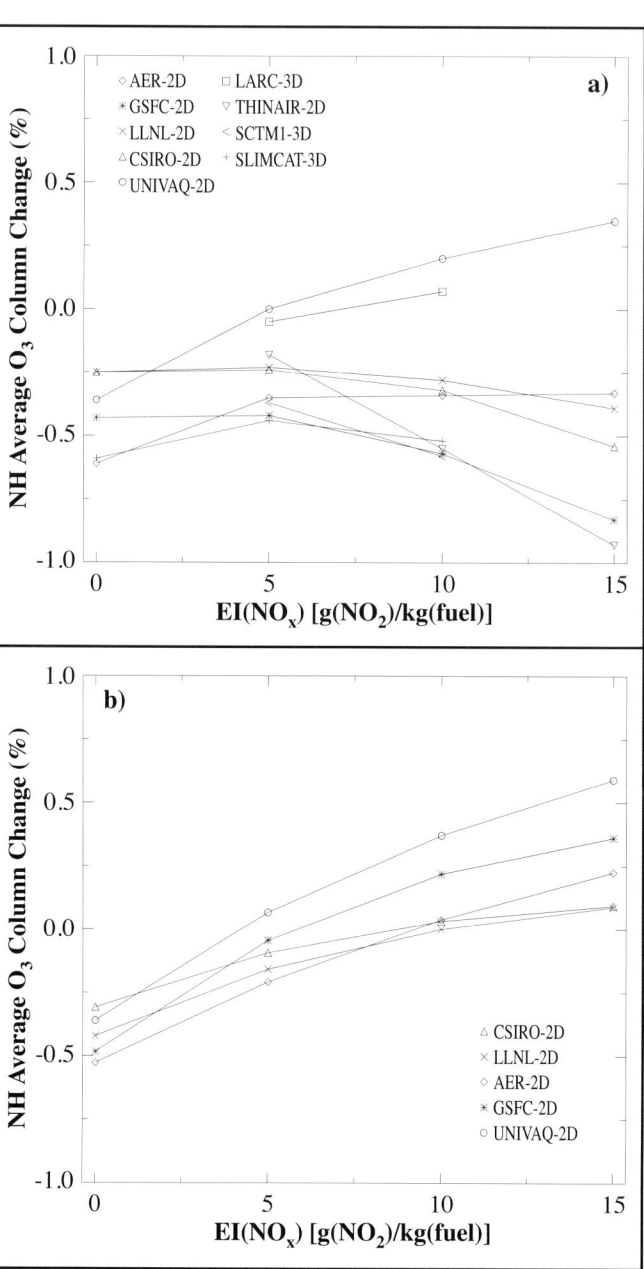

Figure 4-7: Northern Hemisphere total O_3 column change as a function of $EI(NO_x)$ in 2015 for an HSCT fleet size of 500 active aircraft with (a) SA0 sulfate distribution and (b) 4xSA0 sulfate distribution with no sulfur aircraft emissions.

column O_3 depletion (UNIVAQ) down to -0.4% (GSFC and SLIMCAT). At $EI(NO_x)=15$, the overall spread in model-derived total column O_3 change is much larger (+0.4% to -0.9%).

There is also a large spread in interhemispheric gradients in model-derived total column O_3 change (see Tables 4-11 and 4-12). For example, when the emission scenario is S1c [$EI(NO_x)=5$ with 500 supersonic aircraft], the AER, SCTM1, and CSIRO models derived a larger depletion in total column O_3 in the Northern Hemisphere than in the Southern Hemisphere. The UNIVAQ model had very little change in column O_3 in either hemisphere. The LLNL and THINAIR models derived a similar change in total column O_3 in both hemispheres. However, the ratio of Southern Hemisphere to Northern Hemisphere change in total column O_3 is greater than 1.5 in the GSFC, LARC, and SLIMCAT models. This larger depletion in the Southern Hemisphere correlates with increased H_2O and NO_y being transported to the Southern Hemisphere and the sensitivity of cold aerosol processes implemented within the GSFC, LARC, and SLIMCAT models.

4.3.3.4. *Ambient Surface Area Density Sensitivity*

Volcanic activity within the past several decades has greatly modulated ambient SAD. Two major volcanic eruptions—El Chichón (April 1982) and Mt. Pinatubo (June 1991)—are responsible for most of the observed SAD change. Satellite instruments (SAM, SAGE I, SAGE II, and SME; e.g., Thomason *et al.*, 1997) have shown that average SAD in the lower stratosphere between 1979 and 1995 was greater than that during volcanically clean periods (SA0; WMO, 1992) by a factor of 8 in the equatorial region and a factor of 2 to 4 at higher latitudes. These enhancements decay with e-folding time scales on the order of 1 to 2 years (Yue *et al.*, 1991; Thomason *et al.*, 1997).

For this assessment, we examined the impact of supersonic aircraft at a SAD of four times the volcanically clean condition (4xSA0). Results from these higher SAD sensitivity studies are shown in Figure 4-7b. Unlike the volcanically clean ambient atmosphere (Figure 4-7a), when the $EI(NO_x)$ varies between 0 and 15 (S3a-d), all participating models derived a positive slope to the Northern Hemispherical change in total column O_3. All models show a strong interference of the H_2O only [$EI(NO_x)=0$] total O_3 depletion when NO_x emissions are included [$EI(NO_x)=5$]. This response is consistent with the idea that at higher ambient SAD, NO_x catalytic processes are less important, mainly as a result of conversion of active NO_x radicals to nitric acid by heterogeneous reactions on sulfate aerosols. As supersonic aircraft NO_x is enhanced, the additional NO_x interferes with ClO_x, BrO_x, and HO_x odd-oxygen loss processes, reducing the net total odd-oxygen loss.

It should be noted that the model-derived impact on total column O_3 from increasing ambient SAD by natural causes (approximately 2–3% decreases in Northern Hemisphere mid-latitude total column O_3 change; Solomon *et al.*, 1996) exceeds the model-derived impact from supersonic aircraft (typically less than 1% Northern Hemisphere mid-latitude total column O_3 change).

4.3.3.5. *Sulfate Aerosols and Polar Stratospheric Cloud Sensitivities*

Heterogeneous reactions are important in defining the O_3 removal rate in the stratosphere. These reactions occur on sulfate aerosols and PSC particles, though sulfate aerosols have the greater influence because of their global distribution. New sulfate aerosols are produced by aircraft emissions; sulfate particles are generated by plume processing of part of the emitted SO_2/SO_3 (Fahey *et al.*, 1995; Brown *et al.*, 1996; Danilin *et al.*, 1997; Kärcher, 1997; Kärcher and Fahey, 1997). The remaining gas-phase SO_2 is then transported on the atmospheric large/global scale, where it is oxidized to sulfuric acid and condenses onto preexisting sulfate aerosol particles (Bekki and Pyle, 1992; Pitari *et al.*, 1993; Weisenstein *et al.*, 1998). Aircraft also emit H_2O and NO_x and therefore have the potential to increase the saturation ratios of water vapor and HNO_3, so that the size distribution, mass, and SAD of PSCs may be perturbed (Peter *et al.*, 1991; Pitari *et al.*, 1993; see Section 3.3.6).

4.3.3.5.1. *Sulfate aerosol sensitivity—SO_2 gas-to-particle conversion in the supersonic aircraft plume*

Supersonic aircraft consuming sulfur-containing fuel will have some effect on sulfate aerosol amounts in the stratosphere (see

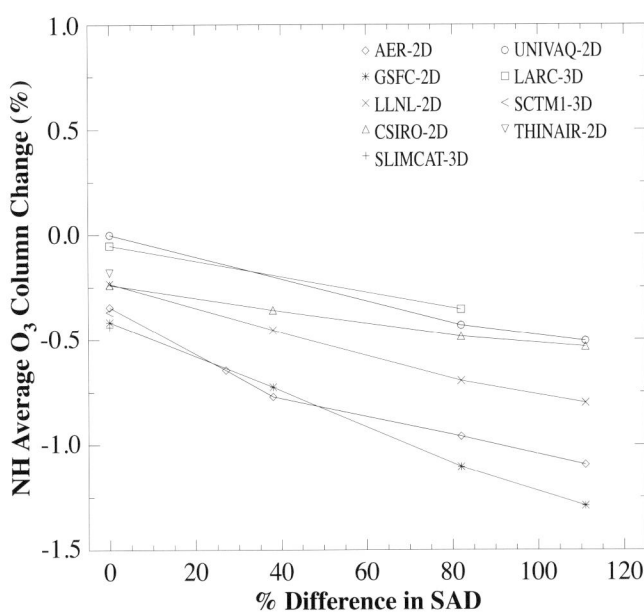

Figure 4-8: Northern Hemisphere total O_3 column change as a function of surface area density (SAD) between 14–21 km and 33–90°N. Calculations are for $EI(NO_x)=5$ and an HSCT fleet size of 500 aircraft in the 2015 atmosphere. The 27, 38, 82, and 111% change in SAD correspond to SA7, SA5, SA1, and SA3 as listed in Table 4-9, respectively.

Section 3.7.3). In this assessment, the increased SAD from sulfur conversion to particles in the plume had a significant impact on stratospheric ozone. Figure 4-8 shows the percentage change in annual average Northern Hemisphere total column O_3 correlated to the percentage increase in SAD for given SO_2 gas-to-particle conversion assumptions. The percentage SAD increase is derived by averaging the SAD change relative to the volcanically clean atmosphere (SA0) within the region between 14–21 km and 33–90°N. The SAD percentage increases for 0, 10, 50, and 100% SO_2 gas-to-particle conversion assumptions in the plume are 27, 38, 82, and 111%, respectively. The 0% SAD increase on the abscissa is taken from supersonic scenario S1c; it represents a perturbation without any consideration of aircraft sulfur emissions. The AER model was the only model to compute the 0% SO_2 gas-to-particle conversion impact on total O_3 change. Here, the additional aircraft-emitted SO_2 gas increased the ambient SAD by 27%, altering the total O_3 decrease for the AER model from 0.3 (SA0) to 0.6% (SA7). When particle conversion was assumed, all participating models derived relatively large decreases in total O_3. In fact, AER and GSFC model results for the Northern Hemisphere showed O_3 depletions of greater than 1% when a 50% SO_2 gas-to-particle conversion efficiency was assumed.

4.3.3.5.2. PSC sensitivity

The change in SAD of PSCs from HSCT aircraft emissions has been calculated by the AER, GSFC and UNIVAQ 2-D models. The surface area increase is a consequence of the stratospheric injection of H_2O and NO_y from aircraft. This process has the effect of increasing the saturation ratios of both water vapor and nitric acid, producing an enhancement of PSC particle mass and SAD. Changes in H_2O and NO_y from HSCT aircraft at 18 km and 60°N and 60°S are listed in Tables 4-13a and 4-13b. The average additional SAD of PSC1+PSC2 aerosols in the 12–24 km altitude layer and poleward of 60°N during December-January-February (DJF) or poleward of 60°S during June-July-August (JJA) are shown in Tables 4-13c and 4-13d. These results should be interpreted with caution because PSC distributions are highly localized.

All three models assume PSC1 to be solid NAT particles, although the treatment for PSC1 and PSC2 formation varies from model to model. AER and GSFC use a thermodynamic equilibrium between gas and condensed phase for HNO_3 and H_2O (Hanson and Mauersberger, 1988), with an imposed size distribution. Supersaturation ratios of 10 and 1.4 are required before PSC1 and PSC2 form in the GSFC model, thus lowering the temperature threshold for particle nucleation by about 3 and 2 K, respectively. Denitrification and dehydration are included in the models by calculating HNO_3 and H_2O loss terms from sedimentation. The UNIVAQ model makes the calculation by integrating on the NAT/ice particle size distribution; the AER and GSFC models assume that all condensed HNO_3 and H_2O form NAT and ice particles of a prescribed size distribution. No supersaturation is assumed in the AER model. In these two models, perturbations of sulfate aerosols from

aircraft emission of SO_2 leave unaffected the PSC. The models do allow H_2O and HNO_3 changes from aircraft to affect PSC. The UNIVAQ model allows interaction between sulfate aerosols and PSC by adopting a microphysical code for both types of particles (Pitari et al., 1993). As the atmosphere cools, PSC1 particles are first formed by nucleating on preexisting sulfate aerosols; below the frost point, PSC1 become coated with ice, thus forming PSC2 particles. With this scheme, a simultaneous presence of sulfate, PSC1, and PSC2 particles is possible, and size distributions are calculated for all particle types.

The spread in absolute change of PSC SADs shown in Tables 4-13c and 4-13d is not surprising because of the different schemes for PSC formation adopted in the models. In particular, the absence of supersaturation assumptions in the AER model consistently produces larger PSC SAD. The spread in relative changes is closely related to different relative increases of H_2O and NO_y. About 50–60% (70–90%) of the calculated PSC perturbation in the Northern Hemisphere (Southern Hemisphere) is caused by HSCT H_2O emissions when EI(NO_x)=5. The more important role of H_2O emissions in the Antarctic winter is a consequence of persistent PSC2 particles there; in the Arctic winter, PSC1 particles contribute to almost the entire available PSC SAD.

4.3.3.5.3.   Column ozone sensitivity to PSCs

Representation of heterogeneous processes on cold aerosols (PSCs) is not consistent among the participating models (Sections 4.3.1.3 and 4.3.3.5.2). These differences are manifest when increased abundances of NO_y and H_2O from supersonic aircraft are considered. As mentioned previously, the transport fields of participating assessment models are significantly different; therefore, the amount of supersonic H_2O and NO_y dispersion between hemispheres and vertically within the same hemisphere is also different. Three models (AER, GSFC, and UNIVAQ) investigated the impact on total O_3 when cold aerosol representation was not included (S4b). These models removed their representation of SAD, dehydration, and denitrification that would be associated with cold aerosol processes. The results were mixed. When cold aerosol processes were included in the model chemistry formulation, the model-derived percentage change in Northern Hemisphere total column O_3 increased by factors of 1.1 and 1.5 for the GSFC and AER models, respectively (S1c vs. S4b). However, in the Southern Hemisphere the GSFC model-derived depletion in total O_3 increased by 2.5 when cold chemistry processes were included. There was minimal change in Southern Hemisphere average O_3 depletion in the AER model. When the EI(NO_x) is set to zero (S4a), a similar conclusion is also drawn (compare to S1b) by the AER and GSFC models. These results are consistent with the dispersion of supersonic H_2O and NO_y between the AER and GSFC models. The UNIVAQ model did show sensitivity to supersonic emissions in the total O_3 hemispheric averages; because the absolute impact was small and centered around zero, however, it is difficult to draw any significant conclusions.

Table 4-13a: H_2O changes from HSCT (18 km, ppmv).

Scenario	#HSCTs	EI(H_2O)	60°N January			60°S July		
			AER	GSFC	UNIVAQ	AER	GSFC	UNIVAQ
S1c	500	1230	0.67	0.84	0.57	0.13	0.26	0.14
S9d	1000	1230	1.19	1.49	1.05	0.24	0.47	0.28

Table 4-13b: NO_y changes from HSCT (18 km, ppbv).

Scenario	#HSCTs	EI(NO_x)	60°N January			60°S July		
			AER	GSFC	UNIVAQ	AER	GSFC	UNIVAQ
S1c	500	5	0.77	1.24	0.61	0.13	0.15	0.22
S1d	500	10	1.57	2.48	1.18	0.28	0.42	0.37
S1e	500	15	2.37	3.71	1.74	0.44	0.69	0.51
S9d	1000	5	1.36	2.14	0.95	0.25	0.15	0.38
S9g	1000	10	2.77	4.33	2.00	0.54	0.50	0.66

Table 4-13c: PSC1+PSC2 SAD changes from HSCT (12–24 km, 60–90°N, DJF).

Scenario	#HSCTs	EI(NO_x)	Absolute Change ($\mu m^2/cm^3$)			% Change		
			AER	GSFC	UNIVAQ	AER	GSFC	UNIVAQ
S1b	500	0	0.06	0.03	0.03	22	39	19
S1c	500	5	0.10	0.05	0.06	33	56	37
S1d	500	10	0.13	0.06	0.08	45	70	50
S1e	500	15	0.16	0.07	0.09	57	82	56
S9d	1000	5	0.19	0.09	0.10	66	102	62
S9g	1000	10	0.26	0.12	0.13	89	137	81

Table 4-13d: PSC1+PSC2 SAD changes from HSCT (12–24 km, 60–90°S, JJA).

Scenario	#HSCTs	EI(NO_x)	Absolute Change ($\mu m^2/cm^3$)			% Change		
			AER	GSFC	UNIVAQ	AER	GSFC	UNIVAQ
S1b	500	0	0.24	0.18	0.17	4.6	11	6.7
S1c	500	5	0.34	0.20	0.19	6.5	12	7.6
S1d	500	10	0.42	0.22	0.25	8.0	13	10
S1e	500	15	0.52	0.24	0.27	10	15	11
S9d	1000	5	0.64	0.37	0.37	12	23	15
S9g	1000	10	0.80	0.42	0.42	15	26	17

4.3.3.6. Ambient Chlorine Sensitivity

Total radical chlorine (Cl_y) abundance in the atmosphere is predicted to decrease in the future as a result of international agreements controlling production of anthropogenic organic source gases such as chlorofluorocarbons (CFCs) and hydrochlorofluorocarbons (HCFCs) (WMO 1995, 1999). Therefore, several sensitivity studies were performed to investigate the impact of ambient Cl_y levels on a given supersonic aircraft perturbation. Here, atmospheric chlorine amounts were varied by changing halocarbon boundary conditions. Four different ground halocarbon amounts were considered, which resulted in reactive Cl_y levels of 1.0, 2.0, 3.0, and 4.0 ppbv in the upper stratosphere. For reference, the 3.0 and 2.0 ppbv Cl_y levels are representative of predicted 2015 and 2050 atmospheric conditions, respectively. In all Cl_y sensitivity studies, only EI(NO_x)=5 and a fleet of 500 supersonic aircraft were considered. Sensitivity studies were conducted under volcanically clean conditions without aircraft sulfur emissions (SA0) and with sulfur emissions assuming 50% SO_2 gas-to-particle conversion in the plume (SA1). Without aircraft sulfur emissions, there was little discernible sensitivity to background Cl_y abundance (S1c, S5a, S6a, and S9a). With sulfur emissions included, however, total ozone depletion for all three participating models was smallest when Cl_y abundance was 1.0 ppbv. As Cl_y increased in abundance, the model-derived ozone depletion increased, maximizing and leveling off between 3.0 and 4.0 ppbv Cl_y (S1f, S5b, S6b, and S9c). Among the participating models, AER showed the most sensitivity to Cl_y abundance, followed by GSFC and UNIVAQ. It should be noted that the $Cl_y=2$ ppbv scenario (S9c) had N_2O and CH_4 boundary conditions consistent with the year 2050. The other three Cl_y sensitivity scenarios (S1f, S5b, and S6b) had N_2O and CH_4 abundances consistent with the year 2015. The AER model did investigate Cl_y sensitivity to N_2O and CH_4 boundary conditions by repeating the $Cl_y=2.0$ ppbv scenario with N_2O and CH_4 values representative of 2015 conditions. The model-derived change in NH total O_3 was -0.6% (instead of -0.7% under 2050 CH_4 and N_2O conditions).

4.3.3.7. Fleet Design

4.3.3.7.1. Supersonic cruise altitude sensitivity

Figure 4-9 (S1c, S2a, and S2b) shows the effect of lowering or raising the supersonic cruise altitude by 2 km relative to the baseline flight cruise altitude band of 17–20 km for an assumed fleet of 500 $EI(NO_x)=5$ aircraft. In most altitude sensitivity studies, ambient SAD was considered to be volcanically clean, and sulfur emission by HSCT aircraft was not included. The model-derived change in annual average Northern Hemisphere total column O_3 had a large dependence on emission altitude in participating assessment models. This cruise altitude impact on O_3 highlights the fact that as NO_x is emitted lower in the stratosphere, the residence time of supersonic aircraft emitted NO_x is significantly shorter. In addition, at a lower cruise altitude the chemical regime is quite different, and past modeling studies have shown that a crossover point exists where an additional NO_x molecule will produce O_3 based on the well-known CH_4-smog mechanism (Johnston and Quitevis, 1975). When the cruise altitude is shifted higher, emissions have a longer stratospheric lifetime and can be transported more readily to regions dominated by NO_x destruction of O_3.

Each of these models showed less total column O_3 depletion when the cruise altitude was shifted down by 2 km. Three models also investigated sensitivity to higher values of $EI(NO_x)$ (10 and 15) at the lower cruise altitude (S2c-d). The model-derived total column O_3 change generally increased in the direction of higher $EI(NO_x)$. These results are consistent with the 1995 NASA supersonic assessment (Stolarski et al., 1995).

Two models (AER and UNIVAQ) did look at sensitivity to cruise altitude with increased SAD from sulfur emissions as 50% particles (S7c). In this scenario, an enhanced SAD was not supplied but derived within the 2-D model, coupled with a sulfate microphysical model. The model-derived annual average Northern Hemisphere depletion in total column O_3 was reduced from -1.0 (S7a) to -0.6% (S7c) in the AER model and from -0.5 (S7a) to -0.2% (S7c) in the UNIVAQ model when the supersonic cruise altitude was lowered by 2 km. The opposite impact was derived by the AER model when the cruise altitude was raised 2 km. Here, the model-derived annual average Northern Hemisphere depletion in total O_3 increased from -1.0 (S7a) to -1.4 % (S7b) in the AER model and from -0.5 to -0.7% in the UNIVAQ model.

4.3.3.7.2. Supersonic fleet size sensitivity

In this assessment, fleet sizes of 500 and 1,000 supersonic aircraft were considered at $EI(NO_x)$ values of 5 and 10. The increase in fleet size from 500 to 1,000 aircraft included additional routes as well as additional aircraft; as a consequence, the geographical distribution of emissions changes with proportionally more flights in the tropics (Baughcum and Henderson, 1998). In addition, by increasing the fleet size, the amount of H_2O emitted is also increased proportionally. For most of the cases considered, doubling the fleet size had a nearly linear effect on the Northern Hemisphere column O_3 reduction. The exception is the UNIVAQ model, which showed almost no change with increasing fleet size. For the 2050 atmosphere with aircraft sulfur emission as 50% particles, increasing the fleet size from 500 to 1,000 aircraft has a less than linear effect, especially with the AER model.

4.4. Uncertainties in Model Results

It is not clear how we can properly represent the true uncertainty associated with atmospheric effects; such an assessment represents an unattainable goal for the present. The model results presented in this chapter have been calculated using state-of-the-art atmospheric models from many groups. As a

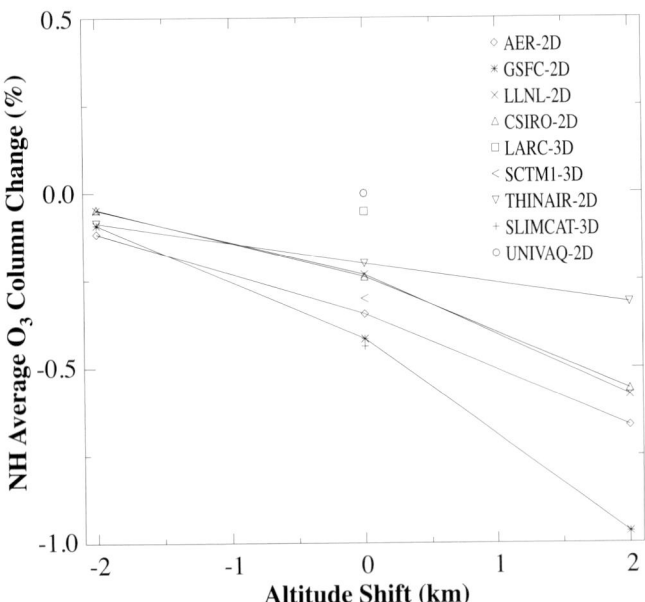

Figure 4-9: Northern Hemisphere total O_3 column change as a function of cruise altitude of the supersonic fleet in 2015 with $EI(NO_x)=5$, SA0 sulfate distribution, and no sulfur aircraft emissions.

result of their different treatment of physical, dynamic, and chemical processes, there is a considerable range in their computed impacts. We have attempted to bracket the results by using a combination of the range of different results between models for the same scenarios and results from sensitivity tests of parameters judged to be important.

Many factors contribute to the uncertainty of model calculations for present and future descriptions of the atmosphere. These factors include deficiencies in the representation of transport processes (e.g., stratosphere/troposphere exchange and tropical and polar region barriers); model resolution; the dimensionality of the models (i.e., 2-D or 3-D); deficiencies in chemical and physiochemical processes; and the state of the future atmosphere. Many of these issues cannot be considered in isolation. For example, the spatial resolution of a model may determine how well it is able to capture the meteorological details required for an accurate simulation of the dispersion of plumes and the formation of PSCs. Furthermore, as noted earlier, the models we used for our calculations can be considered to be of two types: Those that have been formulated for a detailed simulation of the troposphere, and those designed to simulate primarily stratospheric processes. This use of two model types to attack a problem whose resolution requires the ability to model the troposphere and stratosphere adds further uncertainty to the results obtained.

It should also be noted that sensitivity studies from the 3-D tropospheric subsonic scenarios are often based on runs performed with a single model, thus should be viewed with caution. In other cases, such as that of sensitivity to NMHC chemistry, the different models had different sensitivities.

As discussed in Chapter 2, time constants for transport in the stratosphere can be very long, and models must be run for many years to calculate the background state and the perturbed atmosphere. In addition, the parameter space that must be probed for unbuilt HSCTs is much more extensive than for the characteristics of the subsonic fleet. The use of 2-D models allowed for a more detailed analysis of the parameter space associated with supersonic aircraft. However, the results from these runs are also limited by the 2-D nature of the models.

Some of the uncertainty issues noted above have been discussed from a somewhat different perspective in Section 2.3. In addition, several international assessment reports (Stolarski *et al.*, 1995; WMO, 1995; Fabian and Kärcher,1997; Friedl, 1997; Brasseur *et al.*, 1998) have addressed some of these uncertainties. In the following paragraphs, we describe some features related to model uncertainty that we have identified in this study.

4.4.1. Uncertainties in Calculated Effects of Subsonic Aircraft Emissions

4.4.1.1. Transport and the Fuel Study

One means of assessing model uncertainties from transport is by comparing concentrations of inert or passive tracers with specified input and removal rates. For such species, it may be assumed that differences in distribution are solely a result of transport and its representation because chemistry plays no role. The results of such an experiment are described in Section 3.3.4.1 and Danilin *et al.* (1998). Because the emissions in this experiment are based on fuel use by subsonic aircraft, the experiment tests the transport characteristics of models in the UT and LS. Mixing ratios of fuel tracer are roughly 10–100 ng/g between 8–16 km north of 30°N. Southern Hemisphere stratospheric concentrations are about a factor of 4 less than in the Northern Hemisphere. Although the general pattern of the test tracer is qualitatively similar among various models, the maximum tracer concentrations deviate by a factor of 10 among the models. Some of this difference may be an artifact of low vertical resolution, but the results clearly point to uncertainties that need to be resolved.

One of the major problems with model tranport is that stratospheric-tropospheric exchange (STE) is poorly represented in current models of any dimension. Current theory has air entering the stratosphere through the tropics and returning at mid-latitudes. However, recent measurements of chemical tracers such as H_2O and CO, diagnostic of tropospheric air, suggest that the lowest few kilometers of the lowermost stratosphere can be impacted by air being convected from the lower troposphere (Dessler *et al.*, 1995; Lelieveld *et al.*, 1997). These areas require more evaluation.

One of the problems that has not been addressed in these studies is sub-grid-scale parameterizations. This issue includes dispersion of material in the plume to the regional scale, which is typical of the grid resolution of most models. In addition, exchange of air between the troposphere and the stratosphere takes place on a variety of scales, and only the larger scales are modeled with any reliability.

Thus, it is very difficult to assess inherent uncertainties in model results that are products of limitations in modeling transport. Nevertheless, as explained in Section 4.5, we attempt such an assessment by comparing the different models in a semi-quantitative manner.

4.4.1.2. Chemistry

The chemistry of VOCs (NMHC, acetone) and the transport of peroxides to the upper troposphere have recently been found to play important roles in the HO_x radical balance (Singh *et al.*, 1995; Folkins *et al.*, 1997; Jaeglé *et al.*, 1997; McKeen *et al.*, 1997; Prather and Jacob, 1997; Wennberg *et al.*, 1998). An additional source of HO_x in the UT could significantly influence O_3 production efficiencies. This uncertainty would propagate through the 3-D tropospheric models because there are large differences in chemistry modules. To some degree, this effect has been explored and the results tabulated in Table 4-6. Further model studies clearly are needed.

NO_x plays a key role in the O_3 generation process in the UT and LS, and there are significant differences in predicted NO_x

levels among the models. However, because of the limited number of observations and the large temporal and spatial variations of these observations, it is difficult to validate modeled upper tropospheric and lower stratospheric concentrations (see Chapter 2).

Although the process studies referred to in Chapter 2 point to a strong dependence of O_3 production on NO_x levels, the standard simulations and sensitivity studies presented in Chapter 4 show a limited impact of nonlinear chemistry on O_3 perturbations. For example, all of the 3-D CTM standard simulations for 1992, 2015, and 2050 calculated a nearly linear increase in O_3 perturbation as a result of increases in aircraft NO_x emissions despite increases in background NO_x levels. As noted above, there are large differences among models in their calculation of O_3 transport and chemistry between the UT and LS, and it is in this region where O_3 has a significant contribution to the infrared heating budget (Wang and Sze, 1980; Lacis *et al.*, 1990; Fortuin *et al.*, 1995).

Given the limitations of the tropospheric models to estimate O_3 perturbations in the tropopause region—in particular perturbations in the LS—a comparison of fractions of O_3 change that occur in the troposphere is made. Considering different characteristics of each model, meteorological and numerical, the estimated tropopause height may differ considerably among them. In addition, the "real" tropopause height may vary locally by at least 2 km, depending on the season and the local meteorology. Thus, we have applied the following crude estimate of tropopause height to all models: 8 km poleward of 65°, 10 km between 65° and 50°, 12 km between 50° and 35°, 14 km between 35° and 25°, and 16 km in the tropics. These figures are consistent with average tropopause heights used by the Total Ozone Mapping Spectrometer (Fishman *et al.*, 1990).

The fraction of O_3 change from 1992 aircraft emissions that occurs in the troposphere varies between 0.65 and 0.75, with an uncertainty of about 15%. The fraction calculated for the

stratospheric 2-D AER model was 0.81. The ECHAM3/CHEM model predicts the lowest tropospheric fraction (0.65). As noted above, the assumption is that this fraction of 0.65 represents minimum O_3 generation in the troposphere from subsonic emissions by the current (1992) fleet.

However, the model simulations just discussed did not include the generation of sulfate by aircraft, and consideration of this factor may alter the picture somewhat. Because no 3-D tropospheric models were available to investigate sensitivity to sulfate emissions, an attempt to assess their significance was made using the AER and UNIVAQ 2-D microphysical models. Although these models are limited in their applicability to the UT, they do yield results that suggest further work is required to evaluate the significance of calculated O_3 generation by subsonics.

Aircraft sulfur experiments have been carried out for 2015 using an $EI(SO_2)=0.4$ for a fleet flying nominally at 10 km. It was further assumed that the SO_2 emitted was converted in the plume to 10 nm (radius) particles with 50 and 100% efficiency. Table 4-14a summarizes the model predictions of aerosol increases at middle northern latitudes for the 10–14 km height range.

The importance of transport is clear: Although new particles are formed at about 11-km altitude, the models yield significant changes of SAD up to about 14 km, thus creating a potentially important link between subsonic emissions and stratospheric O_3. This conclusion is consistent with the results of the 1992 tracer experiment (see Section 3.3.4), in which 2-D and 3-D models calculate the largest fuel tracer mixing ratios in an altitude layer from about 10 to 14 km.

Table 4-14b gives concomitant O_3 column changes. The AER and UNIVAQ models show that the O_3 column increase from atmospheric NO_x aircraft perturbation tends to be mitigated when sulfate SAD is changed by the aircraft. This result is

Table 4-14a: Mid-latitude lower stratospheric mass density (MD, ng/cm^3) and SAD ($\mu m^2/cm^3$) with/without subsonic aircraft in 2015 [annual average, $EI(SO_2)=0.4$ g/kg, 10–14 km, 30–60°N].

Model	No Aircraft		50% Particles		100% Particles	
	MD	SAD	MD	SAD	MD	SAD
AER	48.3	1.0	55.0	1.6	59.2	2.1
UNIVAQ	59.0	0.7	63.5	2.4	65.7	3.2

Table 4-14b: Global and mid-latitude ozone column changes from subsonic aircraft in 2015 (annual average, %).

Model	No Sulfate Emission		50% Particles		100% Particles	
	Global	45°N	Global	45°N	Global	45°N
AER	0.79	1.29	0.57	0.94	0.51	0.86
UNIVAQ	0.30	0.72	0.14	0.29	0.07	0.15

mainly a consequence of more heterogeneous reactions on sulfate particles, leading to the release of ClO, BrO, and HO_2 in this layer. However, the reduction in magnitude of O_3 generation is much stronger in the UNIVAQ model as compared to the AER model. Clearly, this issue requires more study.

4.4.2. Uncertainties in Calculated Effects of Supersonic Aircraft Emissions

4.4.2.1. Transport

Clearly, an important uncertainty in the evaluation of the effects of aircraft is the representation of transport in the 2-D and 3-D models used to simulate the effect of supersonic aircraft emissions. The research community recognizes this general limitation, and an ongoing exercise (Models and Measurements II, Park *et al.*, 1999) should help to quantify some of the transport differences among the models. Models and Measurements II is an activity supported by NASA to test models through detailed comparisons with measurements. It is a follow-on of the Models and Measurements workshop completed several years ago (Prather and Remsberg, 1993). The results of Models and Measurements II should be available by the end of 1999.

New analysis techniques of models relying on "age of air" (Hall and Plumb, 1994), sulfur hexafluoride (SF_6) (Elkins *et al.*, 1996), and CO_2 (Boering *et al.*, 1996) measurements allow more detailed comparisons to test the transport of models. Most models tend to underestimate the observed "age of air;" therefore, the derived delta NO_y and H_2O for these models is probably a lower limit. As a result, the models will tend to underestimate O_3 depletion from supersonic aircraft.

Another measure of transport uncertainty is illustrated by significant differences in mixing ratios calculated for NO_y accumulation in the stratosphere from supersonic aircraft emissions (Figure 4-6a). This uncertainty reflects differences in lower stratospheric residence times for the different models.

The atmosphere undergoes large variations from one year to the next. These interannual variations will drive a different transport of constituents, which will lead to significant differences in the impact of supersonic emissions on O_3. Jackman *et al.* (1991) simulated very different aircraft-induced O_3 changes using separate dynamical fields; a slow (rapid) circulation simulation caused larger (smaller) O_3 decreases. The SLIMCAT model has been used with a yearly varying transport recently and has computed total O_3 decreases that vary significantly in the same months of different years.

4.4.2.2. Model Resolution/Dimensionality

As noted above, the formidable computational resources required to conduct long-term multiyear 3-D stratospheric simulations for assessment has made it necessary to use 2-D

models. Although the 2-D models have very sophisticated formulations for gas phase chemical processes, their meridianal resolution is typically 5° (550 km) or coarser in latitude and 1 km or more in altitude throughout the stratosphere. Intrinsically, this resolution limits their ability to treat LS transport processes. In addition, because atmospheric species are treated in a zonally averaged manner, effects such as PSC formation and chemical processing must be parameterized more crudely than in 3-D models.

Issues of spatial resolution and dimensionality of models are very pertinent to the treatment of PSCs and the processing of polar air. 2-D models do not adequately parameterize the zonal variation of temperature, thus the formation of PSCs. Even 3-D models do not capture the different horizontal temperature scales within the Arctic stratosphere. For example, even though synoptic conditions would suggest otherwise, low temperatures and PSCs may be induced by wave flow over mountains (Carslaw *et al.*, 1998); this mesoscale phenomenon is not directly addressed by most models to date. This uncertainty in assessing impacts from aircraft could be an even more important issue in trying to assess future impacts when the stratosphere may be cooler because of enhanced CO_2 mixing ratios (Shindell *et al.*, 1998).

Yet another, quite different, resolution problem that may limit our ability to model chemical effects is that of filamentation processes, whereby winter polar air consists of long but narrow (< 5 km) filaments of gas in which species densities are different. O_3 depletion studies performed at very high horizontal resolution for the winter Arctic (Edouard *et al.*, 1996) have underlined the uncertainty in calculating chemical loss of O_3 from differing resolutions.

4.4.2.3. Stratospheric Chemistry

Current models do quite well in reproducing the general behavior of O_3 in the stratosphere. Nevertheless, as is evident from the discussion in Chapter 2 and above, problems remain. For example, there appears to be a discrepancy of about 30% between calculated and measured O_3 levels near the stratopause (Osterman *et al.*, 1997). In addition, related problems emerge with a careful comparison with observations, such as the ratio of active to reservoir species for Cl_y (Michelsen *et al.*, 1996). These problems may be alleviated by the inclusion of new HCl and hydrogen bromide (HBr) formation rates (see Chapter 2).

The models generally assume the same set of reaction rate constants and photodissociation cross-sections based on the DeMore *et al.* (1997) recommendations. Comparisons of results for a subset of models (AER, CSIRO, GSFC, LARC, LLNL, SLIMCAT) in a benchmark at photostationary steady-state (Stolarski *et al.*, 1995) indicate that the differences were less than 1% for CSIRO and LLNL; a few percent for AER, LARC, and SLIMCAT; and a few percent up to 10–15% for GSFC. Similar results are not available for the other models, and it is not possible to conclude that the chemical solvers in all models are equivalent throughout the atmosphere; however, the choice of chemical solver does not appear to be a major uncertainty in our ability to perform assessment calculations.

Uncertainty in photochemical rate constants has been propagated through a 2-D assessment model using a Monte Carlo approach for an $EI(NO_x)=15$ HSCT fleet with background aerosols (SA0) and no sulfur emissions (Stolarski *et al.*, 1995). The effect of uncertainties in gas-phase reaction rate coefficients on perturbations of O_3 are estimated to yield an uncertainty on the order of 1% (1-sigma) in calculated Northern Hemisphere O_3 column perturbations. However, these experiments did not include uncertainties in photolytic cross-sections and heterogeneous processes, which could enhance the uncertainty significantly. Recently, box model sensitivity-uncertainty calculations for O_3 depletion from supersonic aircraft emissions, again with $EI(NO_x)=15$ and background aerosols, were performed at the most perturbed locale using localized outputs from a 2-D model (Dubey *et al.*, 1997). Guided by these sensitivities, 2-D model integrations were completed with nine targeted input parameters altered to 1/3 of their 1-sigma uncertainties to put error bounds on the predicted O_3 change. Results indicated local O_3 loss of 1.5 ± 3% in regions of large NO_x injections. We note, however, that these sensitivity studies were all carried out with relatively high $EI(NO_x)$ values [three times the $EI(NO_x)$ taken as standard for the current assessment], which is likely to further modify the intrinsic sensitivities of the models to perturbations.

4.4.2.4.   *Polar Stratospheric Cloud, High Cirrus Cloud, and Aerosol Processing*

Uncertainty in our knowledge of the present atmosphere affects our ability to calculate formation of and the chemical processing associated with PSCs. This uncertainty is particularly acute in the Arctic winter because early springtime temperatures are higher in the Arctic than in the Antarctic, and are quite close to the threshold temperature for the formation of PSCs. Thus, a temperature error of a few degrees can make a substantial difference in calculating the extent of processing. Objectively analyzed temperatures—those based on measurements and often processed for use in weather forecast models—are frequently used in 3-D CTMs to estimate chemical processing. However, it appears that UKMO and ECMWF objectively analyzed temperatures may be systematically too high by 1 to 2 K in winter (Knudsen, 1996; Pullen and Jones, 1997); this uncertainty will impact our understanding of the microphysics associated with heterogeneous chemistry.

The question of processing of reservoir species such as HNO_3 and HCl on PSCs remains uncertain, and hysteresis effects may be important. That is, the formation of PSCs by condensation and removal by evaporation may not be describable in terms of reversible processes, and the temperature history of the air parcel may be important. Currently, most models avoid this complication.

Conversion of reservoir species may occur as a result of dissolution of HCl in sulfate, forming a ternary solution, or may occur on frozen surfaces of water or NAT ice surfaces. The sensitivity of O_3 depletion to the details of these processes appears to be less important for Antarctic conditions; many models yield relatively complete conversion of reservoir species induced by cold vortex temperatures. However, as noted above, this effect could be more important for the relatively warmer Arctic vortex. This uncertainty associated with chemistry, combined with the dynamic uncertainty, suggests that modeling of the propagation of PSC processing effects onto global scales is very uncertain.

Solomon *et al.* (1997) pointed out that the presence of high cirrus clouds near the tropical tropopause could decrease O_3 in the stratosphere by activating chlorine constituents. Emissions of H_2O and NO_x from supersonic and subsonic aircraft have the potential to increase the occurrence of high cirrus clouds (see Chapter 2). This effect is not evaluated by the model simulations performed in this chapter.

4.4.2.5.   *Underlying Subsonic Fleet*

In all scenarios for 2050 except two (S11a and S12a), the subsonic fleet was as defined for the 2015 scenarios. For scenarios S11a and S12a, a more realistic representation of the subsonic fleet in 2050 was used. For example, scenario S11a was compared with the IS92a predicted subsonic fleet scenario F (scenario Fa1 in Chapter 9), and scenario S12a was compared with the high-demand subsonic fleet scenario G (scenario Fe1 in Chapter 9). As for all other supersonic scenarios, S11a and S12a contain subsonic aircraft emissions as well as HSCT aircraft emissions; the combination accounts for the same passenger demand as in subsonic-only scenarios F and G for 2050. Scenarios S9d, S11a, and S12a have similar supersonic components, but the subsonic components in the perturbed and the base runs (D9, F, and G, respectively) are very different. There are only minor differences in predicted total column O_3 change for a particular model in either hemisphere related to assumptions about the underlying subsonic fleet size (Table 4-12). Therefore, the size of the underlying subsonic fleet is computed to have only a minor impact on predicted O_3 changes from the projected supersonic fleet.

4.4.2.6.   *Future Atmosphere*

An additional uncertainty associated with the calculation of future aircraft impact is related to the increasing trend of greenhouse gases in the atmosphere and the impact on stratospheric temperatures and transport. For example, a CO_2 mixing ratio of about 500 ppmv is the recommendation of WMO (1992) for CO_2 in the year 2050—about 1.5 times the 1980 value. The change in radiative forcing as a result of greenhouse gas increases will affect stratospheric temperature and circulation (see Chapter 12 of WMO, 1999, and references therein); this effect has not been taken into consideration here. The likely cooler future stratosphere would delay the recovery of polar O_3 in Arctic and Antarctic regions, which could affect the future impact of aircraft.

One study of the effect of increasing CO_2 amounts (Pitari and Visconti, 1994) showed that stratospheric cooling mitigated the increase in NO_x. However, this study did not take into consideration the increase in H_2O associated with aircraft; therefore, more complete studies need to be done with a wide range of models.

Another major uncertainty regarding knowledge of the future stratosphere is that associated with source gases, in particular H_2O. This uncertainty is underscored by our lack of understanding of current H_2O trends. Global trends in stratospheric humidity over the period 1992 to 1996 inclusive have been presented recently by Evans *et al.* (1998) using Upper Atmosphere Research Satellite (UARS) measurements of H_2O and CH_4 from the Halogen Occultation Experiment (HALOE) instrument. The authors find that the combined budget of $2CH_4 + H_2O$ is approximately constant with altitude and is increasing with a global mean value of 61 ppbv yr^{-1}. A more local view of water vapor increase in the lower stratosphere was presented by Oltmans and Hofmann (1995) over the period 1981 to 1994. They found a maximum rate of increase in the 18–20 km layer of 32 ppbv yr^{-1}. The rate of increase of H_2O observed in the stratosphere is greater than the increase in tropospheric CH_4. One suggestion is that the increase could be accounted for by a change in tropopause temperature of a few tenths of a degree, and the most recent changes might be a result of changes in the dynamics of the tropics induced by heating of Mt. Pinatubo aerosols (Schauffer and Daniel, 1994). This uncertainty highlights the delicate balance between chemistry and climate.

4.4.3. Soot Emissions by Subsonic and Supersonic Aircraft

The chemistry of soot has been discussed in Chapter 2; the conclusion is that it does not react with O_3 in a catalytic manner. That is, O_3 can be destroyed by soot, but the soot is consumed in the process (see Section 2.1.3.1). However, soot particles may act as condensation nuclei for sulfate or other species, modifying cirrus clouds in the UT or LS, and locally they may be important in the destruction of O_3 (Bekki, 1997; Lary *et al.*, 1997). Thus, we present some estimates of soot densities and SADs for future scenarios to complement those estimated for the current (1992) subsonic fleet already given in Chapter 3. There are large uncertainties with regard to the sources and distribution of soot particles. Emissions from aircraft are particularly uncertain. Few global modeling studies are available for these aerosols (see Chapter 3 for details). Emissions of soot by aircraft [EI(soot)=0.03–0.4 g/kg] (Friedl, 1997) have been studied by Bekki (1997), who used the Cambridge 2-D aerosol/soot model for past and present air traffic, and more recently by Rahmes *et al.* (1998) for current and future aircraft fleets.

For the results presented herein, the aircraft-generated soot distribution was approximated by scaling the distribution of passive tracer calculated according to aircraft fuel burn (see also Chapter 3). A number of models have calculated this distribution for 2015 and 2050 (1992 distribution is shown in Table 3-4); all show the largest mass mixing ratio at about

Table 4-15a: Carbon soot mass mixing ratio from future aircraft [(50°N, annual average, pptm, EI(soot)=0.04 g/kg); approximate conversion factors to ng/m³ are 0.3 at 12 km and 0.1 at 20 km].

Model	2015 Subsonic 12 km	2015 Subsonic 20 km	2015 Subsonic + HSCT 12 km	2015 Subsonic + HSCT 20 km	2050 Subsonic 12 km	2050 Subsonic 20 km
AER	2.0	0.6	4.8	10.8	2.8	0.9
ECHAM3/CHEM	0.9	0.4			1.4	0.6
TM3/KNMI	1.7	0.4				
UiO	2.6	0.7	5.1	9.1	3.8	1.0
UNIVAQ-2D	3.0	0.2	4.1	9.5	4.6	0.3
UNIVAQ-3D	2.9	0.5	3.7	9.2	4.5	0.8

Table 4-15b: Carbon soot column from future aircraft (50°N, annual average, ng/cm², EI(soot)=0.04 g/kg).

Model	2015 Subsonic	2015 Subsonic + HSCT	2050 Subsonic
AER	0.58	2.18	0.84
TM3/KNMI	0.44		
UiO	0.68	1.53	1.00
UNIVAQ-2D	0.70	1.70	1.06
UNIVAQ-3D	0.69	1.52	1.04

12-km altitude and poleward of 40°N. The results for subsonic aircraft in the years 2015 and 2050 (IS92 Scenario 1) are shown in Table 4-15. For 2015, some results are also included for subsonic and HSCT aircraft together. These numbers represent an upper limit because sedimentation is neglected. Soot particles could also be lost by coagulation with larger sulfate aerosols and by condensation of H_2SO_4-H_2O on the particle surface.

Taking 2.0 pptm as a reference mixing ratio for subsonic aircraft soot in 2015 at 12 km and 50°N, an approximate SAD of 0.03 μm^2 cm^{-3} is obtained when assuming (as a first approximation) spherical particles of 2 g cm^{-3} density (Pueschel *et al.*, 1997) and a bimodal size distribution peaked at 20 and 100 nm radii with equal contribution in mass. If the particles are not assumed to be spherical but can be represented by fractal geometry, their SAD could lead to an enhancement by a factor of about 4 (Jennings *et al.*, 1999) over the spherical case. This seems more appropriate than 30, as suggested by Blake and Kato (1995). DeMore *et al.* (1997) recommend reaction efficiencies for O_3 decomposition between 0.003 and 1 x 10^{-5}, with one O_2 molecule produced for each O_3 reacting. As noted above, this O_3 decomposition is not catalytic because reaction sites on soot are rendered inactive when they are used or when the particle becomes coated with H_2SO_4.

4.5. Selection of Model Runs for UV (Chapter 5) and Climate Impact (Chapter 6) Scenario Studies

In this section, the model results that are used as a guide for the ultraviolet B (UV-B) impact (Chapter 5) and the estimation of climate impact (Chapter 6) are selected and uncertainties are assigned in light of the discussion in Section 4.4. A key uncertainty in assessing the impact of subsonic and supersonic exhaust emissions is that all models used to assess the aircraft impact are either tropospheric models with limited representation of stratospheric chemistry and transport (tropospheric 3-D CTMs), or stratospheric 2-D and 3-D CTMs with limited representation of tropospheric processes. In addition, neither the 2-D nor the 3-D models have been designed to deal specifically with transport processes in the tropopause region, where most aircraft emissions take place.

4.5.1. Model Simulations of Subsonic Aircraft

How well can we calculate the impact from subsonic aircraft in the tropopause region using the tropospheric 3-D CTMs or the stratospheric 2-D models? For these aircraft, part of the perturbation occurs in the LS and part in the UT.

As discussed in Chapter 2, based on the limited set of observational data that could be chosen for model validation, none of the 3-D CTMs could be picked as the best assessment model for impact studies of future aircraft emissions. The UiO 3-D model was selected as a representative model for the UV (Chapter 5) and climate impact studies (Chapter 6), because it gave results in the middle range of model results and it was easily available

for sensitivity studies. Height profiles for 30–60°N zonal mean O_3 perturbations for 1992 subsonic emissions (Scenario B-A in Table 4-4) obtained with the AER 2-D model and the UiO 3-D model are shown in Figure 4-10a. There is reasonable agreement between the two models in the 8–12 km region, near the tropopause. The UiO 3-D model, whose chemistry is most suited to the troposphere, shows a smaller O_3 perturbation in the middle and lowest troposphere. The AER 2-D model, whose chemistry is most suited to the stratosphere, shows a larger O_3 perturbation in the LS. An uncertainty range of a factor of 2 was adopted for O_3 perturbation from future subsonic aircraft emissions. This uncertainty range was based on the range of model results obtained by participating models in basic perturbation studies and results from the limited number of sensitivity studies. A "fair" confidence is associated with this uncertainty range for 2015, and a "poor" confidence is associated with this uncertainty range for 2050.

Figure 4-10b shows zonal mean O_3 changes for five calculations from the UiO 3-D model. As illustrated in the figure, predicted changes in O_3 from subsonic aircraft are comparable to changes from surface sources in 2015 and 2050.

4.5.2. Model Simulations of Supersonic Aircraft

The basic assumption in market studies that determine the routing and size of the supersonic fleet is that the supersonic

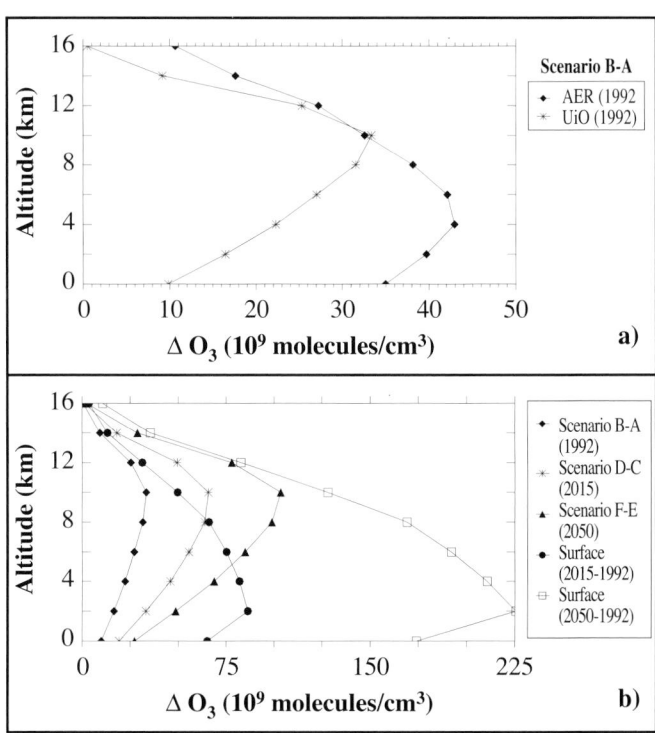

Figure 4-10: Annual average zonal mean change of ozone in the 30–60°N latitude band for various perturbations from UiO-3D and AER-2D models. Predictions are shown for (a) both models in 1992; and (b) UiO-3D model in 1992, 2015, and 2050.

fleet will replace certain routes of the given subsonic fleet, corresponding to about 10% of the subsonic fuel burn. The results of these studies were used to generate the fuel burn for a combined fleet consisting of a supersonic fleet with a modified subsonic fleet. For this reason, it is more appropriate to compare the effect of the combined fleet to the subsonic fleet, rather than to look at the supersonic fleet in isolation. Taking 2015 as an example, numerical results were generated for the 2015 atmosphere without aircraft (scenario C), the 2015 atmosphere with the standard subsonic fleet (scenario D), and the 2015 atmosphere with the combined fleet (scenario S1k).

Recognizing the uncertainties concerning the tropospheric response generated by the stratospheric models, the strategy is to use the change in O_3 computed between S1k and D and the results from the tropospheric models from scenario D minus scenario C for the effect of the subsonic fleet, after adjusting for the 10% difference in fuel burn.

For supersonic aircraft influences on UV effects (Chapter 5) and radiative forcing and climate change (Chapter 6), we have chosen a central or most probable emission scenario (S1k-2015 and S9h-2050), along with a representative assessment model

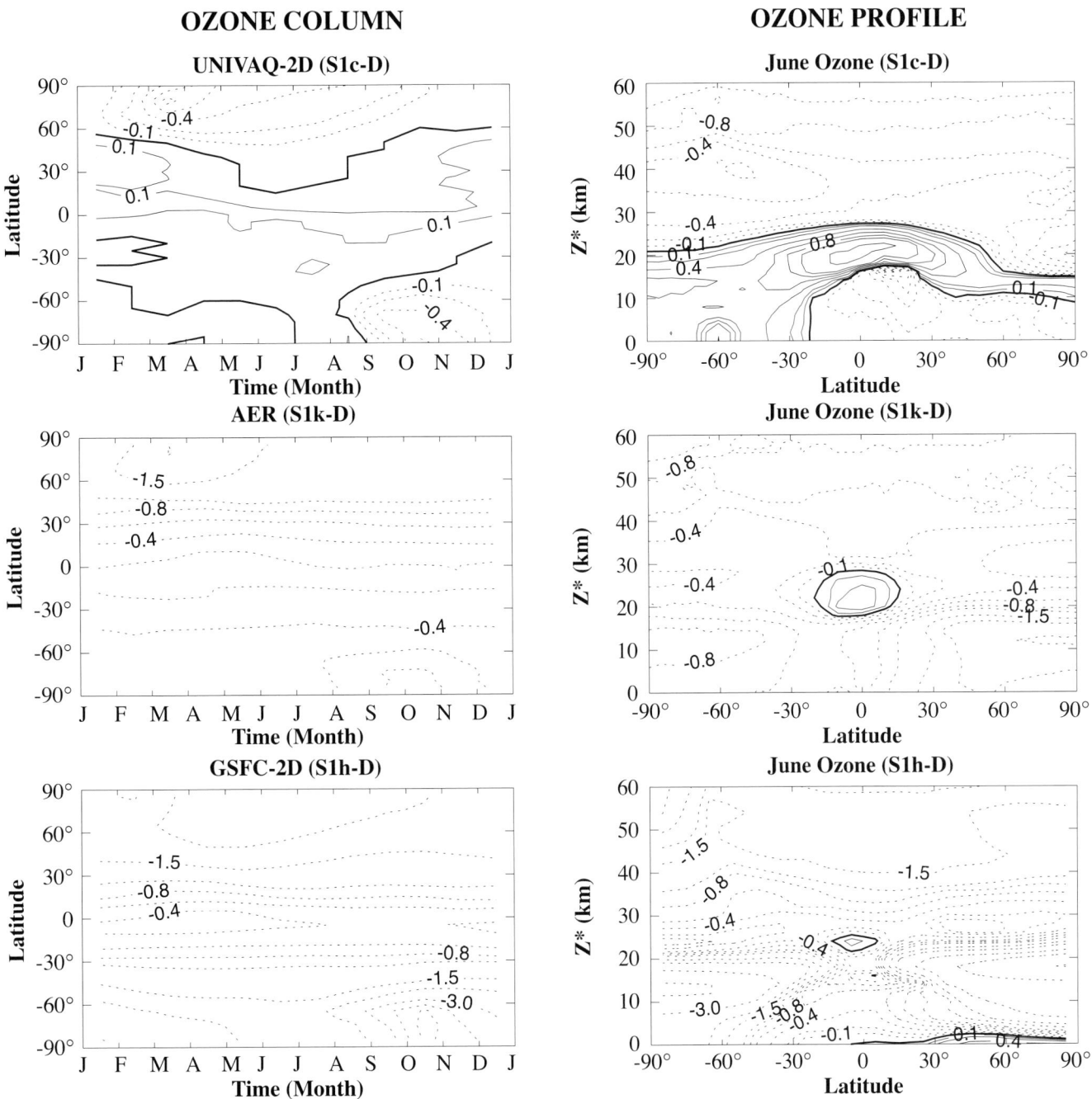

Figure 4-11: Ozone column (left panels) and ozone profile (right panels) percentage difference as a function of latitude and month for three models and three supersonic aircraft scenarios with respect to scenario D: UNIVAQ-2D model for S1c; AER-2D model for S1k; and GSFC-2D model for S1h. This figure mainly illustrates model differences rather than scenario differences.

(AER). This emission scenario assumes a SO_2 gas-to-particle conversion of 10%. The AER model was selected because it was the model that calculated and supplied the enhanced gas-to-particle sulfate aerosol SAD that was used in all participating assessment models. The middle plot in Figure 4-11 shows the percentage change in column O_3 for the AER model using the S1k scenario. At all latitudes and months, the AER model derives a reduction in total column O_3.

Because of coupling between chemistry and transport, there is no simple way to scale O_3 change profiles based on *a priori* estimates of uncertainties. It is possible to get an idea of the uncertainties by looking at the assembly of results from various scenarios performed by different models. Concentrating on the first group of scenarios in Table 4-11 (which represent a 500-plane Mach 2.4 fleet with $EI(NO_x)=5$ flying in a clean sulfate background), we picked S1c from UNIVAQ and S1h from GSFC as representative of the range of possible results. The differences in computed O_3 changes by the AER, UNIVAQ, and GSFC models in northern mid-latitudes are illustrated in Figure 4-12a for 2015. The UNIVAQ and GSFC models were picked because these models represent different ways of treating transport and PSCs that consistently produce the smallest and largest O_3 depletion in most scenarios. The sampling of scenarios also covers the possible range of effects from different

assumptions of gas-to-particle conversion in plume processing of SO_2 emission. For 2050, we focus on a 1,000-plane Mach 2.4 fleet with $EI(NO_x)=5$. The AER S9h scenario is taken as the central case, and the lower and upper extremes were taken to be S9d for UNIVAQ and S9f for GSFC. The differences in O_3 changes computed by the AER, UNIVAQ, and GSFC models in northern mid-latitudes are illustrated in Figure 4-12b for 2050.

It is also important to note which uncertainties, among those discussed in Chapters 2 and 3, are not included in this range. All of the models used rate data from DeMore *et al.* (1997). A previous study (described in Stolarski *et al.*, 1995) showed that uncertainties in rate data could lead to an uncertainty in NH O_3 column change of ±1%. Current studies indicate that most models underestimate the mean age of air as defined by inert tracers that enter the stratosphere via the tropical tropopause. Given that there appears to be a positive correlation between calculated increases in NO_y and H_2O from HSCT and calculated mean age, it has been suggested that models that underestimate age will also underestimate NO_y and H_2O increases from HSCT. It is difficult to quantify the uncertainty given current information. We did not consider the effect of plume processing and possible changes (in temperature and transport circulation) in the future background atmosphere. Finally, the range cited does not include different technology options for different $EI(NO_x)$, different cruise altitudes, and different fleet sizes.

It is possible to arrive at a subjective estimate for uncertainty estimates for changes in column O_3. This value can be used to calculate changes in UV because that depends mostly on the changes in column in the lower stratosphere. It is less obvious whether this value can be used to estimate uncertainties in radiative forcing. Restricting ourselves to the 1,000-plane Mach 2.4 fleet with $EI(NO_x)=5$, the range of model results for annual averaged Northern Hemisphere column O_3 depletion is -0.1% to -1.4% (see Table 4-12). Other model studies lead us to believe that reasonable changes in the background atmosphere (including background sulfate surface area) would not change this range in a significant way. Uncertainties in rate data would expand the range to about +1 to -2.5%. One may also argue that the inability of the models to simulate the correct mean age may add to uncertainties on the negative side by another 1%. Thus, a subjective estimate is that actual atmospheric response would likely lie between +1 and -3.5%.

A 500-plane Mach 2.4 fleet with $EI(NO_x)=5$ would have a similar range of uncertainty. Here, the range of model results for annual averaged Northern Hemisphere column O_3 depletion is 0 to -1.3% (see Table 4.11)—very close to the -0.1 to -1.4% model range for 1,000 planes. Applying the same arguments for this 500-plane fleet, a subjective estimate for the actual atmospheric response would likely be in the range of +1 to -3.5%. We have "fair" confidence in this range for the 500-plane and 1,000-plane fleets.

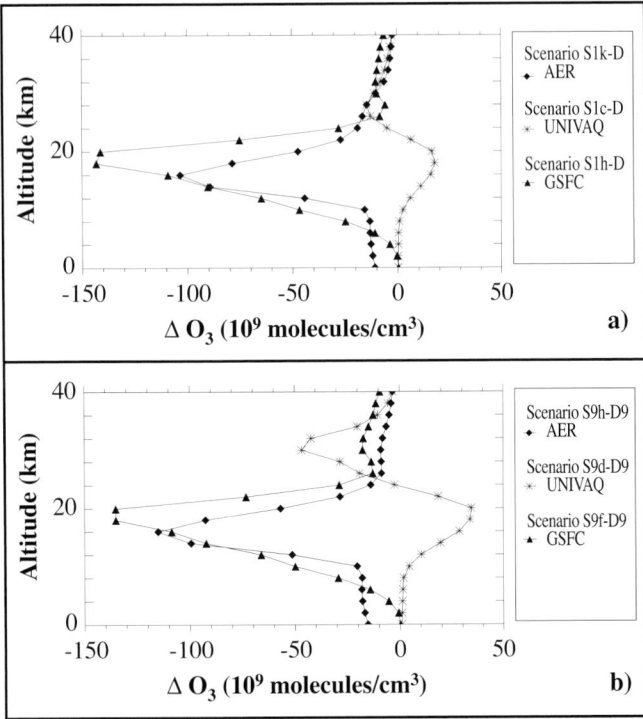

Figure 4-12: Annual average zonal mean change of ozone in the 30–60°N latitude band for supersonic aircraft perturbations from AER-2D, GSFC-2D, and UNIVAQ-2D models in (a) 2015 and (b) 2050. Model scenarios are those passed through to Chapters 5 and 6.

References

Baughcum, S.L. and S.C. Henderson, 1998: *Aircraft Emission Scenarios Projected in Year 2015 for the NASA Technology Concept Aircraft (TCA) High Speed Civil Transport Universal Airline Network*. NASA-CR-1998-207635, National Aeronautics and Space Administration, Langley Research Center, Hampton, VA, USA, 42 pp.

Bekki, S., 1997: On the possible role of aircraft generated soot in the middle latitude ozone depletion. *Journal of Geophysical Research*, **102**, 751–758.

Bekki, S. and J.A. Pyle, 1992: Two-dimensional assessment of the impact of aircraft sulphur emissions on the stratospheric sulfate aerosol layer. *Journal of Geophysical Research*, **97**, 15839–15847.

Berntsen, T. and I.S.A. Isaksen, 1997: A global three-dimensional chemical transport model for the troposphere. 1. Model description and CO and ozone results. *Journal of Geophysical Research*, **102**, 21239–21280.

Berntsen, T.K. and I.S.A. Isaksen, 1999: Effects of lightning and convection on changes in tropospheric ozone due to NO_x emissions from aircraft. *Tellus*, (in press).

Blake, D.F. and K. Kato, 1995: Latitudinal distribution of black carbon soot in the upper troposphere and lower stratosphere. *Journal of Geophysical Research*, **100**, 7195–7202.

Boering, K.A., S.C. Wofsy, B.C. Daube, H.R. Schneider, M. Loewenstein, J.R. Podolske, and T.J. Conway, 1996: Stratospheric mean ages and transport rates from observations of carbon dioxide and nitrous oxide. *Science*, **274**, 1340–1343.

Borrmann, S., S. Solomon, J.E. Dye, D. Baumgardner, K.K. Kelly, and K.R. Chen, 1997: Heterogeneous reactions on stratospheric background aerosols, volcanic sulfuric acid droplets, and type I PSCs: effects of temperature fluctuations and differences in particle phase. *Journal of Geophysical Research*, **102**, 3639–3648.

Brasseur, G.P., J.-F. Müller, and C. Granier, 1996: Atmospheric impact of NO_x emissions by subsonic aircraft: a three-dimensional study. *Journal of Geophysical Research*, **101**, 1423–1428.

Brasseur, G.P., R.A. Cox, D. Hauglustaine, I. Isaksen, J. Lelieveld, D.H. Lister, R. Sausen, U. Schumann, A. Wahner, and P. Wiesen, 1998: European scientific assessment of the atmospheric effects of aircraft emissions. *Atmospheric Environment*, **32**, 2329–2418.

Brown, R.C., R.C. Miake-Lye, M.R. Anderson, C.E. Kolb, and T.J. Resch, 1996: Aerosol dynamics in near-field aircraft plumes. *Journal of Geophysical Research*, **101**, 22939–22953.

Carslaw, K.S., M. Wirth, A. Tsias, B.P. Luo, A. Dornbrack, M. Leutbecher, H. Volkert, W. Renger, J.T. Bacmeister, E. Reimer, and T. Peter, 1998: Increased stratospheric ozone depletion due to mountain-induced atmospheric waves. *Nature*, **391**, 675–678.

Collins, W.J., D.S. Stevenson, C.E. Johnson, and R.G. Derwent, 1997: Tropospheric ozone in a global-scale three-dimensional Lagrangian model and its response to NO_x emission controls. *Journal of Atmospheric Chemistry*, **26**, 223–274.

Considine, D.B., A.R. Douglass, and C.H. Jackman, 1994: Effects of a polar stratospheric cloud parameterization on ozone depletion due to stratospheric aircraft in a 2-D model. *Journal of Geophysical Research*, **99**, 18879–18894.

Dameris, M., V. Grewe, I. Köhler, R. Sausen, C. Brühl, J.-U. Grooß, and B. Steil, 1998: Impact of aircraft NO_x-emissions on tropospheric and stratospheric ozone. Part II: 3-D model results. *Atmospheric Environment*, **32**, 3185–3199.

Danilin, M.Y., J.M. Rodriguez, M.K.W. Ko, D.K. Weisenstein, R.C. Brown, R.C. Miake-Lye, and M.R. Anderson, 1997: Aerosol particle evolution in an aircraft wake: implications for the high-speed civil transport fleet impact on ozone. *Journal of Geophysical Research*, **102**, 21453–21463.

Danilin, M.Y., D.W. Fahey, U. Schumann, M.J. Prather, J.E. Penner, M.K.W. Ko, D.K. Weisenstein, C.H. Jackman, G. Pitari, I. Köhler, R. Sausen, C.J. Weaver, A.R. Douglass, P.S. Connell, D.E. Kinnison, F.J. Dentener, E.L. Fleming, T.K. Berntsen, I.S.A. Isaksen, J.M. Haywood, and B. Kärcher, 1998: Aviation fuel tracer simulation: model intercomparison and implications. *Geophysical Research Letters*, **25**, 3947–3950.

DeMore, W.B., S.P. Sander, D.M. Golden, R.F. Hampson, M.J. Kurylo, C.J. Howard, A.R. Ravishankara, C.E. Kolb, and M.J. Molina, 1997: *Chemical Kinetics and Photochemical Data for Use in Stratospheric Modeling, Evaluation Number 12*. JPL Publication 97-4, Jet Propulsion Laboratory, National Aeronautics and Space Administration, Pasadena, CA, USA, January 15, 1997, 266 pp.

Dessler, A.E., E.J. Hintsa, E.M. Weinstock, J.G. Anderson, and K.R. Chan, 1995: Mechanisms controlling water vapor in the lower stratosphere: a tale of two stratospheres. *Journal of Geophysical Research*, **100**, 23167–23172.

Dlugokencky, E.J., K.A. Masarie, P.M. Lang, and P.P. Tans, 1998: Continuing decline in the growth rate of the atmospheric methane burden. *Nature*, **393**, 447–450.

Dubey, M.K., G.P. Smith, W.S. Hartley, D.E. Kinnison, and P.S. Connell, 1997: Rate parameter uncertainty effects in assessing ozone depletion by supersonic aviation. *Geophysical Research Letters*, **24**, 2737–2740.

Edouard, S., B. Legras, F. Lefevre, and R. Eymard, 1996: The effect of small-scale inhomogeneities on ozone depletion in the Arctic. *Nature*, **384**, 444–447.

Elkins, J.W., D.W. Fahey, J.M. Gilligan, G.S. Dutton, T.J. Baring, C.M. Volk, R.E. Dunn, R.C. Myers, S.A. Montzka, P.R. Wamsley, A.H. Hayden, J.H. Butler, T.M. Thompson, T.H. Swanson, E.J. Dlugokencky, P.C. Novelli, D.F. Hurst, J.M. Lobert, S.J. Ciciora, R.J. McLaughlin, T.L. Thompson, R.H. Winkler, P.J. Fraser, L.P. Steele, and M.P. Lucarelli, 1996: Airborne gas chromatograph for in situ measurements of long-lived species in the upper troposphere and lower stratosphere. *Geophysical Research Letters*, **23**, 347–350.

Eluszkiewicz, J., 1996: A three-dimensional view of the stratosphere-to-troposphere exchange in the GFDL SKYHI model. *Geophysical Research Letters*, **23**, 2489–2492.

Evans, S.J., R. Toumi, J.E. Harries, M.P. Chipperfield, and J.M. Russell III, 1998: Trends in stratospheric humidity and the sensitivity of ozone to these trends. *Journal of Geophysical Research*, **103**, 8715–8725.

Fabian, P. and B. Kärcher, 1997: The impact of aviation upon the atmosphere: an assessment of present knowledge, uncertainties, and research needs. *Physics and Chemistry of the Earth*, **22(6)**, 503–598.

Fahey, D.W., E.R. Keim, K.A. Boering, C.A. Brook, J.C. Wilson, H.H. Jonsson, S. Anthony, T.F. Hanisco, P.O. Wennberg, R.C. Miake-Lye, R.J. Salawitch, N. Louisnard, E.L. Woodbridge, R.S. Gao, S.G. Donnelly, R.C. Wamsley, L.A. Del Negro, S. Solomon, B.C. Daube, S.C. Wofsy, C.R. Webster, R.D. May, K.K. Kelly, M. Loewenstein, J.R. Podolske, and K.R. Chan, 1995: Emissions measurements of the Concorde supersonic aircraft in the lower stratosphere. *Science*, **270**, 70–74.

Fishman, J., C.E. Watson, J.C. Larsen, and J.A. Logan, 1990: The distribution of tropospheric ozone determined from satellite data. *Journal of Geophysical Research*, **95**, 3599–3617.

Folkins, I., P.O. Wennberg, T.F. Hanisco, J.G. Anderson, and R.J. Salawitch, 1997: OH, HO_2 and NO in two biomass burning plumes: sources of HO_x and implication for ozone production. *Geophysical Research Letters*, **24**, 3185–3188.

Fortuin, J.P.F., R. van Dorland, W.M.F. Wauben, and H. Kelder, 1995: Greenhouse effects of aircraft emissions as calculated by a radiative transfer model. *Annales Geophysicae*, **13**, 413–418.

Friedl, R.R., 1997: *Atmospheric Effects of Subsonic Aircraft: Interim Assessment Report of the Advanced Subsonic Technology Program*. NASA Reference Publication 1400, National Aeronautics and Space Administration, Goddard Space Flight Center, Greenbelt, MD, USA, 168 pp.

Fuglestvedt, J.S., I.S.A. Isaksen, and W.-C. Wang, 1996: Estimates of indirect global warming potentials for CH_4, CO, and NO_x. *Climatic Change*, **34**, 405–437.

Grooß, J.-U., T. Peter, C. Brühl, and P.J. Crutzen, 1994: The influence of high flying aircraft on polar heterogeneous chemistry. In: *Impact of Emissions from Aircraft and Spacecraft upon the Atmosphere* [Schumann, U. and D. Wurzel (eds.)]. Proceedings of an international scientific colloquium, 18–20 April 1994, Cologne, Germany. DLR-Mitteilung 94-06, Deutsches Zentrum für Luft- und Raumfahrt (German Aerospace Center), Oberpfaffenhofen and Cologne, Germany, pp. 229–234.

Hall, T.M. and R.A. Plumb, 1994: Age as a diagnostic of stratospheric transport. *Journal of Geophysical Research*, **99**, 1059–1070.

Hanson, D.R. and K. Mauersberger, 1988: Laboratory studies of the nitric acid trihydrate: implications for the south polar stratosphere. *Geophysical Research Letters*, **15**, 855–858.

Hofmann, D.J., 1991: Aircraft sulfur emissions. *Nature*, **349**, 659.

IPCC, 1992: *Climate Change 1992: The Supplementary Report to the IPCC Scientific Assessment* Prepared by IPCC Working Group I [Houghton, J.T., B.A. Callander, and S.K. Varney (eds.)] and WMO/UNEP. Cambridge University Press, Cambridge, United Kingdom, and New York, NY, USA, 200 pp.

IPCC, 1995: *Climate Change 1995: The Science of Climate Change. Contribution of Working Group I to the Second Assessment Report of the Intergovernmental Panel on Climate Change* [Houghton, J.T., L.G. Meira Filho, B.A. Callander, N. Harris, A. Kattenberg, and K. Maskell (eds.)]. Cambridge University Press, Cambridge, United Kingdom and New York, NY, USA, 572 pp.

Jackman, C.H., A.R. Douglass, K.F. Brueske, and S.A. Klein, 1991: The influence of dynamics on two-dimensional model results: simulations of ^{14}C and stratospheric aircraft NO_x injections. *Journal of Geophysical Research,* **96,** 22559–22572.

Jacob, D.J., M.J. Prather, P.J. Rasch, R.L. Shia, Y.J. Balkanski, S.R. Beagley, D.J. Bergmann, W.T. Blackshear, M. Brown, M. Chiba, M.P. Chipperfield, J. de Grandpre, J.E. Dignon, J. Feichter, C. Genthon, W.L. Grose, P.S. Kasibhatla, I. Kohler, M.A. Kritz, K. Law, J.E. Penner, M. Ramonet, C.E. Reeves, D.A. Rotman, D.Z. Stockwell, P.F.J. van Velthoven, G. Verver, O. Wild, H. Yang, and P. Zimmerman, 1997: Evaluation and intercomparison of global atmospheric transport models using ^{222}Rn and other short-lived tracers. *Journal of Geophysical Research,* **102,** 5953–5970.

Jaeglé, L., D.J. Jacob, P.O. Wennberg, C.M. Spivakovsky, T.F. Hanisco, E.J. Lanzendorf, E.J. Hinsta, D.W. Fahey, E.R. Keim, M.H. Proffitt, E.L. Atlas, F. Flocke, S. Schauffler, C.T. McElroy, C, Midwinter, L. Pfister, and J.C. Wilson, 1997: Observed OH and HO_2 in the upper troposphere suggest a major source from convective injection of peroxides. *Geophysical Research Letters,* **24,** 3181–3184.

Jaffe, D., T.K. Berntsen, and I.S.A. Isaksen, 1997: A global three-dimensional chemical transport model for the troposphere, 2, Nitrogen oxides and nonmethane hydrocarbon results. *Journal of Geophysical Research,* **102,** 21281–21296.

Jennings, S.G., K.S. Law, S. Bekki, and M.J. Evans, 1999: Carbonaceous aerosol fractal surface area: implications for heterogeneous chemistry? *Geophysical Research Letters,* (submitted).

Johnston, H.S. and E. Quitevis, 1975: The oxides of nitrogen with respect to urban smog, supersonic transports, and global methane. In: *Radiation Research* [Nygaard, O.F., H.I. Adler, and W.K. Sinclair (eds.)]. Academic Press Inc., New York, NY, USA, pp. 1299–1313.

Johnston, H.S., 1989: Evaluation of excess carbon 14 and strontium 90 data for suitability to test two-dimensional stratospheric models. *Journal of Geophysical Research,* **94,** 18485–18493.

Kärcher, B., 1997: Heterogeneous chemistry in aircraft wakes: constraints for uptake coefficients. *Journal of Geophysical Research,* **102,** 19119–19135.

Kärcher, B. and D.W. Fahey, 1997: The role of sulfur emission in volatile particle formation in jet aircraft exhaust plumes. *Geophysical Research Letters,* **24,** 389–392.

Knudsen, B.M., 1996: Accuracy of Arctic stratospheric temperature analyses and the implications for the prediction of polar stratospheric clouds. *Geophysical Research Letters,* **25,** 3747–3750.

Lacis, A.A., D.J. Wuebbles, and J.A. Logan, 1990: Radiative forcing of climate by changes in the vertical distribution of ozone. *Journal of Geophysical Research,* **95,** 9971–9981.

Lary, D.J., R. Toumi, A.M. Lee, M. Newchurch, M. Pierre, and J.B. Renard, 1997: Carbon aerosols and atmospheric photochemistry. *Journal of Geophysical Research,* **102,** 3671–3682.

Lelieveld, J., B. Bregman, F. Arnold, V. Burger, P.J. Crutzen, H. Fischer, A. Waibel, P. Siegmund, P.F.J. van Velthoven, 1997: Chemical perturbation of the lowermost stratosphere through exchange with the troposphere. *Geophysical Research Letters,* **24,** 603–606.

Lipson, J.B., M.J. Elrod, T.W. Beiderhase, L.T. Molina, and M.J. Molina, 1997: Temperature dependence of the rate constant and branching ratio for the OH+ClO reaction. *Journal of Chemical Society Faraday Transactions,* **93,** 2665–2673.

McKeen, S.A., T. Gierczak, J.B. Burkholder, P.O. Wennberg, T.F. Hanisco, E.R. Keim, R.S. Gao, S.C. Liu, A.R. Ravishankara, and D.W. Fahey, 1997: The photochemistry of acetone in the troposphere: a source of odd-hydrogen radicals. *Geophysical Research Letters,* **24,** 3177–3180.

Michelsen, H.A., C.M. Spivakovsky, and S.C. Wofsy, 1999: Aerosol-mediated partitioning of stratospheric Cl_y and NO_y at temperatures above 200 K. *Geophysical Research Letters,* **26(3),** 299–302.

Michelsen, H.A., R.J. Salawitch, M.R. Gunson, C. Aellig, N. Kampfer, M.M. Abbas, M.C. Abrams, T.L. Brown, A.Y. Chang, A. Goldman, F.W. Irion, M.J. Newchurch, C.P. Rinsland, G.P. Stiller, R. Zander, 1996: Stratospheric chlorine partitioning: constraints from shuttle-borne measurement of HCl, $ClNO_3$, and ClO. *Geophysical Research Letters,* **23,** 2361–2364.

Müller, J.-F. and G. Brasseur, 1995: IMAGES: A three-dimensional chemical transport model of the global troposphere. *Journal of Geophysical Research,* **100,** 16445–16490.

Murphy, D.M. and A.R. Ravishankara, 1994: Temperature averages and rates of stratospheric reactions. *Geophysical Research Letters,* **21,** 2471–2474.

Oltmans, S.J. and D.J. Hofmann, 1995: Increase in lower-stratospheric water vapour at a mid-latitude site from 1981 to 1994. *Nature,* **374,** 146–149.

Osterman, G.B., R.J. Salawitch, B. Sen, G.C. Toon, R.A. Stachnik, H.M. Pickett, J.J. Margitan, J.F. Blavier, and D.B. Peterson, 1997: Balloon-borne measurements of stratospheric radicals and their precursors: implications for the production and loss of ozone. *Geophysical Research Letters,* **24,** 1107–1110.

Park, J.H., M.K.W. Ko, C.H. Jackman, R.A. Plumb, and K.H. Sage (eds.), 1999: *Models and Measurements II.* National Aeronautics and Space Administration, Langley Research Center, Hampton, VA, USA, (in press).

Peter, T., C. Brühl, and P.J. Crutzen, 1991: Increase in the PSC-formation probability caused by high-flying aircraft. *Geophysical Research Letters,* **18,** 1465–1468.

Pitari, G., S. Palermi, G. Visconti, and R.G. Prinn, 1992: Ozone response to a CO_2 doubling: results from a stratospheric circulation model with heterogeneous chemistry. *Journal of Geophysical Research,* **97,** 5953–5962.

Pitari, G., V. Rizi, L. Ricciardulli, and G. Visconti, 1993: High-speed civil transport impact: role of sulfate, nitric acid trihydrate, and ice aerosols studied with a two-dimensional model including aerosol physics. *Journal of Geophysical Research,* **98,** 23141–23164.

Pitari, G. and G. Visconti, 1994: Possible effects of CO_2 increase on the high-speed civil transport impact on ozone. *Journal of Geophysical Research,* **99,** 16879–16896.

Prather, M.J., H.L. Wesoky, R.C. Miake-Lye, A.R. Douglass, R.P. Turco, D.J. Wuebbles, M.K.W. Ko, and A.L. Schmeltekopf, 1992: *The Atmospheric Effects of Stratospheric Aircraft: A First Program Report.* NASA Reference Publication 1272, National Aeronautics and Space Administration, Washington, DC, USA, 244 pp.

Prather, M.J. and E.E. Remsberg (eds.), 1993: *The Atmospheric Effects of Stratospheric Aircraft: Report of the 1992 Models and Measurements Workshop.* NASA Reference Publication 1292, Vols. 1–3, National Aeronautics and Space Administration, Washington, DC, USA, 764 pp.

Prather, M.J. and D.J. Jacob, 1997: A persistent imbalance in HO_x and NO_x photochemistry of the upper troposphere driven by deep tropical convection. *Geophysical Research Letters,* **24,** 3189–3192.

Pueschel, R.F., K.A. Boering, S. Verma, S.D. Howard, G.V. Ferry, J. Goodman, D.A. Allen, and P. Hamill, 1997: Soot aerosol in the lower stratosphere: pole-to-pole variability and contribution by aircraft. *Journal of Geophysical Research,* **102,** 13113–13118.

Pullen, S. and R.L. Jones, 1997: Accuracy of temperature from UKMO analyses of the 1994/95 in the Arctic winter stratosphere. *Geophysical Research Letters,* **24,** 845–848.

Rahmes, T.F., A.H. Omar, and D.J. Wuebbles, 1998: Atmospheric distributions of soot particles by current and future aircraft fleets and resulting radiative forcing on climate. *Journal of Geophysical Research,* **103,** 31657–31667.

Roeckner, E., K. Arpe, L. Bengtsson, S. Brinkop, L. Dümenil, M. Esch, E. Kirk, F. Lunkeit, M. Ponater, B. Rockel, R. Sausen, S. Schubert, and M. Windelband, 1992: *Simulation of the Present-Day Climate with the ECHAM Model: Impact of Model Physics and Resolution.* Report no. 93, Max-Planck-Institut für Meteorologie, Hamburg, Germany, ISSN 0937-1060, 171 pp.

Schauffer, S.M. and J.S. Daniel, 1994: On the effects of stratospheric circulation changes on trace gas trends. *Journal of Geophysical Research,* **99,** 25747–25754.

Schmitt, A. and B. Brunner, 1997: Emissions from aviation and their development over time. In: *Final Report on the BMBF Verbundprogramm, Schadstoffe in der Luftfahrt* [Schumann, U., A. Chlond, A. Ebel, B. Kärcher, H. Pak, H. Schlager, A. Schmitt, and P. Wendling (eds.)]. DLR-Mitteilung 97-04, Deutsches Zentrum für Luft- und Raumfahrt, Oberpfaffenhofen and Cologne, Germany, pp. 37–52.

Shia, R.-L., M.K.W. Ko, M. Zou, and V.R. Kotamarthi, 1993: Cross tropopause transport of excess ^{14}C in a two-dimensional model. *Journal of Geophysical Research,* **98,** 18599–18606.

Shindell, D., D. Rind, and P. Lonergan, 1998: Climate change and the middle atmosphere. Part IV: Ozone response to doubled CO₂. *Journal of Climate,* **11,** 895–918.

Singh, H.B., M. Kanakidou, P.J. Crutzen, and D.J. Jacob, 1995: High concentrations and photochemical fate of oxygenated hydrocarbons in the global troposphere. *Nature,* **378,** 50–54.

Solomon, S., R.W. Portmann, R.R. Garcia, L.W. Thomason, L.R. Poole, and M.P. McCormick, 1996: The role of aerosol variations in anthropogenic ozone depletion at northern midlatitudes. *Journal of Geophysical Research,* **101,** 6713–6727.

Solomon, S., S. Borrmann, R.R. Garcia, R. Portmann, L. Thomason, L.R. Poole, D. Winker, and M.P. McCormick, 1997: Heterogeneous chlorine chemistry in the tropopause region. *Journal of Geophysical Research,* **102,** 21411–21429.

Steil, B., M. Dameris, C. Brühl, P.J. Crutzen, V. Grewe, M. Ponater, and R. Sausen, 1998: Development of a chemistry module for GCMs: first results of a multi-annual integration. *Annales Geophysicae,* **16,** 205–228.

Stevenson, D.S., W.J. Collins, C.A. Johnson, and R.G. Derwent, 1997: The impact of nitrogen oxide emissions on tropospheric ozone studied with a 3-D Lagrangian model including full diurnal chemistry. *Atmospheric Environment,* **31,** 1837–1850.

Stolarski, R.S. and H.L. Wesoky, 1993: *The Atmospheric Effects of Stratospheric Aircraft: A Third Program Report.* NASA Reference Publication 1313, National Aeronautics and Space Administration, Washington, DC, USA, 428 pp.

Stolarski, R.S., S.L. Baughcum, W.H. Brune, A.R. Douglass, D.W. Fahey, R.R. Friedl, S.C. Liu, R.A. Plumb, L.R. Poole, H.L. Wesoky, and D.R. Worsnop, 1995: *1995 Scientific Assessment of the Atmospheric Effects of Stratospheric Aircraft.* NASA Reference Publication 1381, National Aeronautics and Space Administration, Washington, DC, USA, 110 pp.

Thomason, L.W., L.R. Poole, and T. Deshler, 1997: A global climatology of stratospheric aerosol surface area density deduced from Stratospheric Aerosol and Gas Experiments II measurements: 1984–1994. *Journal of Geophysical Research,* **102,** 8967–8976.

Wang, W.-C. and N.D. Sze, 1980: Coupled effects of atmospheric N₂O and O₃ on the earth's climate. *Nature,* **286,** 589–590.

Wang, Y., D.J. Jacob, and J.A. Logan, 1998a: Global simulation of tropospheric O₃-NOₓ-hydrocarbon chemistry. 1. Model formulation. *Journal of Geophysical Research,* **103,** 10713–10725.

Wang, Y., J.A. Logan, and D.J. Jacob, 1998b: Global simulation of tropospheric O₃-NOₓ-hydrocarbon chemistry. 2. Model evaluation and global ozone budget. *Journal of Geophysical Research,* **103,** 10727–10755.

Wauben, W.M.F., P.F.J. van Velthoven, and H. Kelder, 1997: A 3-D chemistry transport model study of changes in atmospheric ozone due to aircraft NOₓ emissions. *Atmospheric Environment,* **31,** 1819–1836.

Wauben, W.M.F., J.P.F. Fortuin, P.F.J. van Velthoven, and H. Kelder, 1998: Comparison of modeled ozone distributions with sonde and satellite observations. *Journal of Geophysical Research,* **103,** 3511–3530.

Weisenstein, D.K., M.K.W. Ko, N-D Sze, and J.M. Rodriguez, 1996: Potential impact of SO₂ emissions from stratospheric aircraft on ozone. *Geophysical Research Letters,* **23,** 161–164.

Weisenstein, D.K., G.K. Yue, M.K.W. Ko, N.-D. Sze, J.M. Rodriguez, and C.J. Scott, 1997: A two-dimensional model of sulfur species and aerosols. *Journal of Geophysical Research,* **102,** 13019–13035.

Weisenstein, D.K., M.K.W. Ko, I.G. Dyominov, G. Pitari, L. Ricciardulli, G. Visconti, and S. Bekki, 1998: The effects of sulfur emissions from HSCT aircraft: a 2-D model intercomparison. *Journal of Geophysical Research,* **103,** 1527–1547.

Wennberg, P.O., T.F. Hanisco, L. Jaeglé, D.J. Jacob, E.J. Hintsa, E.J. Lanzendorf, J.G. Anderson, R.S. Gao, E.R. Keim, S.G. Donnelly, L.A. Del Negro, D.W. Fahey, S.A. McKeen, R.J. Salawitch, C.R. Webster, R.D. May, R.L. Herman, M.H. Proffitt, J.J. Margitan, E.L. Atlas, S.M. Schauffler, F. Flocke, C.T. McElroy, and T.P. Bui, 1998: Hydrogen radicals, nitrogen radicals and the production of ozone in the upper troposphere. *Science,* **279,** 49–53.

WMO, 1992: *Scientific Assessment of Ozone Depletion: 1991.* Report no. 25, Global Ozone Research and Monitoring Project, World Meteorological Organization, Geneva, Switzerland, 366 pp.

WMO, 1995: *Scientific Assessment of Ozone Depletion: 1994.* Report no. 37, Global Ozone Research and Monitoring Project, World Meteorological Organization, Geneva, Switzerland, 578 pp.

WMO, 1999: *Scientific Assessment of Ozone Depletion: 1998.* Report No. 44, Global Ozone Research and Monitoring Project, World Meteorological Organization, Geneva, Switzerland, 732 pp.

Yu, F.Q. and R.P. Turco, 1998: Contrail formation and impacts on aerosol properties in aircraft plumes: effects of fuel sulfur content. *Geophysical Research Letters,* **25,** 313–316.

Yue, G.K., M.P. McCormick, and E.W. Chiou, 1991: Stratospheric aerosol optical depth observed by the Stratospheric Aerosol and Gas Experiment II: decay of the El Chichon and Ruiz volcanic perturbations. *Journal of Geophysical Research,* **96,** 5209–5219.

5

Solar Ultraviolet Irradiance
at the Ground

KEITH R. RYAN AND JOHN E. FREDERICK

Lead Authors:
A.F. Bais, J.B. Kerr, B. Wu

Contributors:
R. Meerkötter, I.C. Plumb

Review Editor:
C. Zerefos

CONTENTS

EXECUTIVE SUMMARY

The solar ultraviolet (UV) irradiance incident on the surface of the Earth is responsible for a variety of biological effects (UNEP, 1994). This radiation varies greatly with local time, altitude, latitude, season, and meteorological conditions. For a given solar elevation, the transmission of UV sunlight through the atmosphere depends on absorption, predominantly by ozone; scattering and absorption by aerosols; and scattering by clouds. Aircraft emissions have the potential to alter each of these processes, and hence to influence the solar radiation field at biologically relevant UV wavelengths. To characterize UV radiation, this chapter adopts the erythemal dose rate (UV_{ery}), which is defined as the irradiance on a horizontal surface, at local solar noon, integrated over wavelength, with a wavelength-dependent weighting factor to account for the sunburning effect of the radiation as a function of wavelength and expressed in $W\ m^{-2}$.

The calculated impacts of present and future fleets of aircraft on atmospheric ozone—hence on the UV_{ery}—are compared with those calculated from other expected changes in the composition of the atmosphere, including changes in bromine and chlorine content and expected increases in emissions of oxides of nitrogen (NO_x) resulting from combustion at the surface. For these calculations, 1970 is taken as the reference year; a combination of results from three-dimensional (3-D) and two-dimensional (2-D) chemical transport models is used to predict changes in ozone for the period 1970 to 2050.

The calculated changes in UV_{ery} show strong dependencies on latitude, season, composition of the background atmosphere, and, in the case of aircraft impacts, whether the aviation fleet is assumed to have a component of supersonic aircraft. For present and projected fleets of subsonic aircraft, the calculations predict a decrease in UV_{ery} relative to the corresponding background atmosphere, which contains no aircraft emissions. The biggest changes are calculated for northern mid-latitudes, where present and expected emissions are greatest. For example, at 45°N in July the change in UV_{ery} relative to the corresponding background atmosphere is predicted to range from -0.5% in 1992 to -1.3% in 2050; the decrease in UV is brought about by the increase in ozone in the upper troposphere resulting from aircraft emissions. The corresponding range at 45°S, where present and predicted air traffic is substantially less, is -0.1% in 1992 and -0.3% in 2050 for January.

Ozone changes for the range of scenarios considered for these calculations have been obtained from Chapter 4. For calculations of the impact of subsonic aircraft on UV_{ery}, Chapter 4 has given a factor of 2 as the uncertainty for calculated ozone changes.

These uncertainties are taken as the 67% likelihood range. We believe that uncertainty in the calculation of ozone changes is by far the greatest uncertainty in the determination of changes in UV_{ery}; accordingly, we have not added additional uncertainties to the range supplied by Chapter 4. In addition, our assessment of the confidence in these calculations is as given by Chapter 4: that is, "fair" for 2015 and "poor" for 2050. Accordingly, the calculated change for 2050 at 45°N in July, for example, can be expressed as -1.3% (-2.5 to -0.7%).

The introduction of mixed supersonic and subsonic fleets in the future will have the potential to modify the impact of aviation on UV_{ery}. Present calculations predict only marginal changes to the values of UV_{ery} in the tropics and increases in mid-latitudes. For example, the predicted changes relative to the corresponding background at 45°N in July are +0.6% in 2015 and +0.3% in 2050. For 45°S in January, the predicted changes in UV_{ery} are +0.4% for 2015 and +0.2% in 2050, relative to the corresponding background atmospheres. Although these estimated changes may be considered small, they do have significantly larger uncertainty limits. The limits for a particular confidence range are difficult to determine. The $\%UV_{ery}$ changes are dominated by changes in ozone; the values for these changes are supplied by Chapter 4, which has also considered three components when assessing the uncertainty for the impact of the hybrid fleet on ozone. These components are the spread obtained by a number of models for a range of plausible scenarios, uncertainties in chemical rate coefficients, and uncertainties introduced by inaccurate treatment of atmospheric circulation in the models. Chapter 4 concluded that the annually averaged impact on ozone for the Northern Hemisphere of a hybrid fleet in 2050, including 1,000 High-Speed Civil Transports (HSCTs), would be in the range of -3.5 to +1% compared with the impact of the subsonic fleet, with a best estimate of -1%. This result again represents the 67% likelihood range with a confidence in this uncertainty range of "fair." As with estimated subsonic impacts, we believe that uncertainties in the changes in ozone caused by the hybrid fleet are much greater than any other uncertainties in the calculation of changes in UV_{ery}. Chapter 4 has considered only the uncertainty estimate for an annually averaged Northern Hemisphere value, whereas reporting of the UV calculations requires estimates of the uncertainties for a range of latitudes and seasons. To achieve this, Chapter 5 has taken the 67% likelihood range for percent change in UV_{ery} to be from (-2% + the best estimate of the percent change) to (+3% + the best estimate of the percent change).

The magnitude of the changes calculated for the impact of aviation may be compared to those calculated for the background atmosphere for the period 1970 to 2050. Expressed relative to 1970, the calculated changes in UV_{ery} at 45°N in July are +8% for 1992, +3% for 2015, and -3% for 2050. The changes calculated for the three background atmospheres reflect the changing levels of halogens and NO_x in the stratosphere and expected increases in NO_x in the troposphere, particularly at northern mid-latitudes, as a result of increased industrial activity. Observed changes in total ozone from 1970 to 1992 imply smaller percentage increases in UV_{ery}, indicating the degree of uncertainty in the model predictions.

Increases in the abundance of atmospheric aerosols or the frequency of cirrus clouds would, in general, lead to a decrease in ground-level UV irradiance, where this change is only weakly dependent on wavelength. The change in aerosol loading expected from increased aircraft operations between the present and 2050 is small relative to the natural aerosol background and to anthropogenic influences other than those related to aviation. The effect of the aircraft-related increase in aerosols is to reduce UV irradiance by less than 0.1%. Calculations indicate that aircraft-related increases in contrails lead to a decrease in UV irradiance of less than 0.2% in an area-averaged sense in regions where 5% of the sky is covered.

5.1. Factors that Determine Ground-Level Ultraviolet Irradiance

This chapter considers the potential impact of aviation on ground-level UV irradiance by using changes in ozone and cloudiness reported in preceding chapters. A series of radiative transfer calculations based on this information yields estimated changes in UV irradiance associated with various aviation scenarios in the years 2015 and 2050 and places these scenarios in context relative to changes expected from other causes during this time period.

Because of strong attenuation by atmospheric ozone, practically no solar radiation reaches the ground at wavelengths shorter than 290 nm. The wavelength range of interest extends from this short wavelength ozone cutoff to 400 nm because biological sensitivities, including the reference action spectrum for erythema (defined as a reddening of human skin in response to irradiation) (McKinlay and Diffey, 1987), extend through both the UV-B (wavelengths 280–315 nm) and the UV-A (315–400 nm). To characterize UV radiation, this chapter adopts the UV_{ery} defined here as the erythemally weighted irradiance, $A(\lambda)E(\lambda)$, integrated over wavelength and expressed in W m^{-2}. The quantity $A(\lambda)$ is the biological weighting function of McKinlay and Diffey (1987); $E(\lambda)$ is the spectral irradiance received on a horizontal surface. This chapter presents calculations of UV_{ery} at local noon. Although action spectra exist for a variety of biological effects (UNEP, 1994), the erythemal weighting has international recognition and is the basis for the widely used UV index. In general, action spectra that are more sharply peaked toward short wavelengths in the UV-B lead to weighted irradiances that are more sensitive to changes in ozone. The discussion in UNEP (1994) addresses this issue; no further detail is required here.

Absorption by ozone is the most important single process that influences the transmission of UV-B radiation through the atmosphere. This absorption leads to a sharp reduction in ground-level spectral irradiance as wavelength decreases from 315 nm. Although the bulk of the absorption occurs at stratospheric altitudes, tropospheric ozone is also important. The dependence of transmission on the geometrical path taken by sunlight, hence on solar zenith angle, leads to a strong dependence of ground-level UV irradiance on latitude, season, and local time. In addition, molecular scattering is significant in the UV and leads to a diffuse irradiance at the surface of the Earth under clear, aerosol-free skies that is comparable to or larger than the direct solar beam, where the relative magnitudes are functions of solar zenith angle. Furthermore, ground-level irradiance increases as surface albedo increases, as a result of backscattering of radiation reflected from the ground. Additional discussion of the factors involved in the transfer of UV radiation appears in Kerr (1997).

Recent studies of radiative transfer in the UV emphasize the roles of clouds and aerosols, primarily sulfates and soot, in altering ground-level irradiance (Seckmeyer *et al.*, 1996; Kerr, 1997). Attenuation of UV sunlight by clouds and aerosols arises primarily from backscattering of radiation to space, although absorption by soot—both freely suspended in the atmosphere and incorporated into cloud drops—can be non-negligible under certain circumstances, particularly in polluted urban areas. The temporal variability inherent in clouds and their geometrical complexity hinder realistic radiative transfer modeling, although simple parameterizations using satellite-based measurements of cloud reflectivities are valuable in assessing the attenuation provided by cloudy skies (Eck *et al.*, 1995; Frederick and Erlick, 1995). The effects of backscattering by aerosols are implicit in such measurements.

Emissions from aircraft could influence the UV irradiance incident on the biosphere by altering the ozone abundance in the stratosphere or troposphere, the abundance and optical properties of atmospheric aerosols, and the geographic extent or optical characteristics of clouds. Our ability to model these three influences decreases in accuracy as one moves from ozone to aerosols to cloudiness. The following sections consider the natural variability in ground-level UV irradiance, comparisons between measurements and calculations, our capability to detect changes, and estimations of the changes that could occur as a consequence of future aviation.

5.2. Comparison of Measured and Calculated Variabilities in UV Irradiance

UV irradiance at the surface of the Earth depends on several variable factors identified in Section 5.1. These factors include scattering (Rayleigh, aerosol, and cloud) and absorption (ozone, aerosol, and pollution) processes that occur in the atmosphere, as well as variations in extraterrestrial solar flux and ground reflectivity. Variability in each of these factors combines to produce large fluctuations in UV irradiance—for example, between corresponding months of different years (Weatherhead *et al.*, 1997). This large variability makes it difficult to quantify systematic decadal changes in UV irradiance and interpret them in terms of cause and effect with instruments other than well-maintained spectroradiometers.

It is important to understand and quantify the individual effects of the many variables affecting UV irradiance at the surface of the Earth. The contributions of the different variables can be studied with the use of radiative transfer models. Several models have been developed for a variety of applications (Dave, 1965; Frederick and Lubin, 1988; Stamnes *et al.*, 1988; Madronich, 1992; Ruggaber *et al.*, 1994; Forster, 1995; Herman *et al.*, 1996). These models use extraterrestrial solar spectral irradiance (Mentall *et al.*, 1981; Neckel and Labs, 1984; Kaye and Miller, 1996; Woods *et al.*, 1996) as input and simulate the physical processes that occur as radiation is scattered and absorbed by the atmosphere and at the surface of the Earth. Model output includes global (direct plus diffuse) and diffuse spectral radiation at the Earth's surface in a form that can be compared with measurements.

The comparison of UV measurements with model simulations is an important exercise for checking both the accuracy of the model and the quality of measurements. Once it is demonstrated that the measurements and model results are in good agreement for a wide range of conditions, a reliable simulation of the transfer of UV radiation through the atmosphere is possible. The model can then be used with confidence to extend a measurement series in time and in space (between ground-based stations), provided measurements of all variables affecting surface UV irradiance are available. The model can also estimate future levels of surface UV irradiance by using predictions of variables such as aerosols or ozone that may change as a result of increased air traffic. Models are used in this context in Section 5.4.

Comparisons between models and measurements are best evaluated by two approaches. One is the comparison of irradiance as a function of wavelength normalized to a certain wavelength—usually in the UV-A, where ozone has a negligible effect. This comparison emphasizes the response of irradiance to variables that are strongly wavelength-dependent (such as ozone absorption or wavelength error) and minimizes effects that are weakly wavelength-dependent, such as clouds, aerosols, or the absolute calibration of instrument responsivity. The second approach is the comparison of absolute irradiances at the wavelength of the normalization to quantify effects that are weakly wavelength-dependent. In any case, comparison of measured irradiances with modeled irradiances suffers from the limited availability and quality of data necessary to describe the atmosphere (Schwander *et al.*, 1997).

Several studies have compared measured irradiances with model simulations. Measurements made under clear sky conditions and no snow cover demonstrate quite convincingly that lower ozone values result in higher UV irradiance levels at the surface of the Earth; these data sets have been used to quantify the dependence statistically (McKenzie *et al.*, 1991; Booth and Madronich, 1994; Kerr *et al.*, 1994). When these measured dependencies are compared with those computed for clear-sky conditions, no aerosols, and low ground reflectivity (no snow cover), reasonable agreement has generally been found (McKenzie *et al.*, 1991; Wang and Lenoble, 1994; Forster *et al.*, 1995).

The availability of aerosol optical depth measurements in the UV has allowed studies of the effects of particulates on ground-level irradiance. Mayer *et al.* (1997) compare clear-sky UV spectral data obtained at Garmish-Partenkirchen, Germany, between 1994–96 with model simulations. The model simulations use ozone and aerosol optical depth measurements as inputs. Systematic differences between measured irradiance spectra and model results were between -11% and +2%. It was necessary to introduce ground-level aerosols into the model to achieve agreement to within 5%. From total ozone, aerosol optical depth, and spectral UV irradiance measurements made under clear-sky conditions at Toronto between 1989–91, Kerr (1997) demonstrates that most of the observed variability of UV irradiance between 300 and 325 nm can be explained by ozone and aerosols. The remaining unexplained variability is 4% at 300 nm and 2% at 325 nm. Comparison of the observed

dependence of UV irradiance on aerosol optical depth with model results suggests that typical aerosols over Toronto are slightly absorbing (Krotkov *et al.*, 1998). The model also shows that a single scattering albedo of about 0.95 for aerosols gives the best agreement with the Toronto data.

Surface UV irradiance is also reduced by atmospheric sulfur dioxide (SO_2), which has strong absorption features at UV wavelengths and occurs both naturally from volcanic emissions and anthropogenically from industrial sources (Zerefos, 1997; Kerr *et al.*, 1998). The presence of SO_2 can interfere with the measurement of ozone and estimates of ozone and UV trends at sites affected by local air pollution (Bais *et al.*, 1993; De Meur and De Backer, 1993). However, measurements made at several sites in less-polluted situations suggest that the effects of SO_2 on UV over wider areas are small (Fioletov *et al.*, 1997).

A method developed recently to calculate surface spectral UV irradiance uses Total Ozone Mapping Spectrometer (TOMS) satellite measurements of ozone and UV reflectivity with a radiative transfer model (Eck *et al.*, 1995; Herman *et al.*, 1996; Krotkov *et al.*, 1998). Comparison of model results with ground-based measurements made at Toronto under clear skies indicates agreement of absolute irradiance to about 2% after correction for the angular response of the ground-based instrument.

The effects of surface albedo have been considered in the UV-A (324 nm), where there is negligible ozone absorption, by observing the difference between measurements made with and without snow cover at several sites (Wardle *et al.*, 1997). The presence of snow was found to enhance irradiance differently from one site to another. The minimum enhancement was 8% at Halifax, Canada; the maximum was 39% at Churchill, Canada. The difference between these two sites is likely to be a result of differences in the surrounding terrain and snow texture. For example, the clean snow on the flat terrain around Churchill would result in a higher average surface albedo than at Halifax, where snow would be dirtier in the suburban areas and not present on nearby open water. Model results show an enhancement of about 50% for an albedo of about 1 (Deguenther *et al.*, 1998; Krotkov *et al.*, 1998). Although there are no direct measurements of albedo available when snow is present, the model gives quite reasonable effective albedo values of about 20% at Halifax and 90% at Churchill. Model results of Deguenther *et al.* (1998) show that irradiance values are affected by surface albedo (snow cover) at distances up to 40 km; most of the dependence is influenced by albedo within a radius of 10 km.

Variability in cloud cover is the largest contributor to short-term changes in surface UV irradiance. It is possible to include the effects of clouds in radiative transfer calculations to various levels of approximation. However, routinely available observational data do not allow a rigorous characterization of cloud optical properties. Measurements show that UV spectral transmittance depends on cloud type, cloud thickness, and whether there are absorbers within the cloud. Although detailed quantification of these dependencies requires further research, some general conclusions can be made. The effects of thin clouds are weakly (<1% per nm) wavelength-dependent, with only broad wavelength features (Seckmeyer, 1989; Seckmeyer *et al.*, 1996; Kylling *et al.*, 1997; Mayer *et al.*, 1998a,b). Under heavy convective clouds—when the amount of radiation is reduced by more than 90%—there is enhanced wavelength dependence as a result of increased absorption due to a longer pathlength through ozone within the cloud (Brewer and Kerr, 1973; Fioletov and Kerr, 1996). The effects of changes in stratospheric ozone on surface UV irradiance through all types of sky conditions (clear and cloudy) have been quantified from statistical analysis of data sets several years in length (Kerr and McElroy, 1993; Wardle *et al.*, 1997). Algorithms that use ozone and reflectivity information from TOMS are able to include the effects of clouds in simulations of surface UV irradiance (Eck *et al.*, 1995; Frederick and Erlick, 1995; Herman *et al.*, 1996), although the results should be interpreted as averages over the large areas covered by the sensor's field of view.

Increased air traffic is expected to lead to changes in the abundances of ozone, NO_x, SO_2, and aerosols, as well as the frequency of cirrus clouds. In general, comparisons of observations with calculations have indicated that radiative transfer models can simulate the effects of gaseous absorbers quite reliably. Greater uncertainties are associated with the treatment of aerosols and cirrus because of the need to specify optical properties and perhaps fractional sky coverage. In the latter case, cirrus can lead to local increases in UV irradiance even though the area-averaged effect is a decrease. Models can simulate both non-absorbing (water or sulfate) and absorbing (carbon) aerosols, although the absorption properties of realistic aerosol types, which consist of mixtures of various chemical components, are not well known.

5.3. Detectability of Changes in Ground-Level Irradiance

Our ability to detect future changes in solar UV irradiance attributable to increases in aircraft emissions depends on two factors. The first factor is the size of the expected change relative to natural variability and long-term changes from other human causes. The second factor is the accuracy and stability with which we are capable of making measurements over long periods of time.

The detection of a long-term change in UV_{ery} depends on the size of the expected change, the length of the data record over which the change occurs, and the variability of UV_{ery}. Weatherhead *et al.* (1998) show that a data record of at least 15 years is necessary to distinguish a statistically significant (2-sigma) long-term change of 5% per decade in UV_{ery} from the natural variability present at most sites. A long-term change of 5% over 50 years (1% per decade to 2050) would just be detectable at the best sites.

The other factor is whether existing instrumentation is able to detect expected long-term changes. Presently, many different instruments devoted to monitoring changes in solar UV during forthcoming decades are deployed over the globe. These instruments are either spectroradiometers or broadband and narrowband detectors operating in the UV region. Spectroradiometers are the most suitable instruments because they provide detailed measurements of global spectral irradiance, which can be used to assess the importance of various UV-controlling factors. Because of the weak intensity of UV irradiance relative to other parts of the solar spectrum, however, it is only in recent years that the quality of such measurements has become sufficient to detect interannual changes in UV irradiance that are attributable to year-to-year changes in ozone over relatively short (< 10 years) measurement records (Kerr and McElroy, 1993; McKenzie *et al.*, 1993; Kerr and McElroy, 1994; Gardiner and Kirsch, 1995; Groebner *et al.*, 1996; Bais *et al.*, 1997; Gurney, 1998). The quality of other types of broadband instruments cannot yet be considered sufficient to detect these small long-term UV changes (Leszczynski *et al.*, 1996; Mayer and Seckmeyer, 1996; Blumthaler, 1997; Weatherhead *et al.*, 1997; WMO, 1997).

Typical uncertainties quoted for UV spectroradiometers range from 5 to 15%, depending on the quality of the instrument (McKenzie *et al.*,1993; Koskela, 1994; Gardiner and Kirsch, 1995; Bais *et al.*, 1997; Webb *et al.*, 1998). When dealing with the shorter wavelength portion of the UV-B region, the uncertainty in measured spectral irradiance increases significantly because of the decreasing signal-to-noise ratio and increasing error related to wavelength calibration in some instruments. Other instrumental factors—such as stray light, temperature sensitivity, angular response, wavelength instability, and degradation of optical components—may further reduce the reliability of data produced by spectroradiometers (Gardiner and Kirsch, 1993, 1995; Seckmeyer and Bernhard, 1993; Slaper *et al.*, 1995; Groebner *et al.*, 1996; Bais, 1997). A large part of this uncertainty appears to arise from the absolute calibration standards and procedures, which can be as high as 6%. For the detection of relative changes or trends, however, instrument stability over time is the most important parameter. Even in this case, the uncertainty of the relative measurements performed by a given instrument operating continuously at the same location can be significantly less than the uncertainty in absolute irradiance. It is possible to achieve a relative uncertainty of 2–5% over a period of several years under special circumstances (Kerr, 1997). Most measurement records fail to achieve this level of stability, however.

Numerous studies during the past decade have established relationships between solar UV radiation and atmospheric parameters identified previously. The effects of changes in ozone on UV irradiance can be detected more easily than the ozone change itself because they are magnified several times by strong absorption, which increases dramatically with decreasing wavelength. Consequently, even the effect of small ozone changes—on the order of a few percent—can be detected in principle (Madronich, 1992; Bais *et al.*, 1993, 1997a,b), provided the absolute irradiance remains above the detection limits of the sensor. The effect of a 1% decrease in ozone on UV irradiance can be detected at wavelengths shorter than about 300 nm, where the corresponding increase in UV irradiance is greater than 4% (Bodhaine *et al.*, 1997; Fioletov *et al.*, 1997). In this spectral region, unfortunately, most single monochromator spectroradiometers suffer from stray light (Gardiner and Kirsch, 1995; Bais *et al.*, 1996). Only a small subset of UV spectroradiometers in the existing worldwide network can achieve the necessary uncertainty limits at these short wavelengths. If changes in column ozone from aviation are larger than 1%, longer wavelengths may be adequate for detection.

In contrast to ozone changes, UV attenuation by aerosols and clouds varies weakly with wavelength, except (occasionally) when clouds contain ozone in significant amounts. The effects of aerosols and clouds on ground-level UV irradiances have been studied with the aid of model simulations as well as by measurements (Frederick and Snell, 1990; Bais *et al.*, 1993; Frederick and Steele, 1995; Blumthaler *et al.*, 1996; Bodeker and McKenzie, 1996; Kylling *et al.*, 1997). Although an increase in aerosols will generally lead to a reduction in UV irradiance via backscattering to space, an increase in radiation is possible, depending on the relative altitudes of the aerosol and ozone layers and the solar zenith angle (Davies, 1993; Tsitas and Yung, 1996). Temporal variations in cloudiness and, to a lesser degree, aerosols lead to large variations in UV irradiance measured at a fixed site. This degree of natural variability will complicate attempts to detect small changes in UV irradiance related to potential changes brought about by future aviation.

Thus, it appears that some existing instrumentation is capable of detecting future changes in UV caused by changes in ozone and perhaps by aerosols and clouds. With the current state of technology, properly maintained instruments should be able to detect UV irradiances that differ from the unperturbed state by 2–5% or more. Whether it would be feasible to distinguish UV changes caused by aircraft emissions from changes caused by the natural background variability in atmospheric parameters depends on the relative magnitude of these variations and the duration of the data sets.

5.4.   Calculated Impact of Aviation on UV at the Surface of the Earth

Present and future fleets of aircraft have the capacity to modify the amount of UV arriving at the surface of the Earth as a result of changes brought about by:

- The amount and distribution of ozone in the upper troposphere and lower stratosphere
- The amount of cloud cover
- The aerosol type, content, and distribution.

This section discusses the expected magnitudes of the effects of each of these changes and compares the calculated impact of

aviation on UV at the ground with that resulting from other changes in the composition of the atmosphere.

Many of the results discussed in this section are expressed in terms of percent change in UV_{ery} between two scenarios being compared. The relationships between UV_{ery}, the erythemal weighting factor, the ground-level irradiance, and the erythemally weighted irradiance are illustrated in Figure 5-1 for 30°N for July and January.

So that percent change in UV_{ery} may be put in context in terms of absolute irradiance, Figure 5-2 shows the calculated spatial and seasonal variation of the absolute value for UV_{ery} with latitude for the 1992 background atmosphere.

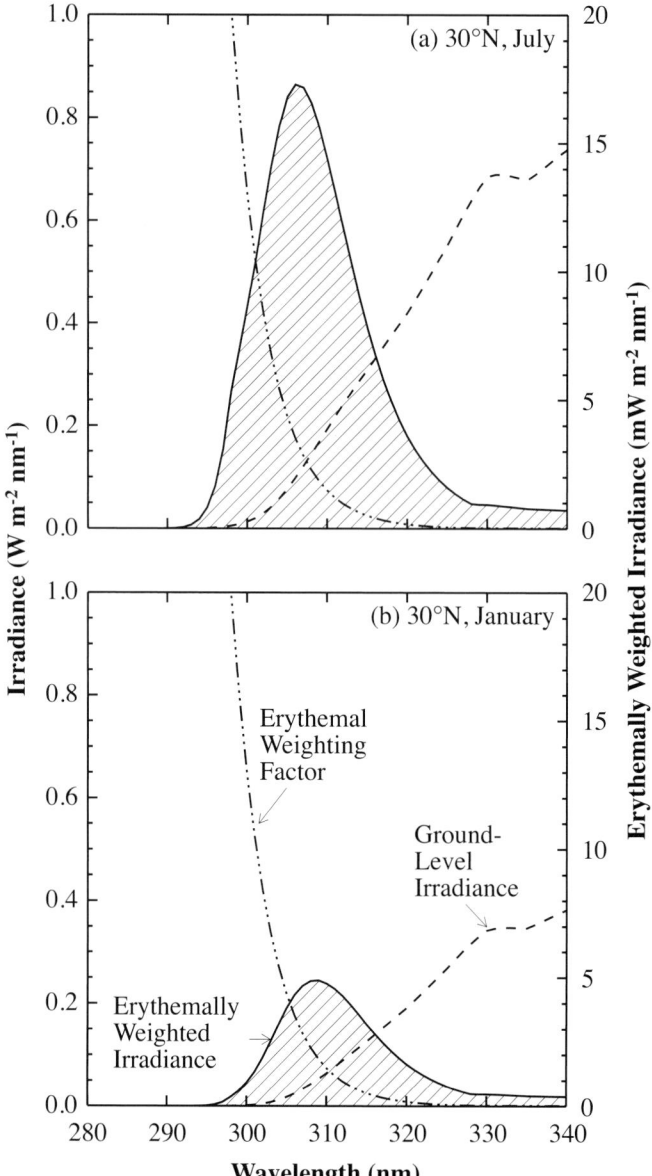

Figure 5-1: Relationships among UV_{ery}, erythemal weighting factor, ground irradiance, and erythemally weighted irradiance at 30°N in July and January.

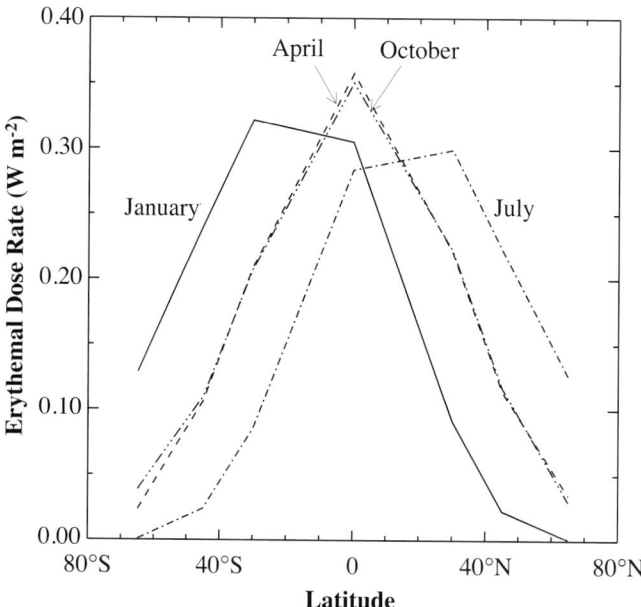

Figure 5-2: UV_{ery} as a function of latitude for the 1992 background atmosphere for January, April, July, and October.

5.4.1. Methodology for Treating Changes in Atmospheric Ozone

The underlying principle for the methodology described below is that calculations of ozone changes derived from chemical transport models should be related to measurements. For this chapter, calculations designed to determine ozone changes for the present and the future have been related to present-day measurements of ozone columns and profiles. Model calculations are then used to estimate the impact of aviation in 1970, 1992, 2015, and 2050. The modeled impacts are compared with other changes that are calculated to have occurred to ozone columns since 1970 and predicted to occur between 1992 and 2050. The radiative transfer calculations reported in this chapter were performed using a stand-alone version of the solar radiation module from the Commonwealth Scientific and Industrial Research Organisation (CSIRO) 2-D chemical transport model (CTM), which has been described in Stolarski *et al.* (1995) and Randeniya *et al.* (1997). The method of Meier *et al.* (1982) is used to account for the effects of multiple scattering, and ray-tracing techniques are used to account for the curvature of the Earth. The CSIRO radiative transfer model participated in the photolysis benchmark intercomparison conducted as part of the 1995 Atmospheric Effects of Stratospheric Aircraft (AESA) assessment (Stolarski *et al.*, 1995), and excellent agreement was obtained with the benchmark calculations. Intercomparisons have also been performed with the Tropospheric Ultraviolet and Visible (TUV) model developed by Madronich using calculations made at the Laboratory of Atmospheric Physics at the University of Thessaloniki. The radiative transfer equation in TUV was solved by using the discrete ordinate radiative transfer (DISORT) code in 16-stream mode (Stamnes *et al.*, 1988). Erythemally weighted fluxes were calculated at the ground under clear-sky conditions for zenith angles from 0 to 80°.

Using the same input parameters, agreement was obtained with the CSIRO model to within 2–3% for zenith angles up to 60° and within 5% for a zenith angle of 80°.

5.4.1.1. Ozone Columns and Profiles

The calculations address the effect of aviation on UV at the surface in 1970, 1992, 2015, and 2050. For the years 2015 and 2050, there will be a range of scenarios that will be influenced primarily by the expected size and composition of the fleets and the amount and composition of emissions. In principle, however, the UV calculations require a representation of a background atmosphere and an atmosphere that includes the effects of aircraft emissions for each of the years 1970, 1992, 2015, and 2050. The term "background" in this context refers to an atmosphere derived from a calculation that includes all expected inputs other than those attributable to aircraft emissions. The calculated impacts of aviation on ozone for 1992, 2015, and 2050 are provided by Chapter 4. The small effect calculated for 1970 is deduced from the 1992 calculations, as discussed in Section 5.4.1.2.

The starting point for subsequent calculations is the representation of ozone columns and profiles for 1992. For this chapter, the columns have been derived from archived TOMS version 7 measurements (Herman *et al.*, 1996). UV calculations have been performed at local noon for January, April, July, and October at the following latitudes: 65°S, 45°S, 30°S, 0, 30°N, 45°N, and 65°N. The monthly averaged and zonally averaged ozone columns derived from the TOMS data for these months and latitudes for the 10-year period 1983 to 1992 are shown in Table 5-1.

Because the values shown in Table 5-1 are averages of measured values, they already include the effects of present-day aircraft emissions on ozone. In other words, to obtain the 1992 background atmosphere, the effects of emissions from aircraft have to be removed from the values shown in Table 5-1. The

Table 5-1: *Ozone columns (Dobson units): Monthly, zonal averages (1983–1992) from TOMS.** *

Latitude	January	April	July	October
65°S	318.8	299.2	(380.0)	312.7
45°S	303.4	286.3	324.3	353.7
30°S	275.7	269.0	290.2	310.1
0°	250.1	260.1	264.9	265.8
30°N	275.3	308.9	294.1	272.5
45°N	352.8	373.0	326.0	295.0
65°N	(380.0)	424.3	329.2	304.8

* There are no TOMS measurements for winter at high latitudes. The values shown in parentheses for winter at 65°S and N are the values given for winter subarctic columns in Anderson *et al.* (1986). The distributions of ozone with altitude for the columns shown in Table 5-1 are derived from Park *et al.* (1999).

manner in which this has been done for these calculations and the extension of the methods applied to obtain the required information for the years 1970, 2015, and 2050 are illustrated in Figure 5-3.

All calculations ultimately have averaged values for ozone given in Table 5-1 as their reference. These 1992 values are represented by 1992_{ss} (where the subscript defines an atmosphere containing emissions from a fleet of subsonic aircraft) in Figure 5-3. Chemical transport calculations are carried out to generate model atmospheres for 1992 that (a) do not include inputs attributable to aviation and (b) do include these inputs. Comparison of the results of these calculations allows one to determine the absolute differences in ozone concentrations predicted by the model.

These absolute differences are applied to 1992_{ss} to produce 1992_{bg}, the 1992 background atmosphere, which is free of inputs from aircraft emissions. A model calculation is then carried out using the prescribed background scenario for 2015. The results of these calculations are compared with the similar calculation for 1992, and the absolute difference obtained from the model calculations is applied to 1992_{bg} to obtain 2015_{bg}, the 2015 background atmosphere. Comparison of a model calculation for a 2015 atmosphere that contains inputs attributable to aviation, assuming only subsonic aircraft will be operating, with the model calculation for the 2015 background atmosphere provides an estimate of absolute changes in ozone concentration between these two calculations. These differences are applied to 2015_{bg} to produce 2015_{ss}, the 2015 atmosphere that includes the effects of emissions from a purely subsonic fleet of aircraft. Clearly, an analogous approach can be adopted to obtain the required information for 2050 and for 1970. Chapter 4 has provided mixing ratio differences calculated for the impact of aviation on ozone profiles for the latitudes and seasons shown in Table 5-1. The authors of Chapter 4 chose results from the Oslo 3-D model and the Atmospheric and Environmental Research, Inc. (AER) 2-D model as representative of the subsonic and supersonic scenario calculations, respectively. Table 5-2 shows the scenarios considered here, with references to the relevant Chapter 4 tables. The method adopted in Chapter 4 for expressing the range in uncertainties for the ozone mixing ratio differences is discussed in Section 5.4.2.3.

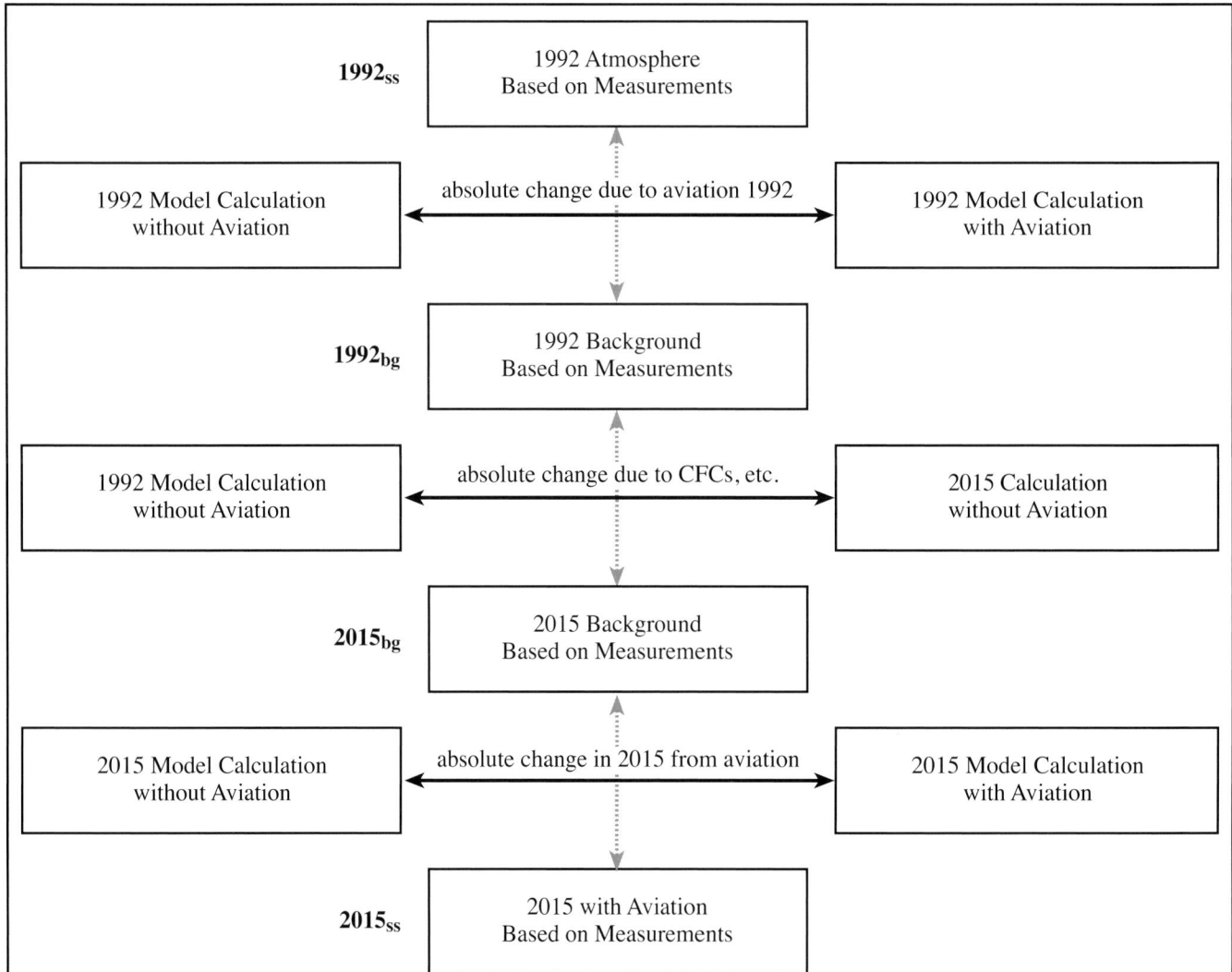

Figure 5-3: Illustration of the derivation of columns used for UV calculations.

Table 5-2: *Brief scenario description.*

Symbol	Description	Chapter 4 Reference
1992_{bg}	1992 background	Scenario A: Table 4-4
1992_{ss}	1992 including subsonics	Scenario B: Table 4-4
2015_{bg}	2015 background	Scenario C: Table 4-4
2015_{ss}	2015 including subsonics	Scenario D: Table 4.4
2015_{hf}	2015 subsonic/supersonic hybrid fleet	Scenario S1k: Table 4-11
2050_{bg}	2050 background	Scenario E: Table 4-4
2050_{ss}	2050 including subsonics	Scenario F: Table 4-4
2050_{hf}	2050 subsonic/supersonic hybrid fleet	Scenario S9h: Table 4-12
1970_{bg}	1970 background	Deduced from 1992_{bg}
1970_{ss}	1970 subsonic	Deduced from 1992_{ss}

Figure 5-3 attempts to highlight the fact that, although 1992_{bg}, 2015_{bg}, and 2015_{ss} are derived from model calculations, they have the measurements of 1992_{ss} as a reference. The implication of this approach is that, although ozone concentrations derived from models clearly have uncertainties, the differences obtained between any two model calculations may be more accurate than the absolute concentrations in either model calculation. This assumption is questionable and must be regarded more as an assumption that allows UV calculations to proceed than one that can be defended strongly. The approach does have the advantage that UV calculations derived from 1992_{bg}, 2015_{bg}, and 2015_{ss} are not based entirely on model calculations. It should be noted that, in assigning differences between model calculations to measured values in 1992_{ss}, absolute differences in model calculations have been chosen rather than percentage differences. Again, the approach taken cannot be defended rigorously. Examination of the 3-D CTM results in Chapter 4 shows a very wide variation in the range of models for estimated ozone differences in the upper troposphere. Arguments can be advanced for adopting either absolute or percentage changes to apply to measured ozone columns; until the reasons for the variations in the range of models are clear, however, neither approach is clearly superior to the other.

The current limitations of multidimensional CTMs generate additional complications that must be addressed in assessing the impacts of aviation on the composition of the atmosphere. The impact of subsonic fleets on ozone, for the present and for 2015 and 2050, has been discussed in Chapter 4, based on a range of 3-D CTMs. In their present state, these predominantly tropospheric models are unable to take into account adequately changes in the chemistry of the stratosphere. Between 1992 and 2050, for example, these changes would be induced predominantly by changes in the concentrations of inorganic chlorine and bromine compounds in the stratosphere. If the principal concern were to evaluate the impact of aviation on ozone for a given year, these limitations would not be too severe. However, if in addition one wishes to compare the effects of aircraft emissions with those that can be attributed to changes in stratospheric processes over the period 1970 to 2050, then one requires more information than current 3-D CTMs can supply. Furthermore, it is expected that future fleets

may contain a supersonic component. These supersonic aircraft will fly in the stratosphere; to calculate the effects of this hybrid subsonic/supersonic fleet, models capable of assessing stratospheric changes must be used.

Calculations of the effects of the hybrid subsonic/supersonic fleets were carried out in Chapter 4 in the following way. For a given year (2015 for example), a 3-D CTM calculation using the Oslo model was performed to determine the change in ozone concentration for the subsonic fleet relative to the background atmosphere for that year. A 2-D calculation was then carried out with the AER 2-D model to determine the change in mixing ratio for ozone between the hybrid fleet and subsonic-only fleet scenarios. For the latitudes and seasons shown in Table 5-1, these mixing ratio differences were added to the mixing ratio differences, calculated by the Oslo 3-D CTM, between the subsonic fleet and the background atmosphere. The scenarios used in Chapter 4 to provide the ozone differences for the subsonic impact were (A,B, 1992), (C,D, 2015), and (E,F, 2050), as defined in Table 4-4. The hybrid fleet impacts were obtained by comparing scenario S1k, defined in Table 4-11, with scenario D of Table 4-10 for 2015 and scenario S9h, defined in Table 4-11, with scenario D9 from Table 4-10 for 2050. Results from these sets of scenarios correspond to the calculated impacts of 500 HSCTs in 2015 and 1,000 HSCTs in 2050.

Calculations designed to show the effects of aviation relative to changes in ozone or UV resulting from changes in the composition of the stratosphere require a slightly more convoluted approach. This approach is illustrated in Figure 5-4. The top panel shows the differences in background atmospheres calculated by the Oslo 3-D CTM for the years 2050 and 1970 (2050_{bg} - 1970_{bg}), 2015 and 1970 (2015_{bg} - 1970_{bg}), and 1992 and 1970 (1992_{bg} - 1970_{bg}). The middle panel shows the results obtained when these calculations are performed with the AER 2-D CTM. The background surface concentrations used in these calculations for 1992, 2015, and 2050 are those shown in Table 4-8. The background surface concentrations used for the 1970 calculation are those given in Table 6.3 of WMO (1994). It is clear from the top panel that the 3-D CTM does not provide any information about changes occurring in the atmosphere above about 18 km for the period from 1970 to 2050. The calculated increases for ozone shown below 18 km and reaching

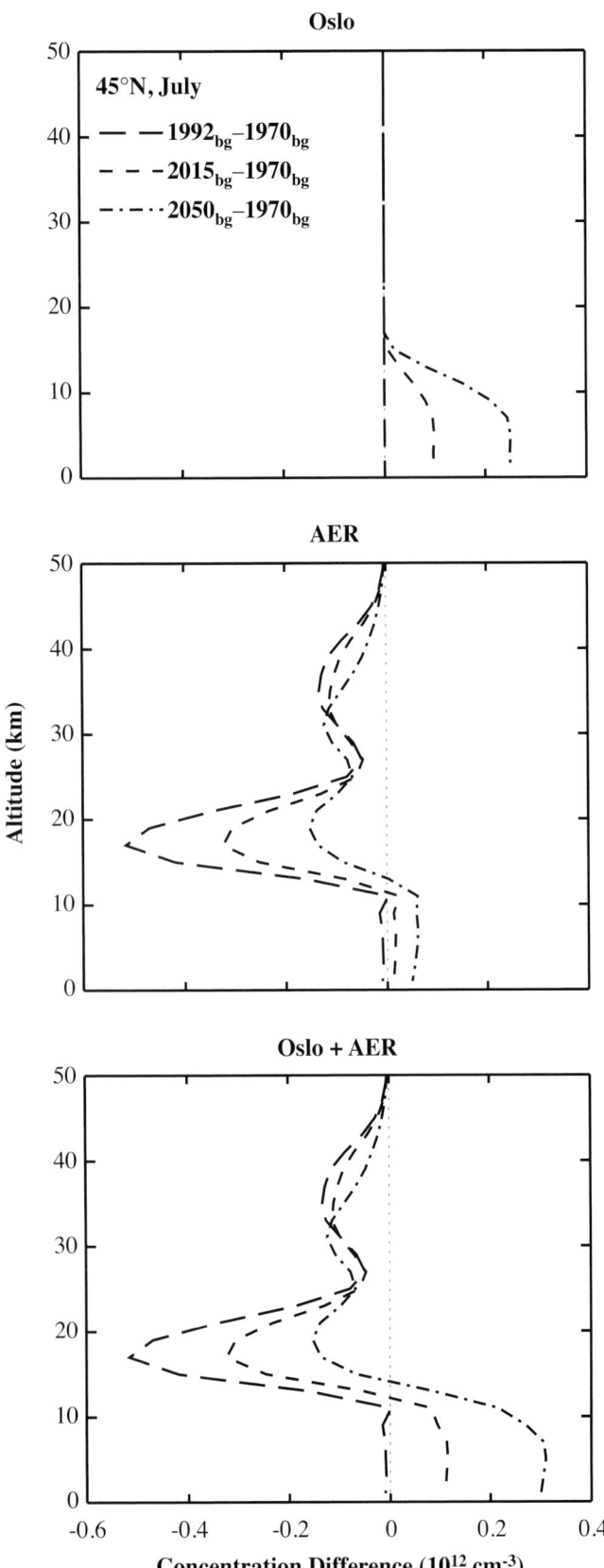

Figure 5-4: Method used to combine 3-D CTM and 2-D CTM results for changes in background atmospheres.

peak values in the lower troposphere result from increased surface emissions, predominantly NO_x, that are expected to occur over this time period. The calculations reported in the middle panel show quite different behavior. In this set of 2-D calculations, surface emissions of NO_x were not increased between 1970 and 2050. The changes in ozone concentrations shown in this panel are directly attributable to expected changes in concentrations of inorganic chlorine and bromine compounds, as well as those for nitrous oxide.

The bottom panel in Figure 5-4 shows the result of the linear combination of the concentration changes calculated in the top and middle panels. Results such as those in the bottom panel are then used to derive the background atmospheres for 1970, 2015, and 2050 from the background atmosphere used for 1992. The procedure of combining the results of 3-D and 2-D CTM calculations in this manner is clearly questionable, but at the present stage of model development this compromise is necessary.

5.4.1.2. *Method for Referring Calculated Changes to 1970*

As discussed elsewhere in this chapter, the calculated impact of aviation on UV is to be compared with impacts on UV resulting from changes in atmospheric composition from other sources. For this purpose, 1970 has been taken as the reference year because it represents a period before the expected onset of substantial changes in the concentration of stratospheric ozone resulting from increases in the concentrations of stratospheric chlorine and bromine. The required ozone columns for 1970 have been obtained in the following manner: It is assumed that the amount of ozone in the lower troposphere of the background atmosphere has not changed between 1970 and 1992. Just as for 2015 and 2050, a 2-D calculation is performed to derive the 1970 background columns from corresponding values for 1992. As explained in Section 5.4.1.1, the background surface concentrations used for the 1970 calculation are those given in Table 6.3 of WMO (1994). It is further assumed that changes in tropospheric ozone due to aviation between 1970 and 1992 are proportional to the amount of NO_x emitted. The amount of NO_x emitted by aviation in 1970 is deduced from a linear extrapolation of the values given in Table 9-4; this amount is estimated to be 0.72 Tg.

5.4.2. **Results**

5.4.2.1. *Ozone Column Changes*

Figure 5-5 shows calculated ozone column changes as a function of latitude for July and October and for a range of scenarios. The top panel shows the change in columns attributable to subsonic fleets over the period 1992–2050 relative to a background atmosphere of the same year. The corresponding calculations for 1970 are not shown, but the effects attributable to aircraft for this period were very much smaller than any shown in the top panel of Figure 5-5 (see Figures 5-6, 5-7, and

5-8). The middle panel shows a corresponding calculation for the hybrid (subsonic+supersonic) fleets for 2015 and 2050, and the bottom panel shows the changes in background between 2050 and 1970, between 2015 and 1970, and between 1992 and 1970. The major features of the aircraft impact on ozone have been discussed in Chapter 4. For the background atmospheres

(bottom panel), the calculations predict a systematic increase in ozone after 1992. For the Southern Hemisphere, this outcome is largely because of the expected decrease in bromine and chlorine. The additional increase in ozone in the Northern Hemisphere relative to the Southern Hemisphere is because the projected release of NO_x at the surface is greater in the Northern Hemisphere.

5.4.2.2. Calculated Changes in Ozone Compared to Changes in UV

An alternative to the method of presentation of results used in Figure 5-5 is to show the evolution of changes in ozone and UV as a function of time, as in Figure 5-6. Here, calculated ozone and UV_{ery} averaged between 65°S and 65°N for July and for a range of scenarios are referred to calculated background values for 1970. When averages are taken over a broad latitude band such as that in Figure 5-6, seasonal behavior is removed, to a large extent. However, a particular latitude may exhibit significant seasonal behavior in the calculated departure from background values for 1970. For example, Figure 5-7 shows calculated changes for 45°N for January, and Figure 5-8 shows the corresponding calculations for July. Comparison of Figures 5-7 and 5-8 shows that, although the calculated ozone and UV_{ery} for the background in 2050 for January are little different from the corresponding values for 1970, there are significant departures in July. This result is almost certainly because of the increased ozone calculated for the upper troposphere in 2050 resulting from the increase in NO_x released from the surface of the Earth relative to 1970.

The method of presentation shown in Figures 5-6, 5-7, and 5-8 has the advantage that the predicted impact of either the hybrid fleet or the subsonic fleet on the background atmosphere at the corresponding time can easily be determined from the diagram; in addition, the departure for any of the scenarios from the calculated value for the background atmosphere in 1970 is readily determined. However, this method of presentation does not show changes as a function of latitude; that information is provided in Figure 5-9, which shows percentage UV_{ery} changes for July and October corresponding to ozone changes in Figure 5-5. The labels on the diagram refer to the atmospheres defined in Table 5-2. The three panels in Figure 5-9 show the calculated changes in UV_{ery} for subsonics (top), subsonics+supersonics (middle), and changes in ground emissions (bottom). Several trends are clear from Figure 5-9. First, the impact of subsonic

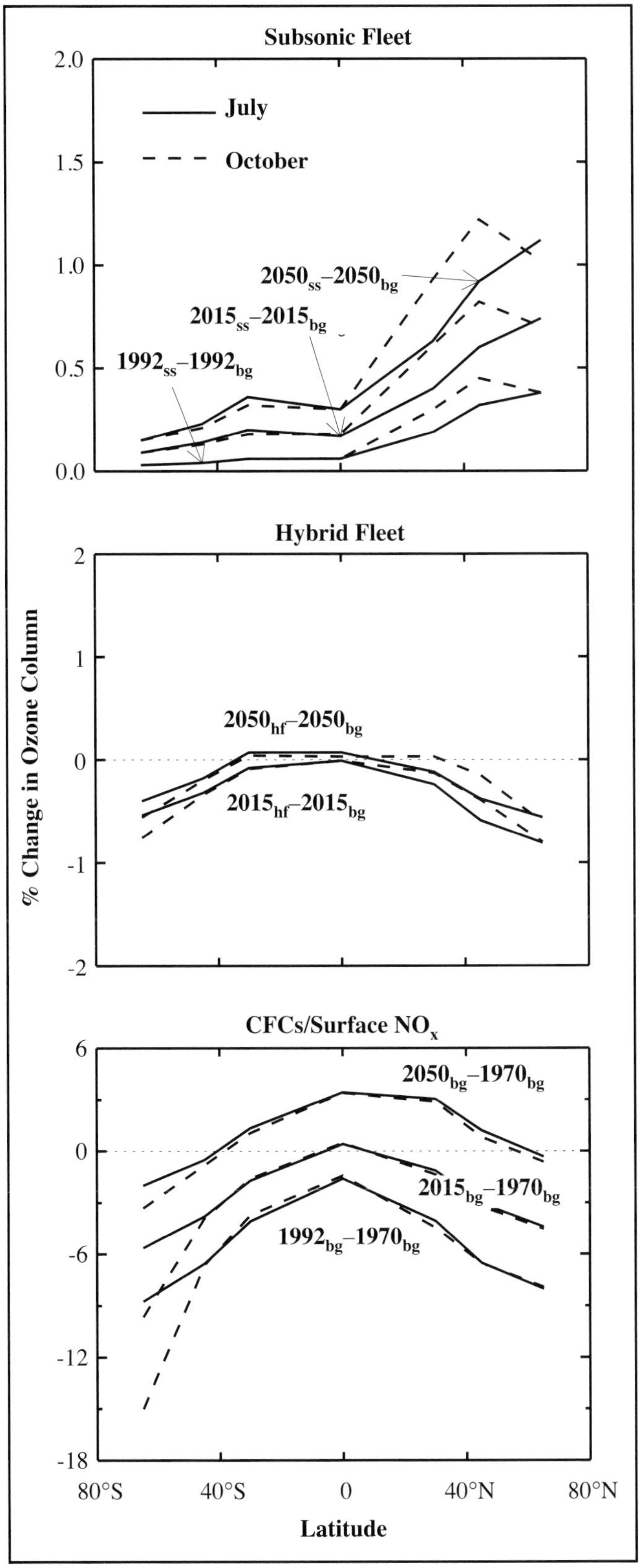

Figure 5-5: *Ozone column changes for July and October.* Top Panel—Subsonic aircraft compared with background atmosphere: 2050 (**2050ss-2050bg**), 2015 (**2015ss-2015bg**), 1992 (**1992ss-1992bg**); Middle Panel—Hybrid (subsonic + supersonic) fleet compared with background: 2050 (**2050hf-2050bg**), 2015 (**2015hf-2015bg**); Bottom Panel—2050 (**2050bg-1970bg**), 2015 (**2015bg-1970bg**), and 1992 (**1992bg-1970bg**) background atmospheres compared with 1970.

aviation is significantly greater in 2015 and 2050 than in 1992 (top panel); at northern mid-latitudes, the impact of the subsonic fleet is roughly proportional to the levels of aircraft emissions assumed. Second, when the effect of the hybrid fleet is compared to the corresponding background, there is an increase in UV_{ery} relative to the case where a pure subsonic fleet is considered (compare the top and middle panels). Third, calculated changes in UV_{ery} relative to 1970 for present and future background atmospheres show an increase in 1992 and a systematic decrease thereafter (bottom panel). This behavior is directly related to

the levels of bromine and chlorine assumed for the stratosphere and the amount of NO_x assumed to be released from the surface of the Earth.

5.4.2.3.   *Treatment of Uncertainties in Ozone Change and UV_{ery} Change Calculations*

The calculations reported in Sections 5.4.2.1 and 5.4.2.2 are based on information provided in Chapter 4. In that chapter,

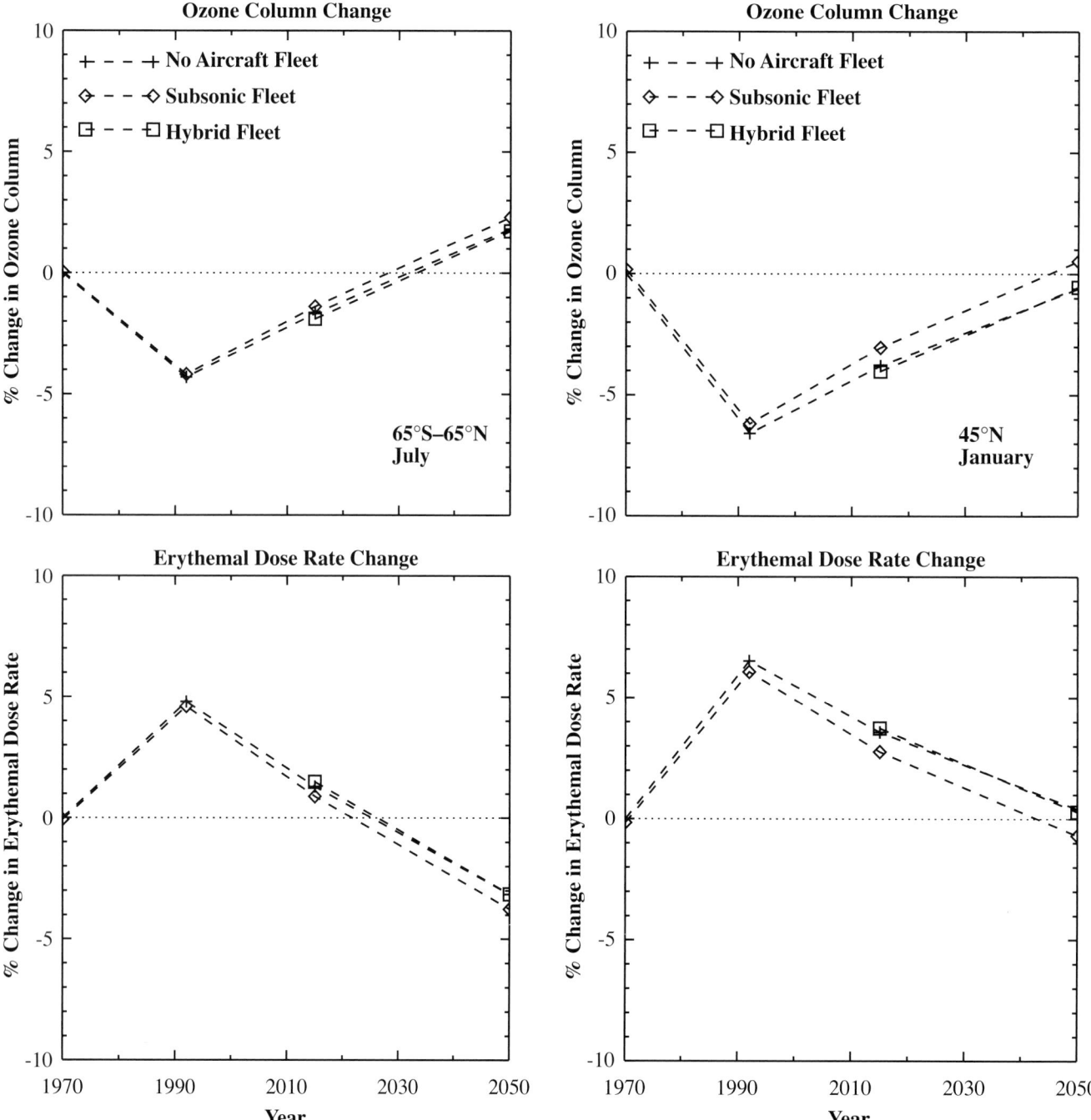

Figure 5-6: Calculated ozone and UV_{ery} averaged between 65°S and 65°N for July and referred to the calculated background values for 1970.

Figure 5-7: Calculated ozone and UV_{ery} at 45°N in January referred to the calculated background values for 1970.

calculations were carried out for a number of scenarios corresponding to the years 1992, 2015, and 2050. For 1992, the calculated impact of the subsonic fleet on atmospheric ozone was estimated using a 3-D CTM. Similar calculations were performed to assess the expected impact of subsonic aircraft in 2015 and 2050. In addition, the impact of hybrid fleets of subsonic and supersonic aircraft for 2015 and 2050 were obtained using a combination of 2-D and 3-D chemical transport calculations. The results for the hybrid fleet in Sections 5.4.2.1 and 5.4.2.2 are based on calculations provided in Chapter 4 for the AER

model for scenarios S1k-D for 2015 and S9h-D9 for 2050. These scenarios are defined in Tables 4-10, 4-11, and 4-12.

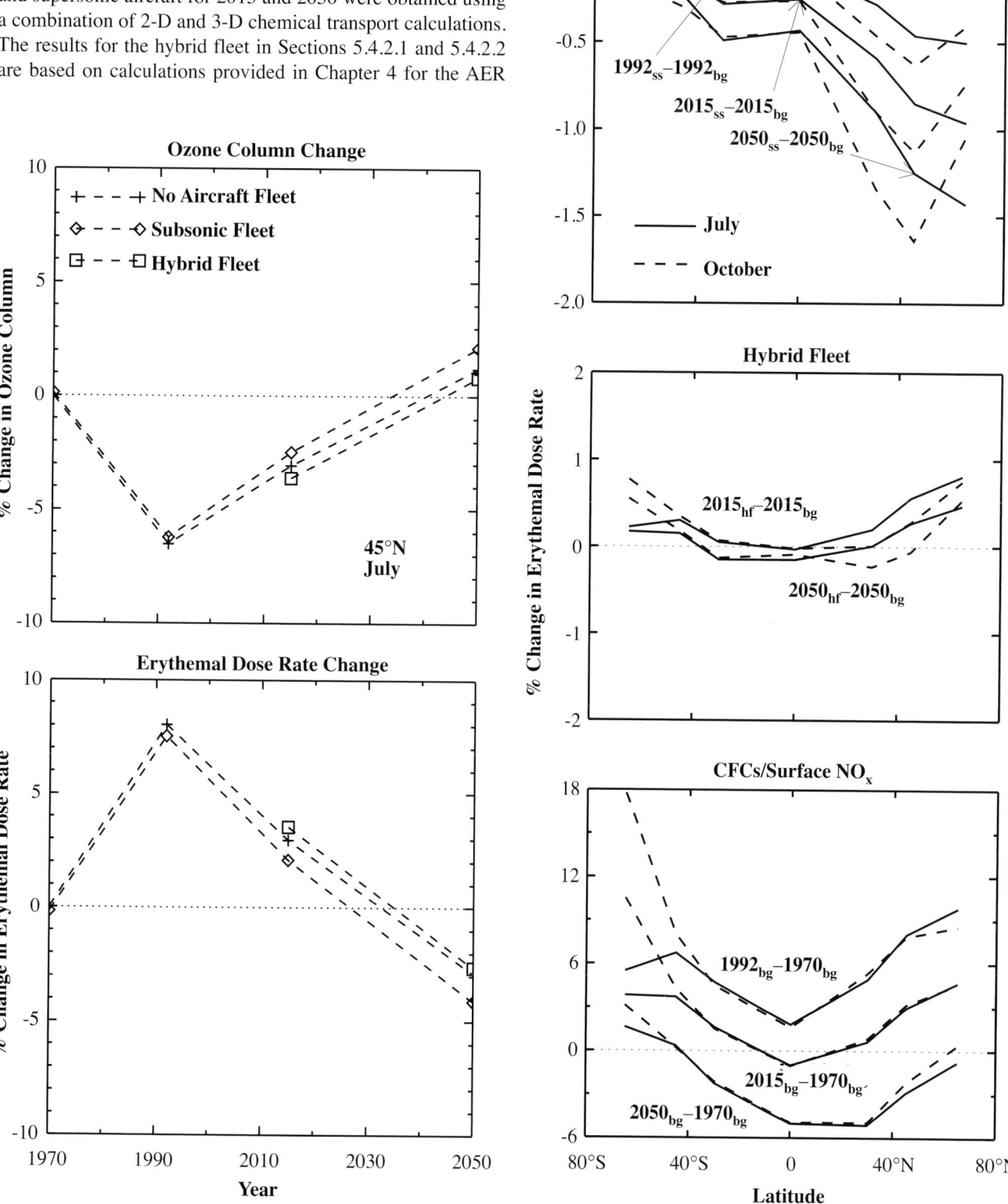

Figure 5-8: Calculated ozone and UV_{ery} at 45°N in July referred to the calculated background values for 1970.

Figure 5-9: Percent changes in UV_{ery} as a function of latitude for July and October for a number of scenarios.

In addition to performing these calculations, the authors of Chapter 4 set uncertainty limits on these results. For ozone changes resulting from the impact of the subsonic fleets, the uncertainty was taken to be a factor of two times the difference obtained by the Oslo 3-D model for scenarios (B-A), (D-C), and (F-E) defined in Table 4-4. The quoted uncertainties are taken as the 67% likelihood range. We believe that uncertainty in the calculation of ozone percentage changes is by far the greatest uncertainty in the determination of percentage changes in UV_{ery}; accordingly, we have not added additional uncertainties to the range supplied by Chapter 4. In addition, our assessment of the confidence in these calculations is as given by Chapter 4—that is, "fair" for 2015 and "poor" for 2050.

Chapter 4 considered three components in assessing the uncertainty for the impact of the hybrid fleet on ozone: The spread obtained by a number of models for a range of plausible scenarios, uncertainties in chemical rate coefficients, and uncertainties introduced by inaccurate treatment of atmospheric circulation in the models. Chapter 4 concluded that the annually averaged impact on ozone for the Northern Hemisphere of a hybrid fleet in 2050 (including 1,000 HSCTs) would be in the range of -3.5 to +1% when compared with the impact of the subsonic fleet and that the best estimate is -1% given by the AER 2-D model. The uncertainty range again represents the 67% likelihood range with a confidence in this uncertainty range of "fair." As with estimated subsonic impacts, we believe that the uncertainty in the change in ozone caused by the hybrid fleet is much greater than any other uncertainties in the calculation of changes in UV_{ery}. Chapter 4 has provided only an annually averaged Northern Hemisphere value because it is not possible at present to assign uncertainty factors as functions of latitude, altitude, and season. The large variations in ozone changes predicted by a range of models as functions of latitude, altitude, and season (see Figures 4-6c, 4-6d, 4-12a, and 4-12b) are clear indications of shortcomings inherent in current models.

Given that the predicted change in UV_{ery} is primarily a function of changes in the ozone column and is less sensitive to the exact altitude dependence of the ozone change, an estimate of the uncertainty for changes in ozone columns as a function of latitude and season should be sufficient. This chapter uses the subjective estimates of uncertainties from Chapter 4 and defines the 67% likelihood range for changes in ozone columns at each latitude and season as follows. If the column change predicted by the AER 2-D model at latitude (Θ) and time of year (T) is $a(\Theta,T)\%$, the uncertainty range in percent column ozone change at that latitude and time of year is given by $[a(\Theta,T)-3]\%$ to $[a(\Theta,T)+2]\%$. The same definition will be adopted for fleet sizes of either 500 or 1,000 HSCTs. These estimates are very subjective; although the confidence attached to these uncertainties for the tropics is "good," outside the tropics the confidence can be considered only "fair." In the absence of a rigorous method of obtaining uncertainties, this chapter also assumes that there is a 5% range in estimated uncertainties for change in UV_{ery}; this range is given by $[b(\Theta,T)-2]\%$ to $[b(\Theta,T)+3]\%$, where $b(\Theta,T)\%$ is the percent change in UV_{ery} corresponding to $a(\Theta,T)\%$ change in ozone column.

In summary, uncertainties in the impact of the subsonic fleet on UV_{ery} may be obtained from any of the diagrams in this chapter that report this change simply by doubling or halving the change shown in the diagram. For example, in the top panel of Figure 5-9, the change in UV_{ery} shown for the subsonic impact for July of 2015 at 45°N is -0.9%. Accordingly, the 67% likelihood range for this particular change is -1.8 to -0.5%; because it is a calculation for 2050, our confidence in this result is "poor" as prescribed by Chapter 4. Similarly, in the middle panel of Figure 5-9, the impact of the hybrid fleet at the equator in July of 2050 is shown as -0.2%. This impact is based on a calculation of ozone changes by the AER 2-D model. From the discussion above, we determined the 67% likelihood range for this impact to be -2.2 to +2.8%, and our estimate of the confidence is "good." At 65°N in July of 2050 (middle panel of Figure 5-9), the calculated impact of the hybrid fleet on UV_{ery} is +0.5%, which gives -1.5 to +3.5% as the 67% likelihood range with an estimated confidence of "fair."

As discussed in Section 5.4.1.1, the calculations shown in Figures 5-6 to 5-9 include percent changes in UV_{ery} for the background atmosphere using 1970 as the reference year. At 45°N in July, these changes are +8, +3, and -3% for 1992, 2015, and 2050, respectively. At 45°S in January, the calculated changes are +9, +4.5, and 0% for 1992, 2015, and 2050, respectively. For comparison, the computed change due to observed ozone depletion at 35–50°N in July is about 4% over the period 1970–1992 (WMO, 1999). The corresponding change for 35–50°S in January is 8%.

5.4.2.4. Contribution of Persistent Contrails, Cirrus Clouds, and Aerosols to the Impact of Aviation on UV

The geographic and temporal variability associated with clouds and aerosols complicates attempts to make general statements concerning their effects on UV irradiance. Nonetheless, this section considers some highly simplified scenarios to place bounds on the influence of altered contrails, cirrus, and aerosol amounts. There are three issues to address: The effect of regional, persistent contrails; the effect of observed trends in cirrus that may result from a variety of processes, including aviation; and the effect of aviation-produced aerosols.

To estimate the effects of persistent contrails on ground-level UV, calculations assume a 5% area coverage, appropriate to the local maximum over the eastern United States of America (see Section 3.4.3), and a contrail optical thickness of 0.3. For latitude 45°N summer, local noon, this scenario leads to a reduction in UV_{ery} of 0.2% relative to clear skies. If persistent contrail coverage were to increase to 10%, the corresponding reduction in UV_{ery} would be 0.4%.

The observed trends in cirrus presented in Chapter 3 include the effects of aviation, as well as any other influences that may be operative. The estimated response of UV_{ery} to trends in the cirrus background is based on the following assumptions: The cirrus have a scattering optical thickness of 0.3, and initially

23% of the land area is covered by cirrus at an altitude of 11 km. Starting from this condition, an increase of 3.5% per decade in the land area covered by cirrus, appropriate to the United States of America in spring (Figure 3-19), leads to a decline in UV_{ery} of approximately 0.1% per decade for local noon at latitude 45°N. If changes in persistent contrails and cirrus, especially on regional scales, differ substantially from the estimates adopted above, the resulting changes in UV_{ery} would have to be modified accordingly. The numbers given here point to the magnitude of reasonable changes, but they should not be viewed as predictions of the future.

The aircraft-related perturbation to atmospheric optical depth in the UV from soot and sulfate aerosols is very small compared to the natural background opacity. The values presented in Table 3-4 imply enhanced aerosol optical depths for soot and sulfates of less than 10^{-3}. The decrease in UV_{ery} associated with such perturbations is less than 0.1%. Calculations indicate that, although an increase in aerosols in the background atmosphere between 1970 and 2050 from sources other than aviation would result in a reduction of UV_{ery} relative to clear skies, the percent change in UV_{ery} attributable to aviation is relatively insensitive to background aerosol loading.

5.5.   Conclusions

Current radiative transfer models are able to predict the influence of changes in ozone on ground-level UV irradiances with a high degree of confidence. The ability of these models to treat aerosols is less well developed because of uncertainties in their optical properties. In addition, the geometrical complexity and temporal variability of clouds still pose challenges to accurate radiative transfer modeling. From the standpoint of assessing the effects of future aviation, it appears that changes in atmospheric ozone abundances are the most important driver of changes in UV irradiance. Outcomes predicted for a range of scenarios are summarized below.

A fleet of subsonic aircraft leads to a latitude-dependent reduction in UV_{ery}, with the greatest percentage changes in the Northern Hemisphere. In the year 2015, the best estimate of the change in UV_{ery} for July associated with a subsonic fleet lies in the range -0.5 to -1.0% poleward of 30°N. When the range of uncertainty is included, the spread in calculated change in irradiance is -1.8 to -0.5% at 45°N. These values are for the year 2015 for a subsonic fleet relative to the year 2015 without this fleet. In the Southern Hemisphere, the corresponding best estimate of change in UV_{ery} for January is less than -0.3% from the equator to 30°S and near 0% at 60°S. For 1992, the calculated impacts for aviation are about half those calculated for 2015. In 2050, the calculated impacts of a pure subsonic fleet are about 50% greater than those for 2015.

Scenarios based on fleets containing subsonic and supersonic aircraft imply aviation-related increases in UV_{ery} in 2015 and 2050 relative to what would otherwise be expected in those years. The best estimates for both years are similar—very little

impact on UV_{ery} in the tropics, rising to increases between about 0.2 and 0.8% around 65° in both hemispheres. Although these estimated changes may be considered small, they do have significantly larger uncertainty limits. For the 67% likelihood range, it is estimated that the percentage change in UV_{ery} relative to the background atmosphere can be expressed as (-2% + best estimate) to (+3% + best estimate)—a 5% range in the change relative to the background column. As can be seen from the middle panel of Figure 5-9, this range always encompasses the possibility of no change in UV_{ery} relative to the corresponding background atmosphere.

The change in aerosol loading expected from increased aircraft operations between the present and 2050 is small relative to both the natural aerosol background and anthropogenic influences other than those related to aviation. The effect of the aircraft-related increase in aerosols is to reduce UV irradiance by less than 0.1%. Calculations indicate that aircraft-related increases in contrails lead to a decrease in UV irradiance of less than 0.5% in regions where 10% of the sky is covered.

The anticipated decline of the chlorine and bromine content of the stratosphere as well as increased emissions of NO_x from combustion between 1992 and 2050 are expected to lead to latitude-dependent increases in ozone amounts and consequent decreases in UV_{ery}, irrespective of future aviation. For a background atmosphere at 45°N in July, the calculated change in UV_{ery} with respect to a 1970 background is +8, +3, and -3% for 1992, 2015, and 2050, respectively. At 45°S in January, the calculated changes are 9, 4.5, and 0% for 1992, 2015, and 2050, respectively. Observed changes in total ozone from 1970 to 1992 imply smaller percentage increases in UV_{ery}, indicating the degree of uncertainty in the model predictions.

References

Anderson, G.P., S.A. Clough, F.X. Kneizys, J.H. Chetwynd, and E.P. Shettle, 1986: *AFGL Atmospheric Constituent Profiles (0–120 km).* AFGL-TR-86-0110, Air Force Geophysics Laboratory, Hanscom Air Force Base, MA, USA, 43 pp.

Bais, A.F., 1997a: Spectrometers: operational errors and uncertainties. In: *Solar Ultraviolet Radiation Modeling, Measurements and Effects* [Zerefos, C. and A. Bais (eds.)]. Series I: Global Environmental Change. NATO-ASI Series, Springer-Verlag, Berlin, Germany, Vol. 1, pp. 163–173.

Bais, A.F., 1997b: Absolute spectral measurements of direct solar ultraviolet irradiance with a Brewer spectrophotometer. *Applied Optics,* **2007,** 5199–5204.

Bais, A.F., C.S. Zerefos, C. Meleti, I.C. Ziomas, and K. Tourpali, 1993: Spectral measurements of solar UV-B radiation and its relations to total ozone, SO_2, and clouds. *Journal of Geophysical Research,* **98,** 5199–5204.

Bais, A.F., C.S. Zerefos, and C.T. McElroy, 1996: Solar UVB measurements with the double- and single-monochromator Brewer ozone spectrophotometers. *Geophysical Research Letters,* **23,** 8, 833–836.

Bais, A.F., M. Blumthaler, A.R. Webb, J. Groebner, P.J. Kirsch, B.G. Gardiner, C.S. Zerefos, T. Svenoe, and T.J. Martin, 1997: Spectral UV measurements over Europe within the SESAME activities. *Journal of Geophysical Research,* **102(D7),** 8731–8736.

Blumthaler, M., J. Groebner, M. Huber, and W. Ambach, 1996: Measuring spectral and spatial variations of UVA and UVB sky radiance. *Geophysical Research Letters,* **23(5),** 547–550.

Blumthaler, M., 1997: Broadband detectors for UV measurements. In: *Solar Ultraviolet Radiation Modeling, Measurements and Effects* [Zerefos, C. and A. Bais (eds.)]. Series I: Global Environmental Change. NATO-ASI Series, Springer-Verlag, Berlin, Germany, pp. 175–185.

Bodhaine, B.A., E.G. Dutton, R.L. McKenzie, and P.V. Johnston, 1997: Spectral UV measurements at Mauna Loa: July 1995–July 1996. *Geophysical Research Letters,* **23**, 2121–2124.

Bodeker, G.E. and R.L. McKenzie, 1996: An algorithm for inferring surface UV irradiance including cloud effects. *Journal of Applied Meteorology,* **35**, 1860–1877.

Booth, C.R. and S. Madronich, 1994: Radiation amplification factors: improved formulation accounts for large increases in ultraviolet radiation associated with Antarctic ozone depletion. In: *Ultraviolet Radiation in Antarctica: Measurements and Biological Research* [Weiler, C.S. and P.A. Penhale (eds.)]. *AGU Antarctic Research Series,* **62**, Washington, DC, USA, 39–42.

Brewer, A.W. and J.B. Kerr, 1973: Total ozone measurements in cloudy weather. *PAGEOPH,* **106–108**, 928–937.

Dave, J.V., 1965: Multiple scattering in a non-homogeneous, Rayleigh atmosphere. *Journal of Atmospheric Science,* **22**, 273–279.

Davies, R., 1993: Increased transmission of ultraviolet radiation to the surface due to stratospheric scattering. *Journal of Geophysical Research,* **98**, 7251–7253.

Deguenther, M., R. Meerkötter, A. Albold, and G. Seckmeyer, 1998: Case study on the influence of inhomogeneous surface albedo on UV irradiance. *Geophysical Research Letters,* **25**, 3587–3590.

De Meur, D. and H. De Backer, 1993: Influence of sulfur dioxide trends on Dobson measurements and on electrochemical ozone soundings. In: *Proceedings of the Society of Photo-Optical Instrumentation Engineers,* **2047**, 18–26.

Eck, T.F., P.K. Bhartia, and J.B. Kerr, 1995: Satellite estimation of spectral UVB irradiance using TOMS derived total ozone and reflectivity. *Geophysical Research Letters,* **22**, 611–614.

Fioletov, V.E. and J.B. Kerr, 1996: Numerical relationship between UV irradiance, total ozone and other variables from analysis of Brewer spectral UV-B measurements archived at the World Ozone and UV Data Center. In: *Proceedings of the XVIII Quadrennial Ozone Symposium, International Ozone Commission, L'Aquila, Italy, 12-21 September 1996* [Bojkov, R.D. and G. Visconti (eds.)]. Edigrafital S.P.A., Italy, pp. 845–848.

Fioletov, V.E., J.B. Kerr, and D.I. Wardle, 1997: The relationship between total ozone and spectral UV irradiance from Brewer spectrophotometer observations and its use for derivation of total ozone from UV measurements. *Geophysical Research Letters,* **24**, 2705–2708.

Fioletov, V.E, E. Griffioen, J.B. Kerr, and D.I. Wardle, 1998: Influence of volcanic sulfur dioxide on spectral UV irradiance as measured by Brewer spectrophotometers. *Geophysical Research Letters,* **25**, 1665–1668.

Forster, P.M.F., 1995: Modeling ultraviolet radiation at the earth's surface. Part I: The sensitivity if ultraviolet irradiances to atmospheric changes. *Journal of Applied Meteorology,* **34**, 2412–2425.

Forster, P.M.F., K.P. Shine, and A.R. Webb, 1995: Modeling ultraviolet radiation at the earth's surface. Part II: Model and instrument comparison. *Journal of Applied Meteorology,* **34**, 2426–2439.

Frederick, J.E. and C. Erlick, 1995: Trends and interannual variations in erythemal sunlight, 1978–1993. *Photochemistry Photobiology,* **62**, 476–484.

Frederick, J.E. and D. Lubin, 1988: The budget of biologically active radiation in the earth-atmosphere system. *Journal of Geophysical Research,* **93**, 3825–3832.

Frederick, J.E. and H.E. Snell, 1990: Tropospheric influence on solar ultraviolet radiation: the role of clouds. *Journal of Climate,* **3**, 373–381.

Frederick, J.E. and H.D. Steele, 1995: The transmission of sunlight through cloudy skies: an analysis based on standard meteorological information. *Journal of Applied Meteorology,* **34**, 2775–2781.

Gardiner, B.G. and P.J. Kirsch (eds.), 1993: *Second European Intercomparison of Ultraviolet Spectroradiometers.* Air Pollution Research Report no. 38, Commission of European Communities, Brussels, Belgium, Vol. 38, 67 pp.

Gardiner, B.G. and P.J. Kirsch (eds.), 1995: *Setting Standards for European Ultraviolet Spectroradiometers.* Air Pollution Research Report no. 53, Commission of European Communities, Luxembourg, 138 pp.

Groebner, J., M. Blumthaler, and W. Ambach, 1996: Experimental investigation of spectral global irradiance measurements errors due to a non ideal cosine response. *Geophysical Research Letters,* **33**, 23, 18, 2493–2496.

Gurney, K., 1998: Evidence for increasing ultraviolet at Point Barrow, Alaska. *Geophysical Research Letters,* **25**, 903–906.

Herman, J.R., P.K. Bhartia, J. Ziemke, Z. Ahmad, and D. Larko, 1996: UV-B increases (1979–1992) from decreases in total ozone. *Geophysical Research Letters,* **23**, 2117–2120.

Kaye, J.A. and T.L. Miller, 1996: The ATLAS series of shuttle missions. *Geophysical Research Letters,* **23**, 2285–2288.

Kerr, J.B., I.A. Asbridge, and W.F.J. Evans, 1988: Intercomparison of total ozone measured by Brewer and Dobson spectrophotometers at Toronto. *Journal of Geophysical Research,* **93**, 11129–11140.

Kerr, J.B., 1997: Observed dependencies of atmospheric UV radiation and trends. In: *Solar Ultraviolet Radiation Modeling, Measurements, and Effects* [Bais, A.F. and C.S. Zerefos (eds.)]. Series I: Global Environmental Change. NATO-ASI Series, Springer-Verlag, Berlin, Germany, Vol. 1, pp. 259–266.

Kerr, J.B. and C.T. McElroy, 1993: Evidence for large upward trends of ultraviolet radiation linked to ozone depletion. *Science,* **262**, 1032–1034.

Kerr, J.B. and C.T. McElroy, 1994: Response (to comment by Michaels *et al.,* 1994). *Science,* **264**, 1342–1343.

Kerr, J.B., C.T. McElroy, D.W. Tarasick, and D.I. Wardle, 1994: The Canadian ozone watch and UV-B advisory programs. In: *Ozone in the Troposphere and Stratosphere, National Aeronautics and Space Administration, Charlottesville, VA, USA, 3-13 June 1992* [Hudson, R.D. (ed.)]. NASA Conference Publication 3266, National Aeronautics and Space Administration, Washington, DC, USA, pp. 794–797.

Koskela, T. (ed.), 1994: *The Nordic Intercomparison of Ultraviolet and Total Ozone Instruments at Izana from 24 October to 5 November 1993.* Final Report. Meteorological publication no. 27, Finnish Meteorological Institute, Helsinki, Finland, 123 pp.

Krotkov, N.A., P.K. Bhartia, J.R. Herman, V. Fioletov, and J. Kerr, 1998: Satellite estimation of spectral surface UV irradiance in the presence of tropospheric aerosols 1: cloud-free case. *Journal of Geophysical Research,* **103**, 8779–8793.

Kylling, A, A. Albold, and G. Seckmeyer, 1997: Transmittance of a cloud is wavelength-dependent in the UV range: physical interpretation. *Geophysical Research Letters,* **24**, 397–400.

Leszczynski, K., K. Jokela, L. Ylianttila, R. Visuri, and M. Blumthaler, 1996: *Report of the WMO/STUK Intercomparison of Erythemally-Weighted Solar UV Radiometers, Spring–Summer 1995, Helsinki, Finland.* Report No. 112, Global Atmosphere Watch, World Meteorological Organization, Geneva, Switzerland, p. 90.

Madronich, S., 1992: Implications of recent total atmospheric ozone measurements for biologically active ultraviolet radiation reaching the earth's surface. *Geophysical Research Letters,* **19(1)**, 37–40.

Mayer, B. and G. Seckmeyer, 1996: All-weather comparison between spectral and broadband (Robertson-Berger) UV measurements. *Photochemistry Photobiolology,* **64(5)**, 792–799.

Mayer, B., G. Seckmeyer, and A. Kylling, 1997: Systematic long-term comparison of spectral UV measurements and UVSPEC modeling results. *Journal of Geophysical Research,* **102**, 8755–8767.

Mayer, B. and G. Seckmeyer, 1998a: Retrieving ozone columns from spectral direct and global UV irradiance measurements. In: *Proceedings of the XVIII Quadrennial Ozone Symposium, International Ozone Commission, L'Aquila, Italy, 12-21 September 1996* [Bojkov, R.D. and G. Visconti (eds.)]. Edigrafital S.P.A., Italy, pp. 935–938.

Mayer, B., A. Kylling, S. Madronich, and G. Seckmeyer, 1998b: Enhanced absorption of UV irradiance due to multiple scattering in clouds: experimental evidence and theoretical explanation. *Journal of Geophysical Research,* **103**, 31241–31254.

McKenzie, R.L., M. Kotkamp, G. Seckmeyer, R. Erb, R. Gies, and S. Toomey, 1993: First southern hemisphere intercomparison of measured solar UV spectra. *Geophysical Research Letters,* **20**, 2223–2226.

McKenzie, R.L., W.A. Matthews, and P.V. Johnston, 1991: The relationship between erythemal UV and ozone, derived from spectral irradiance measurements. *Geophysical Research Letters,* **18**, 2269–2272.

McKinlay, A.F. and B.L. Diffey, 1987: A reference action spectrum for ultraviolet induced erythema in human skin. *CIE Journal,* **6**, 17–22.

Mentall, J.E., J.E. Frederick, and J.R. Herman, 1981: The solar irradiance from 200 to 330 nm. *Journal of Geophysical Research,* **86**, 9881–9884.

Meier, R.R., D.E. Anderson, and M. Nicolet, 1982: Radiation field in the troposphere and stratosphere from 240–1000 nm. I. General analysis. *Planetary and Space Science, 30,* 923–933.

Neckel, H. and D. Labs, 1984: The solar radiation between 3000 and 12500 A^0. *Solar Physics, 90,* 205–358.

Park, J., M.K.W. Ko, C.H. Jackman, R.A. Plumb, and K.H. Sage (eds.), 1999: *Models and Measurements II.* National Aeronautics and Space Administration, Langley Research Center, Hampton, VA, USA, (in press).

Randeniya, L.K., P.F. Vohralik, I.C. Plumb, and K.R. Ryan, 1997: Heterogeneous $BrONO_2$ hydrolysis: effect on NO_2 columns and ozone at high latitudes in summer. *Journal of Geophysical Research, 102,* 23543–23557.

Ruggaber, A., R. Dlugi, and T. Nakajima, 1994: Modelling radiation and photolysis frequencies in the troposphere. *Journal of Atmospheric Chemistry, 18,* 171–210.

Schwander, H., P. Koepke, and A. Ruggaber, 1997: Uncertainties in modelled UV irradiances due to limited accuracy and availability of data inputs. *Journal of Geophysical Research, 102,* 9419–9429.

Seckmeyer, G., 1989: Spectral measurements of the variability of global UV-radiation. *Meteorologische Rundschau, 41(6),* 180–183.

Seckmeyer, G. and G. Bernhard, 1993: Cosine error correction of spectral global irradiances. In: *Atmospheric Radiation* [Stamnes, K.H. (ed.)]. Proceedings of the Society of Photo-Optical Instrumentation Engineers, Tromsoe, Norway, 30 June - 1 July 1993. SPIE, Bellingham, WA, USA, pp. 140–151.

Seckmeyer, G., R. Erb, and A. Albold, 1996: Transmission of a cloud is wavelength-dependent in the UV-range. *Geophysical Research Letters, 23,* 2753–2755.

Slaper, H., H.A.J.M. Reinen, M. Blumthaler, M. Huber, and F. Kuik, 1995: Comparing ground level spectrally resolved solar UV measurements using various instruments: a technique resolving effects of wavelength shift and slit width. *Geophysical Research Letters, 22,* 2721–2724.

Stamnes, K., S.C. Tsay, W. Wiscombe, and K. Jayaweera, 1988: Numerically stable algorithm for discrete ordinate-method radiative transfer in multiple scattering and emitting media. *Applied Optics, 27,* 2502–2509.

Stolarski, R.S., S.L. Baughcum, W.H. Brune, A.R. Douglass, D.W. Fahey, R.R. Friedl, S.C. Liu, R.A. Plumb, L.R. Poole, H.L. Wesoky, and D.R. Worsnop, 1995: *Scientific Assessment of the Atmospheric Effect of Stratospheric Aircraft.* NASA Reference Publication 1381, National Aeronautics and Space Administration, Washington, DC, USA, 64 pp.

Tsitas, S.R. and Y.L. Yung, 1996: The effect of volcanic aerosols on ultraviolet radiation in Antarctica. *Geophysical Research Letters, 23,* 157–160.

UNEP, 1994: *Environmental Effects of Ozone Depletion: 1994 Update.* UNEP Environmental Effects Panel Report, United Nations Environment Programme, Nairobi, Kenya, 108 pp.

Wang, P. and J. Lenoble, 1994: Comparison between measurements and modeling of UV-B irradiance for clear sky: a case study. *Applied Optics, 33,* 3964–3971.

Wardle, D.I., J.B. Kerr, C.T. McElroy, and D.R. Francis (eds.), 1997: *Ozone Science: A Canadian Perspective on the Changing Ozone Layer.* Environment Canada Report no. CARD 97-3, University of Toronto Press, Toronto, Canada, 119 pp.

Weatherhead, E.C., G.C. Tiao, G.C. Reinsel, J.E. Frederick, J.J. DeLuisi, D. Choi, and W. Tam, 1997: Analysis of long-term behavior of ultraviolet radiation measured by Robertson-Berger meters at 14 sites in the United States. *Journal of Geophysical Research, 102(D7),* 8737–8754.

Weatherhead, E.C., G.C. Reinsel, G.C. Tiao, X.-L. Meng, D. Choi, W.-K. Cheang, T. Keller, J. DeLuisi, D.J. Wuebbles. J.B. Kerr, A.J. Miller, S.J. Oltmans, and J.E. Frederick, 1998: Factors affecting the detection of trends: statistical considerations and applications to environmental data. *Journal of Geophysical Research, 103,* 17149–17161.

Webb, A.R., B.G. Gardiner, T.J. Martin, K. Leszczynski, J. Metzdorf, and V.A. Mohnen, 1998: *Guidelines for Site Quality Control of UV Monitoring.* WMO/GAW Publication no. 26, World Meteorological Organization, Geneva, Switzerland, 39 pp.

WMO, 1999: *Scientific Assessment of Ozone Depletion: 1998.* Report No. 44, Global Ozone Research and Monitoring Project, World Meteorological Organization, Geneva, Switzerland, 732 pp.

WMO, 1997: *Report of the WMO/STUK Intercomparison of Erythemally Weighted Solar UV Radiometers* [Leszczynski, K., K. Jokela, L. Ylianttila, R. Visuri, and M. Blumthaler (eds.)]. WMO Report no. 112, World Meteorological Organization, Geneva, Switzerland, 89 pp.

WMO, 1994: *Scientific Assessment of Ozone Depletion.* World Meteorological Organization, Geneva, Switzerland, 508 pp.

Woods, T.N., D.K. Prinz, G.J. Rottman, J. London, P.C. Crane, R.P. Cebula, E. Hilsenrath, G.E. Brueckner, M.D. Andrews, O.R. White, M.E. Van Hoosier, L.E. Floyd, L.C. Herring, B.G. Knapp, C.K. Pankrantz, and P.A. Reiser, 1996: Validation of the UARS solar ultraviolet irradiances: comparison with the ATLAS 1 and 2 measurements. *Journal of Geophysical Research, 101,* 9541–9569.

Zerefos, C., 1997: Factors influencing transmission of solar UV irradiance through the earth's atmosphere. In: *Solar Ultraviolet Radiation Modeling, Measurements and Effects* [Zerefos, C. and A. Bais (eds.)]. Series I: Global Environmental Change. NATO-ASI Series, Springer-Verlag, Berlin, Germany, Vol. 1, pp. 133–141.

6

Potential Climate Change
from Aviation

MICHAEL PRATHER AND ROBERT SAUSEN

Lead Authors:
A.S. Grossman, J.M. Haywood, D. Rind, B.H. Subbaraya

Contributors:
P. Forster, A. Jain, M. Ponater, U. Schumann, W.-C. Wang, T.M.L. Wigley,
D.J. Wuebbles

Review Editor:
D. Yihui

CONTENTS

EXECUTIVE SUMMARY

- Aircraft emissions in conjunction with other anthropogenic sources are expected to modify atmospheric composition (gases and aerosols), hence radiative forcing and climate. Atmospheric changes from aircraft result from three types of processes: direct emission of radiatively active substances (e.g., CO_2 or water vapor); emission of chemical species that produce or destroy radiatively active substances (e.g., NO_x, which modifies O_3 concentration); and emission of substances that trigger the generation of aerosol particles or lead to changes in natural clouds (e.g., contrails).

- Radiative forcing (RF) is the metric used here (and in IPCC) to compare climate perturbations among different aviation scenarios and with total anthropogenic climate change. RF is the global, annual mean radiative imbalance to the Earth's climate system caused by human activities. It predicts changes to the global mean surface temperature: Positive RF leads to global warming. Yet climate does not change uniformly; some regions warm or cool more than others; and mean temperature does not describe vital aspects of climate change such as droughts and severe storms. Aviation's impacts via O_3 and contrails occur predominantly in northern mid-latitudes and the upper troposphere, leading potentially to climate change of a different nature than that from CO_2. Nevertheless, we follow the scientific basis for RF from IPCC's Second Assessment Report and take summed RF as a first-order measure of global mean climate change.

- For the 1992 aviation scenario (NASA-1992*), radiative forcing of climate change from aircraft emissions (gases and aerosols) is estimated to be +0.05 W m-2, which is about 3.5% of total anthropogenic radiative forcing as measured against the pre-industrial atmosphere of +1.4 W m-2 for combined greenhouse gases and aerosols (and +2.7 W m-2 for greenhouse gases alone). The components of aircraft-induced radiative forcing are as follows: CO_2, +0.018 W m-2; NO_x, +0.023 W m-2 (via ozone changes) and -0.014 W m-2 (via methane changes); contrails, +0.02 W m-2; stratospheric H_2O, +0.002 W m-2; sulfate aerosol (direct effect), -0.003 W m-2; and black carbon aerosol (soot), +0.003 W m-2. Changes in "natural" cirrus clouds caused by aircraft may result in negligible or potentially large radiative forcing; an estimate could fall between 0 and 0.04 W m-2. Uncertainty estimates, typically a factor of 2 or 3, have been made for individual components and are intended to represent consistent confidence intervals that the radiative forcing value is likely (2/3 of the time) to fall within the range shown. The uncertainty estimate for

the total radiative forcing (without additional cirrus clouds) is calculated as the square root of the sums of the squares of the upper and lower ranges of the individual components.

- Projection of subsonic fleet growth to 2015 (NASA-2015* scenario) results in a best estimate for total aircraft-induced radiative forcing of +0.11 W m-2 in 2015—about 5% of IS92a projected radiative forcing from all anthropogenic emissions that year.

- Various options for the future development of subsonic air traffic under the International Civil Aviation Organization (ICAO)-developed Forecasting and Economic Subgroup (FESG, or F-type) scenarios for aviation in the year 2050 assume an increase in fuel use by 2050 relative to 1992 by a factor of 1.7 to 4.8. These options result in a range for aircraft-induced total radiative forcing (without additional cirrus clouds) from +0.13 to +0.28 W m-2 in 2050, or 3–7% of IS92a total anthropogenic radiative forcing for that year. However, the upper and lower bounds represent aircraft scenarios that diverge significantly from economic growth assumed for IS92a. Alternative Environmental Defense Fund (EDF, or E-type) scenarios considered here adopt growth in 2050 fuel use by factors of 7 to more than 10 and result in a range of total radiative forcing (without additional cirrus clouds) from +0.4 to +0.6 W m-2.

- For the year 2050, a scenario that matches IS92a economic growth (scenario Fa1) gives total radiative forcing of +0.19 W m-2. Individual contributions to aircraft-induced radiative forcing are as follows: CO_2, +0.074 W m-2; NO_x, +0.060 W m-2 (via ozone changes) and -0.045 W m-2 (via methane changes); contrails, +0.10 W m-2; stratospheric H_2O, +0.004 W m-2; sulfate aerosols (direct effect), -0.009 W m-2; and black carbon aerosols (soot), +0.009 W m-2. The contrail estimate includes an increase in fuel consumption, higher overall efficiency of propulsion (i.e., cooler exhaust), and shifting of routes. An estimate for the radiative forcing from additional cirrus could fall between 0 and 0.16 W m-2.

- As one option for future aviation, we consider the addition of a fleet of high-speed civil transport (HSCT, supersonic) aircraft replacing part of the subsonic air traffic under scenario Fa1. In this example, HSCT aircraft are assumed to begin operation in the year 2015, to grow linearly to a maximum of 1,000 aircraft by the year 2040, and to use new technologies to maintain very low emissions of 5 g NO_2 per kg fuel. By the year 2050, this combined fleet (scenario Fa1H) would add 0.08 W m^{-2} on top of the 0.19 W m^{-2} radiative forcing from scenario F1a. This additional radiative forcing combines direct HSCT effects with the reduction in equivalent subsonic air traffic: +0.006 W m^{-2} from additional CO_2, +0.10 W m^{-2} from increased stratospheric H_2O, -0.012 W m^{-2} from ozone and methane changes resulting from NO_x emissions, and -0.011 W m^{-2} from reduced contrails. In total, the best value for HSCT RF is about 5 times larger than that of displaced subsonic aircraft, although the recognized uncertainty includes a factor as small as zero. The RFs from changes in stratospheric H_2O and O_3 are difficult to simulate in models and remain highly uncertain.

- Although the task of detecting climate change from all human activities is already difficult, detecting the aircraft-specific contribution to global climate change is not possible now and presents a serious challenge for the next century. Aircraft radiative forcing, like forcing from other individual sectors, is a small fraction of the whole anthropogenic climate forcing: about 4% today and by the year 2050 reaching 3–7% for F-type scenarios and 10–15% for E-type scenarios.

- The Radiative Forcing Index (RFI)—the ratio of total radiative forcing to that from CO_2 emissions alone—is a measure of the importance of aircraft-induced climate change other than that from the release of fossil carbon alone. In 1992, the RFI for aircraft is 2.7; it evolves to 2.6 in 2050 for the Fa1 scenario. This index ranges from 2.2 to 3.4 for the year 2050 for various E- and F-type scenarios for subsonic aviation and technical options considered here. The RFI increases from 2.6 to 3.4 with the addition of HSCTs (scenario Fa1H), primarily as a result of the effects of stratospheric water vapor. Thus, aircraft-induced climate change with RFI > 1 points to the need for a more thorough climate assessment for this sector. By comparison, in the IS92a scenario the RFI for all human activities is about 1, although for greenhouse gases alone it is about 1.5, and it is even higher for sectors emitting CH_4 and N_2O without significant fossil fuel use.

- From 1990 to 2050, the global mean surface temperature is expected to increase by 0.9 K following scenario IS92a for all human activity (assuming a climate sensitivity of +2.5 K for doubling of CO_2). Aircraft emissions from subsonic fleet scenario Fa1 are estimated to be responsible for about 0.05 K of this temperature rise.

- At present, the largest aircraft forcings of climate are through CO_2, NO_x, and contrail formation. These components have similar magnitude for subsonic aircraft; for an HSCT fleet, H_2O perturbations in the lower stratosphere, which are the most uncertain, are the most important. The largest areas of scientific uncertainty in predicting aircraft-induced climate effects lie with persistent contrails, with tropospheric ozone increases and consequent changes in methane, with potential particle impacts on "natural" clouds, and with water vapor and ozone perturbations in the lower stratosphere (especially for supersonic transport).

- The advantages of low-NO_x engines (scenario Fa2) are calculated to be modest in terms of total radiative forcing by the year 2050. The climatic impact of technology options, which in these scenarios are phased in linearly between 2015 and 2050, would not be fully felt until after 2050. Although lower NO_x emissions ameliorate the aviation-induced build-up of tropospheric O_3 and its radiative forcing as expected, they also reduce the opposite-sign radiative forcing associated with lowered global CH_4 abundance. The prospect of large canceling RFs, each with large uncertainties, greatly increases the probability of a large residual. Furthermore, RF is a measure only of *global* mean warming. Aircraft O_3 and CH_4 changes have largely different latitudinal contributions to RF; although they partly cancel on average, they may induce *regional* climate change.

6.1. How Do Aircraft Cause Climate Change?

Aircraft perturb the atmosphere by changing background levels of trace gases and particles and by forming condensation trails (contrails). Aircraft emissions include greenhouse gases such as CO_2 and H_2O that trap terrestrial radiation and chemically active gases that alter natural greenhouse gases, such as O_3 and CH_4. Particles may directly interact with the Earth's radiation balance or influence the formation and radiative properties of clouds. Figure 6-1 portrays a causal chain whereby the direct emissions of aircraft accumulate in the atmosphere, change the chemistry and the microphysics, and alter radiatively active substances in the atmosphere, which change radiative forcing and hence the climate.

Chapters 2 and 3 link the direct emissions of aircraft today to changes in radiatively active substances, and Chapter 4 projects these atmospheric changes into the future for a range of aviation scenarios. This chapter presents calculations of radiative forcing from aircraft-related atmospheric changes and discusses implications concerning the role of aircraft in a changing climate. This section begins with the concept of "dangerous climate change," as defined within the mandate of the United Nations Framework Convention on Climate Change (FCCC), then presents the IS92 scenarios for future climate change associated with the Second Assessment Report (IPCC, 1996). Section 6.1 also summarizes aviation's potential role in climate change and its proportion of fossil fuel use. Section 6.2 discusses the concepts of radiative forcing (RF) and global warming potential (GWP). Section 6.3 provides calculations of radiative forcing from aircraft perturbation of greenhouse gases, and Section 6.4 presents calculations of RF from aircraft perturbations of aerosols and contrails. Section 6.5 examines how radiative forcing can be used as a predictor of climate change and presents some case studies of climate change patterns that might be induced by aviation. Finally, Section 6.6 presents the summed radiative forcing, and associated climate change, for a range of projected scenarios and technological options in future aviation.

6.1.1. Anthropogenic Climate Change, Variability, and Detection

What is climate change? The common definition of climate refers to the average of weather, yet the definition of the climate system must reach out to the broader geophysical system that interacts with the atmosphere and our weather. The concept of climate change has acquired a number of different meanings in the scientific literature and in the media. Often, "climate change" denotes variations resulting from human interference, and "climate variability" refers to natural variations. Sometimes "climate change" designates variations longer than a certain period. Finally, "climate change" is often taken to mean climate fluctuations of a global nature, including effects from human activities such as the enhanced greenhouse effect and from natural causes such as volcanic aerosols.

For the purposes of the UNFCCC (and this report), the definition of climate change is: "A change of climate which is attributed directly or indirectly to human activity that alters the composition of the global atmosphere and which is in addition to natural climate variability observed over comparable time periods." This alteration of the global atmosphere includes changes in land use as well as anthropogenic emissions of greenhouse

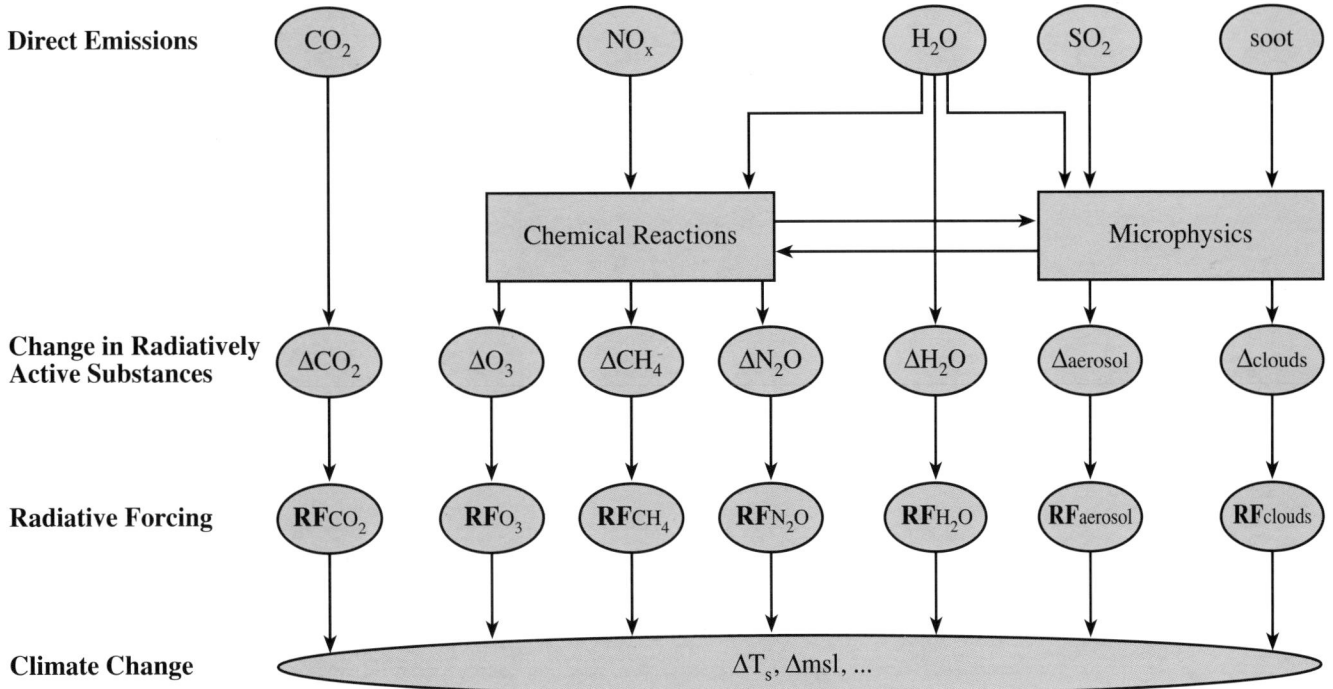

Figure 6-1: Schematic of possible mechanisms whereby aircraft emissions impact climate. Climate impact is represented by changes in global mean surface temperature (ΔT_s) and global mean sea level rise (Δmsl).

gases and particles. This FCCC definition thus introduces the concept of the difference between the effect of human activities (climate change) and climatic effects that would occur without such human interference (climate variability).

What drives changes in climate? The Earth absorbs radiation from the sun, mainly at the surface. This energy is then redistributed by atmospheric and oceanic circulations and radiated to space at longer ("terrestrial" or "infrared") wavelengths. On average, for the Earth as a whole, incoming solar energy is balanced by outgoing terrestrial radiation. Any factor that alters radiation received from the sun or lost to space or the redistribution of energy within the atmosphere and between atmosphere, land, and ocean can affect climate. A change in radiative energy available to the global Earth/atmosphere system is termed here, as in previous IPCC reports, radiative forcing (see Section 6.2 for more details). Radiative forcing (RF) is the global, annual average of radiative imbalance (W m^{-2}) in net heating of the Earth's lower atmosphere as a result of human activities since the beginning of the industrial era almost 2 centuries ago.

Increases in the concentrations of greenhouse gases reduce the efficiency with which the surface of the Earth radiates heat to space: More outgoing terrestrial radiation from the surface is absorbed by the atmosphere and is emitted at higher altitudes and colder temperatures. This process results in positive radiative forcing, which tends to warm the lower atmosphere and the surface. This radiative forcing is the enhanced greenhouse effect—an enhancement of an effect that has operated in the Earth's atmosphere for billions of years as a result of naturally occurring greenhouse gases (i.e., water vapor, carbon dioxide, ozone, methane, and nitrous oxide). The amount of warming depends on the size of the increase in concentration of each greenhouse gas, the radiative properties of the gases involved, their geographical and vertical distribution, and the concentrations of other greenhouse gases already present in the atmosphere.

Anthropogenic aerosols (small particles and droplets) in the troposphere—derived mainly from the emission of sulfur dioxide from fossil fuel burning but also from biomass burning and aircraft—can absorb and reflect solar radiation. In addition, changes in aerosol concentrations may alter cloud amount and cloud reflectivity through their effect on cloud microphysical properties. Often, tropospheric aerosols tend to produce negative radiative forcing and thus to cool climate. They have a much shorter lifetime (days to weeks) than most greenhouse gases (which have lifetimes of decades to centuries), so their concentrations respond much more quickly to changes in emissions.

Other natural changes, such as major volcanic eruptions that produce extensive stratospheric aerosols or variations in the sun's energy output, also drive climate variation by altering the radiative balance of the planet. On time scales of tens of thousands of years, slow variations in the Earth's orbit, which are well understood, have led to changes in the seasonal and latitudinal distribution of solar radiation; these changes have played an important part in controlling variations of climate in the distant past, such as glacial cycles.

Any changes in the radiative balance of the Earth, including those resulting from an increase in greenhouse gases or aerosols, will tend to alter atmospheric and oceanic temperatures and associated circulation and weather patterns. These effects will be accompanied by changes in the hydrological cycle (for example, altered cloud distributions or changes in rainfall and evaporation regimes). Any human-induced changes in climate will also alter climatic variability that otherwise would have occurred. Such variability contains a wide range of space and time scales. Climate variations can also occur in the absence of a change in external forcing, as a result of complex interactions between components of the climate system such as the atmosphere and ocean. The El Niño-Southern Oscillation (ENSO) phenomenon is a prominent example of such natural "internal" variability.

In the observationally based record of global mean surface temperatures shown by the black line on Figure 6-2, both interannual variability and a positive trend are apparent. Year-to-year variations can be interpreted as resulting from internal variability; and the trend, as caused by external forcing mechanisms. For comparison, the yellow line on Figure 6-2 shows a control run from a coupled ocean-atmosphere general circulation model in which concentrations of greenhouse gases and aerosols are held fixed: This indicates that observed natural variability in global mean surface temperatures may be adequately simulated. The red and the blue lines in Figure 6-2 show the surface temperature simulated by two different general circulation models driven by increased greenhouse gas and sulfate aerosol concentrations. Both of the models simulate interannual variability and trends in surface temperature, but differences in model sensitivies (see Section 6.2) lead to differing temperature trends. Figure 6-2 also shows that estimates of the global mean temperature trend resulting from increased greenhouse gas concentrations alone (green line) leads to a larger temperature change than observed.

It is difficult to ascribe climate change to human activities and even harder to identify a particular change with a specific activity. The point at which change is detected in a climate variable is the point at which the observed global mean trend (signal) unambiguously rises above background natural climate variability (noise). Good observational records of climate and sufficiently accurate, reliable models are needed. To simulate climate change, the models require complete representation of all anthropogenic forcing mechanisms (i.e., changes in atmospheric composition). In practice, current climate change is just comparable to natural variability. Therefore, more sophisticated tools have been developed that use the spatial structure of specific climate variables expected to change, which is known as the "fingerprint" method of detection (e.g., Hasselmann, 1993; Santer *et al.*, 1996).

6.1.2. Aircraft-Induced Climate Change

Aircraft emissions are expected to modify the Earth's radiative budget and climate as a result of several processes (see also

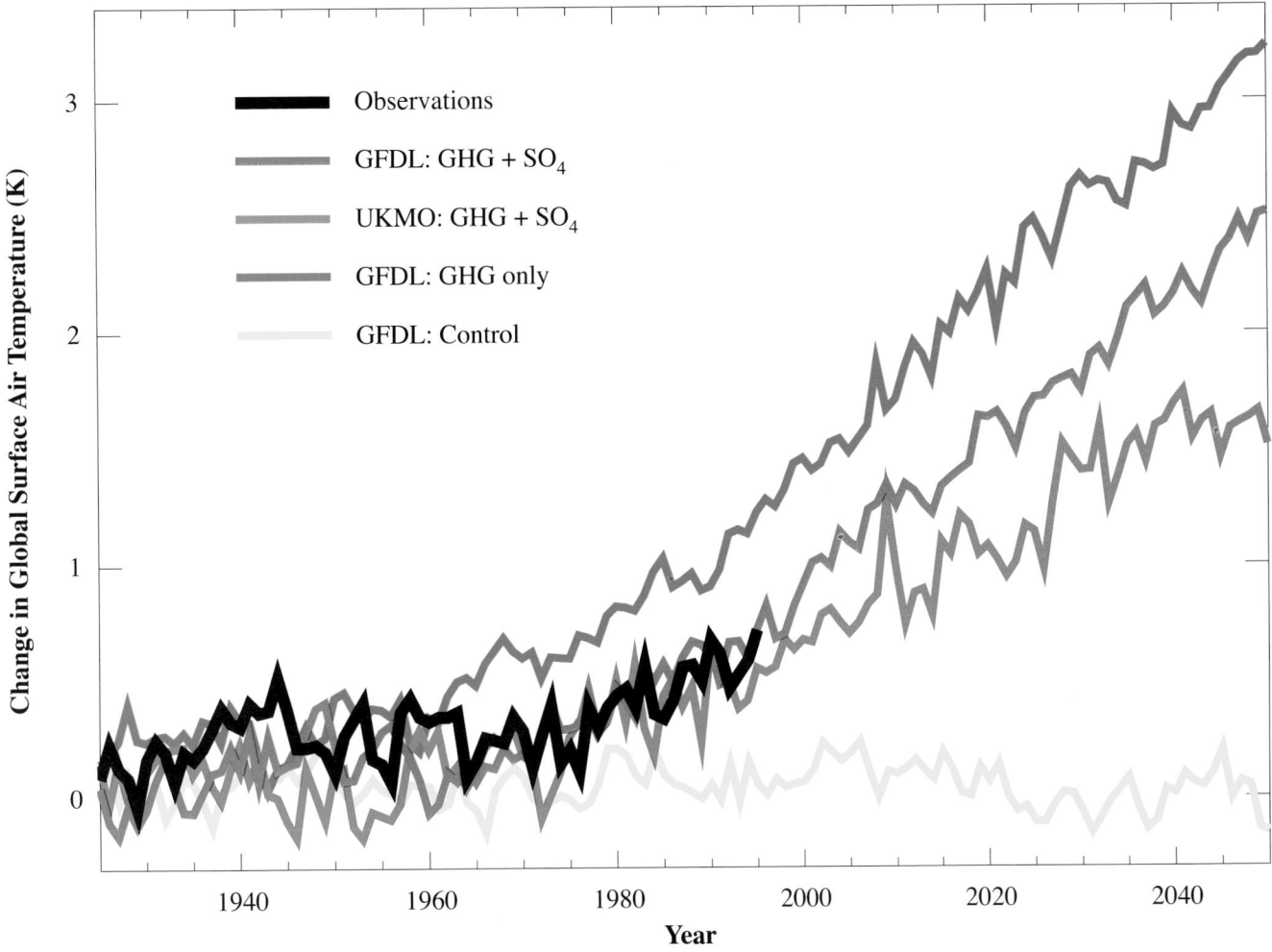

Figure 6-2: Change in global mean surface air temperature (K). Observations are from Jones (1994), modified to include data up to 1995. GFDL data are from modeling studies of Haywood *et al.* (1997b); UKMO data are from modeling studies of Mitchell *et al.* (1995).

Figure 6-1): emission of radiatively active substances (e.g., CO_2 or H_2O); emission of chemical species that produce or destroy radiatively active substances (such as NO_x, which modifies O_3 concentration, or SO_2, which oxidizes to sulfate aerosols); and emission of substances (e.g., H_2O, soot) that trigger the generation of additional clouds (e.g., contrails).

The task of detecting climate change is already difficult; the task of detecting the aircraft contribution to the overall change is more difficult because aircraft forcing is a small fraction of anthropogenic forcing as a whole. However, aircraft perturb the atmosphere in a specific way because their emissions occur in the free troposphere and lower stratosphere, and they trigger contrails, so the aircraft contribution to overall climate change may have a particular signature. At a minimum, the aircraft-induced climate change pattern would have to be significantly different from the overall climate change pattern in order to be detected.

The climatic impact of aircraft emissions is considered on the background of other anthropogenic perturbations of climate. The present assessment is based on the IS92 scenarios used to represent alternative futures in the Second Assessment Report (IPCC, 1996); it does not incorporate recent observed trends in methane (Dlugokencky *et al.*, 1998), nor any implications from the Kyoto Protocol. The scenarios for aircraft flight patterns, fuel burn, and emissions are described in Chapter 9. Atmospheric perturbations are taken from detailed atmospheric models (Chapters 2, 3, and 4), except for CO_2 accumulation (which is discussed in this chapter).

The largest, single, known radiative forcing change over the past century is from the increase in CO_2, driven primarily by the burning of fossil fuel. Figure 6-3 shows the change in CO_2 over the past 1,000 years; the CO_2 concentration increased from about 280 ppmv in 1850 to about 360 ppmv in 1990. Figure 6-4 shows six IPCC projections for anthropogenic emissions of CO_2 (labeled IS92a-f) and predicted atmospheric CO_2 concentrations. We take the central case, IS92a, as the future scenario with which to compare aircraft effects. The central aircraft scenario (Fa1) assessed here matches the economic assumptions of IS92a. Radiative forcing for IS92a is shown in Figure 6-5, including total radiative forcing and individual contributions.

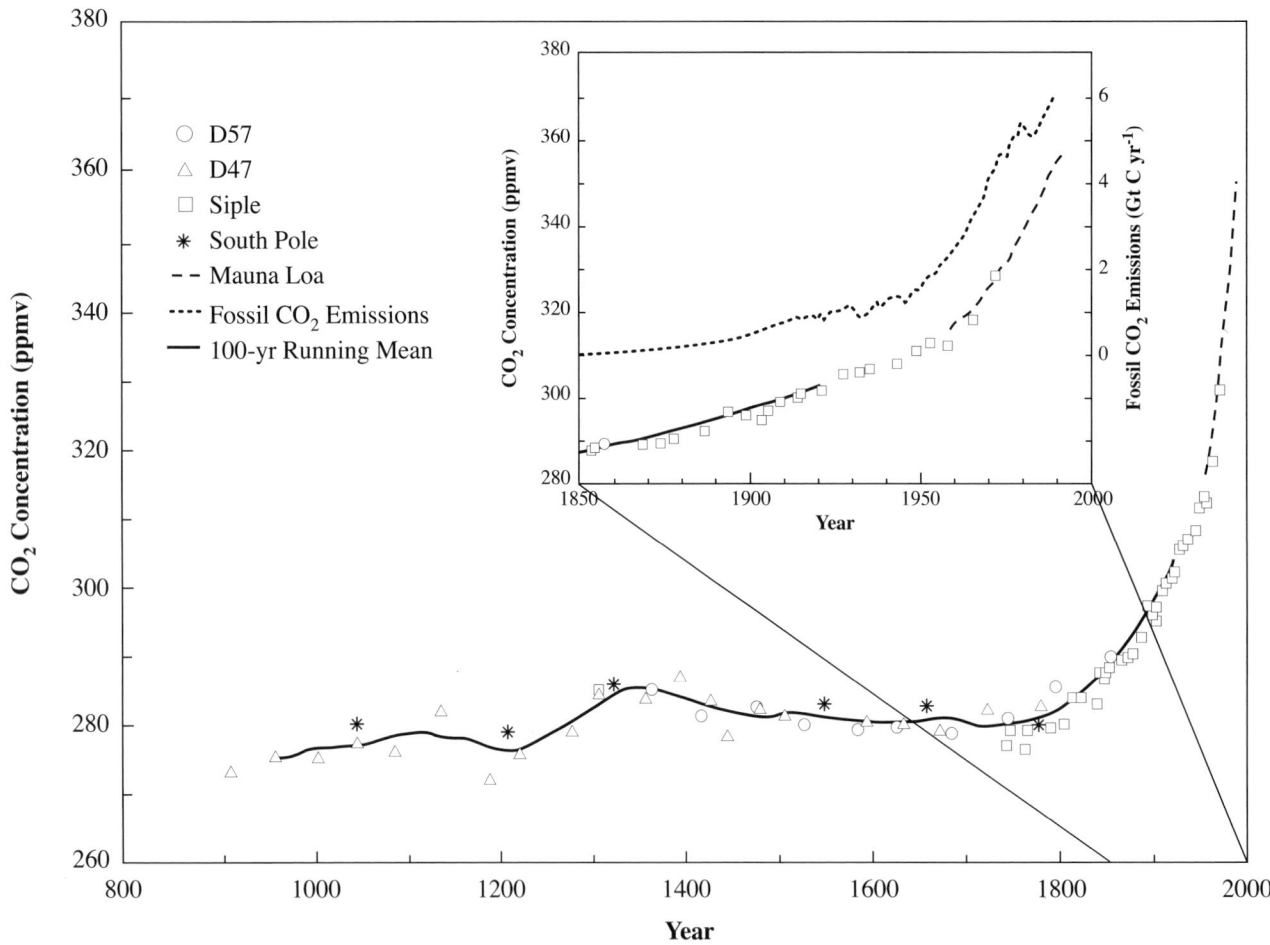

Figure 6-3: CO_2 concentration over the past 1000 years from ice core records (D47, D57, Siple, and South Pole) and (since 1958) from Mauna Loa, Hawaii, measurement site. All ice core measurements were taken in Antarctica. The smooth curve is based on a 100-yr running mean. A rapid increase in CO_2 concentration since the onset of industrialization is evident and has followed the increase in CO_2 emissions from fossil fuels (see inset for period from 1850 onward) (IPCC, 1996, Figure 1a).

6.1.3. Aviation Scenarios Adopted for Climate Assessment

A few, detailed, three-dimensional (3-D) emission inventories for three specific years—1992, 2015, and 2050—are presented in Chapter 9 and are studied with 3-D atmospheric models: NASA-1992, NASA-2015, and the ICAO-developed FESGa and FESGe scenarios for 2050. The FESG scenarios include three economic options, a/c/e, corresponding to economic growth assumed in IS92a/c/e. Each of the FESG scenarios has technology option 1 assuming typical, market-driven advances in engine/airframe technology and technology option 2 with advanced engine technology (i.e., a 25% reduction in NO_x emission index with a 3.5% increase in fuel use; see Chapter 9). Between these fixed-year scenarios, linear interpolation is used to derive continuous scenarios—Fa1, Fa2, Fc1, and Fe1 (see Table 6-3)—that extend from 1990 to 2050. CO_2 increases are derived from carbon-cycle models (see notes to Tables 6-1 and 6-2). Two scenarios based on EDF projections for the years 2015 and 2050 (Vedantham and Oppenheimer, 1998) provide only global CO_2 and NO_x emissions: the EDF-a-base (Eab) and EDF-d-high (Edh) cases. The Edh scenario was not adopted for its relationship to any underlying population or economic

scenario, but because it is a smooth extrapolation of recent growth rates. Atmospheric changes other than CO_2 for Eab and Edh are scaled from the Fa1 and Fe1 scenarios (see notes to Table 6-1). The continuous scenarios are summarized in Table 6-3.

The total fuel in Gt C used by aviation from 1950 to 1992 is shown in Figure 6-6. It also shows two projections to 2050 (Fa1 and Eab; see Tables 6-1 and 6-2), comparing them with projected total fossil carbon emissions for a similar economic scenario (IS92a). Note the logarithmic scale in Figure 6-6. For scenario F1a, fuel use parallels that of IS92a, but for Eab it grows faster than total fossil fuel use. In converting aviation fuel to CO_2 emissions, we adopt a carbon fraction by weight of 86%. Aviation fuel use prior to 1992 is based on International Energy Agency data (IEA, 1991; for table, see Sausen and Schumann, 1999). To account for systematic underestimation of fuel use (see Chapter 9), we have increased NASA-1992 and NASA-2015 emissions by 15% and 5%, respectively, to form the inventories NASA-1992* and NASA-2015*. Figure 6-7 gives an expanded linear scale of aviation fuel use from 1990 to 2050 for scenarios Fc1, Fa1, Fa1H, Fe1, Eab, and Edh, in order of increasing fuel use by 2050.

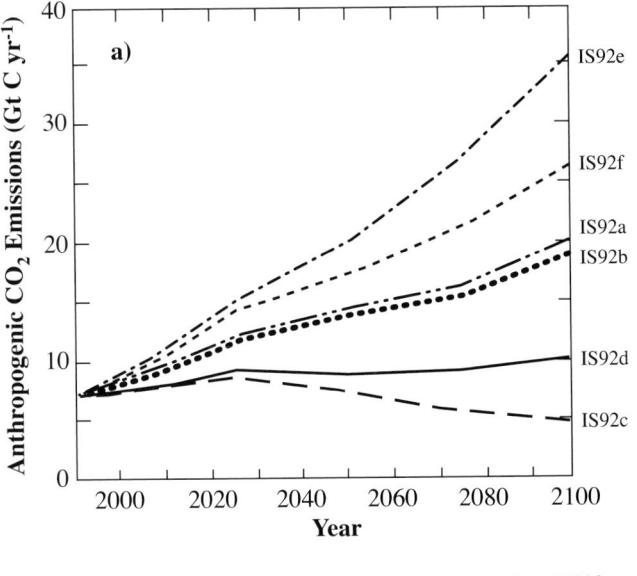

Figure 6-4: a) Total anthropogenic CO_2 emissions (Gt C yr[-1]) under the IS92 emission scenarios, and b) the resulting atmospheric CO_2 concentrations (ppmv) calculated using the "Bern" carbon cycle model (IPCC, 1996, Figure 5).

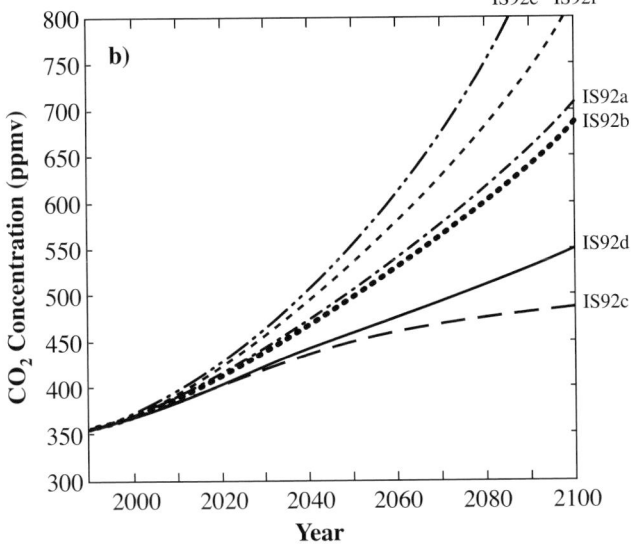

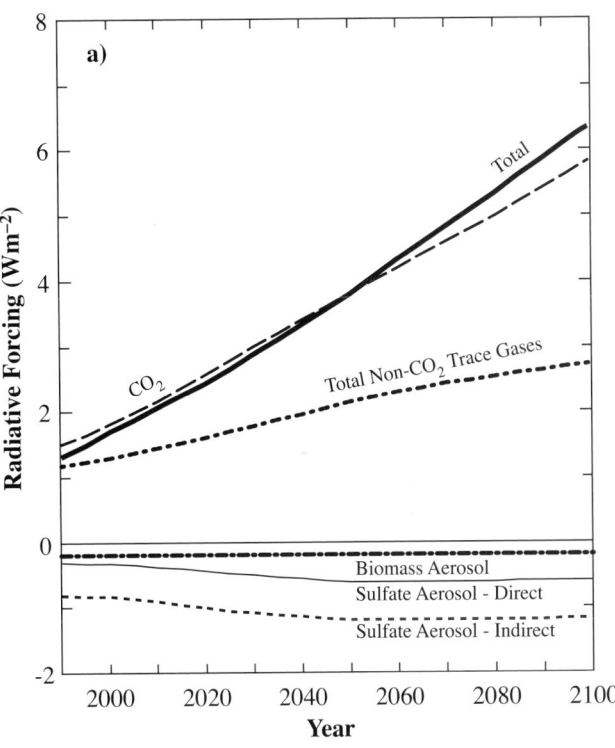

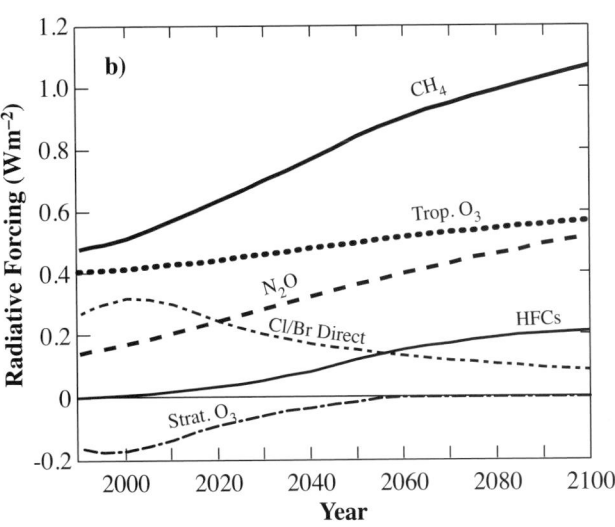

Figure 6-5: a) *Radiative forcing components resulting from the IS92a emission scenario for 1990 to 2100.* The "Total Non-CO_2 Trace Gases" curve includes the radiative forcing from CH_4 (including CH_4-related increases in stratospheric water vapor), N_2O, tropospheric O_3, and the halocarbons (including the negative forcing effect of stratospheric O_3 depletion). Halocarbon emissions have been modified to take account of the Montreal Protocol and its Adjustments and Amendments. The three aerosol components are direct sulfate, indirect sulfate, and direct biomass burning. b) *Non-CO_2 trace gas radiative forcing components.* "Cl/Br direct" is the direct radiative forcing resulting from Cl- and Br-containing halocarbons; emissions are assumed to be controlled under the Montreal Protocol and its Adjustments and Amendments. The indirect forcing from these compounds (through stratospheric O_3 depletion) is shown separately (Strat. O_3). All other emissions follow the IS92a scenario. The tropospheric O_3 forcing (Trop. O_3) takes account of concentration changes resulting only from the indirect effect from CH_4 (IPCC, 1996, Figure 6b-c).

Table 6-1: *Aviation fixed-year (1992, 2015, and 2050) scenarios for emissions and radiative forcing.*

Scenario	Fuel Burn (Mt yr-1)	NOx Emis.[d] (Mt yr-1)	CO2[f] Conc. (ppmv)	Radiative Forcing (W m-2) CO2[f]	O3[g]	CH4[g]	H2O[h]	Contrails[i]	Sulfate[h] Aerosols	BC[h] Aerosols	Total	RFI
NASA-1992*[a]	**160.3**	**1.92**	**1.0**	**+.018**	**+.023**	**-.014**	**+.0015**	**+.020**	**-.003**	**+.003**	**+.048**	**2.7**
Low[b]				+.013	+.011	-.005	+.000	+.005	-.001	+.001		
High				+.023	+.046	-.042	+.005	+.06	-.009	+.009		
NASA-2015*[a]	**324.0**	**4.34**	**2.5**	**+.038**	**+.040**	**-.027**	**+.003**	**+.060**	**-.006**	**+.006**	**+.114**	**3.0**
FESGa (tech1) 2050	**471.0**	**7.15**	**6.0**	**+.074**	**+.060**	**-.045**	**+.004**	**+.100**	**-.009**	**+.009**	**+.193**	**2.6**
Low				+.052	+.030	-.015	+.000	+.03	-.003	+.003		
High				+.096	+.120	-.120	+.015	+.40	-.027	+.027		
FESGa (tech2) 2050[c]	487.6	5.55	6.1	+.075	+.047	-.035	+.005	+.100	-.009	+.009	+.192	
FESGc (tech1) 2050	268.2	4.01	4.9	+.060	+.034	-.025	+.003	+.057	-.005	+.005	+.129	2.2
FESGc (tech2) 2050	277.2	3.14	5.0	+.061	+.026	-.020	+.003	+.057	-.005	+.005	+.127	
FESGe (tech1) 2050	**744.3**	**11.38**	**7.4**	**+.091**	**+.096**	**-.072**	**+.007**	**+.158**	**-.014**	**+.014**	**+.280**	**3.1**
FESGe (tech2) 2050	772.1	8.82	7.6	+.093	+.074	-.055	+.007	+.158	-.015	+.015	+.277	
EDFa-base 2015	297.0[d]	2.85	2.4	+.037	+.026	-.018	+.003	+.055	-.006	+.006	+.103	2.8
EDFa-base 2050	1143.0	7.89	9.4	+.115	+.066	-.050	+.011	+.243	-.022	+.022	+.385	3.3
EDFd-high 2015	448.0[d]	4.30	3.0	+.046	+.040	-.027	+.004	+.083	-.008	+.008	+.146	3.2
EDFd-high 2050	1688.0	11.65	13.4	+.165	+.098	-.073	+.016	+.358	-.032	+.032	+.564	3.4
HSCT (500)	**70.0**	**0.35**			**-.010**		**+.050**					
Low					-.040		+.017					
High					+.010		+.150					
HSCT (1000)	**140.0**	**0.70**			**-.010**		**+.100**					
Low					-.040		+.033					
High					+.010		+.300					
Net HSCT 2050												
+ HSCT	+140.0	+0.70	+0.8	+.010	-.010		+.100				+.100	
- subsonic	-53.6	-0.81	-0.3	-.004	-.007	+.005	-.001	-.011	+.001	-.001	-.018	
FESGa (tech1) + HSCT 2050	557.4	7.04	6.5	+.080	+.043	-.040	+.103	+.089	-.008	+.008	+.275	3.4

[a] The scenarios in boldface were studied in atmospheric models with defined 3-D emission patterns; the others were scaled to these scenarios. The NASA-1992* aviation scenario has been scaled here by 1.15, and the NASA-2015* scenario by 1.05, to account for inefficiencies in flight routing.

[b] Low/High give likely (67% probability) range.

[c] In FESG scenarios, tech 1 is standard, and tech 2 reduces EI(NOx) by 25% with a few percent additional fuel use.

[d] Throughout the table and this report, NOx emissions (Mt yr-1) and indices (EI) use the NO_2 molecular weight.

[e] In the EDF 2015 scenarios, the fuel burns have been revised to 374 (a-base) and 592 (d-high) Mt yr-1, which would increase the added CO2 by 2050 to 10.0 (a-base) and 14.7 (d-high) ppmv.

[f] CO2 is largely cumulative and depends on the assumed previous history of the emissions; CH4 perturbations are decadal in buildup time; all other perturbations reach steady-state balance with emissions in a few years. All except CO2 are assumed here to be instantaneous. Thus, CO2 concentrations are based on complete history of fuel burn—for example, scenario Fa1 = NASA-1992* → NASA-2015* → FESGa (tech1) 2050; and scenario Eab = NASA-1992* → EDFa-base 2015 → EDFa-base 2050, all with linear interpolation between 1992, 2015, and 2050 (see also Section 6.1.3).

[g] The O3 and CH4 RFs are scaled to NOx emissions for non-bold scenarios.

[h] As for note g, stratospheric H_2O, sulfate, and BC aerosols scale with fuel burn.

[i] Contrails do not scale with fuel burn as the fleet and flight routes evolve (see Chapter 3). The contrail RF here is from line-shaped contrail cirrus only. Additional induced cirrus cover RF is positive, and may be of similar magnitude, but no best estimate can be given yet.

Table 6-2*: Emissions, atmospheric concentrations, radiative forcing, and climate change (global mean surface temperature) projected for the years 1990, 2000, 2015, 2025, and 2050 using IPCC's IS92a and the aviation scenarios from Tables 6-1 and 6-3.*

	1990	**2000**	**2015**	**2025**	**2050**
Emissions					
IS92a CO_2 Emissions (Gt C yr^{-1})					
Fossil fuel	6.0	7.0	9.2	10.7	13.2
Total	7.5	8.5	10.7	12.2	14.5
Aviation CO_2 Emissions (Gt C yr^{-1})					
Fa1	0.147	0.187	0.279	0.315	0.405
Fa2	0.147	0.187	0.279	0.319	0.419
Fc1	0.147	0.187	0.279	0.265	0.231
Fe1	0.147	0.187	0.279	0.382	0.640
Eab	0.147	0.179	0.255	0.463	0.983
Edh	0.147	0.224	0.385	0.690	1.452
Fa1H	0.147	0.187	0.279	0.344	0.479
IS92a NO_x Emissions (Mt NO_2 yr^{-1})					
Energy	82	98	122	137	174
Biomass Burn	30	31	32	33	36
Aviation NO_x Emissions (Mt NO_2 yr^{-1})					
Fa1	2.0	2.8	4.3	5.1	7.2
Fa2	2.0	2.8	4.3	4.7	5.6
Fc1	2.0	2.8	4.3	4.2	4.0
Fe1	2.0	2.8	4.3	6.4	11.4
Eab	2.0	2.2	2.9	4.3	7.9
Edh	2.0	2.8	4.3	6.4	11.6
Atmospheric Concentrations					
IS92a Atmosphere					
CO_2 (ppmv)	354	372	405	432	509
CH_4 (ppbv)	1700	1810	2052	2242	2793
N_2O (ppbv)	310	319	333	344	371
Aviation Marginal CO_2 (ppmv)					
Fa1	0.9	1.5	2.5	3.5	6.0
Fa2	0.9	1.5	2.5	3.5	6.1
Fc1	0.9	1.5	2.5	3.2	4.9
Fe1	0.9	1.5	2.5	3.9	7.4
Eab	0.9	1.5	2.4	4.4	9.4
Edh	0.9	1.7	3.0	6.0	13.4
Fa1H	0.9	1.5	2.5	3.5	6.5
Aviation Marginal CH_4 (ppbv)					
Fa1	-31	-49	-75	-97	-152

Notes: The 1990 values are based on IEA data for 1990 fuel use; these values are higher than our estimate for 1992 (see Table 6-1). The scenarios involve linear interpolation of emissions between 1990, 1992, 2015, and 2050 (see Tables 6-1 and 6-3). The projected CH_4 increases are based on the emissions growth in IS92a that do not match the much smaller trends currently observed. These calculations used the methodologies from the IPCC's Second Assessment Report (1996) as contributed by Atul Jain, Michael Prather, Robert Sausen, Ulrich Schumann, Tom Wigley, and Don Wuebbles.

Table 6-2 (continued)

	1990	**2000**	**2015**	**2025**	**2050**
Radiative Forcing					
Differential RF (W m^{-2}/ppmv)					
dRF/dCO$_2$	0.018	0.016	0.015	0.014	0.012
dRF/dCH$_4$	0.38	0.37	0.35	0.33	0.29
IS92a RF (Wm^{-2})					
CO$_2$	1.54	1.84	2.38	2.79	3.83
CH$_4$	0.47	0.51	0.59	0.66	0.83
N$_2$O	0.14	0.17	0.22	0.26	0.36
All greenhouse gases	2.64	3.08	3.81	4.34	5.76
Aerosols (direct and indirect)	-1.26	-1.36	-1.55	-1.66	-1.94
Total	1.38	1.72	2.26	2.68	3.82
Aviation Fa1 Components of RF (Wm^{-2})					
CO$_2$	0.016	0.025	0.038	0.048	0.074
O$_3$	0.024	0.029	0.040	0.046	0.060
CH$_4$	-0.015	-0.018	-0.027	-0.032	-0.045
H$_2$O	0.002	0.002	0.003	0.003	0.004
Contrails	0.021	0.034	0.060	0.071	0.100
Sulfate aerosol	-0.003	-0.004	-0.006	-0.007	-0.009
Soot (BC) aerosol	0.003	0.004	0.006	0.007	0.009
Indirect clouds	no best estimate available				
Total	0.048	0.071	0.114	0.137	0.193
Aviation HSCT (net) Components of RF (Wm^{-2})					
CO$_2$				0.001	0.006
O$_3$				-0.007	-0.017
CH$_4$				0.002	0.005
H$_2$O				0.040	0.099
Contrails				-0.004	-0.011
Total				0.031	0.082
Aviation Scenarios Total RF (Wm^{-2})					
Fa1	0.048	0.071	0.114	0.137	0.193
Fa2	0.048	0.071	0.114	0.136	0.192
Fc1	0.048	0.071	0.114	0.118	0.129
Fe1	0.048	0.071	0.114	0.161	0.280
Eab	0.048	0.068	0.103	0.184	0.385
Edh	0.048	0.083	0.146	0.265	0.564
Fa1H	0.048	0.071	0.114	0.168	0.275
Climate Change					
Global Mean Surface Air Temperature Change (K)					
IS92a	0.000	0.140	0.360	0.510	0.920
Fa1	0.000	0.004	0.015	0.024	0.052
Fc1	0.000	0.004	0.015	0.023	0.039
Fe1	0.000	0.004	0.015	0.026	0.070
Eab	0.000	0.004	0.014	0.026	0.090
Edh	0.000	0.005	0.019	0.038	0.133
Fa1H	0.000	0.004	0.015	0.025	0.066

Table 6-3: *Overview of the scenarios adopted for the climate assessment.*

| Name | Fixed-Year Scenario | | | Comments |
	1992	2015	2050	
Fa1	NASA-1992*	NASA-2015*	FESGa	Technology option 1
Fa2	NASA-1992*	NASA-2015*	FESGa	Technology option 2
Fc1	NASA-1992*	NASA-2015*	FESGc	Technology option 1
Fe1	NASA1992*	NASA-2015*	FESGe	Technology option 1
Eab	NASA-1992*	EDF-a-base	EDF-a-base	
Edh	NASA-1992*	EDF-d-high	EDF-d-high	
Fa1H	NASA-1992*	NASA-2015*	FESGa + HSCT	Fa1 with part of subsonic traffic replaced by HSCT fleet growth from 2015 to 2040

Chapter 4 studies the impact of a fleet of high-speed civil transport (HSCT, i.e., supersonic) aircraft using a range of 3-D emission scenarios with atmospheric chemistry models. These calculations form a parametric range that covers changes in fleet size, NO_x emissions, cruise altitude, sulfate aerosol formation, and future atmospheres. The present chapter combines those results into a continuous scenario for the HSCT fleet, designated Fa1H: On top of the Fa1 scenario it assumes that HSCT aircraft come into service in 2015, grow at 40 planes per year to a final capacity of 1,000 aircraft by 2040, continue operation to 2050, and displace equivalent air traffic from the subsonic fleet (~11% of Fa1 in 2050). This Mach 2.4 HSCT fleet cruises at 18–20 km altitude and deposits most of its emissions in the stratosphere. It has new combustor technology that produces very low emissions of 5 g NO_2 per kg fuel. Table 6-1 gives the breakdown of RF from two specific HSCT studies in Chapter 4: 500 HSCTs in a 2015 background atmosphere and 1,000 HSCTs in a 2050 background atmosphere (e.g., chlorine loading, methane, nitrous oxide). The

likely interval for the RF here combines the uncertainty in calculating the ozone or water vapor perturbation with that from calculating the radiative imbalance.

6.1.4. Aviation's Contribution to the CO_2 Budget

Carbon dioxide released from fossil fuel combustion rapidly equilibrates among atmosphere, surface ocean, and parts of the biosphere, leaving behind excess atmospheric CO_2 that decays slowly over the following century (see carbon cycle discussion in IPCC, 1996). Thus, for CO_2 radiative forcing, it makes no difference whether the fossil fuel is burned in aircraft or other transportation/energy sectors, and the relative role of aircraft can be found by comparing the history of fuel burned by aviation with that of total anthropogenic carbon emissions.

Comparing projected IS92a carbon emissions from fossil fuels in Figure 6-6, CO_2 emissions from aircraft in 1990 account for

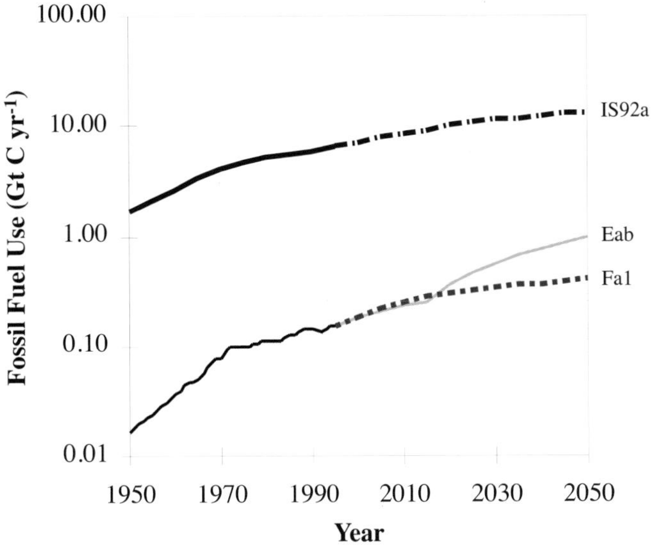

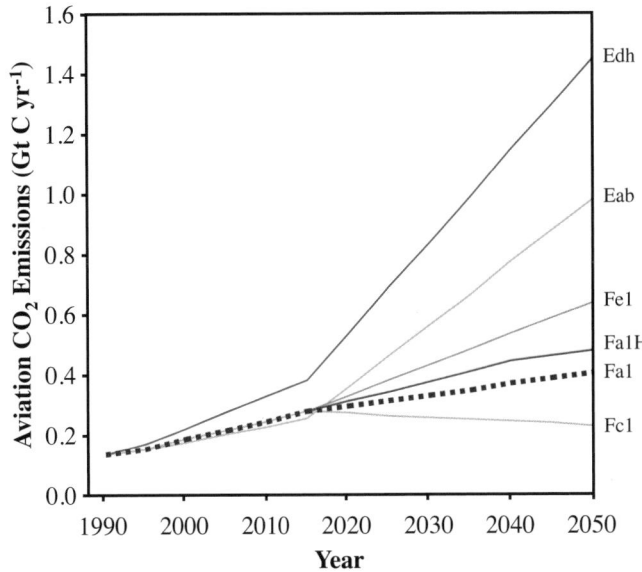

Figure 6-6: Fossil fuel use (Gt C yr^{-1}) shown for historical aviation use (1950–92, solid line) and for projected aviation scenarios Fa1 and Eab. Total historical fossil fuel use and the projection according to scenario IS92a are also shown.

Figure 6-7: Aviation CO_2 emissions (Gt C yr^{-1}) from 1990 to 2050 for the range of scenarios considered here (see Table 6-3).

about 2.4% of the total; they are projected to grow to about 3% (Fa1) or more than 7% (Eab) of all fossil fuel carbon emissions by 2050. Sustained growth in air traffic demand (5%/yr compounded) envisaged in Edh would lead to an aviation fraction of more than 10% by 2050. By comparison, the entire transportation sector is currently about 25% of the total (see discussion in Chapter 8). Clearly, different economic projections, as well as uncertainties in predicting demand for air travel and aviation's ability to meet that demand, can alter this aviation fraction by more than a factor of 2. Technology option 2 (low-NO_x engines, Fa2) increases this fuel fraction slightly to 3.2%, and the HSCT option increases this fraction from 3.1% (Fa1) to 3.6% (scenario Fa1H, assuming no change in air traffic demand) by the year 2050.

The cumulative history of CO_2 emissions allows us to calculate the excess atmospheric CO_2 concentration attributable to aircraft, as shown in Figure 6-8. These calculations are contributed by Jain, Wigley, and Schumann using carbon cycle models consistent with IPCC (1996) and the IS92 scenarios therein. Aviation is estimated to be responsible for about 1 ppmv of the 80 ppmv rise in CO_2 from 1860 to 1990. Uncertainty in the prediction of atmospheric CO_2 is estimated to be ±25%. Resulting CO_2 radiative forcing is only one part of aviation's climate impact. Other changes in greenhouse gases, radiatively important aerosols, and clouds—as noted in Figure 6-1 and broken out in Table 6-1—must be included. The remaining sections of this chapter examine aviation's total role in climate change over the next 50 years for the example scenarios given in Table 6-3.

6.2. Radiative Forcing and GWP Concepts

6.2.1. The Concept of Radiative Forcing

The most useful assessment of the impact of the aircraft fleet on climate would be a comprehensive prediction of changes to

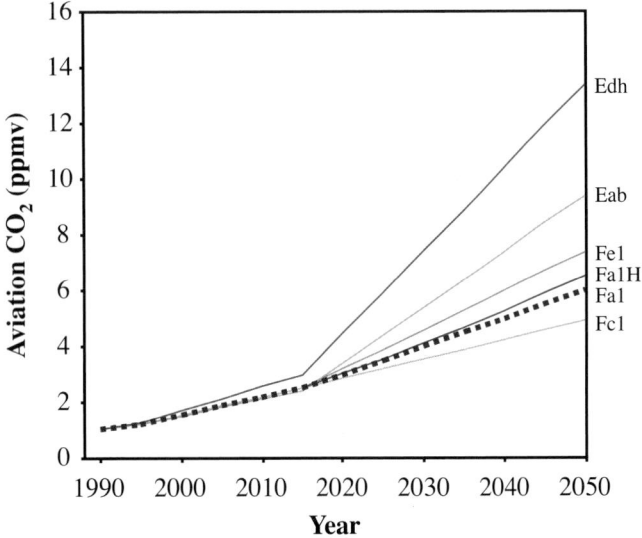

Figure 6-8: Atmospheric CO_2 (ppmv) accumulated from aviation's use of fossil fuel beginning in 1940.

the climate system, including temperature, sea level, frequency of severe weather, and so forth. Such assessment is difficult to achieve given the current state of climate models and the small global forcing of climate attributable to the single sector of aviation chosen for this special report (see discussion in Sections 6.1 and 6.5). Following IPCC (1995, 1996), we choose a single measure of climate change: radiative forcing (RF), which is calculated directly from changes in greenhouse gases, aerosols, and clouds, and which allows ready comparison of the climate impact of different aviation scenarios.

The Earth's climate system is powered by the sun. Our planet intercepts 340 W m^{-2} of solar radiation averaged over the surface of the globe. About 100 W m^{-2} is reflected to space, and the remainder—about 240 W m^{-2}—heats the planet. On a global average, the Earth maintains a radiative balance between this solar heating and the cooling from terrestrial infrared radiation that escapes to space. When a particular human activity alters greenhouse gases, particles, or land albedo, such activity results in radiative imbalance. Such an imbalance cannot be maintained for long, and the climate system—primarily the temperature and clouds of the lower atmosphere—adjusts to restore radiative balance. We calculate the global, annual average of radiative imbalance (W m^{-2}) to the atmosphere-land-ocean system caused by anthropogenic perturbations and designate that change radiative forcing. Thus, by this IPCC definition, the RF of the pre-industrial atmosphere is taken to be zero. (Although the term "radiative forcing" has more general meaning in terms of climate, we restrict its use here to the IPCC definition.)

As an example, burning of fossil fuel adds the greenhouse gas CO_2 to the atmosphere; this burning is responsible for the increase in atmospheric CO_2 from about 280 ppmv in the pre-industrial atmosphere to about 360 ppmv in 1995. Added CO_2 increases the infrared opaqueness of the atmosphere, thereby reducing terrestrial cooling with little impact on solar heating. Thus, the radiative imbalance created by adding a greenhouse gas is a positive RF. A positive RF leads to warming of the lower atmosphere in order to increase the terrestrial radiation and restore radiative balance. Radiative imbalances can also occur naturally, as in the case of the massive perturbation to stratospheric aerosols caused by Mt. Pinatubo (Hansen *et al.*, 1996).

Because most of the troposphere is coupled to the surface through convection, climate models typically predict that the land surface, ocean mixed layer, and troposphere together respond to positive RF in general with a relatively uniform increase in temperature. Global mean surface temperature is a first-order measure of what we consider to be "climate," and its change is roughly proportional to RF. The increase in mean surface temperature per unit RF is termed climate sensitivity; it includes feedbacks within the climate system, such as changes in tropospheric water vapor and clouds in a warmer climate. The RF providing the best metric of climate change is the radiative imbalance of this land-ocean-troposphere climate system—that is, the RF integrated at the tropopause.

When radiative perturbation occurs above the tropopause, in the stratosphere (as for most HSCT impacts), this heating/cooling is not rapidly transported into the troposphere, and the imbalance leads mostly to changes in local temperatures that restore the radiative balance within the stratosphere. Such changes in stratospheric temperature, however, alter the tropospheric cooling; for example, warmer stratospheric temperatures lead to a warmer troposphere and climate system. This adjustment of stratospheric temperatures can be an important factor in calculating RF and is denoted "stratosphere-adjusted."

All RF values used in this report refer to "stratosphere-adjusted, tropopause RF" (*Shine et al.*, 1995). For primarily tropospheric perturbations (e.g., CO_2 from all aviation, O_3 from subsonic aircraft), this quantity can be calculated with reasonable agreement (better than 25%) across models used in this report (see Section 6.3). For specifically stratospheric perturbations (e.g., H_2O and O_3 perturbations from HSCT aircraft), the definition of the tropopause and the calculation of stratospheric adjustment introduce significant sources of uncertainty in calculated RF.

The concept of radiative forcing (IPCC, 1990, 1992, 1995) is based on climate model calculations that show that there is an approximately linear relationship between global-mean RF at the tropopause and the change in equilibrium global mean surface (air) temperature (ΔT_s). In mapping RF to climate change, the complexities of regional and even hemispheric climate change have been compressed into a single quantity— global mean surface temperature. It is clear from climate studies that the climate does not change uniformly: Some regions warm or cool more than others. Furthermore, mean temperature does not provide information about aspects of climate change such as floods, droughts, and severe storms that cause the most damage. In the case of aviation, the radiative imbalance driven by perturbations to contrails, O_3, and stratospheric H_2O occurs predominantly in northern mid-latitudes and is not globally homogeneously distributed (see Chapters 2, 3, and 4), unlike perturbations driven by increases in CO_2 or decreases in CH_4. Does this large north-south gradient in the radiative imbalance lead to climate change of a different nature than for well-mixed gases? IPCC (Kattenberg *et al.*, 1996) considered the issue of whether negative RF from fossil-fuel sulfate aerosols (concentrated in industrial regions) would partly cancel positive RF from increases in CO_2 (global). Studies generally confirmed that global mean surface warming from both perturbations was additive; that is, it could be estimated from the summed RF. Local RF from sulfate in northern industrial regions was felt globally. Nevertheless, the regional patterns in both cases were significantly different, and obvious cooling (in a globally warming climate) occurred in specific regions of the Northern Hemisphere. Such differences in climate change patterns are critical to the detection of anthropogenic climate change, as reported in Santer *et al.* (1996). As a further complication of this assessment, aviation's perturbation occurs primarily in the upper troposphere and lower stratosphere, and thus may alter the vertical profile of any future tropospheric warming. Therefore, the patterns of climate change from individual aviation

perturbations (e.g., CO_2, O_3, contrails) would likely differ, but we take their summed RF as a first-order measure of the global mean climate change (see also the discussion in Section 6.5).

The equilibrium change in mean surface air temperature ($\Delta T_s^{(equil)}$) in response to any particular RF is reached only after more than a century because of the thermal inertia of the climate system (primarily the oceans) and is calculated with long-term integrations of coupled general circulation models (CGCM). A climate sensitivity parameter (λ) relates *RF* to temperature change: $\Delta T_s^{(equil)} = \lambda RF$. Provided that all types of RF produce the same impact on the climate system (in this case measured by mean temperature), the climate sensitivity parameter derived from a doubled-CO_2 calculation can be used to translate other RFs, say from ozone or contrails, into a change in global mean surface air temperature.

For doubled CO_2 relative to pre-industrial conditions (+4 W m^{-2}), surface temperature warming ranges from 1.5 to 4.5 K, depending on the modeling of feedback processes included in the CGCM. The recommended value in IPCC (1996) of 2.5 K gives a climate sensitivity of λ = 0.6 K/(W m^{-2}). With limited feedbacks (e.g., fixing clouds and surface ocean temperatures), the sensitivity parameter is smaller, and most models produce similar responses. In contrast, when all feedbacks are included, model results are quite different, as a result (for instance) of alternative formulation of clouds. The obvious limitation of this approach is that we get no information about regional climate change. The sensitivity parameter for aircraft-like ozone perturbations is discussed in Section 6.5.

In spite of all these caveats, the radiative forcing of an aviation-induced atmospheric perturbation is still a useful index that allows, to first approximation, the different atmospheric perturbations (e.g., aerosols, cloud changes, ozone, stratospheric water, methane) to be summed and compared in terms of global climate impact.

6.2.2. *Global Warming Potential*

Global warming potential (GWP; see Shine *et al.*, 1990, for a formal definition) is an index that attempts to integrate the overall climate impacts of a specific action (e.g., emissions of CH_4, NO_x or aerosols). It relates the impact of emissions of a gas to that of emission of an equivalent mass of CO_2. The duration of the perturbation is included by integrating radiative forcing over a time horizon (e.g., standard horizons for IPCC have been 20, 100, and 500 years). The time horizon thus includes the cumulative climate change and the decay of the perturbation.

GWP has provided a convenient measure for policymakers to compare the relative climate impacts of two different emissions. However, the basic definition of GWP has flaws that make its use questionable, in particular, for aircraft emissions. For example, impacts such as contrails may not be directly related to emissions of a particular greenhouse gas. Also, indirect RF from O_3 produced by NO_x emissions is not linearly proportional to the amount of NO_x emitted but depends also on location and season. Essentially, the buildup and radiative impact of short-lived gases and aerosols will depend on the location and even the timing of their emissions. Furthermore, the GWP does not account for an evolving atmosphere wherein the RF from a 1-ppm increase in CO_2 is larger today than in 2050 and the efficiency of NO_x at producing tropospheric O_3 depends on concurrent pollution of the troposphere.

In summary, GWPs were meant to compare emissions of long-lived, well-mixed gases such as CO_2, CH_4, N_2O, and hydrofluorocarbons (HFC) for the current atmosphere; they are not adequate to describe the climate impacts of aviation.

Nevertheless, some researchers have calculated a GWP, or modified version, for aircraft NO_x emissions via induced ozone perturbation (e.g., Michaelis, 1993; Fuglestvedt *et al.*, 1996; Johnson and Derwent, 1996; Wuebbles, 1996). The results vary widely as a result of model differences, varying scenarios for NO_x emission, and the ambiguous GWP definition for short-lived gases. There is a basic impossibility of defining a GWP for "aircraft NO_x" because emissions during takeoff and landing would have one GWP; those at cruise, another; those in polar winter, another; and those in the upper tropical troposphere, yet another. Different chemical regimes will produce different amounts of ozone for the same injection of NO_x, and the radiative forcing of that ozone perturbation will vary by location (Fuglesvedt *et al.*, 1999). In view of all these problems, we will not attempt to derive GWP indices for aircraft emissions in this study. The history of radiative forcing, calculated for the changing atmosphere, is a far better index of anthropogenic climate change from different gases and aerosols than is GWP.

6.2.3. *Alternative Indexing of Aviation's Climate Impact—RF Index*

A new alternative index to measure the role of aviation in climate change is introduced here: the radiative forcing index (RFI), which is defined as the ratio of total radiative forcing to that from CO_2 emissions alone. Total radiative forcing induced by aircraft is the sum of all forcings, including direct emissions (e.g., CO_2, soot) and indirect atmospheric responses (e.g., CH_4, O_3, sulfate, contrails). RFI is a measure of the importance of aircraft-induced climate change other than that from the release of fossil carbon alone. RFI ranges between 2.2 and 3.4 for the various E- and F-type scenarios for subsonic aviation and technical options considered here (see Section 6.6). Thus, aircraft-induced climate change with RFI > 1 highlights the need for a thorough climate assessment of this sector as performed here. For comparison, in the IS92a scenario the RFI for all human activities is about 1; for greenhouse gases alone, it is about 1.5, and it is even higher for sectors that emit CH_4 and N_2O without significant fossil fuel use.

6.3. Radiative Forcing from Aircraft-Induced Changes in Greenhouse Gases

This section presents RF calculations for perturbations to greenhouse gases attributable to aircraft. Greenhouse gases that have been identified as aircraft-perturbed are CO_2, O_3, CH_4, and H_2O. Each gas presents a special case in terms of predicting its perturbation or deriving radiative forcing. The RF calculations presented here are derived from radiative-balance and comprehensive climate models by subtracting net radiative flux (incoming solar minus outgoing terrestrial infrared) for a control run from that for a run that includes the specified perturbation. In general, these calculations integrate over a full range of latitudinal and seasonal variations that are typical of the Earth's climate, consider the imbalance at the tropopause, and account for the adjustment of stratospheric temperatures. Important factors in deriving representative radiative forcing include realistic temperatures, water vapor, surface albedo, clouds, and tropopause. RF calculations represent the instantaneous imbalance in the troposphere-land-ocean system, thus do not include responses that are considered part of the climate feedback system, such as changes to clouds and tropospheric water vapor.

RF values depend on atmospheric composition as well as temperature, water vapor, and clouds because all of these factors interact with the radiation field. For these calculations, we have adopted the changing composition as specified in IS92a from IPCC (1995) and summarized in Table 6-2. This composition includes substantial increases in CO_2 and CH_4 that alter the Earth's radiation spectrum, thus change the RF for a given unit increase of gas. Although we expect mean global warming of about 1 K by 2050, with concurrent changes in water vapor and possibly cloud cover, there is no consensus in IPCC (1996) regarding what this future atmosphere would be. Thus, these RF values for 2050 are not based on a future climate, and this discrepancy must add to the uncertainty of this assessment. However, all such potential, systematic errors apply equally to the baseline scenario IS92a, and the relative climatic impact of aircraft will have less uncertainty.

6.3.1. *Models for Radiative Forcing*

Calculation of radiative forcing from 3-D models is a relatively new endeavor. The history of this calculation began with one-dimensional models (e.g., Hansen *et al.*, 1984a) that made use of the basic radiative-convective model approach initiated by Manabe and Wetherald (1967). Subsequent researchers expanded this calculation, using GCM output to describe atmospheric lapse rate, water vapor, and cloud cover and including latitudinal and seasonal changes (e.g., Pollack *et al.* 1993). Both the radiative perturbations from aircraft and the background radiative constituents (e.g., clouds, water vapor, albedo) vary with altitude, latitude, longitude, and time.

Three-dimensional modeling of radiative forcing introduces substantial complexity. The different modeling groups cited here have different approaches to calculating RF, involving choices in spatial and temporal domains.

All models report instantaneous values of radiative forcing at the top of the atmosphere and at the tropopause; in other words, these RF values have been calculated with no changes in atmospheric temperature. As discussed in Section 6.2.1, the most appropriate RF value includes allowance for stratospheric temperatures to readjust to radiative perturbation. Only two groups (Forster and Haywood, and Ponater and Sausen—see paragraphs below) have models that allow for such stratospheric adjustment. This correct, adjusted RF usually lies between the instantaneous values at the top of the atmosphere and the tropopause, and we correct the RF reported from other groups so that all RF values here refer to tropopause radiative forcing with stratospheric adjustment.

RF modeling results were contributed by P. Forster and J. Haywood (Forster and Shine, 1997), A. Grossman (Grossman *et al.*, 1997), J. Haywood (Haywood and Ramaswamy, 1998), D. Rind (Rind and Lonergan, 1995), W.-C. Wang (Wang *et al.*, 1995), and M. Ponater and R. Sausen (Ponater *et al.*, 1998).

Forster and Haywood's radiation scheme has been previously used to calculate ozone and water vapor radiative forcings; it is described in Forster and Shine (1997). It employs a 10 cm^{-1} narrowband model (Shine, 1991) in the thermal infrared (IR) and a discrete-ordinate model (Stamnes *et al.*, 1988) at solar wavelengths with 5-nm resolution in the ultraviolet (UV) and 10-nm resolution in the visible. As in Forster and Shine (1997), the fixed dynamic heating approximation (Ramanathan and Dickinson, 1979) is used to calculate stratospheric temperature perturbations. A zonally and annually averaged, 5° latitudinal resolution climatology was used as the basis for the forcing calculations. Temperature and humidity were derived largely from European Centre for Medium-range Weather Forecasts (ECMWF) analyses, averaged over the period 1980–91. In the upper stratosphere, temperatures were derived from Fleming *et al.* (1990). At pressures less than 300 hPA, humidity was based on a combination of Stratospheric Aerosol and Gas Experiment II (SAGE-II) and Halogen Occultation Experiment (HALOE) data. Surface albedos, cloud amounts, and optical depths were 7-year averages from International Satellite Cloud Climatology Project (ISCCP) (Rossow and Schiffer, 1991). Clouds were specified at three levels. The thermal infrared calculations included absorption by nitrous oxide, methane, and carbon dioxide. Ozone climatologies were taken from an observed climatology derived by Li and Shine (1995), a combination of SAGE-II, Solar Backscatter Ultraviolet (SBUV), Total Ozone Mapping Spectometer (TOMS), and ozonesonde data. To calculate forcing, the climatological profiles were perturbed by the absolute annual averages of ozone and water vapor changes.

Grossman (Grossman *et al.*, 1997) uses a set of baseline annual and longitudinal average atmospheric profiles, resolved by 50 layers between 0 and 60 km, at latitudes of 60°N/S, 30°N/S, and at the Equator that are scaled to IS92A (IPCC, 1995) composition. Supersonic and subsonic aircraft O_3 and H_2O

perturbation profiles were added to the baseline atmospheric profiles for RF calculations. The Lawrence Livermore National Laboratory (LLNL) 16-band solar radiation model (Grant and Grossman, 1998) and the LLNL 32-band IR radiation model (Chou and Suarez, 1994) were used to calculate instantaneous tropopause and top-of-atmosphere RFs at each latitude for the global average value for O_3 and H_2O.

Haywood's RF calculations for sulfate and black carbon aerosols were made following the method of Haywood and Ramaswamy (1998). The Geophysical Fluid Dynamics Laboratory (GFDL) R30 GCM incorporates a 26 band delta-Eddington solar radiative code (Ramaswamy and Freidenreich, 1997) and includes the cloud parameterization of Slingo (1989) and aerosol optical properties calculated using Mie theory. RF calculations are performed at the top of the atmosphere every day using mean solar zenith angle. No account is made for stratospheric adjustment, the effects of which are likely to be small for tropospheric aerosol in the solar spectrum.

Ponater and Sausen estimated instantaneous RF using the ECHAM4 GCM (Roeckner *et al.*, 1996). Radiative transfer calculations (one radiative time step only) were performed for each grid point with and without local ozone perturbation, including the actual cloud profile. Several diurnal cycles were calculated for each calendar month, and the radiative flux change was determined for each individual grid point at the top of the atmosphere and at the tropopause. The annual global mean radiative forcing was obtained by averaging over all grid points and over the seasonal cycle. To calculate the stratosphere-adjusted, tropopause RF, a "second atmosphere" was implemented into the ECHAM4 GCM. Whereas the primary atmosphere of the GCM does not "feel" the perturbation of the greenhouse gas, the second atmosphere experiences an additional radiative heating above the tropopause, although dynamic heating is identical to that of the first, unperturbed atmosphere. In the troposphere, the primary and second atmospheres are not allowed to diverge. In this configuration, the model is run for one annual cycle.

Rind used the Goddard Institute for Space Studies (GISS) Global Climate Middle Atmosphere Model (Rind *et al.*, 1988). The radiation scheme in the model is the correlated-k method for modeling non-gray gaseous absorption (Lacis and Oinas, 1991). The procedure involved keeping re-start files and full diagnostics for the first full day of each month from a control run. Radiation and all other routines were called each hour, so a full, diurnal average global response was calculated. Then the first day of each month was re-run with altered atmospheric composition (e.g., changes in ozone). The global net radiation at the top of the atmosphere was compared. The assumption is that with only 1 day of running time, temperatures would not adjust (even in the stratosphere) to the altered composition; hence, the results are instantaneous values.

Wang's RF calculations use the National Center for Atmospheric Research (NCAR) Community Climate Model 3 (CCM3) radiative model with monthly mean, latitude-by-longitude distributions of vertical temperature, moisture, clouds, and surface albedo simulated from the Atmospheric Model Intercomparison Project (AMIP). The year 1992 of the CCM3-AMIP simulations was used because the corresponding year was used to simulate ozone in the Oslo 3-D CTM (Isaksen *et al.*, 1999). Because this State University of New York/ Albany version of CCM3 used the ozone climatology (Wang *et al.*, 1995), CTM-simulated absolute ozone changes for 1992–2015 and 1992–2050 are mapped onto 1992 ozone climatology to calculate the RF. RFs are based on fixed temperature treatment rather than fixed dynamic heating treatment (Wang *et al.*, 1993).

6.3.2.   *Radiative Forcing for CO_2*

Carbon dioxide has a long atmospheric residence time (on the order of many decades); hence, aircraft CO_2 becomes well mixed within the atmosphere and can be treated together with other anthropogenic CO_2 emissions in conventional global warming simulations (e.g., Washington and Meehl, 1989; Cubasch *et al.*, 1992; Murphy and Mitchell, 1995). The aircraft influence depends on the temporal evolution of the amount of the CO_2 increase that can be attributed to aircraft emissions, which is directly proportional to the amount of fuel burned. See Section 6.1.2 and Table 6-2 for the calculation of CO_2 increases attributed to aviation.

Over the period 1990 to 2050, under IS92a we expect an increase in atmospheric CO_2 of about 155 ppmv from burning of fossil fuels, cement production, and other anthropogenic activities that release biospheric carbon. By 2050, F-type aviation scenarios produce a 5–7 ppmv increase, and the high-growth Edh scenario leads to a 13 ppmv increase. Thus, aviation in these scenarios would be responsible for 3–8% of the total anthropogenic increase in CO_2 from 1990 to 2050.

The RF for aviation CO_2 in 1992 is estimated to be +0.018 W m^{-2}, with a likely range of ±30% that includes uncertainties in the carbon cycle and in radiative calculations (see WMO, 1999). Uncertainties and confidence intervals discussed here do not include possible errors in predicting future scenarios. By 2050, the different aviation scenarios have a range of +0.06 to +0.16 W m^{-2}. The technology option 2 scenario (Fa2) leads to a 0.1 ppmv increase in CO_2 by 2050, with only a small increase in CO_2-RF.

The HSCT option, F1aH, has 18% greater fuel use but only 8% greater CO_2 concentrations by 2050, with a corresponding increase in CO_2-RF from +0.074 to +0.080 W m^{-2}. Because the HSCT fleet has just reached maturity in 2040, the extra fuel consumption of the HSCT aircraft is barely felt in terms of the accumulation of CO_2. Similarly, the CO_2 impact of new subsonic technologies that are introduced linearly between 2015 and 2050 is not fully effected by 2050. A fuller evaluation would have to extend the assessment beyond 2050, when the cumulative effects of mature fleets would be felt (e.g., Sausen and Schumann, 1999).

6.3.3.   *Radiative Forcing for O_3*

Ozone is a potent greenhouse gas whose concentration is highly variable and controlled by atmospheric chemistry and dynamics. Aircraft emissions of NO_x accelerate local photochemical production of O_3 in the troposphere; modeling studies suggest that these emissions are responsible today for average O_3 enhancements of 2–5 ppbv in the middle troposphere at northern mid-latitudes, where most aircraft fly (see Chapters 2 and 4). This ozone increase will generally be proportional to the amount of NO_x emitted (Grewe *et al.*, 1999), but evolving atmospheric composition, including increases in surface sources of combustion-related NO_x, will affect the aircraft impact.

Four subsonic ozone perturbations based on detailed 3-D patterns of NO_x emissions were chosen (along with their control atmospheres) for the calculation of RFs: NASA-1992, NASA-2015, FESGa (tech 1), and FESGe (tech 1) at 2050. Results for NASA-1992 were scaled by 1.15 to give NASA-1992*, and those for NASA-2015 by 1.05 to give NASA-2015*. Table 6-1 describes some of the basic properties of these aircraft scenarios, including total NO_x emissions. Chapter 4 supplied the seasonal pattern of O_3 perturbations for these scenarios based on model calculations and reported a factor of 2 uncertainty in this best value. The modeled RFs from these O_3 perturbations agree quite well, and the stratospheric temperature adjustment does not greatly affect the result (as was confirmed by two independent model calculations). Given the predominantly tropospheric perturbation, the uncertainty in modeling RF is small, and the uncertainty in the final result may be a factor of only 3. The ozone RFs are +0.023 W m^{-2} for NASA-1992*, +0.040 W m^{-2} for NASA-2015*, and +0.060 W m^{-2} for FESGa (tech1) 2050 (see also Table 6-1).

The development of atmospheric chemistry models in the past 2 years has allowed a consensus to build such that aircraft ozone perturbations can be calculated with a likely (2/3 probability) range of about a factor of 2 (higher/lower, see Chapter 4). Our estimate of the resulting RF for this predominantly tropospheric perturbation does not significantly enhance that interval.

The NASA-1992, NASA-2015, and FESGa (tech 1) at 2050 scenarios produce global mean column ozone increases (predominantly tropospheric) of 0.5, 1.1, and 1.7 Dobson Units (DU), respectively. Our estimate of the increase in tropospheric ozone associated with all anthropogenic changes (IS92a including aircraft plus surface emissions of NO_x, CO, and hydrocarbons) is about 3 DU from 1992 to 2015 and 7 DU from 1992 to 2050. These results represent an advance in our understanding since the Second Assessment Report (IPCC, 1996), when future ozone changes were scaled only to CH_4 increases and did not include the effects of doubling NO_x emissions from 1990 to 2050.

Two HSCT cases with detailed 3-D emission scenarios—one with 500 aircraft and the other with 1,000 aircraft—were used to calculate RF from stratospheric O_3 and H_2O perturbations (see Table 6-1). Most of the ozone change occurs above the tropopause; thus, there is poorer agreement among RF models and a greater difference in RF values after stratospheric temperatures adjust. The ozone perturbation calculated for 500 supersonic aircraft with the 2015 background atmosphere is substantially different in nature from that calculated for 1,000 aircraft with the 2050 atmosphere, in part because of specified changes in chlorine and methane contents of the stratosphere (see Chapter 4). Nevertheless, the best RF values are about the same: -0.01 W m^{-2}. The uncertainty range in these values is large and changes sign (-0.04 to +0.01 W m^{-2}), reflecting not only the range in O_3 perturbations given by Chapter 4 but also the large uncertainty in deriving RF for stratospheric perturbations. This ozone-related RF for the HSCT fleet is based only on the stratospheric ozone perturbation calculated by the models in Chapter 4; the tropospheric changes are discarded, and a correction for the displacement of about 11% of the subsonic traffic is included, as shown in Table 6-1 (scenario Fa1H).

6.3.4. Radiative Forcing for CH₄

Methane is a long-lived, well-mixed greenhouse gas. It has an atmospheric lifetime of about 9 years. The tropospheric chemical models used to evaluate the impact of the subsonic fleet found unanimously that CH_4 lifetime was reduced by aircraft emissions (see Chapter 4). This instantaneous change (-1.3% in 1992, -2.6% in 2015, and -3.9% in 2050 for scenario FSEGa (tech1)) needs to be increased further by a factor of 1.4 to include the feedback of CH_4 concentrations on lifetime (Prather, 1994; IPCC, 1996). It is then applied to the IS92a CH_4 abundance to calculate the reduction in CH_4 concentration that can be attributed to aircraft. The CH_4 perturbation is assumed to be instantaneous; in reality, however, it takes a couple of decades to appear and will lag the O_3 perturbation. For the purposes of interpolating between RF points in Table 6-1, we assume that the CH_4 perturbation, like the O_3 perturbation, is proportional to NO_x emissions.

6.3.5. Radiative Forcing for H₂O

Water vapor is a potent greenhouse gas that is highly variable in the troposphere, with a short average residence time controlled by the hydrological cycle. In the stratosphere, the slow turnover of air and extreme dryness make precipitation and clouds a rare phenomena, leading to smoothly varying concentrations ranging from 3 to 6 ppmv as CH_4 is oxidized to H_2O (Dessler *et al.*, 1994; Harries *et al.*, 1996). The contribution of aircraft to atmospheric H_2O is directly from the H in the fuel (assumed to be 14% by mass). Most of the subsonic fleet's fuel is burned in the troposphere, where this additional source of water is swamped by the hydrological cycle. A smaller fraction is released in the stratosphere, where longer residence times may lead to greater accumulation. However, because flight routes are close to the tropopause and reach at most into the lowermost stratosphere, this effluent is rapidly returned to the troposphere with little expected accumulation (Holton *et al.*, 1995; see also Section 3.3.4).

Although the uncertainty of predicting the current subsonic RF for water vapor is large—a factor of 3—the absolute number in 1992 is estimated to be sufficiently small, +0.0015 W m^{-2}, making this factor a minor uncertainty in subsonic climate forcing. It is assumed that this value scales linearly with fuel use (see Table 6-1). This value is consistent with earlier studies: Schumann (1994) and Fortuin *et al.* (1995) estimated that present air traffic enhances background H_2O by less than 1.5% for regions most frequently used by aircraft; likewise, Ponater *et al.* (1996) and Rind *et al.* (1996) used GCM studies to conclude that the direct radiative effect on the climate of water vapor emissions from 1992 air traffic is negligibly small.

The projected HSCT fleet, however, would cruise at 20-km altitude and build up much greater H_2O enhancements in the stratosphere. The stratospheric models described in Chapter 4 predicted excess stratospheric water vapor from an HSCT fleet of 500 aircraft (designated HSCT(500) in Table 6-1). This perturbation is difficult to calculate, and the likely (2/3 probability) range includes a factor of 2 higher and lower. Furthermore, RF modeling of this stratospheric H_2O perturbation adds further uncertainty, as indicated in Table 6-1. All results suggest that this effect is the dominant HSCT climate impact, with RF equal to +0.05 (0.017 to 0.15) W m^{-2} for 500 aircraft, increasing to +0.10 (0.03 to 0.30) W m^{-2} for a mature fleet of 1,000 aircraft (HSCT(1000)). Although it takes several years to accumulate this excess stratospheric water vapor, it is assumed that this RF is instantaneously proportional to the HSCT fleet size.

In a GCM study, Rind and Lonergan (1995) looked for climate change caused by H_2O accumulation from a fleet of 500 HSCTs. They found no statistically significant change in surface temperature. Their result is consistent with this assessment and with water vapor as the dominant HSCT climate impact because the magnitude of this radiative forcing from the fleet, +0.05 W m^{-2}, would induce a mean global warming that would be difficult to detect above natural climate variability.

6.3.6. Uncertainties

Assignment of formal uncertainty—or the likely (2/3 probability) interval about the best value—to radiative forcing caused by aircraft perturbations is difficult. For well-mixed gases (e.g., CO_2, CH_4) or for well-defined tropospheric perturbations (subsonic O_3), there is small uncertainty in calculated RF. In these cases, the overall uncertainty interval lies with calculating the perturbation itself: 25% for CO_2, a factor of 2 for O_3, and a factor of 3 for CH_4. For perturbations to stratospheric ozone and water, there is much greater uncertainty in calculating RF, especially because in these cases radiative forcing at the tropopause can be substantially different after stratospheric temperatures adjust. In addition, HSCT-induced ozone and water vapor perturbations—with large variations in the lower stratosphere—present a much more difficult calculation of RF where the placement of the modeled tropopause can lead to additional uncertainty.

As an example of the uncertainty in calculating RF values from an adopted ozone perturbation, the NASA-1992 tropospheric ozone perturbation was calculated by several groups, as shown in Table 6-4. The instantaneous RF at the tropopause is consistent across the models, and the stratospheric adjustment (calculated by two groups) is consistently 0.001 to 0.002 W m^{-2} less. For the HSCT(500) water vapor perturbation, the two groups have significantly greater disagreement, and the correction following stratospheric adjustment is a large fraction of tropopause instantaneous RF. This water vapor perturbation is the result of averaging six model results, and an additional RF is calculated using the water vapor perturbation calculated with a 3-D model that lies at the lower end of this ensemble (Grossman*; see Table 6-4). This table highlights the robustness of calculated RF for tropospheric perturbations and the much greater uncertainty in deriving climate forcing for stratospheric changes.

Uncertainty ranges about the best values are given in Table 6-1 for the NASA-1992* and FESGa (tech 1) 2050 subsonic scenarios and for some components of supersonic scenarios HSCT(500) and HSCT(1000). These intervals are intended to represent the same probability range (67% likelihood), but there is no uniform statistical model (e.g., Gaussian) for all of them, nor are the individual RF contributions fully independent; hence, these ranges cannot be combined into a confidence interval on total RF.

6.4. Radiative Forcing from Aircraft-Induced Changes in Aerosols and Cloudiness

There are two mechanisms by which aerosols may exert radiative forcing: the direct effect, whereby aerosol particles scatter and absorb solar and longwave radiation; and the indirect effect, whereby aerosol particles act as cloud condensation nuclei and modify the physical and radiative properties of clouds. Additionally for aircraft, merely flying through certain meteorological environments can result in formation of contrails (Section 3.4), which affect both solar and longwave radiation budgets. The present-day direct radiative forcing from aircraft emissions of sulfur compounds and black carbon aerosols is investigated in Sections 6.4.1 and 6.4.2; radiative forcing from the formation of contrails and the indirect effect of aerosol emissions is investigated in Section 6.4.3. Section 6.4.4 derives future RF considering our range of scenarios for fuel use. The RF models have been described previously (Section 6.3.1). A summary of radiative forcing calculations and related uncertainties is given in Section 6.4.5.

6.4.1. Direct Radiative Forcing from Sulfate Aerosols

Sulfate aerosol scatters a fraction of incident solar radiation back to space, thereby leading to negative direct radiative forcing. The direct radiative forcing of pure sulfate in the longwave spectrum is likely to be negligible as a result of the size of aerosol particles and the corresponding wavelength dependence of the specific extinction coefficient (e.g., Haywood and Shine, 1997; Haywood *et al.*, 1997a). Myhre *et al.* (1998) summarize 10 detailed studies of the sensitivity of direct radiative forcing from all anthropogenic sources of sulfate. With the exception of one study, sensitivities per unit column mass of anthropogenic sulfate range from -125 to -214 W g^{-1} SO$_4$.

We reexamined these results by inserting a pure ammonium sulfate in a layer between 8 and 13 km in the GFDL R30 GCM and assuming an ambient relative humidity of 45% using the method of Haywood and Ramaswamy (1998). A log-normal

Table 6-4: Results of RF (W m^{-2}) for the 1992 aviation-induced ozone perturbation and for the water vapor from the HSCT(500) fleet as calculated by several models (see section 6.3.1).

NASA-1992 Tropospheric Ozone Perturbation

Type of RF Calculation	Forster & Haywood	Ponater & Sausen	Grossman	Rind	Wang
Top of atmosphere, instantaneous	+0.014	+0.021	+0.013	+0.011	+0.010
Tropopause, instantaneous	+0.020	+0.026	+0.022		+0.020
Tropopause, after stratospheric adjustment	+0.019	+0.024			

HSCT(500) Stratospheric Water Vapor Perturbation

Type of RF Calculation	Forster & Haywood	Ponater & Sausen	Grossman	Grossman*
Top of atmosphere, instantaneous	+0.001	+0.001	-0.002	-0.001
Tropopause, instantaneous	+0.096	+0.049	+0.074	+0.048
Tropopause, after stratospheric adjustment	+0.068	+0.034		

*RF calculation for a lower accumulation rate of H$_2$O (Danilin *et al.*, 1998; Hannegan *et al.*, 1998).

distribution with a dry geometric mean radius of 0.05 μm and a standard deviation of 2.0 was adopted. The resulting global mean sensitivity was found to be approximately -215 W g^{-1} SO$_4$, which is adopted throughout this report. Because the modeled pure sulfate particles scatter incident radiation with no absorption, the RF is not sensitive to their location relative to the tropopause. Thus, the RF is well approximated by the instantaneous radiative forcing at the top of the atmosphere even for aircraft sulfate in the lower stratosphere.

A study of the distribution of aircraft fuel burned and transported as a passive tracer from scenario NASA-1992 involved a range of global models and is presented in Chapter 3 (see also Danilin *et al.*, 1998). The median global mean column burden of sulfate aerosol is derived in this study by adopting an emission index for sulfur EI(S) of 0.4 g kg^{-1} and a 50% effective conversion factor from fuel-sulfur to optically active sulfate aerosols; it is approximately 13.5 μg SO$_4$ m^{-2} (Table 3-4). Thus, global mean radiative forcing from aircraft emissions of sulfate aerosol in 1992 is estimated to be -0.003 W m^{-2}, with a likely range of -0.001 to -0.009 W m^{-2} (see also Table 6-1). This value is much smaller in absolute magnitude than the RF from CO$_2$, O$_3$, CH$_4$, or contrails. We assume that EI(S) remains constant through 2050 and scale the sulfate RF with fuel use.

6.4.2. Direct Radiative Forcing from Black Carbon Aerosols

Tropospheric black carbon (BC) aerosol, also described as soot, primarily absorbs incident solar radiation, which leads to positive radiative forcing. As in the case of sulfate aerosol, the small size of the particles means that radiative forcing in the longwave region of the spectrum is likely to be negligible. The sensitivity of global mean radiative forcing to column loading of total anthropogenic BC is estimated by Haywood *et al.* (1997a), Haywood and Ramaswamy (1998), and Myhre *et al.* (1998) to range from approximately +1100 to +1850 W g^{-1} BC.

We reexamined results by inserting BC aerosol in a layer between 8 and 13 km in the GFDL R30 GCM using the method of Haywood and Ramaswamy (1998). A log-normal distribution with a geometric mean radius of 0.0118 μm and a standard deviation of 2.0 was assumed. The resulting global mean sensitivity was found to be approximately +3000 W g^{-1} BC as a result of the higher sensitivity of the radiative forcing when the BC exists at higher altitudes above a greater proportion of cloudy layers (Haywood and Ramaswamy, 1998). This value is adopted throughout this report because it explicitly takes into account the effect of the elevated altitude of the aerosol.

BC particles primarily absorb sunlight and heat the local air. Thus, unlike BC that resides in the troposphere, BC in the stratosphere contributes negative solar radiative forcing that is countered by induced positive longwave radiative forcing. Thus, radiative forcing from BC aerosols is sensitive to their location relative to the tropopause. We do not have enough information on the location of BC relative to the tropopause,

and thus our use of the instantaneous top-of-atmosphere value overestimates the RF depending on the fraction of aircraft BC in the lower stratosphere.

Using the aircraft fuel-burn scenarios for NASA-1992 noted above and described in Chapter 3, we derive a global mean column burden of BC aerosol from aircraft of 1.0 μg BC m^{-2} (assuming an EI(BC) of 0.04 g kg^{-1}; see also Table 3-4). Thus, we estimate global mean BC aerosol forcing in 1992 to be +0.003 (+0.001 to +0.006) W m^{-2} and assume that it linearly scales with fuel use (see also Table 6-1). This value is much smaller in absolute magnitude than the RF from CO$_2$, O$_3$, CH$_4$, or contrails.

6.4.3. Radiative Forcing from Persistent Contrails and Indirect Effects on Clouds

Aircraft emission of water vapor and particles, as well as the creation of contrails, could lead to a change in global cloudiness. Some atmospheric GCM studies that have looked at the impacts of injecting water vapor or creating contrails (e.g., Ponater *et al.*, 1996; Rind *et al.* 1996) point to the potential importance of these effects on climate, but these pilot studies cannot be used directly in this assessment. Persistent contrails clearly related to aircraft are detectable, however, and their impact on radiative forcing can be evaluated. Section 3.6 (see Table 3-9) estimates direct radiative forcing from persistent contrails to be +0.02 (+0.005 to +0.06) W m^{-2} in 1992 (see also Table 6-1). This estimate is limited to immediately visible, quasi-linear persistent contrails.

Whereas contrail formation and associated radiative forcing is an obvious and visible consequence of aircraft activity, the secondary, indirect effect of aerosols from aircraft on the microphysical and radiative properties of clouds is a very complex issue that has received little attention and is very difficult to quantify (Seinfeld, 1998). Some significant steps in quantifying the indirect effect from anthropogenic aerosols have been made (e.g., Jones *et al.*, 1994; Boucher and Lohmann, 1995). The effects of aerosol particles from aircraft emissions on clouds are more complicated because nucleation and subsequent growth of ice crystals that make up cirrus clouds are more complex and less studied than for water clouds. Cirrus cloud generally exert positive forcing because longwave positive radiative forcing is of a larger magnitude than solar negative radiative forcing. Section 3.6.5 (see Table 3-9) estimates that radiative forcing from aircraft-induced cirrus is positive and may be comparable to contrail RF. The magnitude of this RF remains very uncertain. No best estimate is given in Tables 6-1 and 6-2, but a range for the best estimate could fall between 0 and 0.04 W m^{-2}.

6.4.4. Future Scenarios

Direct RF from sulfate and BC in the future is obtained by scaling the best values for 1992 to future fuel use (see Table 6-1). By 2050, it is projected to increase by factors of about 2–5 for F-type scenarios and 7–10 for E-type scenarios. Atmospheric levels of aircraft sulfate and BC are assumed to respond instantaneously to fuel burn. RF values for these future scenarios given in Table 6-1. For the central FESGa (tech1) 2050 scenario, RF(sulfate) is estimated to be -0.009 (-0.003 to -0.027) W m^{-2}, and RF(BC) is estimated to be +0.009 (+0.003 to +0.027) W m^{-2}. The magnitude of radiative forcing for sulfate and BC aerosol from subsonic aircraft appears to cancel, but this appearance is deceptive because EI(S) and EI(BC) are highly uncertain for the future fleet and are not coupled. For the range of scenarios listed in Table 6-1, each of these RFs remains smaller than the RF from CO_2, O_3, or persistent contrails; these effects still need to be considered, however, especially in the upper limits of the uncertainty range.

We do not evaluate here the climate impact of sulfate and BC aerosols from the projected HSCT fleet (scenario Fa1H). Sulfate released near 20 km would augment the natural Junge layer, adding about 25% to total mass and a smaller fraction to reflectivity (see discussion in Chapter 4). These numbers depend on the sulfur content of HSCT fuel. BC aerosols released from HSCT aircraft would primarily heat the stratosphere and may lead to a small negative value of RF after stratospheric adjustment. Still, the EI(S) and EI(BC) from yet-to-be-developed HSCT aircraft are highly uncertain but are likely to be much smaller than other HSCT-induced RF.

If RF from contrails of +0.02 W m^{-2} in 1992 (see Section 3.6) scales with fuel burn, it would increase to +0.06 W m^{-2} by the year 2050 (Fa1). However, contrails are expected to increase more rapidly than global aviation fuel consumption as a consequence of a number a factors: Air traffic is expected to increase mainly in the upper troposphere, where contrails form preferentially; newer, more efficient engine/airframes will travel greater distances with the same amount of fuel, but larger wide-body aircraft carry more passengers and burn more fuel for the same distance; and more efficient aircraft can trigger contrails at higher atmospheric temperatures, hence at a larger range of altitudes (see discussion in Section 3.7 and Gierens *et al.*, 1999). Thus, global mean RF for persistent contrails is predicted to be larger by an additional factor of 1.6 (+0.10 W m^{-2}). Technology option 2 (scenario Fa2) does not increase contrail RF because the same distances are flown although the fuel burned is greater.

The radiative forcing from aircraft-induced cirrus clouds in 2050 is even more uncertain than for 1992 (Chapter 3). An estimate could fall between 0 and 0.16 W m^{-2} for the 2050 FESGa (tech1) scenario.

Persistent contrails in the stratosphere are not likely because of the low ambient relative humidity. Radiative forcing from contrails from the proposed HSCT fleet may therefore be neglected in future scenarios of radiative forcing (except for the 11% reduction of subsonic RF from air traffic displaced by HSCT aircraft).

6.4.5.   *Uncertainties*

It is difficult to constrain the direct and indirect climate effects of aerosols from aircraft. Measurements at altitude are not adequate to define the aircraft contribution today, so models representing emissions and subsequent chemical and physical transformations have adopted differing parameterizations to estimate atmospheric concentrations. Furthermore, different aerosol size distributions, the hygroscopic nature of some of the aerosol constituents, and mixing of different species of aerosol all lead to uncertainties in the radiative properties of aerosols. These uncertainties in the burden and radiative properties of aerosols emitted by aircraft lead to associated uncertainties in radiative forcing that are much larger than for a well-mixed, directly emitted greenhouse gas such as carbon dioxide.

RF from aircraft emissions of greenhouse gases and aerosols can be constrained when the amount of radiatively important species is directly limited by emissions (e.g., CO_2, H_2O, sulfate, BC). However, it is more difficult to place upper limits on RF when the climate impact occurs indirectly, as in aircraft NO_x production of O_3 or—specifically here—in perturbations to naturally occurring clouds by aircraft aerosols.

The indirect effect of aerosols from aircraft emissions on naturally occurring clouds cannot be quantified at present owing to the complexity of several processes such as ice-cloud nucleation and the dependence of albedo and emissivity on the size of the ice crystals. With our present knowledge, this uncertainty must be considered significant. Radiative forcing from contrails is estimated to grow disproportionately with fuel use, and extrapolation to large values in 2050 is an important uncertainty in the future radiative impact of aviation. Further uncertainty arises from the fact that all estimates of contrail RF were made for the present climate, hence do not account for future climate change (see discussion in Section 3.7.1).

6.5.   **Relation between Radiative Forcing and Climate Change**

6.5.1.   *Radiative Forcing and Limits of the Concept*

In Sections 6.3 and 6.4 we have reported radiative forcing calculations for various aspects of aircraft-induced perturbations to radiatively active substances. One of the basic ideas behind this calculation was the validity of the concept of RF as a quantitative predictor for climate change (see Section 6.2.1). We implicitly assume that contributions from individual perturbations to the change in global mean surface temperature are additive, at least as a first-order approximation. The radiative forcing concept requires a quasi-constant climate sensitivity parameter λ within the same model for different types of RF and different magnitudes of RF. This assumption holds for changes in most of the well-mixed greenhouse gases and for variations in solar irradiance (e.g., Hansen *et al.*, 1984b; Shine *et al.*, 1995; Hansen *et al.*, 1997b; Roeckner *et al.*, 1999). As noted in previous IPCC reports (e.g., IPCC, 1996), there are limitations to the applicability of the radiative forcing concept, especially for RFs that are highly variable geographically or seasonally (i.e., typical of aircraft perturbations).

As a clear example of large spatial differences in radiative forcing, the annual mean radiative imbalance as a function of latitude is shown in Figure 6-9 for the big four RFs for subsonic aviation in 1992. Whereas the radiative imbalances from CO_2 (positive) and CH_4 (negative) are nearly uniform from south pole to north pole, those from O_3 (peaking at 0.065 W m^{-2} at 30°N) and contrails (peaking at 0.10 W m^{-2} at 40°N) are highly concentrated in northern mid-latitudes. The global means of all four contributions are about equal (see Table 6-1). The NO_x-driven perturbations to O_3 and CH_4 produce RFs (+0.23 and -0.14 W m^{-2}, respectively) that are of similar magnitude and in part cancel. However, the latitudinal distribution of the radiative imbalance from these two perturbations does not cancel: The combined O_3+CH_4 forcing is positive in the Northern Hemisphere and negative in the Southern Hemisphere. The response of the climate system to such geographically non-homogeneous forcing is unknown. At least regional differences in climate response can be expected, and there may even be differences in the global mean response (see Section 6.5.2).

The limitations and advantages of the radiative forcing concept have recently been demonstrated by Hansen *et al.* (1997b). A classic test case considers radiative forcing and associated climate change for ozone perturbations occurring at different altitudes. In a series of numerical experiments with a low-resolution GCM (Hansen *et al.*, 1997a), the impact of 100 DU ozone added to the various model layers was investigated. As in the calculations of Lacis *et al.* (1990), Hauglustaine and Granier (1995), Forster and Shine (1997), and Brasseur *et al.* (1998), strong variation of radiative forcing with the altitude of ozone perturbation is found: negative forcing for perturbation in the middle stratosphere, and positive forcing for perturbation in the troposphere and lower stratosphere. RF is a maximum for a perturbation close to the tropopause (Figure 6-10, dotted curve).

Without cloud and water vapor feedbacks, the change in global-mean surface air temperature exhibits a similar dependency on

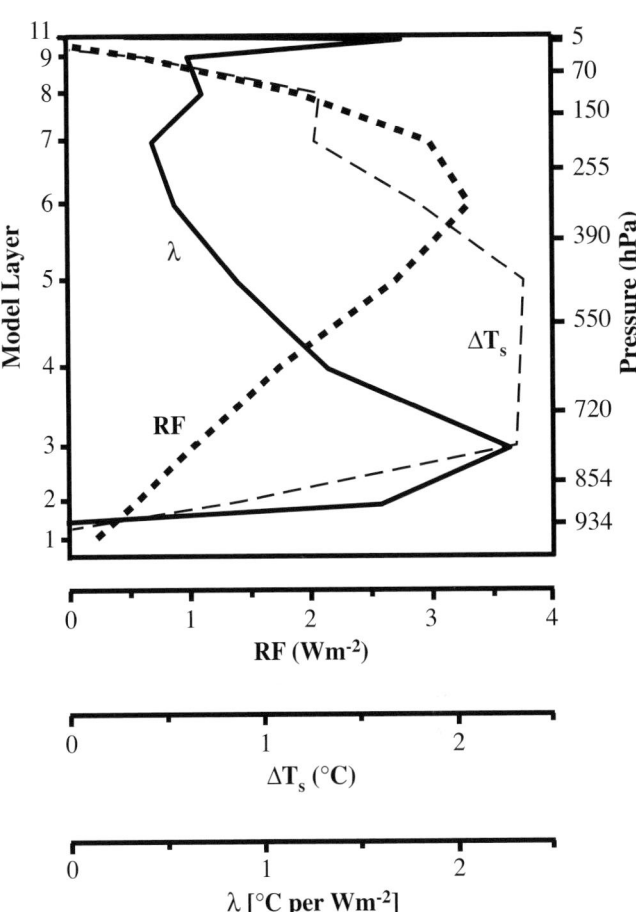

Figure 6-10: Radiative forcing (dotted line), global mean surface temperature change including all internal feedback processes (dashed line), and climate sensitivity parameter λ (solid line) from 100 DU additional ozone as a function of altitude (model layer) to which ozone was added. The plot is based on results of a simplified GCM (data from Table 3 by Hansen *et al.*, 1997b).

the altitude of the ozone perturbation as radiative forcing does (with a maximum close to the tropopause, not shown in figure). However, when all model feedbacks are included, the maximum global-mean surface temperature change in the GISS model occurs for an ozone perturbation in the middle troposphere (Figure 6-10, dashed line). The climate sensitivity parameter λ varies widely as a function of the altitude of the ozone perturbation varies widely from the middle stratosphere to the surface layer (Figure 6-10, solid line). Peak sensitivity in this model occurs in the lower troposphere because of feedbacks on clouds. However, for the altitude range at which aircraft O_3 perturbations contribute significantly to the total RF (e.g., 2–14 km), the value of λ is within a factor of 2 of the climate sensitivity parameter for doubling of CO_2 in the same model, 0.92 K/(W m^{-2}) (Hansen *et al.*, 1997b).

In GCM experiments with a similar model that focused on aircraft-like perturbations, the specific feedbacks that minimized the response for aircraft releases were high-level cloud cover (Rind and Lonergan, 1995—considering ozone impacts) and

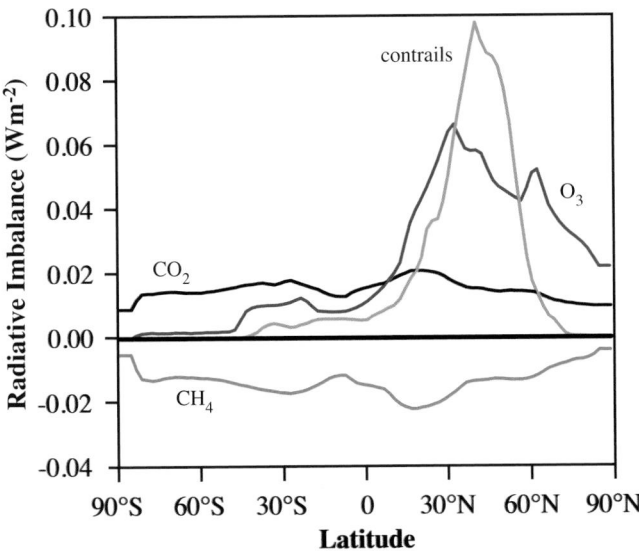

Figure 6-9: Zonal and annual mean radiative imbalance (W m^{-2}) at the tropopause (after adjustment of stratospheric temperature) as a function of latitude as a result of air traffic for 1992.

sea ice (Rind *et al.*, 1996—considering water vapor releases). Sensitivity of λ to layered perturbations is also found by Ponater *et al.* (1998) but with opposite sense: The climate sensitivity parameter for aircraft-induced ozone is *higher* than that for well-mixed greenhouse gases. The difference in these results likely depends on the formulation of clouds within the two models.

With our present understanding of climate modeling and critical feedback processes in the troposphere, we can do no better than adopt the tropopause value of RF after stratospheric adjustment. However, for aircraft-induced climate perturbations, the additivity of RFs across all perturbations cannot be taken for granted and adds further uncertainty.

6.5.2. Climate Signatures of Aircraft-Induced Ozone Perturbations

With uncertainty in relating RF to climate response noted above, it would be useful to compare CGCM-modeled climate responses for aircraft perturbations. Less agreement is expected between models because the magnitude of their feedback responses (i.e., climate sensitivity) can be quite different from one another. A review of different model sensitivities has been given in IPCC (1990, 1992, 1996); the values vary from a sensitivity of about 0.4 K/(Wm^{-2}) to 1.2 K/(Wm^{-2}) for doubled CO_2. How the sensitivity would vary for heterogeneous aircraft perturbations is not known.

The most appropriate tool would be simulations in the transient mode using coupled atmosphere-ocean GCMs. These simulations are computationally very expensive, and it would be very difficult to separate signal from noise. To date, however, all known comprehensive model simulations of aircraft-induced climate change were made in the quasi-stationary mode using either pure atmospheric GCMs (e.g., Sausen *et al.*, 1997) or atmospheric GCMs coupled to mixed layer ocean models (examples below). In other words, these simulations studied quasi-equilibrium response to a stationary or seasonally repeating perturbation. Pure atmospheric models underestimate the response if, for instance, sea surface temperature is fixed to a prescribed value. Equilibrium simulations with coupled models overestimate the aircraft effect in absolute numbers relative to transient simulations (see discussion in Sausen and Schumann, 1999), and the spatial pattern of climate change can serve only as a first estimate (Kattenberg *et al.*, 1996). These simulations must be compared with analogous simulations for well-mixed greenhouse gases (e.g., CO_2 doubling). Then the particular climate sensitivity to aircraft-induced radiative forcing and aircraft signatures of climate changes may possibly be extracted.

Using the GISS 3-D climate/middle atmosphere model, Rind and Lonergan (1995) studied the impact of the combined effect of a stratospheric ozone decrease and a tropospheric ozone increase from an assumed sub- and supersonic aircraft fleet for the year 2015. An equilibrium climate simulation leads to a general stratospheric cooling of a few tenths of a Kelvin,

combined with a warming of the lower stratosphere in northern polar regions as a result of altered atmospheric circulation (Figure 6-11). The globally averaged surface temperature change was found to be not statistically significant compared with climatic variability.

As another example, Figure 6-12 shows the equilibrium annual, zonal mean temperature change from ECHAM4/MLO (Ponater *et al.*, 1998) due to ozone perturbations resulting from 1992 air traffic (Dameris *et al.*, 1998), which differ slightly from the 1992 ozone perturbations reported in this assessment. The overall pattern of temperature change is statistically significant. The RF associated with this case is 0.04 W m^{-2}, and the resulting equilibrium change of global mean surface temperature is 0.06 K, resulting in a climate sensitivity parameter λ of about 1.5 K/(W m^{-2}). Note that a climate sensitivity calculated for such small perturbations is associated with large uncertainty.

For comparison, Figure 6-13 shows the equilibrium climate change associated with the anthropogenic increase of well-mixed greenhouse gases (CO_2, CH_4, etc.) from 1990 to 2015 according to IS92a as simulated with the same model. The associated global mean surface temperature change is 0.9 K, and the RF is 1.1 W m^{-2}. The resulting climate sensitivity parameter, about 0.8 K/(W m^{-2}), is smaller than for the aircraft-induced ozone perturbation. Temperature change patterns are quite different for well-mixed greenhouse gases than for aircraft-induced ozone perturbation.

These and related experiments with the GISS and ECHAM4 models showed that: The aircraft-related climate change pattern and the aircraft-related climate sensitivity parameters are model-dependent, and the climate response for aircraft-induced ozone changes is different from that for conventional greenhouse gases. Internal feedback processes within and between climate models appear to work differently. The exact nature of the climate feedback differences to different pattern of RF remains to be investigated. Thus, the climate factors of human interest, especially the regional climate change for these regional perturbations, await further development of climate models. The details of the climate response to the contrail radiative imbalance, which is spatially heterogeneous on the smallest scales, is likewise unknown.

6.6. The Role of Aircraft in Climate Change— Evaluation of Sample Scenarios

The overall impact of aviation on climate change in the next 50 years is evaluated here for a range of scenarios (see Section 6.1.3 and Table 6-3) in air traffic and potential options in civil aviation (e.g., low-NO$_x$ combustors, high-speed civil transport). Aviation's role is considered within the context of climate change already being forced by greenhouse gases and expected to continue from growth of the world's economies (e.g., IPCC's IS92a scenario; see IPCC, 1992). The impact of air travel and climate change as a whole on society is beyond this scientific assessment of the role of aviation in physical climate change.

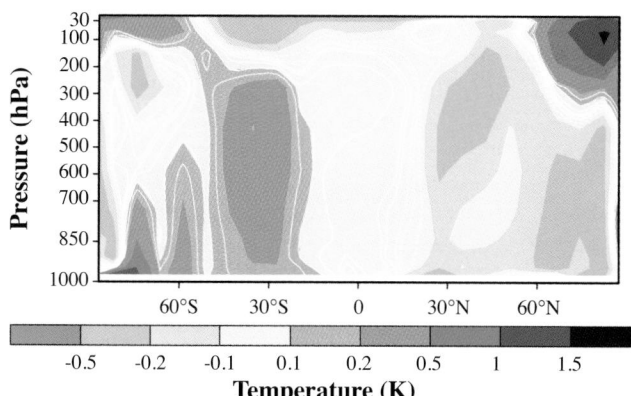

Figure 6-11: Equilibrium change of annual, zonal mean temperature (K) caused by ozone perturbation due to NO_x emissions of a projected sub- and supersonic aircraft fleet for the year 2015, as simulated with the GISS model (Rind and Lonergan, 1995). This calculation is not for any specific scenarios assessed here.

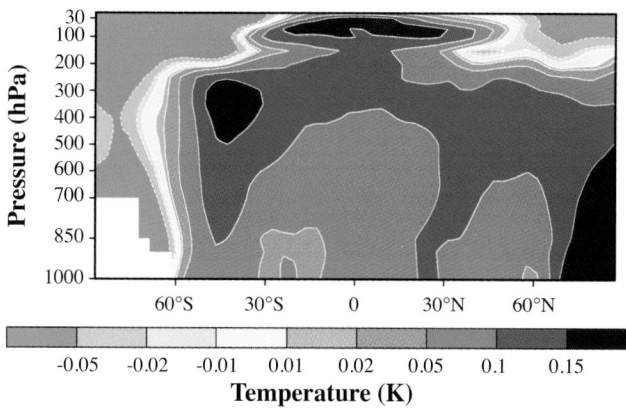

Figure 6-12: Equilibrium change of annual, zonal mean temperature (K) caused by ozone perturbation due to NO_x emissions of 1991–92 air traffic (DLR-2), as simulated with the ECHAM4/MLO model (Ponater *et al.*, 1998). This result is similar to, but not based on, the scenarios analyzed here.

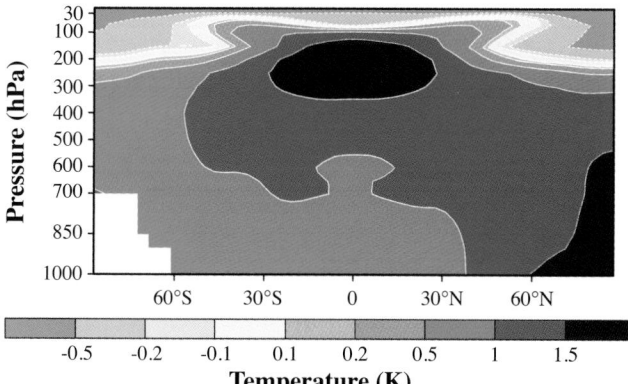

Figure 6-13: Equilibrium change of annual, zonal mean temperature (K) resulting from the anthropogenic increase of well-mixed greenhouse gases from 1990 to 2015, as simulated with the ECHAM4/MLO model (Ponater *et al.*, 1998).

This section combines the aviation impacts from Sections 6.3 and 6.4 evaluated for the fixed-year scenarios (defined in Table 6-1) and places them in a continuous time sequence (defined in Table 6-3) to compare with overall climate changes expected under IS92a. We assume that RFs are additive, so the issue of different climate sensitivities—for example, for aircraft-induced ozone perturbations versus those from CO_2 (as discussed in Section 6.5)—introduces some additional error/ uncertainty when we use summed RF to deduce climate change. We have no alternative but to use RF as a metric of climate change because differences in forcing between the various subsonic options would not be reliably detected above the natural climate variability in many models. The summed RF, however, should provide a relative ranking of the different options discussed below.

6.6.1. Individual Components of Radiative Forcing

Figure 6-14a (IPCC, 1996) shows a bar chart of individual components of RF for all anthropogenic change for the 1990 atmosphere. This figure can be compared with 1992 aircraft RF components from the NASA-1992* scenario in Figure 6-14b (note the change of scale; see also Table 6-1). Aircraft RF is qualitatively different from overall anthropogenic RF: The CH_4 RF is negative, the O_3-RF is greater than the CO_2-RF, and the aerosol-cloud effects are positive (because of persistent contrails) rather than negative. Within the error bars, the effects of CO_2, O_3, and contrails are comparable; that of methane is also comparable, but negative. A caveat here is that we are unable to derive a best estimate for the indirect effects from aircraft on "natural" clouds; these effects might be negligible or may be comparable to that of contrails. Thus, for 1992 our best estimate is that aviation causes RF (without additional cirrus) of approximately +0.05 W m^{-2} and is responsible for about 3.5% of the total anthropogenic RF of about +1.4 W m^{-2} for greenhouse gases plus aerosols (+2.7 W m^{-2} for greenhouse gases alone).

Bar charts for the 2050 atmosphere are presented in Figure 6-15. The same qualitative differences apply. However, in 2050 the contribution by contrails is almost twice as large as the contribution from aircraft CO_2 or O_3. Thus, in 2050 for the Fa1 scenario, our best estimate is that aviation RF (without additional cirrus) grows to about +0.19 W m^{-2}, which is 5% of the total anthropogenic radiative forcing anticipated for IS92a of +3.8 W m^{-2} for greenhouse gases plus aerosols (+5.8 W m^{-2} for greenhouse gases alone). This fraction is more than doubled for E-type scenarios (Table 6-1). There is considerable scientific uncertainty in these estimates—a factor of 2 or more at the 67% confidence level—that is separate from the potential error in projecting future scenarios for aviation and IS92a.

Individual RF components in 2050 attributable to the impact of adding a supersonic fleet are shown in Figure 6-15c (see also Table 6-1). The white bars denote the direct RF of HSCTs; the black bars denote the RF from displaced subsonic air traffic. As already discussed, RF from stratospheric H_2O contributes the

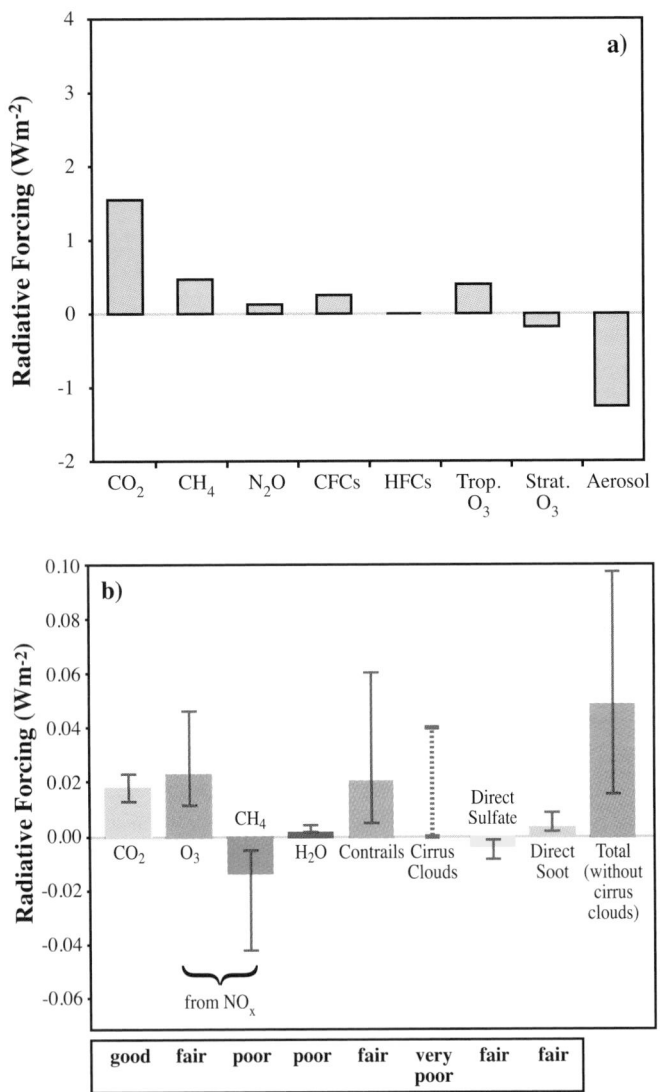

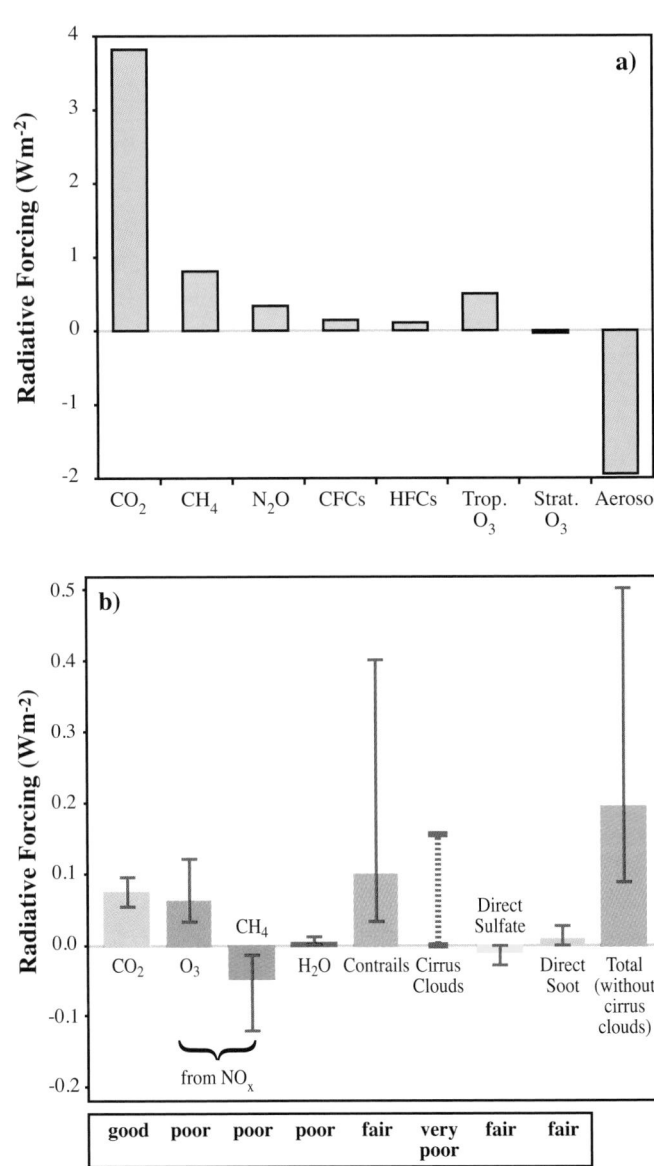

Figure 6-14: Bar charts of radiative forcing from (a) all perturbations in 1990 (from IPCC, 1996) and (b) aviation effects in 1992. Note scale change from (a) to (b). In (b), best estimate (bars) and high-low 67% probability intervals (whiskers) are given. No best estimate is shown for the cirrus clouds; rather, the dashed line indicates a range of possible estimates. The evaluations below the graph are relative appraisals of the level of scientific understanding associated with each component.

Figure 6-15: Bar charts of radiative forcing in 2050 (a) from all perturbations (from IPCC, 1996), (b) from subsonic aviation (Fa1), and (c) from the additional effect due to supersonic air traffic. Note scale change from (a) to (b) and (c). In (b), best estimate (bars) and high-low 67% probability intervals (whiskers) are given. No best estimate is shown for cirrus clouds; rather, the dashed line indicates a range of possible estimates. The evaluations below the graph are relative appraisals of the level of scientific understanding associated with each component. In (c), white bars denote the direct effect of the supersonic fleet (HSCT1000), whereas the black bars display the change resulting from the displaced subsonic air traffic.

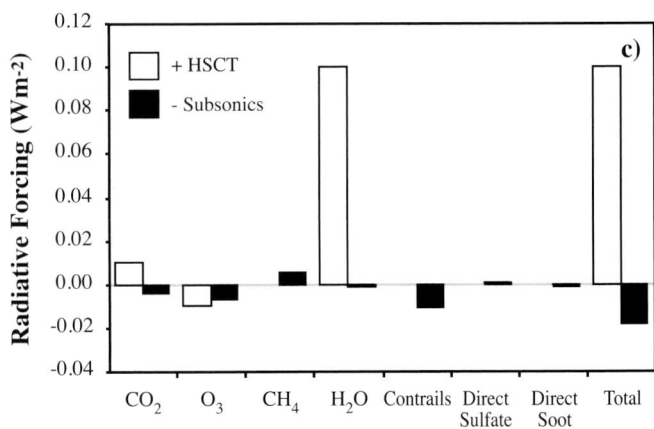

largest component, comparable in magnitude to the big three RFs (CO_2, O_3, contrails) from the subsonic fleet (Figure 6-15b). RFs for the combined fleet (Fa1H) are listed separately in Tables 6-1 and 6-2.

6.6.2.   Uncertainties and Confidence Intervals

Throughout this report, we focus on "best estimates" for each component of atmospheric perturbations caused by aircraft and then of subsequent climate forcing or ultraviolet change. We also try to understand the confidence that we have in these estimates using uncertainty ranges deduced in Chapters 2, 3, and 4 and those from the modeling and combining of RF in this chapter.

Uncertainties in estimating aviation's RF values are addressed with a confidence interval (indicated by error bars or whiskers about each best value) and a description ("good," "fair," "poor," "very poor") of the level of scientific understanding of the physical processes, models, and data on which the calculation is based. The confidence intervals shown in Figures 6-14b and 6-15b define a likelihood range such that the probability that the true value falls within the interval is 2/3. The interval and the quality-of-the-science descriptions are, to a large extent, independent measures covering different aspects of uncertainty.

The likelihood range is defined consistently within this report as the 2/3 or 67% probability range. These probability ranges are meant to be symmetric about the best value; hence, the best value is not always the mean of the upper and lower values. In this case, the probability that the value is less than the lower value is 16%, and the probability that it is less than the upper value is 84%. The range between the low and high values is equivalent to the "1-sigma" range of a normal (i.e., Gaussian) probability distribution. Unfortunately, derivation of these confidence intervals lies with the expert judgment of the scientists contributing to each chapter and may include a combination of objective statistical models and subjective expertise. Thus, the 67% confidence intervals do not imply a specific statistical model and, for example, cannot be used to infer the probability of extreme events beyond the stated confidence interval.

The confidence interval in RF stated here combines uncertainty in calculating atmospheric perturbation to greenhouse gases and aerosols with that of calculating radiative forcing. It includes, but is not based solely on, the range of best values from different research groups. For example, the interval for the HSCT(1000) impact on O_3 was derived from high- and low-end calculations using different combinations of atmospheric models and chemical assumptions. The range in RF from these stratospheric O_3 perturbations was expanded further in this chapter to account for the difficulty in calculating RF for stratospheric perturbations. The tropospheric O_3 perturbation from the subsonic fleet (scenario Fa1) was presented with the 67% confidence interval as a factor of 2 higher and lower than the best value. In this case, the RF calculation did not significantly

add to the uncertainty because tropospheric perturbations can be more accurately calculated. The confidence interval for aviation-induced CH_4 changes is believed to be about 1.5 times larger (log-scale) than that for tropospheric O_3, but potential errors in both are highly correlated. The confidence interval for contrails is taken directly from Chapter 3; the RF from additional cirrus clouds is highly uncertain and no probability range is given.

The RF uncertainties from different perturbations have been determined by different methods; potential errors in individual components may not be independent of one another, and the error bars may not represent Gaussian statistics. The uncertainty ranges for the totals in Figures 6-14b and 6-15b do represent a 2/3 probability range as for the individual components. The uncertainty estimate for the total radiative forcing (without additional cirrus) is calculated directly from the individual components as the square root of the sums of the squares of the upper and lower ranges. There is a further issue on confidence levels that is not quantified here—namely, the accuracy of representing the climate perturbation by the sum of RF values that are global means.

Overall, addition of the best values for RF provides a single best estimate for the total. The uncertainty ranges for individual impacts can be used to assess whether they are potentially major or trivial components and to make a subjective judgment of confidence in the summed RF.

6.6.3.   Aircraft as a Fraction of Total Radiative Forcing

In the aircraft scenarios considered here, the relative amount of aviation fuel burned in 1990 (2.4% of all fossil fuel carbon emissions) would increase by 2050 to about 3% (Fa1) or more than 7% (Eab) depending on projections for air traffic (see Figure 6-16a). For the other scenarios here—with economic assumptions differing from the reference case IS92a, hence a different demand for air transport—these fuel fractions in 2050 may be a factor of 1.5 larger. In comparison, aviation was responsible in 1990 for about 3.5% of total anthropogenic RF. This fraction would grow to about 5% in 2050 for Fa1 and 10% for Eab. Figure 6-16b shows the evolution of aviation RF from 1990 to 2050 for several scenarios as a percentage of total IS92a RF. The climate change forced by aircraft, as measured by the summed RF, is a larger proportion of total RF than indicated by fuel use alone. Thus, aviation is an example of an industry for which the climate impacts of short-lived perturbations such as O_3 and aerosols/clouds must be considered when evaluating the sector as a whole. This effect is also demonstrated by RFI—the ratio of summed RF to CO_2-RF—which is greater than 1 (see Table 6-1).

Each of the demand scenarios has a technology 2 option in which a 25% reduction in $EI(NO_x)$ (corresponding to a 22% reduction in total NO_x emissions) is achieved at an additional fuel cost of about 3.5%. These options are calculated to have very little impact on total RF (Table 6-1) because the reduced O_3-RF is offset by the increased CH_4-RF and the small extra

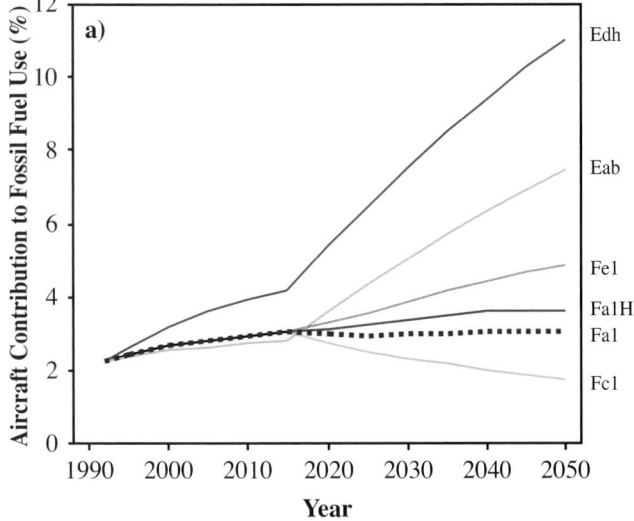

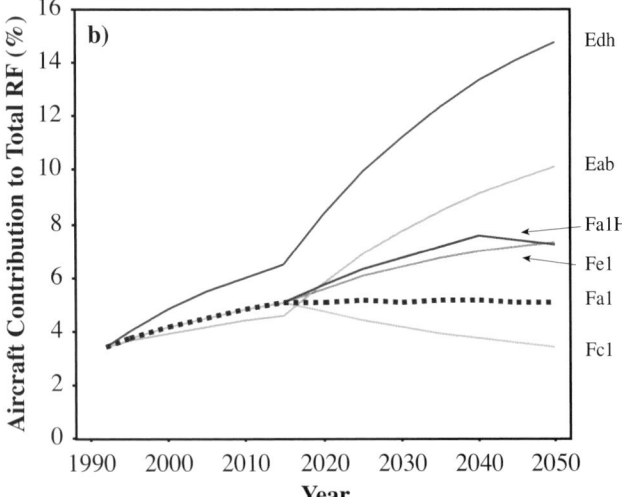

Figure 6-16:(a) Aviation fossil fuel use relative to the IS92a fossil fuel use from 1990 to 2050 for the air traffic scenarios Fc1, Fa1, Fa1H, Fe1, Eab, and Edh; and (b) from aviation as percentage of total radiative forcing (without additional cirrus).

fuel burn (CO_2-RF). If the CH_4 perturbation is tied to the O_3 perturbation as the models now indicate, then the O_3-RF and the CH_4-RF nearly cancel, and efforts at NO_x reduction may not be as effective in reducing RF as previously estimated when only ozone changes were considered. However, these opposing contributions to the radiative imbalance are very non-uniform in geographic extent, and the regional climate response may not simply disappear (see Section 6.5.1). Moreover, the prospect of large and canceling RFs, each with large independent uncertainties, greatly increases the probability of a large residual.

6.6.4. *The HSCT Option and Radiative Forcing*

The HSCT option of building a fleet beginning in 2015 and capping at 1,000 aircraft in 2040 (scenario Fa1H) has significant impact on total aviation RF, as shown in Figure 6-15c. The

addition of HSCT aircraft (+0.08 W m-2) increases RF for scenario Fa1 by a factor of 1.4, becoming comparable to RF for the higher demand scenario Fe1 (see also Figure 6-16b). The HSCT aircraft themselves cause RF of +0.10 W m-2, which is offset only by -0.02 W m-2 from displaced subsonic aircraft. This large RF from HSCT aircraft is driven mainly by RF from increased stratospheric H_2O. This single component has a large uncertainty (at least a factor of 3) associated with the calculation of the stratospheric buildup of water and the value of RF. The differential CO_2-RF penalty for HSCT aircraft is only 3% of total aviation RF. In conclusion, the best estimate of RF from HSCTs is about five times as large as RF from the subsonic fleet they replace; however, with the large confidence intervals on O_3 and H_2O, this factor has a very large uncertainty and may range from 0 to 16.

6.6.5. *Climate Change*

The ranges of aircraft perturbations to climate change examined in this report are, of course, small compared to the overall increase in RF expected throughout the next century. Climate modeling of this transient change over the next 50 years cannot readily evaluate aircraft, or any equally small RF, in terms of specific climate changes because of problems in separating "signal" from "noise." Climate parameters that might be attributed to aircraft, such as the increase in global mean surface air temperature, are calculated here using a simple model consistent with previous IS92a relationships.

The predicted change in global mean surface temperature is shown in Figure 6-17 for total warming in accord with IS92a (solid) and for the case in which all of aviation's contribution to global warming (scenario Fa1) is cut off (dotted line). Of total global warming of 0.9 K anticipated in 2050, about 0.05 K would be attributable to aviation. The Eab case anticipates a larger aviation component (0.09 K). One caveat is that aircraft

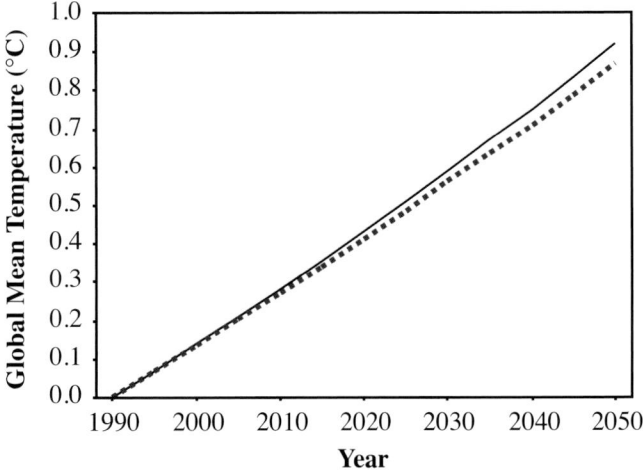

Figure 6-17:Predicted change in global mean surface temperature (K) from 1990 (defined as 0) to 2050 for the IS92a emission scenario (solid line) and for the same scenario without aircraft (Fa1, dotted line).

may produce a different climate signature, one that is not represented by an increase in global mean temperature.

To evaluate individual energy sectors as part of overall climate forcing, it is necessary to compare their summed radiative forcing from all atmospheric perturbations, not just that from their use of fossil-fuel carbon alone. The radiative forcing index—defined here as the ratio of total radiative forcing to that from CO_2 emissions alone—is a measure of the importance of aircraft-induced climate change relative to that from an equivalent sector with the same fossil fuel use but without any effect other than CO_2 (see also Section 6.2.3). In 1992, the RFI for aircraft was about 2.7, with an uncertainty of at least ±1.5. The RFI changes to 3.0 by 2015 then drops to 2.6 for the Fa1 scenario (see Table 6-1). This index ranges from 2.2 to 3.4 in the year 2050 for the various E- and F-type scenarios for subsonic aviation and technical options considered here. The RFI increases from 2.6 to 3.4 with the addition of HSCT aircraft (scenario Fa1H), as a result of the effects of stratospheric water vapor. Thus, aircraft-induced climate change with RFI > 1 points to the need for a complete scientific assessment rather than basing the climate impact on the use of fossil fuel alone. For comparison, in the IS92a scenario, the RFI for all human activities is about 1; for greenhouse gases alone it is about 1.5, and it is much higher for sectors emitting CH_4 and N_2O without significant fossil fuel use.

6.6.6.   *Aviation and Anthropogenic Change*

The overall positive anthropogenic RF today, leading toward global warming, is caused primarily by an increase in anthropogenic emissions of long-lived greenhouse gases, countered in part by short-lived aerosols. Much of this radiative forcing has built up since the industrial revolution. Emissions from aviation are a relatively new contributor to this RF, although they are potentially a growing sector. In evaluating climate change forced by aviation, as well as that by other industrial sectors, the RFI provides a useful indicator. With an RFI of about 3, aviation's role in climate change involves several important climate perturbations beyond that from its release of fossil carbon alone.

This report presents the first thorough IPCC climate assessment of any industrial or agricultural sector. It includes all known climate forcings, many of which are important and not currently represented by indices such as GWP. Comparison of aviation to other industries must await an equally thorough evaluation of the summed effects of human activities by sector that is not available in IPCC (1996) but is anticipated for IPCC's Third Assessment Report in 2001.

The range of technology options that would attempt to reduce the impact of aviation on climate did not significantly change radiative forcing by the year 2050. A lesson taken from the Second Assessment Report (IPCC, 1996, Figure 6a) is that changes in total radiative forcing, even the wide range of emissions for the IS92a-f scenarios, do not appear much before year 2050. Thus, subsequent climate assessments of 21st century

options for civil aviation need to carry out projected changes in greenhouse gases and aerosols, and all chemical and climate feedbacks, to the year 2100.

References

Boucher, O. and U. Lohmann, 1995: The sulfate-CCN-cloud albedo effect. A sensitivity study with two general circulation models. *Tellus,* **47B,** 281–300.

Brasseur, G.P., R.A. Cox, D. Hauglustaine, I. Isaksen, J. Lelieveld, D.H. Lister, R. Sausen, U. Schumann, A. Wahner, and P. Wiesen, 1998: European scientific assessment of the atmospheric effects of aircraft emissions. *Atmospheric Environment,* **32,** 2329–2418.

Chou, M.D. and M.J. Suarez, 1994: *An Efficient Thermal Infrared Radiation Parameterization for Use in General Circulation Models.* NASA Technical Memorandum 104606, Technical Report Series on Global Modeling and Data Assimilation, Vol. 3, National Aeronautics and Space Administration, Greenbelt, MD, USA, 85 pp.

Cubasch, U., K. Hasselmann, H. Höck, E. Maier-Reimer, U. Mikolajewicz, B.D. Santer, and R. Sausen, 1992: Time-dependent greenhouse warming computations with a coupled ocean-atmosphere model. *Climate Dynamics,* **8,** 55–69.

Dameris, M., V. Grewe, I. Köhler, R. Sausen, C. Brühl, J.-U. Grooß, and B. Steil, 1998: Impact of aircraft NO_x-emissions on tropospheric and stratospheric ozone. Part II: 3-D model results. *Atmospheric Environment,* **32,** 3185–3200.

Danilin, M.Y., D.W. Fahey, U. Schumann, M.J. Prather, J.E. Penner, M.K.W. Ko, D.K. Weisenstein, C.H. Jackman, G. Pitari, I. Köhler, R. Sausen, C.J. Weaver, A.R. Douglass, P.S. Connell, D.E. Kinnison, F.J. Dentener, E.L. Fleming, T.K. Berntsen, I.S.A. Isaksen, J.M. Haywood, and B. Kärcher, 1998: Aviation fuel tracer simulation: model intercomparison and implications. *Geophysical Research Letters,* **25,** 3947–3950.

Dessler, A.E., E.M. Weinstock, E.J. Hintsa, J.G. Anderson, C.R. Webster, R.D. May, J.W. Elkins, and G.S. Dutton, 1994: An examination of the total hydrogen budget of the lower stratosphere. *Geophysical Research Letters,* **21,** 2563–2566.

Dlugokencky, E.J., K.A. Masarie, P.M. Lang, and P.P. Tans, 1998: Continuing decline in the growth rate of the atmospheric methane burden. *Nature,* **393,** 447–450.

Fleming, E.L., S. Chandra, J.J. Barnett, and M. Corney, 1990: Zonal mean temperature, pressure, zonal wind, and geopotential heights as a function of latitude: COSPAR international reference atmosphere. *Advanced Space Research,* **10,** 1211–1259.

Forster, P.M. and K.P. Shine, 1997: Radiative forcing and temperature trends from stratospheric ozone changes. *Journal of Geophysical Research,* **102,** 841–855.

Fortuin, J.P.F., R. van Dorland, W.M.F. Wauben, and H. Kelder, 1995: Greenhouse effects of aircraft emissions as calculated by a radiative transfer model. *Annales Geophysicae,* **13,** 413–418.

Fuglestvedt, J.S., T.K. Bernsten, I.S.A. Isaksen, H. Mao, X.Z. Liang, and W.C. Wang, 1999: Climatic forcing of nitrogen oxides through changes in tropospheric ozone and methane Global 3D model studies. *Atmospheric Environment,* **33,** 961–977.

Gierens, K., R. Sausen, and U. Schumann, 1999: A diagnostic study of the global distribution by contrails. Part II: Future air traffic scenarios. *Theoretical Applied Climatology,* (in press).

Grant, K.E. and A.S. Grossman, 1998: *Description of a Solar Radiative Transfer Model for Use in LLNL Climate and Atmospheric Chemistry Studies.* UCRL-ID-129949, Lawrence Livermore National Laboratory, Livermore, CA, USA, 17 pp.

Grewe, V., M. Dameris, R. Hein, I. Köhler, and R. Sausen, 1999: Impact of future subsonic aircraft NO_x emissions on the atmospheric composition. *Geophysical Research Letters,* **26,** 47–50.

Grossman, A.S., D.E. Kinnison, J.E. Penner, K.E. Grant, J. Tamaresis, and P.S. Connell, 1997: O_3 and stratospheric H_2O radiative forcing resulting from a supersonic jet transport emission scenario. In: *IRS '96: Current Problems in Atmospheric Radiation* [Stamnes, K. and W.L. Smith (eds.)]. Proceedings of an international radiation symposium, Fairbanks, Alaska, USA, 19–24 August, 1996. A. Deepak, Hampton, VA, USA, pp. 818–821.

Hannegan, B., S. Olsen, M. Prather, X. Zhu, D. Rind, and J. Lerner, 1998: The dry stratosphere: a limit on cometary water influx. *Geophysical Research Letters,* **25,** 1649–1652.

Hansen, J., A. Lacis, D. Rind, G. Russell, P. Stone, I. Fung, R. Reudy, and J. Lerner, 1984a: Climate sensitivity: analysis of feedback mechanisms. In: *Climate Processes and Climate Sensitivity* [Hansen, J. and T. Takahashi (eds.)]. Geophysical Monograph 29, American Geophysical Union, Washington, DC, USA, pp. 130–163.

Hansen, J., A. Lacis, D. Rind, and G. Russell, 1984b: Climate sensitivity to increasing greenhouse gases. In: *Greenhouse Effect and Sea Level Rise* [Barth, M. and J. Titus (eds.)]. Van Nostrand Reinhold Co., New York, NY, USA, pp. 57–78.

Hansen, J., R. Ruedy, M. Sato, and R. Reynolds, 1996: Global surface air temperature in 1995: Return to pre-Pinatubo level. *Geophysical Research Letters,* **23,** 1665–1668.

Hansen, J., R. Ruedy, A. Lacis, G. Russel, M. Sato, J. Lerner, D. Rind, and P. Stone, 1997a: Wonderland climate model. *Journal of Geophysical Research,* **102,** 6823–6830.

Hansen, J., M. Sato, and R. Ruedy, 1997b: Radiative forcing and climate response. *Journal of Geophysical Research,* **102,** 6831–6864.

Harries, J.E., J.M. Russell III, A.F. Tuck, L.L. Gordley, P. Purcell, K. Stone, R.M. Bevilacqua, M. Gunson, G. Nedoluha, and W.A. Traub, 1996: Validation measurements of water vapor from the Halogen Occultation Experiment (HALOE). *Journal of Geophysical Research,* **101,** 10205–10216.

Hasselmann, K., 1993: Optimal fingerprints for the detection of time dependent climate change. *Journal of Climate,* **6,** 1957–1971.

Hauglustaine, D.A. and C. Granier, 1995: Radiative forcing by tropospheric ozone changes due to increased emissions of CH_4, CO, and NO_x. In: *Atmospheric Ozone as a Climate Gas: General Circulation Model Simulations* [Wang, W.-C. and I.S.A. Isaksen (eds.)]. Springer-Verlag, Berlin, Germany, pp. 189–204.

Haywood, J.M. and K.P. Shine, 1997. Multi-spectral calculations of the radiative forcing of tropospheric sulphate and soot aerosols using a column model. *Quarterly Journal of the Royal Meteorological Society,* **123,** 1907–1930.

Haywood, J.M. and V. Ramaswamy, 1998. Investigations into the direct radiative forcing due to anthropogenic sulfate and black carbon aerosol. *Journal of Geophysical Research,* **103,** 6043–6058.

Haywood, J.M., D.L. Roberts, A. Slingo, J.M. Edwards, and K.P. Shine, 1997a: General circulation model calculations of the direct radiative forcing by anthropogenic sulphate and fossil-fuel soot aerosol. *Journal of Climate,* **10,** 1562–1577.

Haywood, J.M., R.J. Stouffer, R.T. Wetherald, S. Manabe, and V. Ramaswamy, 1997b: Transient response of a coupled model to estimated changes in greenhouse gas and sulfate concentrations. *Geophysical Research Letters,* **24,** 1335–1338.

Holton, J.R., P.H. Haynes, M.E. McIntyre, A.R. Douglass, R.B. Rood, and L. Pfister, 1995: Stratosphere-troposphere exchange. *Review of Geophysics,* **33,** 403–439.

IEA, 1991: Table 4: World Demand by Main Product Groups, World Aviation Fuels. In: *Oil and Gas Information: 1988–1990.* International Energy Agency and Organization for Economic Cooperation and Development, Paris, France, 580 pp.

IPCC, 1990: *Climate Change: The IPCC Scientific Assessment* [Houghton, J.T., G.J. Jenkins, and J.J. Ephraums (eds.)]. Cambridge University Press, Cambridge, United Kingdom and New York, NY, USA, 365 pp.

IPCC, 1992: *Climate Change 1992: The Supplementary Report to the IPCC Scientific Assessment.* Prepared by IPCC Working Group I [Houghton, J.T., B.A. Callander, and S.K. Varney (eds.)]. Cambridge University Press, Cambridge, United Kingdom and New York, NY, USA, 200 pp.

IPCC, 1995: *Climate Change 1994: Radiative Forcing of Climate Change and an Evaluation of the IPCC IS92 Emission Scenarios* [Houghton, J.T., L.G. Meira Filho, J. Bruce, H. Lee, B.A. Callander, N. Harris, and K. Maskell (eds.)]. Cambridge University Press, Cambridge, United Kingdom and New York, NY, USA, 339 pp.

IPCC, 1996: *Climate Change 1995: The Science of Climate Change. Contribution of Working Group I to the Second Assessment Report of the Intergovernmental Panel on Climate Change* [Houghton, J.T., L.G. Meira Filho, B.A. Callander, N. Harris, A. Kattenberg, and K. Maskell (eds.)]. Cambridge University Press, Cambridge, United Kingdom and New York, NY, USA, 572 pp.

Isaksen, I.S.A., W.-C. Wang, T. Berntsen, and X.-Z. Liang, 1999: The impact on radiative forcing from aircraft emission (in preparation).

Johnson, C.E. and R.G. Derwent, 1996: Relative radiative forcing consequences of global emissions of hydrocarbons, carbon monoxide and NO_x from human activities estimated with a zonally averaged two-dimensional model. *Climatic Change,* **34,** 439–462.

Jones, P.D., 1994: Recent warming in the global temperature series. *Geophysical Research Letters,* **21,** 1149–1152.

Jones, A., D.L. Roberts, and A. Slingo, 1994. A climate model study of the indirect radiative forcing by anthropogenic sulphate aerosols. *Nature,* **370,** 450–453.

Kattenberg, A., F. Giorgi, H. Grassl, G.A. Meehl, J.F.B. Mitchell, R.J. Stouffer, T. Tokioka, A.J. Weaver, and T.M.L. Wigley, 1996: Climate models—projections of future climate. In: *Climate Change 1995: The Science of Climate Change* [Houghton, J.T., L.G. Meira Filho, B.A. Callander, N. Harris, A. Kattenberg, and K. Maskell (eds.)]. Cambridge University Press, Cambridge, United Kingdom and New York, NY, USA, pp. 285–357.

Lacis, A.A., D.J. Wuebbles, and J.A. Logan, 1990: Radiative forcing of climate by changes in the vertical distribution of ozone. *Journal of Geophysical Research,* **95,** 9971–9981.

Lacis, A. A. and V. Oinas, 1991: A description of the correlated k distribution method for modeling nongray gaseous absorption, thermal emission, and multiple scattering in vertically inhomogeneous atmospheres. *Journal of Geophysical Research,* **96,** 9027–9063.

Li, D. and K.P. Shine, 1995: A 4-dimensional ozone climatology for UGAMP models. *UGAMP Internal Report, 35.* Center for Global and Atmospheric Modelling, Department of Meteorology, University of Reading, UK, 35 pp.

Manabe, S. and R.T. Wetherald, 1967: Thermal equilibrium of the atmosphere with a given distribution of relative humidity. *Journal of Atmospheric Science,* **24,** 241–259.

Michaelis, L., 1993: Global warming impacts of transport. *The Science of the Total Environment,* **134,** 117–124.

Mitchell, J.F.B., T.J. Jones, J.M. Gregory, and S.B.F. Tett, 1995: Climate response to increasing levels of greenhouse gases and sulphate aerosols. *Nature,* **376,** 510–504.

Murphy, J.M. and J.F.B. Mitchell, 1995: Transient response of the Hadley Centre Coupled Model to increasing carbon dioxide. Part II: temporal and spatial evolution of patterns. *Journal of Climate,* **8,** 57–80.

Myhre, G., F. Stordal, K. Restad, and I. Isaksen, 1998: Estimation of the direct radiative forcing due to sulfate and soot aerosols. *Tellus,* **50B,** 463–477.

Pollack, J.B., D. Rind, A.A. Lacis, and J.E. Hansen, 1993: GCM Simulations of volcanic aerosol forcing. Part I: Climate changes induced by steady state perturbations. *Journal of Climate,* **6,** 1719–1742.

Ponater, M., S. Brinkop, R. Sausen, and U. Schumann, 1996: Simulating the global atmospheric response to aircraft water vapour emissions and contrails—a first approach using a GCM. *Annales Geophysicae,* **14,** 941–960.

Ponater, M. R. Sausen, B. Feneberg, and E. Roeckner, 1998: *Climate Effect of Ozone Changes Caused by Present and Future Air Traffic.* DLR-Mitteilung 103, Deutsches Zentrum für Luft- und Raumfahrt (German Aerospace Center), Institut für Physik der Atmosphäre, Oberpfaffenhofen and Cologne, Germany, ISSN 0943-4771, 26 pp.

Prather, M.J., 1994: Lifetimes and eigenstates in atmospheric chemistry. *Geophysical Research Letters,* **21,** 801–804.

Ramanathan, V. and R.E. Dickinson, 1979: The role of stratospheric ozone in the zonal and seasonal radiative energy balance of the Earth-troposphere system. *Journal of Atmospheric Science,* **36,** 1084–1104.

Ramaswamy, V. and S.M. Freidenreich, 1997: A new multi-band solar radiative parameterization. In: *Proceedings of the Ninth Conference on Atmospheric Radiation, February 2–7, 1997, Long Beach, California.* American Meteorological Society, Boston, MA, USA, pp. 129–130.

Rind, D. and P. Lonergan, 1995: Modeled impacts of stratospheric ozone and water vapor perturbations with implications for high–speed civil transport aircraft. *Journal of Geophysical Research,* **100,** 7381–7396.

Rind, D., P. Lonergan, and K. Shah, 1996: Climatic effect of water vapor release in the upper troposphere. *Journal of Geophysical Research,* **101,** 29395–29406.

Rind, D., R. Suozzo, and N.K. Balachandran, 1988: The GISS global climate/middle atmosphere model. Part II: Model variability due to interactions between planetary waves, the mean circulation and gravity wave drag. *Journal of Atmospheric Science*, **45**, 371–386.

Roeckner, E., K. Arpe, L. Bengtsson, M. Christoph, M. Claussen, L. Dümenil, M. Esch, M. Giorgetta, U. Schlese, and U. Schulzweida, 1996: *The Atmospheric General Circulation Model ECHAM-4: Model Description and Simulation of Present-Day Climate*. Report no. 218, Max-Planck-Institut für Meteorologie, Hamburg, Germany, ISSN 0937-1060, 90 pp.

Roeckner, E., J. Feichter, and M. Esch, 1999: Equilibrium climate responses to historical changes in the concentrations of greenhouse gases and aerosols. *Journal of Geophysical Research*, (in press).

Rossow, W.B. and R.A. Schiffer, 1991: ISCCP cloud data products. *Bulletin of the American Meteorological Society*, **72**, 2–20.

Santer, B.D., T.M.L. Wigley, T.B. Barnett, and E. Anyamba, 1996: Detection of climate change and attribution of causes. In: *Climate Change 1995: The Science of Climate Change* [Houghton, J.T., L.G. Meira Filho, B.A. Callander, N. Harris, A. Kattenberg, and K. Maskell (eds.)]. Cambridge University Press, Cambridge, United Kingdom and New York, NY, USA, pp. 407–443.

Sausen, R. and U. Schumann, 1999: *Estimates of the Climate Response to Aircraft Emissions Scenarios*. Report No. 95, Deutsches Zentrum für Luft- und Raumfahrt, Institut für Physik der Atmosphäre, Oberpfaffenhofen, Germany, ISSN 0943-4771, 26 pp.

Sausen, R., B. Feneberg, and M. Ponater, 1997: Climatic impact of aircraft induced ozone changes. *Geophysical Research Letters*, **24**, 1203–1206.

Schumann, U., 1994: On the effect of emissions from aircraft on the state of the atmosphere. *Annales Geophysicae*, **12**, 365–384.

Seinfeld, J.H., 1998: Clouds, contrails and climate. *Nature*, **391**, 837–838.

Shine, K.P., 1991: On the relative strength of gases such as the halocarbons. *Journal of Atmospheric Science*, **48**, 1513–1518.

Shine, K.P., R.G. Derwent, D.J. Wuebbles, and J.-J. Morcrette, 1990: Radiative forcing of climate. In: *Climate Change: The IPCC Scientific Assessment* [Houghton, J.T., G.J. Jenkins, and J.J. Ephraums (eds.)]. Cambridge University Press, Cambridge, United Kingdom and New York, NY, USA, pp. 41–68.

Shine, K.P., Y. Fouquart, V. Ramaswamy, S. Solomon, and J. Srinivasan, 1995: Radiative forcing. In: *Radiative Forcing of Climate Change and an Evaluation of the IPCC IS92 Emission Scenarios* [Houghton, J.T., L.G. Meira Filho, J. Bruce, H. Lee, B.A. Callander, N. Harris, and K. Maskell (eds.)]. Cambridge University Press, Cambridge, United Kingdom and New York, NY, USA, pp. 163–203.

Slingo, A., 1989: A GCM parameterization for the shortwave radiative properties of water clouds. *Journal of Atmospheric Science*, **46**, 1419–1427.

Stamnes, K., W.J. Wiscombe, and K. Jaraweera, 1988: Numerically stable algorithm for Discrete-Ordinate-Method radiative transfer in multiple scattering and emitting layered media. *Applied Optics*, **27**, 2502–2509.

Vedantham, A. and M. Oppenheimer, 1998: Long-term scenarios for aviation: demand and emissions of CO_2 and NO_x. *Energy Policy*, **26**, 625–641.

Wang, W.-C., Y. Zhuang, and R. Bojkov, 1993: Climate implications of observed changes in ozone vertical distributions at middle and high latitudes of the Northern Hemisphere. *Geophysical Research Letters*, **20**, 1567–1570.

Wang, W.-C., X.-Z. Liang, M.P. Dudek, D. Pollard, and S.L. Thompson, 1995: Atmospheric ozone as a climate gas. *Atmospheric Research*, **37**, 247–256.

Washington, W.M. and G.A. Meehl, 1989: Climate sensitivity due to increases CO_2: experiments with a coupled atmosphere and ocean general circulation model. *Climate Dynamics*, **4**, 1–38.

WMO, 1999: *Scientific Assessment of Ozone Depletion: 1998*. Global Ozone Research and Monitoring Project, World Meteorological Organization, (in press).

Wuebbles, D.J., 1996: Three-dimensional chemistry in the greenhouse – an editorial comment. *Climatic Change*, **34**, 397–404.

7

Aircraft Technology and Its Relation to Emissions

JERRY S. LEWIS AND RICHARD W. NIEDZWIECKI

Lead Authors:
D.W. Bahr, S. Bullock, N. Cumpsty, W. Dodds, D. DuBois, A. Epstein,
W.W. Ferguson, A. Fiorentino, A.A. Gorbatko, D.E. Hagen, P.J. Hart, S. Hayashi,
J.B. Jamieson, J. Kerrebrock, M. Lecht, B. Lowrie, R.C. Miake-Lye, A.K. Mortlock,
C. Moses, K. Renger, S. Sampath, J. Sanborn, B. Simon, A. Sorokin, W. Taylor,
I. Waitz, C.C. Wey, P. Whitefield, C.W. Wilson, S. Wu

Contributors:
S.L. Baughcum, A. Döpelheuer, H.J. Hackstein, H. Mongia, R.R. Nichols,
C. Osonitsch, J. Paladino, M.K. Razdan, M. Roquemore, P.A. Schulte, D.J. Sutkus

Review Editor:
M. Wright

CONTENTS

EXECUTIVE SUMMARY

In a report addressing the effects of aviation on the global atmosphere, the link between emissions and the technological status of aircraft now and in the future is clearly a central issue. The subject is complex. Our approach here, therefore, has been to identify a number of key questions, the answers to which provide an assessment of technical issues, problems, and the prospects of solving them.

The questions and their corresponding answers are as follows:

- *Question*—What are the principal technological factors that determine the nature and scale of emissions from aircraft at altitude?

  Answer—The overriding technological consideration in the design of aircraft today is safety. Given that prerequisite, aircraft are designed to provide an efficient and environmentally acceptable system of transport from ground level to the demanding conditions associated with high-speed flight at high altitudes. To achieve high efficiency, fuel consumption must be minimized by reducing the weight and drag of the aircraft. This requirement also ensures that there is a constant drive toward the highest levels of energy conversion efficiency from the engine. Together, these factors ensure that carbon dioxide (CO_2) and water outputs are minimized.

  The most fuel-efficient engines for today's aircraft are high bypass, high pressure ratio gas turbine engines. No known alternatives are in sight. These engines have high combustion pressures and temperatures; although these features are consistent with fuel efficiency, they increase NO_x formation rates—especially at high power take-off and at altitude cruise conditions.

  Current low-sulfur fuels minimize SO_x emissions. Small amounts of fuel-bound sulfur (400–600 ppm) and associated organic acids provide important lubricity properties for critical fuel system components. Processing to remove all traces of sulfur would remove important organic acids, so sulfur-free fuels are unlikely to be adopted in the short term. Sulfur removal would also result in a small net rise in CO_2.

  At present there is only limited knowledge about the formation and behavior of minor, trace species and aerosols found in the exhaust plumes of engines. Even less is known about how they are influenced by engine features and characteristics.

- *Question*—What progress has been made to date in reducing emissions, and how may new advances in aircraft and engine technology help reduce them further in the future?

  Answer—In the past 40 years, aircraft fuel efficiency has improved by 70% through improvements in airframe design, engine technology, and rising load factors. More than half of this improvement has come from advances in engine technology. These trends are expected to continue, with airframe improvements expected to play a larger role through improvements in aerodynamic efficiency, new materials, and advances in control and handling systems. New, larger aircraft with, for example, a blended-wing body or double-deck cabin offer prospects of further benefits by relaxing some of the design constraints attached to today's large conventional aircraft. Because of the very long total lifetimes of today's aircraft (up to 50 years), however, replacement rates are low, and the fuel efficiency of the whole fleet will improve slowly. Rising market demand will ensure that this trend is maintained, however.

  The intrinsic link between lower CO and rising levels of NO_x is being successfully countered with relatively simple strategies in state-of-the-art combustors. These combustors have achieved 20–40% reductions in NO_x. Consolidation of these improvements to broaden their applicability to newer, even more fuel-efficient engines demands further improvements in combustor technology. Major research programs are underway to do so.

  Although the use of hydrogen as a fuel offers a way to eliminate CO_2 and further reduce NO_x from aircraft, widespread use of hydrogen fuel presents major design problems for aircraft and would entail global changes in supply, ground handling, and storage. Hydrogen would also substantially increase water vapor emissions from aircraft. Thus, kerosene-type fuels are considered to be the only viable option for aircraft within the next 50 years (to 2050).

- *Question*—What data exist about actual emissions from aircraft? What is being done and what needs to be done to improve our understanding of and our ability to predict the scale and nature of these emissions?

  Answer—The International Civil Aviation Organization (ICAO) engine emissions databank is a substantial and growing source of reliable information that is now being used to develop aircraft emissions inventories and to analyze specific emissions. The databank—which includes information on smoke, hydrocarbons (HC), carbon monoxide (CO), and NO_x emitted during a defined landing and take-off cycle—is collated with prescribed correction procedures to guarantee consistency and comparability. There is no comparable source of data relating to sulfur compounds (SO_x) and minor trace species or aerosols from engines. Such data are emerging from individual research programs, but much more must be done to increase the breadth and depth of knowledge about the formation, nature, and scale of these potentially important aircraft emissions.

Important progress has been made in the measurement and observation of the transient behavior of minor and trace species in engines. Again, further work is needed before conclusions can be drawn that might offer clues concerning their control and reduction in future engines.

- *Question*—How are emissions from aircraft currently regulated, and how do these regulations influence emissions at altitude?

Answer—Present aircraft emissions regulations apply only to the landing and take-off cycle up to an altitude of 900 m. However, these regulations exert a controlling influence on emissions from aircraft at cruise altitudes because design changes to achieve lower NO_x at take-off are equally beneficial at medium-power cruise conditions. Methods have been developed to use the ICAO databank to predict aircraft emissions at altitude cruise conditions. These predictions are accurate to within 5–10% for a modern, high bypass ratio engine.

- *Question*—What performance might we expect from new aircraft entering the fleets in 2015 and 2050 (the dates of the scenarios discussed in Chapter 9)?

Answer—The emerging effects of research and technology programs on airframes and engines will influence future fleets of subsonic aircraft. A group of aerospace industry experts has developed some technology projections relating to fuel efficiency and NO_x emissions of aircraft by the years 2015 and 2050. According to these scenarios, average fuel efficiency of new production aircraft in the scheduled commercial fleet may improve by 20% between 1997 and 2015. The corresponding scenarios for improvement between 1997 and 2050 involved two different technology scenarios to take account of tradeoffs between fuel efficiency and low NO_x in aircraft designs. In the first case, with fuel efficiency taking priority, a 40–50% improvement in the fuel efficiency of new production aircraft was projected. In the second case, where NO_x reductions took priority, a 30–40% improvement in fuel efficiency was envisaged.

New commercial supersonic transport aircraft, operating at speeds of Mach 2 to 2.4, have been proposed for introduction into service, though not before 2015. It now seems unlikely that any commercial supersonic transports will exceed flight speeds of Mach 2.5 within the next 50 years as a result of engineering problems, materials limits, fuel efficiency, and other economic considerations. Supersonic aircraft are intrinsically less fuel efficient than subsonic aircraft. They consume about twice as much fuel, on a passenger-kilometer basis, as subsonic aircraft of the same size and range. To minimize stratospheric ozone depletion, the major design criteria for supersonic aircraft focus on flight altitude and low NO_x output. Water vapor emissions may become more important than NO_x emissions. If so, control of H_2O emissions will depend solely on the achievement of greater fuel efficiency. Sulfur aerosols originating in the fuel are an emerging concern in the altitude bands used by supersonic aircraft. As in the subsonic aircraft case, more data are needed to determine their true impact.

- *Question*—What are the likely effects of small aircraft and military aircraft on the environment?

Answer—Small aircraft, including commuter aircraft and general aviation, pose little environmental threat because they consume a very small fraction of the total of aviation fuel. Similarly, military aircraft, which consume less than 20% of the total aviation fuel supply today and, as civil aviation grows, perhaps less than 5% in the next 50 years, are seen as having potentially small environmental impacts.

7.1.   Introduction

This chapter considers how aircraft technologies influence emissions at altitude today and how that may change in the future. This introductory section sets the scene by briefly sketching the development and size of today's industry, outlines the approach the authors have adopted to their task, and describes the structure and content of the rest of the chapter.

Major advances in aircraft technology have been achieved in the past 40–50 years. Over that period, principal methods of propulsion have changed: Propeller aircraft were succeeded by jet-powered aircraft of the 1950s; these jets, in turn, were superseded by today's turbofan-powered aircraft from 1970 onwards. The fleet has expanded rapidly. So too has the capacity of jet aircraft—rising from typical 150-seat versions of the late 1950s to the largest 525-seat variant of the 747-400 aircraft in service today. The performance and capability of aircraft has also changed greatly. Cruise speeds of propeller aircraft have trebled from the 100 knots typical of the 1940s. At the start of the commercial jet age, speeds rose to 450 knots. Today's turbofans cruise at average speeds of around 500 knots, and the Concorde reaches 1350 knots. The search for efficient cruise performance with greater range, particularly for long-haul aircraft, has also resulted in higher flying aircraft—a key factor in determining where most aircraft emissions occur and their resulting impact (as discussed in Chapters 2 to 6). Average cruise altitudes for propeller aircraft rose from about 3 to 7.5 km; today's jets cruise primarily between 10.5 and 11.5 km, with some operating at up to 13 km.

Aviation designers and regulators must address complex challenges, given the operating conditions encountered by aircraft and the correspondingly stringent technical and airworthiness requirements that must be met. Factors that have been and will continue to be considered include the following:

- *Passenger safety* must be assured for all phases of aircraft operations.
- *Aircraft are more severely constrained by volume and weight considerations* than ground-based forms of transportation, placing more stringent limits on available technology choices.
- *Aircraft systems are typically more complex* than other transport modes, and many physical and chemical effects associated with them are closely coupled and interdependent. As such, changes in technology aimed at improving one aspect of performance, (e.g., a particular pollutant, or passenger safety) may have adverse effects on other aspects of performance (e.g., fuel efficiency).
- *Time scales for technology development and product life* are on the order of decades.
- *Costs to develop, purchase, and operate aircraft are high* relative to many other forms of transportation (aviation costs are typically counted in millions and billions of dollars). As with any other commercial product, the impact of technology changes on cost and customer satisfaction must be carefully assessed.

Considering the breadth and complexity of the technology base supporting today's aircraft, this chapter cannot hope to provide more than an overview of the subject. Emphasis has therefore been placed on the following key questions:

- What are the principal technological factors that determine the nature and scale of emissions from aircraft at altitude?
- What progress has been made to date in reducing emissions, and how may new advances in aircraft and engine technology help reduce them further in the future?
- What data exist about actual emissions from aircraft? What is being done, and what needs to be done, to improve our understanding of and our ability to predict the scale and nature of these emissions?
- How are emissions from aircraft currently regulated, and how do these regulations influence emissions at altitude?
- What performance might we expect from fleets operating in 2015 and 2050 and in setting the scenarios discussed in Chapter 9?

The structure and balance of the chapter have been developed to reflect the fact that advances in technology that influence the impact of aircraft on the environment fall broadly into two categories:

- Innovations that improve fuel efficiency, thus reduce the amount of fuel burned (and mass of emissions) *per passenger-km flown*
- Developments that may alter the *percentage concentration* of a particular exhaust gas (e.g., reduce NO_x for a given mass of fuel burned).

Broadly, advances that reduce the weight and drag of the aircraft fall into the first of these two categories. These advances are covered in Sections 7.2 and 7.3, which provide background material and a review of current development themes most relevant to the fuel efficiency of modern aircraft.

Engine technology is more complex. Fuel efficiency is closely linked to engine type (e.g., high bypass ratio) and choice of thermodynamic cycles (e.g., pressure and temperature ratios), but changes in the design of the engine's combustion system can also have a significant effect on the composition of the exhaust plume. These two aspects of engine design are dealt with in Sections 7.4 and 7.5.

Section 7.4 introduces the principal performance and design constraints that designers of new engines face and comments on future trends. Section 7.5 takes account of engine cycle trends on the design requirements of new low-emissions combustors. This section is a key part of the chapter because it deals with the component—the combustor—that has the greatest potential for design changes that may reduce the concentrations of some emissions that are of concern. In particular, this section addresses some of the issues raised in Chapter 2. The challenges are complex, and additional background material is included to describe the many fundamental conflicting characteristics of combustion processes that must be reconciled in emissions reduction technology programs.

Trace species in engine emissions are also considered. Potentially important physical and chemical changes to these species occur in the engine as the gas travels rapidly downstream from the combustor and through the turbine stages, where they undergo sudden changes in pressure and temperature. Section 7.6 discusses the present state of knowledge in this field.

Section 7.7 provides background information about work to date in developing the international engine emissions database. It also reports on progress in using data gathered from ground-based tests to predict corresponding cruise altitude emissions levels. These methods are used in the development of predicted inventories for future scenarios in Chapter 9.

Aircraft fuel continues to be a subject of considerable interest. Kerosene-type fuels are in widespread use today and are likely to remain so in the foreseeable future. This factor inhibits the prospect of further reducing CO_2 by changing fuels. The use of kerosene is addressed in some detail in Section 7.8, as is the question of fuel effects on emissions. Looking further ahead, Section 7.8 also considers briefly the use of alternative fuels in the longer term—beyond 2050.

The later sections of the chapter concentrate on the smaller numbers of special category aircraft in the global fleet. The first of these categories is "small aircraft" as used in the regional sector of air transport. Generally, these aircraft fly at lower altitudes than their larger counterparts and therefore have a much lower potential impact on the climate. However, growth in this sector is expected to continue; for completeness, therefore, Section 7.9 reports on the particular problems and differences of these aircraft now and in the future. Section 7.10 is concerned with the significant technical and operational issues that differentiate supersonic aircraft from the subsonic fleet. These aircraft are small in numbers now and are likely to remain so until 2015 or later. Beyond that time frame, a significant rise in the numbers of such aircraft could present a more challenging environmental problem because of the altitude at which they fly—a matter discussed in Chapter 2. This prospect has spawned research programs addressing the particular problems arising from high-speed, high-altitude operations.

Section 7.11 discusses the effects of military priorities that influence trends in the technology applied to aircraft. Although operational effectiveness in combat will continue to be the key consideration in engine design, Section 7.11 also points out why military operators' interest in the composition of exhaust is similar to that for civil engines. This section has links with Chapter 9, which shows how, in relative terms, the impact of military fleets will fall partly because of the anticipated slight reduction in their numbers but mainly because of the predicted steep rise of global civil fleets over the next 50 years.

7.2. Aircraft Characteristics

Commercial aviation has seen many technology breakthroughs over the past 40 years. Over that period, propeller-driven aircraft were replaced by jet-powered aircraft of the early 1960s, then by turbofan-powered aircraft of the 1970s to 1990s. As more powerful and fuel efficient powerplants were developed, matching baseline airframe improvements in aerodynamics and net weight reductions were also achieved. The driving forces for these improvements were, and continue to be, demand for increased range, better fuel efficiency, greater capacity, and increased speed—all of which have positive impacts on aircraft markets and economics. In many cases, these same characteristics have direct and beneficial influences on the impact of aircraft on the environment.

7.2.1. Aircraft Design: Background

Design of a subsonic transport aircraft begins by establishing its range requirements and the number of passengers it needs to carry. Economic and technical parameters have to be considered with projected market conditions to arrive at design goals. Having established these goals, the aerodynamic design can begin. One of the most important elements is the wing. Wing shape determines that lift is produced in the most efficient and stable manner for each flight mode. During take-off and landing, flaps on the leading and trailing edges of the wing are deployed to generate the extra lift required at the slower speed. As the airflow airspeed increases during the climb, these devices are retracted, and the wing assumes the optimum shape for higher cruise speeds. Forward flight generates "drag," which is manifest in several forms. One is the drag produced by the lift (induced drag). This induced drag varies directly with lift produced. Another is the resistance of the air as it flows over the outer surfaces of the aircraft (termed zero lift drag), which is independent of lift. Sub-components of zero lift drag include skin-friction drag, form drag, roughness or excrescence drag, and interference drag caused by interaction effects of various parts of the aircraft. These drag components, of course, are balanced by the thrust of the engines.

Lift and drag components also create other forces that are controlled by vertical and horizontal tail surfaces, thus enabling the aircraft to be flown accurately. These control surfaces also provide the means to trim the aircraft in level flight, minimizing control inputs during steady parts of the flight profile. In addition to these controls, sections of the trailing edge of the main wing are hinged to form movable control surfaces to control the lateral roll of the aircraft about its longitudinal axis.

All such surfaces and associated maneuvering add, in small measure, to the overall energy required to propel the aircraft forward. A detailed knowledge of the aerodynamic processes involved is therefore called for if this energy is to be minimized.

A well-established equation used in the design process is the Breguet Range Equation (Corning, 1977). The equation provides a basis for comparisons of competing designs by taking into account all of the principal variables—take-off and landing weights, thrust/fuel flow, aerodynamics and speed, as well as mission requirements and passenger load—and providing a figure of merit for the efficiency for each candidate design.

7.2.2   Aircraft Historical and Future Developments

The cruise speed of 1940s propeller-driven aircraft increased from about 100 to 300 knots over a period of 20 years, as shown in Figure 7-1 for Boeing and Douglas aircraft (Condit, 1996). At the start of the commercial jet age, at the end of the 1950s, cruise speeds were about 450 knots. The majority of turbofan-powered aircraft in today's world fleet have average cruise speeds of about 500 knots (Jane's, 1998).

Airworthiness requirements determine the range of safe operational speeds for a given aircraft type. Actual speeds for any given flight will be determined by air traffic control considerations and by individual airline performance management system techniques, the latter taking due account of the need for fuel efficiency. These operational aspects are discussed in Chapter 8.

In the period mentioned above, the cruise altitudes of propeller driven aircraft have risen from about 3 to about 7.5 km, as shown in Figure 7-2. For longer range jet- and turbofan-powered aircraft, average cruise altitudes have remained fairly constant over the past 35 years at 10.5 to 11.5 km, although, since the initial turbofan-powered aircraft were designed, there has been a slow rise in maximum cruise capability. Some aircraft can cruise up to about 13 km. Aircraft will sometimes fly below optimum design cruise altitudes for reasons associated with air traffic control or severe weather conditions (storms and clear air turbulence). Flying at lower cruise altitudes can significantly increase fuel burn (see Chapter 8) depending on range and passenger load.

The civil aircraft fleet average for speed and cruise altitude is not expected to increase significantly beyond 500 knots and 13 km over the next 50 years, as a result of physical and cost limitations.

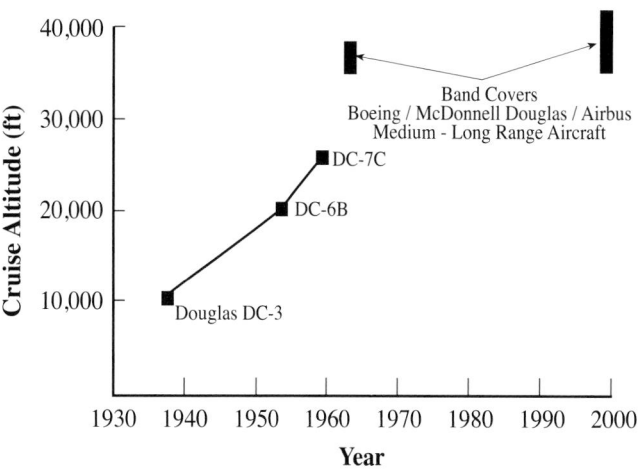

Figure 7-2: Transport aircraft cruise altitude progress.

The high cost of developing new aircraft has led to the adaptation of baseline designs to increase payload and range. Increased weight generally requires more thrust, which can mean using new engines or increasing the performance of existing engines using the "throttle push" approach (described in Section 7.5). Aircraft range has increased from the B707 era (~5,000 nm) to today's long-range (~8,000+ nm) aircraft such as the A340 and the B777. Longer range flights will need to take account of additional passenger amenities such as sleeping facilities. There is little doubt that operational pressures will ensure that existing baseline designs will continue to be developed to achieve further increases in range and payload. Average aircraft size has increased steadily over the transpacific region over the past 20 years, with an associated rise in the numbers of passengers. This route and North Atlantic routes are expected to increase in the future (see Figure 7-3) (Boeing, 1996). For domestic travel, however, shorter range designs with a larger payload can provide benefits in fuel efficiency.

7.2.3.   Time Scales from Technology Development to End of Service Life

The rates at which new aircraft designs and derivatives of current products enter the commercial fleet vary across the range of aircraft size and missions. Progress in airframe and engine technology has an important effect on the time scales over which aircraft mature; so do economic and customer requirements. The implementation of improved technology is limited, however, by the number of opportunities arising for new aircraft projects or major derivatives of existing designs. In some cases, improvements are introduced singularly. In others, combinations of improvements are introduced. Thus, in one case a new airframe might utilize an existing engine (DC-10/CF6-50), whereas in another a new engine might be applied to an existing airframe (A310/JT9D and PW4000). In some cases, changes are at the component level, the B757-300 using the RB211-535E4 LEC (low emissions combustor) engine and the A320 using the CFM 56-5B(DAC) (double annular combustor). The B777/GE90, on the other hand,

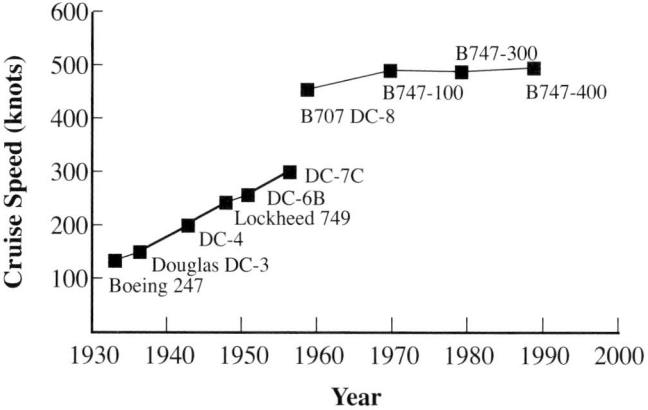

Figure 7-1: Transport aircraft cruise speed progress.

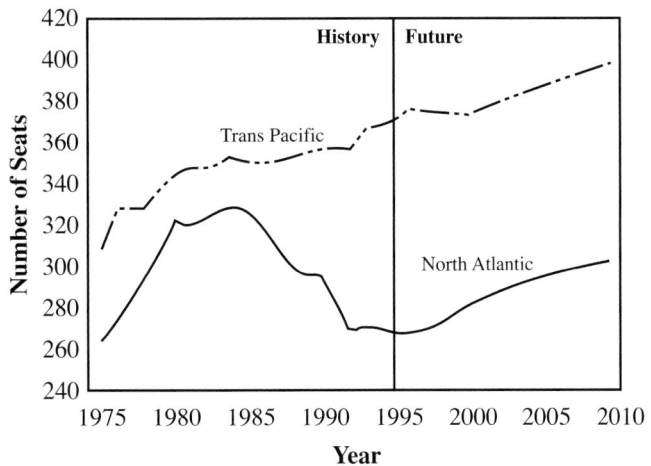

Figure 7-3: Average airplane size.

represents an example of an all new airframe/engine/combustor technology combination.

There is a considerable period of time between the start of a preliminary design of a new or derivative aircraft and the delivery of that aircraft to a "launch" airline. This period includes extensive testing and certification of new equipment. Production rates and aircraft life cycle then determine the time that the aircraft series will remain in the fleet.

A typical time-history for a medium-range commercial aircraft from technology development to the end of airline service life would be as follows:

(i)   Technology development
      preliminary/final design through
      aircraft certification testing   =   5–10 years

(ii) Successful production run   =   15–20 years

(iii) Aircraft lifetime   =   25–35 years

(iv) Total time span (i) through (iii)
      to retirement of aircraft series   =   45–65 years

(v)   Time span (ii) through (iii)
      to retirement of aircraft series   =   40–55 years

7.2.4.   Summary of Aircraft Fuel Efficiency Improvements

Significant improvements in aircraft fuel efficiency have been achieved since the dawn of the jet age in commercial aviation. Historically, these improvements have averaged 1–2% per year for new production aircraft (Koff, 1991; Albritton *et al.*, 1996; Condit, 1996). These advances have been achieved through incorporation of new engine and airframe technology. Changes have included incremental and large-scale improvements. Examined over several decades, however, they represent a

relatively steady and continuous rate of improvement. A similar trend is assumed when fuel efficiency improvements are projected forward to 2050.

Aircraft, airframe, and propulsion production fuel efficiency improvements from the 1950s to today and projected to 2015 and 2050 are summarized in Table 7-1.

Chapter 9 addresses the development of equivalent projected (2015/2050) fleet fuel efficiency improvements from production average fuel efficiencies used in emission scenarios.

Airframe and engine improvements are discussed in detail in Sections 7.3 and 7.4.

7.3.   Airframe Performance and Technology

This section outlines evolving technology of airframe components that has contributed to improvements in fuel efficiency. Corresponding discussions of engine performance and technology appear in Sections 7.4 and 7.5.

7.3.1.   Aerodynamic Improvement

Historically, efforts to improvement aerodynamic efficiency have been aimed mainly at two phases of flight: Take-off/climb and cruise. To this end, significant improvements in lift and drag performance have been achieved (Lynch *et al.*, 1996). A comprehensive range of detailed aerodynamic studies, examining all aspects of the complex flows around the airframe, has been a major part of these efforts. Such work, involving the development and use of high fidelity computational fluid

Table 7-1: *Percentage production fuel-efficiency improvements (ASK kg[-1] fuel).*

Time Period	Airframe	Propulsion	Total Aircraft
1950–1997	30	40	70[a]
1997–2015[b]	10	10	20
1997–2050	25	20	45 (40–50)[c]

[a] To date, approximately 3/7 of the total fuel efficiency improvement of 70% is attributable to advances in airframe technology.

[b] Based on improvement records to date and the discussion in Section 7.3.7, it is reasonable to expect an airframe production average fuel-efficiency improvement of ~10% by 2015. This percentage improvement is further substantiated in other reference material (Greene, 1995). Similarly, a 10% propulsion production average fuel-efficiency improvement is considered feasible in this time frame.

[c] In the longer term (2050) compared to 1997, a total aircraft production average fuel-efficiency improvement of 40–50% is considered feasible (ICCAIA, 1997g). These levels of efficiency improvement are assumed in the 2050 technology scenarios described in Chapter 9. The ratio of airframe to propulsion production average fuel-efficiency improvement over the period 1997 to 2050 is projected to be 55/45 in favor of airframe technology developments. This is equivalent to a 25% airframe fuel-efficiency improvement.

dynamics (CFD) prediction codes (Rubbert, 1994) coupled with improved wind tunnel testing techniques (Lynch and Crites, 1996), has led to much better understanding of the aerodynamic characteristics of new and proposed designs. In turn, this work has led naturally to improved predictions of the effectiveness of measures aimed at improving the performance of aircraft in general and reducing fuel burn rates for future aircraft in particular.

7.3.2. Airframe Weight Reduction

The increasing availability of advanced lighter and stronger materials for use in structural components of the airframe has also been a major factor in the achievement of reduced fuel burn. Of particular note are the greater use of new aluminum alloys, titanium components, and composite materials for secondary (non–load-bearing) structures.

One of the important enabling technologies that has had a major impact on these developments is high-fidelity finite element models (FEMs). FEMs are now extensively used for strength analyses and to obtain better understanding of safety load-factor margins. This work has already contributed to additional reductions in structural weight.

7.3.3. Nacelle Efficiency

As engine bypass ratios (fan bypass airflow divided by engine core flow) have risen over the past 2 decades, so too have the drag and weight of the nacelle (aerodynamic casing surrounding the engine). Furthermore, integration of the engine and the nacelle—which incorporates the air inlet, the engine, and the exhaust nozzle—can be a source of significant interference drag problems. On balance, however, high bypass ratio engines have provided a significant gain for transport aircraft in terms of reduced fuel requirements for a given mission. This development has led to greater performance flexibility for operators wishing to optimize range and payload, hence take-off weights, compared with earlier low bypass ratio engines. Improvements in the aerodynamics of engine-nacelle flows and changes to the shape and length of the inlet section continue to reduce local drag effects and increase efficiency. The current trend is toward higher bypass ratio ducted fan engines having shorter and thinner lip inlets. This approach may be limited in the future, however, by the need to meet more stringent noise regulations. The development of lighter nacelle materials/structures has reduced operating empty weight (OEW). Increasing thrust reverser efficiency for enhanced landing performance can also reduce nacelle package weight.

7.3.4. Propulsion/Airframe Integration (PAI)

Reduction of interference drag caused by flow interactions in the region of the wing-pylon-nacelle during take-off/climb/cruise conditions is a complex design problem (Berry, 1994).

Recent improvements in modeling localized airflow, using CFD, have brought important benefits in terms of reduced interference drag (Lynch and Intemann, 1994). There is an inevitable tradeoff between the higher drag of high bypass ratio engines and the need to minimize interference drag for a given mission fuel burn; a great deal of effort is aimed at achieving an optimum balance. For example, if the nacelle can be located closer to the wing without creating interference penalties, it is possible to reduce pylon weight and drag and reduce landing gear height (and weight). Other tradeoffs, such as noise impacts, also need to be considered.

7.3.5. Control Systems

Older technology aircraft use mechanical, hydraulic, and electrical systems to control flight, propulsion, and environmental systems. Today's modern airframes and airframes under design utilize much lighter fly-by-light (using fiber optics) and fly-by-wire technology, with significant savings in OEW.

Changes to aircraft pressurization and air conditioning systems—particularly increases in the amount of air, which is now recirculated—has reduced engine bleed flow requirements. These measures have significantly reduced engine fuel burn at cruise conditions. Cabin air quality requirements, however, might limit these methods of achieving further fuel savings.

More detailed analysis of PAI/high-lift system interference is regarded as a way to achieve weight reductions in low-speed/take-off drag. Again, CFD techniques are invaluable tools in achieving such improvements. The design of a high-lift system that can provide the same lift versus drag performance at a lower weight is seen as another path towards overall aircraft system improvements that would result in fuel savings.

Increasing use of databus (multiplexing of signals) technology has led to significant reduction in the amount of wiring needed to support the numerous advanced electrical systems in modern aircraft. Although increased wire shielding has become necessary, the overall result has led to further reductions in airframe OEW.

7.3.6. Operational Efficiencies by Design

Avionics improvements have improved navigation accuracy and made more fuel efficient flight paths possible. Chapter 10 deals with this subject in some depth.

Regulatory changes such as the addition of extended twin operations (ETOPS) rules have made it possible for today's highly efficient and reliable twin-engine aircraft to be used on routes that were previously prohibited to them. These routes have larger airfield division distances; hence, a shorter flight distance track can be achieved, which reduces fuel consumption.

7.3.7. *Advanced Future Technologies*

This subsection considers some of the advances being made in aerodynamic-related fields of study. Advances in these areas become candidates for gradual adoption into derivatives of existing production aircraft and the next generation of airliners, as shown in Figure 7-4. Some concepts, such as improved wing tip devices and smoother surface areas, can be considered for derivatives of existing designs. Advanced weight reduction technologies, aircraft control systems, and airframe concepts are also discussed.

7.3.7.1. *Laminar Flow Concepts*

Smooth laminar flow over a body creates less drag than turbulent flow. However, it is difficult to achieve and depends on a number of factors, particularly the shape and surface of the body. Current aircraft designs generate varying degrees of turbulent flow. Passive control concepts that encourage laminar flow are being explored. These concepts include slotted airfoils or actively heated/cooled surfaces, but the benefits still need to be proven. If wing-mounted prop-fan (un-ducted powerplants— see Section 7.4.3.) propulsion technology were to be adopted in the future, laminar flow airfoils that could tolerate the effects of propeller efflux over the wing surface would need to be

developed. Alternative mounting arrangements, such as aft fuselage-mounted prop fans, may also be considered.

Laminar flow suction systems for wing, fuselage, stabilizers, and nacelles have been and continue to be reviewed and evaluated. Development of these systems, which aim to keep the flow attached (laminar) to aerodynamic surfaces by sucking ambient air through porous skins, is a high-risk technical challenge that is likely to require a longer time frame for full development and airline introduction (after 2015). A key consideration is the weight of the laminar flow systems (and their power requirements) compared with savings from drag reduction over the complete mission. Contamination of the porous skin surface by insects/debris can significantly reduce the performance of laminar flow systems and increase maintenance cost. Work in this field to date has not reached the point where these penalties, together with the effects of system failure or other risks, have been fully evaluated and balanced against fuel savings.

7.3.7.2. *Other Aerodynamic Improvements*

Other potential aerodynamic improvements requiring further development and investigation include attachment of riblets (tiny groves in the direction of airflow) to the fuselage,

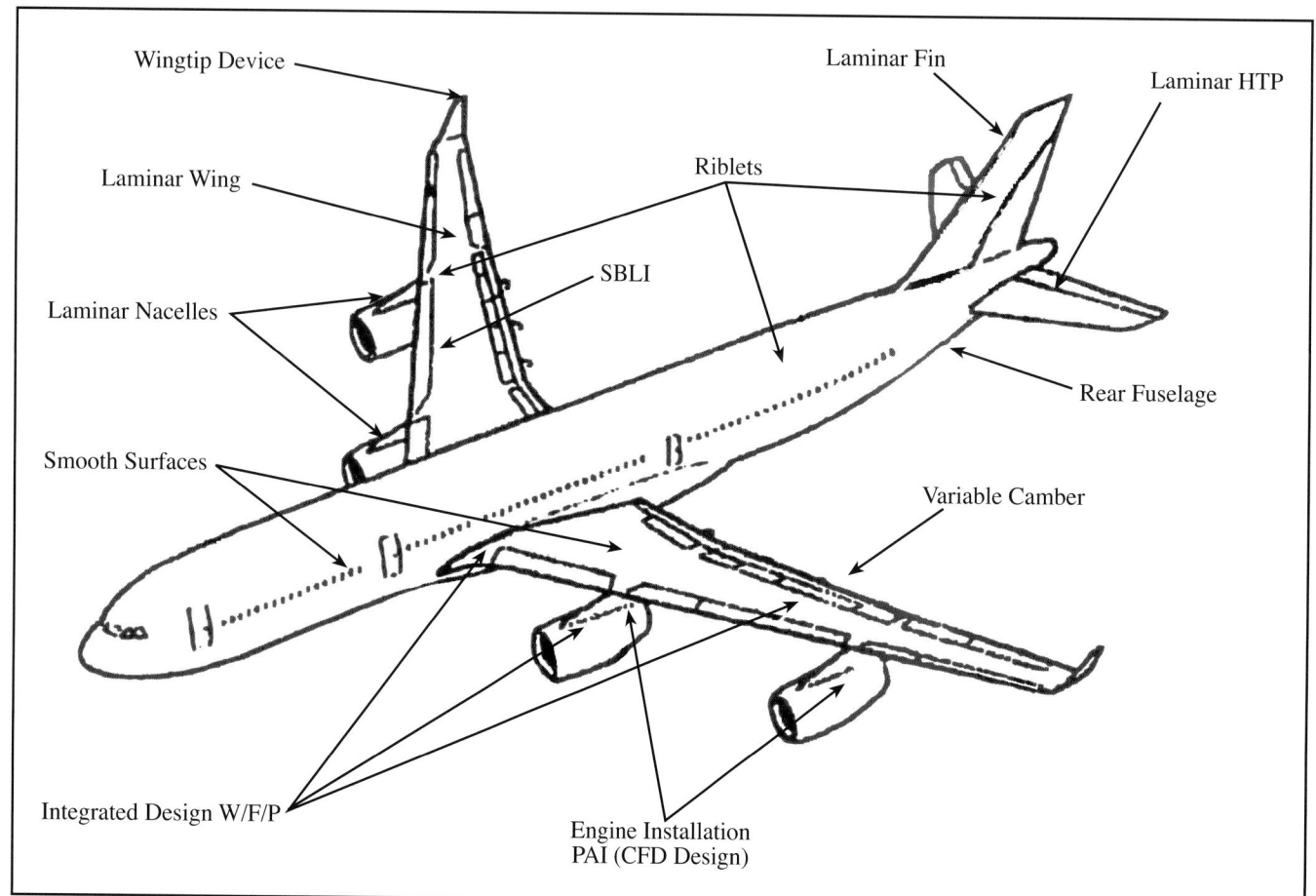

Figure 7-4: Further potential aircraft drag reductions.

wing, and horizontal tail to reduce turbulent flow areas; advanced passive flow control devices (e.g., vortex generators) to enhance lift; advanced winglets on outboard wings; supercritical wing technology to enhance and optimize cruise lift/drag ratio; advanced CFD design methodologies; and advanced manufacturing methods to improve fuselage and wing surface smoothness to reduce drag.

7.3.7.3. Weight Reduction

It is expected that the weight of the airframe structure will continue to decrease through gradual incorporation of improved aluminum alloys and aluminum-lithium composites for sections of primary structures (i.e., fuselage, wing, and empennage), and composites for secondary structures. For primary structures, the process of introduction is slow because of the certification process for structural design, material property characterization, and safety issues, which involve lengthy and costly durability and strength test programs.

Thrust reversers enhance landing performance, especially during wet runway conditions. Significant weight reductions could be obtained by removing them from some aircraft configurations. Estimates indicate that maximum take-off gross weight could be reduced by about 0.3–1%, depending on aircraft configuration and size. Removal of thrust reversers could also improve internal nozzle flow characteristics (e.g., reduced internal thrust losses). This matter needs further study.

Further weight reduction could be achieved via reduced passenger amenities, such as the elimination of windows, in-flight entertainment, and galleys or reduction in seat pitch. These measures may be more applicable to short-range routes. The weight of in-flight entertainment systems is likely to be reduced in the future by technology. However, it is questionable whether such changes would ever be accepted by fare-paying passengers. Increasing demand for passenger comfort items such as flight entertainment systems may also limit changes. Interior cabin furnishings and passive interior noise treatment (e.g., wall bags/environmental control ducts) for cabin noise control may be reduced in the future if active noise control technology is successfully developed for attenuation of broadband and tonal noise sources.

Estimates of weight reductions accruing from successful implementation of these strategies, applied to a medium-range, wide-body aircraft, suggest that 2,000 kg of OEW might be saved. This weight reduction represents approximately a 1% fuel efficiency improvement.

7.3.7.4. Aircraft Systems

"Aircraft systems" is a generic term applied to the large number of subsystems used in a modern aircraft to manage the aircraft in flight. All of these systems offer room for improvements that could reduce fuel burn. It is estimated, for example, that extending the capability of fly-by-wire control systems to include active

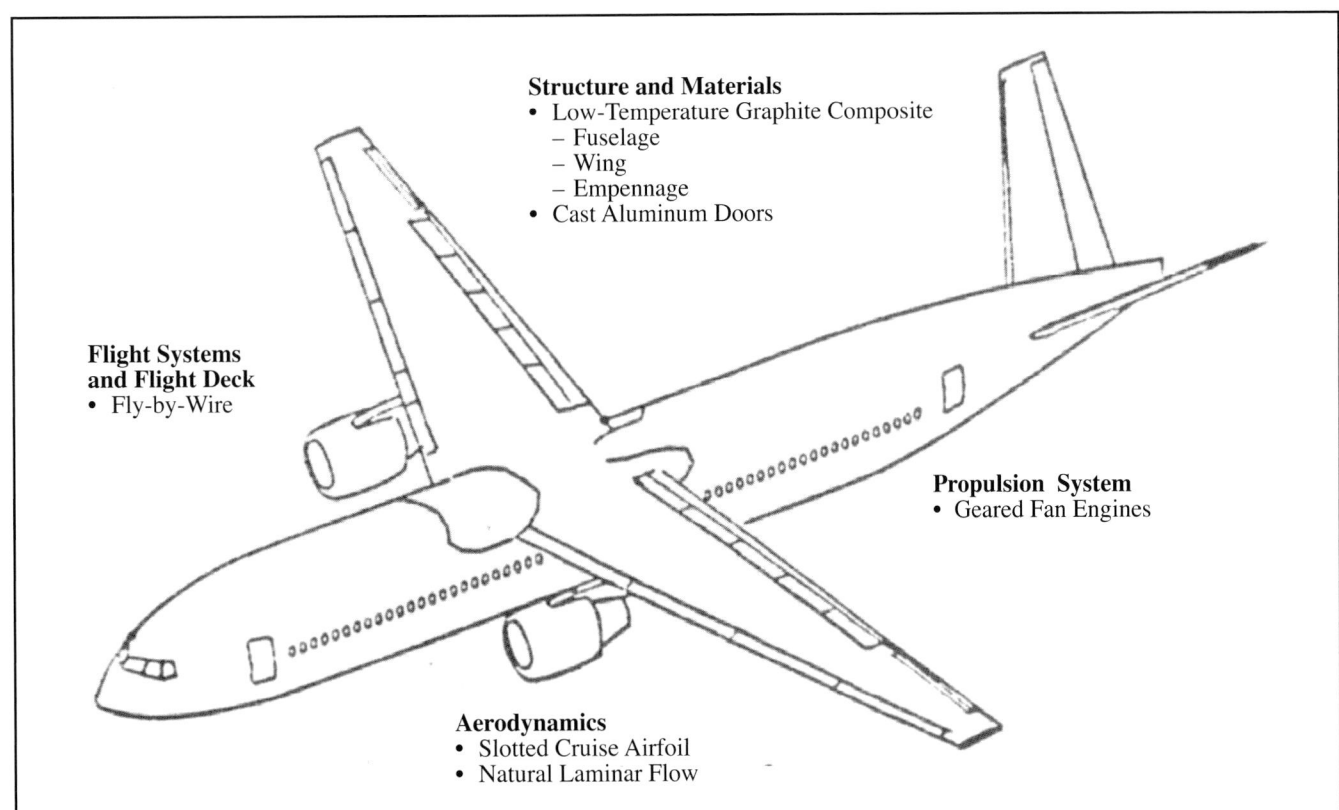

Figure 7-5: 2016 subsonic airplane.

pitch stability augmentation and wing load alleviation offers the potential for a 1–3% improvement in overall fuel efficiency. Development of an "all electric" airplane, which also deletes the current use of air bleeds from the engines for pneumatic and anti-icing secondary power, has the potential to save fuel during cruise. The use of advanced technology fuel cells to replace the auxiliary power unit (APU) could provide savings in overall local/ground-level fuel burn and emissions, with the added benefit of reduced noise near terminals. In some instances, however, the cost and complexity required to overcome failure modes of these systems is high, which may inhibit or delay their use in commercial service. The use of active center-of-gravity control is another potential means to improve fuel efficiencies at cruise conditions. Flight safety improvements could increase OEW.

7.3.7.5. Advanced Airframe Concepts

Aerodynamic efficiency improvements such as higher lift/drag ratio (e.g., slotted cruise airfoil and natural laminar flow), new structural materials, and control system advances (such as fly-by-wire) could collectively improve fuel efficiency by about 10%, compared to current production aircraft. An aircraft representing some of these nearer term (2016) advanced airframe technologies is shown in Figure 7-5 on the previous page (Condit, 1996).

At the upper end of the airframe size scale (> 600 passengers), a more futuristic concept approach such as a blended-wing body (BWB) could be developed. A plan view size comparison between an MD-11, BWB, and conventional 800-passenger aircraft is shown in Figure 7-6 (Liebeck, *et al.*, 1998). Studies have assessed the potential of the BWB design. The advantage of the BWB over conventional or evolutionary designs stem from extending the cabin spanwise, thereby providing structural

and aerodynamic overlap with the wing. This design reduces the total aerodynamic wetted area of the airplane and allows a higher span to be achieved because the deep and stiff centerbody provides "free" structural wingspan. Relaxed static stability allows optimum span loading. If engine and structural material technologies remain the same for the BWB, initial estimates show that fuel burn could be reduced significantly relative to that of conventionally designed large transports (Liebeck *et al.*, 1998). Other large transport configurations are being evaluated (McMasters and Kroo, 1998) and compared to current designs.

In addition to the fuel burn and emissions reduction potential of this concept, the engine installation and airframe can help to minimize exterior noise: Inlets are placed above the wing so fan noise is shielded by the vast centerbody.

Validation of potential fuel burn benefits will require extensive full-scale testing. The principal challenges lie in the overall structural integrity of the oval pressure vessel, integration of propulsion and airframe, emergency egress (evacuation of passengers on land and water), passenger acceptance, and airport compatibility. An initial BWB concept could enter service after the year 2020. However, the passenger size and range of the initial design is not known at this time.

7.4.   Engine Performance and Technology

For the past 50 years, the principal propulsion source for military and civil aircraft has been the gas turbine. For a variety of technical reasons, this situation is likely to continue into the foreseeable future. The earliest military aircraft gas turbines, developed toward the end of World War II, opened the way to high-speed flight by providing high power from low weight, compact engines. For military use, these "jet" engines offered an escape from the aerodynamic limitation of the propeller,

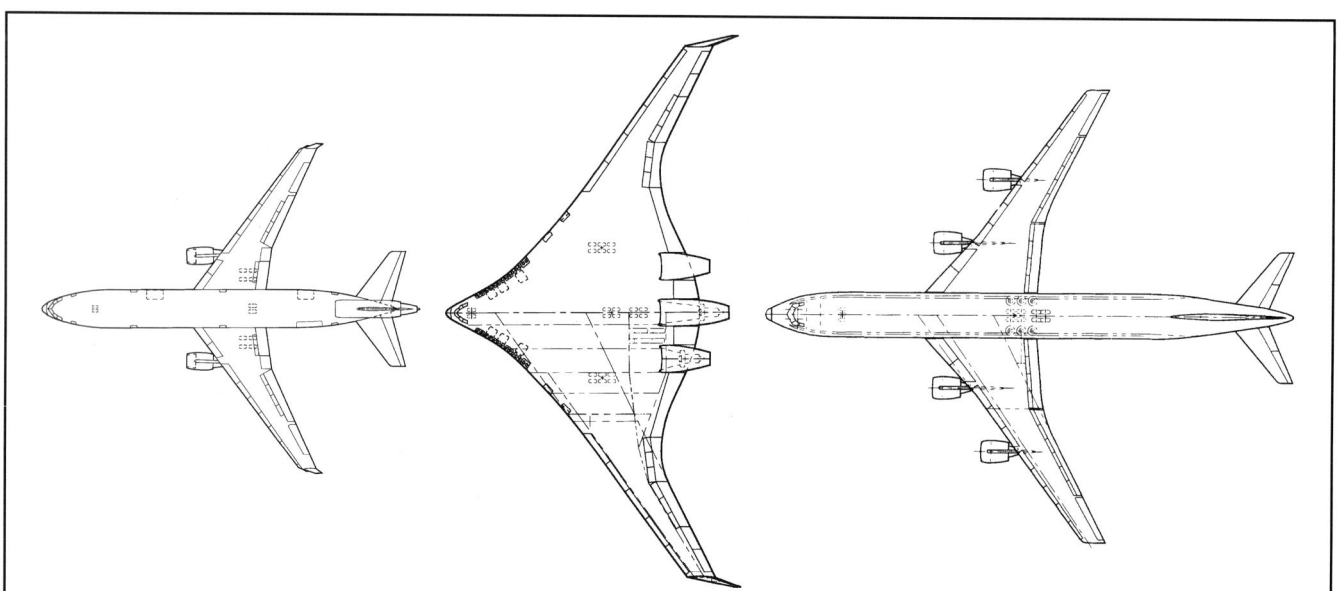

Figure 7-6: MD-11, blended-wing body, and conventional planview size comparison.

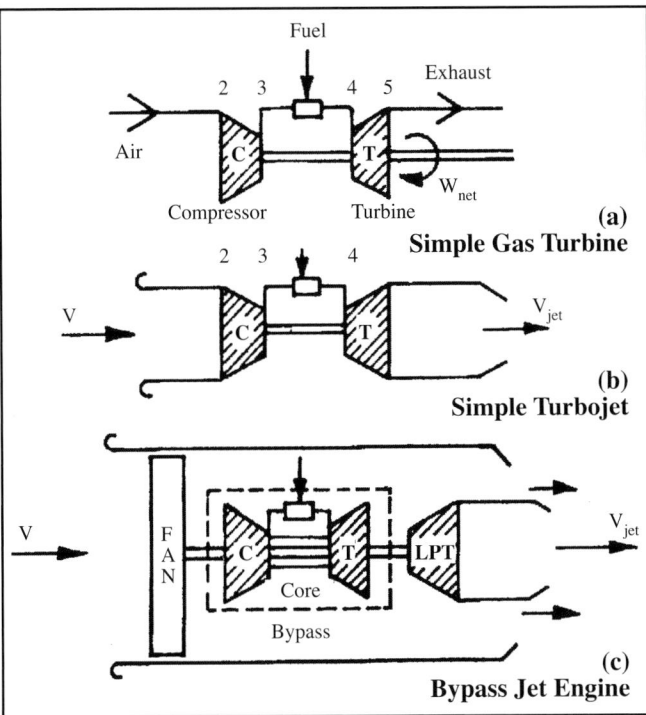

Figure 7-7: Gas turbine schematics.

despite their relatively low thermodynamic efficiency at the time. Early civil aircraft gas turbines continued, at first, to use the engine to drive a propeller in so-called turbo-prop form (engine shown in Figure 7-7a). Several types of turbo-props are still used for short-haul operations where cruise speeds are less important, but larger and faster aircraft have dispensed with the propeller.

Since those early days, huge strides have been made in the critical and pacing technology fields that influence the key design and performance characteristics of engine design. In particular, major advances have been made in the fields of turbo-machinery aerodynamics, combustion, turbine blade cooling, and materials. For military engines, these advances have been realized mainly in increases in the ratio of engine thrust to engine weight. For civil aircraft engines, the benefits have led to high bypass ratio engines with substantially lower fuel consumption, which have contributed to the rapid growth in air transportation over the past 3 decades.

It is common to compare engines in terms of specific fuel consumption (SFC), which is the fuel flow rate per unit thrust at cruise. However, the ultimate goal is to minimize total fuel burned per unit payload, rather than SFC; this computation involves engine weight, installation drag, and their effect on the total fuel required to complete a flight mission. For subsonic transport aircraft, the weight of the engines is on the order of 10–15% of the empty weight of the aircraft; a reduction of one unit of total engine weight translates to a reduction of between 1.5 and 4 units of aircraft empty weight, depending on the design. The relatively larger reduction in aircraft weight derives from concomitant reductions in requirements for supporting structure. The benefits are further magnified by the

fact that the reduction in fuel burn attributable to engine weight savings is proportional to increasing aircraft range.

The following subsections present a brief outline of engine performance issues from an historical perspective (7.4.2) and a look forward into the future (7.4.3). The link between performance considerations and emissions is discussed in Section 7.4.4. Because the gas turbine is expected to remain the principal power source for aircraft propulsion well into the future, however, Section 7.4.1 presents a simplified review of the engine's fundamental principles to explain, among other things, why there is no reasonable alternative to the gas turbine or derivatives thereof in the foreseeable future.

7.4.1. *Fundamental Thermodynamics*

The core of the basic gas turbine consists of three essential elements: The compressor, which mechanically increases the energy of the air (raising the pressure and temperature); the combustor, in which fuel is burned (further raising the temperature of the pressurized air); and the turbine, which mechanically extracts enough energy from the hot compressed gas to drive the compressor (thereby reducing the pressure and temperature of the gas). A fraction of the net energy remaining in the gas after it leaves the turbine is then available to be used in different ways, as shown in Figure 7-7. Case (a) uses an additional turbine stage to mechanically convert the energy to shaft work (such as might be required to drive a propeller or electric generator); case (b) is the turbojet, in which a nozzle is used to accelerate the gas (converting some of the energy into kinetic energy), producing a high-speed jet that can be used to propel the vehicle; and case (c) is the turbofan, in which a further turbine converts most of the energy of the gas into shaft work to drive the bypass air compressor, thereby producing the bypass jet that propels the vehicle. For aircraft applications, the weight of the engine must be low in relation to the power output. This constraint has kept the aircraft type of gas turbine simple—much simpler than those now being built for land-based power generation. (The implications of going to more complicated configurations are considered briefly in Section 7.4.4.)

The overall efficiency of an aircraft engine, η_0, is the mechanical power created by the thrust divided by the energy input rate of the fuel flow. It is convenient to express overall efficiencies by $\eta_0 = \eta_{therm} \times \eta_p$, where η_{therm} is the thermal efficiency and η_p is the propulsion efficiency; these terms are considered below.

7.4.1.1. *Thermal Efficiency*

One of the most fundamental measures needed to assess the effectiveness of a gas turbine is its ability to convert the chemical energy of the fuel into mechanical work—the thermal efficiency of the engine (i.e., η_{therm} = power to the gas stream/ energy input rate). The precise derivation of this parameter requires a knowledge of the details of flows within the engine, including air used inside the engine to cool hot parts. For present purposes, however, a good estimate of the constraints and potential of the engine can be obtained by neglecting such minor flows and by further assuming that the working fluid throughout the engine retains the same properties as air at room temperature throughout the cycle. The compressor and turbine are inevitably less than perfect, and their performance is described by efficiencies (η_{comp} and η_{turb}, respectively). This simple treatment assumes that $\eta_{comp} = \eta_{turb} = 90\%$; these are plausible "state-of-the-art" values. The two properties that affect gas turbine thermal efficiency are T_4/T_2—the ratio of temperature leaving the combustor (i.e., the turbine inlet temperature) to that of the air entering the engine—and the overall pressure ratio between atmospheric pressure and peak pressure within the engine.

Taking first the effects of variations in T_4/T_2 at typical conditions in the cruise phase of a modern civil aircraft (Mach 0.85 at 10.7-km altitude), the inlet temperature, T_2, sensed by the engine for standard atmospheric conditions is 250 K. The upper limit on temperature, T_4, would be reached if sufficient hydrocarbon fuel were burned to consume all the oxygen (stoichiometric combustion; see Figure 7-8). This temperature would be about 2590 K, though this value depends to some extent on the pressure ratio of the engine and the efficiency of the components. These values give $T_4/T_2 \approx 10.3$. For a mid-1990s engine the value of T_4 at cruise is about 1400 K, whereas in the mid-1970s a value of about 1250 K would have been more typical. For these combinations of air inlet temperature, and turbine inlet temperature the ratio T_4/T_2 would be 5.6 and 5.0, respectively. Figure 7-8 shows, for these three examples, the thermal efficiency of a gas turbine as a function of pressure ratio for cruise at Mach 0.85 at 10.7 km. The solid lines are for compressor and turbine efficiencies of 90% and are labeled with the temperature ratio T_4/T_2. The broken line is for ideal compressors and turbines (having $\eta_{comp} = \eta_{turb} = 100\%$); in this case, the thermal efficiency is a function of pressure ratio only.

Figure 7-8 also shows that for high thermal efficiency, the internal pressure ratio of the engine must be moderately high; for new large engines, the value is around 40 at peak throttle settings of take-off and climb. In addition to "internal"

compression of air (by the compressor), the aircraft's forward speed introduces a pressure rise of 1.6 times atmospheric pressure at Mach 0.85, giving an overall pressure ratio of about 64. Increasing the internal pressure ratio much beyond this value raises the temperature at compressor exit to the point where materials are a limitation; furthermore, the higher temperature increases the problem of cooling the turbine and increases levels of NO_x production in the combustor (as discussed in Section 7.5.2). Increasing the internal pressure ratio at current values of T_4/T_2 would offer very little improvement in thermal efficiency.

The increase in turbine inlet temperature over the past 20 years has been the result of research and development, primarily to increase the power-to-weight ratio of the engine. As Figure 7-8 shows, however, a modest increase in T_4/T_2 has had an effect on thermal efficiency. At the theoretical limit, by operating at stoichiometric fuel-air ratios in the combustion system, the temperature ratio would be about double that of current engines. The practical difficulties associated with achieving stoichiometric combustion (largely associated with maintaining structural integrity with materials that lose strength at high temperatures), however, remain formidable. Furthermore, even if these difficulties could be overcome, the thermal efficiency of the engine would still remain below 60% given current levels of turbomachinery component efficiencies. Only when $\eta_{comp} = \eta_{turb} = 100\%$ would thermal efficiency approach 70% for an overall pressure ratio of around 64. Thus, although there are clear gains to be made by further increases in T_4/T_2 and thermal efficiency, the full benefits can be achieved only through parallel efforts aimed at raising propulsive efficiency.

7.4.1.2. Propulsive and Overall Efficiency

The power produced by engine thrust is the product of thrust and flight velocity (V). The ratio of this useful power to the increment in kinetic energy given to the flow in passing through the engine is the propulsive efficiency (η_p); a good approximation of this variable may written in the form $\eta_p = 2V/(V + V_j)$, where V_j is the jet velocity.

High levels of thermal efficiency require the T_4/T_2 ratio to be as high as possible with appropriate pressure ratio and component efficiencies. In the case of a simple turbojet gas turbine engine, this requirement would mean that the jet velocities are also relatively high. For example, for a ratio of $T_4/T_2 = 5.6$ together with an engine pressure ratio of 40, the jet velocity V_j would be about 817 ms[-1], (Cumpsty, 1997). At Mach 0.85 at 10.7 km, when the flight velocity V is 252 ms[-1], the relationship shown above gives a propulsive efficiency of only about 47%. Thermal efficiency at this condition is about 48%, so $\eta_o = \eta_p \times \eta_{therm} = 0.48 \times 0.47 = 23\%$ for the assumptions of this simplified example. For typical aircraft, overall efficiency ranges between 20 and 40%.

The most practical method of raising overall efficiency is to lower the jet velocity and thereby increase propulsive efficiency;

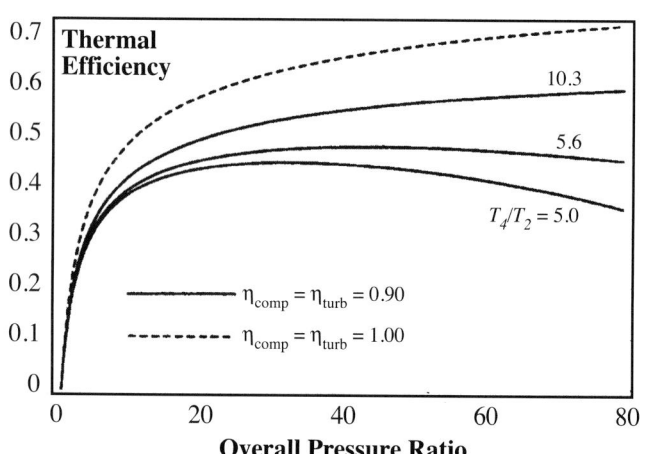

Figure 7-8: Gas turbine thermal efficiency.

this approach has been adopted in the bypass engine used so widely today. In this engine design, hot gases leaving the core turbine expand through further turbine stages, which drive the fan mounted in front of the core compressor. In most modern engines, the pressure ratio across the fan at cruise condition is about 1.6, giving a bypass ratio of about 6 and a jet velocity of about 400 ms^{-1}. At this jet velocity, propulsive efficiency is about 77% at a cruise speed of Mach 0.85 at 10.7 km. Unfortunately, losses associated with the inefficiency of the fan and the turbine driving it inevitably reduce these benefits somewhat, so a typical value for the overall efficiency of such an engine is currently about 30 to 37% at cruise. Increasing the bypass ratio clearly offers the prospect of further increases in propulsive efficiency, but this approach has to be weighed against the penalties of increased size and weight of the installed engine and associated changes in drag (see Section 7.2 for further discussion of airframe efficiency and installation effects). Other prospects for increasing propulsive efficiency (e.g., propellers and unducted fans) are discussed in Section 7.4.3.

This brief review of theoretical considerations summarizes thermodynamic and aerodynamic constraints for engine designers in the continuing drive toward more efficient engines for current and future requirements. Although today's most advanced engines have bypass ratios in the range 5 to 9, there will be efforts to continue to try to increase them. Despite their attractions, however, bypass ratios much in excess of 9 are likely to require a gear box between the power turbine and the fan, or a novel configuration (a topic taken up further in Section 7.4.3), as well as imposing installation and weight problems.

7.4.2. *Historical Trends*

The foregoing discussion briefly reviewed the ways in which basic engine cycle considerations can influence design trends of gas turbines. Practical confirmation of such trends may be obtained by looking at historical trends of principal parameters affecting the performance of gas turbines since they were first introduced. Economic and aircraft range considerations have been uppermost in engine designers' minds. Figure 7-9 clearly

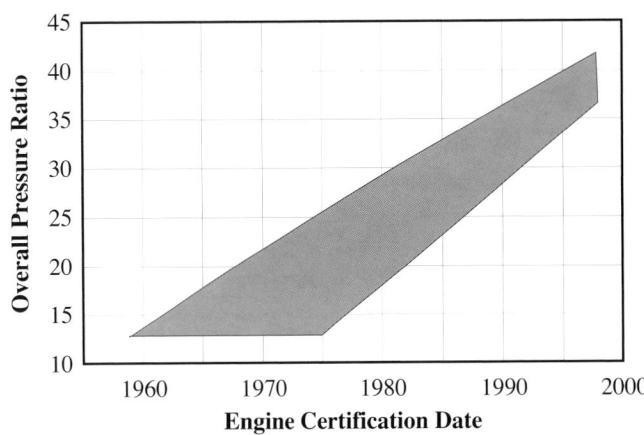

Figure 7-10: Overall pressure ratio trend with time.

shows the impressive progress made in reducing thrust-specific fuel consumption (mass flow rate of fuel burned per unit of thrust) with time. The engines of 1960 to 1970 vintage were either turbojets or first-generation low bypass ratio turbofans (Figure 7-7) with relatively high levels of fuel consumption. The period from 1970 to the mid-1980s saw the introduction of second-generation turbofan engines, which are generally referred to as high bypass ratio engines, which had significantly better fuel consumption than the earlier engines. Improvements in fuel consumption for third generation engines were smaller.

Important improvements in the understanding of complex aerodynamic flows within turbomachinery have been achieved over the past 25 years through mathematical modeling and parallel advances in experimental techniques. Figure 7-10 provides clear evidence of the benefits of that work in the rising trend of overall pressure ratio with time. In line with the fundamental relationships presented in Section 7.4.1, this trend has contributed significantly to the reduced fuel consumption shown in Figure 7-9.

Parallel work in the fields of combustion technology and materials have contributed to increasing levels of peak cycle temperatures—as shown in Figure 7-11, in which the trend of combustor exit temperature is plotted against the same time scale. (Note that this is the maximum temperature, which occurs at take-off. Temperatures are generally lower at cruise.) This increasing temperature trend has contributed not only to the improved fuel consumption trend shown in Figure 7-9—by virtue of improved thermal efficiency—but also to the rise in thrust-to-weight ratio, leading to higher payload for the same overall aircraft weight. The impact of increasing temperatures on various pollutant emissions is described in Section 7.5.

7.4.3. *Future Development Paths for Aircraft Engines*

Figure 7-12 provides a perspective on the future based on historical trends. It complements the points made in Section 7.4.2 and relates to the fundamental thermodynamics of engines described in Section 7.4.1.

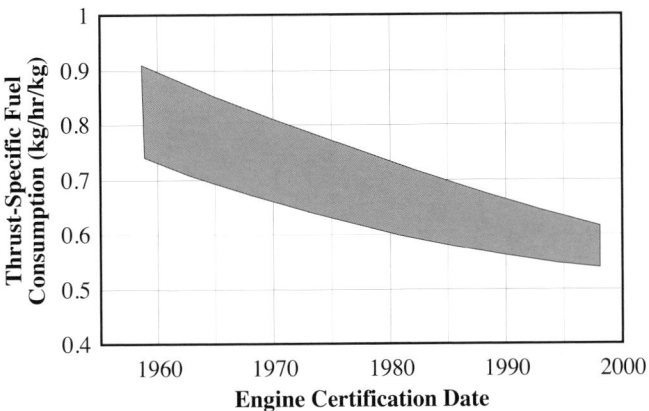

Figure 7-9: Fuel consumption progress with time.

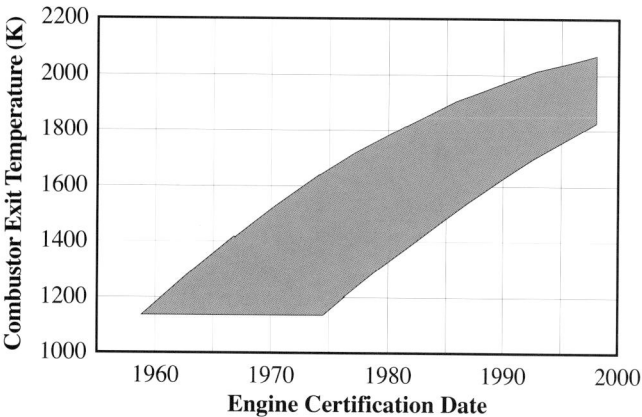

Figure 7-11: Combustor exit temperature trend with time.

Figure 7-12 shows that the future path of engines with rising overall efficiency requires improvements in thermal and propulsive (including transmission) efficiency. In fact, relatively few options are possible for improving the propulsive efficiency of current subsonic aircraft. Of these options, the most practical is to increase the engine bypass ratio, which means enlarging the diameter of the propulsor. Ducted propulsors with bypass ratios up to about 15 have been demonstrated, but they incur drag and installation penalties. Bypass ratios above about 10 generally require the addition of a gearbox to the powertrain, but gearboxes combining required high power capabilities, transmission efficiency, weight, and long-life characteristics have yet to be demonstrated in service. Plans for the introduction of such an engine (with a bypass ratio of 11) have recently been announced, however. Above a bypass ratio of about 15, the duct must be removed to contain weight and drag. Unducted propulsors resembling large, swept propellers, with bypass ratios of more than 30, have been flight tested and offer gains in propulsive efficiency of about 0.15 (i.e., a gain of about 25%) compared to modern transport engines. These engines are labeled as "UDF" (unducted fan) in Figure 7-12. Very high bypass ratio engines, thus very large diameter (above 5–10 m) propulsors, are simply not feasible for installation on existing aircraft because they do not fit within the dimensional limits of the wing and landing gear. Thus, the application of large unducted propulsors is restricted to new aircraft designs for which special accommodation for the propulsion unit can be made. There would be additional costs to adapt airport facilities to handle aircraft of markedly different layout resulting from the use of very large diameter propulsors. Studies of aircraft with ductless propulsors show that aircraft performance tends to optimize at flight speeds 5–10% below that of current transports as a result of aerodynamic efficiency considerations. Community and passenger noise considerations of such engines remain largely unresolved.

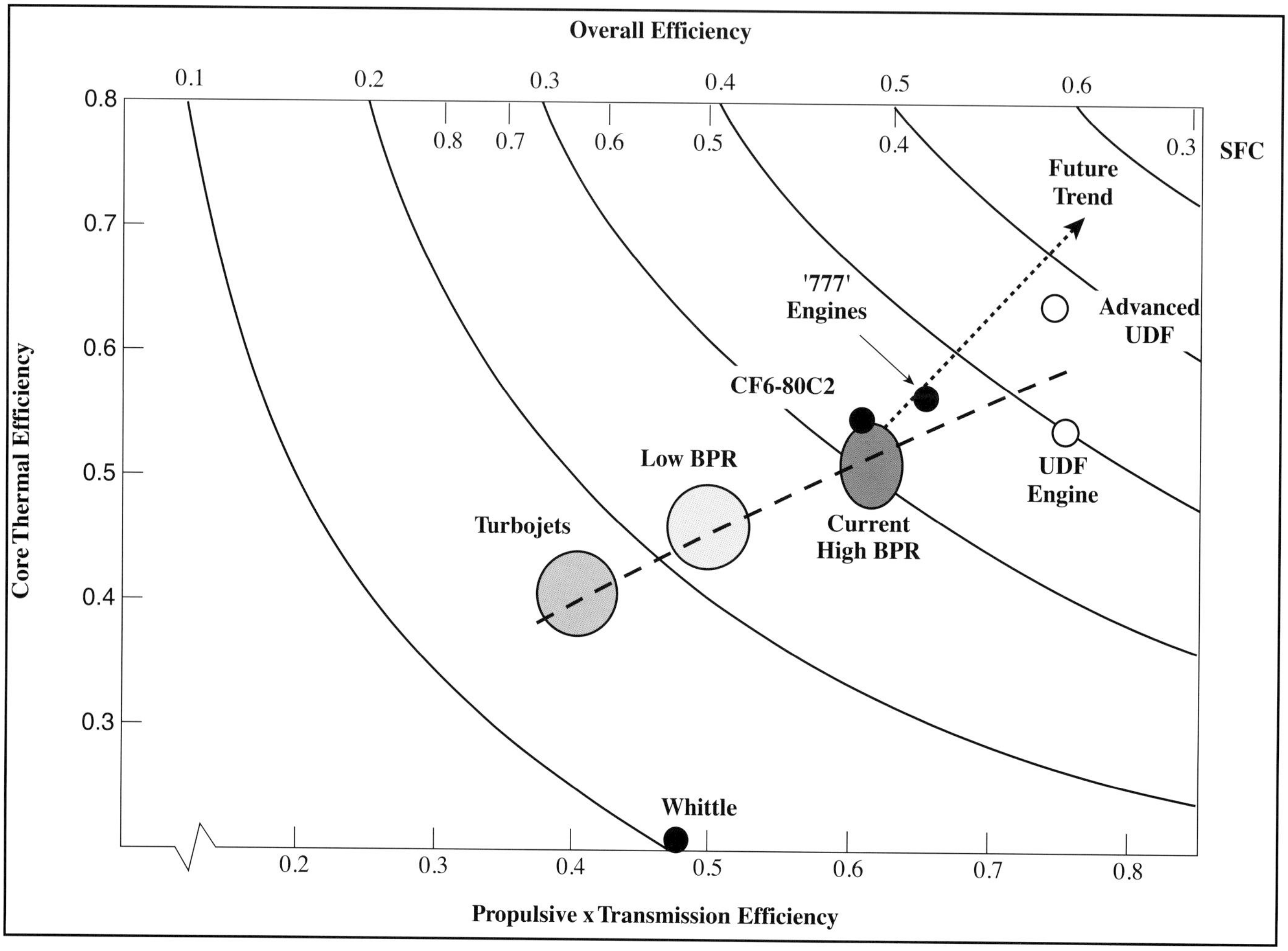

Figure 7-12: Evolution of aircraft gas turbine efficiency (after Koff, 1991).

Studies suggest that propulsive efficiency could be further improved, by as much as 5%, by integrating the propulsion system with the airframe to pass the vehicle boundary layer through the propulsor (thereby canceling the airframe wake). However, this approach would be practical only for very specialized, all-new aircraft designs. Smaller gains (on the order of 1%) may be achieved for a wider range of aircraft through more efficient mixing of propulsor and core flows in the duct before the exit plane of the exhaust nozzle. This principle has already been adopted for certain long-range aircraft for which extensive research has been done to ensure that the weight and drag associated with the mixer did not offset the gains in propulsive efficiency.

Turning to the other option suggested by Figure 7-12—namely thermal efficiency improvements—there are more technology opportunities that can be pursued to effect overall efficiency improvement. These possibilities can be grouped into improvements to current simple cycle bypass designs or new, more complex engine cycles. Improvements to current approaches include the following:

- Further increases in the pressure ratio of compression systems
- Higher temperature hot sections with reduced (or eliminated) cooling requirements
- Improved component efficiencies.

Realization of any or all of the above improvements will require substantial investments in a wide range of research and development fields, including aerodynamics, cooling technology, materials, mechanical design, and engine control. Studies suggest that total gains of 10–20% in the thermal efficiency of the engine might be achievable by pursuing these options (Hill, 1996).

There are, of course, alternative or modified thermodynamic cycle approaches to future engine design, such as incorporation of an inter-cooler and/or a recuperator. Some of these technologies are used in land-based gas turbines, with large potential gains in thermal efficiency. However, they invariably employ heat exchangers, which increase engine weight to an extent that they are currently impractical for aircraft applications.

There is a growing awareness, however, of the potential for reducing the weight of aircraft engines by 20–40%. This general approach offers particular attractions for application to long-range transport aircraft; as mentioned earlier, one unit of engine weight generally saves between 1.5 and 4 units of aircraft empty weight, with a concomitant decrease in fuel burn. Enabling technologies required to achieve significant engine weight reductions include the following:

- Improved materials (composites and high-temperature materials in particular)
- Improved aerodynamics (to reduce the number of turbine and compressor stages)
- Increased turbine entry temperatures (to reduce airflow thus core engine size required for a given power output).

None of these options is new to manufacturers, who already have research and development programs addressing these subjects aimed at civil and military aircraft of the future. As always, progress and success in meeting these objectives is paced by the scale of investment by industry and/or government supporting the work (a subject beyond the scope of this report).

The final question to be asked in reviewing future development paths for aircraft propulsion is: Are there any serious alternatives to the gas turbine or derivatives thereof as the primary source of propulsion for the aircraft of tomorrow? Several other energy conversion technologies have been identified that may have potential for application in commercial aircraft (e.g., solar power, nuclear power, battery/fuel cell power, and hydrogen engines). However, these advanced concepts would require major innovations, development, and changes in infrastructure before they could serve as viable alternatives to hydrocarbon-powered gas turbine engines. We cannot forecast when or if such developments will occur. Given the current state of technology, there are simply no other energy conversion systems identified to date that can offer competitive levels of thermal efficiency and power-to-weight ratio for aircraft propulsion.

7.5. Combustion Technology

7.5.1. Introduction

Section 7.4 explained how major improvements in the fuel efficiency of engines have been achieved through parallel advances in turbomachinery aerodynamics, combustion, cooling, and materials. Significant progress in each of these areas is the direct result of industrial and governmental investment in balanced, interdependent research and technology programs that respond to commercial and environmental pressures. This section focuses on combustion technology, with particular emphasis on achievements to reduce emissions and the prospects of further improvements.

We first describe the modern gas turbine combustion system and the particular requirements placed upon this important engine component. Next we address emissions from combustors. In view of the wide availability of excellent reference material on this subject, however, only a brief assessment of the principal emissions is given here. A new emphasis has been placed on current research aimed at improving the understanding of soot formation and aerosols, a matter of increasing importance (see Chapter 3). We then outline progress made in reducing emissions in recent years. We go on to describe the specific part the combustor plays in controlling NO_x emissions from engines and discuss prospects and plans for further improvements.

7.5.2. The Gas Turbine Combustion System

In the cross-sectional drawing of a modern civil aircraft engine (Figure 7-13), the combustor is shown in its central position between the compressor and turbine. High-pressure air enters

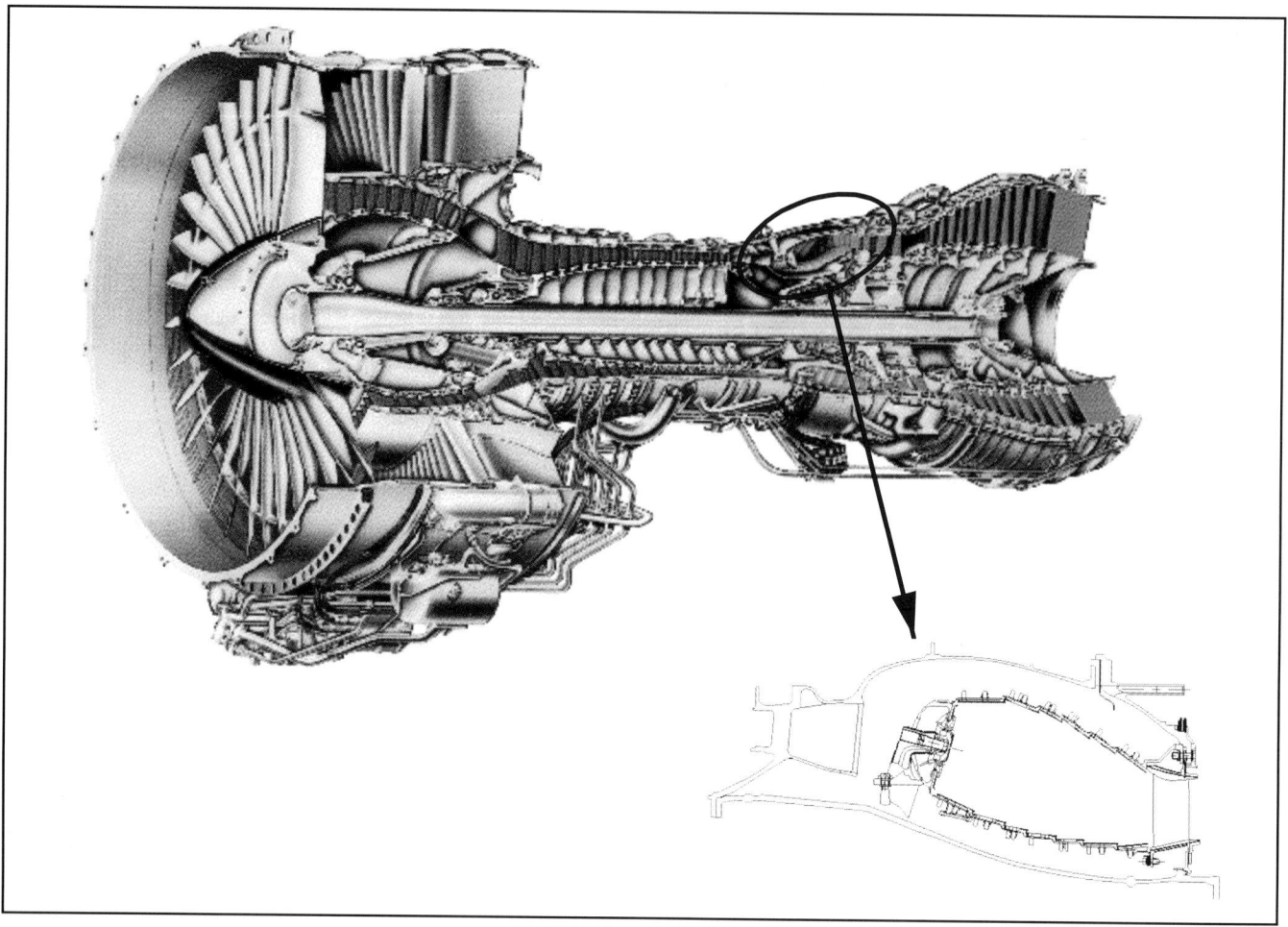

Figure 7-13: Engine cross-section and combustor detail.

the combustor at a relatively high velocity. The air is first carefully decelerated to minimize pressure losses, then forced into the combustion chamber, where fuel is added. The combustion chamber is designed to allow time and space for the fuel and air to mix thoroughly and burn efficiently before entering the turbine stages. Detailed design features of the combustor control the complex burning processes, thus the completeness of the chemical reactions involved and the nature and scale of individual emissions from the engine. Therefore, the combustor has a key role in determining the impact of aircraft on climate.

7.5.2.1. *Combustor Features and Requirements*

Aircraft engine combustors must meet the special requirements of operations over a very wide range of pressures and temperatures. The combustor must be able to ignite and accelerate the engine over a wide operational envelope. For instance, it must be able to ignite at high altitude (up to 9 km) after an unscheduled shutdown when the air is very cold (e.g., 220 K) and pressure is low (e.g., 0.03 MPa). It must also be able to maintain stable burning over a very wide range of air velocities and fuel/air ratios to prevent "flameout" during engine deceleration. At the other end of the power range, when pressures and temperatures are very high, the combustor must be able to burn fuel so that

turbine components are presented with a smooth temperature profile, to minimize damage to the blades and vanes and thereby maximize service life. Together, these requirements present a major engineering challenge because the simplest solutions to meet requirements at one end of the operational envelope often conflict with those required at the other. Relatively recent emissions requirements have added considerably to the time and cost of developing combustors that fully satisfy the operational and environmental requirements placed on today's aircraft.

Future combustors are likely to face even more challenging requirements as manufacturers respond to the continuing need to increase fuel efficiency. As explained in Section 7.4.1, higher cycle pressure ratios lead to improved engine fuel efficiency. However, the compressor delivery/combustor inlet air temperatures rise, reducing the cooling capacity of the air. A greater proportion of total airflow is then needed to cool the hottest parts (liners, blades, etc.). In turn, this requirement reduces airflow available for primary combustion and dilution, making it more difficult to control turbine inlet temperature profiles and emission levels. This cycle of primary and secondary problems, which stem from increases in engine pressure ratios, can be broken only by continuously improving the effectiveness of cooling techniques and devices and/or the use of new and improved materials.

7.5.3. Production of Engine Emissions

Pollutant formation in combustion, regulated aircraft emissions during landing/take-off, and currently unregulated cruise emissions are covered in some depth by Brasseur *et al.* (1998). For convenience, we briefly describe the principal points here, with somewhat greater detail given to more recent findings about soot and particulate emissions.

Under ideal conditions, combustion of kerosene-type fuels produces carbon dioxide (CO_2) and water vapor (H_2O), the proportions of which depend on the specific fuel carbon to hydrogen ratio. Figure 7-14 shows the ideal and "real" combustion processes. Figure 7-14 also illustrates the scale of the combustion products by showing that at cruise conditions they constitute only about 8.5% of total mass flow emerging from the engine. Of these combustion products, only a very small volume (about 0.4%) of residual products arise from non-ideal combustion processes (soot, HC, and CO) and the oxidation of nitrogen (NO_x).

Table 7-2 gives typical emission levels for various operating regimes. The emission values are quoted as an emission index (EI) in units of grams of emittant species per kilogram of fuel burned.

Table 7-2 clearly illustrates the constancy of emissions indices for CO_2, H_2O, and SO_x throughout the flight cycle. These emissions are directly related to the fuel consumption of the

Table 7-2: Typical emission index (g/kg) levels for engine operating regimes.

Species	Operating Condition		
	Idle	Take-Off	Cruise
CO_2	3160	3160	3160
H_2O	1230	1230	1230
CO	25 (10–65)	<1	1–3.5
HC (as CH_4)	4 (0–12)	<0.5	0.2–1.3
NO_x (as NO_2)			
– Short Haul	4.5 (3–6)	32 (20–65)	7.9–11.9
– Long Haul	4.5 (3–6)	27 (10–53)	11.1–15.4
SO_x (as SO_2)	1.0	1.0	1.0

engine in its various flight phases. In contrast, emissions such as NO_x, CO, HC, and soot are strongly influenced by a wide range of variables but particularly engine power setting and ambient engine inlet conditions. CO and HC are products of incomplete combustion. They are highest at low power settings, when the temperature of the air is relatively low and fuel atomization and mixing processes least efficient. This problem area is proving responsive to improvements linked to detailed studies of basic fuel/air mixing processes. NO_x and soot (not shown in the table), on the other hand, are highest at high power settings.

The majority of NO_x emissions are generated in the highest temperature regions of the combustor—usually in the primary combustion zone, before the products are diluted. The fundamental processes of NO_x formation are well known and documented (reviewed and described in detail by Bowman, 1992). They are best expressed as a function of local combustion temperature, pressure, and time. Combustion zone temperature depends on combustor inlet air temperatures and pressure, as well as the fuel/air mass ratio. The dependence of NO_x on fuel/air ratio is shown in Figure 7-15. As illustrated, peak NO_x formation coincides with peak temperature, which occurs close to the stoichiometric fuel/air ratio (or equivalence ratio = 1). In current gas turbine engine combustors, there are always some regions of the flame that burn stoichiometrically, so NO_x formation is very strongly linked to combustor inlet temperature.

"Soot" generally refers to particulates in emissions. These particles are composed primarily of carbonaceous material, the sum of graphite carbon and primary organics resulting from incomplete combustion of carbonaceous material (Novakov, 1982; Chang and Novakov, 1983). "Smoke" refers to combustion emissions particulates that contribute to a visible plume. The formation of soot and its partial oxidation in gas turbine combustors are very complex processes. Soot is produced mainly in the fuel-rich primary zone of the combustor, then

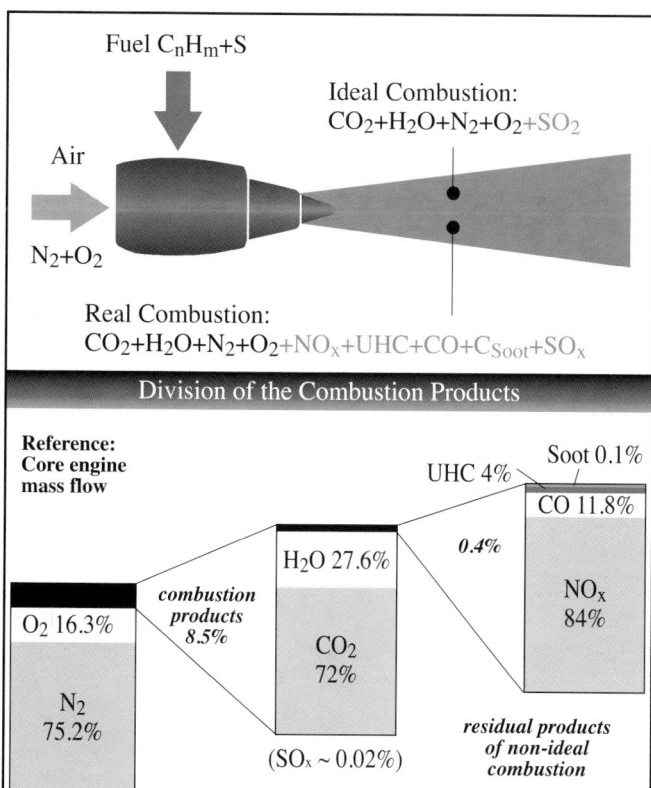

Figure 7-14: Schematic of ideal combustion products (top), and all existing combustion products, showing scale of each.

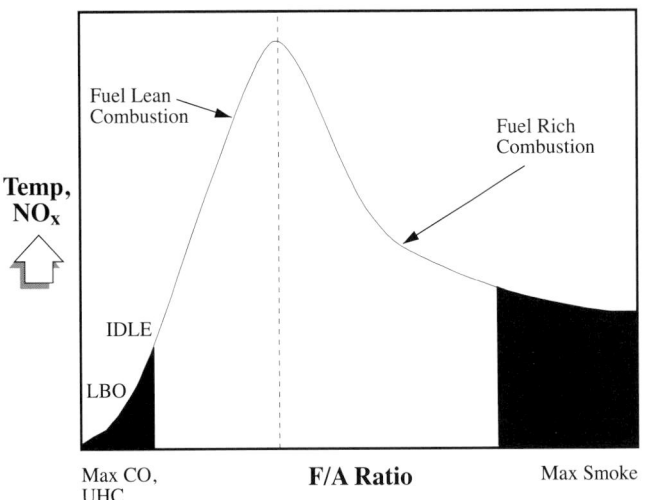

Figure 7-15: Fuel/air ratio dependence on NO_x and temperature.

oxidized in the high-temperature regions of the dilution and intermediate zone. A simplified description of the soot production mechanism is provided in Figure 7-16 (Mullins, 1988).

Estimates of the fleet averaged emission index of soot is 0.04g/kg fuel burned (Döpelheuer, 1997). However, the derivation of this estimate carries considerable uncertainty and at best is only within a factor of 2. The formation of soot in gas turbine combustors and the precise design features and conditions that can influence the process continue to be subjects of important ongoing research. Since 1990, there has been a marked improvement in the engine soot and aerosol emissions database through skilled use of the most modern aerosol measurement techniques (Hagen *et al.*, 1992; Rickey, 1995; Pueschel *et al.*, 1997; Petzold and Schröder, 1998). State-of-the-art instrumentation permits detection of particles as small as 3 nm in diameter (Alofs *et al.*, 1995). These data also show that jet engines emit soot particles, which have log normal-type size distributions

peaking in the 20–30 nm range. Concentrations range between 10^6 and 10^7 particles cm^{-3} (Hagen *et al.*, 1992; Whitefield and Hagen, 1995; Schumann *et al.*, 1996; Pueschel *et al.*, 1997; Anderson *et al.*, 1998; Petzold and Schröder, 1998). Number-based emission indices fall within the range of 10^{12} soot particles per kg of fuel burned for current advanced combustors to 10^{15} for older type engines (Howard *et al.*, 1996; Whitefield *et al.*, 1996). A general range for most engines operating in the current commercial fleet is 10^{14}–10^{15} particles per kg fuel burned (Hagen *et al.*, 1996; Döpelheuer, 1997; Anderson *et al.*, 1998; Petzold and Schröder, 1998; Petzold *et al.*, 1999).

7.5.4. Reduction of Emissions

7.5.4.1. Earlier Developments

Between 1965 and 1975, low-smoke combustors were developed and incorporated into low bypass ratio engines commonly used in early commercial jet aircraft. These changes virtually eliminated visible smoke trails from aircraft. Introduction of higher bypass ratio engines of the late 1960s and early 1970s—with their significantly improved SFCs—marked a new and important step in reducing CO_2 and water vapor emissions from aircraft. These engines also emitted much lower levels of HC and CO at low power ("idle") setting as a result of improved fuel/air mixing and relatively high levels of pressure and temperature in the combustor at this condition. Improved fuel/air mixing in annular combustors of the new engines also reduced take-off smoke. The trend toward even higher bypass engines, with their improved fuel efficiency, continues today, responding not only to the initial and continuing commercial and operational pressures but also to increasing concern about the effect of CO_2 on the environment.

Between 1975 and 1985, new combustor design features that had better fuel atomization and circumferential fuel staging at idle led to further reductions in HC and CO emission levels. A typical example of such improvements in HC and CO emission is presented in Table 7-3, which indicates the scale of some of the benefits that have already been made toward abatement of urban air pollution burdens (Bahr, 1992).

At that time, however, there was little change in the NO_x levels; although, as Table 7-3 shows, they were well within ICAO

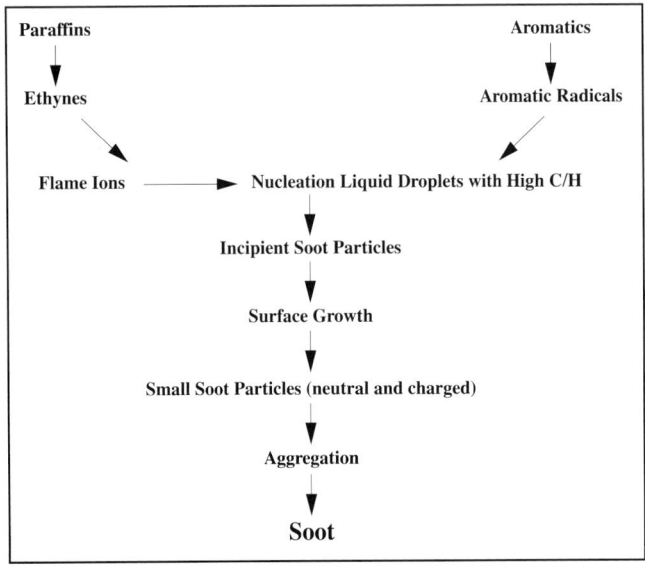

Figure 7-16: Schematic of soot production process.

Table 7-3: *Improvements in emissions levels of GE CF6-50E2 engine.*

Emission	Applicable ICAO Standard	Levels for Original Production Engine	Levels for "Low Emissions" Production Engine
Smoke (SN)	18.8	6.5	12.5
HC (g kN^{-1})	19.6	57.8	3.4
CO (g kN^{-1})	118	97.3	29.8
NO_x (g kN^{-1})	100	58.2	51.6

standards, which applied to landing and take-off operations (ICAO, 1981). Emissions regulation for HC, CO, and NO_x are based on the total mass of each species produced over the ICAO landing/take-off cycle, which is intended to represent typical aircraft operations in the vicinity of the airport. The mass of each species (in grams) is divided by the take-off thrust of the engine (in kilo Newtons) so that different size engines can be reasonably compared; thus, the resulting units are g kN^{-1}. ICAO landing/take-off cycle, measurement procedures, and emissions standards are discussed further in Section 7.7.1.

7.5.4.2   Near-Term Technology

7.5.4.2.1.   General issues

Heightened interest in engine emissions has had a considerable effect on the approach to combustor development over the past 25 years. As with other key fields of gas turbine technology, earlier dependence on semi-empirical and semi-analytical models has been lessened by advances in CFD methods of modeling and analyzing complex flow processes (Mongia, 1994, 1997a). Nevertheless, the complexity of the problems dictates that experimental verification is still required to demonstrate that the combustor meets all emissions, operational performance, and durability requirements. As experience and confidence in these approaches grow, the high costs of combustion system development might be contained—or even reduced slightly. At present, however, modeling is not expected to completely replace experimental testing, especially in the case of more advanced, low-emission concepts for which experience is limited. Thus, the pursuit of advances in combustor technology will remain a major part of the cost of any new engine. A summary of current modeling capabilities, limitations, and potential improvements is given elsewhere (Mongia, 1997a).

It is important to note that the intrinsic complexity of the combustion process and limitations of current analytical tools lead to considerable uncertainty in prediction of emissions from new combustor designs. Mongia (1997a) estimates that even with analytical models that have been "anchored" to measured data—a process to systematically calibrate a model so that it reproduces measured emissions results—prediction accuracy is only about ±15% for NO_x and ±30% for CO and HC. The uncertainty in prediction is even higher for smoke.

One of the clear lessons that emerged from early emissions reduction programs was that changes to reduce NO_x emissions could produce adverse effects on other performance characteristics, leading toward tradeoffs in design. Tradeoffs can be attributed to the engine cycle selected or the combustor design itself. Tradeoffs based on engine cycle reflect changes in overall pressure and bypass ratios, which, in turn, affect fuel burn rates and the conditions at which combustion occurs (see Section 7.4.1.1). Tradeoffs generated by design changes affecting the combustor can influence major combustor performance parameters, including operability, reliability, durability, efficiency, and noise—all of which must be taken into account during engine development.

More fuel efficient, high bypass engines reduce not only CO_2 and H_2O, but also HC and CO emissions. For a given combustor design, however, NO_x formation rates rise as a result of higher air pressures and temperatures. Thus, despite the reduction in the amount of fuel that can form NO_x, the increased formation rate can result in a net rise in the mass output of NO_x. This result is shown clearly in Figure 7-17 (Rudey, 1975). Many earlier researchers have suggested that the only way the more basic tradeoff between CO_2 and NO_x could be altered was by finding satisfactory ways of reducing fuel-rich zones without compromising stability, and reducing the residence time of burning gases in the combustor without compromising exit temperature profiles or pattern factors.

This important NO_x-CO_2 (fuel) tradeoffs issue has been summarized as follows:

"There is no single relationship between NO_x and CO_2 that holds for all engine types. However for the best current aircraft engine and combustor design technologies, there is a direct link between the emissions of NO_x and CO_2. As the temperatures and pressures in the combustors are increased to obtain better fuel efficiency, emissions of NO_x increase, unless there is also a change in combustor technology." (ICCAIA, 1997b)

Improved low-emission combustor technology can alter the precise tradeoff between CO_2 and NO_x. Theoretically, this technology can reduce NO_x levels at any pressure ratio. However, for any currently available aircraft engine combustor technology, there will still be some tradeoff between CO_2 and NO_x, although it will be at lower NO_x levels. Even with very advanced combustor technologies that minimize NO_x formation by premixing fuel and air to control combustion temperatures, there is a tradeoff between CO_2 and NO_x as a result of high

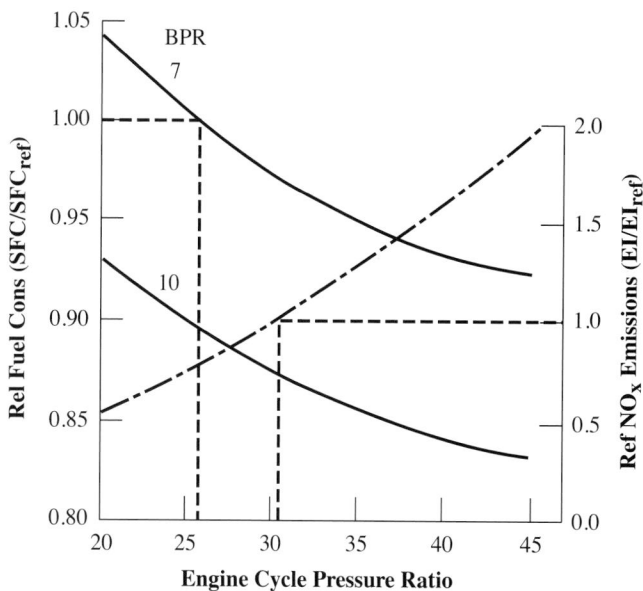

Figure 7-17: Effect of pressure ratio on NO_x-specific fuel consumption.

combustor exit temperatures associated with advanced, highly efficient engines.

A further NO_x-related issue arises from a common practice adopted by industry. In recent years, there have been numerous occasions when operators have sought to increase the capacity (passenger numbers or weight) or range of an existing airplane. A highly cost- and time-effective response to such requirements is to increase the thrust level of an existing engine type. This response has led to a type of development usually referred to as a "throttle push" or "throttle bend" of the core engine. This approach invariably entails increasing overall engine pressure ratio and results in higher combustor inlet pressures and temperatures, thus higher NO_x levels at high power levels. However, analytical studies (ICCAIA, 1997d) of engine growth characteristics have shown that the rate at which the mass formation rate of NO_x increases as the "throttle bend" technique is used to increase thrust is similar to data measured on a wide range of engine families over a significant range of engine operating pressure ratios (ICAO, 1995a). It is also worth noting that, in practice, the required thrust at cruise and measured cruise SFC remain essentially constant for the "throttle push" versions, resulting in increases in fuel efficiency levels as payloads and number of passengers have risen.

7.5.4.2.2. Combustor design considerations

Current combustors can contribute little more to fuel efficiency because at power levels above "idle" the energy conversion level is virtually 100% (see Tables 7-4 and 7-5). However, superimposed on all combustor design considerations is the continuing underlying requirement that low emissions features must not compromise basic combustion requirements or have any significant effect on engine performance. In recent years, this requirement has imposed difficult problems in introducing new emissions reductions features. Table 7-4 provides a list of basic requirements, and Table 7-5 highlights options and compromises that designers have had to face.

Figure 7-18 provides a qualitative indication of the development process as engineers work to reconcile the combustion system constraints listed in Tables 7-4 and 7-5 and the operational requirements of the engine.

7.5.4.2.3. Achievements

Low-emissions combustors currently fall into two categories. The first category is composed of existing combustors, which have incorporated relatively minor changes to liner and/or fuel nozzle designs to improve emissions. Recent examples of changes that have been quite successful in reducing NO_x emissions are shown in Figure 7-19.

The NO_x emissions of state-of-the-art combustors are 20–40% lower than those of older combustors. Figures 7-20, 7-21, and 7-22 provide evidence of some of the progress that has been made in recent years to avoid what had earlier appeared to be an unavoidable link between lower NO_x emissions and relatively high CO and HC emissions. In these examples, relatively minor changes to the combustor's airflow pattern and the location of fuel injection resulted in reduced CO/HC without a NO_x penalty.

Table 7-4: Typical basic performance and operational requirements of a modern aircraft engine combustor.

Item	Requirement	Value	Max/Min
1	Combustion efficiency		
	– At takeoff thrust (%)	99.9	(Min)
	– Idle thrust (%)	99.0	(Min)
2	Low-pressure light-off capability (MPa)	0.03	(Max)
3	Lean blowout fuel/air ratio (at low engine power conditions)	0.005	(Max)
4	Ground light-off fuel/air ratio (with cold air, cold fuel)	0.010	(Max)
5	Total pressure drop—compressor exit to turbine inlet (%)	5.0	(Max)
6	Exit gas temperature distribution		
	– Pattern factor	0.25	(Max)
	– Profile factor	0.11	(Max)
7	Combustion dynamics [dynamic pressure range/inlet air pressure (%)]	3	(Max)
8	Liner metal temperature (K)	1120	(Max)
9	Cyclic life to first repair (cycles)	5000	(Min)

Table 7-5: *Combustion developments linked to emissions performance.*

Requirements*	Emissions Implications and Compromise Required
1, 8, and 9	Lowering of liner cooling flows improves "idle" efficiency (low HC and CO) and ability to reduce NO_x but must be balanced against effect on liner temperatures and life/durability
2, 3, and 4	Wide range of fuel/air ratios and long dwell times in combustor favor good light-off performance but must be balanced against need to control HC and CO emissions at low power and NO_x formation rates at high power
5 and 7	Low overall and dynamic pressure losses required to minimize SFC losses but must be balanced by need to retain good mixing for low emissions and good liner cooling
6	Long mixing lengths improve exit profiles and pattern factors but require more cooling air and increase NO_x formation times

*See Table 7-4 for item requirement numbering scheme.

Although this category of improvement entails relatively minor changes, the development and engine recertification process remains a long one. Because safety is the overriding concern in aviation, the time and cost to introduce a change can be considerable. Safety considerations can also constrain application of new combustor designs to new engines. For example, a new combustor design is often introduced as a package that includes modifications to the combustion chamber, fuel nozzles, and engine control. Positive steps must be in place to prevent intermixing of new and old components during maintenance. Of course, other requirements—such as durability, weight, maintenance, and cost—must also be balanced during product introduction. This first technology "category" can also include improvements in other engine components, such as the engine turbomachinery, with or without concurrent combustor changes. Such improvements alone can lead to better fuel efficiency along with lower peak cycle temperatures, thereby reducing NO_x and CO_2 emissions and improving durability.

The second category of near-term advances involves major changes such as introduction of "staged" combustors (see Figure 7-23). Staging was introduced to improve or provide an additional degree of freedom between operational and emissions requirements. Thus, the high-power stage of a combustor optimized for low NO_x does not have to cope with low-power stability requirements, which are dealt with by bringing in other parts of the staged combustor when needed. Staged combustors, however, do require more complex control systems,

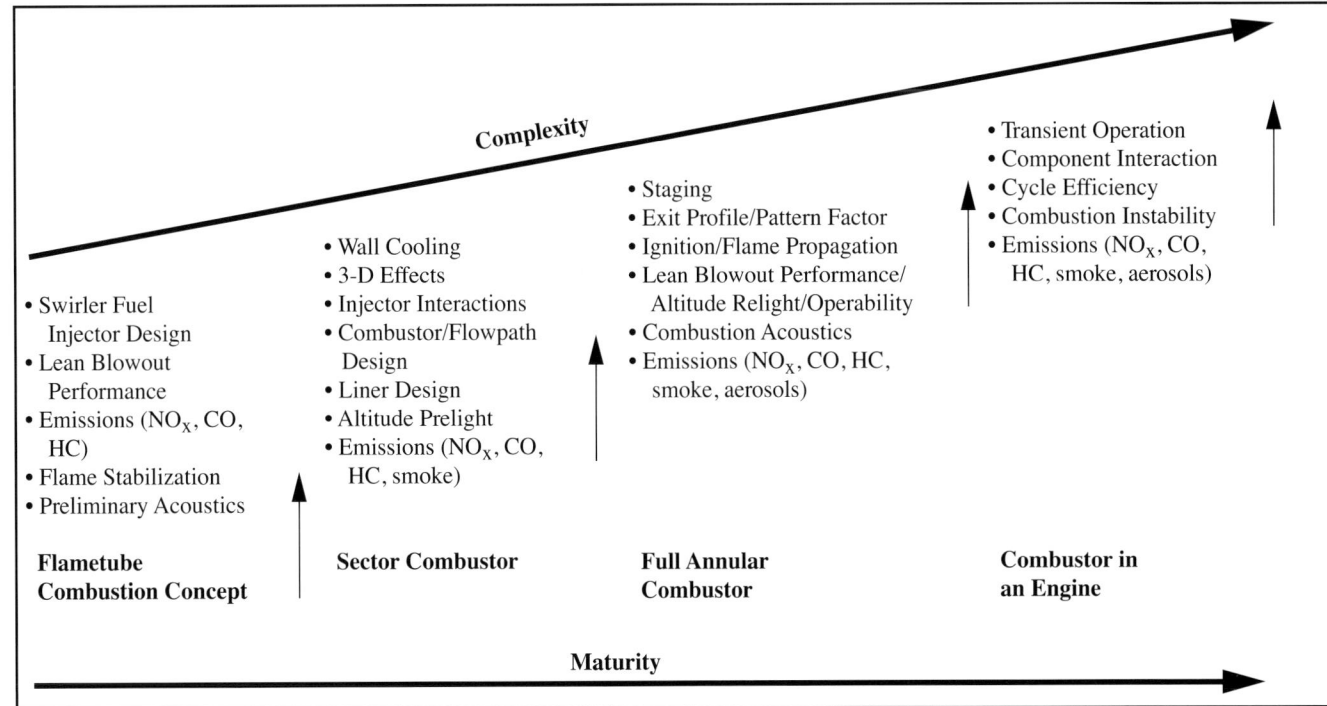

Figure 7-18: Combustor development process.

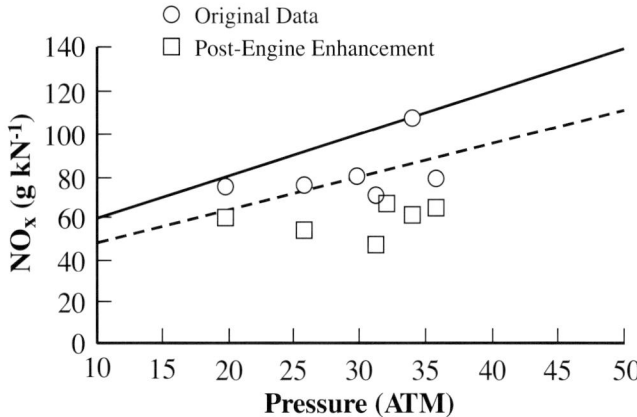

Figure 7-19: Recent combustion system enhancement for aircraft NO_x reduction (ICCAIA, 1997c).

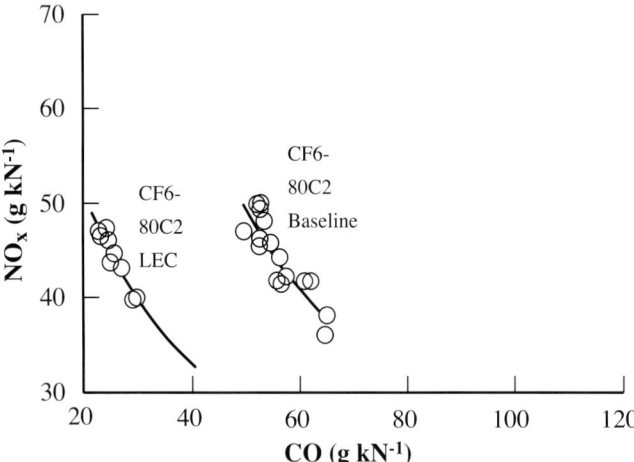

Figure 7-20: Example of combustion system enhancement for CO and HC, reduction EI basis (Mongia, 1997b).

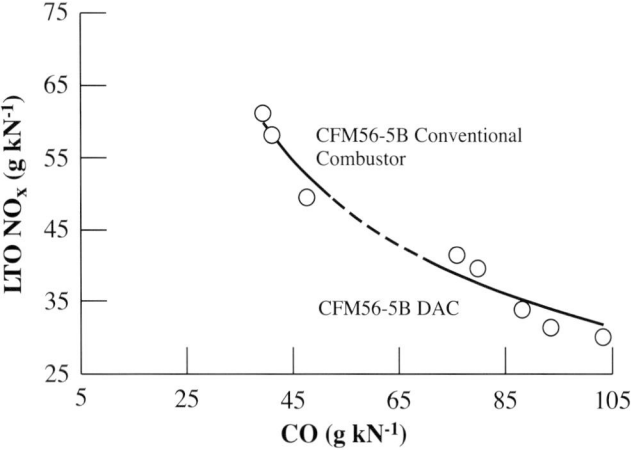

Figure 7-21: Tradeoff between NO_x and CO with a staged combustor (Mongia, 1997b).

incorporating fail-safe operation, to ensure that transient engine performance is not affected. This requirement is especially crucial for the most recent advances in low NO_x technology, which rely on low temperature, lean combustion to achieve low

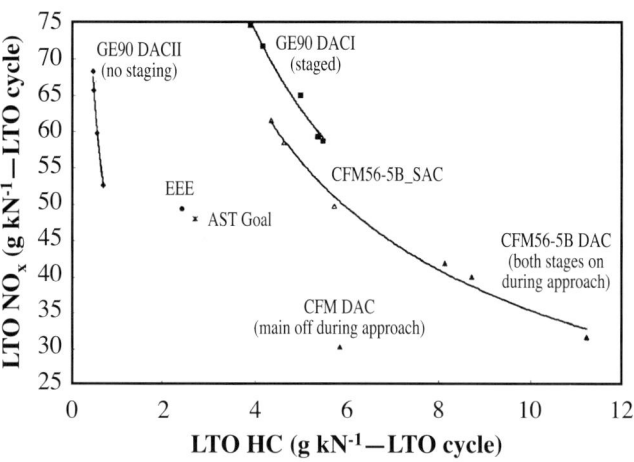

Figure 7-22: NO_x and HC concentrations for a staged combustor.

NO_x. Staged systems can present problems in achieving acceptable combustor exit temperature profiles, with associated losses in turbine efficiency, thus fuel efficiency. They are also heavier. Together, the complex interaction of improvements and penalties translates into a form of tradeoff between NO_x, CO_2, and HC/CO; work in this field continues to define designs that minimize penalties and maximize benefits.

The only example of a staged combustor in aircraft service today is the dual annular combustor (DAC), which is shown in Figure 7-23a. The DAC is a staged system that incorporates two separate combustion zones. The pilot stage provides good operational performance required at low power. The main stage provides low NO_x emissions at high power. Low NO_x emissions are achieved with lean fuel/air mixtures, which reduce flame temperatures, and high throughput velocities, which reduce the residence time available to form NO_x. Relative to current state-of-the-art NO_x levels discussed above, a single annular combustor in an engine having a pressure ratio of approximately 30 achieves about 30% reduction in LTO NO_x emissions, as shown in Figure 7-19, and NO_x levels are about 40% below CAEP/2 standards (ICAO, 1993). However, these improvements do not come without some tradeoffs. For example, Figure 7-20 (Mongia, 1997b) compares CO and NO_x emissions for the conventional baseline combustor and the DAC. Both combustors fall on approximately the same line, indicating an apparent increase in CO with reduced NO_x. The DAC system is also more complex. To obtain requisite staging capability, the engine must be equipped with a full authority digital electronic control (FADEC) system. The FADEC system must deal with increased complexity of engine operation to accommodate the various staging modes and associated engine responses. The added complexity invariably involves additional development effort for any new application to ensure the achievement of acceptable ground-level and altitude starting, combustion efficiency, and turbine inlet temperature patterns. The NO_x benefit is reduced in higher pressure ratio engines because of the increased competition for airflow to meet the conflicting requirements of durability and emissions.

The incorporation of such a staged combustion system into an existing engine type requires changes to several parts of the

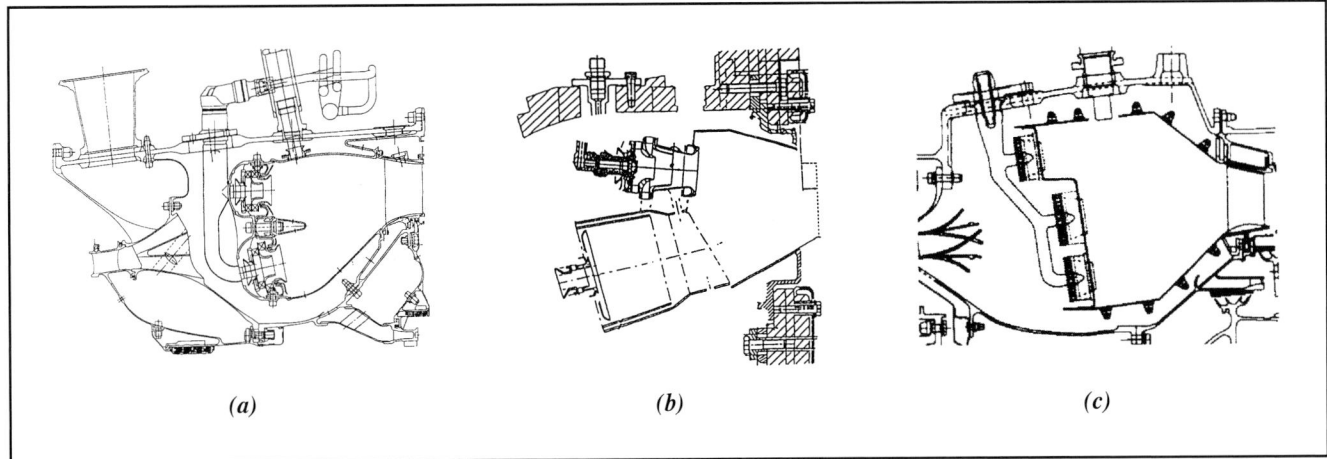

Figure 7-23: Staged combustors: (a) General Electric, (b) Snecma, (c) Pratt and Whitney.

high-pressure section of the engine, including the compressor outlet diffuser, combustor case, and inner structure. Major changes in the fuel control and fuel delivery system are also required. Additional modifications to the turbine may also be needed to accommodate changing temperature patterns during staged operation. Thus, the center and aft sections of the engine, which account for a major fraction of the cost of an engine, may be significantly different from the same engine with a current technology combustor. There may also be increases in weight, maintenance cost, and fuel burn. Several in-service engine models incorporate single burning zone unstaged combustors that employ limited fuel staging to maintain lean blow out performance at low power and operate at reduced NO_x levels at high power.

Retrofitting an older engine model with one of these advanced combustors is technically feasible. However, it could involve not only replacement of the existing combustor but also replacement of almost all other elements of the engine core. Estimates suggest (ICCAIA, 1997c) that retrofit could incur a cost of about one-third the price of a new engine, even if it were accomplished during a standard hot section overhaul. In some cases, aircraft systems and components such as cockpit indication elements, auto throttle, flight management computer, and FADEC interfaces could be affected. It now seems more likely that both categories of combustion system improvement will be considered only for application in new production engine units.

7.5.4.3. Longer-Term Technology

Much of the longer term combustion research is now aimed at what is termed "ultra-low NO_x technology." This technology is defined as technology that produces NO_x levels that are no more than 50% of the ICAO CAEP/2 standards. All major manufacturers of aircraft engines, with the sponsorship of associated government agencies, are currently pursuing combustor technologies aimed at reducing NO_x emissions to these levels, under operating conditions typical of the next generation of engines. This approach entails pursuing technology that is relevant to engines with pressure ratios from current levels

($\approx$30 to 40) to future levels (above 50). The aim is to reduce NO_x production in the vicinity of airports and at subsonic cruise conditions. Two key programs are aimed at large, subsonic, high-pressure ratio aircraft engines:

- U.S. NASA Advanced Subsonic Technology (AST) Program: Sequential NO_x goals of 50 and 70% reduction below CAEP/2 standards, with comparable reductions at subsonic cruise; engine SFC reduction of 8–10% over the most recent production engines by about 2010
- EU-BRITE/EURAM Projects Low NO_x I,II,III: Goal of NO_x reduction more than 60% below CAEP/2; engine SFC reduction of 8–10% over the most recent production engines by about 2010.

These programs specify that there must be no compromise in performance, safety, or any other emissions parameters (smoke, CO, and HC).

Details of combustor technologies are commercially sensitive and are not openly available. However, three parallel strategies are being pursued. They vary in their NO_x reduction potential and their associated increase in complexity, cost, and development challenge:

- Reductions in NO_x levels close to 50% of CAEP/2 standards are being sought through the optimization of current single-stage combustor technology. This approach involves further improvements in fuel injection uniformity, better fuel/air mixing, reduction in combustor liner coolant flow (making more air available for combustion), and decreases in hot gas residence time. Such changes would have minimum impact on engine cost of ownership. The first examples of this type of technology now exist in manufacturers' product plans for the next 5 to 10 years.
- Reductions in NO_x levels to 50–70% below CAEP/2 standards are being explored using multiple burning zones in radial and axial configurations (Figure 7-23). These concepts permit local temperature and residence time in the combustor to be controlled and optimized at each engine operating condition, to minimize NO_x and

other emittants (Bahr, 1992; Segalman *et al.*, 1993). In low-power operations, a single stage is fueled and optimized for stability. At high-power conditions, one or both stages, configured for lean burning, are fueled. Generally, radial staged designs are shorter and lighter but larger in diameter, making it more difficult to achieve a uniform exit temperature profile at off-design conditions. They are also more difficult to cool. Axial staged systems are longer and have a larger number of fuel injectors. In both cases, the increased complexity and weight of these combustors is expected to increase the cost of ownership, as described by DuBell (1995). A major effort is needed to minimize the cost, complexity, weight, and performance of these concepts.

- Reduction in NO_x levels to 85–90% below CAEP/2 standards are focused on emissions from supersonic aircraft at cruise conditions. Such work presently forms part of the U.S. supersonic transport program. Combustors known as lean premixing prevaporizing (LPP) and rich burn quick quench (RBQQ) are being studied. In the former, the burning zone is fed with a lean and homogeneous fuel/air mixture. Premixing and prevaporizing takes place in a premix duct outside the combustor. The RBQQ combustor consists of three zones: A primary rich burning zone; a dilution zone, to rapidly reduce the rich mixture to a lean one without recirculating dilution air into the primary zone; and a lean reburning zone (DuBell, 1995). These combustors uniquely apply to the supersonic engine with its relatively low engine pressure ratio and its requirement for long periods at a single, high-speed cruise operating condition. Application of these concepts to future subsonic engines would pose special problems because of higher pressure ratios. For example, the LPP combustor will have to overcome the greater risk, at very high pressures, of "flashback" or upstream burning—which, if undetected, could damage the combustor. Similarly, the RBQQ combustor—with its high fuel concentration in the primary burning zone—may well result in the generation of large amounts of soot and smoke in high-pressure operations. However, some of the features found in these concepts may be suitable for higher pressure ratio subsonic engines. Partial premixing, coupled with moderately rich sector burning, represents one such concept. Multiple burning zones, together with variable geometry to control local fuel/air ratios, would be another.

A concept incorporating several of these features (shown in Figure 7-23b) is being pursued under the European Union's Targeted Research Action "Efficient and Environmentally Friendly Aero-Engines." A radially staged configuration currently under "main" phase development (LOWNOXIII) combines an RBQQ candidate pilot injector with a premixing main zone injector. This concept might provide low NO_x emissions with acceptable operational capabilities at LTO and cruise conditions of subsonic aircraft (Zarzalis *et al.*, 1995).

7.5.5. *Future Technology Scenarios*

As part of the preliminary work associated with this report, industry was asked to consider what advances in technology might be applicable for aircraft in the year 2050. Numerous projections were made by an expert group from the aeronautical industry (engine, airframe, and aerospace manufacturers). The group provided their best judgments of fuel efficiency and NO_x technology scenarios for the year 2050 (ICCAIA, 1997f) to the ICAO Committee on Aviation Environmental Protection Forecasting and Economic Support Group (FESG) for use in Chapter 9 of this report. The assumptions for the 2050 scenarios were as follows:

- Continued demand for worldwide commercial/regional/ general aviation aircraft
- Technology advancement to be addressed for all engine sizes
- Unrestricted kerosene availability
- Development time and operational life of modern aircraft = 40–50 years
- All airworthiness requirements achievable
- Economically viable
- No impact on noise
- Possible NO_x reduction scenarios.

Chapter 9 of this report considers the conclusions of the group in conjunction with FESG traffic scenario projections. Together, these considerations take account not only of long-term demand but also of fuel burn and emissions (see Chapter 9). The latter two depend on three factors:

- Airframe technology developments (discussed in Sections 7.2 and 7.3)

Table 7-6: *Long-term aircraft technology scenarios.*

Technology Scenario	Fuel-Efficiency Increase by 2050	LTO NO_x Levels
Design for both improved fuel efficiency and NO_x reduction	Average of production aircraft will be 40–50% better than 1997 levels	Fleet average will be 10–30% below current CAEP/2 limit by 2050
Design with much greater emphasis on NO_x reduction	Average of production aircraft will be 30–40% better than 1997 levels	Average of production aircraft will be 30–50% below current CAEP/2 limit by 2020 and 50–70% below current CAEP/2 limit by 2050

- Trends in engine design (discussed in Section 7.4)
- Combustion system development—in particular NO_x reduction scenarios, based in one case on the best near-term combustor technology and in the other on longer term NO_x reduction combustor technology (discussed in Sections 7.5.3 and 7.5.4, respectively).

The benefits arising from the first item influence projections in terms of fuel efficiency alone. The other two items take into account the net effects of cycle changes and emissions reductions strategies. Two potential long-term aircraft technology scenarios emerged from these deliberations. These scenarios are summarized in Table 7-6.

The outcome of these deliberations is that a basis now exists for the development and ongoing monitoring of future research strategies as air transport and concern about its environmental impact continues to grow.

7.5.6. Summary of Key Points Relating to Combustion Technology

Several important conclusions can be drawn from the present assessment of combustion technology in relation to emissions production and control:

- Research and development programs over the past 25 years have provided a basis for vastly improved combustion systems for modern aircraft engines. Current combustion systems achieve virtually 100% energy conversion efficiency at virtually all power settings. Better fuel/air mixing processes, liner cooling techniques, and materials have contributed to this progress. The levels of unburned products, such as CO and HC, from engines are now very low, and visible smoke levels are now under control.
- Reductions in CO_2 depend primarily on engine cycle, not the combustor. The role of the combustor is to ensure that the demands of the more fuel-efficient cycle (low CO_2) are met without compromising engine performance.
- The more fuel-efficient engines, with their high bypass ratios, introduced in the 1970s and 1980s reduced CO_2, HC, and CO emissions but increased NO_x. Although technological improvements have constrained the rise in NO_x, it has become clear that more substantial reductions require more radical solutions (with associated risks and penalties).
- Technology that has reduced NO_x emissions at high power, near the ground, also reduces NO_x at high altitude, though not necessarily by the same amount.
- Current research goals are to achieve NO_x levels of 50% of current standards in 5 to 10 years. Work is also in progress to achieve NO_x levels 50–70% below current NO_x standards using more advanced, staged combustors.
- Although there has been a marked improvement in the measurement and assessment of trace species and

aerosols emerging from engines, knowledge about their formation—hence our ability to control them—is extremely limited.

7.6. Turbine and Nozzle Effects on Emissions

This section describes what is known and not known about changes in various exhaust constituents that occur downstream of the combustor, in the air passages of the turbine and exhaust nozzle. The discussion centers on chemical species that are expected to have the highest potential for impact on the global atmosphere (see Chapters 2 and 3).

The discussion begins in Section 7.6.1 with a review of the factors that have generated interest in chemical processes in the turbine and exhaust nozzle. Section 7.6.2 provides a brief overview of functional requirements and constraints for aircraft turbines and nozzles, and how these requirements have led to particular design choices. Section 7.6.3 describes the relevant chemical and fluid mechanical effects. Section 7.6.4 describes what is known and not known regarding chemical changes in the turbine and exhaust nozzle.

7.6.1. Chemical Processes in Turbine and Exhaust Nozzle

Many engine exhaust species of interest for environmental impact assessment exist in trace amounts, typically tens of ppbv to tens of ppmv. Despite their relatively small concentrations, these trace species emissions can result in perturbations of chemical species in the atmosphere that may induce significant atmospheric effects. Chapters 2 and 3 discuss these issues in relation to atmospheric ozone and cloudiness, respectively. Numerical simulations and a limited number of experiments suggest that several of these trace species can undergo considerable change within the nonuniform, unsteady flow fields of the turbine and nozzle prior to injection into the atmosphere (Hunter, 1982; Harris, 1990; Brown *et al.*, 1996; Lukachko *et al.*, 1998). However, there is a high degree of uncertainty regarding the extent of this change and its dependence on engine technology and operating conditions. Indeed, estimates for the extent of sulfur oxidation ($SO_2 \rightarrow SO_3 + H_2SO_4$) range from 0.4 to 45% or more depending on the modeling assumptions and experimental data considered (Hunter, 1982; Harris, 1990; Arnold *et al.*, 1994, 1999; Frenzel and Arnold, 1994; Miake-Lye *et al.*, 1994; Fahey *et al.*, 1995b; Brown *et al.*, 1996; Kärcher *et al.*, 1996; Hanisco *et al.*, 1997; Lukachko *et al.*, 1998; Miake-Lye *et al.*, 1998). However, most of the predicted conversion efficiencies are less than 10%, with an experimentally derived lower limit of 0.34% (Curtius *et al.*, 1998) (see Section 3.2.2.2). More important, such variations in estimates of the extent of sulfur oxidation within the engine and subsequent changes in aerosol formation lead to changes in predicted column ozone depletion by a fleet of supersonic transport aircraft by as much as a factor of 2 (Weisenstein *et al.*, 1995; Danilin *et al.*, 1997). Large uncertainty regarding trace species processes and the perception that intra-engine changes may be important—along with the desire for a more detailed and complete characterization of engine exhaust emissions to support downstream plume, wake, and atmospheric modeling efforts—have only recently motivated a more detailed study of intra-engine trace species chemistry (Dryer *et al.*, 1993; Brown *et al.*, 1996; NRC, 1997; Lukachko *et al.*, 1998). There is relatively little research yet reported in this area; however, new measurement and modeling capabilities are currently being developed.

7.6.2. *Aircraft Turbine and Nozzle Design*

Sections 7.4 and 7.5 discuss the principal elements and functions of the components of a gas turbine engine. They describe how, after exiting the combustor, the engine core flow passes into the turbine then through the exhaust nozzle (Figure 7-7 *et seq.*). Within these components, the principal engineering constraints are associated with maintenance of the structural integrity of the parts exposed to the high-temperature environment downstream of the combustor and limitations on weight.

Figure 7-24 shows an example of a modern turbine stage. The design of the flow passages ensures maximum operating efficiency and meets engineering integrity requirements of the compressor and combustor. As explained in Section 7.4, the achievement of high thermal efficiency, hence low fuel consumption, means that turbines operate in gas flows that are several hundred Kelvin above the melting point of the materials employed (see, e.g., Kerrebrock, 1992). Cooling of all structures exposed to the flow path is therefore essential. The requirement for cooling adds significant complexity to the blades and vanes because of the need for small internal passages through which

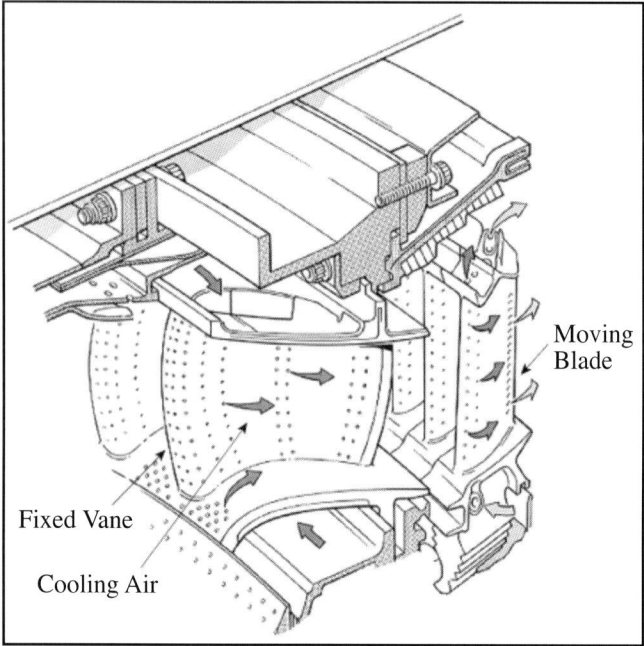

Figure 7-24: Illustration of complex flow passages and blade cooling schemes in a typical turbine stage (after Rolls Royce, 1992).

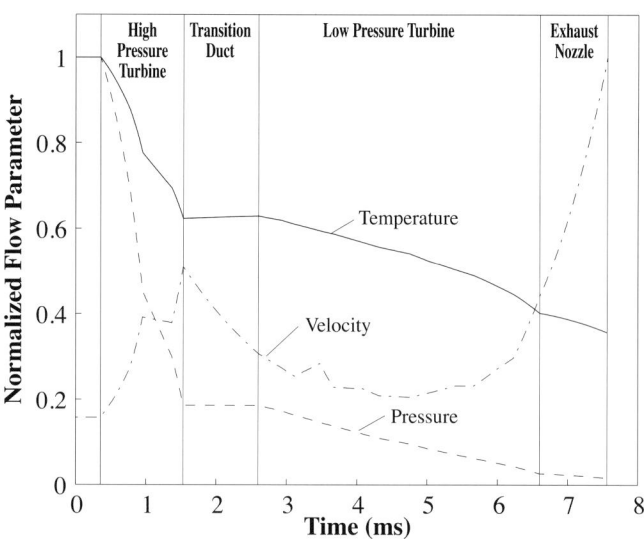

Figure 7-25: Typical static temperature and static pressure histories downstream of the combustor (Lukachko *et al.*, 1998).

air bled from the compressor is channelled. This air is usually injected through small holes in the surfaces of the various components (as shown in Figure 7-24) to form a protective, cooler boundary layer. Air injected in this manner can account for as much as 25% of the flow through the core of the engine. The trend for increased temperature and pressure is expected to continue in the future, as discussed in Section 7.4.

7.6.3. *Chemical and Fluid Mechanical Effects*

Figure 7-25 shows the mean temperature and pressure history as a function of time through the aft end of a typical engine, to illustrate pressure and temperature ranges in which trace species chemistry occurs. Turbine inlet temperatures vary within the flight cycle from 1200 to more than 2000 K, and pressures vary from 0.8–4.5 MPa. The gases remain in residence within the turbine and nozzle for approximately 5–10 ms. Combustor residence times have decreased (currently around 5 ms) as a result of efforts to reduce NO_x, and the time the exhaust gases spend within the turbine and nozzle can thus be longer than that in the combustor. At the engine exit, the temperatures and pressures typically range from 200–600 K and 0.02–0.1 MPa, respectively, depending on the particular engine technology and the operating conditions. Note that some of the temperature change within the turbine results from the addition of cooling air, as discussed above.

Significant variations around the baseline, one-dimensional flow conditions described above exist because of spatial and temporal flow nonuniformities at the combustor exit and throughout the turbine. There are many reasons for this nonuniformity: Combustion turbulence, the combustor exit temperature profile needed to maintain turbine blade life, fuel injector-induced hot spots, introduction of cooling flows, and viscous boundary layers and wakes. Together these effects lead to turbine inlet conditions—which may vary locally at any instant in time—between those associated with combustor inlet conditions and those resulting from stoichiometric combustion.

Downstream of the combustor—in the turbine and exhaust nozzle—the evolution of any particular species within the turbine and exhaust nozzle generally depends on local temperature and pressure, concentrations of other species, and variations of these parameters over time. The multitude of unsteady three-dimensional fluid mechanical effects and the large number of chemical species that interact with one another make this region a complex physical and chemical system. A full understanding and prediction of its behavior has not yet been attained.

Chemical models developed for application to the post-combustion expansion process must apply to a wide range of flow parameters, as discussed previously (Miake-Lye *et al.*, 1993; Brown *et al.*, 1996; Lukachko *et al.*, 1998). The availability of kinetic data applicable to the entire post-combustion range is limited because little research has spanned the wide gap in parameters between combustor and atmospheric conditions for many of the relevant reactions. An example of a chemical mechanism currently employed (Lukachko *et al.*, 1998) consists of 25 species coupled through 74 reactions representing contributions from gaseous SO_x, NO_y, HO_x, and CO_x chemistry (Westley *et al.*, 1983; Tsang and Hampson, 1986; Tsang and Herron, 1991; DeMore *et al.*, 1994; Yetter *et al.*, 1995). Because of gaps in the kinetic data, interpolation of available rates introduces uncertainties into current model results.

Beyond limitations in the range of applicability of basic kinetic data for identified reactions, the overall kinetic mechanism has yet to be validated against experimental data. Thus, the kinetic mechanism may not be complete, and missing mechanisms have yet to be positively identified. In particular, heterogeneous processes (e.g., involving soot and volatile particles) have not been adequately addressed by current models or measurements (see Chapter 3). In addition, recent in-flight (Fahey *et al.*, 1995b; Hanisco *et al.*, 1997; Anderson *et al.*, 1998; Hagen *et al.*, 1998; Miake-Lye *et al.*, 1998; Pueschel *et al.* 1998) and engine test cell measurements (Wey *et al.*, 1998) have indicated that additional SO_x oxidation is occurring that cannot be explained by SO_x reactions currently accounted for in mechanisms used to date.

7.6.4. Current Understanding of Chemical Changes in Turbine and Exhaust Nozzle

Approximately 99.5–99.9% of the molar content of typical commercial engine exhaust consists of N_2, O_2, CO_2, and H_2O. The species that compose the remaining 0.1–0.5% exist in trace amounts. This trace exhaust component consists primarily of NO_x, CO, unburned HC (including soot), the hydroxy family (HO_x, H_2O_x), the sulfur oxide family (SO_x), and elemental species such as O. Figure 7-26 provides a general categorization of chemical processes occurring in the turbine and exhaust nozzle. These processes are discussed in greater detail below.

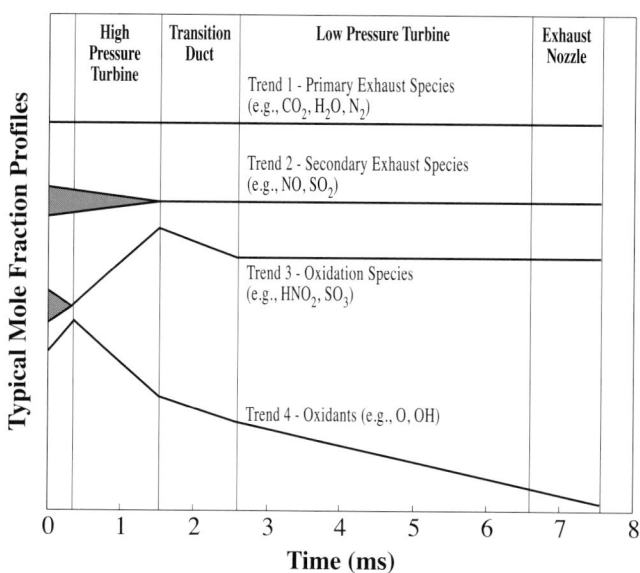

Figure 7-26: General categorization of chemical processes in the turbine and nozzle (Lukachko *et al.*, 1998).

7.6.4.1. Primary Exhaust Constituents (H_2O, CO_2, N_2, O_2)

Apart from the small effect of reactions involving trace species, changes in major species concentration in the turbine and nozzle flow path are caused by the diluting effect of cooling air. CO_2 changes less than a few tenths of a percent as a result of oxidation of CO (increases in H_2O from HO_x recombination are even smaller). This CO_2 fractional increase may grow in the first stages of the high-pressure turbine as more advanced cycles are implemented because associated cycle changes may result in relatively more CO at the entrance to the turbine (Godin *et al.*, 1995, 1997; Leide and Stouffs, 1996). Current small changes and likely future changes in primary exhaust constituents can be predicted with sufficient accuracy (Dryer *et al.*, 1993) for assessment needs, however, and the levels are all relatively easy to derive from measurements.

7.6.4.2. Secondary Combustion Products (NO, NO_2, N_2O, SO_2, CO, stable HC)

Secondary products—such as NO, NO_2, and SO_2, as well as their oxidative products SO_3, HONO, HNO_3, and H_2SO_4—formed via reactions initiated with the reactive radicals OH and O are the principal participants of interest in chemical and microphysical processes occurring soon after emission. Although OH and O are reduced considerably by the engine exit, they continue to play an important role in global atmospheric processes (see Chapters 2 and 3). To understand the processes occurring through the engine, relative and absolute levels of these secondary combustion products, their oxidative products (the acid gases), and the reactive radicals need to be accurately characterized. Emissions indices for NO_x, CO, and HC, as measured by ICAO procedures for stages in a standard LTO cycle, are documented (ICAO, 1995b) for most in-use engines as part of the engine certification process;

these emissions typically correspond to tens to hundreds of ppmv. SO_x emissions are directly proportional to the level of sulfur in the fuel [a 400 ppmm fuel S level corresponds to an $EI(SO_x)$ of 0.8]. Emissions of metals, whether from impurities in the fuel or engine wear, are much smaller than the emissions discussed here but may be of interest in soot activation and condensation processes (Chen *et al.*, 1998; Twohy *et al.*, 1998). NO_x does not change significantly through the turbine and nozzle other than through changes resulting from dilution, although the NO_2/NO ratio may shift as a result of increased oxidation. Oxidation of NO and NO_2 to HONO and HNO_3, respectively, is predicted to be on the order of a few percent or less, occurring largely in the high-pressure turbine (Fahey *et al.*, 1995a; Anderson *et al.*, 1996; Lukachko *et al.*, 1998). Although this change in the NO_x level is not significant, changes in HONO and HNO_3 represent important changes in trace species of NO_y (see below). Ground-based and in-flight measurements indicate that emissions of N_2O are also small relative to NO_x (Kleffmann *et al.*, 1994; Fahey *et al.*, 1995b). Further validation of NO_x chemistry is warranted, but indications are that current models can predict NO_x evolution in the turbine and nozzle with sufficient accuracy for assessment needs (Dryer *et al.*, 1993), and measurements of NO_x with a few percent accuracy are possible.

For typical civil engines, CO and HC are relatively unchanged through the turbine and exhaust nozzle. However, they can be reduced by up to two orders of magnitude in the turbine and exhaust nozzles (Godin *et al.*, 1995, 1997; Leide and Stouffs, 1996; Lukachko *et al.*, 1998) of advanced cycle military engines, where completion of oxidation in the high-temperature regions of the turbine results in the modest increases of CO_2 mentioned above. Measurements of these species (CO_2 and total hydrocarbons) are routine, and it is possible to measure them to several percentage points accuracy (Katzman and Libby, 1975; Spicer *et al.*, 1992, 1994; Dryer *et al.*, 1993; Howard *et al.*, 1996).

7.6.4.3. *Oxidation Products of Secondary Combustion Species (HNO₂, HNO₃, SO₃, H₂SO₄, H₂O₂, HNO)*

Modeling results (Brown *et al.*, 1996; Lukachko *et al.*, 1998) and, for SO_x, initial measurements (Arnold *et al.*, 1992, 1994, 1999; Fahey *et al.*, 1995a,b; Miake-Lye *et al.*, 1998) indicate that significant changes in the levels of these species are possible as their source gases are oxidized. These changes can be important with respect to atmospheric impact (see Chapters 2 and 3). However, most of the important chemical and fluid-mechanical effects remain unexplored. It is known, however, that strong non-linearities in the chemistry (based on as-yet unvalidated chemical models) can be accentuated by complex fluid-mechanical effects such as those associated with viscous boundary layers and wakes and blade cooling (Lukachko *et al.*, 1998). Figure 7-27 shows an example of this process: Two different modeling results are compared for a single blade row such as that shown in Figure 7-24. The first modeling result takes account of spatial temperature nonuniformity associated

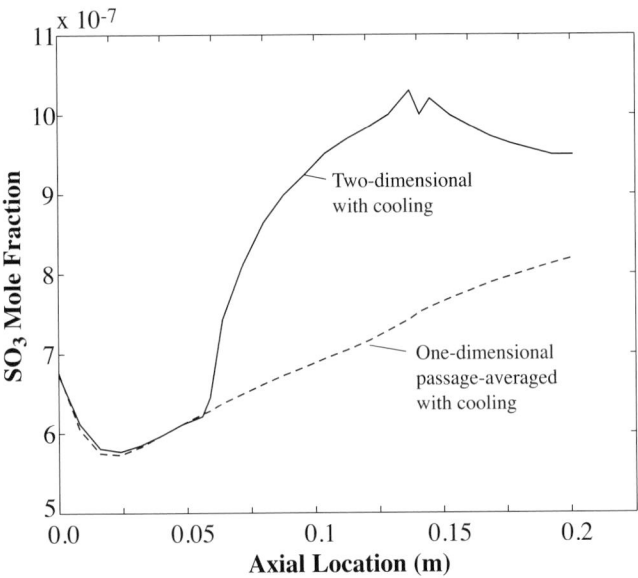

Figure 7-27: Influence of wakes and cold blade surfaces on chemistry for a single blade row in the turbine (Lukachko *et al.*, 1998).

with blade cooling, whereas the second is an averaged, one-dimensional representation of the flow path. SO_x oxidation in the blade row is enhanced when blade cooling is modeled.

Changes in many secondary combustion species from presumably negligible levels after leaving the combustor cannot yet be predicted with sufficient accuracy for assessment needs (Dryer *et al.*, 1993; Miake-Lye *et al.*, 1998). The existing experimental database for these species is insufficient to validate or improve current model predictions. For example, although models and initial measurements suggest that significant oxidation of SO_2 to SO_3 occurs within the engine, the models do not correctly predict the variation in the fraction of sulfur oxidized as fuel sulfur level is changed—suggesting that an unknown oxidative mechanism may be involved (Schumann *et al.*, 1996; Danilin *et al.*, 1997; Kärcher and Fahey, 1997; Miake-Lye *et al.*, 1998). Thus, current understanding is not mature enough to provide accurate initial conditions for aerosol formation models and subsequent atmospheric modeling downstream of the engine exit.

7.6.4.4. *Reactive Species (O, OH, HO₂, SO, H₂, H, N, CH)*

Current chemical models assume the radicals N and CH to be negligible for current technology engines, and their influence on the course of post-combustion chemistry has been ignored. The validity of this assumption has not been assessed, however, and as combustion temperatures increase in future engines these constituents (as well as the others) may become more important. O and OH are considered to be the dominant oxidizing species within current modeling assumptions (HO_2 levels remain less important), and orders of magnitude change are possible in these reactive species throughout the turbine and nozzle. Their evolution is very sensitive to chemical kinetics and species initialization at the combustor exit, and they are influenced by flow nonuniformities in the turbine and nozzle to a similar extent as the oxidation products. Prediction of their levels with sufficient accuracy for assessment needs is beyond currently validated capabilities, and the existing experimental database is insufficient to validate models. Furthermore, recent modeling studies (Yu and Turco, 1997, 1998) have implicated chemi-ions—a product of flame ionization in the combustor—as important in the growth of volatile particles after the exhaust leaves the engine. No measurements of ion levels at the engine exit have been identified; few measurements downstream of the exit exist (Arnold *et al.*, 1994, 1999), and only simple estimates of the recombination inside the engine have been attempted to date, so the uncertainty in the emission levels of chemi-ions is very high.

Pressure and temperature and residence times change over relevant ranges of operating conditions and engine cycles, and the chemical processes are expected to change as a result. Generalizations regarding the exhaust constituents described above are not expected to change significantly over cruise conditions and even over the broader operating conditions representative of the landing/take-off cycle. However, changes in exit speciation are expected for the oxidative products and radical species that depend non-linearly on combustor exit concentrations and local conditions. Only limited research has been carried out on this topic (Hunter, 1982; Brown *et al.*, 1996; Lukachko *et al.*, 1998).

Table 7-7 provides a simplified status summary of the confidence attached to current modeling and measurements of emissions emerging from engines. Reducing uncertainties in the oxidation products of secondary species and in reactive species will require additional new measurements, as well as increased

physical understanding of exhaust species oxidation in the turbine and nozzle.

7.7. Engine Emissions Database and Correlation

Since the early 1970s, when engine manufacturers started measuring emissions to demonstrate compliance with regulations developed for the airport vicinity, there has been a steady increase in data available for the development and refinement of databases associated with subsonic aircraft engine emissions. This section provides a brief overview of current ICAO engine standards and the status of the emissions database. It also discusses methods used to correlate sea-level emissions measurements with in-flight levels. Special emphasis is placed on simplified methods of NO_x prediction and their validation for inventory purposes, as well as emissions variability.

7.7.1. ICAO Engine Standards and Emission Database

To control pollutants from aircraft in the vicinity of airports, ICAO established emissions measurement procedures and compliance standards for soot (measured as smoke number—SN), unburned hydrocarbons, carbon monoxide, and oxides of nitrogen. A landing and take-off cycle was defined to characterize the operational conditions of an aircraft engine within the environs of an airport; this LTO cycle is illustrated in Figure 7-28. The standards are applied to all newly manufactured turbojet and turbofan engines that exceed 26.7 kN rated thrust output at International Standard Atmosphere (ISA) sea level static (SLS) conditions. The smoke standards

Table 7-7: *Summary of confidence attached to current modeling and measurements of emissions emerging from engines.*

Exhaust Products from Engines	Principal Effects of Post-Combustor Reactions and Levels of Confidence Associated with Modeling and Measurements
Primary constituents (e.g., H_2O, CO_2, N_2, O_2)	Present combustors convert almost all of the kerosene to the products of complete combustion. Further CO oxidation in the turbine (a few tenths of a percent or less) slightly increases CO_2 emitted. Prediction capability is good, and levels are easy to derive from basic engine operating conditions or measurements.
Secondary products (e.g., NO, NO_2, N_2O, SO_2, CO, stable HC)	NO_x is little changed by flow through turbine. Oxidation of NO and NO_2 to HONO and HNO_3, occurring mainly in the high-pressure turbine, is a few percent or less. For civil engines, CO and HC are relatively unchanged in the turbine, but significant reductions can occur there in advanced military engines. Accurate NO_x predictions and measurements are now routinely performed for assessment purposes.
Oxidation products of secondary combustion species (e.g., HNO_2, HNO_3, SO_3, H_2SO_4, H_2O_2, HNO)	Chemical mechanisms and reaction rates of trace species are not well known over the range of post-combustor conditions. The impact of fluid mechanics on chemical evolution is not yet fully evaluated by models or measurements.
Reactive species (e.g., O, OH, HO_2, SO, H_2, H, N, CH)	As above, but validation of trace species chemistry mechanisms, via measurements, is also needed over the relevant temperature and pressure range for both classes. Further modeling is required to make the connection with species for which measurements are not available.

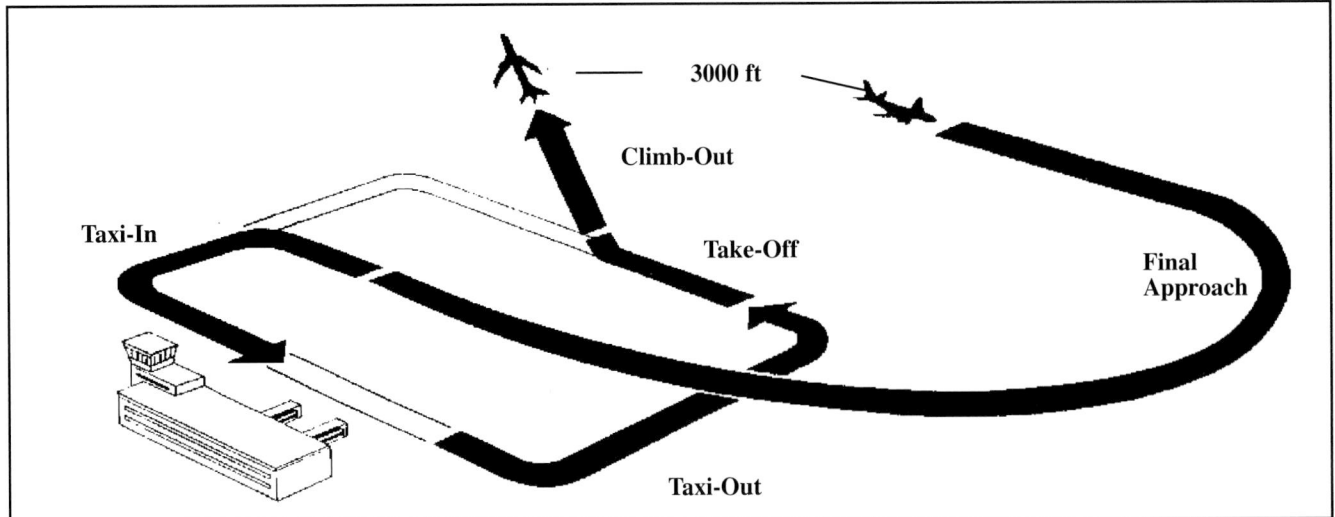

Figure 7-28: The ICAO landing and take-off cycle (LTO).

took effect in 1983, and those for gaseous emissions took effect in 1986. Measurements of the exhaust emissions of a single engine are performed at the manufacturer's test facilities as part of the certification process, in compliance with the requirements of ICAO international standards and recommended practices of Annex 16 to the convention on international aviation (ICAO, 1993).

The data are published in an ICAO exhaust emissions data bank (ICAO 1995b). Engine emissions are given for the standardized LTO cycle represented by an engine power setting of 7 (taxiing), 30 (approach), 85 (climb-out), and 100% (take-off) of rated output and given times in mode (see Figure 7-28 and Table 7-8). Together with fuel flow, emission indices of HC, CO, and NO_x in g per kg of fuel burned and maximum SN are reported. For a variety of engines, the measured SNs for all power settings are provided. Except for smoke, the emissions of each LTO cycle mode (EI x fuel flow x time in mode) are summed (Dp) and expressed in the form Dp/F_{00} (g kN) where (F_{00}) is the maximum thrust of the engine at take-off under ISA SLS conditions.

The emissions measurements are taken at the exit plane of the engine's exhaust nozzle (within 0.5 nozzle diameter). Kerosene-type fuel complying with specified properties—density, heat value, boiling points, aromatics (15–23% volume), sulfur (less 0.3% mass), hydrogen (13.4–14.1% mass)—is used. No additives for smoke suppression are allowed. A set of correction procedures, approved by ICAO bodies, has been developed for gaseous emissions to ensure that the observed emission indices can be compared at reference day conditions [ISA SLS pressure (101.325 kPa) and temperature (288.15 K)]. For NO_x, an additional correction is made to take account of ambient humidity, using as a reference an absolute humidity value of 0.00629 kg water per kg dry air (about 60% relative humidity). For economic reasons, only a small sample of engines of any type is tested. For regulatory purposes, therefore, a statistically based correction is used to account for engine-to-engine variability resulting from manufacturing tolerances. To ensure that the mean value of a population of an engine type will meet the given limits of standards within a confidence level of 90%, an additional factor is applied to the measured mean value of the LTO cycle emission (Dp/F_{00}) to give the so-called characteristic value, which must be in compliance with the regulatory level. The additional statistical factor—derived from a variety of engines measured by all engine manufacturers—depends on the emission species and the number of engines tested. For two engines to be regarded as representative of a type, the increases applied to the mean measured emissions value for each species are as follows: +10.0% for NO_x, +13.9% for CO, +30.1% for HC, and +17.3% for SN. The relative amount to be added to the measured value decreases with the number of engines tested.

Table 7-8: *LTO cycle measurements for a high bypass GE (CF6-80) turbofan engine (ICAO, 1995b).*

Time in Mode	Rated Output (F_{00})	Fuel Flow (kg s⁻¹)	HC	CO	NO_x	SN
0.7 mins. take-off	100%	2.353	0.08	0.52	28.06	7.1
2.2 mins. climb out	85%	1.913	0.09	0.52	21.34	–
4.0 mins. approach	30%	0.632	0.20	2.19	8.97	–
26 mins. idle	7%	0.205	9.68	43.71	3.74	–
Dp/F_{00} (g kN⁻¹) LTO cycle miss./rated output measured avg.			12.43	57.09	42.17	
Dp/F_{00} (g kN⁻¹) characteristic value to be regulated			16.2	65	46.4	8.3
Current regulatory level			19.6	118	80.2	18.3

Table 7-8 shows an example of engine emission data extracted from an ICAO data sheet submitted by the manufacturer with two engines tested.

Ongoing revision of the regulatory level (see last line of Table 7-8), as well as the entire emission certification process, is one of the objectives to be followed by CAEP.

The HC, CO, and SN standards have remained unchanged within the CAEP process. For NO_x, the approach was to tighten NO_x stringency in accordance with technology gains. The baseline (see line CAEP/1 in Figure 7-29) had been introduced to allow NO_x to rise with maximum engine pressure ratio and associated temperature, a parameter that strongly influences the rate of NO_x production (see also Section 7.4.3). In a second stage, the regulatory NO_x level was decreased by 20%. This level is often referred to as the CAEP/2 standard (Figure 7-29); it has been effective for new engine types since 1996 and will apply to newly manufactured engines from the year 2000. From 2004 onward, a further reduction in the LTO regulatory values of NO_x was agreed at a fourth meeting of the CAEP (see line CAEP/4 in Figure 7-29). This value is 16.25% below the CAEP/2 standard at an engine pressure ratio of 30, with some allowances for engines with higher pressure ratios.

Present regulatory procedures based on the LTO cycle were designed to address airport air quality problems, but the CAEP is now pursuing new certification methodologies that take account of the flight mode as well. Although there are links between trends in LTO and flight exhaust emissions, new correlation procedures (see below) are needed to enable ground test measurements to be used to provide quantitative methods for predicting altitude emissions from aircraft. Aircraft engines produce many different emissions. However, only emissions that have been derived for use within the ICAO LTO cycle certification process—such as HC, CO, NO_x, and smoke—or are easily correlated using simple conversion factors related to fuel burn (such as CO_2, H_2O, and SO_x) have been included in significant numbers in today's inventory studies. Apart from the major exhaust emissions, a large variety of minor species are produced. HC emissions, which are typically reported as equivalent mass of CH_4, can be broken down into numerous complex compounds. There has not been sufficient characterization to date for global inventory purposes, with most of the available data acquired from a selection of military engines using kerosene-type fuel with different specifications. Additionally, nitrous oxides (N_2O) have not been rigorously characterized. Other emissions are not currently modeled in emissions databases because of very small quantity or the fact that little data exists. This subject is further discussed in Section 7.6.

7.7.2. Engine Load and Emission Correlation

7.7.2.1. Emission Correlation Methods

Establishment of an aircraft emissions inventory for a given flight traffic scenario requires a knowledge of the engine's emissions

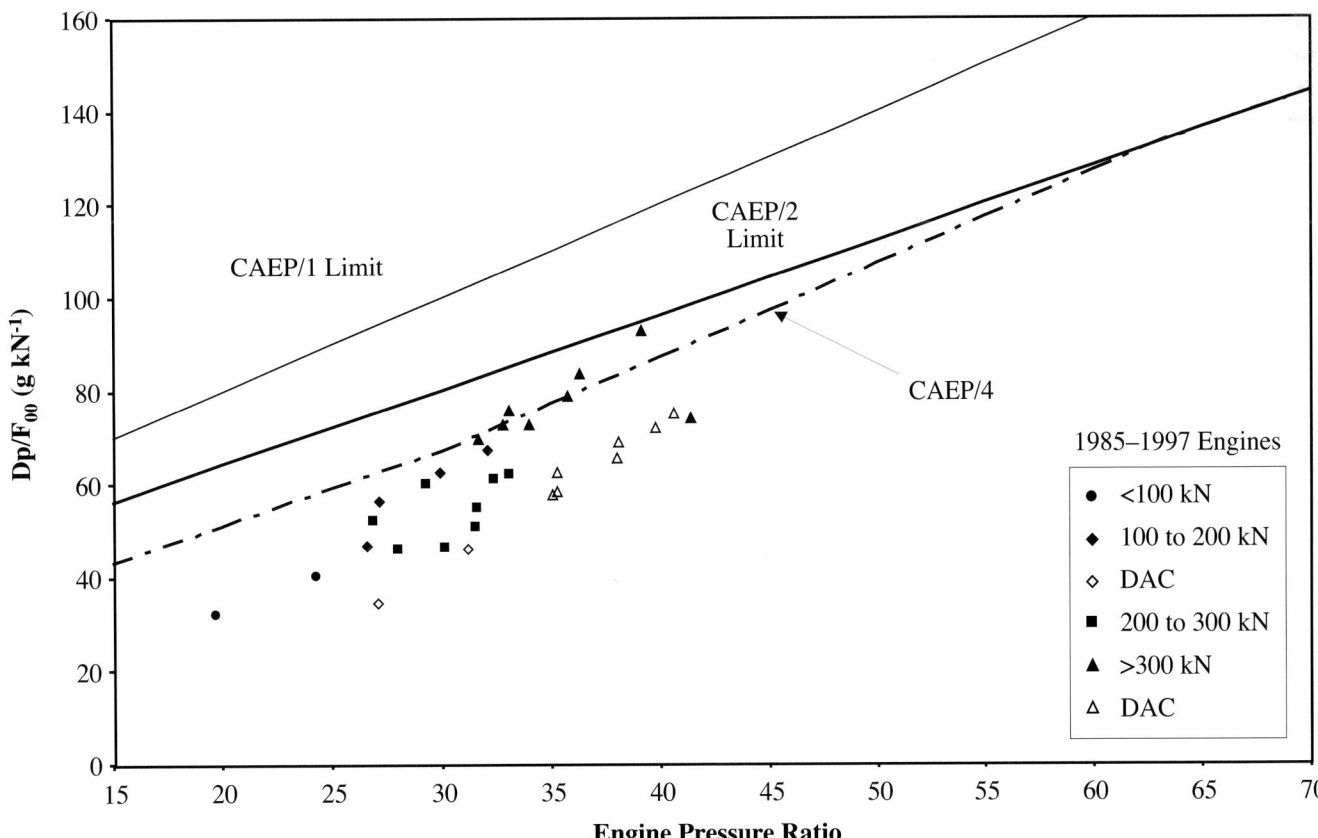

Figure 7-29: Engine NO_x characteristic values of Dp/Foo for the ICAO LTO cycle, and changes in regulatory limits.

over the aircraft's total flight mission. This inventory is relatively straightforward for CO_2, H_2O, and SO_x (in total) emissions because of their direct link with mission fuel burn. For several other emissions that depend on engine power setting, combustor design, and flight condition, correlation methods have been developed to calculate the exhaust emissions for a specific engine type using measured data taken during the engine certification process. This correlation is carried out primarily with species of emissions such as NO_x, CO, HC, or soot by means of a semi-empirical correlation between emissions and principal combustion parameters, using measurement programs involving combustor rigs and engine systems together with theoretical considerations of the main combustion processes. Exhaust emissions production processes generally are complex because they involve unsteady physical processes as well as non-equilibrium chemical processes. A fundamental element in the development of formulas to correlate measured and predicted emission indices is a model of the process based on the relationship of the emission index, chemical kinetic rates, and residence time in the reaction zone. This model is of great value to manufacturers wishing to predict the emissions performance of a new or development combustor. For an existing design, a reference correlation method is used to predict emissions on the basis of measured data. Because combustor inlet pressure (p_3) and temperature (T_3) are the main parameters involved. For EI(NO_x), the "p_3/T_3" method leads to a relationship that, in its simplest form, is as follows:

$$EI(NO_x)/EI(NO_x)_{ref} = (p_3/p_{3ref})^n \, f(Humidity) \text{ for } T_3 = T_{3ref}$$

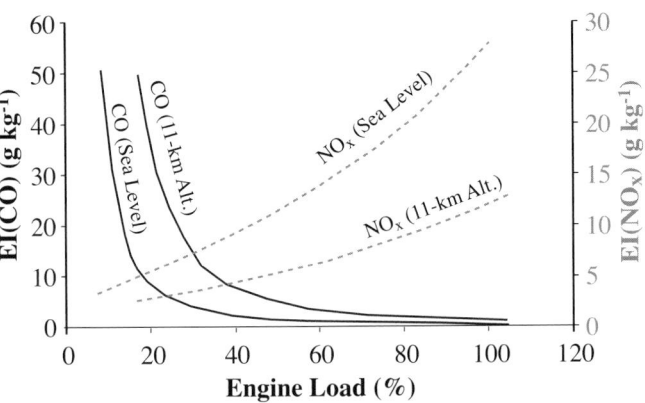

Figure 7-30: Emissions characteristics of a fan engine versus load at sea level and in flight at 11 km.

where n is an exponent in a range between about 0.3 and 0.6 derived from engine and combustor rig testing.

The parameters p_3 and T_3 come either from measurements or from computational engine simulation. Measurement programs, especially for NO_x emission indices, reveal agreement (compared with testing) of better than 5% (AERONOX, 1995; Brasseur *et al.*, 1998). Emissions of HC and CO depend on the completeness or efficiency of the combustion process. Test data are well correlated using a combustor loading parameter, which takes account of the residence time of the burning products and reaction time in the combustor reference volume. As an example,

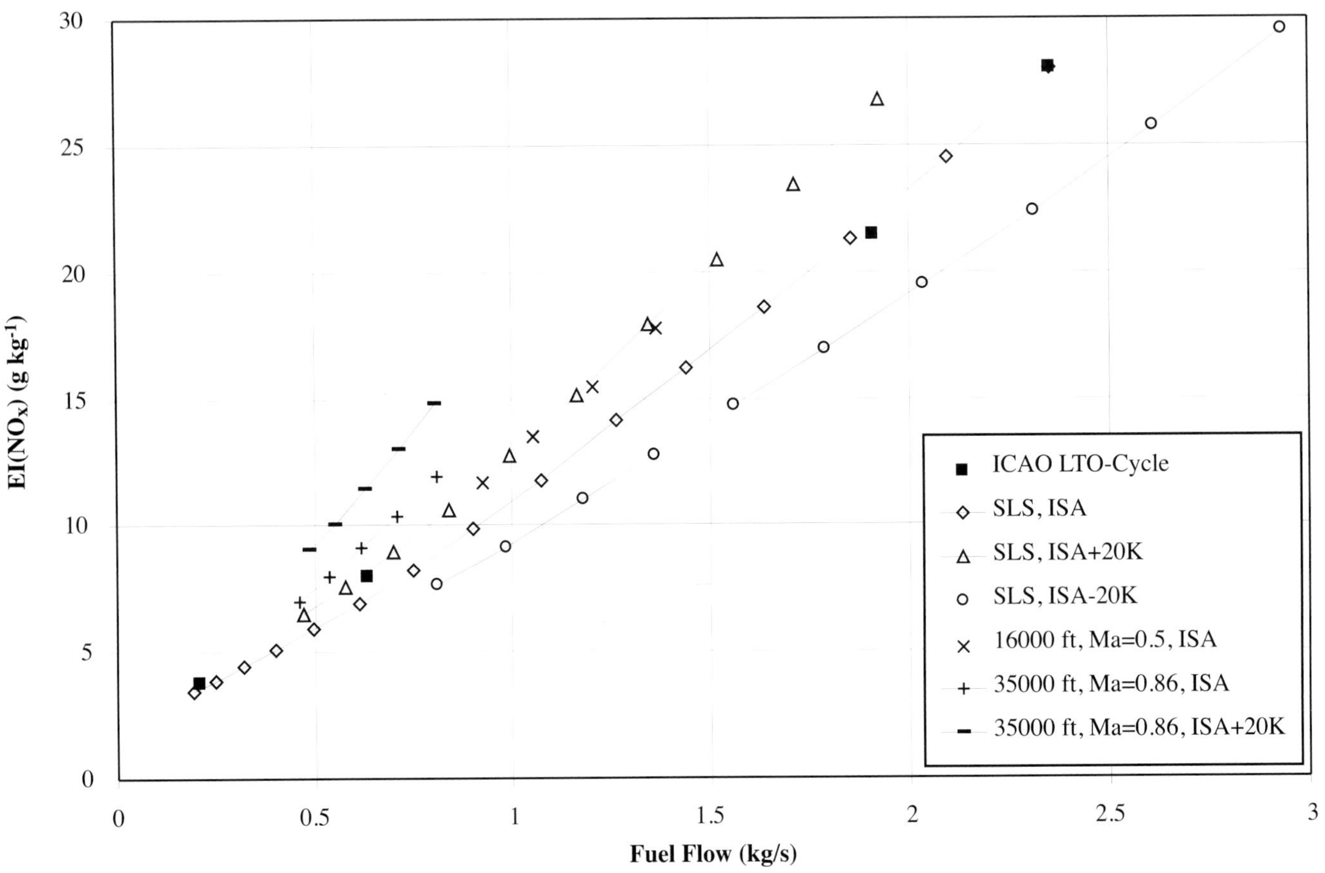

Figure 7-31: Variation in NO_x emission index with engine load and ambient conditions (CF6-80C-type engine modeling).

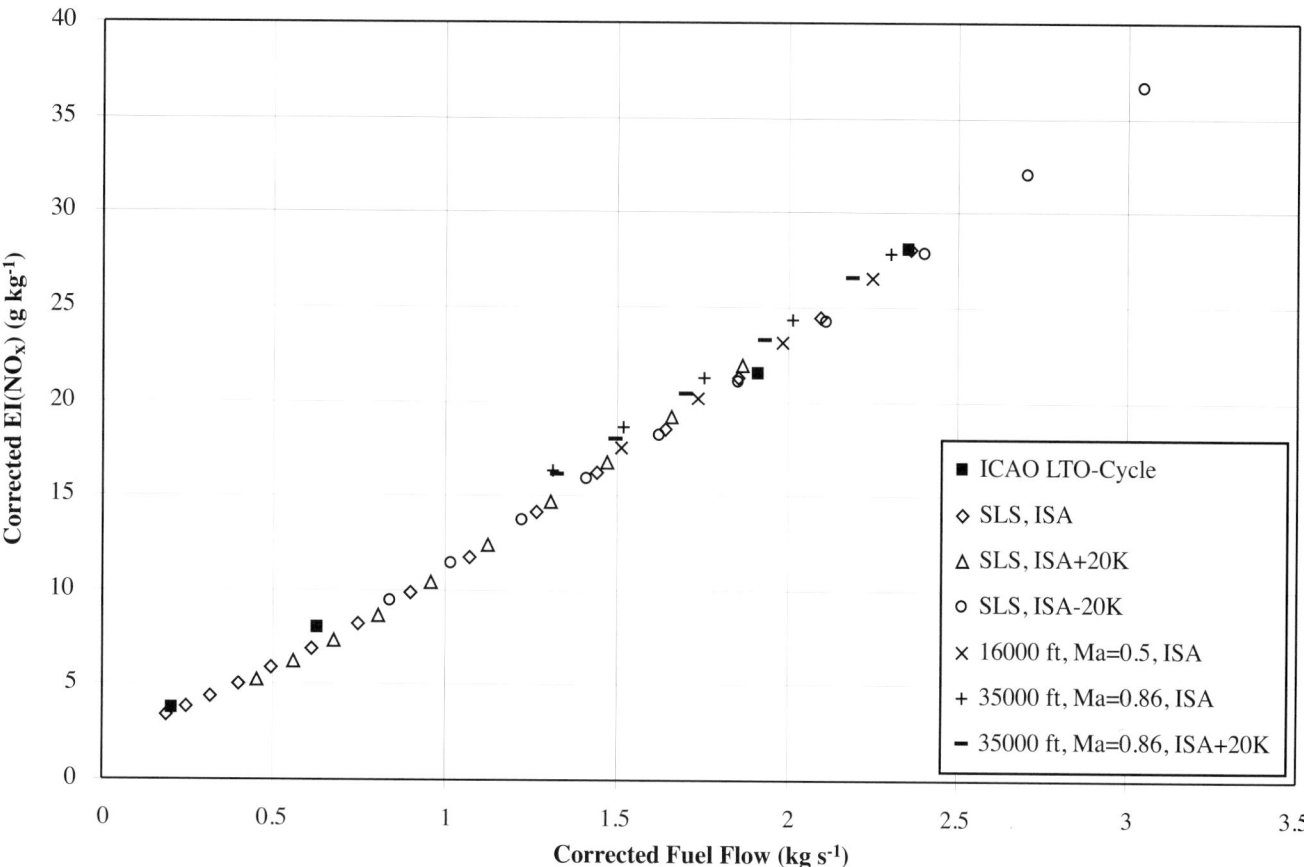

Figure 7-32: Emission indices and fuel flow corrected for ISA SLS conditions.

Figure 7-30 shows typical functions of emission indices of NO_x and CO versus engine load for a turbofan engine at different altitudes on the basis of a thermodynamic engine simulation. Unburned hydrocarbons will follow the same trend as CO, but on a much lower level. Both are products of incomplete combustion; thus, they are controlled by the same mechanism.

Despite the intrinsic complexity of the chemical processes associated with soot production, a semi-empirical correlation between calculated and measured SN has been developed with a 40% standard deviation (De Champlain *et al.*, 1997). By also considering reaction kinetics, Döpelheuer (1997) developed a correlation method based mainly on the concentration of carbon and oxygen in the combustion zone. The concentration of soot is considered to be proportional to the equivalence ratio, to pressure at the combustor inlet (both with exponents to be determined), and to an exponential function controlling the reaction rate. To apply the method as a reference type of correlation, soot density as a function of measured SN (Finch and Eyl, 1976; Whyte, 1982; Champagne, 1988; Hurley, 1993) has been approximated as a first step. Up to a 20% deviation exists between different correlations of soot loading and SN.

Because of the limitations of the standard measuring method for SN (i.e., by collecting soot on a white filter paper and evaluating reflected light intensity) (ICAO, 1993), only the major particles of soot are measured; the remaining wide spectrum of very small particles are not measured (see Section 7.5.3).

7.7.2.2. Simplified Alternate Correlation Methods

The methods described above to correlate engine emissions for different operating conditions all require engine internal gas path data. Such data are normally sensitive from an engine manufacturer's point of view. Alternate correlation methods that do not expose proprietary or sensitive engine data have therefore been sought. These methods are based mainly on emission correlation with the fuel flow of an engine. Thus, SLS engine test data from certification testing, together with relevant methods of correlating emissions with engine performance in flight, is favored as one of the principal sources for the development of aircraft emission inventories. These methods are designed to use unrestricted data from the ICAO engine emission databank coupled with fuel flow data, which can easily be acquired from aircraft missions.

Several methods based on the above principles have been developed and are undergoing continuing improvements. The key element of these simplified methods is the assumption that emission indices at different engine inlet conditions might be correctable to a reference day condition, thus collapsing into a single function of the corrected fuel flow (Lecht and Deidewig, 1994; Martin *et al.*, 1994, 1995; Deidewig *et al.*, 1996). An example of the effectiveness of such a method developed for NO_x emissions is outlined in Figures 7-31 and 7-32. Figure 7-31 shows actual NO_x emission indices according to ambient flight conditions and calculated with a complex correlation formula;

Figure 7-32 gives the result of the same values but corrected for ISA SLS conditions. For comparison, ICAO LTO cycle test data are marked separately.

This method allows ISA SLS measurements to act as a reference function; this function simply has to be re-corrected for actual in-flight engine inlet conditions. These simplified emission correlation methods are highly relevant in terms of their value for the further development of aircraft emission inventories (see Chapter 9). There is, as yet, no fuel flow-based correlation that can be used to predict engine soot for the purposes of inventory preparation because of the use of SN data in the ICAO emissions databank, which cannot readily be converted to soot loading (Döpelheuer and Lecht, 1999).

7.7.3. *Validation of Emission Prediction Methods*

7.7.3.1. *Altitude Chamber Testing*

Cruise-level emission index prediction methods need validation by measurement. Such validation can be carried out at ground-level test chambers in which pressure and temperature can be varied to simulate a wide range of engine operation conditions in flight. Within the AERONOX project, exhaust emissions of two selected engines (Rolls Royce RB211 and Pratt & Whitney PW305) have been analyzed and compared with predictions (Lister *et al.*, 1995). This comparison shows that emission prediction methods developed by engine manufacturers and research institutes can predict the NO_x emission at a flight condition within an error band of 5–10% for a modern high bypass engine. These experiments found a tolerance for fuel-based methods that was nearly as good. Measurements from the AERONOX project also showed that for highest accuracy, the prediction equations must be adapted for specific engine type. A kind of indirect validation has been undertaken by comparing NO_x emissions of fuel flow-based methods with the p_3/T_3 method. Such an evaluation showed an agreement within 13% maximum and 6% standard deviation for a variety of aircraft and flight missions (ICAO, 1995c).

7.7.3.2. *Validation by In-Flight Measurement*

An alternate suitable method is to carry out measurements in the freshly emitted exhaust plume of an aircraft in flight. Several such cruise altitude emissions measurement programs have been conducted. Cruise altitude *in situ* chemical probing was carried out for the first time in December 1991, when the DLR research aircraft "Falcon" flew through the plume of a commercial DC-8 airliner (Arnold *et al.*, 1992). In 1993, the first in-flight measurements of emission indices were accomplished, again using the "Falcon" (Schulte and Schlager, 1996). This measurement involved measuring CO_2 simultaneously with other species of interest (in terms of volume mixing ratios). A detailed discussion of these assumptions and the emission index determination from the measured ratios $\Delta[X]/\Delta[CO_2]$ appears in Schulte and Schlager (1996).

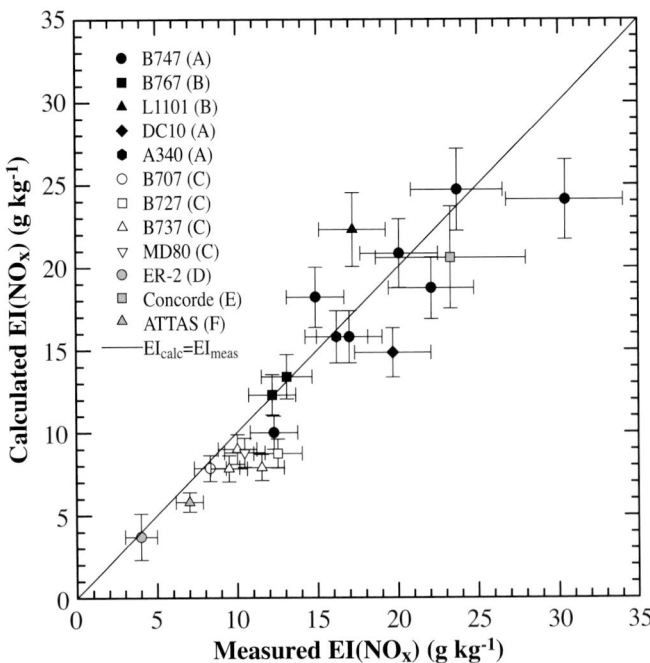

Figure 7-33: Comparison of all available *in situ* measured NO_x emission index values with corresponding predicted values: (A) Schulte *et al.* (1997); (B) Schlager *et al.* (1997); (C) Schulte and Schlager (1996); (D) Fahey *et al.* (1995a); (E) Fahey *et al.* (1995b); and (F) Haschberger and Lindermeir (1996).

In recent years, several *in situ* EI measurements have been carried out. Fahey *et al.* (1995a,b) determined emission indices for NO_x, CO, and N_2O in the exhaust plumes of the NASA ER-2 research aircraft and the Concorde in the lower stratosphere. Haschberger and Lindermeir (1996, 1997) analyzed the exhaust plume of the DLR experimental Advanced Technology Testing Aircraft System of DLR, a two-engine jet aircraft of type VFW 614 (ATTAS) with onboard instruments and derived emission indices for NO_x, CO, and H_2O. The subsonic long-range jet aircraft types that dominate global air traffic were investigated by Schulte *et al.* (1997) to derive NO_x emission indices. Figure 7-33 shows a comparison of all available *in situ* measured $EI(NO_x)$ values with corresponding predicted values. Note that the measurement details were different for each case.

In summary, these measurements revealed good agreement between predicted and measured NO_x emission indices. Special attention was paid to predictions based on fuel flow methods (Lecht and Deidewig, 1994; Deidewig *et al.*, 1996; Martin *et al.*, 1996; Schulte and Schlager, 1996; Schulte *et al.*, 1997) because they are the methods of choice for the building of aircraft NO_x emission inventories. These methods seem to underestimate $EI(NO_x)$ by about 12% on average. However, this 12% deviation is within the uncertainties—indicated as error bars—of the measurements and the predictions.

7.7.4. *Engine Emission Variations*

Engine emissions may vary for a number of reasons. Manufacturing differences, aging characteristics for individual engines, operational and atmospheric conditions, and changes in fuel contents all have parts to play. Data from new engine certification testing reveals a standard deviation for EI(NO_x) of 1–7%. Fuel type, fuel content, and sampling methods have only a slight influence on NO_x emission variation (Lyon *et al.*, 1980; Lukachko and Waitz, 1997). Another major source of variability may be engine deterioration over a long period of time. This deterioration is evident in measurements of engine parameters such as exhaust gas temperatures and fuel consumption rates. Economic considerations recently led to allowable SFC limits of between 2 and 4% because of engine deterioration; exceedances lead to engine overhaul. Measurements of overhauled engines revealed a similar standard deviation for NO_x based on the LTO cycle (Lister and Wedlock, 1978), which implies that in emissions terms their performance is similar to new engines. Lukachko and Waitz (1997) investigated the influence of an ongoing degradation process resulting from aging. In a combined sensitivity study, they found that a 3% SFC increase from deterioration led to a –1 to +4% change of NO_x emission efflux, depending on the engine part mostly affected by the deterioration. This study was conducted under cruise operating conditions. New consideration is being given by ICAO/CAEP to the development of a more appropriate certification methodology in terms of emissions variability over the entire flight cycle.

7.8. Aviation Fuels

This section addresses the major fuel-related issues that have influenced and will continue to influence the development of aircraft into the foreseeable future. Almost all current civil and military aviation around the world uses a kerosene-type fuel. This class of fuel provides a good balance of properties currently required from an aviation fuel, in which energy density, operational issues, cost, and safety all need to be taken into account. This section examines some alternative fuels that will, no doubt, continue to be considered as the demand for air transport continues and its impact on the environment grows.

Fuel specifications that define physical properties, chemical composition, and performance tests have evolved over several decades. These fuel specifications are designed to balance quality, cost, and availability, thus guarantee a product of worldwide consistency. Although many countries have their own fuel specification, by general agreement among governing bodies, fuel suppliers, and aircraft manufacturers, all civil aviation fuel must effectively meet the requirements of American Society of Testing and Materials (ASTM) D1655 (ASTM, 1997) and Defense Evaluation and Research Agency (DERA) DEF STAN 91-91 (DERA, 1998). The ASTM specification contains two relevant fuel designations (Jet A and Jet A-1), which differ only in their freezing points. The DEF STAN specification addresses only Jet A-1. Jet A fuel has a maximum freezing point of -40°C and is used only in the United States, where moderate temperatures combined with short flight times justify a separate specification to increase availability. Jet A-1 has a maximum freezing-point requirement

of -47°C to meet the low-temperature requirements of long, high-altitude flights and is used everywhere in the world except the United States. Other differences between these fuel specifications are relatively minor and immaterial to this report.

The ICAO specification for fuel to be used in emissions testing of aircraft gas turbines is also a kerosene-type fuel (ICAO, 1993). Restrictions are placed only on the 10 properties that potentially affect emissions. Table 7-9 compares the ICAO specification with the ASTM and DEF STAN specifications for the relevant properties. The property limits of ICAO are somewhat more restrictive than the commercial specifications to limit testing concerns, but at the expense of cost and availability.

Also shown in Table 7-9 is the percentage of fuels sold in 1997 in the United Kingdom under DEF STAN 91-91 that meet the ICAO limits. Virtually all DEF STAN 91-91 fuels met the ICAO limits except for naphthalenes and hydrogen content; in both cases, the non-attainment fuels were to the side of lower emissions (i.e., lower naphthalenes and higher hydrogen). This comparison shows that the ICAO fuel specification for emissions testing is relevant to jet fuels being marketed.

The primary military fuels of North America and Western Europe are defined by the identical specifications F-34 (NATO) and JP-8 (United States). These specifications effectively define military fuels throughout much of the world because many countries buy their military aircraft from the same manufacturers. The only significant difference between these military fuels and Jet A-1 is the mandatory use of certain additives in the military fuels; however, some of these same additives may be found in some civilian fuels. For shipboard safety reasons, Navy aircraft use a high-flash-point kerosene fuel that is less volatile, but other relevant properties are similar; in Western Europe and North America, these fuels are designated as F-44 (NATO) and JP-5 (United States). For completeness, we mention here the existence of small volumes of special fuels used by military aircraft that fly at very high altitudes and/or require a higher thermal stability than conventional fuels provide. These fuels are also kerosene-type fuels but may have different volatility/freezing-point requirements and are more highly refined to improve thermal stability.

All jet fuels are composed primarily of hydrocarbons as a blend of saturates, with no more than 25% aromatics. Olefins may be present, but they are effectively kept below about 1% by stability requirements. Additionally, a fuel may contain up to 0.3% sulfur by weight, although the level is generally less than 0.1%. Certain additives may also be present, as mentioned previously. Trace levels of oxygenated organics (e.g., organic acids) may be present but are effectively limited in concentration by the fuel specification to ensure product stability and materials compatibility. Metal contaminants such as iron, copper, and zinc can be picked up from plumbing and storage systems and can be present in the low ppb range. Halogens are not an issue because they are not used in refinery processes for kerosene. Additives currently used in jet fuels are all organic compounds that may also contain a small fraction of sulfur or nitrogen. The

Table 7-9: Comparison of relevant properties of ICAO fuel for emissions testing and commercial jet fuels.

Property	ICAO	Allowable Range of Values		% UK Fuels Meeting ICAO (1997)
		ASTM D 1655 Jet A/Jet A-1	DEF STAN 91-91 Jet A-1	
Density (kg m^{-3}) at 15°C	780–820	775–840	775–840	99.9
Distillation temperature (°C)				
10% boiling point	155–201	≤ 205	≤ 205	100
Final boiling point	235–285	≥ 300	≥ 300	99
Net heat of combustion (MJ kg^{-1})	42.86–43.50	≥ 42.8	≥ 42.8	100
Aromatics (vol %)	15–23	≤ 25	≤ 22[a]	99
Naphthalenes (vol %)	1.0–3.5	(< 3.0)[b]	(< 3.0)[b]	98.5
Smoke point (mm)	20–28	≥ 25[b]	≥ 25[b]	99.3
Hydrogen (mass %)	13.4–14.1	n/a	(Report)[a]	90.4
Sulfur (mass %)	< 0.3	≤ 0.30	≤ 0.30	100
Viscosity at –20°C (mm^2 s^{-1})	2.5–6.5	≤ 8.0	≤ 8.0	100

[a]Aromatics ≤ 25% allowed if hydrogen content is reported
[b]Smoke point ≥ 19mm allowed if naphthalenes < 3.0%.

maximum allowable concentrations of these additives is controlled by relevant fuel specifications. These concentrations vary with the additive but are less than 6 mg L^{-1} (approximately 6 ppm), with the exception of two additives that contain no sulfur or nitrogen. Therefore, these additives presumably will have no measurable impact on emissions and are not an issue for this discussion (based on their constituents and very low concentrations); however, additional testing should be conducted to verify these conclusions.

In summary, differences between jet fuel specifications around the world are relatively minor and have little effect on fleet exhaust emissions. Thus, aviation fuel, in the context of this report, refers to all civilian and military jet fuels unless otherwise specified.

7.8.1.   Databases on Fuel Properties

Only two comprehensive, annual surveys conducted on jet fuel properties are publicly available. Both of these surveys have been carried out annually since the early 1970s—one by the National Institute for Petroleum and Energy Research (NIPER) for U.S. fuels and the other by the Defence Evaluation and Research Agency (DERA) for UK fuels. The 1996 DERA survey covered 1467 batches of fuel representing 14.85 million m^3 (Rickard and Fulker, 1997). The UK survey is thought to be fairly representative of fuel used in the rest of Western Europe as well, because much of the UK fuel comes from refineries

located there. These two surveys are useful in defining trends in fuel properties, although they are both sample averaged, not volume averaged.

Three snapshot surveys have been reported. Bowden *et al.* (1988) reported a survey by the U.S. Army Fuels and Lubricants Research Laboratory of 90 JP-8 fuels from Western Europe, plus one from Singapore and two from Korea. In the late 1980s, Boeing conducted a survey of 39 civil aviation fuels from around the world (Hadaller and Momenthy, 1990). For the two properties that will be discussed here—sulfur and hydrogen contents—the averages and distributions for these two worldwide surveys were similar to the UK results. In 1996, a snapshot survey of U.S. jet fuel properties was conducted jointly by the American Petroleum Institute (API) and the National Petroleum Refiners Association (NPRA); this survey included Jet A fuels from 105 refineries representing 2.47 million m^3. Average sulfur levels in this survey agreed very closely with the average NIPER sampling for that year, lending some credence to the NIPER results on sulfur (API/NRPA, 1997).

7.8.2.   Fuel Composition Effects on Emissions

For the most part, the design of the combustion chamber determines the gaseous and soot emissions from a gas turbine; there are only limited opportunities for fuel properties to influence emissions. Certainly there can be significant effects over the spectrum of hydrocarbon fuels from methane or natural gas to heavy distillate and residual fuels. Within the narrow definition of aviation kerosene, however, there is little opportunity for reducing emissions from current aircraft by fuel modification, with the exception of steps taken to reduce sulfur.

Nitrogen for NO_x comes from the air, not the fuel. Jet fuels contain only trace amounts of fuel-bound nitrogen, which cause storage stability problems and render the fuel unfit for use.

CO_2 and H_2O emissions are influenced by fuel composition. A fuel with a higher H/C ratio will produce lower CO_2 and correspondingly more water; however, only relatively small variations are found in aviation fuel. The NIPER survey of U.S. fuels does not record hydrogen content; however, the 1996 UK survey (Rickard and Fulker, 1997) and the Boeing survey (Hadaller and Momenthy, 1990) are in very good agreement with regard to the range of hydrogen content in jet fuels and the mean value. Figure 7-34 shows the distribution of fuel hydrogen content from the UK survey. The bulk of the data fall between 13.5 and 14.1% with a mean value of 13.84%. More than 90% of data in the 1989 Boeing worldwide survey also fall between these same values, with a mean of 13.85%. Thus, typical emissions indices for CO_2 and H_2O are 3.15 ± 0.01 and 1.25 ± 0.03, respectively, where the variations allow for the range of hydrogen content found in jet fuels.

Increasing the hydrogen content of jet fuel has been considered as a way of reducing CO_2 emissions. Currently, the most efficient and effective means of generating hydrogen is by steam reforming of natural gas; the thermal efficiency of this process is 78.5% (as compared to water electrolysis, with a thermal efficiency of 27.2%) (Encyclopedia of Chemical Technology, 1995). This process is expected to remain the most cost-effective means of producing hydrogen for the foreseeable future (Tindall and King, 1994). Even if pure methane, which has the highest H/C ratio of any fossil fuel, were used as the source, the inefficiencies of hydrogen production would result in about three times as much CO_2 being released as would be saved from combustion of a fuel with higher hydrogen content. Even if hydrogen could be obtained from water without using fossil

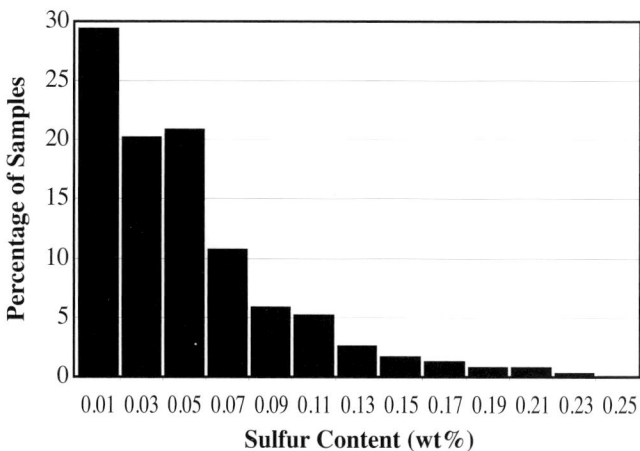

Figure 7-35: 1996 UK survey of sulfur content in jet fuel.

energy, increasing the average hydrogen content by 0.05% (a significant amount) would result in only a 0.6% reduction in the amount of CO_2 produced per pound of jet fuel burned simply based on stoichiometric combustion calculations.

The most significant effect that changes in aviation kerosene can have on emissions is in reduction of SO_x; all of the sulfur in the fuel is converted to sulfur oxides in the exhaust. The sulfur content in aviation fuels is limited to 0.3%, although most aviation kerosene has a sulfur content significantly below this limit. Figure 7-35 shows the distribution of sulfur content for jet fuel in 1996 in the UK. The average sulfur concentration was reported as 0.047% (Rickard and Fulker, 1997). The 1996 average reported in the NIPER survey was slightly higher, at 0.062%. The average value in the Boeing survey was 0.038% (Hadaller and Momenthy, 1990); the U.S. Army survey (Bowden, 1988) had an average value of 0.07%. However, most of the larger values came from two small refineries in southern Europe, whereas larger refineries had much lower values. Thus, average sulfur content around the world is probably in the range of 0.04–0.06%; this results in a typical $EI(SO_2)$ of 0.8–1.2.

Most jet fuel contains a sulfur content much lower than the specifications allow because of the use of low-sulfur crude oils and/or the use of hydro-processing to meet other parts of the jet fuel specification. Sulfur is removed as a part of hydro-processing. Purely from equilibrium calculations (without considering inefficiencies), 1 mole of hydrogen, as H_2, is needed to remove 1 mole of sulfur, reacting to form H_2S. At 78.5% efficiency, steam reforming of methane is currently the most efficient means to produce hydrogen; it produces about 5.1 moles of CO_2 per mole of H_2, hence per mole of sulfur. Assuming 0.06% sulfur in the fuel, this figure translates to an average of 0.0033 kg of CO_2 released to produce 1 kg of sulfur-free fuel. Because burning 1 kg of fuel results in the release of about 3.15 kg of CO_2, mandating zero-sulfur fuel would increase the amount of CO_2 attributable to aircraft by about 0.1%.

Reducing the sulfur level would have no direct impact on engine performance or durability. Although low-sulfur fuels tend to

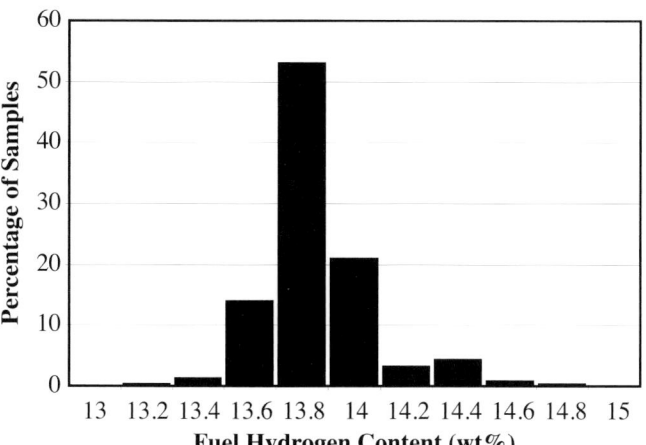

Figure 7-34: 1996 UK survey of hydrogen content in jet fuel.

have low lubricity—which can lead to accelerated wear in fuel pumps and fuel controls—it is not the sulfur that provides the lubricity but organic acids that are removed during the sulfur-removal process (Wei and Spikes, 1986). Military fuels use a lubricity additive to guard against the possibility of low-lubricity fuels. Civilian airlines have not found a need to use such additives, except in a few isolated localities where pump failures have occurred and there was little flexibility in fuel selection.

7.8.3.   *Historical Trends and Forecasts for Sulfur Content*

As mentioned above, surveys show that the sulfur content of most fuels is well below specified limits. All of the surveys show that about 90% of fuels have sulfur content less than 0.1%. Figure 7-36 shows the historical trends for the U.S. and UK surveys. In the UK, average sulfur level has remained relatively constant since 1988 (Rickard and Fulker, 1997); in the United States, however, the average sulfur content in the NIPER survey has been increasing. This trend conflicts with reports (Hadaller and Momenthy, 1993) based on projections of increased hydro-treatment to reduce sulfur in gasoline and diesel fuel. However, changes in gasoline production have not significantly affected jet fuel because there is very little overlap in the boiling range.

The impact of the trend to use low-sulfur diesel fuels is not clear. Many refineries worldwide do not have the hydro-treating capability to make low-sulfur fuels. The API/NPRA survey for 1996 reported that 46% of the jet fuel blendstock in the United States was straight-run material that was not hydro-treated (API/NPRA, 1997). For many of these refineries with limited hydro-treating capability, the most economical approach may be to shift blending stocks with higher sulfur content to jet fuel, saving streams with lower sulfur for diesel fuel.

Without legislation, it is unlikely that average sulfur worldwide will change much from current levels of 0.04–0.06%. The fuel specification certainly could be tightened to allow no more than

0.1% without any apparent increase in cost or availability, and closer to 0.05% might be possible. There will be areas around the world where the sulfur will come down on its own, but there will also be pockets where it will stay relatively high. Specific details on future trends of sulfur content do not exist in the literature; a special survey of worldwide refinery plans would be required to develop a better picture.

Gas-to-liquid conversion processes to produce kerosene (e.g., Fisher-Tropsch processes) yield jet fuel that is almost sulfur-free. Although this approach is attractive in a few regions where there is an abundance of unused natural gas, it is unlikely that this resource will be significant until well into the next century, as the process becomes more economically viable (Singleton, 1997). Biomass gasification could also be used to produce the synthesis-gas feedstock for Fisher-Tropsch conversion.

7.8.4.   *Alternative Fuels to Kerosene*

Current aircraft—along with the airport infrastructure for supply, delivery, and storage of fuel—are specifically optimized for the use of current kerosene fuels; any significant changes in fuel type or specification would require major modifications to all of these elements. These are non-trivial matters involving major perturbations in the existing system, with significant efforts and costs. With this in mind, several alternative fuels have been considered in terms of their environmental impact. These alternatives include alcohols, methane, and hydrogen; more recently, some consideration has been given to using methylated esters of vegetable oils as kerosene extenders. Such fuels must be compatible with the basic capabilities and requirements of existing aircraft. They must have sufficient energy density, for example, to meet payload and range requirements. They must also be compatible with all materials (metallic and non-metallic) used in the engine's fuel system and have adequate lubricity to ensure that current margins and standards of safety-critical items such as fuel pumps are not compromised.

Introduction of an alternative fuel that does not meet the requirements of current aircraft would imply the use of a two-fuel system at all airports until all current aircraft are replaced by new, alternatively fueled aircraft. The prospect of limiting the availability of a new fuel to a few airports does not appear to be viable because of the need to retain full services for aircraft diverted by weather or mechanical problems.

Although alternative fuels may offer some emissions benefits, the major disadvantage is significantly lower energy density compared with kerosene. This density deficit means that the aircraft would have to be designed with larger fuel tanks. Table 7-10 compares the net heats of combustion for several alternative fuels on the basis of mass and volume. For cryogenic fuels, there must also be consideration for the mass and volume of insulation.

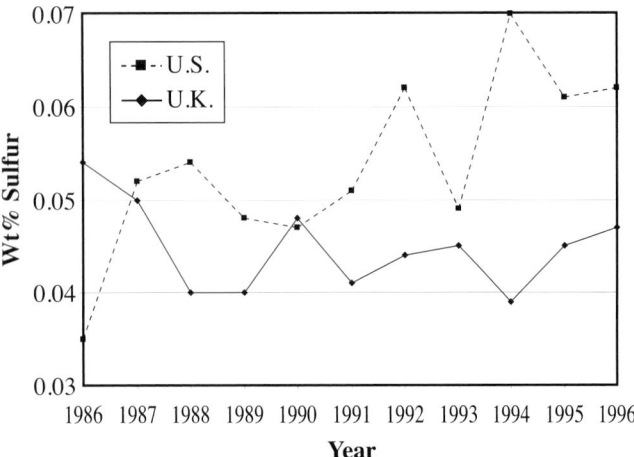

Figure 7-36: Historical trends of sulfur content in U.S. and UK jet fuels.

Ethanol and methanol are liquid fuels that can be pumped and metered in conventional fuel systems, and they can be made

Table 7-10: *Comparison of heats of combustion for candidate alternative aviation fuels.*

	Density (kg m^{-3})	Specific Energy (MJ kg^{-1})	Energy Density (10^3 MJ m^{-3})
Kerosene (typical)	783	43.2	33.8
Ethanol	785	21.8	17.1
Methanol	786	19.6	15.4
Methane (liquid)	421	50.0	21.0
Hydrogen (liquid)	70	119.7	8.4

Table 7-11: *Energy-specific emission indices (kg MJ^{-1}) of CO_2 and H_2O for alternative fuels.*

Fuel	ESEI(CO_2)	ESEI(H_2O)
Jet A/Jet A-1	0.073	0.029
Methane	0.05	0.045
Hydrogen	0	0.075

from renewable energy sources. They are impractical fuels for aviation, however, because of their very low heat content, in mass and volume terms. From a safety standpoint, these alcohols have very low flash points—only 12 and 18°C, respectively—compared with the minimum allowed of 38°C. There are also chemical incompatibilities associated with fuel system materials, although these problems could be remedied with relatively minor changes. Furthermore, the combustion of alcohols produces organic acids and aldehydes in the exhaust at idle conditions on the ground, with attendant health hazards to ground support personnel (Eiff *et al.*, 1992).

There have not been any definitive engine studies using methyl esters of vegetable oils, such as soybean or rapeseed oils, although some evaluations are underway (e.g., Scholes *et al.*, 1998). Adding such a material to jet fuel would not be allowed under any current fuel specifications for jet fuel because of compositional considerations. Furthermore, studies with methyl esters of soybean oil suggest that more than about 2% will raise the freezing point above the specification maximum. Based on the results of Eiff *et al.*, (1992) using ethanol blends with jet fuel, adding methyl esters of vegetable oils to jet fuel would result in lower exhaust smoke/particulates at high-power conditions but increased CO and hydrocarbons at idle conditions, along with the presence of acids and aldehydes. The effect on NO$_x$ is uncertain, but would be directly related to any changes in flame temperature.

Aircraft gas turbines can be designed to operate on cryogenic fuels such as methane or hydrogen; conventional fuel systems, however, cannot handle these fuels. Such fuels would require new aircraft fuel system designs, as well as new ground handling and storage systems. Moreover, cryogenic fuels would have to be stored in the fuselage rather than the wings to reduce heat transfer. Because methane and hydrogen have only 65 and 25%, respectively, of the energy density of jet fuel, fuselages would have to be considerably larger than current designs—increasing drag and fuel consumption. For long-range flights, this penalty would be offset by a reduction in take-off weight because hydrogen and, to a small extent, methane have higher specific energies than kerosene. Design studies for hydrogen-fueled, long-range (10,000 km) aircraft have shown that the lighter fuel weight results in almost a 20% reduction in energy consumption compared to kerosene-fueled

aircraft even accounting for losses (Momenthy, 1996). The same study showed, however, that for medium- and short-range (5500 km and 3200 km) aircraft, there is an energy penalty of 17–38%. For methane, there was only a small benefit for long-range aircraft and penalties of 10–28% for medium- and short-range aircraft.

Of these two cryogenic fuels, hydrogen may be more attractive from an emissions standpoint. CO$_2$ and SO$_x$ emissions would be eliminated. However, water vapor would increase significantly despite the reduction in energy consumption. For the same energy consumption, burning methane would yield about 25% less CO$_2$ and about 60% more H$_2$O than burning jet fuel. Burning hydrogen would result in 2.6 times as much water vapor as burning jet fuel, but no CO$_2$. Table 7-11 compares the energy-specific emissions indices of CO$_2$ and H$_2$O at constant payload. (Effects of weight savings/penalties are not included in Table 7-11 because they depend heavily on the range of the aircraft.)

This basis of comparison could yield a quantitative "greenhouse" comparison of these three fuels if the greenhouse equivalency were known. Figure 7-37 presents the results from such a study for a hydrogen-fueled aircraft derived from the Airbus A310; the analysis takes into account the relative greenhouse effects of H$_2$O, CO$_2$, and NO$_x$ at different altitudes and shows that hydrogen offers a significant reduction in greenhouse effect over kerosene at all altitudes for this aircraft (Klug *et al*, 1996).

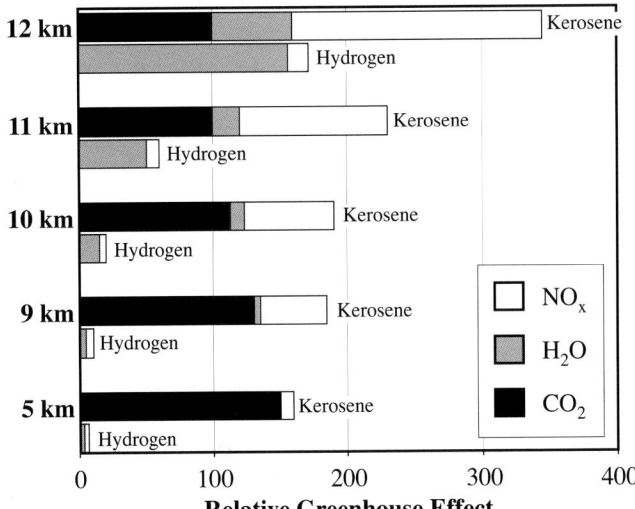

Figure 7-37: Comparison of relative net greenhouse effects for hydrogen and kerosene.

Inefficiencies of production would alter these results somewhat in a comprehensive energy comparison, although water emissions are of environmental concern only at cruise altitudes. Figure 7-38 compares relative CO_2 emissions from the manufacture and use of alternative aviation fuels from different resources (Hadaller *et al.*, 1993). The potential benefits of hydrogen can be realized only if hydrogen can be obtained from water without the use of fossil fuels to provide the energy. Nuclear power is the best method identified in Figure 7-38. The Kvaemer process is being developed as a method for converting hydrocarbons into hydrogen, with carbon as a byproduct (as opposed to CO_2 with the steam reforming process); if the energy requirements are sufficient, this fuel would appear on Figure 7-38 with a value less than 1.0. Kerosene from biomass [via a Fischer-Tropsch (F-T) synthesis process] would also have relative CO_2 emissions less than 1.0 if included in Figure 7-38.

Liquid hydrogen offers an environmental advantage only if this fuel were produced on a renewable energy basis, as explained above. The necessary technology exists, but such liquid hydrogen is not economically competitive with kerosene at current price levels. On the other hand, liquid hydrogen based on renewable energy is the only candidate aviation fuel known today that would completely eliminate CO_2 emission by aviation. Safety issues in the siting of storage and handling systems at airports pose significant challenges, however.

It is very unlikely that military aircraft will ever use any fuel except current kerosene-type fuels, for reasons of logistics and the desire to have one fuel for all aircraft and ground equipment.

7.8.5. *Summary*

On balance, it appears that current types of aviation fuel will continue to be the preferred option for gas turbine powered aircraft. This situation could change if liquid hydrogen could be produced by an environmentally acceptable and economically competitive method or if the need to reduce CO_2 emissions from aviation becomes overwhelming. Aircraft now consume about 2.5% of all fossil fuels burned; therefore, they are not

major contributors to anthropogenic CO_2 discharged into the atmosphere. Future aviation demand growth rates are discussed in Chapter 8. The only emissions that are directly influenced by fuel type are CO_2, H_2O, and SO_x.

The properties of aviation fuels are controlled within fairly narrow limits, and allowable variations can have very little impact on exhaust emissions, with one exception—removal of sulfur. The sulfur content in jet fuel currently averages about 0.05% worldwide, well below the specification allowance of 0.3%. In the United States, sulfur content has been increasing slightly since 1990, but it appears relatively stable in Western Europe and probably the rest of the world. Sulfur will probably remain at this level for the foreseeable future unless it is legislated downward. The penalty for removing sulfur from petroleum-derived jet fuel would be about a 0.1% increase in CO_2 emissions attributable to the aircraft sector as a result of the need to manufacture additional hydrogen; there would be no CO_2 penalty if the hydrogen came from a renewable energy source or if nuclear power were used to extract it from water.

Sulfur-free kerosene can be produced by F-T processes from synthesis gas that could be produced from natural gas, coal, or biomass. The economics of these processes are improving. F-T jet fuel produced from biomass-derived synthesis gas would be essentially CO_2-neutral.

Alternative fuels to kerosene that appear to be environmentally friendly have been identified; even if these benefits can be verified, however, introduction of such fuels will be hindered by significant technical problems in adapting these fuels to current aircraft designs and airport infrastructures. Using current technology, such changes would increase CO_2 released to the atmosphere.

Alternative liquid fuels appear to offer little promise. Alcohols are not compatible with the fuel systems of current aircraft and will suffer significant range penalties as a result of lower heats of combustion. The use of esters of vegetable oils is limited to 2% at most before fuel blends fail freezing-point requirements. Burning alcohols and esters would increase emissions of CO, hydrocarbons, organic acids, and aldehydes.

The introduction of any cryogenic fuel would require the design and development of a new fleet of aircraft, as well as a new supporting infrastructure for the storage and handling of such fuel at airports. Cryogenic fuels are not compatible with the fueling systems of current aircraft, and their lower energy density would require much larger fuel tanks than on current aircraft. Studies have shown that cryogenic hydrogen could be a viable alternative to kerosene with significant reductions in greenhouse effects for long-range commercial aircraft if design and infrastructure problems can be solved.

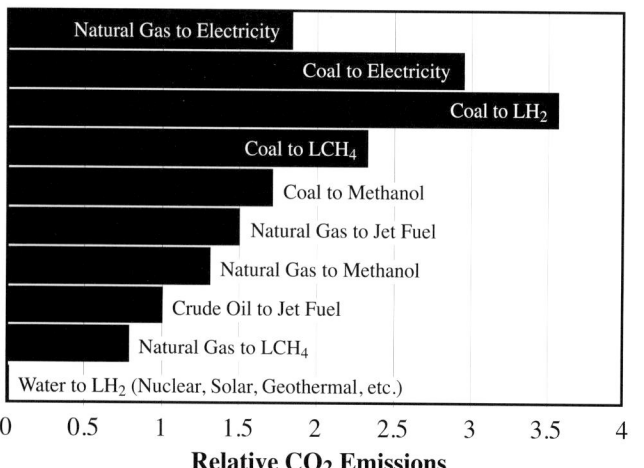

Figure 7-38: CO_2 comparison for the manufacture and use of alternative aviation fuels.

7.9.   **Small Aircraft, Engines, and APUs**

Small aircraft engines are defined here as all turbo engines—excluding piston engines because of the very small amount of fuel burned by these engines (much less than 5% of the world's fleet fuel burn; see Section 7.2) — that are used for regional aircraft, all turbofans with less than 89 kN thrust, all turboprop and turboshaft engines, and all auxiliary power units (APUs) used in civil aircraft.

The small commercial and general aviation segment has been growing rapidly in recent years and is likely to continue to do so. This segment's impact on the environment, however, is unlikely to be significant because of low NO_x emissions levels associated with the generally lower pressure ratio engines they employ and the decreasing percentage of the fleet's fuel burn they represent (see Chapter 9). Furthermore, most of aircraft in this sector fly short missions with lower cruise altitudes and reduced potential for climatic impact. Significant improvements have been made in the idle emissions of small engines in recent years, so that CO, HC, and NO_x emissions from small regional and general aviation aircraft are often comparable, in terms of emissions per kilogram of fuel burned, to those from large engines (Eatock and Sampath, 1993).

This section highlights the key differences between large and small engines in terms of emissions characteristics and control technologies that might apply to small aircraft and engines. A brief overview is presented on small airframe technology, engine performance, engine emission databases, combustor technology, and unique issues related to small engine combustors.

7.9.1.   *Airframe Technology*

The airframe technology applicable to small aircraft parallels that of large aircraft discussed in Section 7.3. Designers and manufacturers continue to strive to reduce drag and increase range/payload performance with a resultant steady improvement in overall fuel-efficiency of new small aircraft. Propeller design, of course, is of much greater importance to this sector of the world's fleets. Detailed aerodynamics research is showing some of the significant performance benefits — thus fuel savings — that can be achieved from relatively small changes in the design of small airframes.

Nacelle problems associated with higher bypass ratio engines for small aircraft are similar to those of large aircraft. Although embedded engines have been considered, there are no plans to adopt them because the small gains in nacelle efficiency are more than offset by losses in wing efficiency.

7.9.2.   *Aircraft Weight Reduction*

In general, simple scale effects mean that reductions in the size and weight of an aircraft's structure and systems have a somewhat greater impact on small aircraft than their larger counterparts. Thus, advances in the power, compactness, and weight of avionics systems are important, as are potential benefits from newer designs of system controls and valves. Ever-increasing uses of composites — not only as aerodynamic

fairings but also in the primary structure — are clearly important, for the same reasons.

Underpinning many of these issues is the emergence of a more centrally computerized product definition database introduced at an early stage in the aircraft/system definition process. This database greatly enhances the design, placement, and routing of environmental control system, hydraulic, fuel, and electrical systems and components.

7.9.3.   *Engine Performance*

Existing small engines operate at overall peak pressure ratios from 8 to about 30, as compared with large engine values (see Section 7.4) of about 20 to > 40. Thus, NO_x production in grams per kilogram of fuel is typically lower for engines designed for regional and general aviation aircraft than it is even for future staged, low NO_x combustors in large engines. However, the low engine cycle pressure ratio makes efficient operation at idle intrinsically more difficult. Combustor inlet pressure at idle power may be less than half that of a modern large engine, with a resulting tendency to cause higher CO and HC emissions. However, technology advances in larger engines are reflected in the latest generation of small engines, with SFC improvements stemming from advanced combustor cooling techniques, high temperature materials, and advanced engine cycles with higher pressures and temperatures. Higher combustor inlet pressures and temperatures in modern small engine combustors now result in even lower fuel burn, which generally balances any tendency toward an increase in NO_x output.

7.9.4.   *Engine Database*

Emissions for small turbofan engines with a thrust of more than 26.7 kN thrust are measured, reported, and certified in exactly the same manner as emissions for large engines and are recorded in the ICAO database (see Section 7.7.1). Engine manufacturers usually have some emissions data on their non-certified engines (turbo-shafts, turboprops, APUs, and turbofans of less than 26.7 kN thrust), but in general these engines are not of certification quality and the data are not readily available. The ICAO LTO cycle is used for turbofan engines, but these points are not well defined for turboshaft or turboprop engines because their thrust depends on the propeller or rotor selected for each application. Shaft power is usually substituted for thrust in emissions calculations to compare turboshaft or turboprop engines, but these emissions cannot be compared directly with turbofan data (which is based on thrust).

Correlations between measured emissions at static sea level conditions and altitude operating conditions, as discussed in Section 7.7.2, are generally applicable to small engines. Exceptions may occur for very small engines in which combustor surface to volume ratios, fuel atomization quality, combustor volumes, and so forth are very different from those in large engines. Small engines may require additional correlation parameters to account for these differences (Rizk, 1994).

7.9.5. Combustor Technology

7.9.5.1. Historical Developments and Current Status

The ability of gas turbine manufacturers to design successful small engine combustors with reduced emissions has improved greatly in the past 20 years. Idle efficiency data, indicative of CO and HC levels, collected by the General Aviation Manufacturers Association (GAMA) (Eatock, 1993) and plotted in Figure 7-39 indicate combustion inefficiencies at idle power of between 5 and 15% in early combustors. The trend, however, shows that modern small engines approach and even match the idle combustion efficiency of modern large turbofan engines. Figure 7-40 shows the basic "Lipfert" correlation (ERAA, 1992; Eatock and Sampath, 1993) of $EI(NO_x)$ in grams of NO_x per kilogram of fuel burned. The correlation relates emission indices of NO_x to compressor discharge pressure (pressure ratio) and covers most engines operating at or near stoichiometric burning at full-power condition. Many small aircraft engines have even lower $EI(NO_x)$ than those anticipated for the future third generation of large turbofan engines.

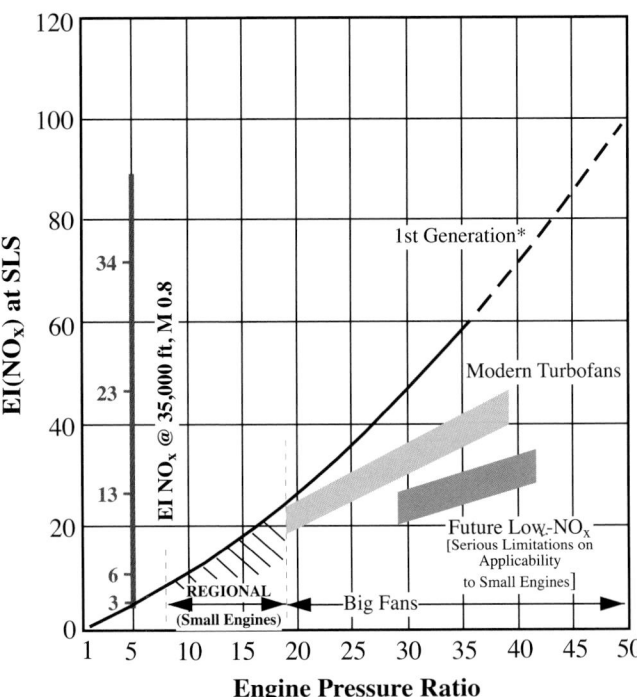

Figure 7-40: Engine pressure ratio versus $EI(NO_x)$ for large and small engines (*Lipfert, 1972; EPA, 1976).

7.9.5.2. Unique Combustors for Small Engines

The major challenge to reduce the emission level of NO_x for small engines is overcoming size-related constraints. NO_x reduction strategies may be restricted, for example, by the smaller passage height between the inner and outer diameters of the combustor, which limits the size and number of fuel injectors that can be used for such purposes. Small combustors are also more sensitive to minor size variations within fixed manufacturing tolerances. For these reasons, and to ensure that all such combustors will meet a given emissions goal, "nominal design" hardware must be able to demonstrate compliance with somewhat larger emissions margins than for larger engines.

Many smaller engines use centrifugal compressors in combination with reverse-flow combustors. These combustors have a higher surface-to-volume ratio. Furthermore, scaling down a design has the effect of rapidly increasing combustor surface area-to-volume ratio (ICCAIA, 1993). Both factors tend to result in the need for a higher percentage of combustor airflow for cooling the combustor, leaving a smaller amount of air available for controlling emissions. Together these effects limit the direct translation of low NO_x combustor technology from large engines into smaller combustion system designs. The NASA-sponsored contract for small turbofan engines (Bruce *et al.*, 1977, 1978, 1981) demonstrated the problems very clearly. This work showed that even a modest 30% reduction in NO_x could be achieved only using a design that featured variable geometry and staged combustion, with a totally impractical increase in the number of fuel injectors.

7.9.5.3. Current and Future Trends

In recent years, manufacturers of small engines have continued to develop new emissions control techniques that have minimum cost and performance impact. These techniques include combustors with optimized stoichiometry in the primary combustion zone, improved fuel/air mixing using efficient swirlers, improved fuel spray quality using piloted air-blast and aerating fuel nozzles, and combustors with optimized liner wall cooling to minimize wall quenching effects on CO and HC emissions.

Several advanced emission control concepts have also been investigated on small engines, including variable geometry and

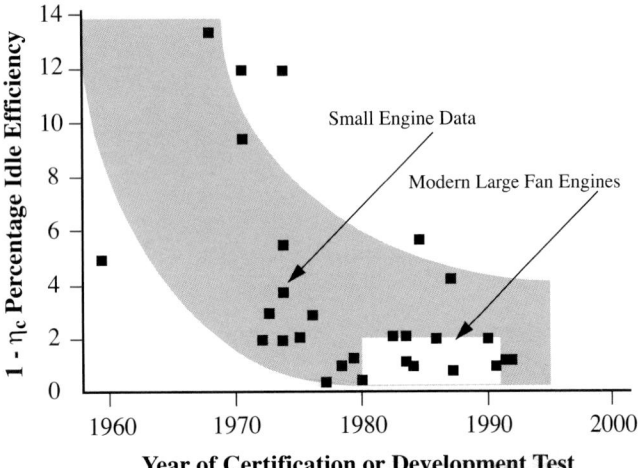

Figure 7-39: Idle efficiency trend for small engines.

staged combustors (Bruce *et al.*, 1977, 1978, 1981). Variable geometry concepts can reduce HC and CO emissions; however, such complexity is particularly unattractive for small engines. An axially staged combustor requires a large number of fuel nozzles, with corresponding fuel passage size reduction and added fuel control system complexity.

In summary, small engine combustor technology has progressed markedly, especially in the past 20 years. In the short term, existing combustion chambers can be modified and optimized, within the constraints defined in Section 7.9.5.2, to obtain the best NO_x versus CO and HC compromise. There are no special fuel requirements for small engines and aircraft, now or in the foreseeable future.

7.10. Supersonic Transport Aircraft

Supersonic transport aircraft are not new. In the 1960s and early 1970s, there were three major supersonic transport projects in Europe, the United States, and the Soviet Union. Only one of those projects, the Concorde, was completed and continues in revenue service—and only between Western Europe (London and Paris) and the eastern U.S. seaboard (New York and, until recently, Washington). Only 13 of these aircraft are in airline service, including charter operations and demonstration flights. The ground and flight operational acceptability for this type of aircraft has been demonstrated worldwide.

The life development of Concorde aircraft now suggests they could continue in service for an additional 10–15 years. The levels of exhaust emissions produced by Concorde's Olympus engines [EI(NO_x) = 18] is higher than could be achieved now but is commensurate with the design knowledge of the 1960s. With a limited small fleet and an average aircraft utilization of 600 hours per year, these aircraft do not appear to constitute a major environmental concern, although they must be considered in aircraft fleet mix scenarios in view of the Concorde's higher cruising altitude.

The Concorde operates for passengers prepared to pay high fares. Improvements in materials, structural, and systems technology that are available or currently being developed could make a second generation of supersonic aircraft more widely affordable. Studies have concluded that an aircraft with between 250 and 300 seats cruising at Mach 2 to 2.4 (at altitudes between 16 and 20 km) is most likely to be successful (Shaw *et al.*, 1997). To make such aircraft effective for the long overseas routes that benefit most from the increased speed and maintain viability with regard to viewpoint of sonic booms, the projected range must be at least 8000 km (and possibly 10400 km). Studies have examined a wide range of speeds and concluded that speeds higher than Mach 2.4 offer little gain in block time, whereas they exacerbate airframe materials and propulsion problems, hence increase technical risk (Zurer, 1995). Prior projections concluded that aircraft with a cruise speed of Mach 2.0 to 2.4 were feasible for entry into service in 2005, and a hypersonic vehicle cruising at Mach 5

might enter service by about 2030 (Zurer, 1995). Events have shown that these projections were optimistic, and it is unlikely that a new Mach 2 to 2.4 vehicle will enter service much before 2015. By the same token, required research for the hypersonic vehicle and its economics would make entry into service of a hypersonic vehicle unlikely before 2050—and possibly later unless scheduling and airport curfews could be accommodated to demonstrate higher cruise speed benefits. Therefore, the focus of the remaining discussion is on vehicles cruising at speeds up to Mach 2.4.

The Concorde has already demonstrated the practicality of Mach 2.05 as an achievable cruise speed with aluminum alloys for the basic structure. For speeds above Mach 2.2, more exotic materials would be required including titanium alloys and organic composites for structural items and more complex air intakes. At speeds between Mach 2 and 2.4, airframe characteristics currently dictate cruise altitudes between 16 and 20 km. Optimization studies are planned to investigate lower cruise altitudes, recognizing the potential benefit of minimized ozone impact. To enable the inclusion of route segments over populated areas without sonic booms, an advanced supersonic airliner must also be capable of cruising efficiently in an environmentally acceptable manner at subsonic speeds and lower cruise altitudes.

7.10.1. Supersonic Transport Characteristics

The characteristics of potential second-generation supersonic transports and their consequent impact on the atmosphere differ substantially from those of subsonic transports. First among these differences is the cruise altitude in the stratosphere, which is near where ozone concentration peaks. For a given level of emissions, the supersonic transport's potential impact on the ozone column is larger than that of subsonic aircraft from an NO_x-ozone depletion perspective. This impact led to strong opposition to U.S. supersonic transport development in the 1970s and, together with the potential airport noise impact and economic considerations, led ultimately to cancellation of the development. There is now consensus that emissions from a second-generation supersonic transport must be limited to levels that will have a "negligible effect" on ozone.

There is also concern regarding the effects of carbon and sulfate-based particulates, and water vapor (Albritton *et al.*, 1993; Stolarski *et al.*, 1995). This subject is discussed in Chapters 2 and 3. Atmospheric modeling has indicated that the effects of NO_x are likely to be small for a fleet of 500 to 1,000 Mach 2–2.4 aircraft if EI(NO_x) is near 5 g kg^{-1}. This issue is discussed in detail in Chapters 4 through 6. This level of emission has become a technology development target for the second generation supersonic transport. As yet there are no such targets for particulates, but such targets may emerge before the decision time for a production development program is reached.

The overall efficiencies of subsonic and supersonic propulsion systems at comparable technology levels are not very different. The lift/drag ratio of supersonic aircraft, however, is substantially lower than that of subsonics—no more than about 9 compared to about 20 for subsonics, as is shown in Figure 7-41. This lower ratio is related to losses caused by shock waves, which can be minimized but not eliminated by sophisticated designs. Thus, for the same range the supersonic transport must carry a larger fraction of its mass as fuel. For a given weight, its engines must also produce more thrust in cruise because of the larger drag. Primarily as a result of these effects, the fuel burn per passenger mile of a supersonic transport is correspondingly large—two or more times that of a subsonic transport. Nevertheless, economic studies that incorporate time zone/ productivity considerations indicate that the greater productivity resulting from shorter block times could enable the supersonic transport to compete with subsonic transports. There is no intent to further quantify this question here. The remaining sections address the technological challenges that set supersonic transport propulsion apart from those faced by advanced subsonic engines.

7.10.2. *Propulsion System Efficiency*

This subject is covered in greater depth in Section 7.4, but the key points are reviewed here briefly with emphasis on supersonic aspects. Overall propulsion system efficiency may be regarded as the product of two factors: Thermal efficiency and propulsive efficiency. Thermal efficiency is the ratio of hot gas power produced by the engine gas generator to the power in the fuel flow. It is ideally controlled by the temperature ratio of the compression process—that is, the ratio of temperature at the discharge of the compressor to ambient temperature. In modern engines, the compressor discharge temperature is limited by the temperature tolerance of materials suitable for use in compressors, to a level of about 850 to 900 K. Ambient temperature is nearly constant at about 220 K (although the temperature varies considerably even at the altitudes flown by supersonic transports). Thus, the temperature ratio is about 4, resulting in an ideal thermal efficiency of about 3/4. The actual efficiency is lower because of various losses but remains quite high.

Propulsion efficiency, which is the ratio of power pushing the airplane to the hot gas power produced by the gas generator, ideally depends only on the ratio of jet velocity to flight velocity. It can be increased toward the limit of unity by increasing the bypass ratio. The bypass ratio that is best for any given application is determined by a balance between the increased drag and weight associated with the larger engine frontal area needed for increased airflow and the increased propulsive efficiency associated with the larger airflow. There is an additional factor in fixing the bypass ratio—the fan pressure ratio. Engines for supersonic applications require a higher fan pressure ratio than subsonics for similar improvements in efficiency. When these compromises are struck for the subsonic and supersonic propulsive systems, the propulsive efficiencies of the two systems at cruise conditions are not very different. Over the past 2 decades, the bypass ratios of subsonic engines have increased from 2 toward 10 as lighter weight and aerodynamically more sophisticated designs have evolved. For supersonic propulsion systems, the cruise bypass ratio of choice is now in the range of 0.5–1.0, whereas for the Concorde and for the U.S. supersonic transport of the 1970s it was zero.

7.10.3. *Supersonic Propulsion System Characteristics*

In contrast to subsonic engines—in which the needs for low cruise fuel consumption and low take-off noise are synergistic in that they both favor high bypass ratio—supersonic engines face a severe conflict between the need for low bypass ratio in transsonic acceleration and cruise, and the need for higher bypass at take-off to limit jet noise. This conflicts leads to supersonic propulsion systems in which the effective bypass ratio can be small at cruise and relatively large at take-off. Such variation can be achieved by several different approaches, characterized at the extremes as variable turbomachinery systems and ejector nozzle systems. Common to all approaches is that the engine operates at a relatively high exhaust velocity at cruise and a low exhaust velocity at take-off.

The engine concept favored by the U.S. program is a low bypass turbofan (0.6) with a mixer-ejector nozzle for noise suppression at take-off. It is projected that a supersonic transport with this propulsion system could meet Federal Aviation Regulations 36 Stage 3 noise requirements. An alternative concept proposed by the European engine consortium has a fan at the midsection of the engine, fed by auxiliary inlets at take-off, to produce a bypass ratio of about 2 at take-off. The auxiliary inlets are closed and the fan airflow is decreased via a geometry change to give a lower bypass in cruise. This engine concept is estimated to meet the Stage 3 noise requirement with a conventional variable area nozzle such as that used on the Concorde.

For designs under current consideration, the compression pressure ratio is about 22 at take-off conditions, and the corresponding temperature rise is about 500 K, producing a compressor outlet temperature of about 800 K. By comparison, a subsonic engine at take-off would have a pressure ratio of 35 with a compressor exit temperature of 900 K. At supersonic cruise, the compressor outlet temperature is limited by the materials to about 950 K. For a subsonic cruise engine with a pressure ratio of 40, the

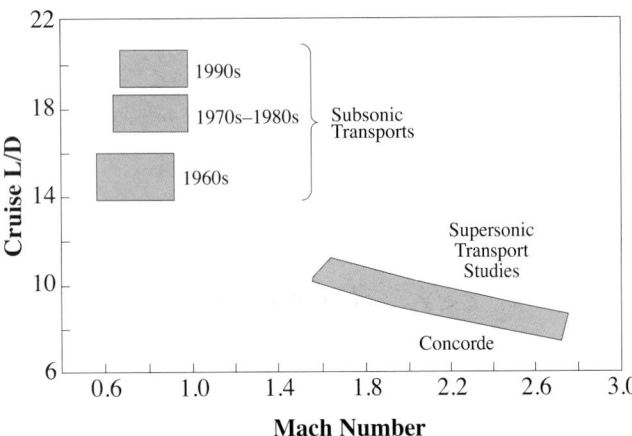

Figure 7-41: Influence of flight Mach number on cruise L/D.

corresponding compressor outlet temperature is about 850 K or lower, limited by practical aerodynamics. The consequence is that the supersonic engine runs hot at cruise and cool at take-off—the opposite of the subsonic engine.

7.10.4. SST Propulsion NO_x Output

At cruise, subsonic engines have an overall pressure ratio, including inlet ram pressure rise, in the range of 35–50. The overall pressure ratio of a supersonic propulsion system, including the intake pressure rise and the remaining turbomachinery pressure ratio, is in the range of 130–140, calculated as a ratio of the compressor exit total pressure divided by the inlet static pressure. Whereas the conventional subsonic system might have $EI(NO_x)$ of 10–15 g kg^{-1} fuel using a conventional combustor, even with the double annular system the advanced supersonic engine would barely achieve $EI(NO_x)$ of 30 (Lowrie, 1993). Combined with the cruise altitude in the high ozone concentration zone, this high NO_x emission would not be acceptable for a supersonic transport. This problem is well recognized, and there are many research activities throughout the western world to develop ultra-low $EI(NO_x)$ systems. Although the combination of the 5g kg^{-1} $EI(NO_x)$ goal and high compressor outlet temperatures may seem formidable, research programs have shown that because of the high altitude, the low absolute pressures in the supersonic transport combustors allow easier control of the combustion mixing process. Contemporary findings therefore support the likely achievement of this target (Shaw *et al.*, 1997).

Supersonic transports will also be subject to LTO cycle emissions rules and potential climb/cruise emissions rules in common with other air traffic. In this respect, the lower combustor inlet temperatures and pressures experienced at take-off will mean that NO_x requirements in the airport neighborhood could be met more easily. The combustion system will be optimized for subsonic climb and cruise as well as supersonic cruise conditions, so attainment of low emissions still requires considerable effort.

7.10.5. Other Contaminants from Supersonic Transport Engines

At this stage, there is no reliable information relating specifically to the design of supersonic propulsion combustion systems, but there is no reason to believe that the degree of inefficiency should be any different from subsonic types. Emissions arising from this form of inefficiency are discussed in Section 7.5. The fundamental nature of the proportion of CO_2 and H_2O emitted with conventional fuels and similar proportions of sulfur in available fuels suggests that these emittants will be produced in the same proportion to fuel usage as in the subsonic fleet, though at supersonic cruise they will be deposited in the stratosphere.

7.10.6. Supersonic Transport Operations

Assessments of advanced supersonic aircraft concentrate on long-range over-water routes but do not assume that such routes will be served exclusively by supersonic services. It is now assumed that supersonic flight will occur only over water. The combination of overland routes where sonic booms are unacceptable and fitting into daily cycles will preserve a place

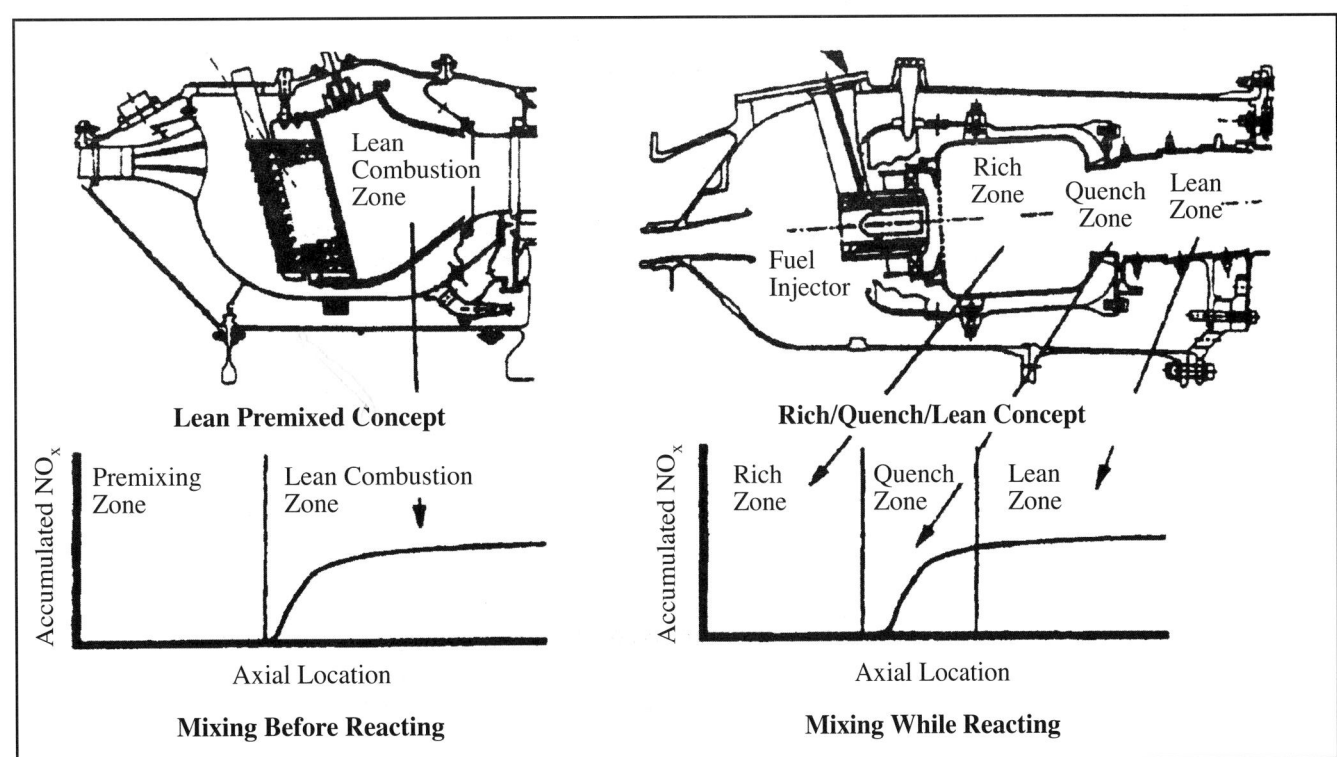

Figure 7-42: Ultra-low NO_x combustors for supersonic commercial transport applications.

for long-range subsonic services to meet airport curfews and provide passengers with comfortable time zone changes.

7.10.7. *Mitigation*

Research programs to develop low NO_x combustion systems for cruise are in place in Europe, the United States, and Japan. Two basic combustion concepts are being researched to produce ultra-low NO_x levels at supersonic cruise conditions. These technologies—the Lean Premixed Prevaporized (LPP) and Rich Burn Quick Quench (RBQQ)—utilize both lean and rich combustion concepts such as those shown in Figure 7-42 on the previous page. As a result of low combustor operating pressures in supersonic transport applications, these concepts appear capable of achieving their full ultra-low NO_x reduction potential while maintaining satisfactory durability and performance. In subsonic transports, the higher pressures dictate compromises in both concepts to avoid incipient flashback in the LPP and excessive soot production in the RBQQ.

The LPP concept has the likely potential of reaching the lowest levels of NO_x. The intent of premixing is to provide the combustion zone with a very lean, uniform fuel/air mixture that is just above the flame extinction limit. This approach results in a low flame temperature with enough residence time to complete combustion and produce low NO_x. Maintaining uniform fuel/air mixtures throughout the combustor is critical because NO_x increases rapidly with any local fuel/air maldistribution. In practice, premixing is achieved with large numbers of small-diameter premixers. Design challenges with this concept include flashback or auto-ignition in the premixer, maintaining combustion near the lean extinction limit over the entire engine cycle operating span, potential fuel clogging of small-diameter fuel injectors, and complexity of the design because of fuel staging requirements.

The RBQQ concept is a derivative of an axially staged combustor and presents the more stable combustion configuration. The fuel/air mixture of the primary combustion zone is fuel-rich, thus producing low flame temperatures and low NO_x. In a second stage, air is quickly introduced to mix with the partially reacted fuel. Combustion is completed in a final stage at lean conditions. Most of the NO_x is produced in the second stage and is a function of the uniformity and time it takes to dilute the reacting mixture. Design challenges with this concept include indirect cooling of the primary combustion zone—which may require high-temperature ceramic materials currently under development—and an advanced second stage that produces nearly instantaneous, uniform mixing of reacting gas and air. Furthermore, this design may require engine power-related control of air and/or fuel staging for practical implementation.

There are three active programs in which this research is being conducted:

- U.S. NASA High Speed Civil Transport (HSCT), with a goal of supersonic cruise NO_x of $EI(NO_x) = 5$

- Japanese Supersonic/Hypersonic Transport (HYPR), with a goal of NO_x emissions below $EI(NO_x) = 5$ at Mach 3 cruise
- EU Low NO_x III, to develop ultra-low NO_x combustor technology for a second-generation civil supersonic transport aircraft.

Under laboratory and component test cell conditions, very low levels of NO_x below $EI(NO_x) = 5$ have been achieved at simulated engine operating conditions of pressure, temperature, and fuel flow with combustor sectors. This result gives credence to the view that $EI(NO_x) = 5$ could well be achieved in engine tests scheduled after the year 2000. The intrinsic outputs from combustion of conventional fuels are more difficult to alleviate. With kerosene as a fuel, for every ton of fuel burned, 3.2 tons of CO_2 and 1.2 tons of water are produced.

7.11. Special Military Considerations

Although today's military and civil aircraft projects clearly respond to radically different national requirements, the underlying engine technology of the two types of aircraft has a significant degree of commonality. This situation is not surprising in the case of the military transport aircraft engine: In common with the civil engine requirements, large payloads must be carried over long distances at the lowest possible costs. The similarities are less obvious, however, with regard to combat aircraft (including fighter/bombers), which have no clear civil parallels. Nevertheless, many of the technical advances that have been developed to meet the military challenge have been adopted, in one form or another, in advanced civil engine applications. Indeed, an increasing number of examples of technical advances derived from civil engine research also are relevant to military engines. This two-way exchange justifies the fact that a significant part of today's basic research underpinning aero-engines for the future is supported directly or indirectly through military and civil engine sources (dual use).

Notwithstanding the parallel development paths and mutual objectives mentioned above, limits must be taken into account. In general terms, increases in the fuel efficiency of combat aircraft are consistent with the logistics and operational needs of the military because aircraft with lower fuel consumption rates would be able to remain engaged for longer periods of time, carry additional payload, reach targets from greater distances, or a combination of the three depending on operational requirements. They would also require less in-flight refueling. However, the prescribed altitude for optimum fuel efficiency may not be not be appropriate for military operations. Similarly, some of the restrictions on civil transport ground operations (e.g., engine start, taxi, and take-off procedures) may not be acceptable to military users. Although military procurement officials and operators are now acutely aware of their responsibilities with regard to environmental effects, operational effectiveness will always be the primary requirement.

The following sections focus on military aero-engines—in particular on aspects of such engines that, for performance and/or operational reasons, differ from their civil counterparts in ways that might influence emissions. It is important to rationalize these differences in terms of their real impact on the environment by taking proper account of current and likely future proportions of the world's military aircraft fleet compared to the global total of all types and operations of aircraft. Chapter 9 shows numerically how the environmental impact of military operations becomes a diminishing percentage of the total of all aircraft operations as the effects of the anticipated strong growth of civil transport takes effect over the next 50 years or so. Differences between military fuels—F-34 (NATO) and JP-8 (US)—and civil AVTUR/Jet A-1 fuel are covered in Section 7.7.

7.11.1. Differences Resulting from Operational and Design Features

7.11.1.1. Combat Aircraft

Aircraft used in combat operations constitute the largest proportion of the various types of aircraft in the military inventory. They outnumber other aircraft types by roughly 3 to 1; for this reason alone they justify a closer look as the most likely sources of divergence between the emissions of military and civil engines in the future. Figure 7-43 is extracted from some of the same data sources used in Chapter 9 dealing with fuel usage and emissions production, in particular the ANCAT/EC2 report (Gardner, 1998) and the recent NASA report (Mortlock and van Alstyne, 1998).

Clearly, combat aircraft will always be built to respond to quite different mission priorities from those applying to civil aircraft. There are, therefore, some differences in the design features of engines to achieve those priorities. Combat aircraft engines will inevitably be designed to extract maximum performance even though this approach entails accepting a shorter life, particularly of key hot-section components such as turbine blades, and shorter periods between maintenance than civil engines. It is also most likely that military aircraft will be the first to adopt the fruits of the most advanced engine technology in a constant drive to achieve superior performance. Thus,

there will always be something of a technology gap between the leading military and civil engines. Of the principal performance requirements, the demand for higher thrust/weight (T/W) ratio engines will continue to be the key driver that will maintain that gap. This consideration inevitably means that T/W targets for new engine cycles will involve higher pressure ratios, higher peak temperatures, and higher fuel/air ratios than the current fleet. The most stringent performance targets today are those of Phase III of the United States' Integrated High Performance Turbine Engine (IHPTET) program (Hill, 1996). This multi-agency/industry initiative has set goals that if fully achieved, will provide important new engine technology levels for adoption in the next century. The principal goal of this third phase of a three-phase program is to achieve +100% in the T/W ratio together with a 40% reduction in fuel burn for a new generation of military engines. These targets are so ambitious that evolutionary improvements in hot-section components of the engine are simply not sufficient. Only radical solutions are likely to achieve the necessary rise in engine cycle temperatures and pressures to meet the T/W targets. Such engine cycle changes imply, as explained in Section 7.4, even higher levels of NO_x emissions if conventional engine technology is retained. This approach, however, is unlikely to be wholly acceptable because of associated rising levels of visibility of the brown-tinted NO_2 component of NO_x gases—which, unchecked, could compromise the stealthiness of the aircraft. It therefore seems inevitable that significant pressure to limit NO_x will remain part of the military aims attached to the IHPTET program and that some if not all of these aims will be relevant to environmental and operational performance. In this respect, therefore, the prospects of total divergence of priorities between military and civil engines seems small.

One important difference between military and civil engines is associated with maneuverability. Under certain conditions, for example, the military engine combustor must be able to accommodate the consequences of high-incidence turns during combat maneuvers. These maneuvers can cause unstable internal flow conditions. The combustor must be able to withstand the consequences of such maneuvers over a wide range of flight speeds and altitudes. It must also relight rapidly in the event of a flame-out. In today's climate, the conventional approach of increasing the fuel/air ratio in the combustor to ensure adequate stability is not acceptable unless high-power emissions can be contained using new designs. The pursuit of effective solutions to these important problems, taking full account of emissions, forms an important part of today's military research and development programs.

Reheat/afterburner operations are an important requirement for combat aircraft engines. Although the majority of fuel used during reheat is burned at low altitude, the extra thrust demands associated with combat maneuvers does mean that a proportion of fuel (typically 8%) is burned at altitudes above 2000 km. Evidence (Seto and Lyon, 1994) suggests, however, that this burning within the jet pipe/exhaust system, where the pressure levels are much lower than in the main combustor, does not increase NO_x production, although the NO_2/NO ratio increases.

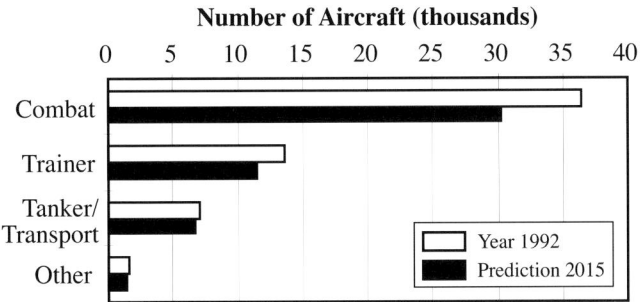

Figure 7-43: Military aircraft inventory (1992 and projection for 2015).

Thus, it appears that although some operational issues can influence the design and therefore the emissions performance of military fighter engines, the broad technology paths aimed at achieving improvements diverge from those of the civil sector only in detail. Military programs will generally continue to lead in providing technology advances that "spin off" into the civil engine sector. Engine cycle advances that improve the thermal efficiency of the core engine will be particularly important in further improving the fuel efficiency of civil engines.

7.11.1.2. Military Transport

Transport requirements for military operations have many parallels with civil operations. This concordance has led to an increasing trend toward development of military transport and tanker aircraft based closely on civil aircraft designs. The links are even closer with regard to engine development for such aircraft, and they are likely to remain so in the future.

7.11.2. Conclusion

The principal conclusion arising from this brief review of the environmental aspects of military engines is that military operators are already well motivated to demand lower emission levels from new engines for operational as well as environmental reasons. Coupled with the evidence presented in Chapter 9 showing that military aircraft will, proportionately, become an even smaller consumer of the world's aviation fuel, a logical extension of this conclusion would be that military operations will have a negligible effect on global emissions from aircraft for the foreseeable future.

References

AERONOX, 1995: *AERONOX. The Impact of NO$_x$ Emissions from Aircraft upon the Atmosphere at Flight Altitudes 8–15 km* [Schumann, U. (ed.)]. EUR-16209-EN, Office for Publications of the European Community, Brussels, Belgium, pp. 236–243.

Albritton, D.L., W.H. Brune, A.R. Douglass, F.L. Dryer, M.K.W. Ko, C.E. Kolb, R.C. Miake-Lye, M.J. Prather, A.R. Ravishankara, R.B. Rood, R.S. Stolarski, R.T. Watson, and D.J. Wuebbles, 1993: *The Atmospheric Effects of Stratospheric Aircraft.* Prepared for the Panel on Atmospheric Effects of Stratospheric Aircraft, Committee on Atmospheric Chemistry, Board on Atmospheric Sciences and Climate. National Research Council, Washington, DC, USA, 156 pp.

Albritton, D., G. Amanaditis, G. Angelettti, J. Crayston, D.H.Lister, M. Macfarlane, J. Miller, A. Ravishankara, N. Sabogal, N. Sundararaman, and H.Wesoky, 1996: *Global Atmospheric Effects of Aviation.* Proceedings of a symposium held in Virginia Beach, Virginia, USA. National Aeronautics and Space Administration, Washington, DC, USA, 96 pp.

Alofs, D.J., C.K. Lutrus, D.E. Hagen, G.J. Sem, and J.L. Blesener, 1995: Intercomparison between commercial condensation nucleus counters and an alternating temperature gradient cloud chamber. *Aerosol Science & Technology,* **23,** 239–249.

Anderson, B.E., W.R. Cofer, D.R. Bagwell, J.W. Barrick, C.H. Hudgins, and K.E. Brunke, 1998: Airborne observations of aircraft aerosol emissions. I: Total nonvolatile particle emission indices. II: Factors controlling non-volatile particle production. *Geophysical Research Letters,* **25,** 1689–1692.

Anderson, M.R., R.C. Miake-Lye, R.C. Brown, and C.E. Kolb, 1996: Calculation of exhaust plume structure and emissions of the ER-2 aircraft in the stratosphere. *Journal of Geophysical Research,* **101,** 4025–4032.

API/NPRA, 1997: *1996 American Petroleum Institute/National Petroleum Refiners Association Survey of Refining Operations and Product Quality.* American Petroleum Institute, New York, NY, USA, July 1997, 12 pp.

Arnold, F., J. Scheid, T. Stilp, H. Schlager, and M.E. Reinhardt, 1992: Measurements of jet aircraft emissions at cruise altitude. 1: The odd nitrogen gases NO, NO$_2$, HNO$_2$, and HNO$_3$. *Geophysical Research Letters,* **19,** 2421–2424.

Arnold, F., J. Schneider, M. Klemm, J. Scheid, T. Stilp, H. Schlager, P. Schulte, and M.E. Reinhardt, 1994: Mass spectrometer measurements of SO$_2$ and reactive nitrogen gases in exhaust plumes of commercial jet airliners at cruise altitude. In: *Impact of Emissions from Aircraft and Spacecraft upon the Atmosphere* [Schumann, U. and D. Wurzel (eds.)]. Proceedings of an international scientific colloquium, Cologne, Germany, 18–20 April 1994. DLR-Mitteilung 94-06, Deutsches Zentrum für Luft- und Raumfahrt (German Aerospace Center), Oberpfaffenhofen and Cologne, Germany, pp. 323–328.

Arnold, F., T. Stilp, R. Busen, and U. Schumann, 1999: Jet engine exhaust chemi-ion measurements: Implications for gaseous SO$_3$ and H$_2$SO$_4$. *Atmospheric Environment,* **32,** 3073–3078.

ASTM, 1997: *D1655 Standard Specification for Aviation Turbine Fuels.* American Society of Testing and Materials, USA, Volume 05.01, pp. 554–563.

Bahr, D.W., 1992: Turbine engine developers explore ways to lower NO$_x$ emission level. *ICAO Journal,* **47(8),** 14–18.

Berry, D.L., 1994: *The Boeing 777 Engine/Aircraft Integration Aerodynamic Design Process.* ICAS-94-6.4.4, International Council on Aeronautical Sciences, BCAG, Anaheim, CA, USA, pp. 1305–1320.

Boeing, 1996: *Current Market Outlook—Commercial Airplane Group Marketing.* Boeing Corporation, Seattle, WA, USA, 36 pp.

Bowden, J.N., S.R. Westbrook, and M.E. LePera, 1988: *A Survey of JP-8 and JP-5 Properties.* BLFRF-253-AD-A207721, Belvoir Fuels and Lubricants Research Laboratory, Southwest Research Institute, San Antonio, TX, USA, September 1988, 109 pp.

Bowman, A.A., 1992: *Control of Combustion-Generated Nitrogen Oxide Emissions: Technology Driven by Regulation.* International Symposium of Combustion, 24, The Combustion Institute, Pittsburgh, PA, USA.

Brasseur, G.P., R.A. Cox, D. Hauglustaine, I. Isaksen, J. Lelieveld, D.H. Lister, R. Sausen, U.Schumann, A. Wahner, and P. Wiesen, 1998: European scientific assessment of the atmospheric effects of aircraft emissions. *Atmospheric Environment,* **32:13,** 2329–2418.

Brown, R.C., R.C. Miake-Lye, M.R. Anderson, C.E. Kolb, A.A. Sorokin, and Y.I. Buriko, 1996: Aircraft exhaust sulfur emissions. *Geophysical Research Letters,* **23(24),** 3603–3606.

Bruce, T.W., F.G. Davis, T.E. Kuhn, and H.C. Mongia, 1977: *Pollution Reduction Technology Program Small Jet Aircraft Engine, Phase I.* Final Report NASA-CR-135214 (September 1977). National Aeronautics and Space Administration, Phoenix, AZ, USA.

Bruce, T.W., F.G. Davis, T.E. Kuhn, and H.C. Mongia, 1978: *Pollution Reduction Technology Program Small Jet Aircraft Engine, Phase II.* Final Report NASA-CR-159415 (September 1978). National Aeronautics and Space Administration, Phoenix, AZ, USA.

Bruce, T.W., F.G. Davis, T.E. Kuhn, and H.C. Mongia, 1981: *Pollution Reduction Technology Program Small Jet Aircraft Engine, Phase III.* Final Report NASA-CR-165386 (December 1981). National Aeronautics and Space Administration, Phoenix, AZ, USA.

Champagne, D.L., 1971: *Standard Measurement of Aircraft Gas Turbine Engine Exhaust Smoke.* ASME 71-GT-88, American Society of Mechanical Engineers, New York, NY, USA.

Chang, S. and T. Novakov, 1983: Role of carbon particles in atmospheric chemistry. In: *Trace Atmospheric Constituents* [Schwartz, S.E. (ed.)]. Wiley & Sons, New York, NY, USA.

Chen, Y., S.M. Kreidenweis, L.M. McInnes, D.C. Rogers, and P.J. DeMott, 1998: Single particles analyses of ice nucleating aerosols in the upper troposphere and lower stratosphere. *Geophysical Research Letters,* **25,** 1391–1394.

Condit, P.: 1996: *Performance, Process, and Value: Commercial Aircraft Design in the 21st Century*. Paper presented at the World Aviation Congress and Exposition, Los Angeles, CA, USA. The Boeing Company, Los Angeles, CA, USA, 18 pp.

Corning, G., 1977: *Supersonic and Subsonic CTOL and VTOL Airplane Design*. University of Maryland Press, College Park, MD, USA, 4th ed., pp. 2:37–2:38.

Cumpsty, N.A., 1997: *Jet Propulsion*. Cambridge University Press, Cambridge, United Kingdom and New York, NY, USA.

Curtius, J., B. Sierau, F. Arnold, R. Bauman, R. Busen, P. Schulte, and U. Schumann, 1998: First direct sulfuric acid detection in the exhaust of a jet aircraft in flight. *Geophysical Research Letters*, **25**, 923–926.

Danilin, M.Y., J.M. Rodriguez, M.K.W. Ko, D.K. Weisenstein, R.C. Brown, R.C. Miake-Lye, and M.R. Anderson, 1997: Aerosol particle evolution in an aircraft wake—implications for the high-speed civil transport fleet impact on ozone. *Journal of Geophysical Research*, **102(D17)**, 21453–21463.

De Champlain, A., D. Kretschmer, J. Tsogo, and G.F. Pearce, 1997: Prediction of soot emissions in gas turbine combustors. *Journal of Propulsion and Power*, **13:1**, 117–122.

Deidewig, F., A. Döpelheuer, and M. Lecht, 1996: *Methods to Assess Aircraft Engine Emissions in Flight*. Proceedings of the 20th Congress of the International Council of the Aeronautical Sciences (ICAS-96-4.1.2), International Council on Aeronautical Sciences, Sorrento, Italy, ISBN 1-56347-219-8, pp. 131–141.

Deidewig, F. and M. Lecht, 1994: NO_x emissions from aircraft/engine combinations in flight. In: *Impact of Emissions from Aircraft and Spacecraft upon the Atmosphere* [Schumann, U. and D. Wurzel (eds.)]. Proceedings of an international scientific colloquium, Cologne, Germany, 18–20 April 1994. DLR-Mitteilung 94-06, Deutsches Zentrum für Luft- und Raumfahrt, Oberpfaffenhofen and Cologne, Germany, pp. 44–49.

DeMore, W.B., S.P. Sander, D.M. Golden, R.F. Hampson, M.J. Kurylo, C.J. Howard, A.R. Ravishankara, C.E. Kolb, and M.J. Molina, *Chemical Kinetics and Photochemical Data for Use in Stratospheric Modeling, Evaluation 11*. JPL-94-26, Jet Propulsion Laboratory, Pasadena, CA, USA, August 1994, 273 pp.

DERA, 1998: *Aviation Turbine Fuel F-35*. Specification DEF STAN 91-91, Issue 2, Defence Evaluation and Research Agency, Directorate of Standardisation, Ministry of Defence, Kentigern House, Glasgow, United Kingdom.

Döpelheuer, A., 1997: *Berechnung der Produkte unvollständiger Verbrennung aus Luftfahrttriebwerken*. IB-325-09-97, Deutsches Zentrum für Luft- und Raumfahrt, Institut für Antriebstechnik, Oberpfaffenhofen and Cologne, Germany, 38 pp.

Döpelheuer, A. and M. Lecht, 1999: *Influence of Engine Performance on Emission Characteristics*. Symposium on Gas Turbine Engine Combustion, Emissions and Alternative Fuels, Lisbon, Portugal, 12–16 October 1998, 11 pp. (in press).

Dryer, F.L., C.E. Kolb, R.C. Miake-Lye, W.J. Dodds, D.W. Fahey, and S.R. Langhoff, 1993: Engine exhaust trace chemistry committee report. In: *The Atmospheric Effects of Stratospheric Aircraft: A Third Program Report* [Stolarski, R.S. and H.L. Wesoky (eds.)]. NASA Reference Publication 1313, National Aeronautics and Space Administration, Office of Space Science and Applications, Washington, DC, USA, November 1993, pp. 245–316.

DuBell, T.L. and S.A. Syed, 1995: *Control of Aircraft Engine Emissions: Status and Future Directions*. International Gas Turbine Conference, Yokohama, Japan, October 27, 1995, pp. I195–I201.

Eatock, C., 1993: Advanced techniques being used to develop low emission combustors for small turbine engines. *ICAO Journal*, **June**, 30–33.

Eatock, H.C. and P. Sampath, 1993: *Low Emissions Combustor Technology for Small Aircraft Gas Turbines*. Paper presented at the 82nd Symposium, Technology Requirements for Small Gas Turbines, Advisory Group for Aerospace Research and Development, October 1993. AGARD Propulsion and Energetics Panel, Specialised Printing, Sussex Ltd., Laughton, Essex, United Kingdom, 15 pp.

Eiff, G., S. Putz, and C. Moses, 1992: *Combustion Properties of Ethanol Blended Turbine Fuels*. Paper presented at the 2nd Annual FAA/AIAA Symposium on General Aviation Systems, Wichita, KS, USA, March 1992. College of Aviation Technologies, Southern Illinois University, Carbondale, IL., USA, 65 pp.

Encyclopedia of Chemical Terminology, 1995: *4th Edition*, Vol. 13. John Wiley and Sons, New York, NY, USA, p. 853.

EPA, 1976: *Aircraft Technology Assessment—Status of the Gas Turbine Program* [Munt, R. and E. Danielson (eds.)]. U.S. Environmental Protection Agency, Washington, DC, USA, 150 pp.

ERAA, 1992: *The Vital Earth—Regional Operation and the Atmosphere* [Ambrose, B. (ed.)]. European Regional Airline Association, October 1992, 21 pp.

Fahey, D.W., E.R. Keim, E.L. Woodbridge, R.S. Gao, K.A. Boering, B.C. Danube, S.C. Wofsy, R.P. Lohmann, E.J. Hintsa, A.E. Dessler, C.R. Webster, R.D. May, C.A. Brock, J.C. Wilson, R.C. Miake-Lye , R.C. Brown, J.M. Rodriguez, M. Loewenstein, M.H. Proffitt, R.M. Stimpfle, S.W. Bowen, and K.R. Chan, 1995a: In situ observations in aircraft exhaust plumes in the lower stratosphere at midlatitudes. *Journal of Geophysical Research*, **100**, 3065–3074.

Fahey, D.W., E.R. Keim, K.A. Boering, C.A. Brock, J.C. Wilson, H.H. Jonsson, S. Anthony, T.F. Hanisco, P.O. Wennberg, R.C. Miake-Lye, R.J. Salawitch, N. Louisnard, E.L. Woodbridge, R.S. Gao, S.G. Donnelly, R.C. Wamsley, L.A. Del Negro, S. Solomon, B.C. Danube, S.C. Wofsy, C.R. Webster, R.D. May, K.K. Kelly, M. Loewenstein, J.R. Podolske, and K.R. Chan, 1995b: Emission measurements of the Concorde supersonic aircraft in the lower stratosphere. *Science*, **270**, 70–74.

Finch, S.P. and A.W. Eyl, Jr., 1976: *Prediction of Test Cell Visible Emissions*. U.S. Air Force Civil Engineering Center, Report No. AF CEC/TR76-47, AD-A037694. Tyndall Air Force Base, Florida, USA, pp. 3–8.

Frenzel, A. and F. Arnold, 1994: Sulfuric acid cluster ion formation by jet engines: implications for sulfur acid formation and nucleation. In: *Impact of Emissions from Aircraft and Spacecraft upon the Atmosphere* [Schumann, U. and D. Wurzel (eds.)]. Proceedings of an international scientific colloquium, Cologne, Germany, 18–20 April 1994. DLR-Mitteilung 94-06, Deutsches Zentrum für Luft- und Raumfahrt, Oberpfaffenhofen and Cologne, Germany, pp. 106–112.

Gardner, R.M. (ed.), 1998: *ANCAT/EC Aircraft Emissions Inventory for 1991/92 and 2015*. Final Report EUR-18179, ANCAT/EC Working Group. Defence Evaluation and Research Agency, Farnborough, UK, 108 pp.

Godin, T., S. Harvey, and P. Stouffs, 1995: *High Temperature Reactive Flow of Combustion Gases in an Expansion Turbine*. Paper presented at the ASME Cogen-Turbo Power Conference, Vienna, Austria, August 23–25, 1995. ASME-95-CTP-7, American Society of Mechanical Engineers, New York, NY, USA.

Godin, T., S. Harvey, and P. Stouffs, 1997: *A Numerical Study of Chemically Reactive Flow of Hot Combustion Gases in a First-Stage Turbine Nozzle*. Paper presented at the International Gas Turbine Institute Gas Turbine and Aeroengine Congress and Exhibition, Orlando, FL, USA, June 2–5, 1997. ASME-97-GT-147, American Society of Mechanical Engineers, New York, NY, USA.

Greene, D.L., 1995: *Commercial Air Transport Energy Use and Emissions: Is Technology Enough?* Paper presented at the 1995 Conference on Sustainable Transportation Energy Sources, 31 July – 3 August 1995. Pacific Grove, CA, USA, 26 pp.

Hadaller, O.J. and A.M. Momenthy, 1993: Characteristics of future aviation fuels. In: *Transportation and Global Climate Change* [Greene, D.L. and D.J. Santini (eds.)]. American Council for an Energy-Efficient Environment, Washington, DC, USA, ISBN 0-918249-17-1, Chapter 10.

Hadaller, O.J. and A.M. Momenthy, 1990: *Characteristics of Future Fuels*. D6-54940, Boeing Commercial Airplane Company, Seattle, WA, USA, September 1990, 12 pp.

Hagen, D.E., M.B. Trueblood, and P.D. Whitefield, 1992: A field sampling of jet exhaust aerosols. *Particle Science Technology*, **10**, 53.

Hagen, D.E., P.D. Whitefield, J. Paladino, M. Trueblood, and H. Lilenfeld, 1998: Particulate sizing and emission indices for a jet engine exhaust sampled at cruise. *Geophysical Research Letters*, **25:10**, 1681–1684.

Hagen, D.E., P.D. Whitefield, and H. Schlager, 1996: Particulate emissions in the exhaust plume from commercial jet aircraft under cruise conditions. *Journal of Geophysical Research*, **101**, 19551–19557.

Hanisco, T.F., P.O. Wennberg, R.C. Cohen, J.G. Anderson, D.W. Fahey, E.R. Keim, R.S. Gao, R.C. Wamsley, S.G. Donnelly, L.A. Del Negro, R.J. Salawitch, K.K. Kelly, and M.H. Proffitt, 1997: The role of HO_x in super- and subsonic aircraft exhaust plumes. *Geophysical Research Letters*, **24(1)**, 65–68.

Harris, B.W., 1990: Conversion of sulfur dioxide to sulfur trioxide in gas turbine exhaust. *Journal of Engineering for Gas Turbines and Power,* **112,** 585–589.

Haschberger, P. and E. Lindermeir, 1997: Observation of NO and NO_2 in the young plume of an aircraft jet engine. *Geophysical Research Letters,* **24,** 1083–1086.

Haschberger, P. and E. Lindermeir, 1996: Spectrometric in-flight measurement of aircraft exhaust emissions: first results of the June 1995 campaign. *Journal of Geophysical Research,* **101(25),** 995–26,006.

Hill, R.J., 1996: The purpose and status of IHPTET—1995. In: *Advanced Aero-Engine Concepts and Controls.* AGARD Conference Proceedings. Advisory Group for Aerospace Research and Development, Neuilly-sur-Seine, France, June 1996, vol. 572, p. 18.

Howard, R.P., R.S. Hiers, P.D. Whitefield, D.E. Hagen, J.C. Wormhoudt, R.C. Miake-Lye, and R. Strange, 1996: *Experimental Characterization of Gas Turbine Emissions at Simulated Flight Altitude Conditions.* AEDC-TR-96-3, Arnold Engineering Development Center, Manchester, TN, USA, September 1996.

Hunter, S.C., 1982: Formation of SO_3 in gas turbines. *Transactions of the ASME,* **104,** 44–50.

Hurley, C.D., 1993: *Smoke Measurements Inside a Gas Turbine Combustor.* Paper presented at the 29th Joint Propulsion Conference and Exhibit, Monterey, CA, USA. AIAA-93-2070, American Institute of Aeronautics and Astronautics (AIAA), USA, 9 pp.

ICAO, 1981: *International Standards and Recommended Practices, Environmental Protection. Annex 16 to the Convention on International Civil Aviation, Volume II, Aircraft Engine Emissions.* International Civil Aviation Organization, Montreal, Canada, 1st ed., 60 pp.

ICAO, 1993: *International Standards and Recommended Practices, Environmental Protection. Annex 16 to the Convention on International Civil Aviation, Volume II, Aircraft Engine Emissions.* International Civil Aviation Organization, Montreal, Canada, 2nd ed., 55 pp.

ICAO, 1995a: *ICCAIA Position on Stringency of NO_x Emissions Regulations.* Working Paper 28, International Civil Aviation Organization Working Group 3 (Emissions), Montreal, Canada, December 1995.

ICAO, 1995b: *ICAO Engine Exhaust Emissions Data Bank.* ICAO-9646-AN/943, International Civil Aviation Organization, Montreal, Canada, 1st ed.

ICAO, 1995c: *Report of the Emissions Inventory Sub-Group at Working Group 3.* International Civil Aviation Organization, Committee on Aviation Environmental Protection, Working Group 3, Bonn, Germany, June 1995, 45 pp.

ICCAIA, 1993: *Feasibility of Reducing Oxides of Nitrogen Levels of Engines in the 6,000 to 15,000 Pounds Thrust Turbofan Engine Class.* Working Paper 2/6 presented at the second meeting of the Technology Subgroup of ICAO/CAEP, International Civil Aviation Organization Committee on Aviation Environmental Protection Working Group 3 (Emissions), Amsterdam, The Netherlands, 10 pp.

ICCAIA, 1997a: *Emissions Technology Review.* Working Paper 2/4 presented at the 2nd meeting of the International Civil Aviation Organization, Committee on Aviation Environmental Protection, Working Group 3 (Emissions). Seville, Spain, January 1997, 38 pp.

ICCAIA, 1997b: *Emissions Technology Review.* Working Paper 3/11 presented at the 3rd meeting of the International Civil Aviation Organization, Committee on Aviation Environmental Protection, Working Group 3 (Emissions), Savannnah, GA, USA, May 1997, 24 pp.

ICCAIA, 1997c: *Engine Enhancements for Lower Emissions.* Working Paper 3/12 presented at the 3rd meeting of the International Civil Aviation Organization, Committee on Aviation Environmental Protection, Working Group 3 (Emissions), Savannnah, GA, USA, May 1997, 5 pp.

ICCAIA, 1997d: *Responses to Comments on WG3 Working Paper 3/11.* Working Paper 4/16 presented at the 4th meeting of the International Civil Aviation Organization, Committee on Aviation Environmental Protection, Working Group 3 (Emissions), Bern, Switzerland, November 1997, 8 pp.

ICCAIA, 1997e: *Responses to Questions Pertaining to Emissions Trade-Offs.* Working Paper 4/17 presented at the 4th meeting of the International Civil Aviation Organization, Committee on Aviation Environmental Protection, Working Group 3 (Emissions), Bern, Switzerland, November 1997, 9 pp.

ICCAIA, 1997f: *2050 Fuel Efficiency and NO_x Technology Scenarios.* Working Paper 4/22 presented at the 4th meeting of the International Civil Aviation Organization, Committee on Aviation Environmental Protection, Working Group 3 (Emissions), Bern, Switzerland, November 1997, 14 pp.

Jane's, 1998: *All the World's Aircraft, 1997–98.* Jane's Publishing Co., London, United Kingdom, 1029 pp.

Kärcher, B. and D.W. Fahey, 1997: The role of sulfur emission in volatile particle formation in jet aircraft exhaust plumes. *Geophysical Research Letters,* **24,** 389–392.

Kärcher, B., M.M. Hirschberg, and P. Fabian, 1996: Small-scale chemical evolution of aircraft exhaust species at cruising altitudes. *Geophysical Research Letters,* **101(15),** 15169–15190.

Katzman, H., and W.F. Libby, 1975: Hydrocarbon emissions from jet engines operated at simulated high-altitude supersonic flight conditions. *Atmospheric Environment,* **9,** 839–842.

Kerrebrock, J.L., 1992: *Aircraft Engines and Gas Turbines.* MIT Press, Cambridge, MA, USA, 2nd ed.

Kleffmann, J., R. Kurtenbach, and P. Wiesen, 1994: Emissions of nitrous oxide and methane from aero engines: monitoring by tunable diode laser spectroscopy. In: *Impact of Emissions from Aircraft and Spacecraft upon the Atmosphere* [Schumann, U. and D. Wurzel (eds.)]. Proceedings of an international scientific colloquium, Cologne, Germany, 18–20 April 1994. DLR-Mitteilung 94-06, Deutsches Zentrum für Luft- und Raumfahrt, Oberpfaffenhofen and Cologne, Germany, pp. 82–87.

Klug, H.G., S. Bakan, and V. Gaylor, 1996: *Cryoplane—Quantitative Comparison of Contribution to Anthropogenic Greenhouse Effect of Liquid Hydrogen Aircraft versus Conventional Kerosene Aircraft.* European Geophysical Society, XXI General Assembly, The Hague, The Netherlands, May 1996, 22 pp.

Koff, B.L., 1991: *Spanning the Globe with Jet Propulsion.* AIAA-2987, American Institute of Aeronautics and Astronautics, Arlington, VA, USA, p. 2.

Lecht, M. and F. Deidewig, 1994: *No_x Emission Correlation for Varying Engine Inlet Conditions.* Presented at the 5th meeting of the Certification Sub-Group of the International Civil Aviation Organization, Committee on Aviation Environmental Protection, Working Group 3 (Emissions), Seattle, WA, USA, 7 pp.

Leide, B. and P. Stouffs, 1996: Residual reactivity of burned gases in the early expansion process of future gas turbines. *Journal of Engineering for Gas Turbines and Power,* **118,** 54–60.

Liebeck, R.H., M.A. Page, and B.K. Rawdon, 1998: *Blended Wing Body Subsonic Commercial Transport.* AIAA-98-0438, American Institute of Aeronautics and Astronautics, 36th Aerospace Sciences Meeting, 12-15 January 1998, Reno, Nevada, USA.

Lipfert, F.W., 1972: *Correlation of Gas Turbine Emissions Data.* ASME 72-GT-60, American Society of Mechanical Engineers, New York, NY, USA, 16 pp.

Lister, D.H. and M.I. Wedlock, 1978: *Measurement of Emissions Variability of a Large Turbofan Aero-Engine.* ASME-78-GT-75, American Society of Mechanical Engineers, New York, NY, USA, 11 pp.

Lister, D.H., E. Eyeh, C. Baudoin, J. Burbank, F. Deidewig, R.S. Falk, M. Kapernaum, J. Kleffmann, V. Kluge, R. Kurtenbach, M. Lecht, M. Metcalf, A. Sami, C. Wahl, P. Wiesen, and N. Zarzalis, 1995: Engine exhaust emissions. In: *AERONOX. The Impact of NO_x Emissions From Aircraft Upon the Atmosphere at Flight Altitudes 8–15 km* [Schumann, U. (ed.)]. EUR-16209-EN, Office for Publications of the European Communities, Brussels, Belgium, pp. 33–127.

Lowrie, B., 1993: *Combustors for Supersonic Transport Propulsion.* Presentation at the AGARD meeting of Fuels and Combustion Technology for Advanced Engines. AGARD 536, 81st Symposium, 10–14 May 1993, 2-1 – 2-6.

Lukachko, S.P. and I.A. Waitz, 1997: *Effects of Engine Aging on Aircraft NO_x Emissions.* Presented at the International Gas Turbine and Aeroengine Congress and Exhibition, Orlando, FL, USA, 2-5, June 1997. ASME 97-GT-386, American Society of Mechanical Engineers, New York, NY, USA.

Lukachko, S.P., I.A. Waitz, R.C. Miake-Lye, R.C. Brown, and M.R. Anderson, 1998: Production of sulfate aerosol precursors in the turbine and exhaust nozzle of an aircraft engine. *Journal of Geophysical Research,* **103(D13),** 16159–16174.

Lynch, F.T. and R.T. Crites, 1996: *Some Constraints Imposed on the Aerodynamic Development Process by Wind-Tunnel Circuit Design Characteristics.* Advisory Group for Aerospace Research and Development, Moscow, Russia, September 1996.

Lynch, F.T., R.C. Potter, and F.W. Spaid, 1996: *Requirements for Effective High Lift.* CFD-ICAS 96-2.7.1, International Council on Aeronautical Sciences, Sorrento, Italy.

Lynch, F.T. and G.A. Intemann, 1994: *The Modern Role of CFD in Addressing Airframe/Engine Integration Issues for Subsonic Transports.* ICAS 94-6.4.3, International Council on Aeronautical Sciences.

Lyon, T.F., W.J. Dodds, and D.W. Bahr, 1980: *Determination of Pollutant Emissions Characteristics of General Electric CF6-6 and CF6-50 Model Engines, Final Report.* FAA-EE80-27, Federal Aviation Administration, Washington, DC, USA.

Martin, R.L., C.H. Oncina, and P.J. Zeeben, 1994: *A Simplified Method for Estimating Aircraft Engine Emissions.* Oral presentation at ASME, The Hague, 1994. Reported as "Boeing Method 1 fuel flow methodology description" in appendix C of "Scheduled Civil Aircraft Emission Inventories for 1992" [Baughcum, S. (ed.)]. NASA Contractor Report 4700, April 1996.

Martin, R.L., C.H. Oncina, and P.J. Zeeben, 1995: A Simplified Method for Estimating Aircraft Engine Emissions. ICAO/CAEP/Working Group 3, Certification Subgroup, March 1995. Reported as "Boeing Method 2 fuel flow methodology description in appendix D of "Scheduled Civil Aircraft Emission Inventories for 1992" [Baughcum, S. (ed.)]. NASA Contractor Report 4700, April 1996.

Martin, R.L., C.H. Oncina, and P.J. Zeeben, 1996: A simplified method for estimating aircraft engine emissions. In: *Scheduled Civil Aircraft Emission Inventories for 1992.* NASA-CR-4700, National Aeronautics and Space Administration, Washington, DC, USA, April 1996, pp. 13–15.

McMasters, J.H. and I.M. Kroo, 1998: *Advanced Configurations for Very Large Transport Airplanes.* Presented at 36th Aerospace Sciences Meeting, 12-15 January 1998, Reno, NV, USA. AIAA-98-0439, American Institute for Aeronautics and Astronautics.

Miake-Lye, R.C., B.E. Anderson, W.R. Cofer, H.A. Wallio, G.D. Nowicki, J.O. Ballethin, D.E. Hunton, W.B. Knighton, T.M. Miller, J.V. Seeley, and A.A. Viggiano, 1998: SO_x oxidation and volatile aerosol in aircraft exhaust plumes depend on fuel sulfur content. *Geophysical Research Letters,* **25,** 1677–1680.

Miake-Lye, R.C., M. Martinez-Sanchez, R.C. Brown, and C.E. Kolb, 1993: Plume and wake dynamics, mixing, and chemistry behind a high speed civil transport aircraft. *Journal of Aircraft,* **30,** 467–479.

Miake-Lye, R.C., R.C. Brown, M.R. Anderson, and C.E. Kolb, 1994: Calculations of condensation and chemistry in an aircraft contrail, in impact of emissions from aircraft and spacecraft upon the atmosphere. In: *Impact of Emissions from Aircraft and Spacecraft upon the Atmosphere* [Schumann, U. and D. Wurzel (eds.)]. Proceedings of an international scientific colloquium, Cologne, Germany, 18–20 April 1994. DLR-Mitteilung 94-06, Deutsches Zentrum für Luft- und Raumfahrt, Oberpfaffenhofen and Cologne, Germany, pp. 280–285.

Momenthy, A.M., 1996: *Jet Fuel Data Status and Importance.* Presentation at International Air Transport Association Fuel Trade Forum, Johannesburg, South Africa, May 1996. Boeing Airplane Group, Seattle, WA, USA.

Mongia, H.C., 1994: *Combustor Modeling in Design Process: Applications and Future Direction.* AIAA-94-0466, American Institute for Aeronautics and Astronautics.

Mongia, H.C., 1997a: *Gas Turbine Combustion Design, Technology and Research: Current Status and Future Direction.* AIAA-97-3369, American Institute for Aeronautics and Astronautics, 17 pp.

Mongia, H.C., 1997b: *Recent Progress in Low-Emissions Gas Turbine Combustors.* Presented at the XIII International Symposium on Airbreathing Engines, Chattanooga, TN, USA, 7-12 September 1997, 7 pp.

Mortlock, A., and R. van Alstyne, 1988: *Military, Charter, Unreported, Domestic Traffic and General Aviation 1976, 1984, 1992 and 2015.* NASA-CR-1998-207639, National Aeronautics and Space Administration, Langley Research Center, Hampton, VA, USA, March 1998, pp. 1–58.

Mullins, J., A. Simmons, and A. Williams, 1988: *Rates of Formation of Soot from Hydrocarbon Flames and Its Destruction.* AGARD-CP-422, Advisory Group for Aerospace Research and Development, 23 pp.

Novakov, T., 1982: Soot in the atmosphere. In: *Particulate Carbon Atmospheric Life Cycle* [Wolff, G.T. and R.L. Klimisch (eds.)]. Plenum Press, New York, NY, USA, pp. 19–37.

NRC, 1997: *An Interim Assessment of AEAP's Emissions Characterization and Near-Field Interactions Elements.* National Research Council Panel on the Atmospheric Effects of Aviation, National Academy Press, Washington, DC, USA, 16 pp.

Petzold, A. and F.P. Schröder, 1998: Jet engine exhaust aerosol characterization. *Aerosol Science and Technology,* **28,** 62–76.

Petzold, A., J. Ström, F.P. Schröder, and B. Kärcher, 1999: Carbonaceous aerosol in jet engine exhaust: emission characteristics and implications for heterogeneous chemical reactions. *Atmosphere and Environment,* (in press).

Pueschel, R.F., *et al.,* 1997: Soot aerosol in the lower stratosphere: pole-to-pole variability and contribution by aircraft. *Journal of Geophysical Research,* **102,** 13113–13118.

Pueschel, R.F., S. Verma, G.V. Ferry, S.D. Howard, S. Vay, S.A. Kinne, J. Goodman, and A.W. Strawa, 1998: Sulfuric acid and soot particle formation in aircraft exhaust. *Geophysical Research Letters,* **25,** 1685–1688.

Rickard, G.K. and R. Fulker, 1997: *The Quality of Aviation Fuel Available in the United Kingdom, Annual Survey 1996.* Technical Report DERA/SMC/SM1/TR970039, Defence Evaluation and Research Agency, Farnborough, Hampshire, United Kingdom, May 1997, 35 pp.

Rickey, J.E., 1995: *The Effect of Altitude Conditions on the Particle Emissions of a J85-GE-5L Turbojet Engine.* NASA-TM-106669, National Aeronautics and Space Administration, Lewis Research Center, Cleveland, OH, USA, February 1995, pp. 1–50.

Rizk, N.K. and D.A. Smith, 1994: *Regional and Business Aircraft Mission Emissions.* ASME-94-GT-300, American Society of Mechanical Engineers, New York, NY, USA, June 1994, 10 pp.

Rolls-Royce, 1992: *The Jet Engine.* Rolls-Royce plc, Derby, United Kingdom, 4th ed., 187 pp.

Rubbert, P.E., 1994: *CFD and the Changing World of Airplane Design.* ICAS 94-0.2, International Council on Aeronautical Sciences, BCAG, Anaheim, CA, USA, September 1994.

Rudey, R.A., 1975: *Status of Technological Advancements for Reducing Aircraft Gas Turbine Pollutant Emissions.* NASA-TMX-71846, National Aeronautics and Space Administration, Washington, DC, USA, December 1975.

Schlager, H.P., P. Schulte, and H. Ziereis, 1997: In situ measurements in aircraft exhaust plumes and in the North Atlantic flight corridor. In: *Pollutants from Air Traffic: Results of Atmospheric Research 1992-1997* [Schumann, U. (ed.)]. DLR Mitteilung 97-04, ISBN 1434-8462, pp. 57–66.

Scholes, J., D. Stanley, and M. Kimble-Thom, 1998: Evaluation of Soy Diesel in Jet A as a Commercial Aircraft Fuel. Purdue University, Aviation Technology Department, Lafayette, IN, USA.

Schulte, P. and H. Schlager, 1996: Flight measurements of cruise altitude nitric oxide emissions indices of commercial jet aircraft. *Geophysical Research Letters,* **23,** 165–168.

Schulte, P., H. Schlager, H. Ziereis, U. Schumann, S.L. Baughcum, and F., Deidewig, 1997: NO_x emission indices of subsonic long-range jet aircraft at cruise altitude: in situ measurements and predictions. *Journal of Geophysical Research,* **102,** 21431–21442.

Schumann, U., 1997: *Pollution From Aircraft Emissions in the North Atlantic Flight Corridor (POLINAT).* EUR 16978. Office for Publications of the European Communities, Luxembourg, ISBN 92-827-8569-6, 304 pp.

Schumann, U., J. Ström, R. Busen, R. Baumann, K. Gierens, M. Krautstrunk, F.P. Schröder, and J. Stingl, 1996: In situ observations of particles in jet aircraft exhausts and contrails for different sulfur-containing fuels. *Journal of Geophysical Research,* **101,** 6853–6869.

Segalman, I., R.G. McKinney, G.J. Sturgess, and L. Huang, 1993: Reduction of NO_x by fuel-staging in gas turbine engines—a commitment to the future. In: *Fuel and Combustion Technology for Advanced Aircraft Engines.* Paper No. 29, Advisory Group for Aerospace Research and Development Conference Proceedings, Propulsion and Energetics Panel, 81st Symposium, AGARD CP-536, Fiuggi, Italy, May 1993.

Seto, S.P. and T.F. Lyon, 1994: Nitrogen oxide emission characteristics of augmented turbofan engines. *Journal of Engineering for Gas Turbines and Power,* **116.**

Shaw, R.J., L. Koops, and R. Hines, 1997: *Progress Toward Meeting the Propulsion Technology Challenges for a 21st Century High-Speed Civil Transport.* ISABE-97-7048, International Symposium on Airbreathing Engines, September 7–12, 1997.

Singleton, A.H., 1997: Advances make gas-to-liquids process competitive for remote locations. *Oil & Gas Journal,* 4 August 1997.

Spicer, C.W., M.W. Holdren, R.M. Riggin, and T.F. Lyon, 1994: Chemical composition and photochemical reactivity of exhaust from aircraft turbine engines. *Annales Geophysicae,* **12,** 944–955.

Spicer, C.W., M.W. Holdren, D.L. Smith, D.P. Hughes, and M.D. Smith, 1992: Chemical composition of exhaust from aircraft turbine engines. *Journal of Engineering for Gas Turbines and Power,* **114,** 111–117.

Stolarski, S.S., S.L. Baughcum, W.H. Brune, A.R. Douglass, D.W. Fahey, R.R. Friedl, S.C. Lui, R.A. Plumb, L.R. Poole, H.L. Wesoky, and D.R. Worsnop, 1995: *1995 Scientific Assessment of the Atmospheric Effects of Stratospheric Aircraft.* NASA reference publication number 1381. National Aeronautics and Space Administration, Washington, DC, USA, 110 pp.

Tindall, B.M. and D.L. King, 1994: Designing steam reformers for hydrogen production. *Hydrocarbon Processing,* **73:7,** 69–75.

Tsang, W. and J.T. Herron, 1991: Chemical kinetic data base for propellant combustion I: reactions involving NO, NO_2, HNO, HNO_2, HCN and N_2O. *Journal of Physical and Chemical Reference Data,* **20(4),** 609–663.

Tsang, W. and R.F. Hampson, 1986: Chemical kinetic data base for combustion chemistry: Part I—methane and related compounds. *Journal of Physical and Chemical Reference Data,* **15,** 1087.

Twohy, C. and B.W. Gandrud, 1998: Electron microscope analysis of residual particles from aircraft contrails. *Geophysical Research Letters,* **25,** 1359–1362.

Wei, D. and H.A. Spikes, 1986: The lubricity of diesel fuels. *Wear,* **111,** 217–235.

Weisenstein, D.K., M.K.W. Ko, N-D. Sze, and J.M. Rodriguez, 1995: Potential impact of SO_2 emissions from stratospheric aircraft on ozone. *Geophysical Research Letters,* **23(2),** 161–164.

Westley, F., J.T. Herron, and R.J. Cvetanovic, 1983: *Compilation of Chemical Kinetic Data for the Combustion Chemistry: Part I—Non-Aromatic C, H, O, N, And S Containing Compounds (1971–1982).* NSRDS-NBS 73, National Bureau of Standards, Washington, DC, USA, Part 1.

Wey, C.C., C. Wey, D.J. Dicki, K.H. Loos, D.E. Noss, D.E. Hagen, P.D. Whitefield, M.B. Trueblood, M.E. Wilson, D. Olson, J.O. Ballenthin, T.M. Miller, A.A. Viggiano, J. Wormhoudt, T. Berkoff, and R.C. Miake-Lye, 1998: *Engine Gaseous, Aerosol Precursor, and Particulate at Simulated Flight Altitude Conditions.* NASA/TM-1998-208509/ARL-TR-1804, National Aeronautics and Space Administration, Lewis Research Center, Cleveland, OH, USA, 176 pp.

Whitefield, P.D. and D.E. Hagen, 1995: *Particulates and Aerosols Sampling From Combustor Rigs Using the UMR MASS.* Paper presented at the 33rd Aerospace Sciences Meeting, Reno, NV, USA, January 1995. AIAA-95-0111, American Institute for Aeronautics and Astronautics.

Whitefield, P.D., D.E. Hagen, and H.V. Lilenfeld, 1996: Ground-based measurements of particulate emissions from sub-sonic and super-sonic transports. In: *Proceedings of the 30th Section Anniversary Technical Meeting of the Central Start Section of the Combustion Institute, St. Louis, MO, USA.* Paper No. 39, Combustion Fundamentals and Applications, American Institute of Aeronautics and Astronautics, Combustion Institute, Pittsburgh, PA.

Whyte, R.G., 1982: *Alternative Jet Engine Fuels.* AGARD Advisory Report No 181, Vol. 2, ISBN 92-835-1423-8, 168 pp.

Yetter, R.A., F.L. Dryer, M.T. Allen, and J.L. Gatto, 1995: Development of gas-phase reaction mechanisms for nitramine combustion. *Journal of Propulsion and Power,* **11,** 683–697.

Yu, F. and R.P. Turco, 1998: Contrail formation and impacts on aerosol properties in aircraft plumes: effects of fuel sulfur content. *Geophysical Research Letters,* **25,** 313–316.

Yu, F. and R.P. Turco, 1997: The role of ions in the formation and evolution of particles in aircraft plumes. *Geophysical Research Letters,* **24,** 1927–1930.

Zarzalis, N., G. Tellischek, G. Meikis, B. Glaeser, and G. Huster, 1995: *NO_x-Reduction in Aero Engine Combustors by Application of the Rich Lean Combustion Concept.* ASME-IMECE-WAM-95-6, American Society of Mechanical Engineers, San Francisco, CA, USA, November 1995, 12 pp.

Zurer, P.S., 1995: NASA cultivating basic technology for supersonic passenger aircraft. *Chemical Engineering News,* **24 April,** 10–16.

8

Air Transport Operations and Relation to Emissions

GERARD BEKEBREDE

Lead Authors:
D. Dimitriu, L. Dobbie, V. Galotti, A. Lieuwen, S. Nakao, D. Raper, H. Somerville, R.L. Wayson, S. Webb

Contributors:
A. Gil, D.R. Marchi, B. Miaillier, B. Nas, J. Templeman

Review Editor:
C.V. Oster, Jr.

CONTENTS

EXECUTIVE SUMMARY

This chapter assesses what is known about measures already being taken or that might be taken to improve the fuel efficiency of aviation through changes in aircraft operations and changes in the way air traffic is managed against a background of continuing growth in air transport activity. The objective of such improvement in fuel efficiency is to reduce the amount of fuel consumed for a given demand for air transportation—which would have the effect of reducing emissions.

The existing air navigation system and its subsystems suffer from technical, operational, political, procedural, economic, social, and implementation shortcomings. The International Civil Aviation Organization (ICAO), national airworthiness authorities, regional bodies, and private-sector stakeholders have identified the need for radical improvement of the air traffic management system to accommodate the continuing growth of aviation and to promote efficient airspace use. Implementation of a new concept for air traffic management that includes enhanced communications, navigation, and surveillance in support of an improved air traffic management system has already begun. Communications, navigation, and surveillance/air traffic management (CNS/ATM) systems benefit the air transportation sector by reducing delays, increasing the capacity of existing infrastructure, and improving operational efficiency. This system results in fuel savings, hence reduced emissions for a given demand.

Several studies associated with the implementation of CNS/ATM systems have been carried out. Although some of these studies provide results in terms of cost/benefit and associated fuel savings and do not specifically address environmental benefits, there is an obvious correlation with reductions in gaseous emissions. These studies suggest that improvements in air traffic management could help to improve overall fuel efficiency by 6–12%.

Other strategies that exist for mitigating the environmental impact of emissions from aviation could achieve environmental benefits through reduced fuel burn. These strategies include: optimizing aircraft speed, reducing additional weight, increasing the load factor, reducing nonessential fuel on board, limiting the use of auxiliary power units, and reducing taxiing. Airlines are already under strong pressure to optimize these parameters, largely because of economic considerations and requirement within the industry to minimize operational costs. The potential reduction in fuel burn by further optimization of these operational measures is in the range of 2–6%.

To answer the question of whether emissions reductions could be achieved by substituting the use of air transport by other modes, several studies have compared fuel burn and carbon dioxide emissions from different modes of transport. The amount of carbon dioxide emitted per passenger-km for the different modes of transport is very dependent on the distance travelled; the type of aircraft, train, or car; the load factor; and the source of energy used. Substitution of rail and coach for air travel could result in the reduction of emissions per passenger-km. However, the potential reduction would be achieved only for passengers travelling relatively short distances and only on high-density routes that have rail or coach links. Finally, substituting other transport modes for air transport could have environmental impacts on, for example, local air quality and noise exposure (which are outside the scope of this report).

8.1. Introduction

The air transport industry has grown rapidly over recent years, and growth is expected to continue. An effect of this growth is an increasing amount of emitted exhaust gases and consequential environmental impact. Chapter 9 discusses aspects of this growth.

Chapter 8 is the second of two chapters that assess what is known about means of addressing adverse environmental impacts associated with aircraft engine emissions. Whereas Chapter 7 considers how aircraft and their engines might be improved, this chapter assesses what is known about measures that might be, or have been, taken to change aircraft operations and improve air traffic management (ATM) systems and procedures to reduce the amount of fuel consumed, which would have the effect of reducing emissions. Aircraft operations and measures that could be taken to improve efficiency are well understood, but only limited knowledge exists on the potential environmental impact of many of these measures.

Section 8.2 describes the constraints and limitations of the conventional ATM system and changes that are presently being implemented through new technology and improved procedures, as well as changes anticipated for the future designed to create a more integrated, global system of air traffic management. Whereas inefficient routings, cruising at less than optimum flight levels, and airborne holdings that characterize the present system result in unnecessary amounts of emissions injected into the atmosphere, new systems and procedures could result in fuel savings, hence reductions of emissions for a given demand for air transportation.

Section 8.3 discusses potential fuel reductions from other (non-ATM) operational factors, including improvement of aircraft utilization, optimization of speed, reductions in weight and nonessential fuel on board, limiting the use of auxiliary power units, and reduced taxiing. The airlines' need to contain fuel costs already provide a powerful incentive for progressive improvements in these areas. Application of these measures is restricted to some degree, however, by safety and regulatory aspects.

To answer the question of whether emissions reductions could be achieved by substituting other modes of travel for air transport, several studies have compared fuel burn and carbon dioxide emissions from different modes of transport. This comparison between carbon dioxide emissions for different modes of transport is discussed in Section 8.3.

Throughout this chapter, the assumption is made that improvements in operations and fuel efficiency will lead to reductions in emissions. Improved efficiency however, may result in attracting additional air traffic, although no studies providing evidence on the existence of this so-called rebound effect have been identified. In this respect, the scenarios presented in Chapter 9 have assumed an optimized air traffic management system. As a consequence, Chapter 9 includes the effects of an optimized ATM system for air transport demand and related emissions. Chapter 8 discusses the relation between improved operational efficiency and emissions for a given demand for air transport.

8.2. Air Traffic Management System—Present and Future

8.2.1. Introduction

Demand for air travel is growing rapidly. Shortfalls in capacity and other constraints on the efficiency of airport and aircraft operations have negative effects on airline costs, passenger convenience, and the environment. Making efficient use of finite airspace and airport resources while ensuring high levels of safety is the primary mission of ATM which involves considerable coordination of planning and operations among regulators, service providers, and users at the global, regional, and national levels.

This section discusses the conventional ATM system in the context of operational phases of flight, highlighting constraints and limitations and the negative effects they have on airport and aircraft operations—including unnecessary fuel burn and, consequently, excessive emissions. The section then describes changes anticipated for the future based on new technologies and improved procedures that are expected to lead to the creation of a more efficient and integrated global ATM system.

Improved ATM as envisaged will encompass traditional elements of air traffic services (ATS)—air traffic control (ATC), air traffic flow management (ATFM), and airspace management (ASM)—but will also functionally integrate these elements with ATM-related aspects of flight operations into a total system. Today, ATC accounts for the greatest percentage of ATS on a global basis; ATC serves primarily to prevent collisions between aircraft and between aircraft and obstructions in the airport maneuvering area and to expedite and maintain an orderly flow of air traffic. For current (i.e., 1998–99), worldwide aircraft fleet operations, improvements to the ATM system alone could reduce fuel burn per trip by 6–12% (EUROCONTROL, 1997b; FAA, 1998a; ICAO, 1998b).

Improving ATM requires that advanced technological and management systems and procedures be adopted more rapidly and on a broader scale than is presently the case. Specific improvements related to ATM and the operation of aircraft that could reduce fuel burn are covered in this chapter; institutional, regulatory, and economic policy measures that could also have an important influence on future traffic growth and associated fuel burn are covered in Chapter 10.

8.2.2. Limitations of Current ATM System

8.2.2.1. General

In 1983, the ICAO Council established a committee to identify and assess new technologies and make recommendations for the future development of air navigation. After close analysis, the special committee on the future air navigation system (FANS) recognized that the existing air navigation system and its subsystems suffered from technical, operational, procedural, economic, and implementation shortcomings. In addition to infrastructure constraints, conventional airspace organization of flight information regions and their supporting infrastructure of routes and ground-based facilities and services are based largely on national rather than international requirements. For these reasons, aircraft must plan their flights along strictly defined routes and be channelled, to a certain degree, so that air traffic controllers can keep aircraft safely separated from each other.

8.2.2.2.   Airport and Terminal Maneuvering Area (TMA) Operations and Capacity

In some regions, limited airport capacity is one of the main constraints on continued growth in air transport; this limited capacity results in congestion and delays. There is also a lack of adequate awareness and shared decisionmaking among ATC, ramp, and taxi areas. In low-visibility conditions, movements are severely restricted, and there is increased risk of runway incursion. Insufficiently developed taxiways and aprons also limit runway and airport capacity. Operational limitations for noise control may also have a negative effect on access to and from key airports. Automated ground-based systems to manage departures and arrivals efficiently are not available in most cases, and onboard automation is therefore underutilized. Published arrival and departure procedures—created to ease controller workload and ensure separation between departures and arrivals—are often inflexible, indirect, and less than optimum.

8.2.2.3.   En Route and Oceanic Operations

The existing worldwide route structure often imposes mileage penalties compared to the most economic routes (generally great-circle routes); it also takes into account wind, temperature, and other factors such as aircraft weight, charges, and safety. Use of a fixed-route network often results in concentration of traffic flows at major intersections, which can lead to a reduction in the number of routes and flight levels that are available. Studies on penalties to air traffic associated with the European ATS Route Network alone suggest that ATM-related problems add an average of about 9–10% to the flight track distance of all European flights en route and in terminal maneuvering areas (TMA) (EUROCONTROL, 1992). Lack of international coordination in the development of ground ATC systems exacerbates these problems. Examples include inconsistent separation standards in radar and non-radar airspace and operation at less than optimum flight levels in oceanic airspace as a result of communication deficiencies.

8.2.2.4.   Meteorological Information

Currently, three main areas can be distinguished in which improvements need to be made in the way meteorological

information is provided to international civil aviation: Timeliness, presentation, and accuracy.

Timeliness problems are largely related to the inability of telecommunications channels in some regions to cope with increasing message traffic. As a result, tight restrictions have been established concerning the exchange of operational meteorological information, which now does not fully meet flight planning requirements for increasingly long-range aircraft operations. The presentation of meteorological information has also been largely dictated by the telecommunications channels used, which have imposed a predominance of alphanumeric messages over graphical information, especially in the cockpit.

Finally, the accuracy of meteorological information needs improvement. For the en route phase of flight, the information provided is not always based on output from the most advanced numerical weather prediction models. In the terminal area, up-to-date and accurate meteorological information may not be available to the pilot because of congestion of voice channels and/or lack of modern observing systems.

8.2.2.5.   Restricted and Military Airspace

The fundamental premise that every state has complete and exclusive sovereignty over the airspace above its territory can be traced to the Convention on International Civil Aviation (ICAO, 1997). States implement restrictions on the use of airspace for a variety of reasons, including technological limitations, political considerations, security, and environmental concerns. However, by far the most important reason for restricted airspace is to accommodate the needs of states' military forces. Restricted airspace does not allow aircraft to minimize their emissions by direct routing between two points. Significant regions of airspace are permanently reserved or restricted, thereby forcing civil air transport to circumnavigate these areas.

The extent of the problem varies by region. In the European region, for example, 24 states are applying the flexible use of airspace (FUA) concept (EUROCONTROL, 1998a). The basis for FUA is that airspace should no longer be considered as either military or civil airspace but should be considered as a continuum, shared in accordance with user needs and used flexibly on a day-to-day basis. Although national security requirements must be key factors in revising a nation's restricted airspace allocation, problems related to restricted and military airspace could partly be solved by modernization of the ATM system. Negotiation of overflying rights to shorten routes would also contribute to solving the problems.

8.2.3.   Improvements to ATM System

8.2.3.1.   Introduction

The description, goals, and objectives of an improved global air navigation infrastructure emanate from the concept known as communications, navigation, and surveillance/air traffic management (CNS/ATM) systems. CNS/ATM was formally endorsed by the worldwide civil aviation community at the Tenth Air Navigation Conference in Montreal in 1991 (ICAO, 1991).

The primary goal of an integrated ATM system is to enable aircraft operators to meet planned times of departure and arrival and adhere to preferred flight profiles with minimum constraints and no compromise on safety. Therefore, although an integrated, global ATM system grew out of a need to meet growing demand, the ultimate effect will be enhanced operations and improved efficiency, primarily through less fuel burn for a given level of demand. ATM systems will therefore be developed and organized to overcome shortcomings discussed in this chapter and to accommodate future growth to offer the best possible service to all airspace users and to provide adequate economic benefits to the civil aviation community, with due regard for environmental concerns.

CNS/ATM has been defined as a system employing digital technologies, including satellite systems together with various levels of automation, applied in support of a seamless global air traffic management system. The main elements of CNS/ATM systems are described in detail in the ICAO Global Air Navigation Plan for CNS/ATM Systems (ICAO, 1998a). CNS/ATM systems will use very high frequency (VHF) and high frequency (HF) communication channels to transmit digital data between aircraft and between aircraft and ground stations. Satellite data and voice communications capable of global coverage are also being introduced. Improvements in navigation include progressive introduction of area navigation (RNAV) capabilities based on a global navigation satellite system (GNSS). Improvements in surveillance techniques will allow aircraft to automatically transmit their positions using data link technology.

It is generally agreed that these newer technologies will optimize the worldwide route structure. Planners will be less confined by the location of ground aids and more direct tracks will be used, allowing substantial savings in fuel. Ultimately, rigid route structures will be gradually eliminated or redesigned at a number of critical intersections. This flexibility has the potential to relieve congestion in very high density traffic areas.

8.2.3.2. Airport Operations and Capacity

In some regions, the increasing gap between traffic demand and capacity provided by the physical infrastructure at many key airports is a critical limiting factor. CNS/ATM systems can contribute to increasing capacity. Sophisticated automation and digital data links will help to make maximum use of available capacity and meet throughput requirements by improving the identification and predicted movement of aircraft and vehicles in the airport movement area. Additionally, increasing levels of collaboration and information-sharing between aircraft operators and ATM providers will create a more realistic picture of airport departure and arrival demand, allowing operators to make dynamic scheduling and flight planning decisions based on the ATM situation at any given time.

8.2.3.3. TMA Operations and Capacity

Enhanced instrument approach techniques will improve the flexibility of approach operations, thereby reducing noise and emissions levels. Parallel runways spaced as closely as 760 m or less are expected to routinely accommodate independent instrument flight rule (IFR) approaches based on high-data-rate secondary surveillance radars, data link technologies, improved cockpit and air traffic controller displays, and advanced automation. This technology will provide capacity increases in instrument meteorological conditions (IMC) at locations with such closely spaced runways. Also, automation tools will assist air traffic managers in establishing efficient flows of approaching aircraft for parallel and converging runway configurations.

8.2.3.4. En Route Operations

The flow management process will monitor capacity resources and demand at airports and in terminal and en route airspace and will implement strategies, where required, to protect ATC from overloads and to provide an optimal flow of traffic by making best use of available airspace capacity. Clearances involving position and time, using an ATM data link interface with flight management computers, will be principal tools in assuring that ATM constraints are met with minimum deviation from user-preferred trajectories. The ability to predict optimum trajectory and monitor conformance of aircraft along these trajectories will allow the most efficient flight profiles and routes—resulting in an increase in overall efficiency, reductions in average fuel consumption per flight, and, consequently, reduced emissions levels for a given demand.

8.2.3.5. Oceanic Operations

Future oceanic ATM operations will make extensive use of data link technologies, GNSS, HF and satellite-based digital communications, aviation weather system improvements, and collaborative decisionmaking techniques. Planned implementation of reduced longitudinal and lateral separation minima and reduced vertical separation (RVSM), along with more flexible handling of flights, is expected to lead to fuel savings. Implementation of RVSM above 8,850 m in the North Atlantic region has already resulted in increased capacity and reduced fuel consumption.

8.2.3.6. Meteorological Information

More timely weather information will be available as a result of two developments: Implementation of the final phase of the world area forecast system (WAFS), which uses direct satellite communications to deliver information to states, and increasing use of air-to-ground data link communications to uplink operational meteorological information. With these developments, the restrictions placed on the exchange of operational meteorological information are gradually being lifted.

With regard to the presentation of meteorological information, the increasing use of graphical information will be made possible by the introduction of air-to-ground data links.

With regard to the accuracy of meteorological information, introduction of the final phase of WAFS will increase the quality of meteorological information provided. However, future improvements in accuracy will depend significantly on the availability of frequent automatic dependent surveillance (ADS) reports, which include a meteorological information data block. Improved observing and forecasting techniques for volcanic ash and clear-air turbulence will eliminate overprediction of airspace affected by these phenomena, which can restrict the use of airspace. In the terminal area, integrated terminal weather systems will improve the accuracy of information provided regarding hazardous weather phenomena. Furthermore, real-time wind models run by ATC computers, based on ADS reports obtained during the climb-out phase, will be used to monitor the evolution of the wind field, which is required by ATC for sequencing of approaching aircraft.

8.2.4. *Implementation*

8.2.4.1. *Emerging Concepts*

To attain the benefits of an improved ATM system, the United States and Europe have designed future ATM concepts as a logical progression of ICAO's work on CNS/ATM systems. The improvements described in Section 8.2.3 are expected to be implemented over the next 20 years. For a discussion of institutional and financial arrangements related to the implementation of CNS/ATM systems, see Chapter 10.

The U.S. effort, known as Free Flight, would enable optimum and dynamically determined flight paths for all airspace users through CNS/ATM technologies and the establishment of ATM procedures that maximize flexibility while ensuring positive separation of aircraft (RTCA, 1995). The European Organisation for Safety and Air Navigation (EUROCONTROL) has defined the future uniform European ATM System (EATMS) and an associated ATM Strategy for 2000+ as a system whereby flights will be managed from a total system perspective from gate to gate (EUROCONTROL, 1998b). This approach is expected to reduce fragmentation and improve efficiency considerably. The ATM 2000+ strategy and EATMS are taking environmental considerations into account, as a consequence of the environmental provision of the revised EUROCONTROL Convention (EUROCONTROL, 1997a).

Free Flight and the ATM Strategy for 2000+ were developed to accommodate the significant requirements of these two regions

of the world within the context of ICAO concepts and policies. Remaining regions will be guided more specifically by ATM concepts being developed by ICAO. Harmonization of concepts under the CNS/ATM umbrella will lead, in theory, to an integrated global ATM system. CNS/ATM systems, along with efforts aimed at overcoming national limitations associated with sovereignty issues and funding of new systems, should result in a reduction in restrictions and a more optimized route system.

8.2.4.2. *Regional Variations*

There are a number of differences between air navigation infrastructure requirements in the various regions of the world. The United States and Europe embody the differences between two complex and highly developed regions. However, these two regions also share some similarities; increased capacity, for example, is the primary driver of implementation planning in both regions. On the other hand, developing regions are more concerned with improving safety, efficiency, and accessibility.

Integration and harmonization of various concepts and needs are required to achieve consistency in terms of safety and regularity and to attain the seamlessness required for efficient operation. Planning for implementation of improved air navigation systems based on CNS/ATM therefore cannot be accomplished in isolation. Regional air navigation plans, comprising listings of facilities and services required to serve civil aviation, are coordinated through the global mechanisms established by ICAO, and the ICAO standards and recommended practices provide a common framework for implementation and planning of systems. Regional bodies such as the African Civil Aviation Commission (AFCAC), the Arab Civil Aviation Commission (ACAC), EUROCONTROL, and the Latin American Civil Aviation Commission (LACAC) also provide common platforms for planning and implementation while ensuring commonality and coherency. Regions also establish bilateral relationships—such as between EUROCONTROL and the U.S. Federal Aviation Administration (FAA)—to address their unique and specific needs and to work more rapidly than traditional mechanisms allow.

8.2.5. *Environmental Benefits Associated with CNS/ATM Implementation*

8.2.5.1. *Potential Environmental Benefits*

Several studies associated with implementation of CNS/ATM systems have been carried out. Although some of these studies state their results in terms of cost/benefit and associated fuel savings and do not specifically address the environment, obvious correlations to reduced gaseous emissions (including NO_x and carbon dioxide) can be drawn. Table 8-1 provides a summary of results from several case studies; this summary is taken from a paper presented by ICAO during the worldwide CNS/ATM systems implementation conference held in Rio de Janeiro in May 1998 (ICAO, 1998b).

Table 8-1: Summary of fuel-saving case studies.

Case Study	Period	Fuel Saving	Remarks
India	1997–2001	1.0–1.2 tons of fuel per flight	Study examined potential savings to airlines from CNS/ATM implementation within Calcutta flight information region (FIR)
Spain	1997–2016	1.3–1.5% reduction in total fuel burn	Cost/benefit study with the objective of assessing the economic feasibility and financial implications of implementing CNS/ATM systems in Spain
U.S. airlines	1997–2010	6–9% fuel burn savings in medium to long term	Study estimated costs of inadequate ATC infrastructure
Mitre	1996–2010	12% fuel burn savings worldwide	Preliminary study that presented an overview of the analysis of emissions reductions achievable through CNS/ATM

Source: ICAO, 1998b.

A preliminary EUROCONTROL study considered various ATM strategies and concepts that have the potential to reduce fuel burn. For two particular scenarios—free route airspace above 10,200 m and unconstrained "direct flight gate-to-gate"—potential savings of 1–2 and 7–8% were estimated, respectively (EUROCONTROL, 1997b). Table 8-2 presents the results of a preliminary assessment of potential fuel savings and resulting environmental benefits resulting from introduction of CNS/ATM in the national airspace system of the United States for the year 2015 (FAA, 1998b).

8.2.5.2. Rebound Effects

Introduction of capacity- and efficiency-enhancing measures such as those associated with CNS/ATM systems may attract additional air traffic. This phenomenon is referred to as the

Table 8-2: Potential annual fuel savings and resulting environmental benefits from the introduction of CNS/ATM in U.S. national airspace system.*

Phase of Flight	Fuel	NO_x	CO	HC
Above 3000 m	9,683	204.3	197.1	56.7
Below 3000 m	219	4.0	1.1	0.1
Surface	358	1.2	13.2	3.1
Total	10,259	209.5	211.4	59.9
% Savings	6.1	9.9	12.7	18.0

*Annual savings in millions of pounds.

rebound effect. Introducing efficiencies hitherto not available to operators—efficiencies that result in lower fuel consumption and subsequently lower operating costs—may reduce fares and stimulate traffic and growth beyond that already anticipated based on the forecast demand for air travel. This eventuality cannot be overlooked because the net effect could be an increase in air traffic and consequently an increase in fuel burn and emissions. It is unclear to what extent any additional emissions caused by a rebound effect might offset anticipated reductions in emissions arising from CNS/ATM systems implementation. No studies providing evidence on the existence or size of the rebound effect have been carried out.

8.3. Other Operational Factors to Reduce Emissions

From an environmental viewpoint, efficiency of transportation ("mobility efficiency") can be described as the fuel required to transport one person over a distance of 1 kilometer—the energy required per passenger-km. Comparison of freight hauling by aircraft to other modes of freight transportation is more complex because of differences in modes, such as weight restrictions. If various restrictions to the analysis are assumed and stated, freight may be compared on the basis of energy use per tonne-km. Because of these complexities and the requirements of various assumptions, this chapter concentrates on comparisons of passenger mobility.

The potential to improve the mobility efficiency of air transport is discussed in Section 8.3.1. The consequences for the aviation transportation system of avoiding specific areas at particular times and at particular altitudes and the resultant impact on fuel consumption are described in Section 8.3.2. Section 8.3.3 compares CO_2 emissions from various transport modes. Section 8.3.4 then discusses ground-based, aircraft-related emissions. Section 8.3.5 provides concluding remarks.

8.3.1. Aircraft Performance

This section describes potential reduction of fuel consumption by optimizing the operational use of an aircraft. Issues discussed include improving aircraft capacity utilization, reducing the operational weight of an aircraft, and optimizing the speed of an aircraft. In general, however, these variables have already been optimized by airlines, largely because of economic pressures and the requirement within the industry to minimize operational costs.

8.3.1.1. Aircraft Capacity Utilization

The energy required per passenger-km depends on the distance travelled and the load factor, where the load factor is defined as the ratio between the transported payload and the maximum payload. Increasing payload improves fuel efficiency, as illustrated by the effect of payload on fuel burn for a Tupolev-154 (see Table 8-3). The Tupolev-154 is an older aircraft with lower fuel efficiency than modern aircraft that is frequently used in the Eastern part of Europe and the former Soviet Union. The effect of load factor on fuel efficiency for modern aircraft is similar (see Table 8-4); fuel efficiency of modern aircraft is greater at all payloads (Table 8-4 and Section 8.3.3). For a discussion of trends in fuel efficiency over different generations of aircraft and levels of technology, see Chapter 7.

Apart from load factor, aircraft type, and level of technology, the configuration of an aircraft (number of seats, distribution between seating and cargo capacity) also has an important influence on fuel burned per passenger-km. The configuration is determined by the airline in consultation with the aircraft manufacturer and can be altered during the aircraft's lifetime. It differs between airlines and is based on market considerations. An example is provided by the way Japan Airlines configures its Boeing 747-400 aircraft. The 747-400 in long-range full passenger configuration has 262 seats, whereas the 747-400D used in high-density local Asian service has 568 seats, even when used in similar lengths of flights (see Table 8-4). Seat configuration can also vary in smaller aircraft. For example,

Table 8-3: *Effect of load factor on fuel burned per passenger-km—Tupolev-154.**

Load Factor (%)	Passengers	Fuel Burned per Passenger-km (kg)	MJ per Passenger-km
19	32	0.21	8.9
51	84	0.08	3.6
70	115	0.065	2.9
82	135	0.053	2.3
100	164	0.051	2.0

* Based on 164 seats, a maximum payload of 19,100 kg, and the Bucharest–London Heathrow route (2,235 km). Information provided by Tarom Romanian Air Transport.

Tarom Romanian Air Transport operates BAC 1-11 aircraft in configurations with 104 economy passengers or a combination of 12 business and 77 economy passengers. Tarom also operates Boeing 737-300s with the same operating weights with 12 business and 120 economy passengers or 20 business and 102 economy passengers.

Most calculations of the mobility efficiency of passenger aircraft do not take into account the cargo that is carried in aircraft used for scheduled passenger services. For example, in 1995 about 28% of tonne-km flown by British Airways operations consisted of freight (British Airways, 1998a), much of which was carried on passenger aircraft. The United States reported 33.9% for all types of passenger aircraft in 1996 (Bureau of Transportation Statistics, 1997). The 1996 U.S. cargo load factor for freight carriers was reported to be 61.8% for all aircraft types. Although relatively little freight is carried on short-haul passenger service, the carriage of freight might be adding a significant penalty to the perceived mobility efficiency of passenger air transport for long-haul flights.

Optimization of aircraft configuration and load factors can result in reduced emissions. This conclusion is supported by

Table 8-4: *Effect of configuration on fuel consumption in Boeing 747-400 aircraft.**

Length of Flight (km)	B747-400 Long-Range Configuration (262 Seats)		B747-400D High-Density Configuration (568 Seats)	
	Fuel Consumption	MJ per Seat-km	Fuel Consumption	MJ per Seat-km
1000	0.049	2.1	0.023	1.0
2000	0.042	1.8	0.020	0.9
4000	0.040	1.7	0.019	0.8
6000	0.040	1.7		
8000	0.042	1.8		
10000	0.043	1.8		
12000	0.045	1.9		

*Information provided by Japan Airlines.

data presented in Tables 8-3 and 8-4. Airline economics dictate that costs, including fuel costs, be minimized, therefore that the load factor be optimized. As an illustration of the development of load factors for all domestic and international scheduled services, International Air Transport Association (IATA) statistics show an average 0.4% per annum increase in load factor over the past decade (IATA, 1996). These data are consistent with data in ICAO (1996), which showed an increase in load factor of 4% in a period of 10 years. However, it is unclear whether the load factor can continue to grow indefinitely without consequences to passenger service. Moreover, load factor and aircraft configuration are driven not only by the need for improved efficiency but also by market considerations.

The increase in demand for air travel is an important factor in aircraft capacity utilization. (Chapter 9 describes the main drivers for this increase.) A second important factor is the (further) development of the "hub and spoke" system in some parts of the world, such as the United States and Europe. Because transport efficiency is not always optimized by direct point-to-point travel, air travel has evolved from direct connections between major city pairs to a complex network that includes consolidation hubs. Costs are minimized by maximal use of single-sector, out-and-back operations. By combining these sectors in a hub-and-spoke network, an efficient multiplication of potential point-to-point markets served can be achieved. Use of the hub-and-spoke system results in greater distances than direct point-to-point flights; it can also result in fewer aircraft operations for the entire system. Any area of the world with, for example, 500 airports will have about 250,000 potential point-to-point flights. For many pairs of airports, however, a direct flight will never occur except where demand makes direct routes economically viable.

Recent studies (e.g., Peterse and Boering, 1997) have suggested that apart from further intensification of the hub-and-spoke system, more point-to-point connections are likely to be introduced. This "hybrid" development arises for several reasons:

- Two-engine aircraft built for extended twin operations (ETOPS) over water are available.
- Further segmentation of the air transport market. Business passengers are prepared to pay more for a direct connection instead of flying via hubs.
- Capacity problems at main hubs are likely to lead to growth at smaller, secondary hubs.

With the introduction of ETOPS aircraft, a hybrid system will not necessarily result in a decrease of transport efficiency from an economic point of view.

8.3.1.2. *Aircraft Cruise Speed*

Historically, the speed of jet transport operations was set at a constant Mach number. The Mach number is defined as the ratio of an aircraft's speed to the local speed of sound. The local speed of sound varies with ambient static temperature,

which generally decreases with altitude. The selected Mach number is based on overall time-variable costs, as well as fuel economy, which has always been a major component of operating costs. Thus, typical cruise speeds were Mach 0.82 to 0.84 for first-generation jet aircraft such as the Boeing 707, Boeing 727, or McDonnell Douglas DC-8 and 0.84 to 0.86 for early Boeing 747s.

Subsequently, a number of fuel-conscious airlines developed the concept of a long-range cruise (LRC) speed schedule, usually based on Mach number. LRC was introduced as a compromise between maximum speed and the speed that provides the highest mileage in terms of km per kg of fuel burned in cruise (maximum range cruise, or MRC speed), taking some account of costs associated with flight time. LRC is defined as the fastest speed at which cruise fuel mileage is 99% of fuel mileage at MRC. At the time LRC was introduced, it was not possible to fly at lower speeds, closer to MRC, because of the stability needs of the autothrottle and/or the autopilot. At speeds close to MRC, the autothrottle would continuously "hunt" which could give rise to an increase in fuel burn.

In the mid-1970s, fuel conservation was further enhanced by development of the performance management system (PMS) for aircraft such as the Boeing 737 and 747 and the McDonnell Douglas DC-10. Later aircraft (such as most Airbus aircraft; the Boeing 757, 767, 747-400, and 777; and the McDonnell Douglas MD-11) included the flight management system (FMS) as a built-in feature. PMS or FMS computer systems can be used to minimize overall trip cost, which is a balance between fuel- and time-related costs. The most efficient cruise speed (ECON) can be calculated on a real-time basis by using the cost index facility on the system. With full-time autothrottle, late-model aircraft can fly between MRC and LRC to optimize fuel savings.

Figure 8-1 shows the relationship between the difference in block time and the difference in fuel consumption for various cruise speed schedules—such as constant Mach number, LRC,

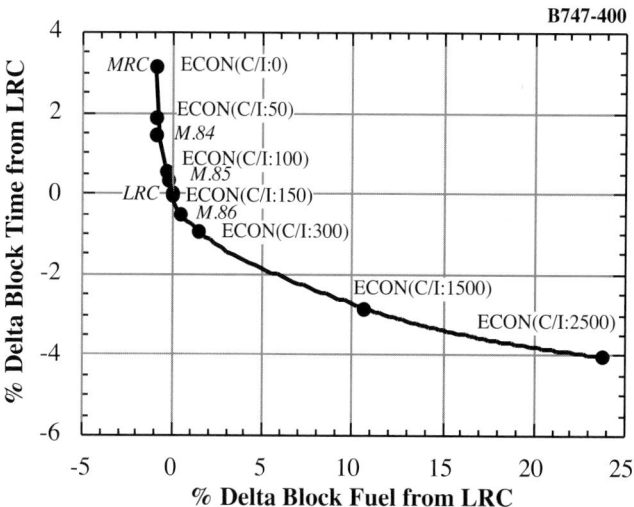

Figure 8-1: Effect of cruise speed on block fuel and block time.

MRC, or ECON—for the Boeing 747-400 (block time is the time between engine start at the airport of origin and engine stop at the airport of destination; block fuel is the fuel burned in this time). Data in Figure 8-1 are consistent with those in Fransen and Peper (1993). The data presented in Figure 8-1 suggest that reduction of fuel use by further speed optimization is likely to be small.

Bradshaw (1994) investigated the variation in block fuel and block time for Airbus aircraft. The report shows higher figures (up to 10%) for the potential reduction in fuel burn from reduction of cruise speed. However, the report also concludes that reduction of Mach number will involve penalties such as significant increase in block time and the subsequent effect on direct operating costs.

8.3.1.3. Aircraft Weight Reduction

A negative effect on the mobility efficiency of aircraft can be caused by additional non-essential weight. Additional weight is introduced if an aircraft takes more fuel onboard than that required by the fuel flight plan. Tankering is the term for loading of fuel used for subsequent flight segments. The main reasons for tankering of fuel are commercial—for example, in cases where the cost of fuel consumed in carrying additional fuel is more than offset by the difference in the price of fuel at the departure point and a destination where fuel could be loaded. Factors that can affect fuel costs and decisions on tankering include the following:

- Genuine high fuel costs because of expensive distribution infrastructure and local taxes
- Fuel availability at some remote airports
- Government-imposed fuel pricing (at some Eastern European airports, the price of aviation fuel is more than 50% more than at Western European airports)
- Monopoly distribution of fuel, which can involve cross-subsidies from large to small airports and expensive manpower practices
- Concern over fuel quality (e.g., water content) at particular locations
- Slot availability (where limited aircraft turnaround time allows insufficient time for refueling, an aircraft may have to tanker to minimize the risk of losing slots; problems in this area are exacerbated at congested airports, where there may be limitations in runway and/or terminal capacity).

Estimates from British Airways suggest that additional fuel burn as a result of tankering is on the order of 0.5% of total aircraft fuel consumption.

Apart from tankering, all commercial flights must carry a certain amount of additional fuel, often mandated by national legislation, for safety reasons. The minimum amount of fuel required for the planned route is calculated by taking into consideration the weather forecast, the route, the weight of the aircraft, and other

factors (including an allowance for diversion to secondary destinations). In most cases, this calculation is carried out with a computerized database. The pilot is presented with a flight plan and the required amount of fuel, although the captain decides how much fuel is finally carried. The pilot may adjust the amount to be loaded in light of any exceptional circumstances and his or her own interpretation of the risk of diversion or likelihood of weather changes down the route. The amount of excess fuel carried for safety reasons is likely to be one to several thousand kg per flight.

Other factors that introduce additional weight are potable water and emergency equipment. Adjustment of the amount carried to the anticipated requirement, with a contingency allowance, could accomplish some fuel savings. The extent to which emergency equipment is carried varies from airline to airline and the nature of the flight. In addition aircraft weight is increased by customer service considerations such as quality of seating, in-flight entertainment, and duty-free goods.

Although little quantitative data are available, such weight reduction could result in a potential fuel savings on the order of 1–2%. As described, only part of this potential reduction can be achieved. Hence, the potential reduction of total fuel burn from weight reduction is probably less than 1%.

8.3.1.4. Other Operational Issues

During very rare emergency situations, it may be necessary to jettison fuel into the atmosphere to reduce the overall weight of an aircraft to a safe landing weight. The potential effects of this (relatively small amount of) jettisoned fuel are not described here. Chapters 2, 3, and 6 discuss the possible effects of emitting unburned aviation fuel into the atmosphere.

Emergencies requiring jettisoning of fuel are mainly mechanical in nature, such as serious engine malfunction, or airframe structural failure. Severe illness of passenger(s) is also a major cause of emergency landings. Jettisoning of fuel is largely confined to larger aircraft flying long-haul routes. For these aircraft, the maximum landing weight may be significantly lower than the maximum take-off weight (both specified through certification by the manufacturer). In the event of an emergency that requires fuel to be jettisoned, airline instructions, as specified in aircraft operating manuals, and local operating procedures call for the aircraft to climb to a specified altitude or to fly to designated fuel dumping areas away from centers of population. British Airways estimates that only a very small percentage (on the order of 0.01%) of fuel used by the aviation industry each year is jettisoned. This estimate is subject to major uncertainty, particularly with reference to the former Soviet Union, and does not take into account military operations.

Other potential fuel-reduction measures include reducing non-revenue flights and changing the loading distribution of passengers, cargo, and fuel to change the aircraft's center of gravity. However, such measures are not expected to have a significant impact.

8.3.1.5. Tradeoff between Noise and Emissions

Apart from aircraft emissions, noise exposure around airports is also an environmental issue. The relationship between noise exposure and fuel efficiency sometimes involves a tradeoff. For example, the Aircraft Noise Design Effects Study (ANDES) concluded that "...a general rule of thumb is that a 3-decibel noise reduction at flyover (where the noise rewards are greatest) would, on average, increase fuel burn and hence emissions by some 5%." (ICCAIA, 1994). This figure applies to a new aircraft design; to achieve this reduction in noise at the other measuring points (approach and sideline), a higher fuel burn penalty is involved. For the same result to be achieved by modifying an existing design, the ANDES study concluded that the penalty would be greater.

One major area of potential impact is in retrofitting of engine equipment on older aircraft to conform with current aircraft noise standards. For example, it is possible to convert older, more noisy "Chapter 2" aircraft to comply with the tighter "Chapter 3" noise standards that will be a requirement of all civil subsonic jet aircraft operating at airports in the United States in 2000 and in Europe and some other countries by 2002. Such a change is not possible for all "Chapter 2" aircraft. However, where it is possible, the increased weight of noise abatement equipment ("hushkits") can lead to an increase in fuel consumption of up to 5%. The fuel increase depends on the type of hushkit used. A lightweight hushkit, with smaller improvements in noise reduction, may have a negligible effect on fuel consumption. In addition, for a very limited number of aircraft, there is a possibility of fitting modern, quiet, fuel efficient engines—but at higher cost because new engines generally are more costly than hushkits.

In addition, noise restrictions may cause flight paths, arrival paths, and departure paths to be longer than the shortest routings—resulting in increased fuel burn. Airport neighbors continue to lobby for noise reductions. Such noise mitigation measures that increase flight time and distance may include departures over bodies of water, selective runway use, routing around populated areas, and so forth. Agencies worldwide have implemented these types of noise control strategies. For example, in the United States, the FAA has defined 37 noise control strategy categories (Cline, 1986). Of these 37 categories, 25 could have a direct effect on aircraft operations at the airport. Some of the remaining 12 categories could also have an indirect effect on aircraft operations.

8.3.2. Ambient Factors

In the past decade, discussions of daylight flying restrictions and on the potential sensitivity to aviation emissions of specific geographical areas and flight levels have resulted in research into consequences for the aviation transportation system of avoiding these areas (Fransen and Peper, 1993; ICAO, 1995; Sausen *et al.*, 1996). Chapters 2, 3, and 6 of this report describe evolving knowledge on the possible atmospheric effects of these issues. In general, this scientific assessment is unable to make clear-cut recommendations regarding possible reductions in adverse environmental impacts that might be achieved with such flight restrictions. The description in this section is confined to consequences for the aviation transportation system.

Aircraft operate most efficiently at specific cruise altitude; generally, less fuel is consumed at higher altitudes, but more engine thrust and equipment may be required to reach those altitudes. Any requirement to stay, for example, below the tropopause would undoubtedly increase fuel consumption. In that case, flights travelling close to the North Pole during the winter would have to descend to levels that could be 2,500–5,000 m below the most fuel efficient level. Consequently, fuel consumption would increase for many flights (ICAO, 1995). A study by Fransen and Peper (1993) showed that total fuel burned by aviation in the North Atlantic corridor would increase by 4–5% using flight level 310 (approximately 9,500 m) as the maximum allowed level; the increase for an individual flight could be as much as 20% (Lecht, 1994). This restriction would lead to payload and range limitations because some current aircraft operate at or close to maximum range. Such limitations, in turn, could lead to requirements for intermediate stops—resulting in less direct routes, increased flight times, and increased fuel burn. Another effect would be concentration of flights, hence increased congestion, which would contribute to additional fuel burn.

Limiting flights to specific parts of the day is practically impossible for a number of reasons, not least because long-range commercial flights can last for up to 14 hours and during that time can cross a number of time zones. Confining travel to the hours of darkness would require intermediate stops, which would result in an increase in the number of landing and take-offs and an associated increase in the amount of fuel used. An additional problem concerning night flights is public pressure to reduce these flights for noise reasons.

8.3.3. Intermodality

This section compares CO_2 emissions from various transport modes. Only CO_2 emissions are discussed, primarily because other pollutant emissions are not comparable at all altitudes. Climate forcing from CO_2 is independent of the altitude and geographical location of its release. Potential atmospheric effects of other gaseous particle emissions from aircraft will likely depend on where and when they occur. For example, the potential climate forcing and ozone perturbation effects of NO_x emitted at altitude will be significantly different from the effects of NO_x emitted by ground transport sources. Finally, substituting other transport modes for air transport can have environmental impacts on, for example, local air quality and noise exposure (which are outside the scope of this report).

8.3.3.1. Introduction

Emissions of CO_2 from all transport sectors currently account for about 22% of all global emissions of CO_2 from fossil fuel use (IPCC, 1996a). In 1990, aviation was responsible for about 12% of CO_2 emissions from the transport sector (see Figure 8-2) (Faiz *et al.*, 1996; IPCC, 1996b; OECD, 1997a,b). Regional variations also occur, as shown in Figure 8-3 for North America. Consequently, aviation is currently responsible for about 2% of total global emissions of CO_2 from the use of fossil fuels (Sprinkle and Macleod, 1993; WMO, 1995; Gardner *et al.*, 1996).

Civil aviation can be divided into domestic and international flights, and their respective levels of CO_2 emissions can be determined. Balashov and Smith (1992) provided an estimate of this breakdown (see Table 8-5).

Chapter 9 provides a detailed assessment of the future growth of the aviation industry, fleet scenarios, and projected fuel burn.

Rail and other forms of transport, such as bus or motorcar, have been suggested as substitutes for short-haul air travel (there is little alternative for longer distances). Flight stages of 800 km or less are estimated to represent only about 15–20% of all scheduled passenger operations (expressed in terms of available seat-km), according to an estimate provided by the ICAO Secretariat based on an analysis of 1996 airline schedules. Bearing in mind that a significant proportion of passenger journeys include more than one flight stage and that some short-haul flights bypass physical obstacles such as water, mountains, or inadequate ground infrastructure, the potential scope for replacing air transport with other modes seems unlikely to exceed about 10%.

Fare, trip time, and frequency of service rather than environmental considerations influence the choice a passenger makes for a particular mode of transport. A paper presented to a European Air Traffic Forecasting Forum (ECAC, 1996) estimates that "...even under favorable assumptions for rail, less than 10% of the European air passengers could be substituted by high-speed train."

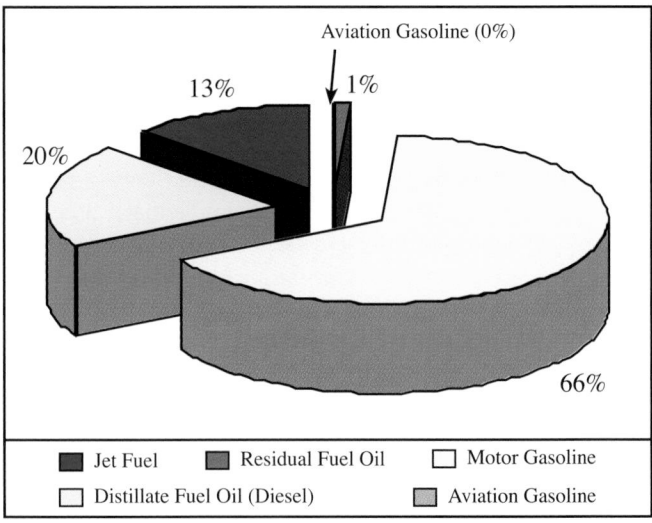

Figure 8-3: CO_2 emissions by transportation mode: United States – 1996.

8.3.3.2.   Comparison of Carbon Dioxide Emissions from Different Forms of Passenger Transport

Figure 8-4 compares CO_2 emissions from major passenger transport modes. The wide ranges in CO_2 intensity of passenger transport reflect many differences between countries and regions, including the availability of renewable and nuclear energy, the extent of road and rail infrastructure, and culture. For example, in the United States, greater importance is placed on aircraft and automobile modes of travel than on rail. The automobile mode serves mainly shorter urban and commuter needs, whereas aircraft serve longer intercity needs. This pattern is illustrated by the number of domestic flights—more than five times the number of international flights—which demonstrates greater U.S. dependence on short-haul aviation than in other geographic regions. In addition, U.S. cars and light trucks tend to be higher emitters because of increased vehicle and engine size (right of the range shown in Figure 8-4).

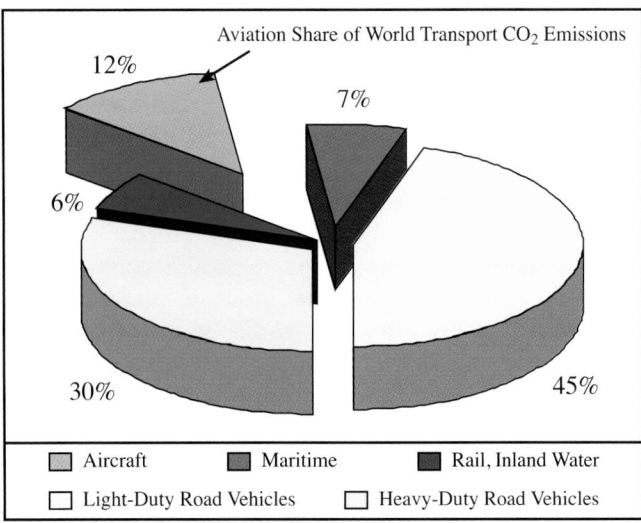

Figure 8-2: Aviation share of world transport CO_2 emissions.

Table 8-5: *Fuel consumption and CO_2 production of civil aviation.*

	Total Fuel Consumption (Mt)	Total Fuel Consumption (% World)	CO_2 Emissions (Mt C)
International Services			
Scheduled	53	40.0	45
Non-scheduled	10	7.5	9
Domestic Services			
United States	38	28.6	33
Russian Fed.	15	11.3	13
Other	17	12.8	15
World Total	133	100	114

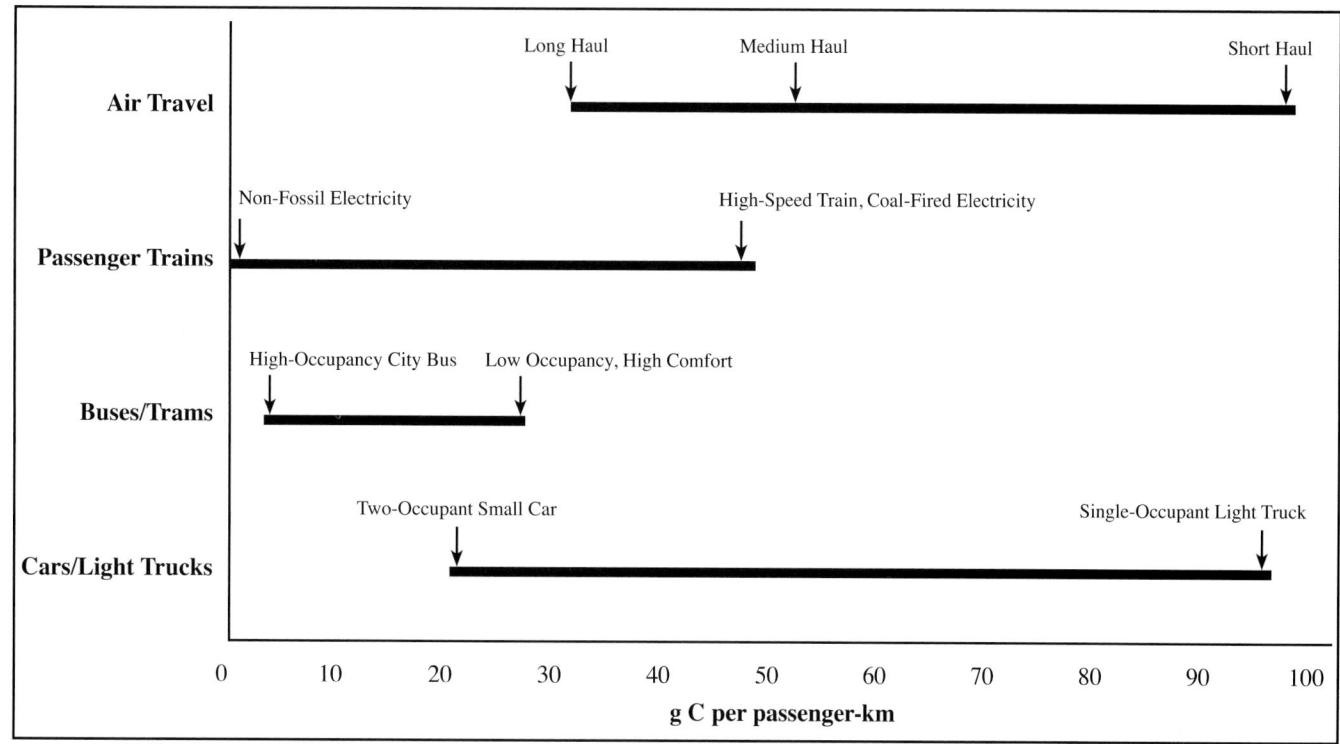

Figure 8-4: CO_2 intensity of passenger transport (TEST, 1991; Whitelegg, 1993; Faiz *et al.*, 1996; Centre for Energy Conservation and Environmental Technology, 1997a; OECD, 1997a).

Care must be taken in interpreting such data to appreciate underlying assumptions and statistics from which they have been drawn. The load factor of transportation modes is critical to the analysis. Car occupancy, in particular, can vary between 1 and 4. For example, in Europe the average is 1.65 (Centre for Energy Conservation and Environmental Technology, 1997b), but in the United States this value is generally less than 1.2 (Institute of the Association for Commuter Transportation, 1997)—which implies a significant margin in specific emissions (per passenger-km) relative to average occupancy. Occupancy levels for air, rail, and bus also vary significantly, but because of commercial pressures they are more likely to operate at higher levels than private road vehicles. European scheduled airlines typically operate at a load factor of about 70% (AEA, 1997) and charter airlines at about 90%. These figures are comparable to those in the United States, where in 1996 the average passenger load factor ranged from 48.6 to 75.4% for various passenger aircraft types and was 69.4% for all air carrier aircraft types (Bureau of Transportation Statistics, 1997).

Energy consumption and CO_2 emissions from electrically powered vehicles, particularly trains, are very dependent on the mode of electrical power generation. In countries that have a large dependency on hydroelectric or nuclear power generation, emission of CO_2 per passenger-km by rail may be very low (Figure 8-4). Conversely, emissions of CO_2 per passenger-km from high-speed locomotives with power derived from coal-fired electricity are considerably higher. In the case of aviation, flight distance is very important. On a short flight (250 km), energy consumption and CO_2 emissions are significantly higher than they are for medium- or long-haul flights (see Figures 8-4 and 8-5), because a greater proportion of

the flight is at take-off power (with a relatively higher fuel consumption). Also, available data do not differentiate between aviation fuel used for passenger transport and that for freight. The Organisation for Economic Cooperation and Development (OECD) has calculated that passengers roughly account for 71% of the load carried (OECD, 1997a), although on short-haul routes freight may account for less than 10% of the weight (Centre for Energy Conservation and Environmental Technology, 1997b).

Figure 8-5 provides a more detailed analysis of CO_2 emissions (g C per passenger-km) from a range of aircraft based on flight and fuel use data collected by British Airways and studies from Germany, Switzerland, The Netherlands, and the European Commission (Hofstetter and Meienberg, 1992; Prognos AG, 1995; Centre for Energy Conservation and Environmental Technology, 1997a,b; British Airways, 1998b; European Commission, 1998). New aircraft, particularly long-haul aircraft (e.g., B747-400), are significantly more fuel efficient and emit less CO_2 per passenger-km than older aircraft (e.g., DC-10). Over short-haul routes, advanced turboprops (ATP) emit about 20% less CO_2 (as carbon) per passenger-km than new jet aircraft (B737-400) and up to three-fold less CO_2 than older jet aircraft (MD81/F-100). Chapter 7 provides a detailed assessment of aircraft fuel efficiencies.

8.3.3.3. Comparison of Carbon Dioxide Emissions from Different Forms of Freight Transport

Figure 8-6 provides a comparison of CO_2 emissions from major freight transport modes. In terms of CO_2 (g C per tonne-km),

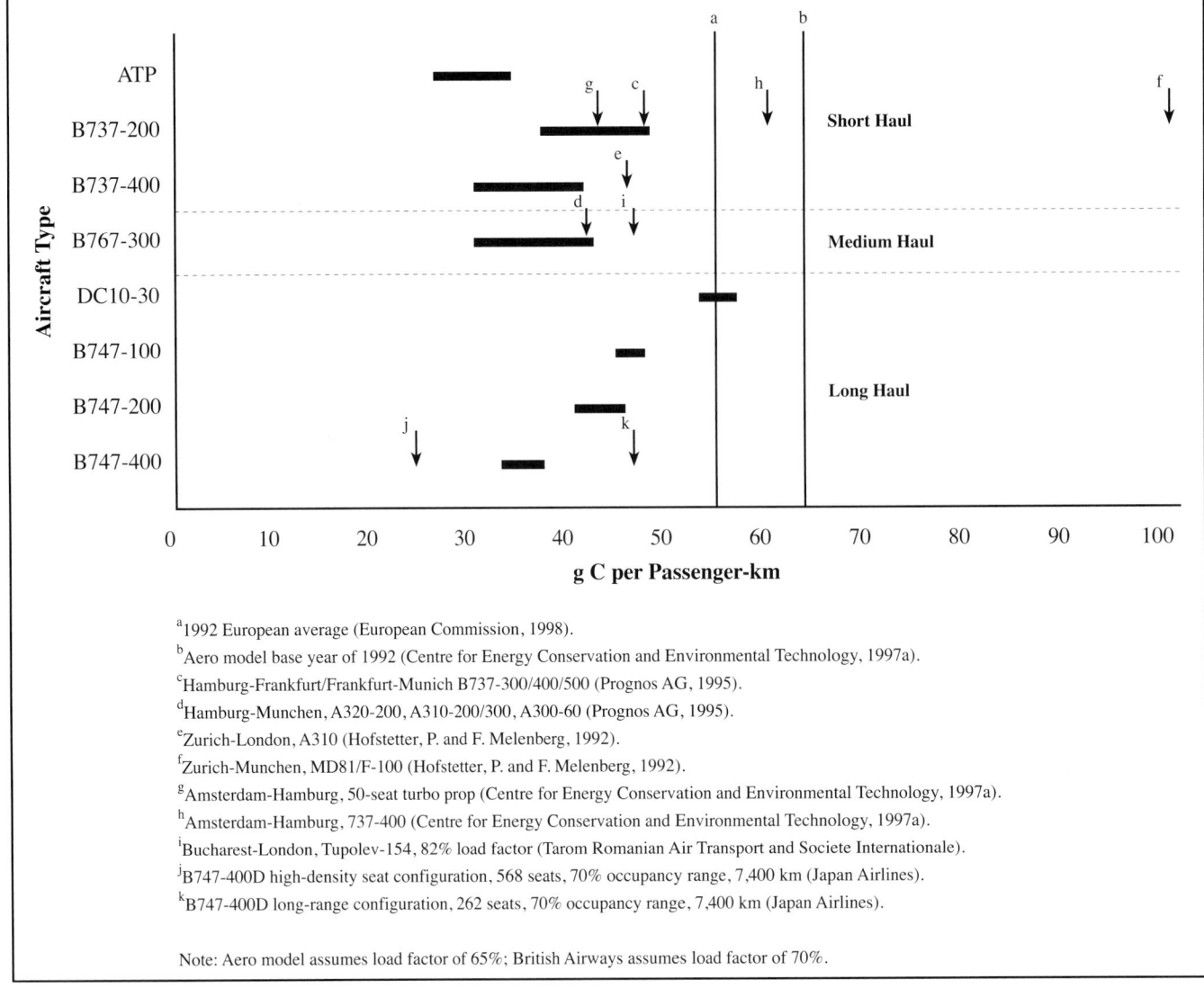

Figure 8-5: CO_2 emissions for different aircraft types, based on British Airways fleet of 1997–98.

aviation emits 1 to 2 orders of magnitude more carbon than other forms of transport. Cost and weight limitations do not allow aircraft to compete in the transport of heavy goods. For perishable freight and high-value goods, however, there may be no other suitable form of transportation.

Freight may be carried to improve capacity utilization on combination freight-passenger aircraft or passenger aircraft with appropriate space. This approach improves efficiencies, though such improvements are not necessarily reflected in all statistics.

8.3.3.4.   *Comparison between Different Modes of Transport*

The amount of CO_2 emitted per passenger-km for different modes of transport is very dependent on the type of aircraft, train, or car and on the load factor. Typical CO_2 emissions for air transport are in the range of 30 to 110 g C per passenger-km, which is comparable with passengers travelling by car or light truck.

Emission of CO_2 per passenger-km from bus or coach transport is significantly lower (< 20 g C per passenger-km). For rail travel, CO_2 emissions per passenger-km depend on several factors, such as source of primary energy, type of locomotive, and load factor; emissions vary between < 5 and 50 g C per passenger-km.

8.3.4.   *Ground-Based Aircraft Emissions*

It is not within the scope of this report to consider all of the ground-based activities associated with aviation. Overall emissions resulting directly from aircraft flights do include emissions associated with taxiing and the use of auxiliary power units at the gate. These emissions are considered briefly here.

8.3.4.1.   *Auxiliary Power Units*

Auxiliary power units (APUs) are engine-driven generators contained in the aircraft (usually in the tail) that provide the

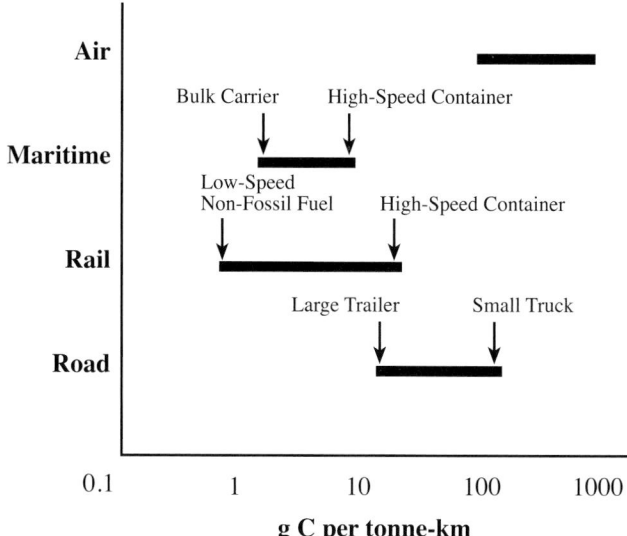

Figure 8-6: CO_2 intensity of freight (Whitelegg, 1993; IPCC, 1996a; OECD, 1997a).

aircraft with necessary energy during the time the aircraft is at the gate. Part of the generated energy is used for air conditioning. As an alternative, the required energy can be supplied by ground-based equipment that delivers electrical power at 400 Hz and preconditioned air to the aircraft. Herau (1992) investigated the financial and energy savings of such ground-based installations at Brussels Zaventum Airport's 2000 terminal and concluded that a significant net saving of carbon emissions could be achieved. At Zurich Airport in 1997, about 68,000 aircraft used terminal A and B, which has the facility to provide preconditioned air and 400 Hz power. Provision of these services achieved estimated savings of about 95% in APU fuel consumption and emissions. However, fuel used by APUs is only a relatively small part of the total fuel use of an aircraft. For example, for B737, B747, A310, MD81, and F100 aircraft, average APU fuel use is only 2.6, 0.8, 1.4, 2.5, and 3.5% of the fuel use at cruise per hour of operation, respectively (USFAA, 1982, 1995). British Airways estimates that the amount of fuel used by an APU is less than 1% of the total fuel used by an aircraft.

8.3.4.2. Taxiing

Although standard landing and take-off (LTO) cycles have been used for simplicity, there are large variations in LTO cycles from airport to airport. Although landing and take-off times are quite similar for similar fleet mixes, the amount of taxi/idle varies significantly from airport to airport (Wayson and Bowlby, 1988). This variance in taxi/idle time is the key factor in the variability of emissions from aircraft during airport operations (USFAA, 1982, 1988). Various scenarios have been discussed and attempted to reduce these emissions. These scenarios include the use of high-speed taxiways, towing of aircraft to runways (Fleuti, 1992), last-minute start-up, improvements in engine design, restriction of certain aircraft

types, realignment of taxiways, improvements at the gate area, and taxiing in with minimal engines running.

8.3.5. Summary of Other Operational Factors

Aviation passenger mobility efficiency is very dependent on the type of aircraft, the configuration, the load factor, and the distance flown. Old aircraft use much more fuel per passenger-km than new aircraft of similar size. The fuel efficiency of different aircraft is examined in Chapter 7. The required energy per passenger-km is in the range of 1.0 to 3.0 MJ per passenger-km, or about 30 to 110 g C per passenger-km. Airlines have generally optimized energy use per passenger-km, largely because of economic pressures and the requirement within the industry to minimize operational costs. Thus, with or without environmental considerations, market and cost considerations are drivers for airlines to optimize the utilization of an aircraft as much as possible.

CO_2 intensity for rail transport also depends on factors such as energy source, type of locomotive, and load factor, and emissions of CO_2 range between < 5 and 50 g C per passenger-km. However, a passenger's choice of mode of transport is based on fares, total trip time, and frequency—not just environmental considerations. ECAC (1996) estimates that less than 10% of the European air passenger travel could be replaced by high-speed train. Yet the scope for substitution is greater in Europe than in many other parts of the world.

Other operational factors to reduce aircraft fuel burn include optimization of cruise speed, reduction of tankering, reduction of additional weight, and energy savings at the airport such as limitations on the use of APUs and reduced taxi times. The total potential reduction in fuel burn by further optimization of these factors is in the range of 2–6%. The relative contribution of each factor is indicated in Table 8-6.

Finally, aircraft noise mitigation measures such as operational changes and retrofitting of engine equipment on older aircraft to conform with current aircraft noise standards could have an adverse effect on fuel use. Application of hushkits could lead to an increase in fuel consumption of up to 5%. However, lightweight hushkits may have a negligible effect on fuel use.

Table 8-6: *Reduction in fuel burn by other operational factors.*

Operational Factor	Relative Contribution
Improvement in load factor	+++
Optimization of aircraft speed	++
Reduction of tankering	++
Limitation of APU use	++
Reduction of additional weight	++
Reduction of taxi times	++
Reduction of jettisoning fuel	+

References

AEA, 1997: *Association of European Airlines Yearbook.* Association of European Airlines, Brussels, Belgium.

Balashov, B. and A. Smith, 1992: ICAO analyses—trends in fuel consumption by world's airlines. *ICAO Journal,* August, 18–21.

Bradshaw, P.J.T., 1994: *Effect of Non-Optimum Cruise Altitudes on the NO$_x$ Emissions, Performance and Economics of Typical Modern Aircraft.* Defence Evaluation and Research Agency, Farnborough, United Kingdom, 53 pp.

British Airways, 1998a: *Reports and Accounts 1997–1998.* British Airways, London, United Kingdom, 68 pp.

British Airways, 1998b: *Annual Environmental Report.* British Airways, London, United Kingdom, pp. 21–26.

Bureau of Transportation and Statistics, 1997: *Form 41 Traffic and Financial Data.* Bureau of Transportation and Statistics, Washington, DC, USA.

Centre for Energy Conservation and Environmental Technology, 1997a: *European Aviation Emissions: Trends and Attainable Reductions* [Dings, J.M.W., W.J. Dijkstra, and R.C.N. de Wit (eds.)]. Centre for Energy Conservation and Environmental Technology, Delft, The Netherlands, 71 pp.

Centre for Energy Conservation and Environmental Technology, 1997b: *Energy and Emission Profiles of Aircraft and Other Modes of Passenger Transport over European Distances* [Roos, J.H.J., A.N. Bleijenberg, and W.J. Dijkstra (eds.)]. Centre for Energy Conservation and Environmental Technology, Delft, The Netherlands, 106 pp.

Cline, P.A., 1986: *Airport Noise Control Strategies.* FAA-EE-86-2, Federal Aviation Administration, Washington, DC, USA, 126 pp.

ECAC, 1996: *Potential Influence of High Speed Train Substitution on Airport Capacity.* Working paper EURAFFOR/2 presented at the European Civil Aviation Conference, Paris, France, 7–8 March 1996. European Civil Aviation Conference, Paris, France, 12 pp.

EUROCONTROL, 1992: *Penalties to Air Traffic Associated with the ATS Route Network in the Continental ECAC States Area.* Document no. 921016, EUROCONTROL, Brussels, Belgium, 26 pp. + appendices.

EUROCONTROL, 1997a: *Revised EUROCONTROL Convention,* June 1997. EUROCONTROL, Brussels, Belgium, 60 pp. + appendices.

EUROCONTROL, 1997b: *Potential Environmental Impact on ATM Capacity, a Preliminary Assessment.* Third progress report, 24 October 1997. EUROCONTROL, Brussels, Belgium, 22 pp.

EUROCONTROL, 1998a: *Guidance Document for the Implementation of the Concept of the Flexible Use of Airspace.* ASM.ET1.ST08.5000-GUI-01-00, Edition 1.0, February 1996, Amendment 1, July 1996, and Amendment 2, March 1998. EUROCONTROL, Brussels, Belgium, 54 pp. + appendices.

EUROCONTROL, 1998b: *ATM Strategy for 2000+.* November 1998. EUROCONTROL, Brussels, Belgium, Volume 1, 48 pp.

European Commission, 1998: *Transport and CO$_2$.* European Commission, Brussels, Belgium, 34 pp.

FAA, 1998a: *Impact of CNS/ATM Improvements on Aviation Emissions.* Information paper WW/IMP-WP71 presented by the United States at the World-wide CNS/ATM Systems Implementation Conference, Rio de Janeiro, Brazil, 11–15 May 1998, 5 pp.

FAA, 1998b: *The Impact of National Airspace Systems (NAS) Modernization on Aircraft Emissions.* DOT/FAA/SD-400-98/1, Federal Aviation Administration, Washington, DC, USA, September 1998, 38 pp.

Faiz, A., C.S. Weaver, and M.P. Walsh, 1996: *Air Pollution from Motor Vehicles.* The World Bank, Washington, DC, USA, 346 pp.

Fleuti, E., May, 1992: *Operational Towing at Zurich Airport.* Zurich Airport Authority, Zurich, Switzerland, May 1992, 6 pp.

Fransen, W. and J. Peper, 1993: *Atmospheric Effects of Aircraft Emissions.* Royal Meteorological Institute in The Netherlands, National Aerospace Laboratory, De Bilt, The Netherlands, 129 pp.

Gardner, R.M., K. Adams, T. Cook, S. Ernedal, R. Falk, E. Fleuti, E. Herms, C.E. Johnson, M. Lecht, D.S. Lee, M. Leech, D. Lister, B. Massé, M. Metcalfe, P. Newton, A. Schmitt, C. Vandenbergh, and R. Van Drimmelen, 1996: The ANCAT/EC global inventory of NO$_x$ emissions from aircraft. *Atmospheric Environment,* **31(12),** 1751, 1756.

Herau, G., 1992: Saving fuel and emissions during ground operations. *Airport Technology International,* Sterling International Publishers.

Hofstetter, P. and F. Meienberg, 1992: *Ein ökologischer und ökonomischer Vergleich verschiedener Verkehrsträger anhand von Städtereisen in Europa.* Traffic Club Switzerland (VCS) and Swiss Student Travel (SSR), Zurich, Switzerland, 78 pp.

IATA, 1996: *World Air Transportation Statistics.* International Air Transport Association, Geneva, Switzerland, 40th ed., 111 pp.

ICAO, 1991: *Report of the Tenth Air Navigation Conference.* Document no. 9583, AN-CONF/10, International Civil Aviation Organization, Montreal, Canada, 241 pp.

ICAO, 1995: *Report of the Emissions Inventory Sub-Group at Working Group 3.* International Civil Aviation Organization, Committee on Aviation Environmental Protection, Working Group 3, Bonn, Germany, June 1995, 45 pp.

ICAO, 1996: *Civil Aviation Statistics of the World 1995.* Document no. 9180/21. International Civil Aviation Organization, Montreal, Canada, 206 pp.

ICAO, 1997: *Convention of International Civil Aviation.* Document no. 7300/7, International Civil Aviation Organization, Montreal, Canada, 52 pp.

ICAO, 1998a: *ICAO's Global Plan for CNS/ATM.* Information paper WW/IMP-WP6, presented by the ICAO Secretariat at the World-wide CNS/ATM Systems Implementation Conference, Rio de Janeiro, 11-15 May 1998. International Civil Aviation Organization, Montreal, Canada, 5 pp.

ICAO, 1998b: *Costs and Benefits for Providers and Users.* Information paper WW/IMP-WP20 presented by the ICAO Secretariat at the World-wide CNS/ATM Systems Implementation Conference, Rio de Janeiro, Brazil, 11–15 May 1998. International Civil Aviation Organization, Montreal, Canada, 11 pp.

ICCAIA, 1994: *ANDES Aircraft Noise Design Effects Study.* International Coordinating Council of Aerospace Industries Association, Washington, DC, USA, 72 pp.

Institute of the Association for Commuter Transportation, 1997: *TDM Case Studies and Commuter Testimonials.* Transportation Demand Management Institute of the Association for Commuter Transportation, Washington, DC, USA.

IPCC, 1996a: *Climate Change 1995. Impacts, Adaptations and Mitigation of Climate Change: Scientific Technical Analysis. Contribution of Working Group II to the Second Assessment Report of the Intergovernmental Panel on Climate Change* [Watson, R.T., N.C. Zinyowera, and R.H. Moss (eds.)].Cambridge University Press, Cambridge, United Kingdom, and New York, NY, USA, 880 pp.

IPCC, 1996b: *Technologies, Policies and Measures for Mitigating Climate Change: IPCC Technical Paper I.* Intergovernmental Panel on Climate Change Working Group II [Watson, R.T., M.C. Zinyowera, and R.H. Moss (eds.)]. Intergovernmental Panel on Climate Change, Geneva, Switzerland, 85 pp.

Lecht, M., 1994: *Schadstoffe des Luftverkehrs: Maßnahme gegen die zunehmende Verunreinigung der Atmosphäre durch den Luftverkehr.* DLR-Nachrichten, Heft 75 (May 1994), Deutsches Zentrum für Luft- und Raumfahrt (German Aerospace Center), Cologne, Germany, 5 pp.

OECD, 1997a: *Special Issues in Carbon/Energy Taxation: Marine Bunker Fuel Charges.* Working Paper no. 11, Annex I expert group on the United Nations Framework Convention on Climate Change. Organization for Economic Cooperation and Development, Paris, France, 79 pp.

OECD, 1997b: *Special Issues in Carbon/Energy Taxation: Carbon Charges on Aviation Fuel.* Working Paper no. 12, Annex I expert group on the United Nations Framework Convention on Climate Change. Organization for Economic Cooperation and Development, Paris, France, 67 pp.

Peterse, A. and H.J. Boering, 1997: *A Future Better and Less Environmentally Impacting Aviation.* National Aerospace Laboratory, Amsterdam, The Netherlands.

Prognos AG, 1995: *Bedeutung und Umweltwirkungen von Schienen-und Luftverkehr in Deutschland.* ADV (German Airports Association), Deutsche Bahn, Deutsches Verkehrsforum, Lufthansa, Basel, Germany, 201 pp.

RTCA, 1995: *Report of the Board of Directors' Select Committee on Free Flight.* RTCA Incorporated, Washington, DC, USA, 33 pp.

Sausen, R., D. Nodorp, C. Land, and F. Deidewig, 1996: *Ermittlung optimaler Flughöhe und Flugrouten unter dem Aspekt minimaler Klimawirksamkeit.* DLR-Forschungsbericht 96-13, Deutsches Zentrum für Luft- und Raumfahrt, Cologne, Germany, 105 pp.

Sprinkle, C.H. and K.J. Macleod, 1993: *Aviation and the Environment.* Presented at the 5th International Conference on Aviation Weather Systems in Vienna. Published by the American Meteorological Society, Boston, pp. 89–91.

TEST , 1991: *Wrong Side of the Tracks? Impacts of Road Transport on the Environment: A Basis for Discussion.* Transport and Environmental Studies, London, United Kingdom, 286 pp.

USFAA, 1982: *Air Quality Procedures for Civilian Airports and Air Force Bases.* FAA-EE-82-21, Federal Aviation Administration, Washington, DC, USA.

USFAA, 1988: *A Microcomputer Pollution Model for Civilian Airports and Air Force Bases. Model Application and Background.* FAA-EE-88-5, Federal Aviation Administration, Washington, DC, USA, 94 pp.

USFAA, 1995: *Federal Aviation Administration Emission Database, Version 2.1.* Federal Aviation Administration, Washington, DC, USA.

Wayson, R.L. and W. Bowlby, 1988: Inventorying airport air pollutant emissions. *Journal of Transportation Engineering,* **114(1),** 1–20.

Whitelegg, J., 1993: *Transport for a Sustainable Future.* Belhaven Press, London, United Kingdom, 156 pp.

WMO, 1995: *Scientific Assessment of Ozone Depletion: 1994.* Report no. 37, Global Ozone Research and Monitoring Project, World Meteorological Organization, Geneva, Switzerland, 578 pp.

9

Aircraft Emissions: Current Inventories and Future Scenarios

STEPHEN C. HENDERSON AND UPALI K. WICKRAMA

Lead Authors:
S.L. Baughcum, J.J. Begin, F. Franco, D.L. Greene, D.S. Lee, M.-L. McLaren,
A.K. Mortlock, P.J. Newton, A. Schmitt, D.J. Sutkus, A. Vedantham, D.J. Wuebbles

Contributors:
R.M. Gardner, L. Meisenheimer

Review Editor:
O. Davidson

CONTENTS

EXECUTIVE SUMMARY

- Three-dimensional (latitude, longitude, altitude) global inventories of civil and military aircraft fuel burned and emissions have been developed for the United States National Aeronautics and Space Administration (NASA) for the years 1976, 1984, and 1992, and by the European Abatement of Nuisances Caused by Air Transport (ANCAT)/European Commission (EC) Working Group and the Deutsches Zentrum für Luft- und Raumfahrt (DLR) for 1991/92. For 1992, the results of the inventory calculations are in good agreement, with total fuel used by aviation calculated to be 129.3 Tg (DLR), 131.2 Tg (ANCAT), and 139.4 Tg (NASA). Total emissions of NO_x (as NO_2) in 1992 were calculated to range from 1.7 Tg (NASA) to 1.8 Tg (ANCAT and DLR).

- Forecasts of air travel demand and technology developed by NASA and ANCAT for 2015 have been used to create three-dimensional (3-D) data sets of fuel burn and NO_x emissions for purposes of modeling the near-term effects of aircraft. The NASA 2015 forecast results in a global fuel burn of 309 Tg, with a NO_x emission of 4.1 Tg (as NO_2); the global emission index, $EI(NO_x)$ (g NO_x/kg fuel), is 13.4. In contrast, the ANCAT 2015 forecast results in lower values—a global fuel burn of 287 Tg, an emission of 3.5 Tg of NO_x, and a global emission index of 12.3. The differences arise from the distribution of air travel demand and technology assumptions.

- Long-term emission scenarios for CO_2 and NO_x from subsonic aviation in 2050 have been constructed by the International Civil Aviation Organization (ICAO) Forecasting and Economic Support Group (FESG); the United Kingdom Department of Trade and Industry (DTI); and the Environmental Defense Fund (EDF), whose projections extend to 2100. The FESG and EDF scenarios used the Intergovernmental Panel on Climate Change (IPCC) IS92 scenarios for economic growth to project future air traffic demand, though with different approaches to the relative importance of gross domestic product (GDP) and population. Each group also makes different assumptions about projected improvements in fleet fuel efficiency and NO_x reduction technology. In addition, the Massachusetts Institute of Technology (MIT) has projected emissions from a "high speed" sector that includes aviation, and the World Wide Fund for Nature (WWF) has published a projection of aviation emissions for the year 2041.

- All future scenarios were constructed by assuming that the necessary infrastructure (e.g., airports, air traffic control) will be developed as needed and that fuel supplies will be available. System capacity constraints, if any, have not been evaluated.

- Future scenarios predict fuel use and NO_x emissions that vary over a wide range, depending on the economic growth scenario and model used for the calculations. Although none of the scenarios are considered impossible as outcomes for 2050, some of the EDF high-growth scenarios are believed to be less plausible. The FESG low-growth scenarios, though plausible in terms of achievability, use traffic estimates that are very likely to be exceeded given the present state of the industry and planned developments.

- The 3-D gridded outputs from all of the FESG 2050 scenarios and from the DTI 2050 scenario are suitable for use as input to chemical transport models and may also be used to calculate the effect of aviation CO_2 emissions. The FESG scenarios project aviation fuel use in 2050 to be in the range of 471–488 Tg, with corresponding NO_x emissions of 7.2 and 5.5 Tg (as NO_2) for IS92a, depending on the technology scenario; 268–277 Tg fuel and NO_x of 4.0 and 3.1 Tg for IS92c; and 744–772 Tg fuel and NO_x of 11.4 and 8.8 Tg for IS92e. (For all of the individual FESG IS92-based scenarios, higher fuel usage—thus CO_2 emissions—were a result of the more aggressive NO_x reduction technology assumed). The DTI scenario projects aviation fuel use in 2050 to be 633 Tg, with NO_x emissions at 4.5 Tg.

- As a result of higher projected fuel usage, EDF projections of CO_2 emissions are all higher than those of FESG by factors of approximately 2.4 to 4.3 for IS92a, 3.1 to 5.7 for IS92c, and 1.7 to 3.1 for IS92e. Results from EDF scenarios based on IS92a and IS92d are suitable for use in calculating the effect of CO_2 emissions as sensitivity analyses; the latter scenario projects CO_2 emissions levels from aviation 2.2 times greater in 2050 than the highest of the FESG scenarios.

- The effects of a fleet of high-speed civil transport (HSCT) aircraft on fuel burned and NO_x emissions in the year 2050 were calculated using the FESG year 2050 subsonic inventories as a base. A fleet of 1,000 HSCTs operating with a program goal $EI(NO_x)$ of 5 in 2050 was calculated to increase global fuel burned by 12–18% and reduce global NO_x by 1–2% (depending on the scenario chosen), assuming that low-NO_x HSCTs displace traffic from the higher NO_x subsonic fleet. A fleet of 1,000 HSCT aircraft was chosen to evaluate the effect of a large fleet; it does not constitute a forecast of the size of an HSCT fleet in 2050.

- The simplifying assumptions used in calculating all of the historical and present-day 3-D inventories (1976 through 1992)—great circle routing, no winds, standard temperatures, no cargo payload—cause a systematic underestimate of fuel burned (therefore emissions produced) by aviation on the order of 15%, so calculated values were scaled up accordingly. By 2015, we assume that the introduction of advanced air traffic management systems will reduce this underestimate to approximately 5%. Full implementation of these systems by 2050 should reduce the error somewhat further, but given the wide range of year 2050 scenario projections, adjustments to calculated fuel values in 2050 were not considered to be necessary.

9.1. Introduction

The nature and composition of aircraft emissions has been described in Chapter 1, and their effects on the composition of the atmosphere are described in Chapters 2 and 3. Chapter 4 uses aircraft emissions data in modeling studies to provide chemical perturbations that feed into the ultraviolet (UV) irradiance and radiative forcing calculations presented in Chapters 5 and 6, respectively. In this chapter, the aircraft emissions data that were used in calculations described in Chapters 4 and 6 are presented and discussed.

Compilation of global inventories of aircraft NO_x emissions has been driven by requirements for global modeling studies of the effects of these emissions on stratospheric and tropospheric ozone (O_3). Aircraft carbon dioxide (CO_2) emissions are easily calculated from total fuel burned. Early studies used one- (1-D) and two-dimensional (2-D) models of the atmosphere (see Section 2.2.1). Most of these early studies considered effects on the stratosphere (e.g., COMESA, 1975), but some also included assessments of the (then) current subsonic fleet on the upper troposphere and lower stratosphere (e.g., Hidalgo and Crutzen, 1977; Derwent, 1982). An early height- and latitude-dependent emissions inventory of aircraft NO_x was given by Bauer (1979), based on earlier work by A.D. Little (1975). This work was used by Derwent (1982) in a 2-D modeling study of aircraft NO_x emissions in the troposphere.

Later estimations of global aircraft emissions of NO_x were still made by relatively simple methods, using fuel usage and assumed $EI(NO_x)$ (e.g., Nüßer and Schmitt, 1990; Beck et al., 1992). Concerted efforts were subsequently made by a number of groups to construct high-quality global 3-D inventories of aircraft emissions. Such work was undertaken for a variety of programs and purposes: United Kingdom input to ICAO Technical Working Groups (McInnes and Walker, 1992); the U.S. Atmospheric Effects of Stratospheric Aircraft (AESA) Program (Wuebbles et al., 1993); the German "Schadstoffe in der Luftfahrt" Program (Schmitt and Brunner, 1997); and the ANCAT/EC Emissions Database Group (ANCAT/EC, 1995), which combined European efforts to produce an aircraft NO_x inventory for the AERONOX Program (Gardner et al., 1997). Subsequently, methodologies for the production of global 3-D inventories of present-day aircraft NO_x emissions (based on 1991–92) have been refined and have produced results that have largely superseded earlier work. These inventories cover the 1976–92 time period and have been extended to the 2015 forecast period. These gridded inventories—which calculate aviation emissions as distributed around the Earth in terms of latitude, longitude, and altitude—have been produced by NASA, DLR, and ANCAT/EC for national and international work programs (Baughcum et al., 1996a,b; Schmitt and Brunner, 1997; Gardner, 1998).

This chapter is not the first attempt to synthesize information on aircraft emissions inventories; earlier assessments were made by the World Meteorological Organization (WMO)/ United National Environment Programme (UNEP) (1995) Scientific Assessment of Ozone Depletion, ICAO's Committee on Aviation Environmental Protection (CAEP) Working Group 3 (CAEP/WG3, 1995), the NASA Advanced Subsonic Technology Program (Friedl, 1997), and the European Scientific Assessment of the Atmospheric Effects of Aircraft Emissions (Brasseur et al., 1998).

Any assessment of present and potential future effects of subsonic and supersonic air transport emissions relies heavily on input emissions data. Thus, considerable effort has been expended on understanding the accuracy of present-day inventories and the construction of forecasts and scenarios. Forecasts are quite distinct from scenarios, as noted in Chapter 1. Forecasts of aviation emissions for a 20–25 year time frame are generally considered possible, whereas such confidence is not the case for longer time frames. Thus, scenarios generally rely on many more assumptions and are less specific than forecasts.

In planning this Special Report, it was clear that there were no gridded emission scenarios of NO_x emissions from subsonic aircraft for the year 2050 that could be used as input to 3-D chemical transport models (see Chapters 2 and 4). The IPCC made a request to ICAO to prepare 3-D NO_x scenarios, which was carried out under the auspices of ICAO's FESG (CAEP/4-FESG, 1998). The UK DTI also responded to this requirement, producing an independent 3-D NO_x scenario for 2050 (Newton and Falk, 1997). The EDF had also published scenarios of aircraft emissions of NO_x and CO_2 extending to 2100 (Vedantham and Oppenheimer, 1994, 1998), but these scenarios were not gridded; thus, although the aviation CO_2 scenarios could be used in radiative forcing calculations (see Chapter 6), the NO_x scenarios could not be used to calculate O_3 perturbations and subsequent radiative forcing. Other scenario data exist for aircraft emissions, including those from WWF (Barrett, 1994) and MIT (Schafer and Victor, 1997). As with the EDF data, these scenarios were not gridded for NO_x emissions, therefore could not be used in O_3 perturbation calculations in Chapter 4. Furthermore, the MIT data do not explicitly represent aircraft emissions; instead, they cover high-speed transport modes, including some surface transportation modes.

HSCT scenarios prepared for NASA's AESA Program are considered distinct from subsonic scenarios; these HSCT scenarios represent a technology that does not yet exist but might be developed. Therefore, the HSCT scenarios represent a quite different set of assumptions from other long-term scenarios, which only consider continued development of a subsonic fleet. The HSCT scenarios were used in modeling studies (Chapters 4 and 6) as sensitivity analyses for studying the effects of their emissions on stratospheric O_3.

In this chapter, methodologies of inventory and forecast construction are compared, and a review and assessment of long-term scenarios and their implicit assumptions provided. This is the first detailed consideration of long-term scenarios and their implications.

By way of background, Section 9.2 provides an overview of factors that affect aircraft emissions, such as market demand for air travel and developments in the technology. The aircraft emissions data discussed in this chapter are of four distinct types: Historical inventories (e.g., for 1976 and 1984); inventories that represent the "present day" (i.e., 1991–92); forecasts for 2015; and long-term scenarios for 2050 and beyond. The methodologies and a comparison of historical, present-day, and forecast inventories are presented in Section 9.3. Section 9.4 describes and comments on available long-term scenarios for 2050 and beyond. Scenarios of high-speed civil transport (HSCT) that incorporate certain assumptions about the development of a supersonic fleet and its impact on the subsonic fleet are presented separately in Section 9.5. Finally, Section 9.6 discusses underlying assumptions and drivers of long-term subsonic scenarios. The plausibility of the assumptions are also considered in terms of implications for fleet size, infrastructure requirements, and global fossil-fuel availability.

9.2.   Factors Affecting Aircraft Emissions

9.2.1.   Demand for Air Travel

In the past 50 years, the air transport industry has experienced rapid expansion as the world economy has grown and the technology of air transport has developed to its present state. The result has been a steady decline in costs and fares, which has further stimulated traffic growth. As an example of this growth, the output of the industry (measured in terms of tonne-km performed) has increased by a factor of 23 since 1960; total GDP, which is the broadest available measure of world output, increased by a factor of 3.8 over the same period (ICAO, 1997a).

Although growth in world air traffic has been much greater than world economic growth, economic theory and analytical studies indicate that there is a high correlation between the two, and most forecasts of aviation demand are based on the premise that the demand for air transport is determined primarily by economic development. Statistical analyses have shown that growth in GDP now explains about two-thirds of air travel growth, reflecting increasing commercial and business activity and increasing personal income and propensity to travel. Demand for air freight service is also primarily a function of economic growth. Air travel growth in excess of GDP growth is usually explained by other economic and structural factors:

- Improvement in service offerings as routes and frequencies and infrastructure are added, stimulation from reductions in airline fares as costs decline, and increasing trade and the globalization of business (Boeing, 1998)
- Population and income distribution (Vedantham and Oppenheimer, 1998)
- Travel behavior, including travel time budgets and travel costs (Zahavi, 1981; Schafer and Victor, 1997).

Changes in technology and in the regulatory environment have also had great effects on the growth in air travel demand. The

modern era of air transportation began in the 1960s, driven by the replacement of piston-engined aircraft with jet aircraft that increased the speed, reliability, and comfort of air travel while reducing the cost of operation. The continuing trend of declining fares (as measured in constant dollars) began in this period. In real terms, fares have declined by almost 2% per year since 1960. Deregulation of airline services in the United States in 1978 allowed airlines to improve services by expanding their route systems and reduce average costs by greatly increasing the efficiency of scheduling and aircraft use. Trends toward liberalization of airline services in Europe and elsewhere will continue to increase airline efficiency.

Sharp increases in oil prices have had important (though temporary) effects on traffic demand. In addition to an adverse effect on the world economy, the 10-fold increase in crude oil prices in 1973–74 and further escalation in 1979–81 (since ameliorated) greatly increased aviation fuel prices. Air fares increased in response to higher costs, with a resulting decline in demand growth rates.

Figure 9-1 provides evidence of the relationship between the economy and traffic demand by illustrating fluctuations in the rate of growth of each from 1960 to the present. The economic recessions of 1974–75, 1979–82 (largely caused by the increase in oil prices), and 1990–91 (the Gulf War) and their impact on air traffic are clearly visible.

The growth rate in global passenger demand over the past 35 years is shown in Figure 9-2. Freight traffic, approximately 80% of which is carried in the bellies of passenger airplanes, has also grown over the same time period. The declining trend in the rate of growth as the size of the industry has increased by more than 20-fold is a natural result of the total size of the industry (it is difficult to sustain an "infant industry" growth rate as size increases) and a maturing of certain markets—primarily those in the developed world—that dominate the statistics.

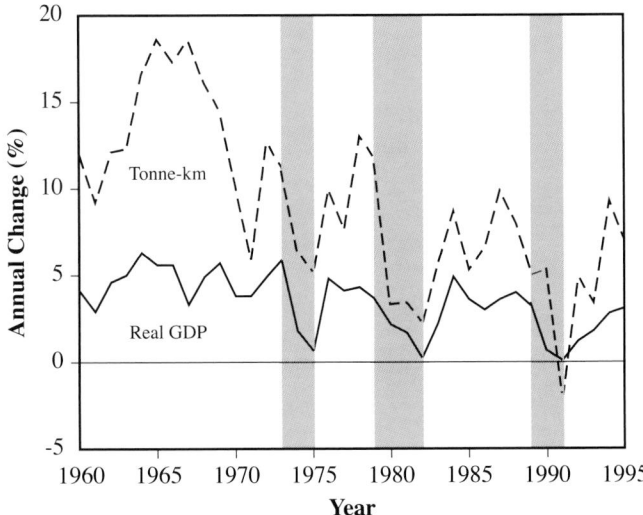

Figure 9-1: Relationship between economic growth and traffic demand growth (IMF, WEFA, ICAO Reporting Form A-1).

Changes in demand in regional markets are given in Table 9-1 for the period 1970–95. Over this period, global traffic measured in revenue passenger kilometers (RPK) increased by a factor of 4.6 (Boeing, 1996). Table 9-1 is ordered by 1995 regional RPK value.

9.2.2. Developments in Technology

The trend in fuel efficiency of jet aircraft over time has been one of almost continuous improvement; fuel burned per seat in today's new aircraft is 70% less than that of early jets. About 40% of the improvement has come from engine efficiency improvements and 30% from airframe efficiency improvements (Figure 9-3, after Figure III-A-1 in Albritton *et. al*, 1997).

The growth rate of fuel consumed by aviation therefore has been lower than the growth in demand. Improvement in engine fuel efficiency has come mainly from the increasing use of modern high-bypass engine technology that relies on increasing engine pressure ratios and higher temperature combustors as a means to increase engine efficiency. These trends have resulted in drastic decreases in emissions of carbon monoxide (CO) and unburned hydrocarbons (HC), though they tend to increase emissions of oxides of nitrogen (NO_x). As a result, total NO_x emissions from aircraft are growing faster than fuel consumption (see Figure 9-4, from NASA emissions inventories discussed in

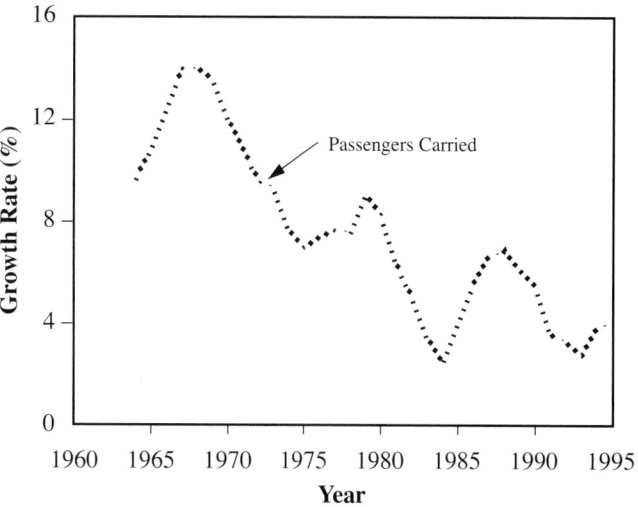

Figure 9-2: Growth rate of passengers carried (ICAO Reporting Form A-1). Note the assumption of 5-year moving average of annual growth rates, excluding operations in the Commonwealth of Independent States (CIS).

Section 9.3). A discussion of the technology required to reduce NO_x emissions while continuing to improve engine efficiency appears in Chapter 7.

Table 9-1: Regional share of total demand.

Regional Traffic Flow	1970 RPK x 10⁹	1995 RPK x 10⁹	1970–95 Growth Factor	1970 Market Share	1995 Market Share	1970–95 Change in Share
Intra North America	190.897	697.880	3.7	34.6%	27.5%	-7.1%
Intra Europe	61.275	317.099	5.2	11.1%	12.5%	1.4%
North America – Europe	72.143	277.909	3.9	13.1%	11.0%	-2.1%
China Domestic/Intra Asia/Intra Oceania	10.234	207.405	20.3	1.9%	8.2%	6.3%
North America – Asia/Oceania	14.760	188.799	12.8	2.7%	7.4%	4.8%
Europe – Asia	6.732	134.343	20.0	1.2%	5.3%	4.1%
Asia – India/Africa/Middle East	13.959	115.204	8.3	2.5%	4.5%	2.0%
North America – Latin America	16.087	75.538	4.7	2.9%	3.0%	0.1%
Europe – Latin America	7.124	73.090	10.3	1.3%	2.9%	1.6%
Domestic Former Soviet Union	75.496	67.603	0.9	13.7%	2.7%	-11.0%
Japan Domestic	8.181	61.607	7.5	1.5%	2.4%	0.9%
Europe – Africa	18.478	61.045	3.3	3.4%	2.4%	-0.9%
Intra/Domestic Latin America	13.432	55.331	4.1	2.4%	2.2%	-0.3%
Europe – Middle East	9.838	41.224	4.2	1.8%	1.6%	-0.2%
Intra/Domestic Middle East – Africa	5.065	39.213	7.7	0.9%	1.5%	0.6%
International Former Soviet Union	3.677	29.508	8.0	0.7%	1.2%	0.5%
Indian Subcontinent – Asia/Middle East/Oceania	3.249	29.500	9.1	0.6%	1.2%	0.6%
Europe – Indian Subcontinent	2.333	19.858	8.5	0.4%	0.8%	0.4%
Intra/Domestic Africa	5.826	16.808	2.9	1.1%	0.7%	-0.4%
Intra Indian Subcontinent	3.215	13.218	4.1	0.6%	0.5%	-0.1%
North America – Africa/Middle East	1.149	10.777	9.4	0.2%	0.4%	0.2%
U.S. Military Airlift	8.112	3.605	0.4	1.5%	0.1%	-1.3%
Total	**551.262**	**2536.561**	**4.6**	**100.0%**	**100.0%**	**0.0%**

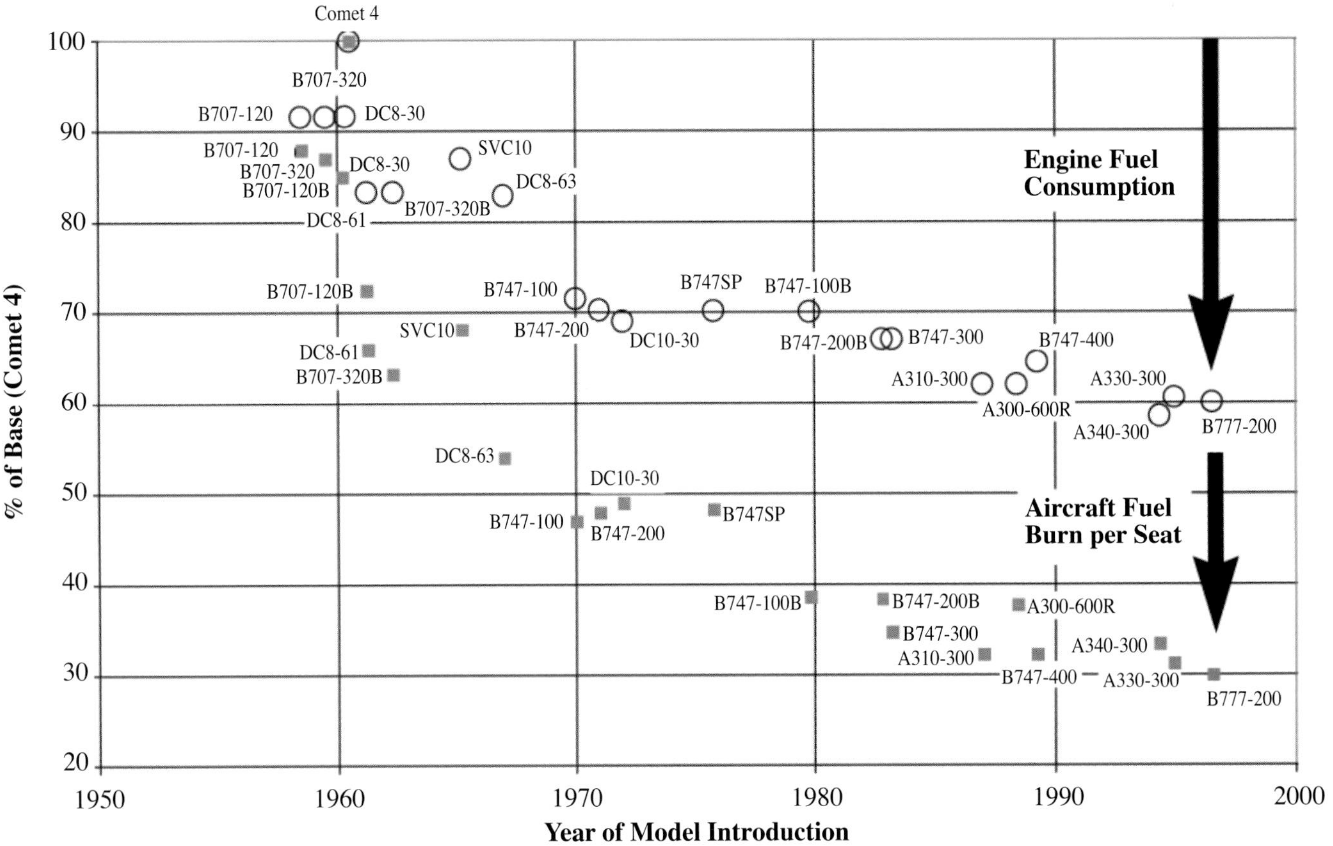

Figure 9-3: Trend in transport aircraft fuel efficiency.

9.3.   Historical, Present-Day, and 2015 Forecast Emissions Inventories

Studies on the effects of CO_2 emissions from aircraft on radiative forcing require only a knowledge of total emissions. However, to examine the potential effects of other emissions from aviation (e.g., those considered in Chapter 4), estimates of the amount and the distribution of emissions are required. Such 3-D inventories for present and projected future aviation operations have been produced under the aegis of NASA's Atmospheric Effects of Aviation Project (AEAP), the European Civil Aviation Conference's ANCAT and EC Emissions Inventory Database Group (EIDG), and DLR.

These inventories consist of calculated aircraft emissions distributed over the world's airspace by latitude, longitude, and altitude. Historical inventories of aviation emissions have been produced for 1976 and 1984 by NASA. Present-day and 2015 forecast inventories (where present-day is taken to be the most recent available—1991–92) have been produced by NASA, ANCAT, and DLR. DLR has also produced emissions inventories of scheduled international aviation only for each year from 1982 through 1992, and for total scheduled aviation for 1986 and 1989. DLR has also constructed a four-dimensional (4-D) inventory with diurnal cycles for scheduled aviation in March 1992.

All of the aforementioned 3-D emissions inventories have a common approach of combining a database of global air traffic

(fleet mix, city-pairs served, and flight frequencies) with a set of assumptions about flight operations (flight profiles and routing) and a method to calculate altitude-dependent emissions of aircraft/engine combinations in the fleet. Figure 9-5 shows how these processes are combined.

All of the historical, present-day, and 2015 forecast inventories considered in this section assume idealized flight routings and profiles, with no winds or system delays. Thus, minimum

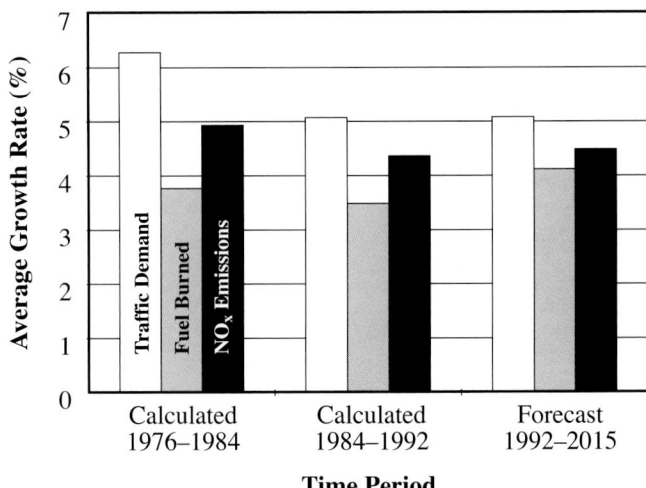

Figure 9-4: Comparison of growth rates for civil traffic, fuel consumption, and NO_x emissions.

fuel burn and emissions possible for each flight operation are implicit, given the onboard load assumed. Simplifying assumptions for military operations vary according to aircraft type.

9.3.1. NASA, ANCAT/EC2, and DLR Historical and Present-Day Emissions Inventories

The NASA, ANCAT, and DLR 3-D inventories adopt a similar overall approach but differ in some of the components and data used. This section describes the common approaches and explains the differences. More detailed information appears in the source material for these inventories (Baughcum *et al.*, 1996a,b; Schmitt and Brunner, 1997; Gardner, 1998).

All of the inventories use a "bottom-up" approach in which an aircraft movement database was compiled, aircraft/engine combinations in operation were identified (to differing levels of detail), and calculations of fuel burned and emissions along great-circle paths between cities were made. Flight operation data were calculated as the number of departures for each city pair by aircraft and engine type—which, combined with performance and emissions data, gave fuel burned and emissions by altitude along each route. This approach resulted in data on fuel burned and emissions of NO_x (as NO_2) on a 3-D grid for each flight. In addition, the NASA inventories provide 3-D distributions of CO and total HC. NASA and ANCAT inventories were calculated on a 1° longitude x 1° latitude x 1-km altitude resolution, whereas the DLR inventory used a 2.8° longitude x 2.8° latitude horizontal resolution.

Different approaches were taken for constructing underlying traffic movements databases. The NASA inventories use scheduled jet and turboprop aviation operations for the years 1976, 1984, and 1992 (Baughcum *et al.*, 1996a,b). Movements for charter carriers, military operations, general aviation, and the domestic fleets of the former Soviet Union (FSU) and the People's Republic of China were estimated separately (Landau *et al.*, 1994; Metwally, 1995; Mortlock and Van Alstyne, 1998). Military aircraft contributions to emissions were calculated by estimating the flight activity of each type of military aircraft by country. The 1976 and 1984 NASA inventories were based on operations for 1 month in each quarter of the year, whereas the 1992 inventory compiled movements on a monthly basis to reflect the seasonality of aviation operations.

The ANCAT approach used a combination of air traffic control (ATC) data and scheduled movements, favoring ATC data where available (Gardner, 1998). Where ATC data were unavailable, scheduled data were taken from the ABC Travel Guide (ABC), the Official Airline Guide (OAG), the Aeroflot time table, and a German study of Chinese domestic aircraft movements. Only jet aircraft were represented in the ANCAT/EC2 inventory. The most significant omission of ATC data was the United States, for which data were unavailable for security reasons. Thus, only time table data were used for the United States; so nonscheduled U.S. domestic charters and other flights were not recorded. To compensate for this problem, fuel usage data were factored up by 10% (Gardner, 1998). ATC data accounted for half of the non-U.S. aircraft movements in the database. Military movements were estimated by allocating fuel and emissions to countries' boundaries from an analysis of the world's military fleet composition.

The DLR inventory for 1991/92 (Schmitt and Brunner, 1997) used the ANCAT/EC2 civil movements database. Emissions inventories for 1986, 1989, and 1992 were based on scheduled air traffic only; a 4-D inventory with diurnal cycles for March 1992 was based on ABC data. ICAO data (ICAO, 1997b) were used for emissions inventories for international (only) scheduled air traffic in the years 1982 to 1992.

Calculation of fuel burned and emissions for aircraft differs between the three inventories. NASA used detailed manufacturers' proprietary performance information on each aircraft-engine combination and the flight profile shown

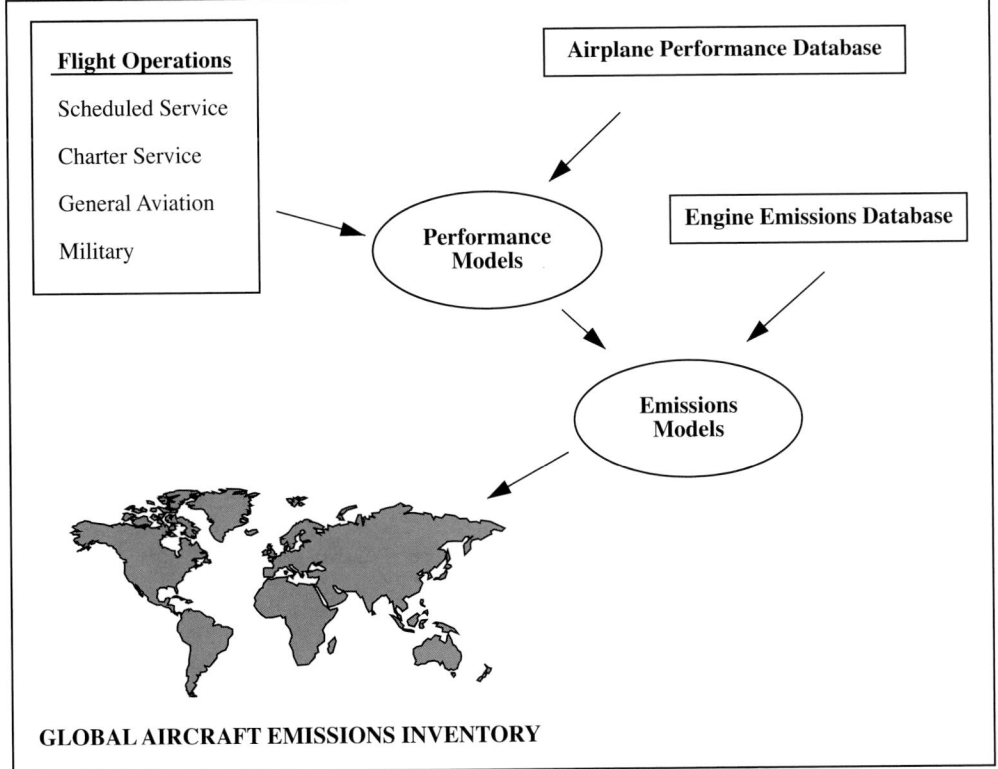

GLOBAL AIRCRAFT EMISSIONS INVENTORY

Figure 9-5: Aircraft emissions inventory calculation schematic.

in Figure 9-6. Emissions were calculated from the information in the ICAO Engine Exhaust Emissions Data Bank (ICAO, 1995), through the use of Boeing "Method 2" procedures (Baughcum *et al.*, 1996b, Appendix D), which allow extrapolation of sea-level data in the ICAO data bank to the operating altitudes and temperatures encountered throughout the aircraft flight profile.

The ANCAT/EC2 inventory used commercial software for flight and fuel profiling, along with Project Interactive Analysis and Optimization (PIANO), a parametric aircraft design model. The global civil fleet was modeled with a selection of 20 representative aircraft types. These representative aircraft were assumed to be fitted with generic engines typical of the technology and thrust requirements of each type. PIANO generated fuel profiles covering the entire flight cycle, including steps in cruise for each aircraft. Fuel use during ground operations was estimated from ICAO certification timings (ICAO, 1993).

The DLR inventory used airline data and an in-house flight and fuel profile model (Deidewig *et al.*, 1996). The DLR approach also used different aircraft/engine combinations from those utilized by ANCAT. The aircraft mission was simulated by using a simplified flight modeling code as point-to-point missions with no step cruise. Although the climb was calculated in iterative steps, the cruise segment was treated as one section, applying the Breguet formula to calculate the cruise fuel. Descent was assumed to be a gliding path with minimum engine load; no separate approach procedure was used. A thermodynamic model for design and off-design operation of a two-shaft fan engine was applied. Constant efficiencies and constant relative pressure losses for main engine components were assumed for simplicity.

The ANCAT/EC2 and DLR inventories calculated NO_x emissions from the fuel using the DLR fuel flow method. This method has been tested and correlated with information from airlines, flight measurements, and altitude chamber measurements (Deidewig *et al.*, 1996; Schulte *et al.*, 1997).

9.3.2. NASA, ANCAT/EC2, and DLR 2015 Emissions Forecasts

The first NASA subsonic aircraft emissions inventory for 2015 was created as part of an assessment of the effects of a future HSCT (Baughcum *et al.*, 1994); it has now been superseded by a new study (Baughcum *et al.*, 1998; Mortlock and Van Alstyne, 1998) that includes new emissions technology assumptions and more detailed fleet mix and route system calculations. The NASA 2015 forecast inventory was calculated using methods similar to those used for NASA's historical and present-day inventories. Separate forecasts were created for scheduled operations (flights shown in the OAG database), charter operations, cargo operations, domestic operations in the FSU and China, military operations, and general aviation.

The forecast for scheduled traffic was based on the 1996 Boeing Current Market Outlook (Boeing, 1996), which projects separate traffic growth rates by region. Growth in worldwide demand for air travel was expected to average about 5% per

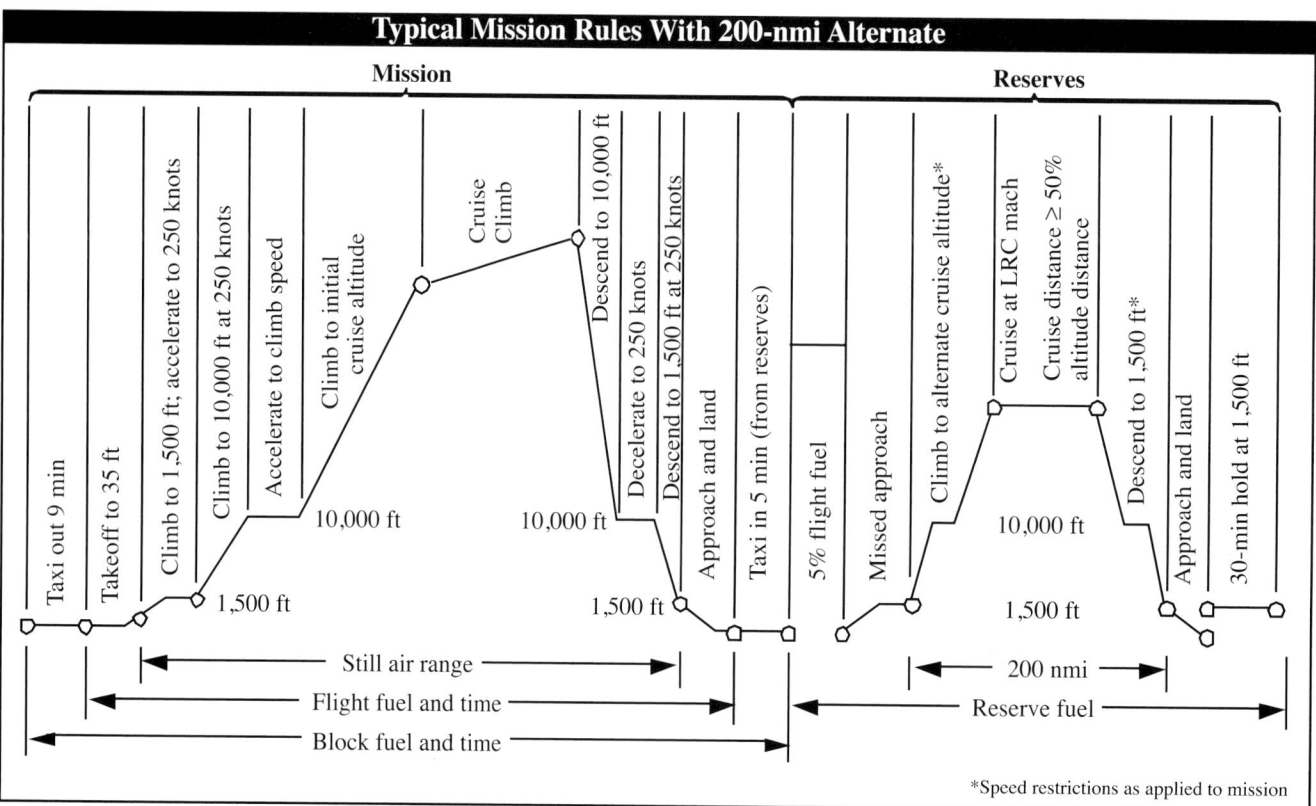

Figure 9-6: Scheduled aircraft mission profile.

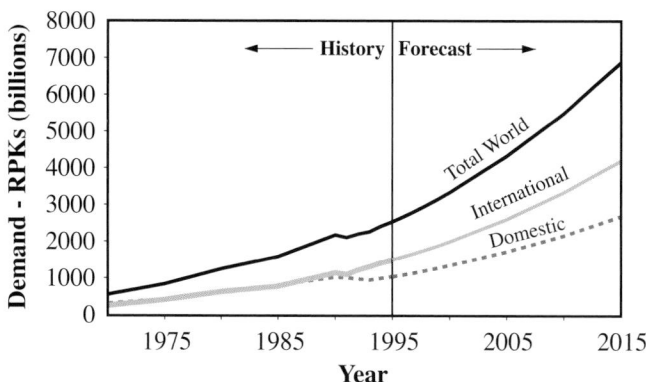

Figure 9-7: Passenger traffic demand growth to 2015.

year to the year 2015, with international travel growing at a slightly faster rate than domestic travel (Figure 9-7). By 2015, demand for air travel is projected to be 2.5 times greater than in 1996.

The total projected demand for scheduled air travel in the year 2015 was assigned to actual aircraft on a projected city-pair schedule derived from the schedules for 1995 published in the OAG. Individual city-pair service schedules for 1995 within each of the traffic flow regions were grown to 2015 by using the consolidated regional growth rate applicable for that region. Aircraft types were assigned to routes by using a market share forecast model. The turboprop market (for which there was no detailed forecast) was projected for 2015 by assuming that city pairs not served by the smallest turbojet category (50–90 seats) after demand growth to 2015 will continue to be served by small, medium, or large turboprops.

The result of the fleet assignment task was a detailed city-pair flight schedule by aircraft type required to satisfy forecast scheduled passenger demand in 2015. This schedule was used to calculate the 3-D emissions inventory for scheduled passenger service. Simplifying assumptions were the same as those used in calculating the historical and present-day inventories.

Projections of engine and aircraft technology levels for the 2015 scheduled fleet with regard to fuel efficiency and NO_x emissions were made by assuming a continuation of present trends. In general, engines in the 2015 scheduled fleet represent the state-of-the-art in engine technology available either in production or in the final stages of development at the time the assignments were made (1997). These engines include low-emissions derivatives of previously existing engines. It is unlikely that any radical changes in airframe or engine design—even if such designs were acceptable—would have much of an effect on the 2015 fleet, given the time required to bring new designs into service. The combined effects of 2015 fleet mix and technology projections on the NO_x technology level of the projected 2015 fleet appear in Figure 9-8, which shows the percentage of total fleet fuel burned by aircraft having landing/take-off cycle (LTO) emissions at a given level relative to the CAEP/2 NO_x limit. (CAEP is chartered to propose worldwide certification standards for aircraft emissions and

noise. The CAEP/2 designation refers to emissions certification standards adopted at the second meeting of the CAEP in December 1991.) Much more of the fleet consists of low-NO_x aircraft-engine combinations in 2015, with ~70% of fuel burned in engines with NO_x emission levels between 20 and 40% below the CAEP/2 certification limit.

DTI has developed a traffic and fleet forecast model for civil aviation, which was adapted under the direction of ANCAT and EIDG to produce an estimate of fuel burned and NO_x emitted by civil aviation for the forecast year of 2015 (Gardner, 1998). Fuel and NO_x growth factors—base to forecast—were calculated and applied to the ANCAT/EC2 city-pair gridded 1992 base year inventory to produce a gridded 2015 forecast.

DTI's top-down regional traffic demand forecasting model has a horizon of 25 years. Traffic coverage in the model includes all scheduled civil operations but excludes the former Soviet Union, Eastern Europe, freight, military, non-European charter traffic, business jets, and general aviation. Factors were developed to account for these traffic sectors in the forecast. The traffic forecast assumes a relationship between traffic [available seat-kilometers (ASK)] and GDP growth, and is assessed on a regional and flow basis (i.e., traffic flow between specific regions). The relationship is modified by assumptions on airline yields—a surrogate for fares price—and by a market maturity term that modifies demand as a function of time. Future fleets are estimated from traffic forecasts in terms of size and composition.

The concept of "traffic efficiency" was used to estimate fuel consumption from traffic values. Traffic efficiency is defined as the amount of traffic or capacity (ASK) per unit of fuel consumed. Aircraft manufacturers' traffic efficiency data for current aircraft types and projections for future aircraft types were used to develop efficiency trends for the eight categories of generic aircraft adopted for forecasting purposes, over a range of flight sector lengths. This approach permitted estimation of fuel consumption on the basis of regional and global traffic forecasts.

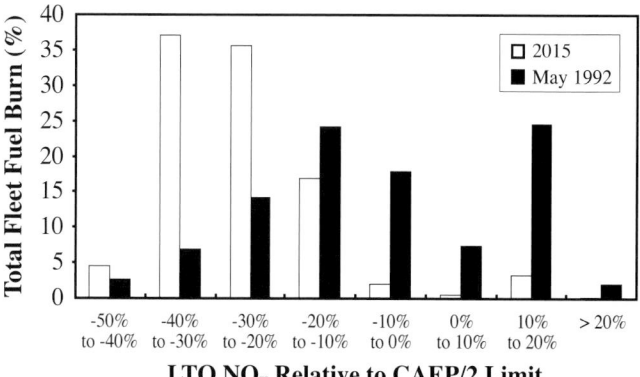

Figure 9-8: Percentage of total scheduled fleet fuel burned by aircraft in specific LTO NO_x emissions categories for May 1992 (Baughcum *et al.*, 1996b) and the year 2015 projection (Baughcum *et al.*, 1998).

Average efficiency figures were also calculated for the eight generic aircraft types in the 1992 base year fleet; a fleet average value of about 24.0 seat-km per liter was found. This figure compares well with those in Greene (1992) and Balashov and Smith (1992) for the years 1989 and 1990, respectively, which gave traffic efficiencies of 20.5 seat-km per liter.

Greene (1992) and Balashov and Smith (1992) forecast an annual improvement in commercial air fleet fuel efficiency (see Table 9-2). These efficiencies include improvements arising from the introduction of new aircraft into the fleet and changes to operating conditions and passenger management. For the DTI work, the Greene (1992) forecasts were used to 2010. Annual improvements in fuel efficiency was assumed to decrease to 1% per year beyond 2010.

Using this efficiency trend, traffic efficiencies were calculated for the future aircraft fleet. The base year fleet average was estimated to increase to 31.8 seat-km per liter by 2015.

The same trends in fuel efficiency were applied to all size and technology classes. This approach represents a simplification because improvement figures are really a fleet average and would be influenced strongly by the rate of introduction of new aircraft. Given the much smaller contribution of older aircraft to global traffic performance, however, this factor will be only a second-order effect.

The emission performance of the forecast fleet was determined in part by the assumed response of the engine manufacturing industry to an assumed regulatory scenario. An emissions certification stringency regime was proposed for the forecast period, and compliance with the tighter limits was achieved by modifying the emissions performance of engines as they became noncompliant. This calculation was assessed from a base year engine fleet, comprising engines typical of and representing those found in the fleet (and compatible with the aircraft generic types described above). Performance improvements were applied only to new fleet entrants and were appropriate for staged and ultra-low NO_x control technology in some cases.

This process results in an estimate of fuel burn and NO_x emissions for the base year and forecast fleet using the same methodology; 1992–2015 fuel and NO_x growth factors are thereby calculated. The growth factors were applied to the ANCAT/EC2 base year gridded fuel and NO_x estimates to provide a 2015 gridded forecast.

The methods used to project civil aviation traffic demand for the DLR 2015 inventory were based on regional growth factors calculated by DTI. Thus, the DLR 2015 forecast differs from the ANCAT/EC2 forecast only in that the base year inventory is slightly different because of the different fuel and profiling methodology and the aircraft generic types. Thus, in the comparison of results, ANCAT and DLR 2015 forecasts are not assumed to be different because the DLR forecast is essentially an application of the DTI/ANCAT forecast.

Table 9-2: *Future trends in fuel efficiency improvement.*

Time Period	Fuel Efficiency Improvement
1993–2000	1.3% yr^{-1}
2000–2010	1.3% yr^{-1}
2010–2015 (extrapolation)	1.0% yr^{-1}

9.3.3.   Other Emissions Inventories

Studies of atmospheric effects of aviation were conducted using the global inventory of McInnes and Walker (1992) and emissions data sets produced by the Dutch Institute of Public Health and Environmental Protection for 1990, 2003, and 2015 (Olivier, 1995) based on the McInnes and Walker (1992) data. Other emissions estimates are predominantly made on a national level (e.g., in Austria and Sweden).

The Dutch Aviation Emissions and Evaluation of Reduction Options (AERO) Project was initiated in 1994 by the Dutch Civil Aviation Department to estimate economic and environmental impacts of possible measures to reduce aviation emissions (see Chapter 10). Within this project, the flights and emissions model (FLEM) was developed for the calculation of worldwide fuel use and emissions per grid cell (ten Have and de Witte, 1997). The base year traffic movements database is a combination of data from ANCAT/EC2, International Air Transport Association, the ABC schedule, ICAO, and the U.S. Department of Transportation (DOT). Global volumes for aircraft kilometer, fuel consumption, and emissions (CO_2, NO_x) resulting from computations of the AERO modeling system for civil aviation for base year 1992 and forecast for 2015 (called FPC-2015 scenario) are listed in Table 9-3. Further details appear in Pulles (1998).

9.3.4.   Comparisons of Present-Day and 2015 Forecast Emissions Inventories (NASA, ANCAT/EC2, and DLR)

Table 9-4 lists the totals for calculated fuel burned and emissions from the NASA, ANCAT, and DLR inventories for 1976, 1984, 1992, and 2015. Because these inventories consisted of 3-D

Table 9-3: *Results from AERO modeling analysis.**

	1992	2015	Annual Change
Aircraft kilometers (km yr^{-1})	20.7 x 10^9	49.6 x 10^9	3.9%
Fuel consumption (Tg yr^{-1})	144	278	2.9%
CO_2 emissions (Tg yr^{-1})	453	877	2.9%
NO_x emissions (Tg yr^{-1})	1.84	3.86	3.3%

*AERO results for base and datum (FPC-2015 scenario).

Table 9-4: Calculated fuel and emissions from NASA, ANCAT, and DLR inventories.

	NASA 1976	NASA 1984	NASA 1992	ANCAT 1992	DLR 1992	NASA 2015	ANCAT 2015	DLR 2015
Calculated Fuel Burned (Tg)								
Scheduled	45.83	64.17	94.84			252.73		
Charter	8.47	9.34	6.57			13.50		
FSU/China	6.05	7.43	8.77			15.79		
General Aviation	4.04	5.62	3.68			6.03		
Civil Subtotal	64.38	86.56	113.85	114.20	112.24	288.05	272.32	270.50
Military	35.66	29.76	25.55	17.08	17.10	20.59	14.54	14.50
Global Total	100.04	116.31	139.41	131.3	129.34	308.64	287.86	285.00
Calculated CO$_2$ Emissions (Tg C)								
Scheduled	39.41	55.18	81.56			217.35		
Charter	7.28	8.03	5.65			11.61		
FSU/China	5.20	6.39	7.54			13.58		
General Aviation	3.47	4.83	3.16			5.18		
Civil Subtotal	55.36	74.44	97.91	98.22	96.52	247.72	234.21	232.63
Military	30.67	25.59	21.98	14.68	14.71	17.71	12.50	12.47
Global Total	86.03	100.03	119.89	112.92	111.23	265.43	246.71	245.10
Calculated NO$_x$ Emission (Tg as NO$_2$)								
Scheduled	0.50	0.79	1.23			3.57		
Charter	0.09	0.11	0.09			0.19		
FSU/China	0.04	0.06	0.06			0.12		
General Aviation	0.06	0.07	0.05			0.07		
Civil Subtotal	0.70	1.02	1.44	1.60	1.60	3.95	3.37	3.41
Military	0.28	0.25	0.23	0.20	0.20	0.18	0.16	0.16
Global Total	0.98	1.28	1.67	1.81	1.80	4.12	3.53	3.57
Calculated Fleet Average NO$_x$ Emission Index [g NO$_x$ (as NO$_2$) kg^{-1} fuel burned]								
Scheduled	10.9	12.3	13.0			14.1		
Charter	10.8	11.3	13.3			13.8		
FSU/China	7.4	7.4	7.4			7.4		
General Aviation	14.5	12.6	14.4			11.3		
Civil Subtotal	10.8	11.8	12.6	14.0	14.2	13.7	12.4	12.6
Military	8.0	8.5	8.9	11.9	11.8	8.7	10.7	10.8
Global Total	9.8	11.0	12.0	13.8	13.9	13.4	12.3	12.5

data sets, the differences in spatial distributions as well as totals are compared. The NASA inventories also included emissions of CO and HC, which are summarized in Table 9-5.

The NASA inventories include piston-powered aircraft in the general aviation fleet. This category of aircraft is excluded from the ANCAT and DLR inventories, but the contribution to total fuel burned from these aircraft is small (2.6% of fuel burned in 1992). Piston-powered aircraft are large contributors to CO and HC emissions relative to the amount of fuel they burn (39% of CO and 13% of HC emissions in 1992). This large relative contribution is reflected in the emissions indices of these two pollutants in the general aviation category.

A comparison of calculated global total values for fuel burned and NO$_x$ emissions from the NASA, ANCAT, and DLR inventories

for 1992 and 2015 is shown in Figure 9-9. All three inventories for 1992 have approximately the same calculated values for total fuel burned in the civil air fleet; the difference in total fuel (7% maximum) arises almost entirely from different calculated contributions for military aviation operations, for which the ANCAT inventory calculates 33% lower fuel burned. Because military fuel is estimated to be between 13 and 18% of total fuel in 1992, the effect of this large difference in estimates between military sectors on the total is small. Use of the NASA inventories as a base is arbitrary and does not imply that differences from the NASA results are errors. Exclusion of turboprop operations from the ANCAT inventory results in about a 2% underestimate (if data from the NASA inventory are used).

Calculated values for total NO$_x$ emissions from the three inventories for 1992 are within 9% of each other. The ANCAT

Table 9-5: *Emissions of CO and HC from NASA inventories.*

	NASA 1976	NASA 1984	NASA 1992	NASA 2015
Calculated CO Emissions (Tg)				
Scheduled	0.41	0.41	0.50	1.12
Charter	0.03	0.04	0.02	0.05
FSU/China	0.10	0.12	0.15	0.26
General Aviation	0.73	0.75	0.62	0.60
Civil Subtotal	1.27	1.32	1.29	2.04
Military	0.43	0.35	0.29	0.23
Global Total	1.70	1.67	1.57	2.27
Calculated Fleet Average CO Emissions Index (g CO kg^{-1} fuel burned)				
Scheduled	8.9	6.3	5.3	4.5
Charter	4.0	4.0	3.7	3.9
FSU/China	16.6	16.6	16.6	16.6
General Aviation	180.1	133.0	167.6	99.4
Civil Subtotal	19.7	15.2	11.3	7.1
Military	12.0	11.9	11.2	11.3
Global Total	17.0	14.4	11.3	7.4
Calculated HC Emissions (Tg)				
Scheduled	0.27	0.20	0.20	0.17
Charter	0.01	0.01	0.00	0.01
FSU/China	0.02	0.02	0.03	0.05
General Aviation	0.03	0.05	0.04	0.05
Civil Subtotal	0.33	0.28	0.26	0.28
Military	0.09	0.07	0.06	0.05
Global Total	0.42	0.35	0.32	0.33
Calculated Fleet Average HC Emissions Index (g HC kg^{-1} fuel burned)				
Scheduled	5.8	3.2	2.1	0.7
Charter	0.9	0.9	0.5	0.6
FSU/China	3.2	3.2	3.2	3.2
General Aviation	8.2	8.5	9.9	8.6
Civil Subtotal	5.1	3.3	2.3	1.0
Military	2.5	2.3	2.4	2.5
Global Total	4.2	3.0	2.3	1.1

and DLR values are higher than those from NASA—a result of a combination of differing fleet mixes, a different method of calculating NO$_x$ emissions, and the offsetting effects of civil and military calculations. This variation is also reflected in the calculated EI(NO$_x$) for the fleet components: The ANCAT and DLR inventories have a total fleet emission index that is 15% higher than that of the NASA inventory.

Differences between inventory totals widen for the 2015 case, although total fuel burned is still within 8%. Total NO$_x$ emissions in the NASA 2015 forecast are almost 15% greater than those in the ANCAT forecast, a result of different assumptions about the direction of NO$_x$ reduction technology (the NASA assumptions result in an increase in NO$_x$ emissions index in the civil sector, whereas the ANCAT forecasts assume a reduction).

Other differences between the NASA, ANCAT, and DLR inventories relate to the distribution of calculated fuel burned and emissions, geographically (latitude and longitude) and with altitude. Although all three inventories place more than 90% of global fuel burned and emissions in the Northern Hemisphere, there are differences between inventories in the details of the distribution. Figure 9-10 shows the distribution of fuel burned as calculated in 1 month (May) of the 1992 NASA inventory. The most heavily trafficked areas are clearly visible (United States, Europe, North Atlantic, North Asia).

For geographical comparison purposes, data in the files of the NASA and ANCAT 1992 inventories were divided into 36 regions, defined by 60° spans of longitude and 20° spans of latitude. Figure 9-11 shows the differences between the ANCAT and NASA 1992 inventories with regard to geographical distribution.

The major differences between the NASA and ANCAT inventories (on a geographical basis) lie in the estimate of fuel burned and NO_x emissions in the regions covering North America and Europe. The ANCAT inventory places 32% of total fuel burned and 30% of total NO_x over North America, whereas the NASA inventory places 27% of fuel burned and 27% of NO_x over that region. ANCAT places 16% of the fuel and 15% of the NO_x over Europe, whereas the NASA inventory places 21% of the fuel and 19% of total NO_x over that region. Part of this difference may be explained by the 10% scaling of U.S. traffic assumed by ANCAT as a method of approximating the U.S. charter market.

NASA and ANCAT fuel and NO_x emissions projections for 2015 are similar to the respective 1992 inventories in that no new city pairs were used in the 2015 traffic projections. Growth rates from 1992 to 2015 vary with region, so the geographical distribution of emissions changes over time.

The altitudinal distributions of fuel burned in the present-day NASA, ANCAT, and DLR inventories are shown for civil aviation in Figure 9-12 and for military aviation in Figure 9-13. The civil aviation distributions are similar, with the NASA inventory showing more fuel burned at higher altitudes. The military distributions are quite different, with fuel burned in the NASA inventory concentrated at the higher altitudes and fuel burned in the ANCAT inventory at lower altitudes. This difference may be because of a higher proportion of transport operations in the NASA inventory. The altitudinal distribution of NO_x emissions follows closely that of fuel burned. The three inventories show that more than 60% of the fuel burned and NO_x emissions occur above 8 km, whereas a major fraction of CO and HC are emitted near the ground.

Although the three inventories show comparably low variations for total global monthly figures over the year, the seasonal dependency can be quite large for some regions (Figure 9-14). Operations in the North Atlantic and North Pacific show a clear yearly cycle, with a maximum in the northern summer and a minimum during winter. In contrast, Southern Hemisphere operations show little seasonal variation overall, with small peaks in February and November.

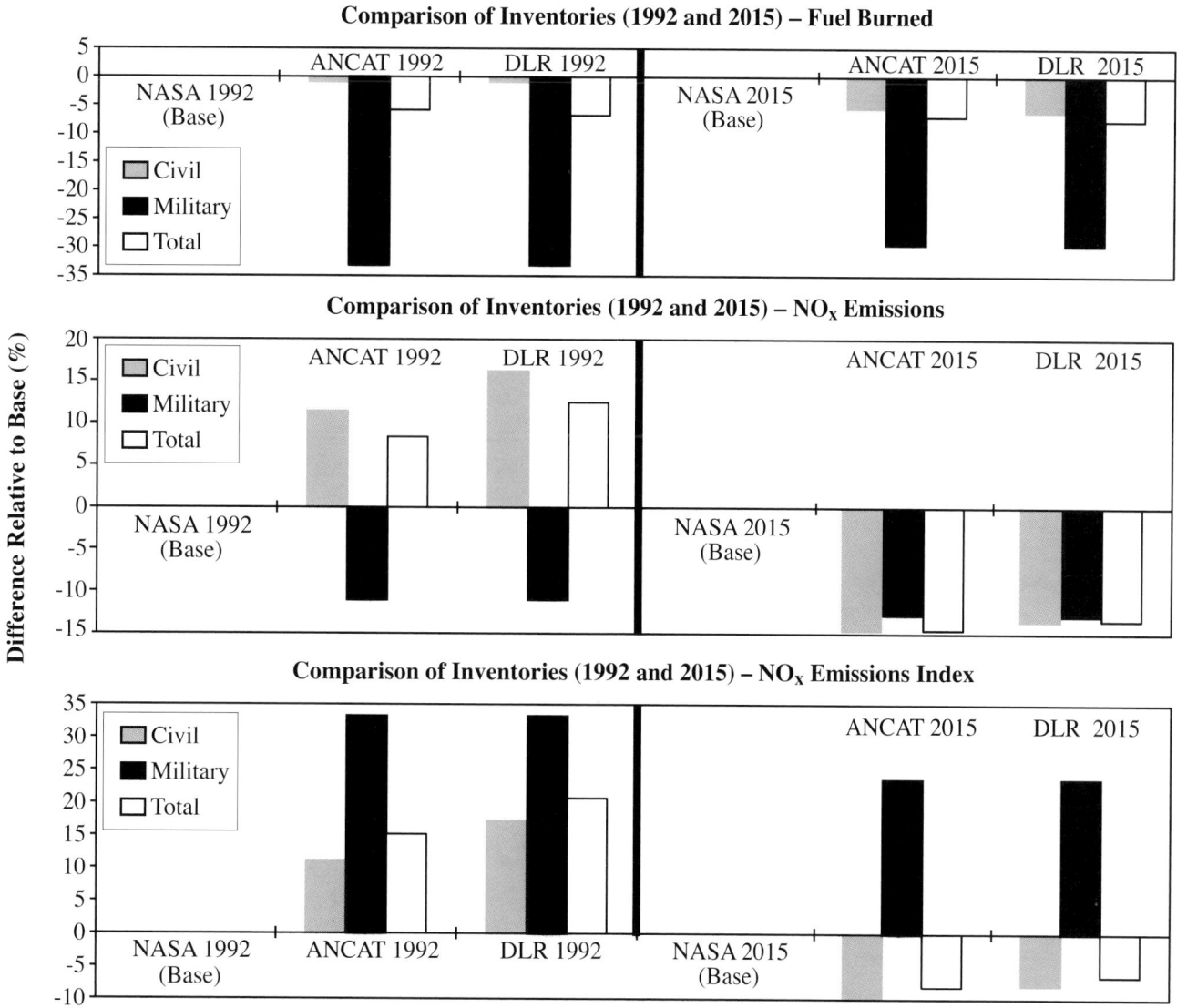

Figure 9-9: Comparison of inventories.

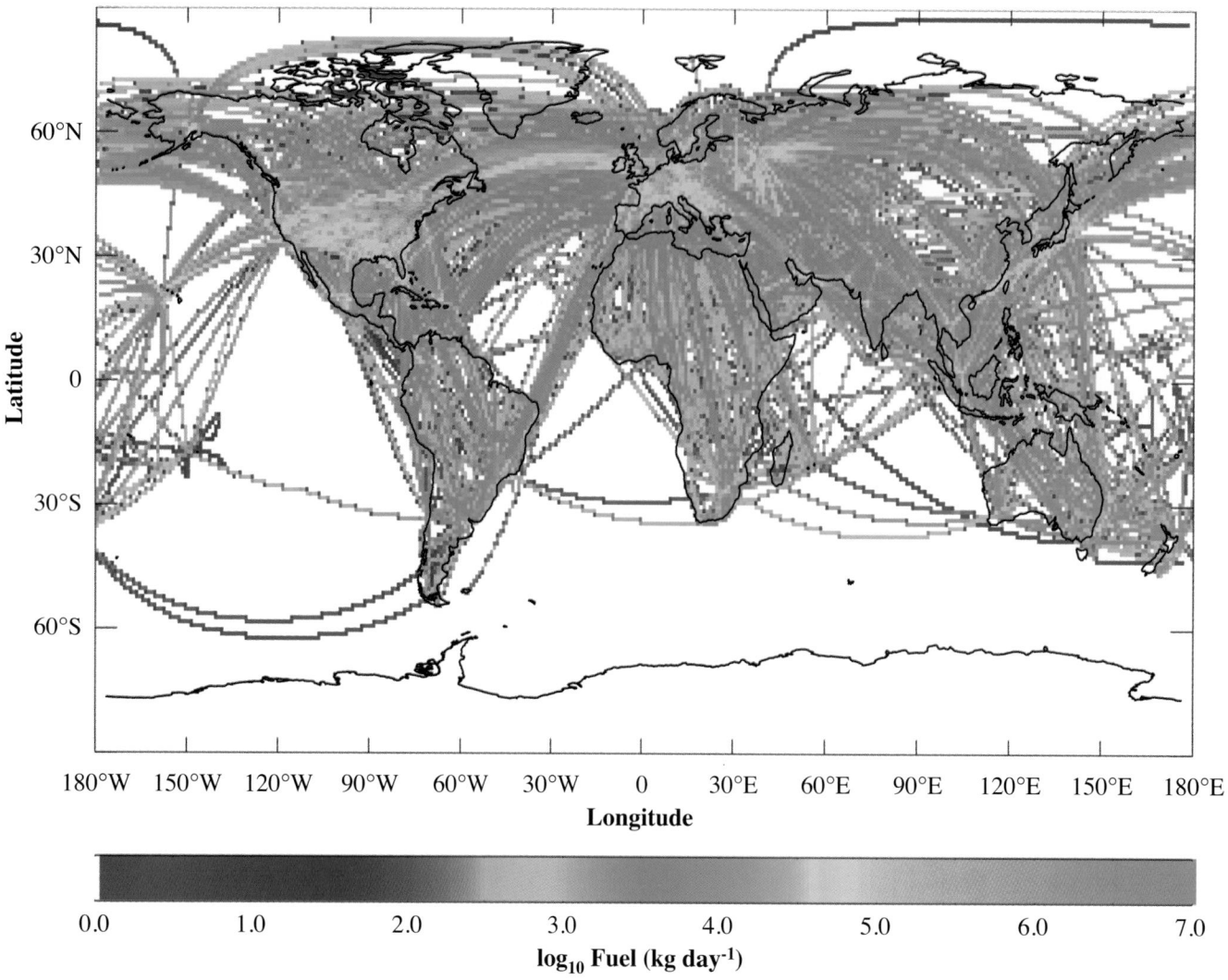

Figure 9-10: Geographical distribution of fuel burned by civil aviation (May 1992).

DLR has also examined longer trends in fuel burned and emissions for air traffic (Schmitt and Brunner, 1997). 3-D gridded inventories of fuel burned and emissions were calculated for 1982 through 1992 using ICAO statistics on annual values for international scheduled air traffic and ABC time table data of all scheduled air traffic for the same week of September in 1986, 1989, and 1992. Emissions inventories were produced for each of these data sets using the same methods as in the 1992 DLR inventory described above. These inventories concentrate on scheduled services because reasonably accurate calculations are possible for this segment of aviation. Because these data do not include non-scheduled flights, military traffic, general aviation, or former Soviet Union/China traffic, they are of limited use in global modeling studies. However, they do provide a consistent set of data to track the growth of the international and domestic scheduled sector. Table 9-6 gives the totals for the yearly inventories.

9.3.5. *Error Analysis and Assessment of Inventories*

Simplifying assumptions used in creating all of the 3-D emissions inventories have introduced systematic errors in the

calculations. An analysis of the effects of the simplifying assumptions on fuel burned used in the 1992 NASA inventory has been performed by Baughcum *et al.* (1996b). All of the assumptions have the effect of biasing the calculation toward an underestimate of fuel burned and emissions produced, as detailed in Table 9-7. The effects of the assumptions on the ANCAT and DLR inventories may be expected to be similar, because most of the simplifying assumptions used in those inventory calculations were similar to those in the NASA inventory.

The assumption of great-circle flight paths results in an underestimate of distance flown, although the practice of routing to take advantage of winds may result in lower fuel consumption than a great-circle path for a given flight. A study of international and domestic flights from German airports showed an average increase in flight distance of 10% for medium- and long-haul flights above 700 km, with larger deviations from great-circle routes for shorter flights (Schmitt and Brunner, 1997). Ground delays and in-flight holding at relatively low altitudes caused by congestion in the air traffic control system also adds to fuel consumption. Aircraft in service are subject to factors that may increase fuel consumption by up to 3% (e.g., engine deterioration, added weight from added systems, and increased surface roughness).

Factors that cause underestimates of fuel burned do not necessarily operate at the same time, so they are not additive. Sutkus *et al.* (1999) compared fuel burned for certain carriers

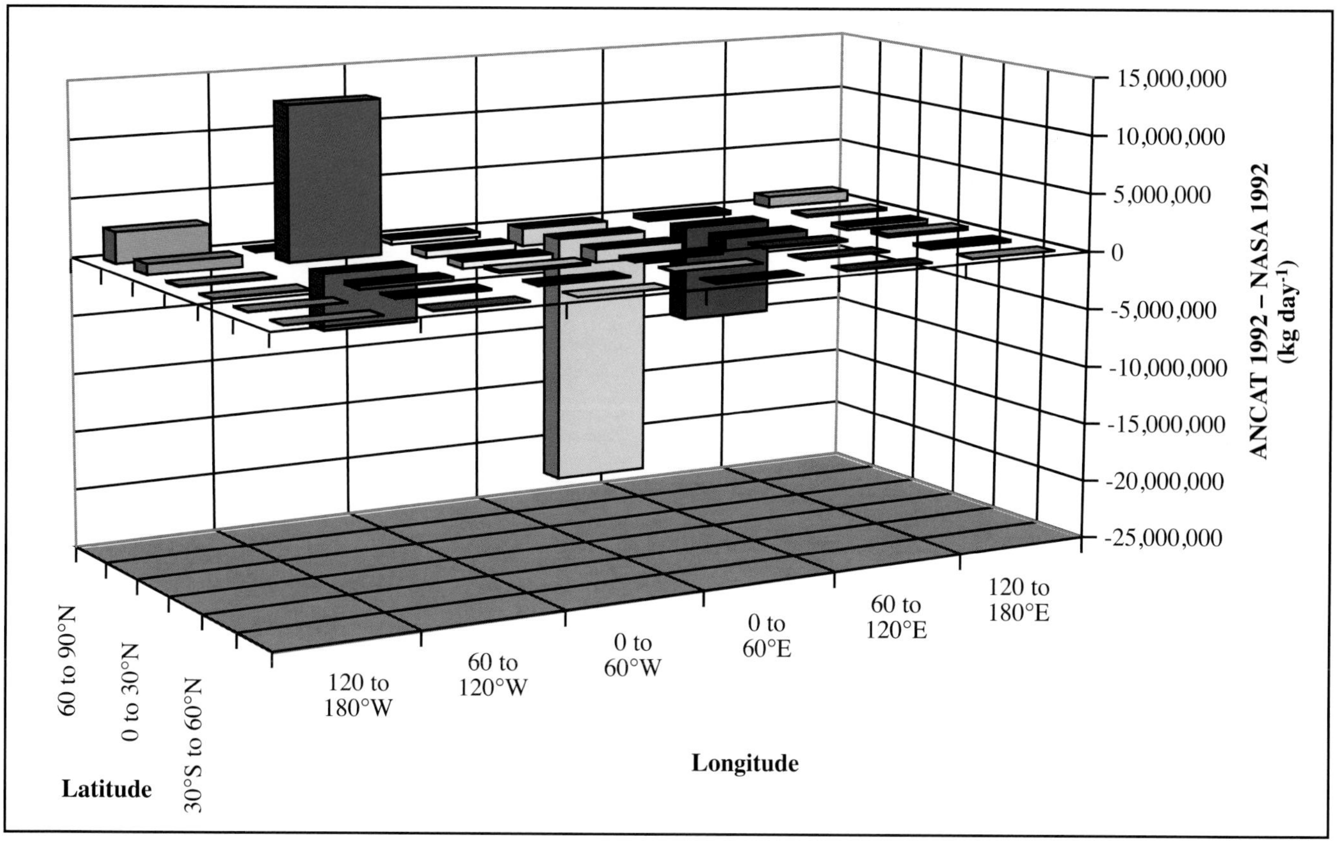

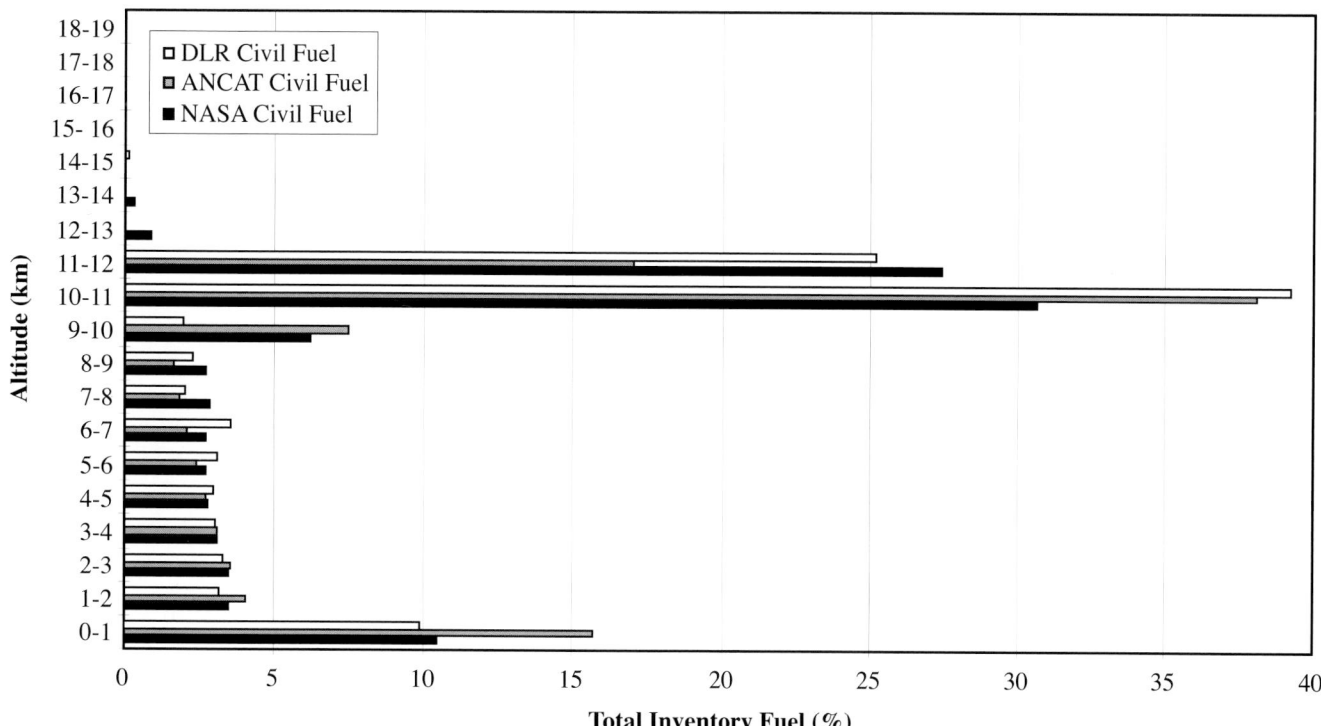

Figure 9-11: Differences in geographical distribution of fuel burned.

Figure 9-12: Comparison of altitude distribution of 1992 inventories for civil aviation fuel burned.

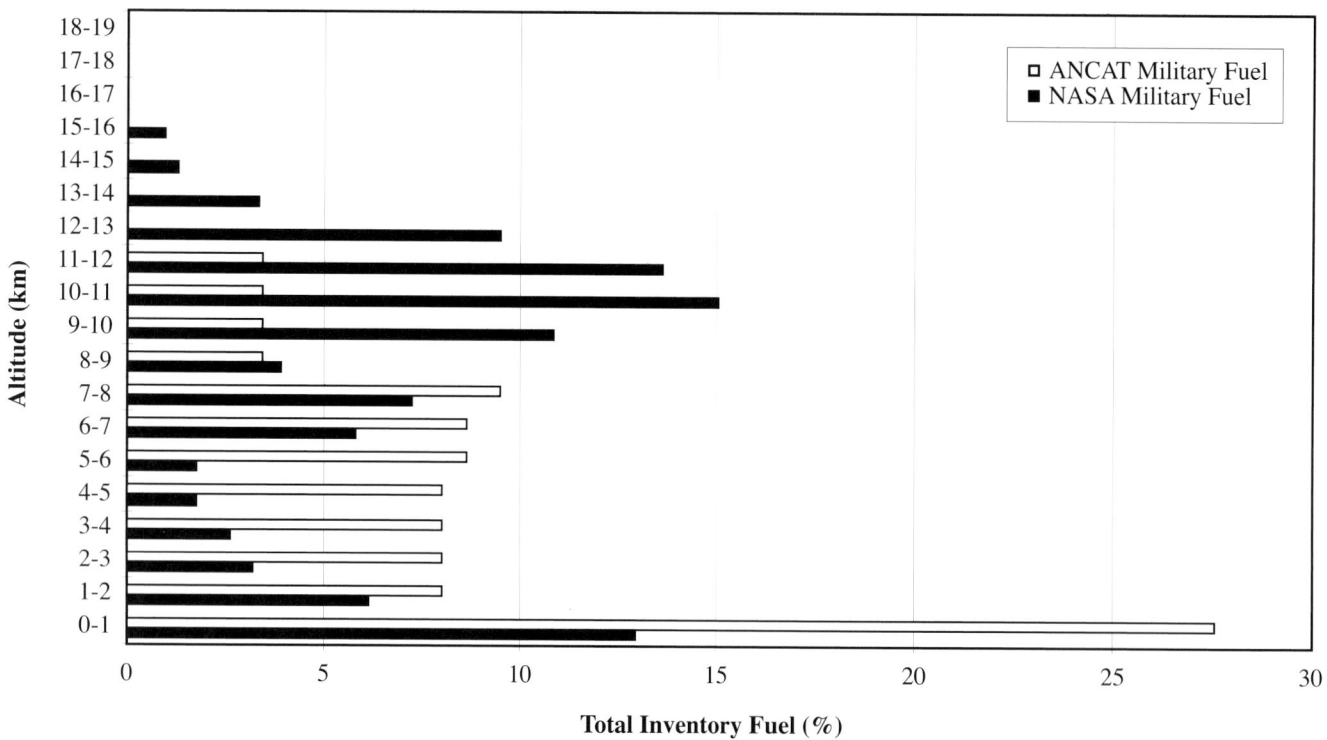

Figure 9-13: Comparison of altitude distribution of 1992 inventories for military aviation fuel burned.

and certain specific aircraft types reported to DOT by U.S. air carriers, with the value for fuel burned calculated for these carriers and aircraft types in the 1992 NASA inventory. The comparison shows that a combination of factors outlined above results in systematic underestimation of total fleet fuel burned by 15–20% for domestic operations. The assumptions in the foregoing analysis apply to the civil aviation fleet. An error analysis of the calculation of fuel burned and emissions from military operations is not possible, given the nature of the estimates used in the calculations.

The present-day inventories described above have reported global fuel consumption values for 1992 ranging from 129 to 139 Tg. However, reported aviation fuel production was somewhat larger, at 177 Tg (OECD, 1998a,b). Calculated fuel consumption therefore accounts for 73–80% of total fuel reported produced in 1992. Simplifying assumptions used in calculating the inventories probably account for most of the

difference. Reported fuel production values are not an ideal reference, however, because they do not necessarily represent fuel delivered to airports for use in aircraft. Jet fuel, in particular, is a fungible product; it can be reclassified and sold as kerosene or mixed with fuel oils or diesel fuel, depending on market requirements (e.g., when low freezing point fuel oil is needed in winter). Other distillate fuels from refineries may satisfy jet fuel requirements and could be purchased and used as jet fuel. As a consequence, reported jet fuel production data do not provide a rigorous upper or lower limit to jet fuel use. Fuel production data represent a compilation of reports of varying accuracy from many (not all) countries, whose overall accuracy has not been evaluated (Baughcum *et al.*, 1996b; Friedl, 1997).

OECD data on aviation fuel production from 1971 (the first year the data includes the former Soviet Union) to 1996 are shown in Figure 9-15. These data shown are the sum of OECD and non-OECD country production data. Reported data include production of aviation gasoline, naphtha-type jet fuel (mostly JP-4, used for military aircraft), and kerosene-type jet fuel (Jet A, the most common transport aircraft jet fuel). Also shown are calculated values of aviation fuel burned from the NASA, ANCAT, and DLR present-day inventories. (NASA values have been increased by 15% as a rough estimate of systemic underestimate of civil fuel burned.)

Aviation gasoline has declined as a percentage of total aviation fuel—from 4% of production in 1971, to just over 1% in 1995. Production of naphtha-type jet fuel reached just over 10% of total fuel in 1983, but has since declined to less than 1% as military aviation has phased out its use in favor of kerosene-type fuels. Prior to 1978, production of naphtha-type fuel was

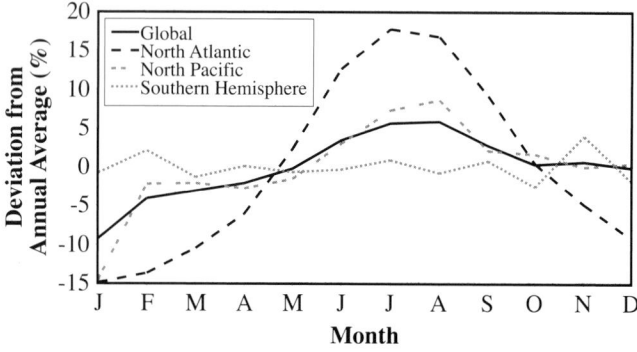

Figure 9-14: Regional seasonality of traffic.

Table 9-6*: Fuel burned and emissions from scheduled air traffic, 1982–92 (DLR).*

Year	Fuel [Int'l] (Tg)	NO$_x$ [Int'l] (Tg)	CO [Int'l] (Tg)	HC [Int'l] (Tg)	Fuel [Total] (Tg)	NO$_x$ [Total] (Tg)	CO [Total] (Tg)	HC [Total] (Tg)
1982	19.2	0.31	0.05	0.02				
1983	20.9	0.35	0.06	0.02				
1984	24.7	0.41	0.06	0.03				
1985	24.9	0.41	0.06	0.02				
1986	26.7	0.44	0.07	0.03	72.2	1.03	0.24	0.10
1987	30.0	0.51	0.08	0.03				
1988	32.6	0.55	0.09	0.03				
1989	35.8	0.61	0.10	0.03	76.5	1.14	0.27	0.09
1990	37.2	0.62	0.11	0.04				
1991	36.3	0.59	0.11	0.04				
1992	39.3	0.62	0.10	0.03	93.0	1.31	0.34	0.10

Table 9-7*: Analysis of the underestimate of fuel burned caused by simplifying assumptions (Baughcum et al., 1996b).*

Changes to Simplifying Assumptions	Maximum % Fuel Burned Increase
No winds to actual winds	2.6 (Autumn winds, North Pacific route)
Standard temperatures to actual temperatures	0.7 (Summer temperature, North Pacific route)
Combined wind and temperature effects	3.1 (Autumn winds, ISA+5°C temperature, North Pacific)
Payload: increase load factor to 75%	0.8 (747-400, North Pacific)
Payload: increase load factor to 75%	2.5 (737-300, Los Angeles–San Francisco)
Payload: volume limited cargo	7.7 (747-400, North Pacific)
No fuel tankering to actual practice	4.0 (737-300, Los Angeles–San Francisco, four leg mission)

not reported as a separate item in the OECD database; it was included in the kerosene-type production data.

The three inventories are in good agreement; given the different approaches and data sources used, the inventory results (particularly for the present day) are remarkably consistent. Assumptions regarding the state of NO$_x$ reduction technology in 2015 cause the biggest difference in the results of the three forecasts. The 1992 and 2015 inventories of NASA, ANCAT, and DLR are all suitable for calculating the effects of aircraft emissions on the atmosphere, taking account of differences in the details of the inventories and systematic underestimates examined above. To correct for the systematic underestimation of fuel burned in the inventories when calculating the effects of aviation CO$_2$ emissions, fuel burned values for 1992 should be increased by 15% and those for 2015 should be increased by 5%, based on the assumption that inefficiencies in the air traffic control system responsible for extra fuel burned will be much reduced by 2015. A summary of the results from these inventories is given in Table 9-8. The DLR "trend" inventories (1982–92) include only a portion of total

aviation operations (scheduled international service for all years and total scheduled service in 1986, 1989, and 1992); as such, they are valuable for historical growth analysis and for comparisons with the NASA and ANCAT/EC2 scheduled traffic segments.

9.4. Long-Term Emissions Scenarios

Long-term projections to the year 2050 producing 3-D emissions data have been made by the Forecasting and Economic Analysis Subgroup of CAEP, using the NASA studies as a base (CAEP/4-FESG, 1998), and by DTI using the ANCAT studies as a base (Newton and Falk, 1997). Long-term projections of total demand, fuel consumption, and emissions (but not providing 3-D data) have also been made by EDF (Vedantham and Oppenheimer, 1994, 1998), WWF (Barrett, 1994), and MIT (Schafer and Victor, 1997).

Predictions of traffic demand and resulting emissions beyond 2015 become increasingly uncertain because the probability

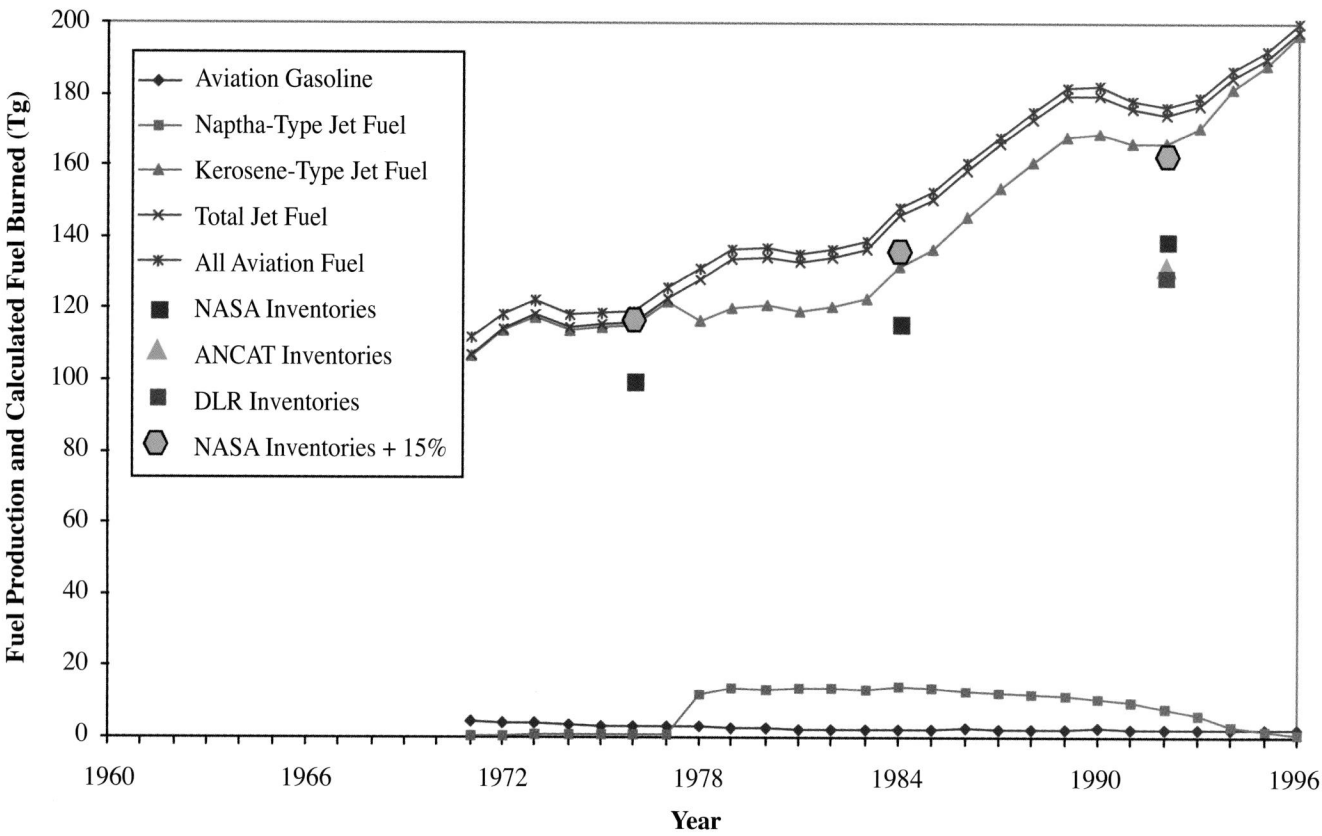

Figure 9-15: Comparison of calculated fuel burned by aviation with reported production.

for unforeseeable major changes in key factors influencing the results steadily increases. The best approach for insight into the evolution of long-term futures is the application of scenarios. A scenario is simply a set of assumptions devised to reflect the possible development of a particular situation over time. These assumptions are used as inputs to a model that describes the manner in which an activity might develop over time. A range of possible futures can be described by a set of independent scenarios. The results of the scenario are difficult to judge in terms of confidence level: They are simply the outcome of input assumptions. However, scenarios can be objectively judged as implausible by showing that their assumptions or outcomes conflict with industry trends or with invariant rules and laws that might reasonably be expected to remain

unchanged during the scenario time period or by revealing internal inconsistencies or incompatibilities with other dominating external developments. Investigation of the consequences and implications of scenarios can be used to support a subjective assessment regarding which of the remaining possible scenarios might be more plausible than others.

9.4.1.   FESG 2050 Scenarios

9.4.1.1.   Development of Traffic Projection Model

In developing long-term traffic scenarios, various models of traffic demand were considered (CAEP/4-FESG, 1998), particularly

Table 9-8: *Summary comparison of historical, present-day, and 2015 forecast 3-D emissions inventories.*

Inventory Year	Inventory Source	Calculated Fuel Burned (Tg)	Calculated CO_2 (as C) (Tg)	Calculated NO_x (as NO_2) (Tg)	Calculated Fleet $EI(NO_x)$ (g NO_2 kg^{-1} fuel)
1976	NASA	100.0	86.0	1.0	9.8
1984	NASA	116.3	100.0	1.3	11.0
1992	NASA	139.4	119.9	1.7	12.0
1992	ANCAT	131.2	112.9	1.8	13.8
1992	DLR	129.3	111.2	1.8	13.9
2015	NASA	308.6	265.4	4.1	13.4
2015	ANCAT	287.1	246.9	3.5	12.3
2015	DLR	285.0	245.1	3.6	12.5

those incorporating a market maturity concept. Under this concept, historical traffic growth rates in excess of economic growth are considered unlikely to continue indefinitely, and traffic growth will eventually approach a rate equal to GDP growth as the various global markets approached maturity. Based on this assumption, a single global model of traffic demand per unit of GDP was developed, based on a logistics growth curve function:

$$\frac{RPK}{GDP} = \left(\frac{26.24}{1 + 9.04\exp(-.073t)} \right)$$

t = time
RPK = revenue passenger-km
GDP = gross domestic product

The parameters in the model equation were estimated from historic values of RPK/GDP for the period 1960 through 1995. No constraints were imposed on the values the parameters could take. Further details of the modeling process appear in CAEP/4-FESG (1998). Table 9-9 lists the GDP growth assumptions used in developing these scenarios (Leggett *et al.*, 1992). The key assumptions of this approach follow:

• The world can be treated as a single, gradually maturing aviation market that is the sum of regional markets at various stages of maturity.
• Historical values of world demand and GDP over time provide sufficient information about the stage of development of the industry to provide reliable estimates of market maturity.
• Business and personal travel sectors can be combined.
• Global traffic growth is driven primarily by global GDP; as markets mature, overall passenger growth rates will eventually grow in line with GDP growth.
• Fuel will be available, and fuel prices will not increase greatly relative to other costs.
• Whatever aviation technological or regulatory changes occur, they will have no significant impact on ticket prices, demand, or service availability.
• Infrastructure will be sufficient to handle demand.
• There will be no significant impacts from other travel modes (e.g., high-speed rail) or alternative technologies (e.g., telecommunications).

Perhaps the most critical assumption of this methodology was that historical global traffic totals contained sufficient information about the maturity of the industry as a whole to provide a

reasonable basis upon which long-term aviation trends could be projected. There is a question of whether the signals of recent years (i.e., that overall traffic growth is slowing) are sufficiently robust to provide a reliable indication of future long-term growth. A related concern is that historical world traffic totals are dominated by OECD experience, thus may not adequately capture the potential for growth in other, less-developed regions (CAEP/4-FESG, 1998). To a large extent, the FESG scenarios for 2050 reflect assumptions of no fundamental change in overall revenue/cost structure trends of the aviation industry and no fundamental changes in the trends in technology or society. They also assume that the growth of air traffic demand will not be significantly constrained by other limiting factors. Sections 9.6.5 and 9.6.6 examine the availability of infrastructure and fuel with regard to the plausibility of all of the long-term scenario projections.

Growth rates from the model were applied to 1995 reported world traffic demand (Boeing, 1996)—together with GDP growth rates from the IPCC IS92a, IS92e, and IS92c scenarios—to produce FESG base case (Fa), high (Fe), and low (Fc) scenarios of scheduled traffic demand. The high case (Fe) was adjusted slightly to match the NASA traffic forecast for 2015 on which the NASA 2015 emission inventory was based. The basis for the NASA 2015 traffic forecasts were GDP forecasts that were similar to the IS92e GDP scenario (Boeing, 1996). The resulting traffic demand and average growth rate for the three 2050 scenarios are illustrated in Figures 9-16 and 9-17 and listed in Table 9-10. The traffic demand scenarios have

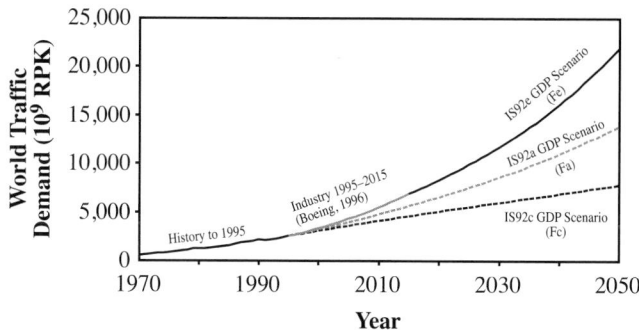

Figure 9-16: ICAO/FESG traffic demand scenarios to 2050 (based on IPCC IS92a, IS92c, and IS92e).

Table 9-9: *Summary of IPCC GDP scenarios used in FESG model.*

| | **Average Annual Global GDP Growth Rate** | |
Scenario	1990–2025	1990–2100
IS92a	2.9%	2.3%
IS92c	2.0%	1.2%
IS92e	3.5%	3.0%

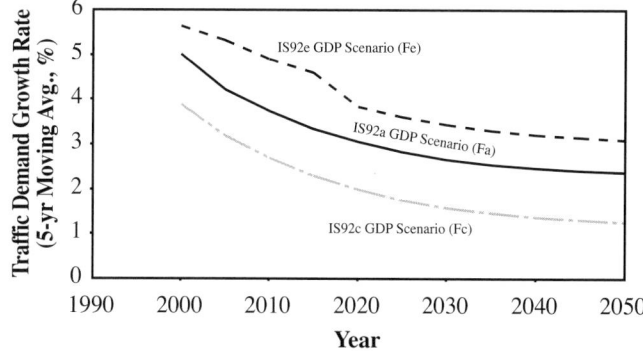

Figure 9-17: Average traffic growth rates from FESG model.

Table 9-10: Traffic projections and 5-year average growth rates from FESG (CAEP/4 – FESG Report 4, 1998).

Year	Fa Demand (10^9 RPK)	Fa Growth Rate (%)	Fc Demand (10^9 RPK)	Fc Growth Rate (%)	Fe Demand (10^9 RPK)	Fe Growth Rate (%)
1995	2,536.6[a]		2,536.6[a]		2,536.6[a]	
2000	3,238.0	5.0	3,068.8	3.9	3,336.1	5.6
2005	3,981.4	4.2	3,591.9	3.2	4,322.4	5.3
2010	4,782.6	3.7	4,103.0	2.7	5,491.7	4.9
2015	5,638.6	3.3	4,596.1	2.3	6,876.2	4.6
2020	6,552.9	3.1	5,070.7	2.0	8,302.4	3.8
2025	7,533.6	2.8	5,530.7	1.8	9,908.5	3.6
2030	8,592.7	2.7	5,981.4	1.6	11,727.0	3.4
2035	9,744.9	2.5	6,429.8	1.5	13,794.9	3.3
2040	11,006.8	2.5	6,881.5	1.4	16,155.8	3.2
2045	12,396.5	2.4	7,342.5	1.3	18,864.2	3.1
2050	13,933.5	2.4	7,817.2	1.3	21,978.2	3.1

[a]Actual reported traffic for 1995.

been labeled Fa through Fe for brevity; these labels, when combined with the appropriate technology assumption designator (1 or 2; see Section 9.4.1.2), form the complete designator for the FESG scenarios used throughout the rest of this report.

Global traffic from the model projections was apportioned over 45 regional traffic flows with a separate market share model because certain regions grow faster than others, and the correct distribution of traffic is important in the calculation of the effects of emissions on the atmosphere. In this procedure, regional traffic flows were expressed as a share of the global market; using the market share and historical growth patterns ensures consistency between regional flows and the global forecast. The underlying assumption of this procedure is that each regional share approaches its ultimate share of the total market asymptotically. Mature markets tend to have declining shares approaching an asymptotic value, whereas developing markets tend to increase their shares. Adjustments of traffic flows were made so that the the "top-down" traffic projections of the FESG global model were matched by a reasonable "bottom-up" distribution of regional traffic flows. These traffic flows include all traffic in all regions, and regional variations in growth rates are highlighted. Factors that affect the operations of military and general aviation aircraft were also estimated, and projections were made of the growth of these sectors (CAEP/4-FESG, 1998).

9.4.1.2. FESG Technology Projections

Calculations of fuel burned and NO_x emissions produced by the 2050 scheduled fleet were made by applying projections of overall improvement in fleet fuel efficiency and emission characteristics to regional traffic flows and summing the results. These projections were created from technology-level estimates for new aircraft over time made by a working group

of the International Coordinating Council of Aerospace Industries Associations (ICCAIA) (Sutkus, 1997); they are discussed in Section 7.5.5. A "fleet rollover" model was used to project a fleet average fuel efficiency trend, using characteristics of the present-day fleet and traffic demand from the FESG scenarios (Greene and Meisenheimer, 1997). The ICCAIA projections were made for two technology scenarios. The first scenario assumes that fuel efficiency and NO_x reduction will be considered in the design of future aircraft in a manner similar to the current design philosophy. The second technology scenario assumes a more aggressive NO_x reduction design strategy that will result in smaller improvements in fuel efficiency. The assumptions associated with the two technology scenarios are given in Table 9-11. The basis for projections of aircraft emissions made by FESG for the year 2050 was the 3-D NASA emissions scenarios for the year 2015 discussed in Section 9.3.2. The NASA 2015 emissions inventory was factored on the basis of the product of the ratios of regional traffic (as departures), fleet fuel efficiency, and fleet EI(NO_x) as calculated for 2050 over the same values in 2015. For all flights in a given region:

$$NO_x \text{ Emissions}_{2050} = NO_x \text{ Emissions}_{2015} \times (\text{regional traffic}_{2050}/ \text{regional traffic}_{2015}) \times (\text{fleet fuel efficiency}_{2050}/ \text{fleet fuel efficiency}_{2015}) \times (\text{fleet EI}(NO_x)_{2050}/ \text{fleet EI}(NO_x)_{2015})$$

Figure 9-18 shows the trend for average new production and fleet average fuel efficiency as a function of time, derived from ICCAIA inputs and the fleet rollover model for the FESG high-demand traffic growth scenario. The average NO_x emission index for the scheduled fleet over the same time period is shown in Figure 9-19. The 2050 fleet average values used in the calculation of emissions from scheduled traffic as well as the baseline 2015 value are given in Table 9-12 (Sutkus, 1997). Fleet fuel efficiency is predicted to improve by about 30% between 2015 and 2050.

Table 9-11: *ICCAIA NO$_x$ and fuel-efficiency technology assumptions for 2050.*

Technology Scenario	Fuel Efficiency Increase by 2050	LTO NO$_x$ Levels
Design for fuel efficiency and NO$_x$ reduction	Average of production aircraft will be 40–50% better relative to 1997 levels	Fleet average will be 10–30% below CAEP/2 limit by 2050; fleet average EI(NO$_x$) = 15.5 in 2050
Design for aggressive NO$_x$ reduction	Average of production aircraft will be 30–40% better relative to 1997 levels	Average of production aircraft will be 30–50% below CAEP/2 limit by 2020 and 50–70% below CAEP/2 limit by 2050; fleet average EI(NO$_x$) = 11.5 in 2050

Traffic in the FSU and the People's Republic of China has not historically been reported in airline schedule databases such as the OAG. Fuel burned and emissions from aviation in these regions were estimated individually and projected to 2015 (Mortlock and Van Alystyne, 1998), then extended to 2050 (CAEP/4-FESG, 1998).

9.4.1.3. FESG Emissions Scenario Results

Results of calculations of fuel burned and NO$_x$ emissions for the year 2050 based on the long-term scenarios described above are given in Table 9-13. The FESG complete scenarios are identified below and in the remainder of this chapter by combining the demand scenario (e.g., Fa) with the technology scenario number (e.g., Fa1, Fe2).

9.4.2. DTI 2050 Scenarios

The DTI projection for air traffic and emissions for 2050 (Newton and Falk, 1997) has been developed from the DTI traffic and fleet forecast demand model, in conjunction with data from the ANCAT/EC2 inventory. The forecast model was developed from DTI's global and regional traffic forecast models for passenger and freight traffic. Fuel consumption trends were estimated with a fleet fuel efficiency model, and fleet emissions performance were estimated on the basis of assumed regulatory change. Finally, appropriate fuel and emissions factors were calculated to estimate 2050 figures

from the base year; these factors were then applied to the 1992 ANCAT/EC2 emissions inventory to produce gridded results for the 2050 scenario.

The DTI model relates air traffic demand in RPKs with regional and global economic performance as reflected in GDP trends, as was the case with the ANCAT/EC2 2015 forecast. Generally, a load factor of 70% is assumed to estimate ASKs (capacity) from traffic demand. Long-term traffic demand is also assumed to be modified by the same assumptions on fares pricing, market maturity, and so forth that the ANCAT/EC2 2015 forecast used. Capacity estimates are converted to fuel consumption

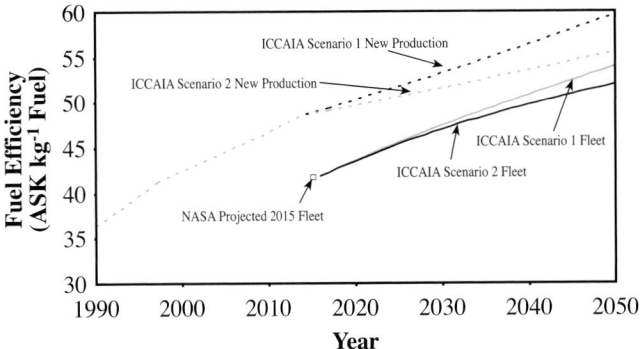

Figure 9-18: Fuel efficiency trends to 2050 corresponding to the two ICCAIA technology scenarios for the FESG high traffic demand case.

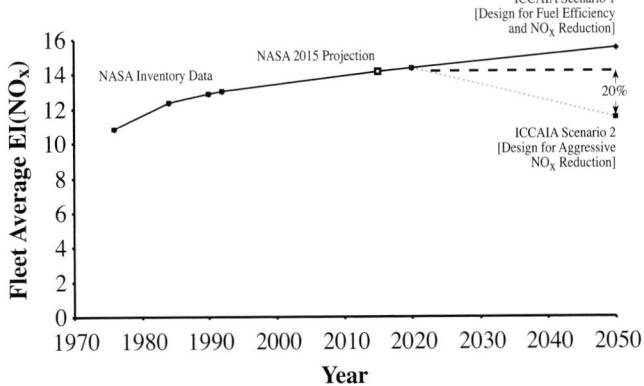

Figure 9-19: Fleet average trends in EI(NO$_x$) showing projections for the two ICCAIA technology scenarios.

Table 9-12: *Projected scheduled fleet fuel efficiency (Sutkus, 1997).*

	Scheduled Fleet Fuel Efficiency (ASK kg⁻¹ Fuel)	
2015 NASA Inventory	41.8	
Traffic Scenario	Technology Scenario 1	Technology Scenario 2
Demand scenario Fa	53.6	51.8
Demand scenario Fc	53.1	51.4
Demand scenario Fe	54.0	52.0

Table 9-13: Results of FESG year 2050 scenarios calculations.

Sector	Fa1	Fa2	Fc1	Fc2	Fe1	Fe2
Calculated Fuel Burned (Tg)						
Scheduled	396.1	410.8	224.0	232.3	620.0	643.9
Charter	21.4	22.2	12.1	12.6	33.5	34.8
FSU/China	30.3	31.4	8.8	9.1	67.5	70.1
General Aviation	8.8	8.8	8.8	8.8	8.8	8.8
Civil Subtotal	456.6	473.2	253.8	262.8	729.8	757.7
Military	14.4	14.4	14.4	14.4	14.4	14.4
Global Total	471.0	487.6	268.2	277.2	744.3	772.1
Calculated CO$_2$ Emissions (Tg C)						
Scheduled	340.7	353.3	192.7	199.7	533.2	553.7
Charter	18.4	19.1	10.4	10.8	28.8	29.9
FSU/China	26.0	27.0	7.5	7.8	58.1	60.3
General Aviation	7.6	7.6	7.6	7.6	7.6	7.6
Civil Subtotal	392.7	407.0	218.2	226.0	627.7	651.6
Military	12.4	12.4	12.4	12.4	12.4	12.4
Global Total	405.1	419.4	230.6	238.4	640.1	664.0
Calculated NO$_x$ Emissions (Tg as NO$_2$)						
Scheduled	6.1	4.7	3.5	2.7	9.6	7.4
Charter	0.4	0.3	0.2	0.2	0.6	0.4
FSU/China	0.5	0.3	0.1	0.1	1.0	0.8
General Aviation	0.1	0.1	0.1	0.1	0.1	0.1
Civil Subtotal	7.0	5.4	3.9	3.0	11.3	8.7
Military	0.1	0.1	0.1	0.1	0.1	0.1
Global Total	7.2	5.5	4.0	3.1	11.4	8.8
Calculated Fleet Average EI(NO$_x$) [g NO$_x$ (as NO$_2$) kg^{-1} fuel burned]						
Scheduled	15.5	11.5	15.5	11.5	15.5	11.5
Charter	16.7	12.4	16.7	12.4	16.8	12.4
FSU/China	14.9	11.1	14.9	11.1	14.9	11.0
General Aviation	9.0	9.0	9.0	9.0	9.0	9.0
Civil Subtotal	15.4	11.5	15.3	11.4	15.4	11.5
Military	8.7	8.7	8.7	8.7	8.7	8.7
Global Total	15.2	11.4	15.0	11.3	15.3	11.4

estimates by using the concept of traffic efficiency as described in Section 9.3.2 and a fuel efficiency trend for the scenario period. Model coverage includes all global aviation markets, but separate fuel consumption estimates are made for freight and for the FSU on the basis of aligning growth with global civil passenger market trends.

The scenario modeled for 2050 assumes that sufficient aviation infrastructure would be available to accommodate the forecast increase in traffic. No new city pairs are introduced during the scenario period, and aircraft flight profiles remain unaltered from the present day; altitude, speed, and method of operation are assumed to be the same as present-day values, even for larger aircraft types (600+ seats) that are assumed to enter service beginning in about 2005. All traffic is assumed to be carried by a subsonic aircraft fleet (i.e., no HSCT would be operating by 2050). The model forecasts traffic growth to be

positive throughout the scenario, but growth rate declines during the period. Decadal capacity growth rates—actual and forecast—are given in Table 9-14. The traffic forecast includes civil and freight operations as well as civil charter and business jet traffic but excludes military aviation activity and possible future supersonic operations.

Fuel usage was determined for the base year fleet from the capacity offered in that year (ASKs) and the fleet's traffic efficiency (ASK per kg fuel). A fuel efficiency trend suggested by Greene (1992) and modified by DTI was included as a scenario parameter, as given in Table 9-15.

The traffic efficiency of the fleet over the scenario period was estimated to range from 30 ASK kg^{-1} in the base year 1992 to 48 ASK kg^{-1} in 2050 (a 60% improvement). This estimate was based on the performance of existing aircraft types and forecasts of the type and number of aircraft (categorized by seat band and technology level) that might be flying in 2050. Future aircraft types included size developments to 799 seats.

A major scenario element was the NO_x reduction technology assumption. Current technology will allow engines to achieve reductions of around 30% below the current certification level (CAEP/2 standards). The basis of the technology scenario was that NO_x regulations would be made considerably more stringent than today and that the manufacturing industry would develop appropriate technology solutions. This development was modeled by assuming that from 1992:

- CAEP/2 certification standard applies to all new production from 2000
- 30% reduction in ICAO recommended limits from CAEP/2 in 2005
- 60% reduction from CAEP/2 phased in equally over 8 years from 2035.

With a fleet development trend determined by the capacity forecast, the rate of introduction of the scenario above implies a global fleet emissions index trend that is as compatible with the relatively modest fuel efficiency assumption given in Table 9-16. The fleet $EI(NO_x)$ of 7.0 implies widespread use of ultra-low NO_x technology (Section 7.5). The total calculated fuel burned and emissions for 2050 under the DTI/ANCAT scenario are given in Table 9-17.

9.4.3. Environmental Defense Fund Long-Term Scenarios

EDF has produced projections of total traffic demand, fuel use, and emissions through 2100 (Vedantham and Oppenheimer, 1994, 1998). The EDF projections use a logistic model to simulate the stages of demand growth in aviation markets, focusing particularly on demand growth in developing countries (where aviation has only recently become a commonplace travel mode). Two sets of aviation demand scenarios—base-level and high-level—describe traffic under each of the six IPCC 1992 scenarios (IS92a through IS92f) for global expectations of gross national product (GNP), population, and emissions (Leggett *et al.*, 1992). Data produced are regional and global totals.

The model logic incorporates the assumption (based on observation) that latent demand in a region previously not served by airlines will result in an initial period of rapid growth; once an airport network is in place, business and personal habits will incorporate the new transport option, causing a period of continuing strong growth rates. Barring unforeseen developments, the experience of some OECD nations suggests that aviation demand will eventually reach maturity, and relative growth rates will slow as the market approaches saturation. Continued growth of GNP and population imply continuing, albeit slow, growth in demand, even over the very long term.

EDF uses a logistic model with a time-varying capacity to model the dynamics in several sectors of rapid expansion,

Table 9-14: *Actual and forecast global capacity growth rates used in the DTI model.*

Year	ASK Annual Global Growth Rate (%)
1994	5.36
2000	5.16
2010	4.82
2020	3.62
2030	3.01
2040	2.49
2050	1.72

Table 9-15: *Assumed annual improvements in fuel efficiency in DTI model.*

Year	Annual Improvement in Fuel Efficiency (%)
1991–2000	1.3 (Greene, 1992)
2001–2010	1.3 (Greene, 1992)
2011–2020	1.0 (DTI extrapolation)
2021–2030	0.5 (DTI extrapolation)
2031–2040	0.5 (DTI extrapolation)
2041 on	0.5 (DTI extrapolation)

Table 9-16: *Trend of civil fleet $EI(NO_x)$ in DTI projections.*

Year	$EI(NO_x)$
1992	11.1
2010	10.73
2020	10.43
2030	10.3
2040	9.5
2050	7.0

Table 9-17: *Results of DTI 2050 projections (military operations not included).*

Scenario	Traffic (10^9 RPK)	Fuel (Tg)	NO_x (Tg NO_2)	$EI(NO_x)$
DTI	18106	633.2	4.45	7.0

continued growth, and eventual slowdown in growth rates without imposing a zero growth-rate ceiling. Growth rates and market capacities for different regions of the world were chosen after a review of economic and aviation market history in industrial nations. The demand model is consistent with the history of the U.S. domestic market.

The EDF model sorts the nations of the world into five economic groups (see Table 9-18). For each of the five economic groups, the three sectors of civil business passenger, civil personal passenger, and civil freight are modeled as logistics with

Table 9-18: Definition of regional economic groups in the EDF model.

Group	Members
1	OECD members, except Japan
2	Asian newly industrialized countries (NICs), Japan
3	China and the rest of Asia
4	Africa, Latin America, Middle East
5	Former Soviet Union (FSU), Eastern Europe

time-varying market capacities. The civil business passenger and civil freight sectors experience logistic expansion toward a time-varying capacity level that is proportional to the nation's GNP.

The model assumes that expansion in business travel is accompanied by expansion in personal travel, which includes tourism and leisure visits. Personal travel by air has high income elasticity, and aviation demand will increase rapidly when a poor nation experiences an economic boom and per capita income increases. Depending on the income distribution, there can be significant demand for aviation even in countries with very low per capita incomes (Atkinson, 1975). As incomes rise and seat prices (as well as cargo costs) fall, growth in aviation demand will result from the penetration of aviation services into lower income brackets (Boeing, 1993). The civil personal passenger sector experiences logistic expansion toward a time-varying capacity level proportional to the nation's population (the model does not account for possible feedback relationships between GNP and population). The military and general aviation sectors do not experience logistic expansion; both sectors grow nominally, at the same rate as global GNP. The mathematical basis of the model and further details on the assumptions are given by Vedantham and Oppenheimer (1994, 1998).

The base-demand and high-demand sets include expected start date for market expansion, market capacity levels, and maturity period length. These assumptions for the two demand sets reflect implicit assumptions about diverse social factors, including travel trends in developing countries (Gould, 1996), penetration of future telecommunications technologies, and development of competing modes of transportation. Assumptions on start dates of aviation market expansion for rapidly developing economies, slowly developing economies, and post-Communist economies reflect EDF's own assessment of near-term economic expectations and were not made in relation to IPCC scenarios. Prior to the start date, demand is assumed to grow nominally, at the same rate as global GNP. The base-demand and high-demand sets include assumptions on market capacity levels based on multiples of 2 (base-demand) and 3 (high-demand) relative to the 1990 demand levels for Economic Group 1 (OECD less Japan), because these markets are closest to maturity today.

EDF's analysis of the history of the U.S. domestic market concluded that there was approximately a 70-year period from

start of market expansion to maturity. The model assumes that nations that are building their airport infrastructure today may well attain market maturity faster because they will benefit from technological improvements and some fraction of their populace will be more familiar with lifestyle and business habits that incorporate aviation. Another region-specific assumption was that markets in the post-Communist economies may mature faster because they have undergone industrialization.

The six IPCC scenarios for GNP and population, combined with the two demand sets described above, provide a total of 10 demand projections (because the IS92a and IS92b scenarios share the same GNP and population expectations). Figure 9-20 shows five of the global demand scenarios; sharp upswings when different regions start expansion are clearly visible. Annotations attached to the curves are shorthand nomenclatures for the scenarios used in this report.

Under the IS92a scenario (the IPCC base case), the base-demand level in 2050 is higher than the 1990 level by a factor of 10.7 and has an average annual demand growth rate of 4.03% over the 60-year forecast period (forecasts to 2100 are given by Vedantham and Oppenheimer, 1998). For the base-demand set, the range of traffic demand expected for different population and GNP estimates spans a factor of almost 5 in 2050; the full range across all 10 scenarios spans a factor of more than 20. Assumptions about rates of expansion and maturity have a sizable impact: The high-demand projection for the IS92a scenario in 2050 is 78% higher than the base-demand value.

The 10 demand scenarios produced by the EDF model are synthesized with expectations for fuel efficiency improvement and changes in emissions indices to produce fuel use, CO_2 emissions, and NO_x emissions scenarios.

Although fuel efficiency has increased steadily over the past few decades, improvements in fuel efficiency are becoming less dramatic over time. The technology projections of the EDF model use a constant-capacity logistic that extrapolates Greene's (1992) forecast for a base-case annual increase

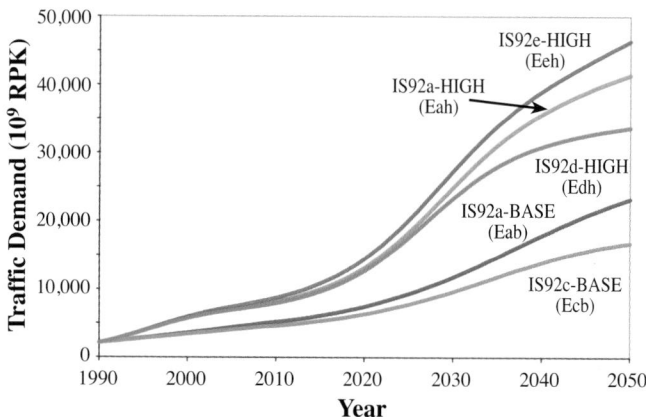

Figure 9-20: EDF global aviation demand projections.

Table 9-19: *Excerpt of EDF results—demand, fuel use, CO_2, % of global CO_2, and NO_x.*

IPCC Scenario	Factor	Year				
		1990	2000	2015	2025	2050
IS92a Base (Eab)	Demand (10^9 RPK)	2,171	3,629	6,115	9,339	23,256
	Fuel Use (Tg)	179	258	374	544	1,143
	CO_2 (Tg C)	154	222	322	468	983
	Percentage of Global CO_2	2.1%	2.6%		3.8%	6.8%
	NO_x (Tg)	1.96	2.57	3.28	4.42	7.88
IS92a High (Eah)	Demand (10^9 RPK)	2,171	5,801	9,954	18,332	41,392
	Fuel Use (Tg)	179	395	610	1,123	2,086
	CO_2 (Tg C)	154	340	525	966	1794
	Percentage of Global CO_2	2.1%	4.1%		7.9%	12.4%
	NO_x (Tg)	1.96	3.92	5.34	9.12	14.39
IS92c Base (Ecb)	Demand (10^9 RPK)	2,171	3,447	5,337	7,802	16,762
	Fuel Use (Tg)	179	243	325	455	837
	CO_2 (Tg C)	154	209	280	391	720
	Percentage of Global CO_2	2.1%	2.8%		4.5%	9.6%
	NO_x (Tg)	1.96	2.42	2.85	3.70	5.77
IS92d High (Edh)	Demand (10^9 RPK)	2,171	5,729	9,647	17,619	33,655
	Fuel Use (Tg)	179	390	592	1,082	1,689
	CO_2 (Tg C)	154	336	510	932	1,453
	Percentage of Global CO_2	2.1%	4.5%		10.0%	16.2%
	NO_x (Tg)	1.96	3.88	5.19	8.79	11.64
IS92e High (Eeh)	Demand (10^9 RPK)	2,171	5,964	10,850	20,202	46,362
	Fuel Use (Tg)	179	408	668	1,234	2,297
	CO_2 (Tg C)	154	351	574	1,061	1,975
	Percentage of Global CO_2	2.1%	3.9%		7.0%	9.8%
	NO_x (Tg)	1.96	4.05	5.85	10.02	15.84

of 1.3% in fleet-wide fuel efficiency from 1989 to 2010. Significant differences in fuel efficiency exist today across regions, and there may be a tendency toward higher fuel efficiency in wealthier regions. The EDF model assumes differences in fuel efficiency across economic groups and builds projections on the assumption that the technology gap between wealthier and poorer nations will close over time.

The NO_x emissions scenarios reflect changes in EI(NO_x) based on a constant-capacity logistic that extrapolates a best-fit approximation to the 1993 NASA numbers for EI(NO_x) in 1990 and 2015 (Stolarski and Wesoky, 1993). The model does not reflect specific technology choices for fuel efficiency or changes in EI(NO_x), although the fleet EI(NO_x) of 6.9 that results from the extrapolation is in the ultra-low technology regime. Results for all scenarios are summarized in Table 9-19.

Figure 9-21 shows CO_2 emissions scenarios [which assume a constant EI(CO_2) of 3.16]. Under the base IS92a scenario, CO_2 emissions grow at an annual rate of 3.2% to reach 983 Tg C in 2050—an increase of a factor of 6.6. For all scenarios, projected

CO_2 emissions climb rapidly after 2015. For the IS92c scenario (which reflects low population and GNP growth) under both demand sets, the level of CO_2 emissions in 2100 is lower than that in 2050, reflecting a successful catch-up effect whereby technological improvements have compensated for demand growth (Vedantham and Oppenheimer, 1998).

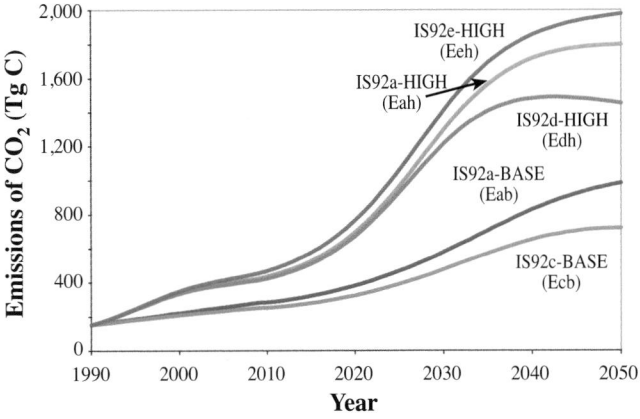

Figure 9-21: EDF CO_2 emissions projections.

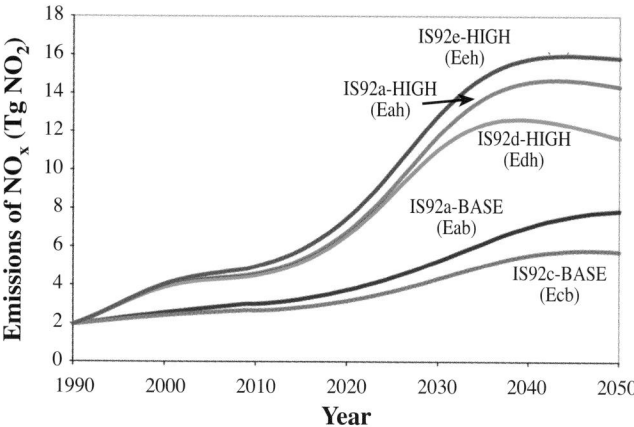

Figure 9-22: EDF NO$_x$ emissions projections.

Comparing the EDF scenarios for aviation's CO$_2$ emissions projections with the IPCC scenarios for total anthropogenic CO$_2$ emissions (including emissions from energy consumption and deforestation) provides a benchmark measure of the environmental importance of the aviation sector. For the base-demand IS92a scenario, aviation's share of global CO$_2$ emissions rises from its current value of 2.1% to a level of 3.8% in 2025 and 6.8% in 2050. Across all scenarios, aviation's share of global CO$_2$ emissions ranges between 3.3 and 10% in 2025 and between 5.6 and 17.6% in 2050. These scenarios imply that aviation may become a significant contributor to global CO$_2$ emissions.

Figure 9-22 shows the NO$_x$ emissions scenarios; these scenarios incorporate the effects of fuel efficiency improvements as well as changes in EI(NO$_x$). For the base-demand IS92a scenario, NO$_x$ emissions rise sharply from almost 2 Tg (as NO$_2$) in 1990 to 7.9 Tg in 2050. Because total NO$_x$ emissions are reduced as a result of fuel efficiency improvements and EI(NO$_x$) reduction, technological improvement can compensate for a greater fraction of demand growth than in the case of CO$_2$ emissions.

Table 9-19 presents an excerpt of EDF model results of traffic demand, fuel burned, and emissions of CO$_2$ and NO$_x$ through the year 2050 for the several sets of assumptions. The three-letter designators for the EDF scenarios (e.g., Eab, Eeh) are used throughout this report.

9.4.4.   World Wide Fund for Nature Long-Term Scenario

A study by WWF addresses future aviation demand by analyzing load factors and capacity constraints, particularly in the freight market (Barrett, 1994). Analysis of historical data shows that increases in the number of seats per aircraft have begun to level off. The study examines the effects of pollution control strategies such as phasing out of air freight and policies to encourage intermodal shifts to road and rail. Technological options for reducing the environmental impact of aviation (such as operational improvements, changes in cruise altitude and alternative fuel sources) are examined. In particular, these models consider the feasibility that increases in load factors

(percentage of total passenger seats that are occupied) could increase fuel efficiency per seat-km for aviation. The model evaluates a wide range of policy and operational choices, including a 100% load factor and a 100% fuel tax.

The model includes explicit assumptions of fixed growth rates in leisure travel, business travel, average trip length for passenger and freight traffic, and freight tonnage. It assumes that passenger load factors rise to 75% by 2020 in the base case. Constant rates of improvement are assumed for aircraft size, airframe efficiency, and EI(NO$_x$).

With an annual growth rate of 5.2%, demand rises by a factor of more than 12 between 1991 and 2041 in the "business-as-usual" case. Proposed policies, including changes in load factor, and technological improvements result in a forecast for demand increase of about a factor of 3 in the "demand management" case. Carbon emissions in 2041 constitute 550 Tg C, and aviation's share of global carbon emissions rises to 15% by 2041.

9.4.5.   Massachusetts Institute of Technology Long-Term Scenarios

A study of the long-term future mobility of the world population has been undertaken at MIT. This study constructed scenarios based on the simple yet powerful assumption that time spent and share of expenditures on travel remain constant (Zahavi, 1981), on average, over time and across regions of the globe (Schafer and Victor, 1997). Stability of average time budgets for travel (motorized and nonmotorized) is substantiated by a considerable amount of aggregate historical data. Although there is some variability in travel budgets from poorer to richer nations, within each society travel budgets have generally followed a predictable pattern—rising with income and motorization and stabilizing at 10–15%.

Using the constant travel budget hypothesis, Schafer and Victor (1997) produced global passenger mobility scenarios for 11 world regions and four transport modes for the period 1990–2050. Adding estimates of changes in the energy intensity of transportation modes, they also generated scenarios of CO$_2$ emissions from passenger transport (see Table 9-20).

The high-speed travel category includes aviation, but the aviation portion of high-speed travel is not explicitly characterized. Results of this model projection therefore cannot be used directly in evaluations of the effect of aviation on the atmosphere, nor can they be directly compared to other long-term projections of emissions from aviation.

9.5.   High-Speed Civil Transport (HSCT) Scenarios

The technology for commercial (supersonic) HSCT is being developed in the United States, Europe, and Japan. The goal is to develop an aircraft that can carry approximately 300 passengers, with a 9,260-km range, cruising at Mach 2.0–2.4 at altitudes of 18–20 km. As described in Chapter 7, NASA has an aggressive technology program to develop combustors with NO_x emission levels of 5 g NO_x (as NO_2) per kg fuel burned at supersonic cruise conditions. The HSCT is expected to fly supersonically only over water because of the need to mitigate sonic booms over populated land masses. The potential market for the HSCT is limited by economic and environmental considerations.

9.5.1. Description of Methods

3-D emissions inventories of fuel burned, NO_x, CO, and unburned HC for fleets of 500 and 1,000 active (high utilization) HSCTs have been developed based on market penetration models and forecasts of air traffic in 2015 (Baughcum *et al.*, 1994; Baughcum and Henderson, 1995, 1998). Although such large fleets clearly will not be in operation by 2015, the year was chosen as a base year because detailed industry projections of air traffic on a route-by-route basis are available only to that time period. Although the introduction of an economical HSCT may stimulate total traffic growth by an unknown amount, the HSCT will certainly displace some traffic from the subsonic fleet on major long-range intercontinental routes. For this study, possible stimulative effects were ignored to reduce the number of variables, and HSCT-generated RPKs were explicitly substituted for subsonic RPKs on a route-by-route basis.

The most recent set of scenarios based on the NASA technology concept aircraft (TCA) HSCT were used for most of the atmospheric impact calculations presented in Chapter 4. It is not clear when HSCT technology will be mature enough for viable commercial service, so fleet sizes and technology levels are treated parametrically.

The projected flight tracks for a fleet of 500 HSCTs above 13-km altitude are shown in Figure 9-23. Because of its speed advantage over subsonic aircraft, the HSCT would likely be used primarily on long intercontinental routes, where that advantage can best be utilized. Because of the sonic boom that trails below the aircraft, the best HSCT routes have a large portion of the flight path over water. These conditions combine to put a majority of HSCT routes at northern mid-latitudes over the North Atlantic and North Pacific.

To project the HSCT fleets and their displacement of subsonic aircraft in the scenarios to 2050, the following procedure was used:

1) 3-D displacement scenarios of subsonic traffic by a fleet of 1,000 active HSCTs was calculated for the year 2015 using differences in the 3-D scenarios calculated for the NASA all-subsonic fleet (Baughcum *et al.*, 1998) and the NASA subsonic fleet in the presence of an HSCT fleet (Baughcum and Henderson, 1998).

2) This subsonic displacement scenario was then scaled for the technology growth factors described in the discussion of the FESG scenario and combined with the HSCT only-scenario (assuming the TCA technology level) and 2050 all-subsonic scenarios.

The 1,000-unit fleet should not be considered a forecast of the actual number of HSCTs that might be in the fleet in 2050. For this sensitivity study, the 1,000-unit value was chosen to represent a fleet that would be the result of a successful HSCT program; this fleet size also was chosen so that previous fleet projections could be used (Baughcum and Henderson, 1998). No changes in fuel efficiency or NO_x emissions technology relative to the assumptions used in the reference were assumed

Table 9-20: Results of MIT reference scenario–passenger travel and carbon emissions.

	1990 (10^12 pkm[a])	2050 (10^12 pkm)	1990 (10^12 mt C)	2050 (10^12 mt C)
Industrialized				
High-Speed	1.5	32.7	0.09	0.66
Total	12.4	44.4	0.52	1.12
Reforming				
High-Speed	0.3	2.1	0.02	0.04
Total	2.3	7.1	0.07	0.20
Developing				
High-Speed	0.4	7.2	0.02	0.14
Total	8.6	53.8	0.18	1.29
World				
High-Speed	2.2	42.0	0.13	0.84
Total	23.3	105.3	0.77	2.61

Source: Schafer and Victor (1997); additional data supplied by David Victor (June 1998).
[a]pkm=passenger kilometers.

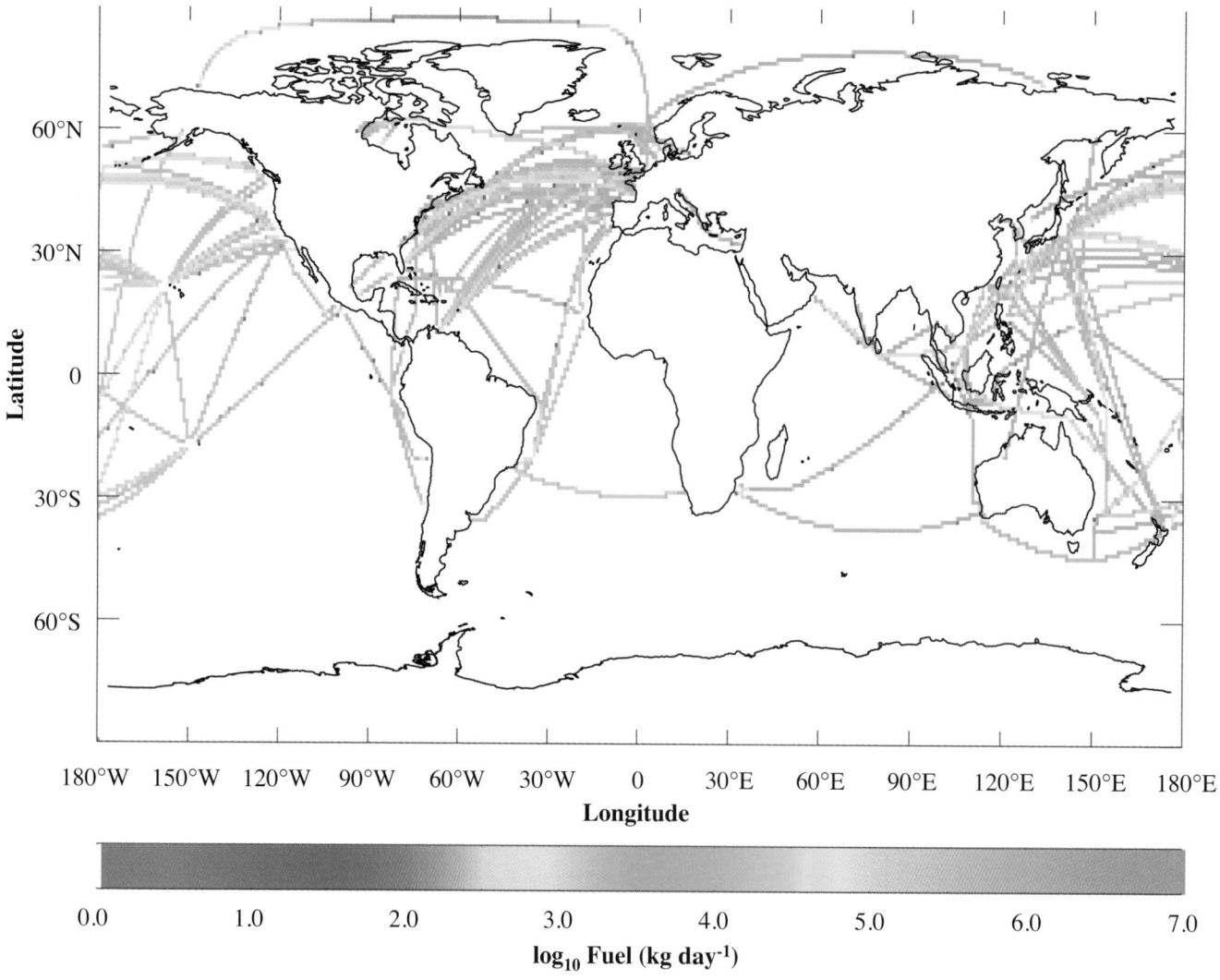

Figure 9-23: Flight tracks above 13-km altitude for a fleet of 500 high-speed civil transports (Baughcum and Henderson, 1998).

for the 2050 HSCT. A detailed description of the route system flown by the 1,000 HSCTs is given by Baughcum and Henderson (1998).

9.5.2. Description of Results

Fleet fuel burned with the HSCT was calculated by assuming that the fuel efficiency and NO_x emissions of the subsonic fleet were described by NO_x technology scenario 1, the "fuel efficiency" scenario. Table 9-21 gives the total fleet fuel burned and NO_x emissions with and without the assumed 1,000-unit HSCT fleet. Fleet fuel burned increases as a result of the substitution of less fuel efficient HSCTs for subsonic airplanes (present HSCT designs have about half the fuel efficiency, measured as RPK per fuel burned, of present subsonic airplanes). However, fleet NO_x emissions decrease in spite of the increase in fuel burned because the HSCT is assumed to be designed for very low NO_x emissions [cruise $EI(NO_x)$ of 5].

A comparison of the altitudinal distributions of fuel use and NO_x emissions between the all-subsonic fleet and a fleet

containing subsonic and HSCT aircraft is shown in Figures 9-24 and 9-25 for the FESG year 2050 IS92a scenario. The introduction of an HSCT fleet with $EI(NO_x)=5$ combustors would be expected to increase emissions above 12-km altitude and lead to a decrease of NO_x emissions below 12, particularly in the 10-12 km band, assuming that the introduction of an HSCT will cause a displacement of subsonic traffic.

9.6.   Evaluation and Assessment of Long-Term Subsonic Scenarios

9.6.1.   Difficulties in Constructing Long-Term Scenarios

Long-term (beyond 20 years) projections of aviation traffic demand, fleet fuel burned, and fleet emissions are inevitably speculative. Difficulty in forecasting technological developments that might be appropriate for the long term, possible shifts in traffic demand, and myriad uncertainties resulting from human society's development over the period in question all conspire to make long-term projections unreliable—sometimes astoundingly so. Given the state of the aviation industry 50

Table 9-21: Results of substitution of 1,000-unit parametric HSCT fleet in 2050.

Scenario	Fuel (Tg)	CO₂ (Tg as C)	% Change (Fuel)	NOₓ (Tg as NO₂)	% Change (NOₓ)	Fleet EI(NOₓ)
Fa1—All Subsonic	471	405	Base	7.2	Base	15.2
Fa1H—With 1,000 active HSCTs	557	479	+18	7.0	-2	12.6
Fe1—All Subsonic	744	641	Base	11.4	Base	15.3
Fe1H—With 1,000 active HSCTs	831	715	+12	11.3	-1	13.6

years ago (in 1947), it is doubtful that either the technology or the scope of the industry in 1997 could have been forecast. However, because the transport aviation market and aviation technology seem to be maturing, a plausible way of making projections far into the future is to make reasonable extrapolations based on our knowledge of present trends in the world and in the aviation industry. These extrapolations are termed scenarios, rather than forecasts, as outlined in Section 9.1.

9.6.2. Structure and Assumptions

Before we review the outcomes of the scenario studies in the following section, we consider some differences and similarities between the models. This comparison is restricted to the EDF, DTI, and FESG models. Although the MIT model provides an interesting insight into future travel options based on the thesis of invariant travel time and travel expenditure budgets, it is

excluded from this comparison because it provides only a highly aggregated scenario for the future mobility of total motorized passenger traffic; air traffic is only one—albeit important—portion of this picture, and the aircraft component cannot be identified. The WWF aviation scenario for 2041 provides aggregated fuel burned and CO₂ emissions projections but does not provide regionally distributed NOₓ emissions estimates.

Of the long-term scenarios considered, the EDF, FESG, and DTI studies allow assessment of the impacts of CO₂ from aviation. However, only the results from the DTI and the FESG models are suitable for use in chemical transport models for modeling other emissions (see Chapters 2 and 4) and their effects on radiative forcing (see Chapter 6) because they provide gridded data that include a consideration of the potential changes in the spatial distribution of emissions. Only the EDF study provides scenarios for demand from the aviation sector and subsequent global CO₂ and NOₓ emissions to 2100.

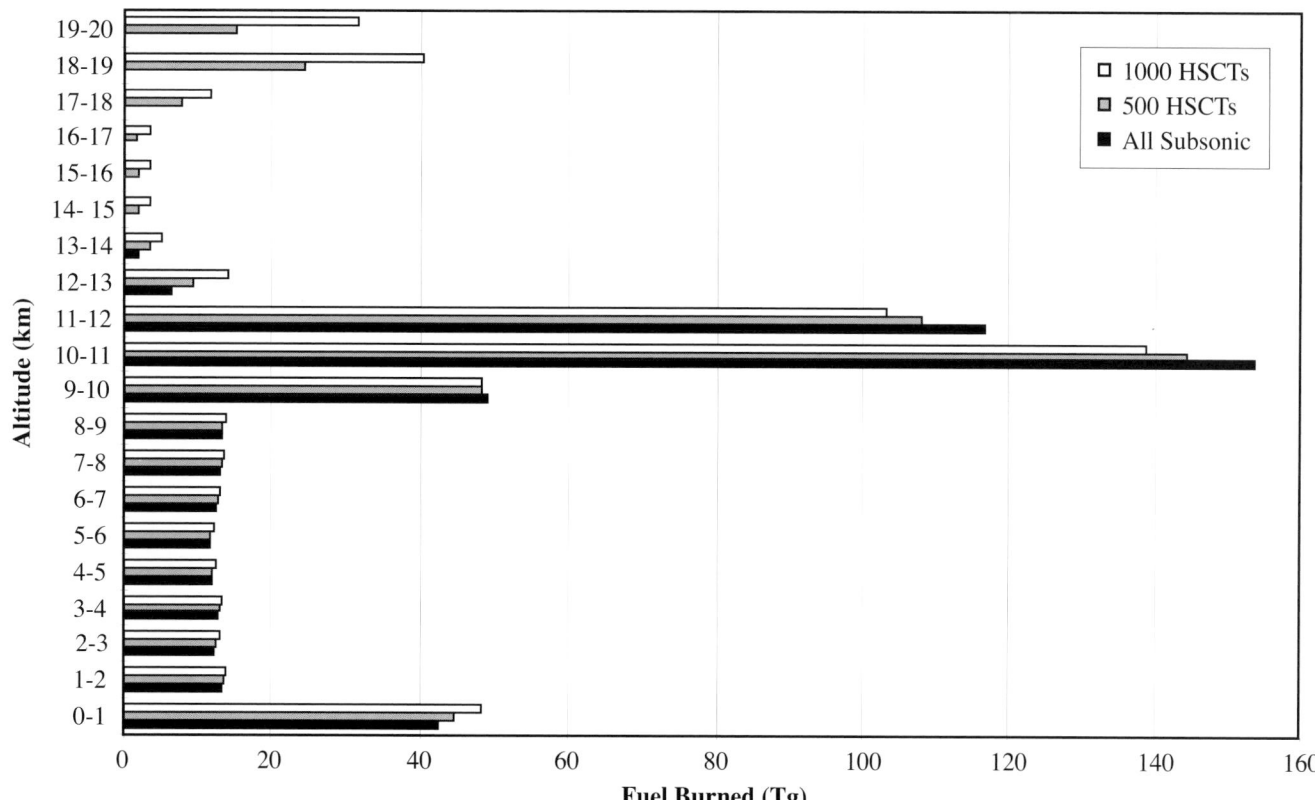

Figure 9-24: Altitude distribution of fuel burned—with and without HSCT fleet—based on IS92a scenario (Fa1,2).

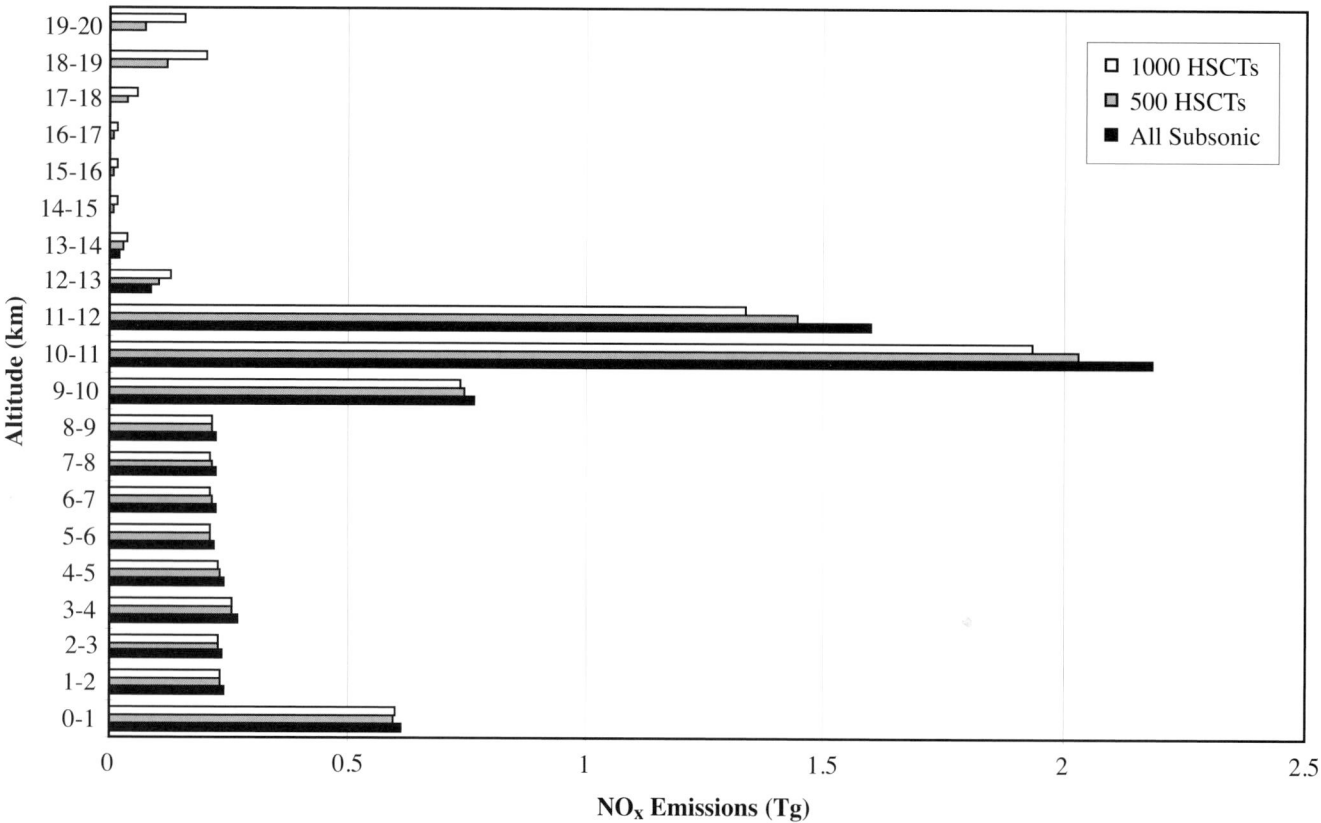

Figure 9-25: Altitude distribution of NO_x emissions—with and without HSCT fleet—based on IS92a scenario (Fa1, FaH).

The EDF study provided 10 scenarios based on five different IPCC IS92 world scenarios for the long-term development of world economy and population and two air traffic demand scenarios (base case and high case). The FESG study calculated three air traffic demand scenarios based on the IPCC IS92a, IS92c, and IS92e world scenarios, which were combined with two engine technology scenarios to produce six different emissions inventories.

The FESG scenarios of regional and global air traffic were based on a logistic regression model of traffic demand since 1960 using global GDP as a predictor. The FESG model used a combination of top-down and bottom-up approaches, in which global volumes of civil aircraft flight kilometers were predicted using the regression model for different GDP scenarios. All available information on regions, including regional variation in growth, was then used to disaggregate these global values in a consistent way over 45 traffic flows within and between the regions of the world by using a market share allocation model. Year 2050 values of fuel burned and NO_x emissions for military traffic were estimated separately.

The EDF scenarios also were based on the use of logistic growth curves to model air traffic growth for business and personal travel (plus military and freight traffic). Model parameters were chosen through observation of historical traffic trends in the United States. Regional population was used as a predictor of personal passenger travel, and regional GNP was used as a predictor of business passenger travel and

freight demand. Both the FESG and EDF models incorporate the underlying assumption that the chosen parameters are satisfactory predictors of aviation demand and that aviation markets eventually mature.

There are large differences between the EDF and FESG models with respect to the development of emissions scenarios. The EDF model uses a constant capacity logistic to describe fuel efficiency improvements, which extrapolates Greene's (1992) forecast to 2010 with varied rates for five geographic world regions and the military/freight aviation sector. For the trend in fleet $EI(NO_x)$ a single global logistic model extrapolates from the 1990 and 2015 values. The FESG scenarios are based on two engine technology scenarios developed by ICCAIA for ICAO/FESG and IPCC (see Chapter 7). These scenarios represent an industry perspective on likely future developments in fuel efficiency and NO_x reduction technologies, as well as further potentials and limitations. The fuel efficiency technology element of the DTI scenario was similar in this respect, but a NO_x technology scenario appropriate to stricter emissions regulations was assumed, in which subsonic engine research programs would deliver emissions levels similar to those targeted in the NASA HSCT program.

Additional assumptions are also important to the results of the scenario models. In the EDF model, assumptions about the dates of market expansion and maturity and the ultimate capacity levels chosen for the economic regions strongly influence the outcomes.

The EDF, FESG, and DTI models all use statistics of traffic/air traffic from international organizations and OECD countries, as well as numerous other recently published sources, and adopt one or more of the IPCC IS92 scenarios to describe the long-term development of worldwide economic growth and population. The FESG, EDF, and DTI models also use information from the NASA and ANCAT/EC gridded inventories of traffic flows and related emissions. The FESG models used new, partly proprietary, information from industry as a base to project emissions in the year 2050.

None of the 3-D gridded inventories for 2050 assume any changes in design that would alter the cruise altitudes of subsonic aircraft. Furthermore, no consideration was given in any of the 2050 scenarios reported here to the possible stimulative (or otherwise) effect of HSCT introductions on traffic.

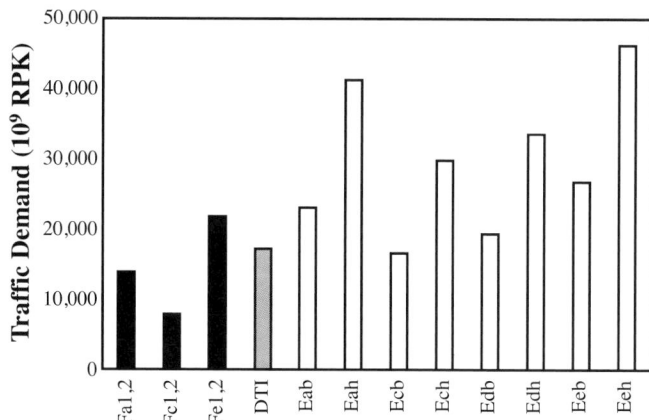

Figure 9-26: Comparison of traffic demand in 2050.

9.6.3. Traffic Demand

Total traffic demand projected for the year 2050 for three of the long-range scenarios is shown in Figure 9-26. The values shown for the FESG model projection do not include military or general aviation traffic. Military and general aviation fuel burned and emissions were estimated separately for the year 2050; they were 3.1 and 1.8% of total fuel burned, respectively (Fa1,2). The demand values shown for the EDF model include military as well as freight demand, with projected billion tonne-km values converted to RPK. The DTI model includes passenger, freight, and business jet traffic but excludes military operations. The WWF model includes passenger and freight only, but a demand value for 2041 was not published.

Although the FESG and EDF models use the same IS92 economic scenarios (IS92 population scenarios also are inputs to the EDF model), the traffic demand projections for 2050 from the EDF model are higher than those of the FESG model by a factor of 1.2 to almost 4, depending on the scenario. The DTI model, which does not directly depend on the IS92 scenarios, projects a traffic demand about 80% that of the FESG high case (Fe1,2).

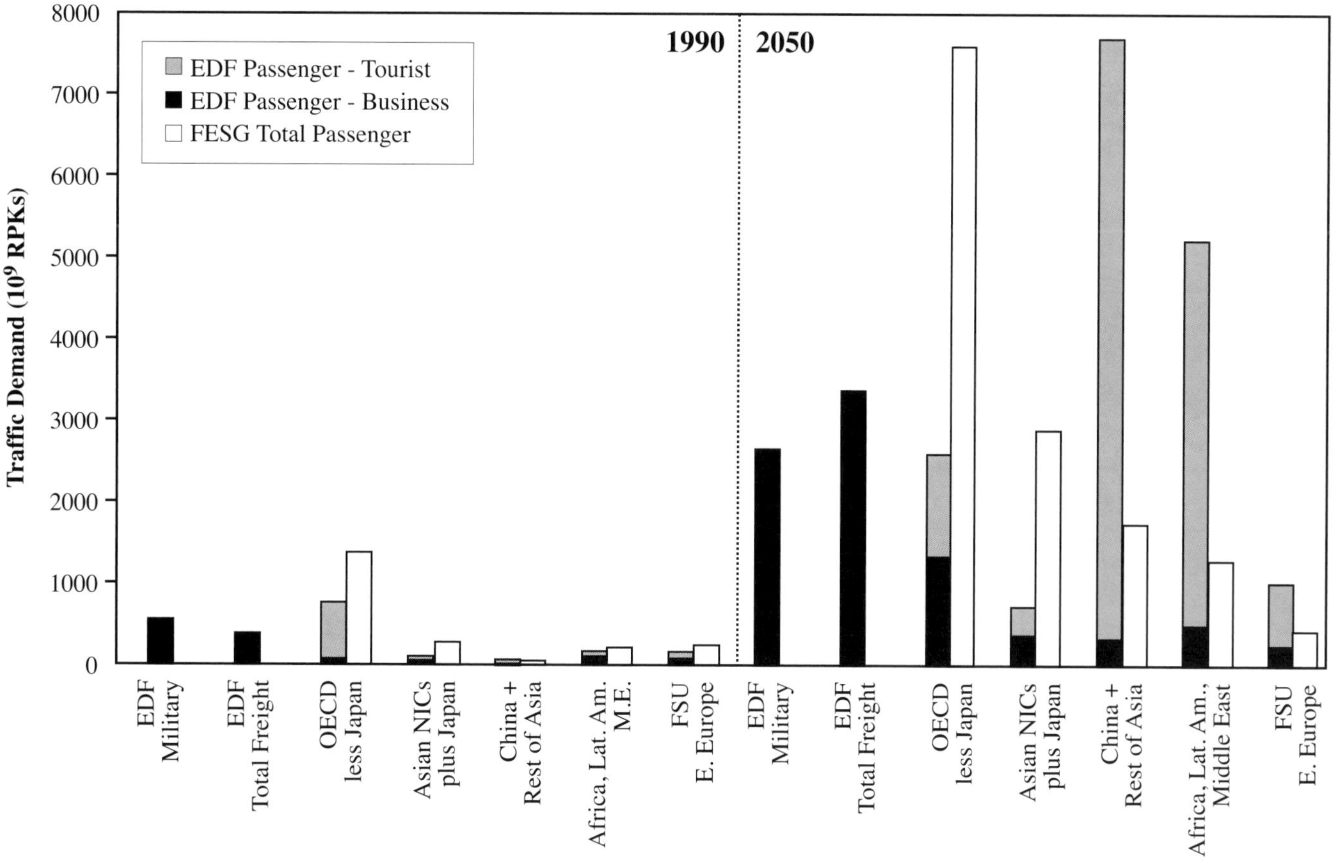

Figure 9-27: Comparison of 1990 and 2050 regional demand values based on EDF and FESG models (IS92a scenario).

Table 9-22: Comparison of FESG and EDF model results for year 2050 based on IS92a.

Region	1990 % World GNP	1990 % World Population	1990 FESG % Demand	1990 EDF % Demand	2050 % World GNP	2050 % World Population	2050 FESG % Demand	2050 EDF % Demand
1) OECD, less Japan	57	12	63	62	45	8	55	15
2) Asian NICs + Japan	16	3	13	5	13	2	21	4
3) China, Rest of Asia	6	52	2	5	15	49	12	45
4) Africa, Latin America, Middle East	9	25	10	14	18	37	9	30
5) FSU, Eastern Europe	12	7	11	14	9	5	3	6

Clues to the reasons behind the large differences in projected traffic demand between the FESG and EDF models can be found by examining the details of the results of each model. The EDF model projects passenger business and personal traffic in five world regions, plus military and freight traffic. To make comparisons between the two models, the 45 traffic demand flows (allocated from the global growth projection) in the FESG model were assigned to the five regions used in the EDF model. Demand flows between two of the EDF-defined regions were allocated by assigning 50% of the FESG traffic demand to each region.

On the previous page, Figure 9-27 shows a comparison of traffic demand in 1990 and 2050 from the FESG and EDF models, with demand sorted by region and/or type. The EDF base case demand (Eab) is compared with the Fa1,2 demand scenario. Large differences in the distribution of demand between the two models are apparent: The FESG model assigns the largest share of passenger traffic in 2050 to the OECD area, whereas the EDF model assigns the largest share to the China-Africa area (with personal travel making up the bulk of the demand).

Table 9-22 provides data on passenger demand from the EDF and FESG models by region in 1990 and 2050 and regional

distribution of world GNP and population over the same time periods. The basis of both models in this comparison is the IS92a GNP and population scenario. In 1990, the demand distributions of both models are roughly the same and reflect to a great extent the regional distribution of GNP. In 2050, the regional demand distribution from the FESG model reflects the shift in GNP distribution, demonstrating the economics-driven basis of the FESG model. The 2050 FESG values also show that the market share tool has probably underestimated the share of demand in region 4; percentage of GNP has increased from 1990 to 2050, but percentage of demand has decreased.

In contrast, the 2050 demand distribution from the EDF model differs greatly from the distribution of GNP in 2050 and reflects the population-driven basis of much of the EDF model.

The differences between the FESG and EDF models are further illustrated in Figures 9-28 and 9-29, which show the cumulative distribution of traffic growth over time for the IS92a scenario. Figure 9-28 shows the growth and regional proportions of traffic demand as projected by the FESG model. The shares of demand reflect the GDP of each region. Figure 9-29 shows the cumulative distribution of demand for the five regions as projected by the EDF model. The EDF model, unlike the FESG model, projects business and personal passenger demand separately (business demand is a function of GNP; personal demand is a function of population); both sectors of demand are shown in the figure. Notable is the lack of projected growth in personal demand in region 1 (OECD less Japan). Driven by projected slow growth and eventual decline in OECD population, demand growth in this sector is projected to be less than 1% per year after 2005 and negative after 2035. Notable also is the relative lack of growth projected for region 2 (Asian newly industrialized countries + Japan). The effect of the population-driven personal demand sector is shown by the rapid growth in regions 3 (China + rest of Asia) and 4 (Africa, Latin America, Middle East). Personal demand in these two regions is projected by the EDF model to grow at rates exceeding 12% per year for 25 years (region 3) and 10% per year for 20 years (region 4) to create 75% of total passenger demand in 2050 (up from 19% in

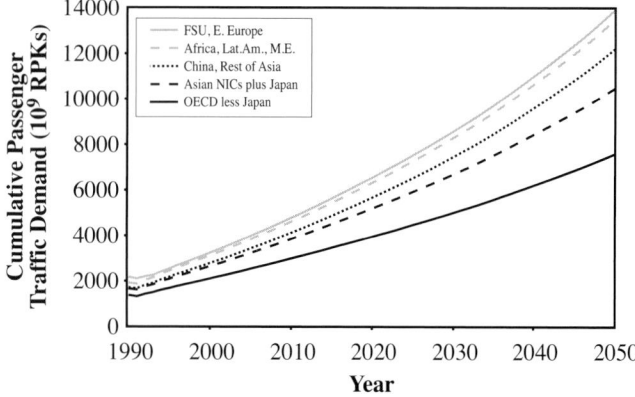

Figure 9-28: Cumulative traffic demand (IS92a scenario, Fa1,2).

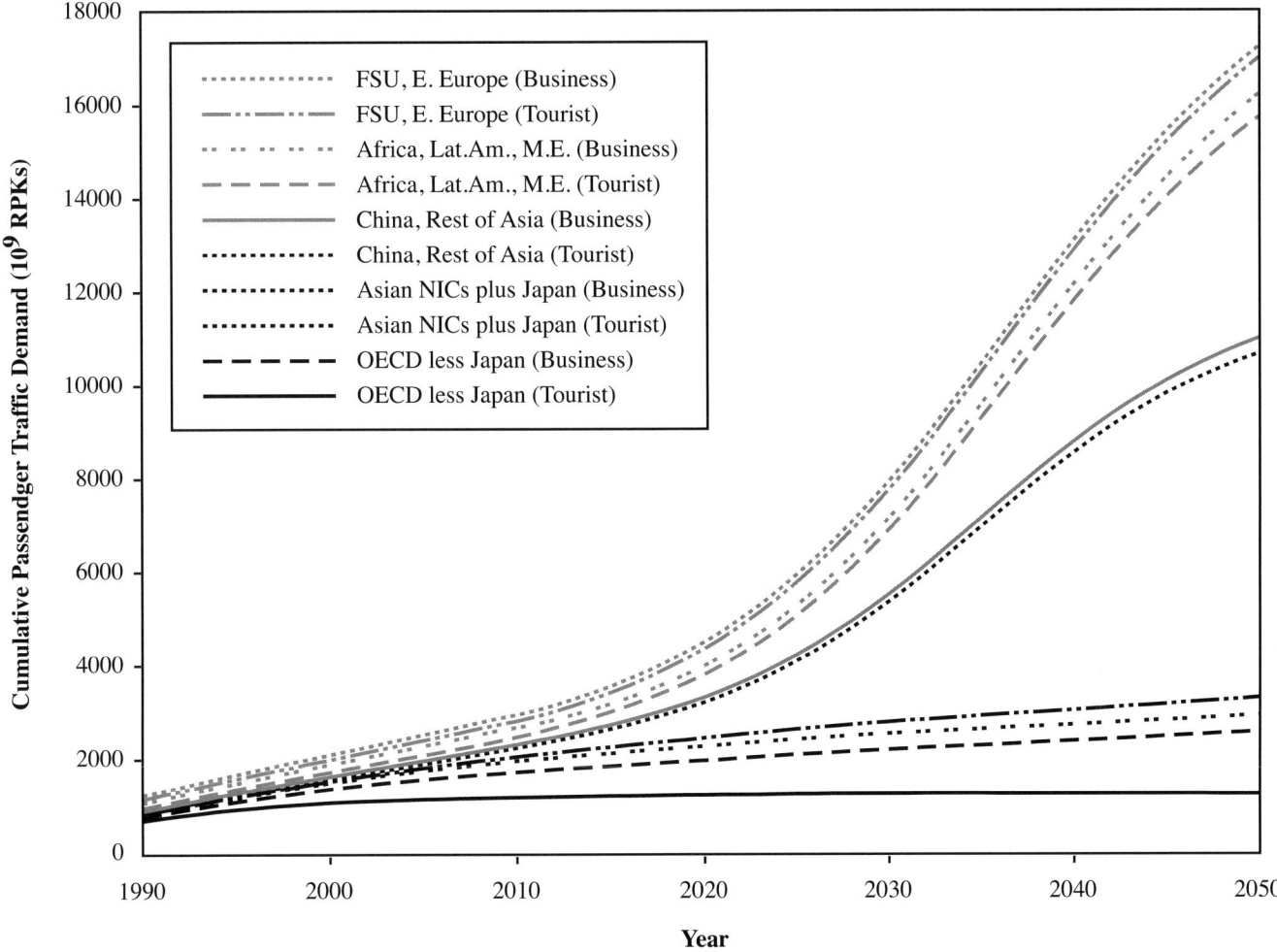

Figure 9-29: Distribution of passenger demand (IS92a scenario, Eab).

1990). This value contrasts with the 21% of total passenger demand projected for these two regions in the FESG model, based on the two regions' 33% share of GNP.

9.6.4. NOₓ Technology Projections

A list of NO_x emissions index projections is given in Table 9-23 for the three long-term models (IS92a scenarios).

The fleet $EI(NO_x)$ in 1992 was calculated as 12.0 (NASA), 13.8 (ANCAT/EC2), and 13.9 (DLR).

Expectations for the development of NO_x technology are quite different among the models. The two FESG model NO_x technology estimates were based on ICCAIA technology projections for new aircraft (Sutkus, 1997) and estimates of how quickly such new technology would enter the fleet (Greene and Meisenheimer, 1997). The assumptions in the DTI model were that regulatory pressures would require reductions in NO_x emissions, and the fleet emissions index would be forced down as the engine industry responded with specific technology developments through 2035. These developments assumed the introduction of emission control

technology that would produce engine emissions indices appropriate to those anticipated for staged combustor and ultra-low NO_x combustor technology—the latter of the type being developed for HSCT applications [$EI(NO_x)$ =5]. Ultra-low NO_x technology concepts now being developed may not be suitable for future high pressure ratio subsonic engine designs, so achieving fleet NO_x emission levels assumed in the DTI and EDF models may be very difficult (see Section 7.5).

The EDF model used a logistic extrapolation of NO_x trends from NASA work (Stolarski and Wesoky, 1993), but no changes in technology were explicitly specified.

Table 9-23: Comparison of fleet $EI(NO_x)$ from technology projections to 2050.

Scenario	Fleet $EI(NO_x)$
Fa1	15.5
Fa2	11.5
DTI	7.0
Eab	6.9

9.6.5. *Infrastructure and Fuel Availability Assumptions*

All of the long-term scenarios reviewed in this chapter were developed with the implicit assumption that sufficient system infrastructure and capacity will be available to handle the demand in an unconstrained fashion (infrastructure and capacity are defined for airports as runways, terminals, gates and aprons, roads, etc., and for airways as air navigation services, air traffic control, etc.). However, lack of infrastructure development may well impede future aviation growth. Lack of infrastructure will result in congestion and delay, additional fuel burn (in the air and on the ground), higher operating costs, higher ticket prices, and reduced service.

In some parts of the world, particularly in North America and Europe, the airway and airports system is currently operating under constraints that limit its ability to provide service. These constraints are likely to become more acute in the future as the demand for aviation services continues to grow. Congestion resulting from capacity constraints impairs the economic and environmental performance of airlines and the entire aviation system. To accommodate future demand, physical and technological infrastructure must be upgraded and expanded. In many areas, however, strong local pressures (especially related to noise created by aircraft movements) have constrained development of new airports and capacity improvements at existing airports. It is therefore important to note that the traffic forecasts reviewed in this chapter are all unconstrained forecasts that do not evaluate system capacity constraints when estimating future traffic growth.

Aviation also depends on petroleum fuels. For the past 50 years, known reserves of petroleum have continued to expand to satisfy 20–30 years of predicted demand. Over the short-term future, little change in the demand/supply situation is expected. Oil companies predict continued supply of their raw material, and kerosene supplies should have similar availability as the present day. Despite the forecast for increasing demand, oil prices are projected to rise only moderately over the next 20 years (Hutzler and Andersen, 1997). Over the period of these scenarios (to 2050), estimates of availability are less clear, but there is a general view that the oil industry will continue to meet demand (Rogner, 1997). There are, however, less optimistic views for oil production, with some predictions of a production decline occurring within the next decade (Campbell and Laherrere, 1998). The long-term scenarios assessed for this report implicitly assume continued availability of fuel at moderate prices. This is a key assumption for all scenarios because large increases in the price of fuel and/or shortages in supply would act to restrain demand for passenger and cargo air transport.

All of the scenarios ignore (in their baseline assumptions) possible changes in service patterns or infrastructure that a future HSCT might require. The effects of an HSCT fleet are considered in Section 9.5.

9.6.6. *Plausibility Checks*

Although none of the long-term scenarios reviewed here is considered impossible, some may be more plausible than others. We devised three simple checks to assess plausibility. The first estimated the fleet size required to carry projected traffic in 2050; the second examined implications for airport and infrastructure; and the third examined implications for kerosene demand. These plausibility checks represent an initial examination of the implications of the scenarios and are intended to illustrate possible consequences of traffic estimates resulting from the different scenarios. It must be emphasized that the fleet numbers produced by this analysis are approximate and are provided for comparative purposes only.

9.6.6.1. *Fleet Size*

The fleet sizes implied by five of the scenarios were determined from the DTI traffic and fleet forecast model (see Section 9.3.2), which was developed primarily to project demand for new aircraft implied by 25-year traffic forecasts. The DTI model requires an annual traffic growth rate as an input; for this assessment purpose, this value was assumed to be a constant annual rate calculated from the base year traffic and the model's projection for 2050. The model assumes the fleet to comprise a range of jet aircraft types, described by seat capacity as follows: 80–99, 100–124, 125–159, 160–199, 200–249, 250–314, 315–399, 400–499, 500–624, and 625–799. The larger aircraft sizes have yet to be produced but are assumed to enter service beginning about 2005. Regional variations in fleet composition are reflected in the global fleet, based on current trends. This analysis does not capture the effects of compositional change that could be created as new markets develop. Average aircraft size growth is assumed (reflecting the historical trend of greater seating capacity for individual aircraft types over time). The future fleet required to satisfy the scenario demand estimates is derived through an iterative process by matching capacity to traffic demand, based on assumptions regarding aircraft unit productivity in capacity terms. Other model assumptions are as follows:

* Subsonic aircraft supply-projected demand (no supersonics)
* A short- and long-haul market share
* Future market functions as does the present day (i.e., no assumptions regarding wider deregulation are made)
* Unconstrained demand
* Aircraft retired at an average age of 25 years (reduced productivity from 20–30 years)
* Aircraft productivity to improve by an average of 0.75% annually.

The assumption regarding lack of constraints requires comment. Today's civil aviation market is constrained only by the practical limitations of airport capacity and access restrictions, airspace restrictions, and economic restraint resulting from taxation, charges, and so forth that affect ticket price. Any constraints in the future, whether to address environmental problems or as a result of government policy, will affect or limit demand and therefore affect the emissions burden from civil aviation. In contrast, measures such as the introduction of advanced air traffic control systems may improve the efficiency of traffic management (see Chapter 8) but could lead to a traffic increase, with the consequence of increasing emissions from aircraft. Neither the scenarios nor the analysis of their impact have examined such possibilities because there would be too many permutations of possibilities to define a scenario acceptable to all.

Table 9-24 summarizes the estimated traffic in RPK x 10⁹ for five of the scenarios. The global fleets (numbers of aircraft of all types) appropriate for each traffic estimate are also given.

In addition, it is necessary to consider the extent to which the freighter fleet might grow. An independent study was performed using figures for the current inventory of freighters and extrapolating the Boeing freighter forecast from 2015 at two growth rates—5.1% (high) and 2.5% (low). Assuming the high growth rate, the freighter fleet could grow to approximately 19,000 aircraft by 2050. The low-growth rate would require 8,000 freighter aircraft (Campbell-Hill Aviation Group, 1998). This calculation results in the adjusted commercial fleet profile given in Table 9-25.

9.6.6.2. Airport and Infrastructure Implications

From the implied 2050 fleet sizes, it is likely that more airports will be required to support the growth in traffic and flight operations. The number of new airports required will depend on the total fleet, the number of airports now capable of handling jet operations, and the number of gates available at each airport. Table 9-26 shows the number of new airports required, assuming the lowest and highest growth cases for the passenger fleet and total fleet (from model by Campbell-Hill Aviation Group, 1998).

Table 9-24*: Projected traffic and size of global fleet.*

| | Traffic (RPK x 10⁹) and Global Fleet | | |
Scenario	Year 1990 (*1995)	Year 2050	Fleet[a]
Eab	2,171	23,256	35,000
Eeh	2,171	46,362	69,000
Fa1,2	2,536*	13,934	21,000
Fc1,2	2,536*	7,800	15,000
DTI	2,553	18,106	30,000

[a]Passenger fleet rounded to nearest 1000.

Table 9-25*: Effect of freighter fleet on 2050 total fleet size.*

| Sector | Current Fleet | 2050 Fleet Forecast[a] | |
		Lowest Growth	Highest Growth
Passenger	10,000	15,000 (Fc1,2)	69,000 (Eeh)
Freighter	1,347	8,000 (2.5% growth)	19,000 (5.1% growth)
Total Fleet	11,347	23,000	88,000

[a]Rounded to nearest 1000.

The present inventory of airports was taken as the number of airports now having one or more jet departures per day, thereby obviously capable of handling large jet transport aircraft, and the total number of airports in the OAG—that is, airports with scheduled air service (many not presently capable of handling large jet transport aircraft). For example, if all of the airports now having one or more jet departures per day had 15 or more gates, no new airports would be needed to handle the fleet in

Table 9-26*: Number of new airports required to accommodate year 2050 fleet.*

	New Airports Required (Lowest Growth Case – Passenger Fleet)			New Airports Required (Highest Growth Case – Passenger Fleet)		
Present Airport Inventory	10 gates/airport	15 gates/airport	20 gates/airport	10 gates/airport	15 gates/airport	20 gates/airport
1,490[a]	103	0	0	3,026	1,779	1,155
3,750[b]	0	0	0	1,941	6,94	70

	New Airports Required (Lowest Growth Case – Total Fleet)			New Airports Required (Highest Growth Case – Total Fleet)		
Present Airport Inventory	10 gates/airport	15 gates/airport	20 gates/airport	10 gates/airport	15 gates/airport	20 gates/airport
1,490[a]	521	0	0	4,041	2,455	1,663
3,750[b]	0	0	0	2,956	1,371	578

[a]Now having 1 or more jet departure/day.
[b]All airports in OAG.

the lowest growth case. Conversely, the highest growth case would require more than 1,300 new airports of 15 gates each (two new airports per month for 60 years) even if all 3,750 airports now listed in the OAG had 15 gates and were capable of handling large jet transport aircraft (which they do not and are not). This analysis ignores infrastructure location and the problems associated with its provision. In populous parts of the world, where civil aviation is established, the addition of airport capacity is often difficult given local environmental pressures such developments create. However, in developing countries, where much of the future traffic growth is anticipated, new infrastructure might encounter less environmental sensitivity and therefore be more readily provided. Nonetheless, the infrastructure projects required to satisfy the highest growth scenarios are unprecedented in scope.

9.6.6.3. Fuel Availability

All of the 2050 scenarios imply large increases in fuel consumption by aircraft. In the highest FESG scenario (Fe2), aircraft fuel consumption increases from 139 to 772 Tg yr[-1] over the period 1992 to 2050. In the highest EDF scenario (Eeh), the increase is from 179 Tg yr[-1] in 1990 to 2,297 Tg yr[-1] in 2050. Because both scenarios are based on the IS92e scenario, it is appropriate to compare these figures to total energy use in the IS92e scenario. According to scenario Fe2, aircraft will account for 13% of the total transportation energy usage in 2050 and require 15% of the world's liquid fossil fuel production. The EDF scenario (Eeh) implies that aircraft account for 39% of the transportation energy usage and require 45% of the world's liquid fuel production. These comparisons assume that aircraft do not use biomass fuels or fuels derived from natural gas.

Under either scenario, the world will be straining the limits of conventional oil resources by 2050. Total remaining resources of conventional petroleum, discovered and undiscovered, have been estimated at between one trillion (Campbell and Laherrere, 1998) and two trillion barrels (Masters *et al.*, 1994—based on the optimistic 5% probability estimate of undiscovered oil). The IS92e scenario implies cumulative production of liquid fuels of 1 trillion barrels by 2025 and 2 trillion barrels by 2050. Cumulative consumption by 2050 by aircraft alone amounts to 0.15 trillion barrels in the Fe2 scenario and 0.35 trillion barrels in the Eeh scenario. However, production of liquid fuels is not necessarily limited by conventional oil resources. Liquid fuels can be produced from heavy oil, tar sands, oil shale, or even coal, albeit with significantly greater environmental consequences and at higher costs. High fuel prices would violate the explicit assumptions used in developing the scenarios.

9.6.6.4. Manufacturing Capability and Trends in Aircraft Capacity

In 1997, the global aircraft manufacturing capability delivered 634 passenger jet aircraft, bringing the global jet passenger fleet to approximately 10,000 aircraft. The rate of new aircraft

Table 9-27: *Required yearly delivery rates of aircraft implied by scenarios.*

Year	Total Aircraft Deliveries (in these years)			
	Eab	Eeh	Fa1,2	DTI
2020	1,106	1,667	813	1,012
2030	1,072	1,933	686	945
2040	1,566	3,058	931	1,354
2050	1,714	3,831	948	1,440

deliveries has followed a generally increasing trend since the mid-1950s, and this trend must continue over the scenario period to satisfy predicted demand for new and replacement aircraft. For the demand cases examined above, deliveries of new aircraft are estimated to reflect the schedule given in Table 9-27.

The delivery rate for the Fa1 and Fa2 scenarios would be achievable with existing manufacturing capacity. The delivery rate required by the highest scenario, Eeh, implies a considerable increase in manufacturing capacity—approximately six times that existing today. Although this level is not impossible, such an expansion of aircraft manufacturing capability is likely to be difficult to achieve and sustain during the period. The Eab scenario implies a delivery rate that is approximately three times the level existing today, which is not implausible for 2050.

One assumption intrinsic to the fleet size analysis was that the average number of seats per aircraft will increase by 1% each year, reflecting current trends. This assumption has a large effect on fleet size estimates, particularly for high-demand cases. As a sensitivity analysis, the factor was changed to 2% per year for the Eeh high scenario and for the Fa1 and Fa2 scenarios. Such a change may reflect potential market pressures for larger aircraft, which is not inappropriate for a high traffic growth scenario. The results are given in Table 9-28.

As this analysis shows, a different assumption in aircraft size growth has a significant effect on the estimated future fleet. The projected numbers of the largest aircraft types (between 625 and 799 seats) in future fleets are particularly sensitive in this analysis, which suggests that there might be more than 7,000 such aircraft in the fleet by 2050 in the Eeh scenario (compared with about 10,000 passenger aircraft of all sizes today) or about 4,000 additional aircraft for the more conservative Fa1 and Fa2 scenarios.

Increased capacity can be supplied by additional aircraft, increased flying hours (i.e., more efficient use of the fleet),

Table 9-28: *Sensitivity of fleet size to aircraft capacity.*

Aircraft Size Growth Assumption	Fleet at 2050 – Total Aircraft	
	1% yr[-1]	2% yr[-1]
Eeh scenario	69,275	42,448
Fa1,2 scenario	21,209	11,913

Table 9-29: Summary data from long-term scenarios.

Scenario Year	Scenario Name	Traffic Demand (10^9 RPK)	Calculated Fuel Burned (Tg yr^{-1})	Calculated CO_2 (as C) (Tg yr^{-1})	Calculated NO_x (as NO_2) (Tg yr^{-1})	Calculated Fleet EI(NO_x) (g NO_2/kg fuel)
2041	WWF	n/a	639.5[a]	550.0	n/a	n/a
2050	Fa1	13,934	471.0	405.1	7.2	15.2
2050	Fa1H	13,934	557.0	479.0	7.0	12.6
2050	Fa2	13,934	487.6	419.4	5.5	11.4
2050	Fc1	7,817	268.2	230.6	4.0	15.0
2050	Fc2	7,817	277.2	238.4	3.1	11.3
2050	Fe1	21,978	744.3	640.1	11.4	15.3
2050	Fe1H	21,978	831.0	714.7	11.3	13.6
2050	Fe2	21,978	772.1	664.0	8.8	11.4
2050	Eab	23,257	1,143.0	983.0	7.9	6.9
2050	Eah	41,392	2,086.0	1,794.0	14.4	6.9
2050	Ecb	16,762	837.0	720.0	5.8	6.9
2050	Ech	29,934	1,528.0	1,314.1	10.5	6.9
2050	Edb	19,555	959.0	824.8	6.6	6.9
2050	Edh	33,655	1,689.4	1,452.8	11.6	6.9
2050	Eeb	26,886	1,298.0	1,116.3	9.0	6.9
2050	Eeh	46,363	2,297.0	1,975.4	15.8	6.9
2050	DTI	18,106	633.2	544.6	4.5	7.0
2050	MIT	n/a[b]	977.0[a]	840.0[b]	n/a	n/a

[a]Fuel burned calculated from published CO_2 data.
[b]Contains unspecified fraction from high-speed rail.

larger aircraft, or a combination of these factors. The high aircraft growth assumption used as a sensitivity analysis here suggests that about 70% of future capacity growth will be supplied by an increase in aircraft size. Although such an industry trend is not impossible, it is unlikely to occur in such a prescriptive manner if the industry remains relatively deregulated. Deregulation tends to favor increased frequency and direct flights with smaller aircraft between departure and destination. However, it is likely that some markets would favor the proliferation of very large aircraft, especially those with dense traffic flows. The size of the fleets suggested by the 2% per year aircraft size growth assumption must therefore be regarded as toward the low end of the range.

9.6.6.5. Synthesis of Plausibility Analyses

Given the range of estimates for traffic, fuel consumption, and emissions from the 2050 aircraft scenarios available to this assessment, it is necessary to comment on the plausibility of the results—not least to demonstrate that results used in subsequent analyses are bounded by sensible limits within which the aviation industry is currently envisaged to develop.

The foregoing analyses suggest that although none of the scenarios considered for 2050 is impossible, some of the high-growth scenarios (e.g., Eah and Eeh) are probably less plausible. The fleet size and infrastructure implications suggest radical developments that are likely to be beyond the scope of changes observed in the industry thus far (or anticipated for the future). Similarly, the low-growth scenarios—though plausible in terms of achievability—give traffic estimates that are likely to be exceeded given the present state of the industry and planned developments. Although all of the FESG scenarios discount the possibility of truly radical developments in technology over the next 50 years, they are considered to fall within a plausible range of outcomes and suggest achievable developments for the industry.

The 3-D gridded output from scenarios Fa–Fe (with T1 and T2 technology scenarios) and from DTI are suitable for use as input to chemical transport models and may be used to calculate the effect of aviation CO_2 emissions. Scenarios Eab and Edh are suitable for use in Chapter 6 to calculate the effect of CO_2 emissions as sensitivity analyses because the latter scenario projects CO_2 emissions levels from aviation that are 2.2 times greater in 2050 than the highest of the FESG scenarios. Table 9-29 provides a summary of all of the long-term scenarios examined in the chapter.

References

ANCAT/EC, 1995: *A Global Inventory of Aircraft NO_x Emissions.* A first version (April 1994) prepared for the AERONOX programme. Abatement of Nuisances Caused by Air Transport (ANCAT) and European Community Working Group, European Civil Aviation Conference, London, United Kingdom, 20 pp.

Albritton, D.L, G.T. Amanatidis, G. Angeletti, J. Crayston, D.H. Lister, M. McFarland, J.M. Miller, A.R. Ravishankara, N. Sabogal, N. Sundararaman, and H.L. Wesoky, 1997: *Global Atmospheric Effects of Aviation: Report of the Proceedings of the Symposium.* NASA-CP-3351, National Aeronautics and Space Administration, Washington, DC, USA, 52 pp.

Atkinson, A.B., 1975: *The Economics of Inequality.* Clarendon Press, Oxford, United Kingdom, pp. 22, 246.

Balashov, B. and A. Smith, 1992: ICAO analyses—trends in fuel consumption by world's airlines. *ICAO Journal, 47(8),* 18–21.

Barrett, M., 1994: *Pollution Control Strategies for Aircraft.* World Wide Fund for Nature, Gland, Switzerland, 60 pp.

Bauer, E., 1979: A catalog of perturbing influences on stratospheric ozone, 1955–1975. *Journal of Geophysical Research, 84,* 6929–6940.

Baughcum, S.L., S.C. Henderson, P.S. Hertel, D.R. Maggiora, and C.A. Oncina, 1994: *Stratospheric Emissions Effects Database Development.* NASA-CR-4592, National Aeronautics and Space Administration, Langley Research Center, Hampton, VA, USA, 156 pp.

Baughcum, S.L., and S.C. Henderson, 1995: *Aircraft Emission Inventories Projected in Year 2015 for a High Speed Civil Transport (HSCT) Universal Airline Network.* NASA-CR-4659, National Aeronautics and Space Administration, Langley Research Center, Hampton, VA, USA, 117 pp.

Baughcum, S.L., S.C. Henderson, and T.G. Tritz, 1996a: *Scheduled Civil Aircraft Emission Inventories for 1976 and 1984: Database Development and Analysis.* NASA-CR-4722, National Aeronautics and Space Administration, Langley Research Center, Hampton, VA, USA, 144 pp.

Baughcum, S.L., S.C. Henderson, T.G. Tritz, and D.C. Pickett, 1996b: *Scheduled Civil Aircraft Emission Inventories for 1992: Database Development and Analysis.* NASA-CR-4700, National Aeronautics and Space Administration, Langley Research Center, Hampton, VA, USA, 196 pp.

Baughcum, S.L., D.J. Sutkus, Jr., and S.C. Henderson, 1998: *Year 2015 Aircraft Emission Scenario for Scheduled Air Traffic.* NASA-CR-1998-207638, National Aeronautics and Space Administration, Langley Research Center, Hampton, VA, USA, 44 pp.

Baughcum, S.L., and S.C. Henderson, 1998: *Aircraft Emission Scenarios Projected in Year 2015 for the NASA Technology Concept Aircraft (TCA) High Speed Civil Transport.* NASA CR-1998-207635, National Aeronautics and Space Administration, Langley Research Center, Hampton, VA, USA, 31 pp.

Beck, J.P., C.E. Reeves, F.A.A.M. de Leeuw, and S.A. Penkett, 1992: The effect of aircraft emissions on tropospheric ozone in the Northern Hemisphere. *Atmospheric Environment, 26A,* 17–29.

Boeing Commercial Airplane Group, 1993: *1993 Current Market Outlook* [Sepanen, D. (ed.)]. Boeing Commercial Airplane Group, Seattle, WA, USA, March 1993, 58 pp.

Boeing Commercial Airplane Group, 1996: *1996 Current Market Outlook* [Meskill, T. (ed.)]. Boeing Commercial Airplane Group, Seattle, WA, USA, March 1996, 38 pp.

Boeing Commercial Airplane Group, 1998: *1998 Current Market Outlook* [Meskill, T. (ed.)]. Boeing Commercial Airplane Group, Seattle, WA, USA, March 1998, 60 pp.

Brasseur, G.P., R.A. Cox, D. Hauglustaine, I. Isaksen, J. Lelieveld, D.H. Lister, R. Sausen, U. Schumann, A. Wahner, and P. Wiesen, 1998: European scientific assessment of the atmospheric effects of aircraft emissions. *Atmospheric Environment, 32,* 2329–2418.

CAEP/4-FESG, 1998: *Report 4. Report of the Forecasting and Economic Analysis Sub-Group (FESG): Long-Range Scenarios.* International Civil Aviation Organization Committee on Aviation Environmental Protection Steering Group Meeting, Canberra, Australia, January 1998, 131 pp.

CAEP/WG3, 1995: *Report of the Emissions Inventory Sub-group.* International Civil Aviation Organization Committee on Aviation Environmental Protection Working Group 3 [Ralph, M. (ed.)], London, United Kingdom, 45 pp.

Campbell-Hill Aviation Group, 1998: *Airport and Infrastructure Implications,* private communication, Brian M. Campbell to Anu Vedantham, December 1998.

Campbell, C.J. and J.H. Laherrere, 1998: The end of cheap oil. *Scientific American,* **278(3).**

COMESA, 1975: *Report of the Committee on Meteorological Effects of Stratospheric Aircraft.* UK Department of Defence, Bracknell, Berks, United Kingdom

Deidewig, F., A. Döpelheuer, and M. Lecht, 1996: *Methods to Assess Aircraft Engine Emissions in Flight.* Proceedings of the 20th International Council of Aeronautical Sciences Congress, Sorrento, Italy, Sept. 1996. American Institute of Aeronautics and Astronautics, Reston, Virginia, USA, Vol. I, pp 131–141.

Derwent, R.G., 1982: Two-dimensional model studies of the impact of aircraft exhaust emissions on tropospheric ozone. *Atmospheric Environment, 16,* 1997–2007.

Friedl, R.R. (ed.), 1997: *Atmospheric Effect of Subsonic Aircraft: Interim Assessment Report of the Advanced Subsonic Technology Program.* NASA Reference Publication 1400, National Aeronautics and Space Administration, Goddard Space Flight Center, Greenbelt, MD, USA, 154 pp.

Gardner, R.M., K. Adams, T. Cook, F. Deidewig, S. Ernedal, R. Falk, E. Fleuti, E. Herms, C.E. Johnson, M. Lecht, D.S. Lee, M. Leech, D. Lister, B. Masse, M. Metcalfe, P. Newton, A. Schmitt, C. Vandenbergh, and R. Van Drimmelen, 1997: The ANCAT/EC global inventory of NO_x emissions from aircraft. *Atmospheric Environment, 31(12),* 1751 and 1766.

Gardner, R.M. (ed.), 1998: *ANCAT/EC2 Global Aircraft Emissions Inventories for 1991/1992 and 2015: Final Report.* EUR-18179, ANCAT/EC Working Group, ISBN-92-828-2914-6, 84 pp. + appendices.

Greene, D.L., 1992: Energy-efficiency improvement potential of commercial aircraft. In: *Annual Review of Energy and the Environment* [Hollander, J., J. Harte, and R. H. Socolow (eds.)]. Annual Reviews, Inc., Palo Alto, CA, USA, Vol. 17, pp. 537–574.

Greene, D.L. and L. Meisenheimer, 1997: *Commercial Air Transport Emission Scenario Model.* Oak Ridge National Laboratory Center for Transportation Analysis, Oak Ridge, TN, USA, October 6, 1997, 6 pp.

Gould, R., 1996: *Environmental Management for Airports and Aviation: Jane's Special Report.* Jane's Information Group, Surrey, United Kingdom, January 1996, 253 pp.

Hidalgo, H. and P.J. Crutzen, 1977: The tropospheric and stratospheric composition perturbed by NO_x emissions of high-altitude aircraft. *Journal of Geophysical Research, 82,* 5833–5866.

Hutzler, M.J. and A.T. Andersen, 1997: *International Energy Outlook 1997.* Energy Information Agency, Washington, DC, USA.

ICAO, 1993: *Environmental Protection: Annex 16 to the Convention on International Civil Aviation – Volume II, Aircraft Engine Emissions.* International Civil Aviation Organization, Montreal, Quebec, Canada, 63 pp.

ICAO, 1995: *ICAO Engine Exhaust Emissions Databank, 1st Edition—1995.* ICAO-9646-AN/943, International Civil Aviation Organization, Montreal, Quebec, Canada, 173 pp.

ICAO, 1997a: *Outlook for Air Transport to the Year 2005.* ICAO Circular 270-AT/111, International Civil Aviation Organization, Montreal, Quebec, Canada, 53 pp.

ICAO, 1997b: *Traffic: Commercial Air Carriers 1982–96, Digest of Statistics No. 337.* International Civil Aviation Organization, Montreal, Quebec, Canada, 353 pp.

Landau, Z.H., M. Metwally, R. Van Alstyne, and C.A. Ward, 1994: *Jet Aircraft Engine Exhaust Emissions Database Development—Year 1990 and 2015 Scenarios.* NASA-CR-4613, National Aeronautics and Space Administration, Langley Research Center, Hampton, VA, USA, 88 pp.

Leggett, J., W.J. Pepper, and R.J. Swart, 1992: Emissions scenarios for the IPCC: an update. In: *Climate Change 1992: The Supplementary Report to the IPCC Scientific Assessment.* Prepared by IPCC Working Group I [Houghton, J.T., B.A. Callander, and S.K. Varney (eds.)] and WMO/UNEP. Cambridge University Press, Cambridge, United Kingdom, and New York, NY, USA, pp. 68–95.

Little, A.D., 1975: *Preliminary Economic Impact Assessment of Possible Regulatory Action to Control Atmospheric Emission of Selected Halocarbons.* Report C77327-10, U.S. Environmental Protection Agency, Menlo Park, CA, USA.

McInnes, G. and C.T. Walker, 1992: *The Global Distribution of Aircraft Air Pollutant Emissions.* Warren Spring Laboratory Report LR 872, Department of Trade and Industry, Warren Spring Laboratory, Stevenage, Hertfordshire, United Kingdom, 41 pp.

Metwally, M., 1995: *Jet Aircraft Engine Emissions Database Development—1992 Military, Charter, and Nonscheduled Traffic.* NASA-CR-4684, National Aeronautics and Space Administration, Langley Research Center, Hampton, VA, USA, 61 pp.

Mortlock, A.M. and R. van Alstyne, 1998: *Military, Charter, Unreported Domestic Traffic and General Aviation: 1976, 1984, 1992, and 2015 Emission Scenarios.* NASA-CR-1998-207639, National Aeronautics and Space Administration, Langley Research Center, Hampton, VA, USA, 118 pp.

Newton, P.J. and R.S. Falk, 1997: *DTI Forecast of Fuel Consumption and Emissions from Civil Aircraft in 2050 Based on ANCAT/EC2 1992 Data.* DTI/ADI3c/199701/1.0, Department of Trade and Industry, London, United Kingdom, 13 November 1997, 15 pp.

Nüßer, H.-G. and A. Schmitt, 1990: The global distribution of air traffic at high altitudes, related fuel consumption and trends. In: *Air Traffic and the Environment, Lecture Notes in Engineering* [Schumann, U. (ed.)]. Springer-Verlag, Berlin, Germany, pp. 1–11.

OECD, 1998a: *Energy Balances of OECD Countries* (database software). Organization for Economic Cooperation and Development, OECD Publications, Paris, France.

OECD, 1998b: *Energy Balances of Non-OECD Countries* (database software). Organization for Economic Cooperation and Development, OECD Publications, Paris, France.

Olivier, J.G.J., 1995: *Scenarios for Global Emissions from Air Traffic. The Development of Regional and Gridded (5° x 5°) Emissions Scenarios for Aircraft and for Surface Sources, Based on CPB Scenarios and Existing Emission Inventories for Aircraft and Surface Sources.* RIVM-773002003, Rijksinstituut voor Volksgezondheid en Milieu, Bilthoven, The Netherlands, July 1995, 113 pp.

Pulles, H., 1998: *Report of the (ICAO/CAEP) Focal Point on Charges (FPC) Prepared for CAEP/4. Emission Charges and Taxes in Aviation.* International Civil Aviation Organization Committee on Aviation Environmental Protection, The Hague, The Netherlands, March 1998, 79 pp. + appendices.

Rogner, H.H., 1997: An assessment of world hydrocarbon resources. In: *The Annual Review of Energy and the Environment.* Annual Reviews, Inc., Palo Alto, CA, USA, Vol. 22, pp. 217–262.

Schafer, A. and D.G. Victor, 1997: *The Future Mobility of the World Population.* Discussion Paper 97-6-4, Center for Technology, Policy and Industrial Development and the MIT Joint Program on the Science and Policy of Global Change, Massachusetts Institute of Technology, Cambridge, MA, USA, Rev. 2, September 1997, 39 pp.

Schmitt, A. and B. Brunner, 1997: Emissions from aviation and their development over time. In: *Final Report on the BMBF Verbundprogramm, Schadstoffe in der Luftfahrt* [Schumann, U., A. Chlond, A. Ebel, B. Kärcher, H. Pak, H. Schlager, A. Schmitt, and P. Wendling (eds.)]. DLR-Mitteilung 97-04, Deutsches Zentrum für Luft- und Raumfahrt, Oberpfaffenhofen and Cologne, Germany, pp. 37–52.

Schulte, P., H. Schlager, H. Ziereis, U. Schumann, S.L. Baughcum, and F. Deidewig, 1997: NOx emission indices of subsonic long-range jet aircraft at cruise altitude: *In situ* measurements and predictions. *Journal of Geophysical Research,* **102,** 21431–21442.

Stolarski, R., and H. Wesoky, 1993: *The Atmospheric Effects of Stratospheric Aircraft: A Third Program Report.* NASA Reference Publication 1313, National Aeronautics and Space Administration, Washington, DC, USA, November 1993, 422 pp.

Sutkus, D.J., Jr., 1997: *2050 Fuel Efficiency and NOx Technology Scenarios.* International Coordinating Council of Aerospace Industries Association Working Paper WG3-WP4/22 presented at the Fourth Meeting of International Civil Aviation Organization Committee on Aviation Environmental Protection Working Group 3 (Emissions), 12–14 November 1997, Bern, Switzerland, 22 pp.

Sutkus, D.J. Jr., D.L. Daggett, D.P. DuBois, and S.L. Baughcum, 1999: *An Evaluation of Aircraft Emissions Inventory Methodology by Comparisons with Reported Airline Data.* National Aeronautics and Space Administration, Langley Research Center, Hampton, VA, USA, (in press).

ten Have, H.B.G. and T.D. de Witte, 1997: *User Manual of the Flights and Emissions Model—Model Version 3.11.* NLR-CR 97063-L, National Aerospace Laboratory, Amsterdam, The Netherlands.

Vedantham, A. and M. Oppenheimer, 1994: *Aircraft Emissions and the Global Atmosphere.* Environmental Defense Fund, New York, NY, USA, 77 pp.

Vedantham, A. and M. Oppenheimer, 1998: Long-term scenarios for aviation: demand and emissions of CO_2 and NO_x. *Energy Policy,* **26(8),** 625–641.

WMO, 1995: *Scientific Assessment of Ozone Depletion: 1994.* Report no. 37, Global Ozone Research and Monitoring Project, World Meteorological Organization, Geneva, Switzerland, 578 pp.

Wuebbles, D.J., D. Maiden, R.K. Seals, S.L. Baughcum, M. Metwally, and A. Mortlock, 1993: Emissions scenarios development: report of the emissions scenarios committee. In: *The Atmospheric Effects of Stratospheric Aircraft: A Third Program Report* [Stolarski, R.S. and H.L. Wesoky (eds.)]. NASA Reference Publication 1313, National Aeronautics and Space Administration, Washington, DC, USA, pp. 63–85.

Zahavi, Y., 1981: *The UMOT-Urban Interactions.* DOT-RSPA-DPB 10/7, U.S. Department of Transportation, Washington, DC, USA, 151 pp.

10

Regulatory and Market-Based Mitigation Measures

JOHN F. HENNIGAN

Lead Authors:
H. Aylesworth, J. Crayston, L. Dobbie, E. Fleuti, M. Mann, H. Somerville

Review Editor:
C.V. Oster, Jr.

CONTENTS

EXECUTIVE SUMMARY

- Although improvements in aircraft engine technology and air traffic management technology will bring environmental benefits, these benefits are not expected to be sufficient to fully offset the projected growth of aviation emissions arising from increased demand for air transportation services (as discussed in the scenarios in Chapter 9). These scenarios adopt simplifying assumptions of full implementation of communication, navigation, and surveillance/air traffic management (CNS/ATM) and no infrastructure constraints. As a result, they are thought to show unrealistically high projected growth of air traffic and emissions. Policy measures aimed at addressing the growth of aviation emissions will need to take this factor into account.

- The International Civil Aviation Organization (ICAO) is the United Nations (UN) specialized agency that has global responsibility for the establishment of standards, recommended practices, and guidance on various aspects of international civil aviation, including the environment. The Kyoto Protocol committed countries to work through ICAO in limiting or reducing emissions of greenhouse gases from aviation bunker fuels, but international aviation emissions are not covered by emissions reduction targets in the Kyoto Protocol.

- Policy measures have the potential to limit the growth of aircraft emissions by encouraging technological innovation, effecting greater operating efficiencies in the aviation industry, and affecting demand. Such measures include initiatives to ensure timely implementation of CNS/ATM; development of regulatory standards for aircraft emissions, operational measures, and avoidance of delay; and market-based instruments such as environmental levies and emissions trading. Voluntary and mandated policies to reduce emissions internalize to the producer or consumer many of the associated environmental costs.

- Cost-benefit analysis (CBA) has been used as a tool to conduct economic analysis of environmental mitigation measures across many areas of economic activity, but some difficulties have arisen in its application to aviation, primarily because of measurement and information problems. CBA provides a framework for balancing economic and environmental impacts associated with different policy options. The precautionary and "polluter pays" principles set out in the 1992 Rio Declaration are also relevant to economic analysis of policies aimed at limiting the growth of aviation emissions.

- In 1981, ICAO established aircraft engine emission standards for oxides of nitrogen (NO_x). Since then, the NO_x standard has been made more stringent (by 20% in 1993). A further 16% change to newly certified engines after 2003 was recommended in 1998. ICAO has established an approach that requires that any actions to mitigate the environmental effects of aviation show environmental need, technical feasibility, and economic reasonableness. The current method of regulating NO_x based on the landing/take-off (LTO) cycle does not fully address emissions at altitude. ICAO is developing a new parameter for emissions certification during climb and at cruise altitude to complement the existing LTO cycle-based parameter.

- Timely implementation of CNS/ATM is important to success in leading to reduced fuel burn and emissions; CNS/ATM is critically dependent on the development of necessary institutional and financial arrangements.

- Existing studies indicate that levies on fuel and en route charges are viewed as the most environmentally effective levies. If passed through to consumers, environmental levies could reduce consumption of aviation fuel, hence aircraft emissions, by providing incentives to develop and use more energy-efficient aircraft, by optimizing operations, and by reducing the growth in demand for air transport. However, a practical difficulty is that most bilateral air service agreements exempt fuel used internationally from tax. Levies at a regional level may avoid this and other problems. There is concern, however, that this approach would not have sufficient environmental impact, in some cases could increase other emissions, and could lead to economic distortions.

- Emissions trading is a market-based approach that enables participants from all industries to cooperatively minimize the costs of reducing emissions. The role of governments would be to set a cap on emissions and rules for trading reductions under such a cap. Initial experience at the national and international levels in other industries suggests that emissions trading can provide an effective incentive for firms to innovate and achieve low-cost emissions reduction.

- Other potential mitigation measures are voluntary agreements to meet environmental targets and funding of research to better understand the environmental impact of aircraft emissions.

- There are uncertainties over future trends in traffic and emissions and the impact of mitigation measures.

10.1. Introduction

Although the introduction of aircraft emissions reduction technology (see Chapter 7), ATM improvements, and other aircraft operational measures (see Chapter 8) highlight the technological potential to reduce future aircraft emissions, the question has been raised about whether these technologies will be adopted by industry and implemented by governments at sufficient levels and soon enough to appropriately address projected scenarios of growth in aviation traffic and emissions (Chapter 9).

This chapter reviews policy measures that are viewed as ways of accelerating emission mitigation trends, if needed (IPCC, 1996b). Such measures for policymakers include initiatives to ensure timely implementation of CNS/ATM, development of regulatory standards and recommended practices, market-based measures such as environmental levies (taxes and charges), and emissions trading. This chapter also examines analytical frameworks to evaluate alternative policy options.

When evaluating potential mitigation measures for aviation, policymakers probably would benefit from having reliable information on the relative costs of mitigating aviation's greenhouse gas or other emissions compared with the costs of mitigating emissions from other sources. At present, however, it is not known whether addressing aviation is likely to be more or less costly than achieving comparable reductions from other sources.

In terms of guidance, the 1995 IPCC assessment states, "Achieving the economic and policy potential of greenhouse gas reductions ... depends on the economic and other priorities of the providers and users of transport services. The extent to which these potentials are achieved depends on a complex interaction among technology, the economy, and choices made by consumers, producers, and policymakers. It is this interaction that most needs to be understood and addressed if policymakers wish to limit greenhouse gas emissions from the transport sector." (IPCC, 1996a).

Because most developed and developing country economies depend on aviation services, balancing the individual interests of aviation stakeholders with the broader public interest, including environmental interests, is critical. It is also important to understand the institutional and global context in which the industry operates, the drivers of aviation economic activity, and the incentives and motivations of the parties.

In addition, the literature on this subject suggests that flexible, cost-effective policies that rely on economic incentives, as well as the coordinated use of policy instruments, can considerably reduce mitigation or adaptation costs and increase the cost-effectiveness of emission reduction measures. Appropriate long-run signals are required to allow producers and consumers to adapt cost effectively to constraints on greenhouse gas emissions and to encourage investment, research, development, and demonstration (IPCC, 1996c). To attain the proper balance

of interests, international standardization of regulations and economic instruments and uniformity in application are necessary. Other criteria in assessing the effectiveness of policy tools are environmental need, technical feasibility of proposed solutions, and ease of implementation.

10.2. Institutional Framework

10.2.1. *International Civil Aviation Organization*

Because of the international nature of aviation and the need to promote international standards of safety, regulation of air transport is harmonized on a worldwide basis as much as possible through ICAO, a UN specialized agency established in 1944 under the Convention on International Civil Aviation (the "Chicago Convention") (ICAO, 1997a). With the technical support of states and the aviation community, ICAO has developed international standards and recommended practices for aviation. These standards and practices are attached as annexes to the Chicago Convention and are considered binding. However, if any state finds it impossible to comply with them, the state is required to inform ICAO of any differences. ICAO also provides guidance material for states on various aspects of international civil aviation.

In addition to operating under the harmonization achieved through ICAO, international flights are also subject to bilateral air service agreements between states. Although ICAO's mandate relates to international civil aviation, most states also take ICAO's international standards and recommended practices and guidance material into account in regulating their domestic aviation.

The ICAO strategic action plan acknowledged "a perception that the aviation sector may be contributing unduly to both existing and future environmental problems." The ICAO program of implementation action on the strategic action plan for the 1996–98 triennium lists several activities that could be relevant to environmental issues (ICAO, 1997c).

In the environmental field, ICAO has established standards for aircraft noise and aircraft engine emissions (Annex 16), as well as standards and procedures for operational measures such as landing and departure (ICAO, 1993a,b). It has also developed broader policy guidance on fuel taxation and charging principles that have relevance in an emissions context (ICAO, 1994, 1997b).

Much of ICAO's work in the environmental field is undertaken by its Committee on Aviation Environmental Protection (CAEP). CAEP examines the effectiveness and reliability of proposed aircraft certification schemes from the viewpoint of technical feasibility, economic reasonableness, and environmental need. CAEP also examines other issues related to emissions from aircraft engines—including international and national research into the impact of emissions and possible means of controlling emissions, such as operational measures and emissions-related levies. CAEP works closely with regional bodies and national airworthiness authorities to discuss and propose changes in recommended environmental standards.

10.2.2. United Nations Framework Convention on Climate Change

Although the United Nations Framework Convention on Climate Change (UNFCCC) does not specifically refer to emissions from aviation, its coverage includes emissions from all sources. One of the commitments in the Convention is that parties to the Convention compile national inventories of their emissions sources (UNFCCC, 1992).

IPCC guidelines provide advice on the quantification of aviation emissions (IPCC, 1996e). Even after aviation emissions have been quantified, an important distinction is whether such emissions are domestic or international. For domestic flights, emissions are considered to be part of the national inventory of the country within which the flights occur. The IPCC guidelines require international aviation emissions to be estimated by the country where the fuel is sold, although such emissions are not to be included in that country's total emissions. For international flights, the problem is how to allocate the emissions (referred to as "emissions from international aviation bunker fuels" in UNFCCC terminology, although "international" is not always specified) to national inventories. A similar problem exists for shipping.

In an attempt to resolve this problem, the UNFCCC Subsidiary Body for Scientific and Technological Advice (SBSTA) has been presented with the following allocation options for emissions from international aviation and marine bunker fuels (UNFCCC, 1996):

- *Option 1*—No allocation
- *Option 2*—Allocation of global bunker sales and associated emissions to parties in proportion to their national emissions
- *Option 3*—Allocation according to the country where the bunker fuel is sold
- *Option 4*—Allocation according to the nationality of the transporting company, or to the country where an aircraft or ship is registered, or to the country of the operator
- *Option 5*—Allocation according to the country of departure or destination of an aircraft or vessel; alternatively, emissions related to the journey of an aircraft or vessel shared by the country of departure and the country of arrival
- *Option 6*—Allocation according to the country of departure or destination of passengers or cargo; alternatively, emissions related to the journey of passengers or cargo shared by the country of departure and the country of arrival
- *Option 7*—Allocation according to the country of origin of passengers or owner of cargo
- *Option 8*—Allocation to a party of all emissions generated in its national space.

SBSTA subsequently noted that there were three separate issues: adequate and consistent inventories, allocation of emissions, and control options. Appropriate allocation of responsibility for emissions from international bunker fuels would be connected to inventory and control issues. Reviewing the foregoing eight options, SBSTA decided that Options 1, 3, 4, 5, and 6 should be the basis for further work on this issue (UNFCCC, 1997).

To date, however, there has been no agreement among parties to the Convention on which option to choose. At Kyoto in December 1997, the Conference of Parties to the UNFCCC urged SBSTA to elaborate further on the inclusion of emissions from fuel sold to aircraft and ships engaged in international transport in overall greenhouse gas inventories (UNFCCC, 1998a).

The Kyoto Protocol to the UNFCCC, which has not yet entered into force, requires Annex I (industrialized) countries to reduce their collective emissions of greenhouse gases by approximately 5% by 2008–2012 compared with 1990 levels, with the reduction varying from country to country. The agreed-on targets apply to national totals of greenhouse gases. Consequently, each Annex I country can determine how the various emission-producing sectors in its economy should be called upon to assist in achieving the country's national target.

Because international aviation and marine emissions are not included in national inventories, they are currently excluded from the agreed-on targets. However, the treatment of aviation and marine emissions was considered in Kyoto in the context of discussions on policies and measures to be pursued by Annex I countries, and a provision was included in the Kyoto Protocol. The relevant text (Article 2, paragraph 2) reads as follows:

> "The Parties included in Annex I shall pursue limitation or reduction of emissions of greenhouse gases not controlled by the Montreal Protocol from aviation and marine bunker fuels, working through the International Civil Aviation Organization and the International Maritime Organization, respectively." (UNFCCC, 1998b)

A question has since arisen in the aviation community regarding whether this provision covers emissions from international aviation only or emissions from both international and domestic aviation. Although this issue has not been resolved, ICAO commented in a statement to the Conference of Parties to the UNFCCC in November 1998 that ICAO's mandate (from the 1944 Convention on International Civil Aviation) does not extend to domestic aviation but that ICAO's standards, recommended practices, and procedures in many circumstances have a *de facto* application domestically (ICAO, 1998c).

At the ICAO assembly in September/October 1998, the ICAO Council was asked to study policy options to limit or reduce greenhouse gas emissions from civil aviation—taking into account the findings of this IPCC Special Report and the requirements of the Kyoto Protocol—and to report to the

Assembly at its next ordinary session in 2001. ICAO was also asked to work with SBSTA to consider the various options for allocation of emissions from international aviation (ICAO, 1998b).

10.3. Framework for Evaluation

10.3.1. Methodology for Economic Analysis

Economic appraisal provides a conceptual framework for assessing benefits arising from environmental mitigation measures and the costs they impose. This appraisal involves defining objectives, examining options, and summing costs and benefits (and uncertainties) associated with each option. All environmental regulations or market-based options aim to internalize external costs and are likely to raise the costs of production, leading to higher prices along with benefits resulting from reduced amounts of pollution and associated damage costs. In conducting an economic appraisal of an environmental mitigation measure, the loss of material benefits to those who purchase goods or services needs to be weighed against the benefits from reduced pollution.

CBA is a common method for assessing benefits arising from an investment project, regulation, or policy and the costs that it imposes. It seeks to quantify—in monetary terms, where possible—as many of the costs and benefits as possible, taking into account their timing. Other techniques include risk analysis, cost-effectiveness, and environmental impact analysis; these approaches are generally not mutually exclusive, and in many cases they are used complementarily as part of a CBA framework. CBA has a long history as a useful aid to decisionmaking across many areas of economic activity. There are important obstacles, however, to its complete application—notably the lack of transparent information, where information is closely held for competitive reasons by private-sector companies; the difficulty of putting monetary values on environmental benefits; and the challenge of applying it to proposals that are implemented over a long time. Where benefits cannot be valued, it may be necessary to present them in physical units or to rely on some form of cost-effectiveness analysis (EASG, 1995; U.S. Department of Transportation and Federal Aviation Administration, 1998).

The following principles, set out in the 1992 Rio Declaration on Environment and Development (UNCED, 1992) regarding the precautionary and "polluter pays" principles, are relevant to the economic analysis framework:

> "In order to protect the environment, the precautionary approach shall be widely applied by states according to their capabilities. Where there are threats of serious or irreversible damage, lack of full scientific certainty shall not be used as a reason for postponing cost-effective measures to prevent environmental degradation." (Principle 15)

> "National authorities should endeavor to promote the internalization of environmental costs and the use of economic instruments, taking into account the approach that the polluter should, in principle, bear the cost of pollution...." (Principle 16)

The UNFCCC provides further guidance on the thinking underlying the precautionary principle (UNFCCC, 1992). Article 3 states:

> "The Parties should take precautionary measures to anticipate, prevent, or minimize the causes of climate change and [to] mitigate its adverse effects. Where there are threats of serious or irreversible damage, lack of full scientific certainty should not be used as a reason for postponing such measures, taking into account that policies and measures to deal with climate change should be cost-effective so as to ensure global benefits at the lowest possible cost. To achieve this, such policies and measures should take into account different socio-economic contexts, be comprehensive, cover all relevant sources, sinks, and reservoirs of greenhouse gases and adaptation, and comprise all economic sectors. Efforts to address climate change may be carried out cooperatively by interested Parties."

The precautionary principle is a guideline that indicates when mitigation action should be taken. CBA is a useful analytical tool to guide policy decisions. The "polluter pays" principle establishes upon whom the cost should fall.

10.3.2. The Application of Economic Analysis

Studies examining the economic impact of environmental measures affecting aviation have included the ICAO policy framework for imposing operating restrictions on certain noisier aircraft, global implementation of CNS/ATM, and potential noise and emissions stringency reduction options available to ICAO. Although the precise methodologies have differed according to the circumstances and available information, the common approach in all of these studies has been the use of a cost-benefit framework.

The ICAO study, "The Economic Implications of Future Noise Restrictions on Subsonic Jet Aircraft," estimated operating cost and revenue loss implications of early replacement of Chapter 2 aircraft as a result of certain states' implementation of operating restrictions on these aircraft for noise reasons (ICAO, 1989). The benefit of the proposed noise restrictions was analyzed by estimating the land area and population around affected airports with and without the proposed restrictions on noisy aircraft. The cost of the phaseout to air carriers would be acceleration of the fleet modernization process, reducing the value of the operators' assets and requiring them to make capital commitments earlier than might be justified from purely commercial considerations. A similar study was conducted for Chapter 2 phaseout of noisy aircraft in the United States (U.S. Department of Transportation and Federal Aviation Administration, 1991). These types of issues would need to be examined if proposals for operating restrictions to limit emissions were put forward.

Studies on the economic impact of implementation of CNS/ATM are discussed in Chapter 8 and summarized in Table 8-1. Similarly, a report conducted for CAEP's third meeting (CAEP/3) (ICAO, 1996a) described steps required to carry out an economic analysis of environmental policy measures for aviation (EASG, 1995). It considered alternative methodologies for appraising policy options and judged that CBA was the most appropriate economic appraisal methodology to apply to aviation environmental issues. This methodology was applied to noise and emissions stringency options under consideration at CAEP/3. CBA methodology was also applied to the emissions stringency option considered at CAEP's fourth meeting (CAEP/4) (FESG, 1998).

A number of issues and uncertainties have arisen in the application of CBA to potential environmental issues affecting aviation:

- The absence of transparent cost data capable of independent validation (related to industry commercial confidentiality)
- The need to forecast over a sufficiently long appraisal period to reflect the gradual buildup of benefits from emissions stringency measures
- Scientific uncertainties regarding the impact of aviation emissions, leading to the use of estimated reductions in the amount of pollutant emitted as a surrogate measure for benefits
- Uncertainties about future trends in technology, underlying trends in traffic growth, and aircraft fleet mix. This uncertainty is illustrated by recently revised projections for fleet NO_x emission index (EI) made by Boeing for 2015, showing more than a 50% increase from its previous estimate, caused primarily by reassessment of technology likely to be present in the 2015 fleet (Sutkus, 1997).

In addition, uncertainty has arisen regarding whether new worldwide standards for aircraft engine emissions would encourage additional local rules by airports and governments, leading to a reduction in the value of the existing aircraft fleet. Two studies reported in the economic analysis subgroup (EASG) report to CAEP/3 (EASG, 1995) concluded that emissions stringency options under consideration would lead to a reduction in existing aircraft fleet values for this reason. Some subsequent studies commissioned by the forecasting and economic support group (FESG) and included in its report to CAEP/4 found that the limited quantitative data available provide inconclusive evidence of any material impact on aircraft fleet values arising from an earlier change in the worldwide emissions stringency standard agreed at CAEP/2 for NO_x (ICAO, 1998a).

CBA may have to be applied somewhat differently in the context of environmental levies. All environmental mitigation measures aim to internalize external costs, but environmental levies achieve this internalization more explicitly and to a greater degree. External costs arise where, for example, the cost of pollution from an economic activity does not accrue to the party responsible for it. Placing monetary values on external costs may eliminate

the need for CBA, so the focus of economic analysis can be compliance costs, economic distortion, and the environmental effectiveness of alternatives. With aircraft emissions, however, the monetary value of these external costs is subject to a wide range of uncertainty, as discussed in Section 10.4.3. In these circumstances—and in particular under a regime of emissions targets introduced on the basis of the precautionary principle— CBA techniques will be necessary in providing assistance to decisionmakers on the most economically efficient means of meeting such targets.

10.3.3. Elasticity of Demand Considerations

The extent to which cost increases arising from environmental mitigation policies are passed on to consumers and affect demand for aviation services depends on elasticity of demand (the responsiveness of consumers to fare increases).

Evidence regarding fare elasticities is particularly relevant in considering the impact of environmental levies (charges or taxes) that directly increase the cost of air travel. If demand is relatively unresponsive to fare increases (i.e., price is inelastic), such measures will have little impact in reducing emissions through dampening of demand. Conversely, if passenger responsiveness is high (i.e., price is elastic), policy measures will have a greater impact. This analysis can be illustrated as follows: A fare elasticity of -0.5 would mean that a 10% increase in fares would result in a 5% reduction in traffic volume; a fare elasticity of -2.0 would mean that a 10% fare increase would reduce traffic by 20%.

Estimates of fare elasticities differ by journey purpose; leisure travel generally is significantly more price elastic than business travel. Fare elasticities also exhibit some regional variation; for example, studies show comparatively higher price elasticities in North America than in other regions of the world. A study commissioned by the U.S. Federal Aviation Administration (FAA) (U.S. Department of Transportation, 1995) reviewed more than 25 price elasticity studies and found that most presented elasticity of demand estimates in the range of -0.8 to -2.7.

Studies examining the effectiveness of environmental levies have had to make assumptions about the reaction of airlines and how the demand for air transport changes in response to a change in fare. Many commentators agree that in the long term, in addition to encouraging more energy-efficient technological alternatives, an increase in the price of fuel will lead to an increase in the cost of travel and consequent reduction in demand. However, airlines operate in a highly competitive environment, and in the short term many may absorb fare increases at a cost to profitability rather than pass them on to passengers.

10.4. Mitigation Measures

Aviation is a technologically intensive industry, both as a user and as a driver of advanced technology. Until quite recently, business units within the air service industry were often characterized as a series of local monopolies connected by a set of protected routes, which contributed to economic and environmental inefficiencies. Continued aviation liberalization has not brought decreased regulation in the areas of safety, security, and the environment however, and continued cooperation among national authorities is necessary to ensure that such regulations are globally and competitively neutral.

10.4.1. Underlying Trends Relevant to Mitigation

The general trend in aircraft engine technology development over the past few decades has been to reduce specific fuel consumption (SFC). This trend has resulted in lower emissions of carbon dioxide (CO_2) and water vapor (H_2O) and most other exhaust gases per unit of thrust. Advances in combustor technology have resulted in considerable reduction of NO_x emissions at a given pressure ratio. Future developments in engine technology are discussed in Chapter 7.

Developments in communication, navigation, and surveillance technology, as well as air traffic management systems (see Chapter 8), have enabled more efficient use of the air traffic system. This trend has resulted in considerable fuel savings. The complete transformation—that is, modernization of the air traffic system—is expected to generate significant safety, operational, and environmental benefits. Air traffic measures for present and future systems offer potential for reduced fuel consumption, hence emissions, through improvement in the overall capacity and efficiency of the air traffic system. Potential environmental benefits to be gained from operational measures within the current air traffic system, though important, are thought to be smaller than those that may be gained through modernization of the air traffic system. The environmental effect will depend on the rate at which these measures are adopted.

Chapter 9 examines the implications of alternative growth and technology scenarios on aviation emissions. Mid-term projections of aviation growth have meaningful margins of error, and those beyond 20 years are highly uncertain. Accordingly, rather than use projections, we develop a range of future industry growth and technology scenarios to the year 2050. Comparison of scenarios gives order-of-magnitude estimates of emissions changes resulting from variation in the rate of aviation growth and different technology options. Although these scenarios produce a wide range of outcomes, they all show an increase in aircraft emissions as growth outpaces engine technology improvements.

The scenarios assume minimal fuel consumption resulting from the use of optimal flight paths and absence of delay. Unlimited development of the aviation infrastructure projects unconstrained growth of aviation. Chapter 8 addresses current policies, which are expected to diminish the environmental effects of additional fuel consumption resulting from present air traffic management technology and systems. Unconstrained growth—an assumption that is not representative of conditions today or in the future—thereby becomes the dominant factor. Because the scenarios assume complete modernization of the air traffic system and no infrastructure constraints, growth of traffic and emissions are thought to be overstated in the modeling exercises.

10.4.1.1. Deregulation and Subsidies

An important factor for promoting the sustainable development of aviation is ensuring that producers and consumers receive appropriate signals about pollution costs and natural resource scarcities. The focus of aviation policy over the past 15 years has begun to shift from command-and-control policies to enhancing efficiency and responding to changing demand patterns. Two broad categories of issues—the operation of markets, and government supports and ownership—bear on environmental mitigation measures (OECD, 1997b).

Government regulation of competition in international aviation has restricted market entry and exit, fares, and capacity. Internationally, liberalization of bilateral air service agreements is aimed at increasing competition within "the home market" and promoting greater efficiency. Although aviation remains a quasi-protected industry in some countries, deregulation is becoming the predominant trend within national markets.

Economic deregulation of the U.S. airline industry began in 1976, and was completed in 1983 when all regulations on domestic fares, entry, and exit were eliminated. This deregulation resulted in a 33% lower fare structure and higher average industry load factors, which have increased from 52% (1960–69) to more than 70% (1995–98). Following deregulation, increases in traffic volume, fuel use, airport and air traffic congestion, and noise that some observers consider to be above "normal" have been attributed to the hub-and-spoke method of operation that ensued. Studies comparing the effect of direct versus hub-and-spoke routing on fuel consumption, traffic levels, and fleet mix have not been undertaken (Winston, 1998).

When the environmental cost of a supported activity is placed outside the market transaction, many subsidy or tax treatments are thought to have adverse environmental consequences to the extent that they generate negative effects. The most important examples of explicit government supports to aviation are nonmarket pricing of infrastructure services, below-market financing, tax and depreciation preferences for oil, and direct government ownership. These and other issues are being studied (OECD, 1997b). Lack of inclusion of environmental costs is currently thought to be an implicit subsidy to industry and consumers. All of the mitigation measures discussed below internalize or include external costs to the service provider to some degree. An important issue for further research is whether the full costs of externalities can be quantified in advance.

10.4.1.2. Cost Structure

After 1976, costs of operation fell rapidly for U.S. operators. In real terms, costs per revenue-tonne-km have declined more than 25%. Similar declines have been achieved internationally. Lower passenger fares and cargo rates have paced this reduction in costs. As a result, airline profits have remained low relative to other industries. For the industry, airline profits are a significant source of funds for purchasing capital assets. Timely acquisition of new technologies has been viewed as necessary for environmental mitigation. Recent volatility in industry profits is related primarily to short-term changes in demand and short-term overcapacity of aircraft (Winston, 1998).

Fuel cost and consumption are important to the mitigation measures discussed below. Examination of the historical record indicates that aviation fuel productivity has increased considerably. ICAO estimates that, while aggregate fuel consumption for the 1976–90 period grew approximately 60%, world civil air traffic (passengers, freight, and mail combined) increased about 150%. These figures are explained by an increased average load factor, a more optimal aircraft fleet mix, greater engine efficiency, and improved system capacity (Balashov and Smith, 1992). The amount of fuel consumed and fuel productivity affect airline profits and ticket prices.

Of interest to the policymaker is the effect of a change in fuel price on airline cost and fare structures and on passenger and cargo demand. Figure 10-1 shows the percentage of total annual

airline expense attributable to fuel cost for U.S. and world airlines over time and compares changes in this ratio against changes in the index of fuel prices for the United States and the world. Data for the figure were extracted from industry and ICAO published sources (ICAO, 1996b; Aerospace Industries Association of America, 1998). The data indicate that average fuel cost across the industry has recently represented about 11–15% of operating costs, but has been higher during times of high fuel prices (e.g., nearly 30% in the early 1980s). Further analyses of the impact of fuel prices on demand and industry profitability are needed.

Against the background of these underlying trends, the potential role of mitigation measures is considered in further reducing the growth of aircraft emissions by forcing technology, changing industry practices, and dampening growth in traffic.

10.4.2. Regulatory Measures

Atmospheric environmental impacts associated with aviation emissions include greenhouse gas emissions, ozone depletion, acidification, and impact on local air quality. Cruise altitude (above 900 m) aircraft emissions of interest are CO_2, NO_x, particulates and aerosols, sulfur compounds, and H_2O. Ground-level (altitude below 900 m) aircraft emissions of interest to local air quality issues are NO_x, carbon monoxide (CO), unburned hydrocarbons (HC), and other volatile organic compounds. Ground-level CO_2 emissions from aircraft are of interest because of their effects on climate. This section examines more fully two technology-based means for mitigating the adverse effects of aviation emissions.

10.4.2.1. Operational Means to Reduce Emissions

The environmental aim of air traffic system modernization and operational measures is to reduce fuel consumption, hence emissions, through improvement in the overall efficiency of the air traffic system. For 1995, U.S. airlines reported a 79% "on-time" arrival rate for all domestic flights (i.e., flights arriving within 15 minutes of schedule), including canceled flights and those with mechanical delays. For 1996, this figure was 75%. This decline indicates the scope of current congestion and delay problems (U.S. Department of Transportation, 1998).

U.S. Federal Aviation Administration (FAA) data show that annual system delays of 15 minutes or longer (as a percent of total operations) have declined steadily from 2.2% in 1989 to 1.2% in 1997. Over that period, 55–70% of these delays were caused by weather; 25–35% were caused by terminal volume and closed runways and taxiways (U.S. Federal Aviation Administration, 1977). FAA estimates of the total cost for air carrier delay (operating plus passenger time costs) grew from $6.5 billion in 1987 to $9.5 billion in 1994. Except for the cost of fuel, environmental costs were not considered (U.S. Federal Aviation Administration, 1995).

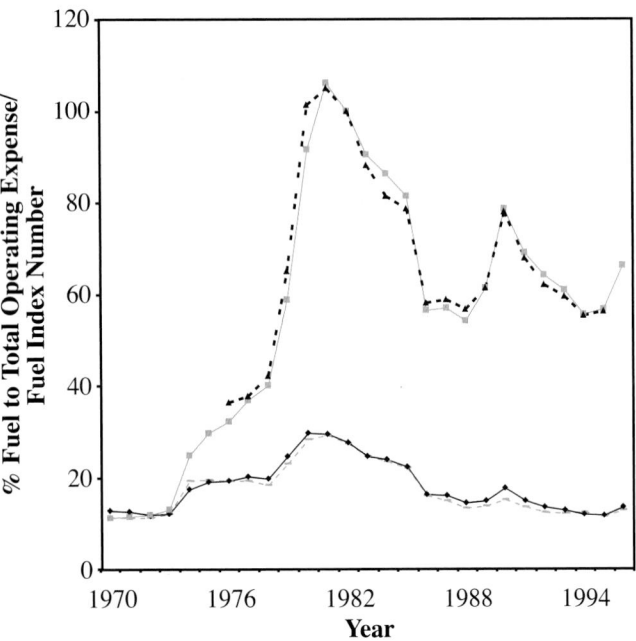

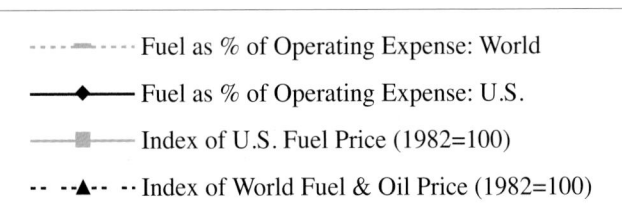

Figure 10-1: Comparison of annual fuel expense as percentage of total annual operating expense for world and U.S. airlines against world fuel and oil and U.S. fuel cost indicies, 1970–96.

The European Organisation for Safety and Navigation (EUROCONTROL) reports that air traffic delays in Europe are primarily the result of national institutional factors, capacity overloads, and air traffic control inefficiencies. Association of European Airlines (AEA) data show that in 1989, the number of departures for short- and medium-term flights delayed by more than 15 minutes was 23.8%. This figure dropped to 12.7% in 1993, but has risen steadily since then: The rate was 19.5% in 1997 and 20.1% for the January–June 1998 period (Association of European Airlines, 1998). These data are consistent with EUROCONTROL data for the same years (EUROCONTROL, 1998).

ICAO air traffic system policies, plans, standards, and recommended practices have not been regarded as measures that might be used to achieve environmental gains. Studies assessing the costs and benefits of CNS/ATM modernization conclude that the benefits to airlines and passengers from reduced delays and fuel savings far outweigh the incremental costs. Recent studies have focused on estimating the environmental benefits of a more efficient air traffic system. These studies indicate that emissions from aviation may be reduced significantly. Although states are working within ICAO toward global modernization of the air traffic system, this activity has only recently been placed in the context of addressing environmental issues (EISG, 1995; Aylesworth, 1997; U.S. Federal Aviation Administration, 1998a,b).

Studies and working papers prepared for CAEP/3 Working Group 3 (emissions) and the ICAO Worldwide CNS/ATM Systems Implementation Conference (see Chapter 8) indicate that improvements to the air traffic system could reduce annual fuel consumption 6–12%. For the total flight regime, 94% of fuel saving would occur at cruise altitude and 6% below 900 m. NO_x reductions are estimated to be 10–16%. This discussion of measures for mitigation leads to the following conclusions:

- Potential gains from operational measures within the current air traffic system are much smaller than those that may be gained through modernization of the air traffic system.
- Although modernizing the air traffic system and improved operational measures are important environmental policy instruments in their own right, delayed or early realization of their potential would have adverse or beneficial environmental consequences, respectively.

These conclusions have important environmental policy implications. Successful implementation of CNS/ATM is a daunting task that requires worldwide and regional collaboration and cooperation. Impediments to be overcome include restricted airspace, sovereignty issues, institutional development, and finance, particularly for developing nations and countries in transition.

10.4.2.2. Engine Emissions Stringency

ICAO Annex 16 emission standards are certification requirements for individual engines (ICAO, 1993b). Because of extensive regulation of the industry to maintain high levels of safety and to maintain and promote competition, as well as comparatively long technology development and product economic lifetimes (Chapter 7), ICAO has set limitations on emissions at their sources (stringency) on the basis of the best achievable technology by all manufacturers.

These ICAO standards define emissions levels for unburned HC, CO, NO_x, and smoke. They do not address aircraft emissions of CO_2, sulfur dioxide (SO_2), or H_2O. Neither are there any ICAO standards for trace compounds such as particulates, aerosols, certain types of HC, and other nitrogen compounds. Aviation is not a source of nitrous oxide (N_2O), a greenhouse gas. ICAO Annex 16 standards are related to engine emissions during the LTO cycle; they do not specifically address engine emission levels for altitudes above 900 m (cruise), although these emissions are related to emissions during the LTO cycle.

In 1981, the ICAO Council adopted standards and recommended practices for aircraft engine emissions by establishing regulatory emissions maxima for engines manufactured after 1985. The regulatory conformance criteria for covered emissions are based on characteristic engine performance during the LTO phases of flight, which represents on average of 15% of the total flight segment. LTO cycle operating requirements are characterized by very low and very high power settings, which are not representative of aircraft performance at cruise conditions. For purposes of estimation, LTO cycle parameters are transformed into an emissions index (EI) for a given airframe-engine combination to characterize emissions under cruise conditions.

On the recommendation of CAEP/2, in 1993 the ICAO Council amended the emissions standards for NO_x, reducing permitted levels by 20% at a representative engine pressure ratio of 30. The certification regime follows the LTO cycle basis established under the original stringency standard. The CAEP/2 standard was adopted by ICAO member states with few, if any, exceptions as the universally recognized international standard for aviation. It is to be applied to all engines produced after 2000, to all new and derivative engines for which certification has been or is to be applied for after 1995, and, as a practical matter, to currently certified in-production engines that are to be altered to meet the standard. CAEP has since reviewed the NO_x stringency issue in light of local air quality and atmospheric concerns and in 1998 recommended that the standard be further reduced by 16% at an engine pressure ratio of 30 (ICAO, 1998a). For pressure ratios above 30, the slope of the NO_x maxima was returned to that of the 1981 standard to permit greater fuel efficiency and therefore greater reductions in CO_2, H_2O, and oxides of sulfur (SO_x). Recognizing the need to protect the asset value of the existing fleet, the new standard is to be applied only to new or derivative engines certified after 31 December 2003. In-production engines are thereby left unaffected by the stringency requirement. The ICAO Council is scheduled to review the proposed standard and recommended practice (SARP) by the first quarter of 1999. Adoption by the Council is expected, after which the SARP will be incorporated into ICAO Annex 16 for adoption by member states.

Although there is a wide range of views concerning the scale of costs and benefits resulting from increased NO_x stringency, an analysis of the environmental benefits obtained to date under current ICAO engine stringency standards has not been initiated.

The three ICAO SARPs are shown in Figure 10-2 (ICAO, 1995a). It suggests that ICAO stringency requirements have "pushed" engine emissions reduction technology or, at a minimum, have ensured its incorporation into new and derivative engine designs. The relatively wide variation in technology levels among individual engine types and whole engine families represented in Figure 10-2 raises three important issues. For purposes of regulating aviation engine emissions, the first policy matter is whether use of the existing "best achievable" standard is preferable to the "best available" standard used for other industries. The second question concerns the optimal level at which the standard should be set to promote sustained and rapid progress on further emissions reduction. The third policy matter pertains to the means to evaluate tradeoffs concerning technical limitations in minimizing various emissions species and those pertaining to safety, performance, and environmental objectives, in particular noise control.

ICAO practice has inferred that the "best technology" concept has a specific meaning for aviation. The airworthiness concern is that setting standards based on unproved, anticipated, or nonexistent technology might result in untenable solutions, as might requirements that all engines meet a single, extremely low emissions threshold. A matter yet to be addressed is that selection of a "best available" technology or technology level for aircraft engines has inherent competition and market structure implications. Accordingly, as part of the work program adopted at CAEP/4, members called for looking into "establishing long-term and forward-looking CAEP goals for aircraft emission reductions" to meet the industry need for extended planning horizons while more aggressively "pushing" technology development (ICAO, 1998a).

ICAO certification standards are being regarded as a means for reducing overall emissions levels. Since CAEP/4, work has begun on developing parameters for aircraft emissions certification during climb and cruise phases of flight (above 900 m) to complement the existing LTO-based standard.

10.4.3. *Economic and Market-Based Measures*

IPCC (1996b), OECD (1989), and others have examined other potential economic instruments for mitigating aviation environmental effects:

- Fuel taxes and charges (levies) to promote fuel efficiency and reduce demand

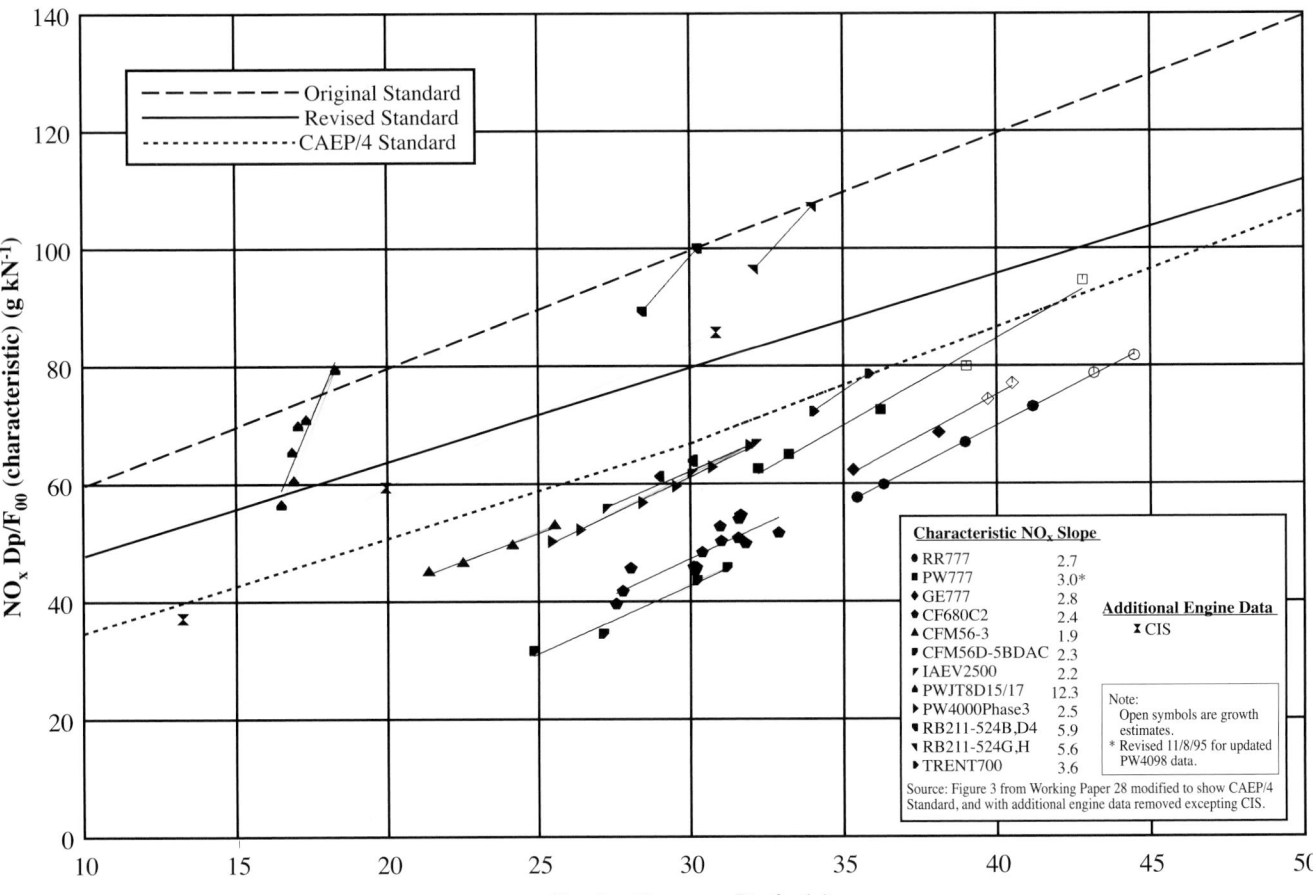

Figure 10-2: Engine certification to ICAO NO_x standards by engine type and engine.

- Emissions charges aimed at encouraging adoption of lower emitting technology
- Emissions trading to encourage emissions reductions through market forces
- Ticket taxes or charges
- Levies on empty aircraft seats to promote improvement in seat load factor
- Levies on excessive traffic per destination, destinations served, or type of equipment serving a destination
- Levies on route length to reduce the number of flights that are less than a minimum distance
- Subsidies or rebates to act as an incentive for polluters to change their behavior, such as grants, soft loans, tax allowances or differentiation, and instruments similar to effluent, product, or administrative charges.

Some of these measures are currently being considered by ICAO. Although examples of some of these and other measures exist in some countries, however, they have not been actively pursued by governments or ICAO.

10.4.3.1. Environmental Levies

Aviation is subject to various taxes and charges (grouped for convenience and referred to as levies in this report), most of which are related to the recovery of infrastructure costs and the provision of services. These levies consist primarily of landing charges imposed by airports on air service providers, route facility charges imposed by providers of air navigation services on operators, and passenger and cargo taxes based on the value of the ticket or waybill. The term environmental levy or emissions-related levy refers to charges and taxes. There are, however, relatively few examples of environmental charges or taxes related to emissions. The primary goal of such charges is to ensure that the external costs of the service provided are fully reflected in the charges paid. By internalizing external costs, environmental levies have the potential to reduce aircraft emissions by providing further incentives to develop and purchase low-emission technology, improve operational efficiency, and reduce demand via higher fares.

ICAO makes a distinction between charges and taxes and, because charges are based on cost of service and the revenue is retained by the aviation sector, has expressed a strong preference for the former. It regards charges as levies to defray the costs of providing facilities and services for civil aviation and taxes as levies to raise government revenues, which are applied to non-aviation purposes. In applying this principle to environmental levies, the ICAO Council Resolution on Environmental Charges and Taxes recommends that, where introduced, levies should be guided by the following principles:

- There should be no fiscal aims behind the charges.
- Charges should be related to costs.
- Charges should not discriminate against air transport compared with other modes of transport.

The Resolution was adopted in December 1996, and endorsed by the 32nd ICAO Assembly (ICAO, 1998b).

One of the recommendations at the CAEP/4 meeting in April 1998 (ICAO, 1998a) was to "identify and evaluate the potential role of market-based options, including emissions charges, fuel taxes, carbon offset, and emissions trading regimes." This recommendation was approved by the ICAO Council, and the 32nd ICAO Assembly subsequently requested that a conclusion be reached on guidance to be given to states on emissions-related levies in time for the next ICAO Assembly in 2001 (ICAO, 1998b). In its future work, CAEP will identify a range of market-based options, assess their limitations, identify their environmental effects, consider the application of revenues that might be accrued, and consider implementation mechanisms that might be employed.

ICAO does not, however, cover member states' tax policies for domestic aviation. The United States, for example, applies a fuel tax on domestic air carriers and general aviation aircraft; the receipts are held in the Airport and Airways Trust Fund and are prohibited by law from being used for non-aviation purposes. Similarly, some domestic aviation fuel is already taxed in Europe, where the status of the mandatory exemption of international aviation fuel from taxation has been under discussion for some time. In March 1997, the European Commission proposed in a draft directive that this aviation fuel tax exemption should remain until the international legal situation permits imposition of such a tax (European Commission, 1997).

Zurich Airport has added an emissions surcharge to the landing fee based on engine certification information contained in the ICAO Engine Exhaust Emissions Data Bank (ICAO, 1995b; Zurich Airport Authority, 1997). This charge was intended primarily to provide an incentive to encourage operators to use their lowest emissions aircraft into Zurich and to accelerate the use of best available technology. Revenues are used to finance emissions reduction measures at the airport. A similar emissions-related charge was applied at 10 Swedish airports beginning 1 January 1998. These charges are revenue neutral and do not affect consumer demand. They do, however, provide an incentive to airlines to purchase and operate aircraft with lower engine emissions. These charges are considered consistent with ICAO principles.

Concern has been expressed that uncoordinated introduction of emissions charges at the national, sub-federal, or regional level or by airports will have international repercussions because they are discriminatory to some aircraft that comply with internationally recognized standards, and the original intention of the NO_x certification standards was not to set local environmental restrictions, including levies. One implementation issue that would need to be addressed with fuel taxation is that international taxation of aviation fuel is currently precluded by provisions contained in many bilateral air service agreements between countries, the main legal framework underlying the operation of international civil aviation. Three studies assessing the economic and environmental impact of environmental levies are considered below.

A 1997 OECD study (OECD, 1997a) found that although fuel price increases have had little impact on the demand for air travel, the rate of energy intensity reduction in civil aviation has been very responsive to fuel price. On the basis of historical data, the study concluded that, when implemented on an international basis, a meaningful rise in fuel prices introduced at a moderate rate each year could have a large impact on increasing the rate of energy intensity reduction. The study considered fuel charge options equivalent to 2, 10, and 50% of the price of aviation fuel, and considered the use of revenues to reduce the general level of taxation and to fund research and development within the aviation industry.

Although the OECD study did not attempt to identify a direct relationship between fuel price and energy intensity over the scenario period, it did suggest that cumulative fuel levies resulting from an increase in the price of aviation fuel of up to 5% per year could result in at least a 30% reduction in aviation energy use in 2020 relative to the reference scenario. However, if charges were not applied at a uniform global level, the study concluded that the increase in fuel price would lead to a distortion of competition and a weakening of incentives to develop and adopt energy-efficient aircraft. It further concluded that a fuel charge was unlikely to have any substantial long-term effect on air traffic growth, particularly if it were introduced gradually.

In its report to CAEP/4, the Focal Point on Charges (FPC) (ICAO, 1998a) considered the potential economic and environmental impacts of various environmental levies. It concluded that, among the options studied, the most effective options for addressing global emissions were a fuel levy and en route charges. The report showed that any resulting cost increases passed on to consumers would result in a reduction in emissions. This result would occur primarily through lower traffic demand, but airline and manufacturer supply-side responses also would be stimulated, with limited impact on airline operating results. The portion of the fuel price increase not passed on to customers would be borne by the airlines, affecting their profitability, cash flow, and retained earnings—which, in turn, could affect the ability of airlines to purchase more environmentally beneficial equipment.

The report considered alternative applications of the proceeds of environmental levies. The revenue-neutral approach did not cause a problem with the redistribution of revenues, but the environmental benefits were found to be limited. Although general taxes are feasible on implementation grounds, they raise serious problems of equity and acceptability. A prevention cost approach using revenues to fund future technology improvement was regarded as better than general taxation in reducing total emissions; however, rechanneling of revenues to the aviation industry would give rise to administrative complexities and raises equity problems and risks of distortion of competition.

A feasibility study by the Centre for Energy Conservation and Environmental Technology (CE) (Bleijenberg and Wit, 1998)

considered the feasibility of introducing an environmental charge on civil aviation in Europe. Using a scenario-based approach and an aviation fuel price to include the environmental cost at 125% of the cost of the fuel, the study found that the rate of growth in aviation emissions in Europe would be approximately halved. A charge based on calculated emissions was found to have similar environmental effectiveness to a fuel charge, but the former had smaller economic distortions and fewer legal obstacles than the latter. The study found that an emissions charge in European airspace would have little impact on competition between domestic and non-European carriers. A revenue-neutral emissions charge was judged to be the most feasible option, with high environmental effectiveness and relatively few economic distortions.

Several comparisons may be made of the main features of these three studies. In the FPC study, most of the benefits arose from a reduction in the growth of demand, whereas in the OECD and CE studies, the supply-side influence was predominant. Partly as a consequence of the strong supply-side effect, the OECD and CE studies estimated a larger reduction in total emissions than did the FPC report. Both the FPC and the OECD considered options to rechannel revenues for technology improvement within the aviation sector. The FPC study found greater distortions to competition between airlines arising from a European emissions or fuel charge than the CE study.

Two areas require further study. The first is the applicability of environmental levies to the circumstances of countries other than those in OECD and countries in transition; the other is how revenues generated from a levy would be used.

10.4.3.2. Other Market Approaches

Emissions trading of greenhouse gases has been adopted as part of the Kyoto Protocol as a potential means of achieving reductions in these gases at the lowest possible costs. Emissions trading allows market forces to operate to achieve the lowest possible costs of achieving an environmental goal. It can provide companies such as airlines with the flexibility to reduce their own emissions or to purchase equivalent reductions from others, if doing so would be less expensive. It gives firms the incentive to employ innovative technologies and reduce emissions beyond what any standard would require.

Markets play a central role in the efficient exchange of commodities, shares, bonds, and financial instruments. Existing international markets have well-established practices for contracting, delivery, and settlement that could be applicable to emissions trading. Emissions trading differs considerably from the more traditional standards approach that, for example, has been used by ICAO. Under that approach, a specific emissions limit has been established, and each aircraft must meet the standard. Under emissions trading, an overall level of emissions production is set, and firms are allowed flexibility to jointly meet that standard. Firms that can achieve low-cost reductions to meet their requirements have an incentive to reduce below the required levels and sell their excess emissions reductions to other firms. Firms facing higher control costs can purchase these reductions to comply with their requirements at lower costs than by using alternative means.

A credible system of monitoring and verifying emissions reductions that allows for trading with minimal transaction costs would be needed to achieve the cost savings potential of these mechanisms. As in other areas of environmental policy, an emissions trading regime would be likely to meet environmental objectives at the lowest cost because it sets overall environmental goals, provides geographic and temporal flexibility, would allow for flexible trading across industry boundaries, and would offer incentives for meeting the goals (Dudek and Goffman, 1997).

Among OECD countries, the efficacy of emissions trading in meeting the objectives of the Kyoto Protocol has been compared to direct market and regulatory intervention. Some observers have suggested emissions trading is difficult to enforce and raises potentially difficult liability issues. Others have expressed concern about the possibility that companies with excess reductions will not sell them, thus creating a barrier to new entrants in the market. However, successful allowance-based SO_2 emissions trading among competitive electric power generation companies in the United States favored emissions trading over other options. Monitoring, reporting, verification, and certification systems need to be explicitly defined and developed.

Emissions trading has been used in the United States in the control of SO_2 emissions responsible for acid rain. Flexibility in this program has resulted in pollution permit prices of about $100 per ton, compared to estimated prices before the program was implemented on the order of $250–400 per ton (Council of Economic Advisors, 1998a). In addition, modeling indicates that programs using tradable permits could enable more cost-effective control of local and global atmospheric pollutants than other regulatory options under consideration (IPCC, 1996d; GAO, 1997). Emissions trading has been incorporated in the Montreal Protocol, the international treaty that limited ozone-depleting substances. Under that treaty, restrictions were placed on several specific chlorofluorocarbon (CFC) compounds as a group, instead of separate restrictions on each compound. Countries could permit firms to trade among CFCs, reducing those that cost the least first. In addition, in a separate provision, production permits were allowed to be traded across nations to allow for a lower cost, more orderly phaseout of manufacturing facilities (Montreal Protocol, 1987).

The Kyoto Protocol (UNFCCC, 1998b) contains emissions trading provisions that provide substantial flexibility for nations to reduce their costs of meeting agreed emissions goals. As with the Montreal Protocol, targets for greenhouse gases were not set for individual compounds but rather as a single comprehensive goal combining six major greenhouse gases (sources and sinks). Nations can adopt plans that minimize the costs of meeting this target based on the relative costs of controls among these different categories of greenhouse gases.

The Kyoto Protocol provisions allow developed nations to trade emissions on a project-by-project basis (Article 6) and through an emissions trading system (Article 17). A clean development mechanism (Article 12) was established to allow developed countries to support and get emissions credit for actions in conjunction with developing countries that reduce emissions in those countries.

Relevant principles, modalities, rules, guidelines for emissions trading, and the role of governments are being discussed by the Conference of the Parties to the UNFCCC and its subsidiary bodies, with the goal of reaching agreement before the end of the year 2000. A credible system of monitoring and verifying reductions that allows for trading with minimum transaction costs will be needed to achieve the cost savings potential of these mechanisms. International aviation emissions are not covered by the emissions-related targets in the Kyoto Protocol. The prerequisite for emissions trading is adoption of emissions reduction targets or caps. In principle, the aviation sector could be included in the emissions targets agreed in the Kyoto protocol, but the feasibility of applying an emissions trading regime depends on establishing a method to allocate international aviation bunker fuels. Emissions trading would likely be available across all industries, allowing progress in emissions reduction at the lowest cost. High-cost compliance industries with limited compliance options could purchase rights from lower-cost producers of other commodities.

An analysis of the potential for reducing the costs of meeting the Kyoto targets from these flexible, market-based approaches was conducted within the United States. This study suggests that reductions in permit prices on the order of 70–90% were possible because of these provisions (Council of Economic Advisors, 1998b).

10.4.4. Other Measures (Communication, Voluntary Agreements, Intermodality, Research)

In contrast with some other sectors, voluntary agreements within the aviation industry to meet environmental targets without recourse to regulatory standards have been rare. In its consultation paper on air transport and the environment, the European Commission has recommended that the industry examine options for establishing voluntary environmental agreements. A further recommendation was that better information should be provided to consumers about the environmental performance of alternative air travel options (European Commission, 1998). A similar consultative process is occurring in the United States, where the White House Climate Change Task Force is seeking voluntary reductions in greenhouse gas emissions from many industries, including aviation.

In addition, the Kyoto Protocol and the following Conferences of Parties have stimulated interest in the possible role of other voluntary measures (such as carbon sequestration) as a further way to mitigate the effects of CO_2 emissions. At present, there is no clear picture of the role such measures could play in joint implementation and other mechanisms that may have international significance in the limitation of CO_2 emissions.

One additional potential mitigation measure is encouragement of surface transport—principally rail—in place of air travel, through the use of economic instruments or restrictions on the use of air travel on certain routes. Intermodal substitution is considered in Section 8.3.3 of this report, where it is noted that the scope for substituting other modes for air is up to 10%. Moreover, comparison of carbon emissions among different modes of transport is highly dependent on the type of aircraft, train, or car and the type of service.

Interactions between emissions and climate effects are not fully understood; research efforts on the environmental impact of aircraft emissions can assist the decisionmaking process. Research relating to the atmospheric effects of aircraft emissions and their mitigation through technological and operational measures is taking place throughout the world. The European Commission and the U.S. government sponsor substantial programs. These efforts are also complemented with aviation-focused scientific assessments that are cited in other chapters of this report.

10.4.5.   *Open Questions For Future Research and Policy Considerations*

Based on full consideration of mitigation options available to policymakers to influence aviation emissions, if needed, and underlying information requirements, the main areas that require future research follow:

- Uncertainties over future trends in traffic, technology, and therefore emissions, depending on the scenarios chosen. Factors underlying this uncertainty include uncertainties about the pace of introduction of CNS/ATM and the impact of infrastructure constraints in limiting growth in demand. These future projections and scenarios form the base case against which policy measures are considered.
- The impact of aviation emissions on the environment and the monetary value associated with the benefits of mitigating those impacts. There is uncertainty about the appropriate level at which any environmental levy should be set.
- The impact of deregulation and subsidies to airlines on fuel consumption and emissions.
- The environmental benefits of ICAO engine emissions stringency standards.
- How the "best available" technology concept used in other sectors might be applied in aviation.
- The impact of various options on encouraging adoption of technological changes.
- The applicability of various market options to the circumstances of countries not included in Annex 1 to the UNFCCC.
- Practical experience with emissions trading at the global level.

References

Aerospace Industries Association of America, 1998: *Aerospace Facts and Figures,* Washington, DC, USA, various years.

Association of European Airlines, 1998: *Press Release* dated October 19, 1998, and other dates.

Aylesworth, H. Jr., 1997: *Use of the Global Warming Commitment Concept to Measure the Benefits of Differing Policy Approaches to Reduce the Atmospheric Effects of Aviation.* Working Paper WG3-WP3/20, International Civil Aviation Organization Committee on Aviation Environmental Protection Working Group 3 (Emissions), Third Meeting, Savannah, GA, USA, 20–23 May 1997, 14 pp. + appendix.

Balashov, B. and A. Smith, 1992: ICAO analyses—trends in fuel consumption by world's airlines. *ICAO Journal,* **47(6),** 18–21.

Bleijenberg, A.N. and R.C.N. Wit, 1998: *A European Environmental Aviation Charge, Feasibility Study.* Final Report, Center for Energy Conservation and Environmental Technology, Delft University, March 1998, 195 pp.

Council of Economic Advisors, 1998a: *Economic Report of the President.* United States Government Printing Office, Washington, DC, USA, 449 pp.

Council of Economic Advisors, 1998b: *The Kyoto Protocol and the President's Policies to Address Climate Change.* U.S. Council of Economic Advisors, Washington, DC, USA, July 1998, 80 pp. + appendices.

Dudek, D. and J. Goffman, 1997: *Emissions Budgets: Building an Effective International Greenhouse Gas Control System.* Paper presented by invitation to the United Nations Framework Convention on Climate Change (UNFCCC), Ad Hoc Group on the Berlin Mandate (AGBM), Geneva, Switzerland, February 1997, 26 pp.

EASG, 1995: *Final Report of the Economic Analysis Subgroup Regarding Economic Analysis of CAEP/3 Noise and Emissions.* Working Paper 4.4, Economic Analysis Subgroup of the Committee on Aviation Environmental Protection Coordination Working Group (WG4), International Civil Aviation Organization Committee on Aviation Environmental Protection, Bonn, Germany, June 1995, 51 pp. + appendices.

EISG, 1995: *Report of the Emissions Inventory Sub-Group.* Emissions Inventory Subgroup of the Committee on Aviation Environmental Protection Emissions Working Group (WG3), International Civil Aviation Organization Committee on Aviation Environmental Protection, Bonn, Germany, June 1995, 45 pp.

EUROCONTROL, 1998: *Delays to Air Transport in Europe (CODA Report).* EUROCONTROL, Brussels, Belgium, various years.

European Commission, 1997: *Proposal for a Council Directive Restructuring the Community Framework for the Taxation of Energy Products.* COM (97) 0030, European Commission, Brussels, Belgium, 17 March 1997.

European Commission, 1998: *Consultation Paper on Air Transport and the Environment.* European Commission, Directorate General for Transport, Brussels, Belgium, July 1998.

FESG, 1998: *Report of the Forecasting and Economic Support Group (FESG) to CAEP/4.* International Civil Aviation Organization Committee on Aviation Environmental Protection, Montreal, Canada, 6–8 April 1998, 132 pp.

GAO, 1997: *Air Pollution: Overview and Issues on Emissions Allowance Trading Programs.* Testimony Before the Joint Economic Committee, Congress of the United States, July 1997 by P. Guerrero, Director, Environmental Protection Issues, Resources, Community and Economic Development Division. GAO/T-RCED-97-183, U.S. General Accounting Office, USA, 14 pp.

ICAO, 1989: *Economic Implications of Future Noise Restrictions on Subsonic Jet Aircraft.* Circular 218-AT/86, International Civil Aviation Organization, Montreal, Canada, 60 pp.

ICAO, 1993a: *International Standards and Recommended Practices— Environmental Protection.* Annex 16 to the Convention on International Civil Aviation, Volume I: Aircraft Noise, 3rd ed. International Civil Aviation Organization, Montreal, Canada, July 1993, 130 pp.

ICAO, 1993b: *International Standards and Recommended Practices— Environmental Protection.* Annex 16 to the Convention on International Civil Aviation, Volume II: Aircraft Engine Emissions, 2nd ed. International Civil Aviation Organization, Montreal, Canada, July 1993, 63 pp.

ICAO, 1994: *ICAO's Policies on Taxation in the Field of International Air Transport.* Document no. 8632-C/968, International Civil Aviation Organization, Montreal, Canada, 2nd ed., 19 pp.

ICAO, 1995a: *ICCAIA Position on Stringency of No$_x$ Emissions Regulations.* Working Paper 28, Committee on Aviation Environmental Protection Working Group 3, December 5–15, 1995, Montreal, Canada, Figure 10.2 adapted from underlying document.

ICAO, 1995b: *ICAO Engine Emissions Data Bank.* Document no. 9646, International Civil Aviation Organization, Montreal, Canada, 177 pp.

ICAO, 1996a: *Report of the Committee on Aviation Environmental Protection.* Committee on Aviation Environmental Protection Working Group 3, Third Meeting, Montreal, Canada, 5–15 December 1995. Document no. 9675, International Civil Aviation Organization, Montreal, Canada, 149 pp.

ICAO, 1996b: *Civil Aviation Statistics of the World.* Document no. 9180, International Civil Aviation Organization, Montreal, Canada, various years, 200 pp.

ICAO, 1997a: *Convention of International Civil Aviation.* Document no. 7300/7, International Civil Aviation Organization, Montreal, Canada, 7th ed., 52 pp.

ICAO, 1997b: *Statements of the Council to Contracting States on Charges for Airports and Air Navigation Services.* Document no. 9082/5, International Civil Aviation Organization, Montreal, Canada, 5th ed., 23 pp.

ICAO, 1997c: *International Civil Aviation Organization Strategic Action Plan.* State Letter M 2/5-97/20 (with Appendices A and B), 14 March 1997. International Civil Aviation Organization, Montreal, Canada, 20 pp.

ICAO, 1998a: *Report of the Committee on Aviation Environmental Protection.* Committee on Aviation Environmental Protection Working Group 4, Fourth Meeting, Montreal, Canada, 6–8 April 1998. Document no. 9720, International Civil Aviation Organization, Montreal, Canada, 99 pp.

ICAO, 1998b: *Report of the Executive Committee on Agenda Item 21 to the 32nd Session of the ICAO Assembly.* A32-WP/243, International Civil Aviation Organization, Montreal, Canada, October 1998, 19 pp.

ICAO, 1998c: *Statement from the International Civil Aviation Organization (ICAO) to the Fourth Session of the Conference of the Parties to the United Nations Framework Convention on Climate Change.* International Civil Aviation Organization, Buenos Aires, Argentina.

IPCC, 1996a: *Climate Change 1995: Impacts, Adaptations, and Mitigation of Climate Change: Scientific-Technical Analyses. Contribution of Working Group II to the Second Assessment Report of the Intergovernmental Panel on Climate Change* [Watson, R.T., N.C. Zinyowera, and R.H. Moss (eds.)]. Cambridge University Press, Cambridge, United Kingdom and New York, NY, USA, 880 pp.

IPCC, 1996b: *Technologies, Policies, and Measures for Mitigating Climate Change: IPCC Technical Paper 1.* Intergovernmental Panel on Climate Change Working Group II [Watson, R.T., M.C. Zinyowera, and R.H. Moss (eds.)]. World Meteorological Organization, Geneva, Switzerland, 85 pp.

IPCC, 1996c: *The IPCC Second Assessment, Synthesis of Scientific-Technical Information Relevant to Interpreting Article 2 of the UN Framework Convention on Climate Change.* World Meteorological Organization, Geneva, Switzerland, 64 pp.

IPCC, 1996d: *Climate Change 1995: Economic and Social Dimensions of Climate Change. Contribution of Working Group III to the Second Assessment Report of the Intergovernmental Panel on Climate Change* [Bruce, J.P., H. Lee, and E.F. Haites (eds.)] Cambridge University Press, Cambridge, United Kingdom and New York, NY, USA, 448 pp.

IPCC, 1996e: *Revised 1996 IPCC Guidelines for National Greenhouse Gas Inventories: Reference Manual.* Intergovernmental Panel on Climate Change.

Montreal Protocol, 1987: *Protocol on Substances that Deplete the Ozone Layer.* Article 1, Paragraph 8, Definition of "industrial rationalization." 16 September 1987, Montreal, Canada.

OECD, 1989: *Economic Instruments for Environmental Protection.* Organization for Economic Cooperation and Development, Paris, France, 132 pp.

OECD, 1997a: *Special Issues in Carbon/Energy Taxation: Carbon Charges on Aviation Fuels.* OECD/GD(97)78, Policies and Measures for Common Action, Working Paper 12, Organization for Economic Cooperation and Development, Annex 1 Expert Group on the United Nations Framework Convention for Climate Change, Paris, France, March 1997, 67 pp.

OECD, 1997b: *Sustainable Development, OECD Policy Approaches for the 21st Century.* Organization for Economic Cooperation and Development, Paris, France, 1997, 190 pp.

Sutkus, Jr., D.J., 1997: *2050 Fuel Efficiency and NO$_x$ Technology Scenarios.* Paper submitted to a meeting of Working Group 3 of the Committee on Aviation Environmental Protection, Bern, Switzerland, November 1997.

UNCED, 1992: *Report of the United Nations Conference on Environment and Development - Annex I: Rio Declaration on Environment and Development.* A/CONF.151/26 (vol. 1), Rio de Janeiro, 3-14 June 1992. 12 August 1992, United Nations, New York, NY, USA.

UNFCCC, 1992: *United Nations Framework Convention on Climate Change.* Adopted 9 May 1992 in New York, NY, USA.

UNFCCC, 1996: *Communications from Parties included in Annex I to the Convention: Guidelines, Schedules, and Process for Consideration.* Secretariat note FCCC/SBSTA/ 1996/9/Add. 1, United Nations Framework Convention on Climate Change, 8 July 1996, Geneva, Switzerland, 22 pp.

UNFCCC, 1997: *Report of the Subsidiary Body for Scientific and Technological Advice on the Work of its Fourth Session, Geneva, Switzerland, 16–18 December 1996.* FCCC/SBSTA/1996/20, 27 January 1997, United Nations Framework Convention on Climate Change, 16 pp. + annex.

UNFCCC, 1998a: *Addendum to the Report of the Conference of the Parties on Its Third Session, Kyoto, Japan, 1–11 December 1997.* FCCC/CP/1997/ 7/Add. 1, United Nations Framework Convention on Climate Change, 18 March 1998, Kyoto, Japan, Decision 2/CP.3, 60 pp.

UNFCCC, 1998b: *Addendum to the Report of the Conference of the Parties on Its Third Session, Kyoto, Japan, 1–11 December 1997.* FCCC/CP/1997/ 7/Add. 1, United Nations Framework Convention on Climate Change, 18 March 1998, Kyoto, Japan, Decision 1/CP.3, 60 pp.

U.S. Department of Transportation, 1995: *Report to Congress, Child Restraint Systems.* Report of the Secretary of Transportation to the United States Congress pursuant to Section 522 of the Federal Aviation Administration Authorization Act of 1994, Public Law 103-305, Volume II, Appendix G. Department of Transportation, Washington, DC, USA, 14 pp.

U.S. Department of Transportation, 1998: *Air Travel Consumer Report.* Office of Aviation Enforcement and Proceedings, Department of Transportation, Washington, DC, USA, March 1999 and various editions, Flight Delay section, 29 pp.

U.S. Department of Transportation and Federal Aviation Administration, 1991: *Final Regulatory Impact Analysis, Final Regulatory Flexibility Determination, and Trade Impact Determination, Transition to an All Stage 3 Fleet Operating in the 49 Contiguous United States of America and the District of Columbia.* Office of Aviation Policy and Plans and Office of Environment and Energy, Federal Aviation Administration, Washington, DC, USA, 23 September 1991, 127 pp. + appendix.

U.S. Department of Transportation and Federal Aviation Administration, 1998: *Economic Analysis of Investment and Regulatory Decision—Revised Guide.* FAA-APO-98-4, Federal Aviation Administration, Washington, DC, USA, January 1998, 133 pp.

U.S. Federal Aviation Administration, 1995: *Total Cost for Air Carrier Delay for the Years 1987–1994.* Office of Aviation Policy and Plans, Federal Aviation Administration, Washington, DC, USA, December 1995, 4 pp. + appendices.

U.S. Federal Aviation Administration, 1997: Air Traffic Activity and Delay Report. December 1997 and various dates, 13 pp.

U.S. Federal Aviation Administration, 1998a: *Impact of CNS/ATM Improvements on Aviation Emissions.* WW/IMP-IP7, Paper presented at the International Civil Aviation Organization World-Wide CNS/ATM Systems Implementation Conference, Rio de Janeiro, Brazil, 11–15 May 1998, 4 pp.

U.S. Federal Aviation Administration, 1998b: *The Impact of National Airspace Systems (NAS) Modernization on Aircraft Emissions.* FAA DOT/FAA/SD-400-98/1, Federal Aviation Administration, Washington, DC, USA, September 1998, 38 pp. + appendices.

Winston, C., 1998: U.S. industry adjustment to economic regulation. *The Journal of Economic Perspectives,* **12(3),** 89–110.

Zurich Airport Authority, 1997: *Aircraft Engine Emission Charges at Zurich Airport.* Zurich Airport Authority, Environmental Protection, Zurich, Switzerland, April 1997, 10 pp.

AVIATION AND THE GLOBAL ATMOSPHERE: ANNEXES

Prepared by IPCC Working Groups I and II

A

Authors, Contributors,
and Expert Reviewers

AUTHORS AND CONTRIBUTORS

Chapter 1. Introduction

Lead Authors

J.H. Ellis	Federal Express Corporation, USA
N.R.P. Harris	European Ozone Research Coordination Unit, United Kingdom
D.H. Lister	DERA, United Kingdom
J.E. Penner	University of Michigan, USA

Review Editor

B.S. Nyenzi	Directorate of Meteorology, Tanzania (United Republic of)

Chapter 2. Impacts of Aircraft Emissions on Atmospheric Ozone

Coordinating Lead Authors

R.G. Derwent	Meteorological Office, United Kingdom
R.R. Friedl	Jet Propulsion Laboratory, USA

Lead Authors

I.L. Karol	Main Geophysical Observatory, Russia
H. Kelder	KNMI, The Netherlands
V.W.J.H. Kirchhoff	Instituto Nacional de Pesquisas Espaciais, Brazil
T. Ogawa	National Space Development Agency of Japan, Japan
M.J. Rossi	Ecole Polytechnique Federale de Lausanne, Switzerland
P. Wennberg	California Institute of Technology, USA

Contributors

T.K. Berntsen	University of Oslo, Norway
C.H. Brühl	Max-Planck Institut für Chemie, Germany
D. Brunner	KNMI, The Netherlands
P. Crutzen	Max-Planck Institut für Chemie, Germany
M.Y. Danilin	Atmospheric and Environmental Research Inc., USA
F.J. Dentener	Institute for Marine and Atmospheric Research, The Netherlands
L. Emmons	Space Physics Research Laboratory, USA
F. Flatoy	Geophysics Institute, Norway
J.S. Fuglestvedt	University of Oslo, Norway
T. Gerz	DLR Institut für Physik der Atmosphäre, Germany
V. Grewe	DLR Institut für Physik der Atmosphäre, Germany
D. Hauglustaine	CNRS Service Aeronomie, France
G. Hayman	AEA Technology, United Kingdom
Ø. Hov	Ministry of Environment, Norway
D. Jacob	Harvard University, USA
C.E. Johnson	Meteorological Office, United Kingdom
M. Kanakidou	Centre des Faibles Radioactivités, France
B. Kärcher	DLR Institut für Physik der Atmosphäre, Germany
D.E. Kinnison	Lawrence Livermore National Laboratory, USA
A.A. Kiselev	Main Geophysical Observatory, Russia
I. Köhler	DLR Institut für Physik der Atmosphäre, Germany
J. Lelieveld	Institute of Marine and Atmospheric Research, The Netherlands
J. Logan	Harvard University, USA
J.-F. Müller	Belgian Institute for Space Aeronomy, Belgium
J.E. Penner	University of Michigan, USA
H. Petry	University of Köln, Germany,
G. Pitari	Università Degli Studi dell' Aquila, Italy
R. Ramaroson	ONERA, France
F. Rohrer	Forschungszentrum Jülich, Germany
E.V. Rozanov	University of Illinois, USA
K. Ryan	CSIRO, Australia
R.J. Salawitch	Jet Propulsion Laboratory, USA
R. Sausen	DLR Institut für Physik der Atmosphäre, Germany
U. Schumann	DLR Institut für Physik der Atmosphäre, Germany
R.F. Slemr	Fraunhofer-Gesellschaft, Institut für Atmosphärische, Umweltforschung, Germany
D. Stevenson	Meteorological Office, United Kingdom
F. Stordal	Norwegian Institute for Air Research, Norway
A. Strand	Geophysics Institute, Norway
A. Thompson	NASA Goddard Space Flight Center, USA
G. Velders	RIVM, The Netherland
P.F.J. van Velthoven	KNMI, The Netherlands
P. Valks	RIVM, The Netherlands
Y. Wang	Georgia Institute of Technology, USA
X.W.M.F. Wauben	KNMI, The Netherlands
D.K. Weisenstein	Atmospheric and Environmental Research Inc., USA

Review Editor

A. Wahner	Forschungszentrum Jülich, Germany

Chapter 3. Aviation-Produced Aerosols and Cloudiness

Coordinating Lead Authors

D.W. Fahey	NOAA Aeronomy Laboratory, USA
U. Schumann	DLR Institut für Physik der Atmosphäre, Germany

Lead Authors

S. Ackerman	University of Wisconsin, USA
P. Artaxo	University of Sao Paulo, Instituto de Fisica, Brazil
O. Boucher	Université de Lille, France
M.Y. Danilin	Atmospheric and Environmental Research Inc., USA
B. Kärcher	DLR Institut für Physik der Atmosphäre, Germany
P. Minnis	NASA Langley Research Center, USA
T. Nakajima	Center for Climate System Research, University of Tokyo, Japan
O.B. Toon	University of Colorado, USA

Contributors

J.K. Ayers	Analytical Services and Materials Inc., USA
T.K. Berntsen	Center for International Climate and Environmental Research, Norway
P.S. Connell	Lawrence Livermore National Laboratory, USA
F.J. Dentener	Utrecht University, Institute for Marine and Atmospheric Research, The Netherlands
D.R. Doelling	NASA Langley Research Center, USA
A. Döpelheuer	DLR Institut für Antriebstechnik, Germany
E.L. Fleming	NASA Goddard Space Flight Center, USA
K. Gierens	DLR Institut für Physik der Atmosphäre, Germany
C.H. Jackman	NASA Goddard Space Flight Center, USA
H. Jäger	Fraunhofer-Gesellschaft, Institut für Atmosphärische Umweltforschung, Germany
E.J Jensen	NASA Ames Research Center, USA
G.S. Kent	NASA Langley Research Center, USA
I. Köhler	DLR Institut für Physik der Atmosphäre, Germany
R. Meerkötter	DLR Institut für Physik der Atmosphäre, Germany
J.E. Penner	University of Michigan, USA
G. Pitari	Università Degli Studi dell' Aquila, Italy
M.J. Prather	University of California at Irvine, USA
J. Ström	University of Stockholm, Sweden

Y. Tsushima	Center for Climate System Research, University of Tokyo, Japan,
C.J. Weaver	NASA Goddard Space Flight Center, USA
D.K. Weisenstein	Atmospheric and Environmental Research Inc., USA

Review Editor

K.-N. Liou	UCLA, Department of Atmospheric Sciences, USA

Chapter 4. Modeling the Chemical Composition of the Future Atmosphere

Coordinating Lead Authors

I. Isaksen	University of Oslo, Norway
C.H. Jackman	NASA Goddard Space Flight Center, USA

Lead Authors

S.L. Baughcum	The Boeing Company, USA
F.J. Dentener	Institute for Marine and Atmospheric Research, The Netherlands
W.L. Grose	NASA, USA
P. Kasibhatla	Duke University, USA
D.E. Kinnison	Lawrence Livermore National Laboratory, USA
M.K.W. Ko	Atmospheric and Environmental Research Inc., USA
J.C. McConnell	York University, Canada
G. Pitari	Università Degli Studi dell' Aquila, Italy
D. Wuebbles	University of Illinois, USA

Contributors

T.K. Berntsen	University of Oslo, Norway
M.Y. Danilin	Atmospheric and Environmental Research Inc., USA
R.S. Eckman	NASA Langley Research Center, USA
E.L. Fleming	NASA Goddard Space Flight Center, USA
M. Gauss	University of Oslo, Norway
V. Grewe	DLR Institut für Physik der Atmosphäre, Germany
R.S. Harwood	University of Edinburgh, United Kingdom
D. Jacob	Harvard University, USA
H. Kelder	KNMI, The Netherlands
J.-F. Muller	Belgian Institute for Space Aeronomy, Belgium
M.J. Prather	University of California at Irvine, USA
H. Rogers	University of Cambridge, United Kingdom

R. Sausen	DLR Institut für Physik der Atmosphäre, Germany
D. Stevenson	Meteorological Office, United Kingdom
P.F.J. van Velthoven	KNMI, The Netherlands
M. van Weele	KNMI, The Netherlands
P. Vohralik	CSIRO, Australia
Y. Wang	Georgia Institute of Technology, USA
D.K. Weisenstein	Atmospheric and Environmental Research Inc., USA

Review Editor
| J. Austin | Meteorological Office, United Kingdom |

Chapter 5. Solar Ultraviolet Irradiance at the Ground

Coordinating Lead Authors
| J.E. Frederick | University of Chicago, USA |
| K.R. Ryan | CSIRO, Australia |

Lead Authors
A.F. Bais	Aristotle University of Thessaloniki, Greece
J.B. Kerr	Atmospheric Environment Service, Canada
B. Wu	Institute of Atmospheric Physics, China (Deceased)

Contributors
| R. Meerkötter | DLR Institut für Physik der Atmosphäre, Germany |
| I.C. Plumb | CSIRO, Australia |

Review Editor
| C. Zerefos | Aristotle University of Thessaloniki, Greece |

Chapter 6. Potential Climate Change from Aviation

Coordinating Lead Authors
| M.J. Prather | University of California at Irvine, USA |
| R. Sausen | DLR Institut für Physik der Atmosphäre, Germany |

Lead Authors
A.S. Grossman	Lawrence Livermore National Laboratory, USA
J.M. Haywood	Meteorological Office, United Kingdom
D. Rind	NASA Goddard Institute of Space Studies, USA
B.H. Subbaraya	ISRO HQ, India

Contributors
P.M. Forster	University of Reading, United Kingdom
A.K. Jain	University of Illinois, USA
M. Ponater	DLR Institut für Physik der Atmosphäre, Germany
U. Schumann	DLR Institut für Physik der Atmosphäre, Germany
W.-C. Wang	State University of New York at Albany, USA
T.M.L. Wigley	NCAR, USA
D.J. Wuebbles	University of Illinois, USA

Review Editor
| D. Yihui | China Meteorological Administration, China |

Chapter 7. Aircraft Technology and Its Relation to Emissions

Coordinating Lead Authors
| J.S. Lewis | Rolls Royce Plc., United Kingdom |
| R. Niedzwiecki | NASA Lewis Research Center |

Lead Authors
D.W. Bahr	G.E. Aircraft Engines, USA (Retired)
S. Bullock	Rolls-Royce Plc, United Kingdom
N. Cumpsty	Cambridge University, United Kingdom
W. Dodds	General Electric Company, USA
D. DuBois	The Boeing Company, USA
A. Epstein	Massachusetts Institute of Technology, USA
W.W. Ferguson	Pratt & Whitney Aircraft, USA
A. Fiorentino	Pratt & Whitney Aircraft, USA
A.A. Gorbatko	CIAM, Russian Federation
D.E. Hagen	University of Missouri-Rolla, USA
P.J. Hart	Allison Engine Company, USA
S. Hayashi	National Aerospace Laboratory, Japan
J.B. Jamieson	Consultant, United Kingdom
J. Kerrebrock	MIT, USA
M. Lecht	DLR Institut für Antriebstechnik Germany
B. Lowrie	Rolls-Royce Plc, United Kingdom
R.C. Miake-Lye	Aerodyne Research Inc., USA
A.K. Mortlock	The Boeing Company, USA
C. Moses	Southwest Research Institute, USA
K. Renger	Airbus Industrie, France
S. Sampath	Pratt & Whitney Canada Inc., Canada
J. Sanborn	Allied Signal Engines, USA
B. Simon	Motoren-und Turbinen-Union, Germany
A. Sorokin	CLAM ECOLEN, Russia
W. Taylor	Exxon Research and Energy Co, USA

I. Waitz	MIT Gas Turbine Laboratory, USA
C.C. Wey	Army Vehicle Propulsion Directorate, USA
P. Whitefield	University of Missouri-Rolla, USA
C.W. Wilson	Combustion and Emissions Group Propulsion, United Kingdom
S. Wu	Beijing University of Aeronautics and Astronaytics, China

Contributors

S.L. Baughcum	The Boeing Company, USA
A. Döpelheuer	DLR Institut für Antriebstechnik, Germany
H.J. Hackstein	AI/EE-T Engineering Directorate, France
H. Mongia	General Electric Company, USA
R.R. Nichols	The Boeing Company, USA
C. Osonitsch	Gulfstream, USA
R. Paladino	University of Missouri-Rolla, USA
M.K. Razdan	Allison Engine Company, USA
M. Roquemore	WL/POSF, USA
P.A. Schulte	DLR Institut für Physik der Atmosphäre, Germany
D.J. Sutkus	The Boeing Company, USA

Review Editor

| M. Wright | Consultant, United Kingdom |

Chapter 8. Air Transport Operations and Relation to Emissions

Coordinating Lead Author

| G. Bekebrede | NLR, The Netherlands |

Lead Authors

D. Dimitriu	Romanian Air Transport, Romania
L. Dobbie	IATA
V. Galotti	ICAO
A. Lieuwen	Eurocontrol Agency, Belgium
S. Nakao	Japan Airlines, Japan
D. Raper	Manchester Metropolitan University, United Kingdom
H. Somerville	British Airways, United Kingdom
R.L. Wayson	University of Central Florida, USA
S. Webb	The Weinberg Group Inc., USA

Contributors

A. Gil	Airports Council International, Switzerland
D.R. Marchi	Airports Council International, USA
B. Miaillier	Eurocontrol Airspace and Navigation, Belgium
B.O. Nas	SAS, Sweden
J. Templeman	The Boeing Company, USA

Review Editor

| C.V. Oster, Jr. | Indiana University, USA |

Chapter 9. Aircraft Emissions: Current Inventories and Future Scenarios

Coordinating Lead Authors

| S.C. Henderson | The Boeing Company, USA |
| U.K. Wickrama | ICAO |

Lead Authors

S.L. Baughcum	The Boeing Company, USA
J.J. Begin	Northwest Airlines Inc., USA
F. Franco	_____, Canada
D.L. Greene	Center for Transportation Analysis, USA
D.S. Lee	DERA, United Kingdom
M.-L. McLaren	Consulting and Audit Canada, Canada
A.K. Mortlock	The Boeing Company, USA
P.J. Newton	DTI, United Kingdom
A. Schmitt	DLR Verkehrsforschung, Germany
D.J. Sutkus	The Boeing Company, USA
A. Vedantham	Richard Stockton College of New Jersey, USA
D.J. Wuebbles	University of Illinois, USA

Contributors

| R.M. Gardner | DERA, United Kingdom |
| L. Meisenheimer | Furman University, USA |

Review Editor

| O. Davidson | University of Sierra Leone, Sierra Leone |

Chapter 10. Regulatory and Market-Based Mitigation Measures

Coordinating Lead Author

| J.F. Hennigan | Federal Aviation Administration, USA |

Lead Authors

H. Aylesworth, Jr.	Aerospace Industries Association, USA
J. Crayston	ICAO
L. Dobbie	IATA
E. Fleuti	Zurich Airport Authority, Switzerland
M. Mann	DETR, United Kingdom
H. Somerville	British Airways, United Kingdom

Review Editor

| C.V. Oster, Jr. | Indiana University, USA |

EXPERT REVIEWERS

Albania
E. Demiraj Bruci   Hydrometeorological Institute

Australia
E. Curran   Bureau of Meteorology
K. Ryan   CSIRO
J. Zillman   Bureau of Meteorology

Austria
M. Blumthaler   Innsbruck University
T. Glöckel   Federal Environment Agency
K. Radunsky   Federal Environment Agency

Belgium
G.T. Amanatidis   EC/DGX11-DT
D. Brockhagen   The Greens in the European Parliament
A. Lieuwen   Eurocontrol Agency
K. Smeekens   VITO
J.-P. van Ypersele   Université Catholique De Louvain
M. Vanderstraeten   OSTC

Brazil
V.W.J.H. Kirchhoff   Instituto Nacional de Pesquisas Espaciais

Canada
J. Masterton   Atmospheric Environment Service
T. McElroy   Atmospheric Environment Service

Chile
L. Gallardo Klenner   CONAMA
M. Fiebig-Wittmack   Universidad de La Serena

China
D. Yihui   China Meteorological Administration
Z. Jingmeng   China Meteorological Administration (Deceased)
B. Wu   Institute of Atmospheric Physics (Deceased)
S. Wu   Beijing University of Aeronautics and Astronautics

Czech Republic
J. Pretel   Hydrometeorological Institute

Denmark
T.S. Jørgensen   Danish Meteorological Institute
B. Kuemmel   KVL
N. Larsen   Danish Meteorological Institute
H. Lyse Nielsen   Danish Environment Protection Agency

European Commission
G. Angeletti
R. Dunker
M. Raquet

Finland
R. Korhonen   VTT-Energy
M. Kulmala   University of Helsinki
I. Savolainen   VTT-Energy

France
J.-J. Becker   Mission Interministérielle de l'Effet de Serre
O. Boucher   Laboratoire d'Optique Atmospherique
D. Cariolle   Meteo-France
M. Desaulty   SNECMA
M. Gillet   Mission Interministérielle de l'Effet de Serre
M. Kanakidou   Centre des Faibles Radioactivités
L. Michaelis   OECD/International Energy Agency
P. Vesseron   Mission Interministérielle de l'Effet de Serre
R. von Wrede   Airbus Industrie

Germany
C.H. Brühl   Max-Planck Institut für Chemie
P. Crutzen   Max-Planck Institut für Chemie
M. Ernst   Federal Ministry for the Environment
P. Fabian   LBI
N. Gorisson   Federal Environmental Agency
B. Kärcher   DLR Institut für Physik der Atmosphäre
D. Kley   Institut für Chemie der KFA Jülich GmbH
H.G. Klug   Daimler-Benz Aerospace Airbus GmbH
P. Koepke   Meteorologisches Institut Munchen
A. Petzold   DLR Institut für Physik der Atmosphäre
R. Sausen   DLR Institut fur Physik der Atmosphäre
H. Schlager   DLR Institut für Physik der Atmosphäre
U. Schumann   DLR Institut für Physik der Atmosphäre
G. Seckmeyer   Fraunhofer-Institut für Atmosphäre
B. Simon   Motoren-und Turbinen-Union
W. Stockwell   Fraunhofer-Institut für Atmosphäre
J. Szodruch   Deutsche Aerospace Airbus GmbH
M. Treber   Germanwatch
F. Walle   Leiter Umweltfragen
H. Weyer   DLR Institut für Antriebstechnik

IATA
B. Bourke
J. de la Camera
L. Dobbie
A. Hardeman
R. Walder
G. Zaccagnini

ICAO
A. Costaguta
J. Crayston
G. Finnsson

Israel
Y. Rudich — Weizmann Institute of Science

Italy
V. Santacesaria — IROE-CNR
G. Visconti — Università Degli Studi dell' Aquila

Japan
Y. Kondo — Nagoya University

Kenya
C.K. Gatebe — University of Nairobi
P.M. Mbuthi — Ministry of Energy
J.K. Njihia — Kenya Meteorological Department

New Zealand
V. Gray — Coal Research Association of New Zealand
M.R. Manning — NIWA
R. McKenzie — National Institute of Water and Atmospheric Research Ltd

Norway
K. Alfsen — CICERO
M. Asen — Ministry of Environment
T.K. Berntsen — University of Oslo
O. Christophersen — Ministry of Environment
J.S. Fuglestvedt — University of Oslo
A. Gaustad — Norwegian Civil Aviation Administration
O. Hov — Ministry of Environment
I. Isaksen — University of Oslo
S. Lindahl — Norwegian Ministry of Transport
M.V. Pettersen — State Pollution Control Authority
K. Rypdal — Central Bureau of Statistics

Russia
A.A. Gorbatko — CIAM
I.L. Karol — Main Geophysical Observatory
A. Sorokin — CLAM, ECOLEN

Sierra Leone
O. Davidson — Sierra Leone University

Slovenia
A. Kranjc — Hydrometeorological Institute

Spain
B. Artinano — CIEMAT
J.M. Ciseneros — National Meteorological Institute
E. Ferdinandez — INTA
M. Gil — INTA
J.C. Rodriguez Murillo — Centro de Ciencias Medioambientales
I. Sastre — DGAC

Sweden
N.E. Nertun — Scandanavian Airlines
K.J. Noone — University of Stockholm
K Pleijel — Swedish Environmental Research Institute
J. Ström — University of Stockholm
L. Westermark — Swedish Commission on Climate Change

Switzerland
W. Bula — Federal Office for Civil Aviation
B.C. Cohen — UNECE Committee on Energy
C. Frohlich — World Radiation Center
J. Romero — Federal Office of Environment, Forest, and Landscape
J. Staehelin — ETH Atmospheric Physics Laboratory
S. Wenger — Federal Office for Civil Aviation

The Netherlands
H. Kelder — KNMI
J.W. Nieuwenhuis — Ministry of Housing
M. Oosterman — Royal Netherlands Meteorological Institute
H.J. Pulles — Netherlands Civil Aviation Authority
G. Velders — Air Research Lab, RIVM

Uganda
B. Apuuli — Department of Meteorology

UNEP
N. Sabogal

UNFCCC
N. Höhne
D. Tirpak

United Kindom
I. Colbeck — University of Essex
P.M. Forster — University of Reading
B. Gardiner — British Antarctic Survey
J. Houghton — IPCC WGI Co-Chairman
C. Hulme — British Airways
C. Hume — British Aerospace
L.R. Jenkinson — Department of Aeronautical and Automotive Engineering and Transport Studies
C. Johnson — DETR
P.R. Jonas — UMIST
D. Lary — University of Cambridge
D.S. Lee — DERA
D.H. Lister — DERA
A. McCulloch — AFEAS/ICI C&P
N.J. Mills — DERA
K. Morris — British Airways
P. Newton — DTI
K. Pearson — DETR
A. Pyle — University of Cambridge

M. Rossell	DETR	B. Manning	Environmental Protection Agency
K. Shine	University of Reading	R. Marshall	Pratt & Whitney Aircraft
J. Tilston	DERA	J. Montgomery	American Airlines
		A. Mortlock	The Boeing Company

United States of America

H. Aylesworth, Jr.	Aerospace Industries Association	C. Newberg	Environmental Protection Agency
S.L. Baughcum	The Boeing Company	R. Nichols	The Boeing Company
J. Begin	Northwest Airlines Inc.	R. Niedzwiecki	NASA Lewis Research Center
F.J. Berardino	Gellman Research Associates Inc.	R.C. Oliver	Institute for Defence Analyses
K. Boering	University of California	M. Oppenheimer	Environmental Defense Fund
T. Carmichael	Coalition for Clean Air	J. Pershing	US Department of State
L. Chang	Environmental Protection Agency	A. Petsonk	Environmental Defense Fund
U. Covert	Massachusettes Institute of Technology	B. Pierce	NASA Langley Research Center
R. Cuthbertson	The Boeing Company	L.R. Poole	NASA Langley Research Center
P.J. De Mott	Colorado State University	R. Pueschel	Ames Research Center
S. Dollyhigh	NASA Langley Research Center	V. Ramaswamy	Geophysical Fluid Dynamics Laboratory
D. DuBois	The Boeing Company	A.R. Ravishankara	NOAA Aeronomy Laboratory
J.H. Ellis	Federal Express Corporation	D. Rind	NASA Goddard Institute of Space Studies
D. Ercegovic	NASA Lewis Research Center		
A. Fiorentino	Pratt & Whitney Aircraft	R. Rubenstein	Environmental Protection Agency
N. Fitzroy	GE (Retired)	C. Russo	NASA Lewis Research Center
S. Gander	Center for Clean Air Policy	A. Schafer	Massachusetts Institute of Technology
M. Geller	New York State University		
A. Gettelman	US Climate Action Network	K. Shah	Goddard Institute for Space Studies
T.E. Graedel	IGBP/IGAC, AT&T Bell Laboratories	S.F. Singer	Science and Environmental Policy Project
K. Green	Consultant		
W.L. Grose	NASA Langley Research Center	R. Stolarski	Goddard Space Flight Centre
M. Guynn	NASA Langley Research Center	W. Strack	Modern Technologies Corp
P. Hamill	San Jose State University	L. Thomason	NASA Langley Research Center
S. Henderson	The Boeing Company	A. Thompson	NASA Goddard Space Flight Center
J. Holton	University of Washington	K.E. Trenberth	National Center for Atmospheric Research
C.H. Jackman	NASA Goddard Space Flight Center		
D. Jacob	Harvard University	A. Vedantham	Richard Stockton College of New Jersey
R. Kassel	Natural Resources Defense Council		
G. Kent	NASA Langley Research Center	D. Victor	Council for For Foreign Relations
H. Kheshgi	EXXON	W-C. Wang	State University of New York at Albany
E. Levina	Center for Clean Air Policy		
J. Levy	NOAA	S.G. Warren	University of Washington
J. Logan	Harvard University	S. Webb	The Weinberg Group Inc.
S.P. Lukachko	Massachusetts Institute of Technology	H.L. Wesoky	Federal Aviation Administration

B

Glossary of Terms

AERONOX
EU project to study impact of NO_x emissions from aircraft at altitudes between 8 to 15 km.

Aerosols
Airborne suspension of small particles.

Aerosol Precursors
Gases or chemi-ions that may undergo gas to particle conversion.

Aerosol Size Distribution
Particle concentration per unit size interval.

AEROTRACE
Project funded by the EU to measure trace species in the exhaust of aero engines.

Albedo
The ratio between reflected and incident solar flux.

Anthropogenic
Caused or produced by humans.

Background Atmosphere
The atmosphere remote from anthropogenic or volcanic influences.

Binary Nucleation
Nucleation from two gas phase species.

Black Carbon
Graphitic carbon, sometimes referred to as elemental or free carbon.

Block Time
The time elapsed from start of taxi out at origin to the end of taxi in at destination.

Bunker Fuels (International)
Fuels consumed for international marine and air transportation.

Catalytic Cycle
A cycle of chemical reactions, involving several chemical compounds, that depends on the presence of a specific compound which remains unchanged during these reactions.

Charged Particles
Particles carrying a positive or negative electric charge.

Chemi-ion
Charged cluster of a few molecules.

Cirrus
High, thin clouds composed of mainly ice particles.

Climate Model
A numerical representation of the climate system. Climate models are of two basic types: (1) static, in which atmospheric motions are neglected or are represented with a simple parameterization scheme such as diffusion; and (2) dynamic, in which atmospheric motions are explicitly represented with equations. The latter category includes general circulation models (GCMs).

Cluster
A set of molecules forming an entity.

Coagulation
Collision between two (or more) particles resulting in one larger particle.

Combustion Efficiency
Ratio of the heat released in combustion to the heat available from the fuel.

Condensation
The process of phase transition from gas to liquid.

Condensation Nucleus
A particle that can be activated to continual growth through the condensation of water by exposure to a high supersaturation with respect to water.

Contrail
Condensation trail (i.e., white line-cloud often visible behind aircraft).

Differential Mobility Analysis
A technique for measuring a particle's size by putting an electric charge on it, and measuring its electric mobility in an electric field.

Direct Radiative Impact
Radiative forcing of aerosols or gases by scattering and absorption of solar and terrestrial radiation.

Dp/F_{00}
The ICAO regulatory parameter for gaseous emissions, expressed as the mass of the pollutant emitted during the landing/take-off (LTO) cycle divided by the rated thrust (maximum take-off power) of the engine.

Economies in Transition
National economies that are moving from a period of heavy government control toward lessened intervention, increased privatization, and greater use of competition.

Emission Index
The mass of material or number of particles emitted per burnt mass of fuel (for NO_x in g of equivalent NO_2 per kg of fuel; for hydrocarbons in g of CH_4 per kg of fuel).

Energy Efficiency
Ratio of energy output of a conversion process or of a system to its energy input; also known as first-law efficiency.

Engine Pressure Ratio
The ratio of the mean total pressure at the last compressor discharge plane of the compressor to the mean total pressure at the compressor entry plane, when the engine is developing its take-off thrust rating (in ISA sea-level static conditions).

Equivalence Ratio
Ratio of actual fuel-air ratio to stoichiometric fuel-air ratio.

Feedback
When one variable in a system triggers changes in a second variable that in turn ultimately affects the original; a positive feedback intensifies the effect, and a negative reduces the effect.

Freezing
The process of phase transition from liquid to solid state.

Freezing Nucleus
Any particle that, when present within a mass of supercooled water, will initiate growth of an ice crystal about itself.

Greenhouse Gas
A gas that absorbs radiation at specific wavelengths within the spectrum of radiation (infrared) emitted by the Earth's surface and by clouds. The gas in turn emits infrared radiation from a level where the temperature is colder than the surface. The net effect is a local trapping of part of the absorbed energy and a tendency to warm the planetary surface. Water vapor (H_2O), carbon dioxide (CO_2), nitrous oxide (N_2O), methane (CH_4), and ozone (O_3) are the primary greenhouse gases in the Earth's atmosphere.

Heterogeneous Chemistry
Chemical reactions that involve both gaseous and liquid/solid ingredients.

Heterogeneous Nucleation
Formation of liquid or solid particles on the surface of other material.

Homogeneous Chemistry
Chemistry in the gas phase.

Homogeneous Nucleation
Formation of particles from gas-phase species.

Indirect Radiative Impact
Radiative forcing induced not directly but by changing other scattering or absorbing components of the atmosphere (clouds or gases).

Jet
The continuous strong stream of exhaust gases leaving the engine exit.

Kerosene
Hydrocarbon fuel for jet aircraft.

Landing/Take-Off (LTO) Cycle
A reference cycle for the calculation and reporting of emissions, composed of four power settings and related operating times for subsonic aircraft engines [Take-Off - 100% power, 0.7 minutes; Climb - 85%, 2.2 minutes; Approach - 30%, 4.0 minutes; Taxi/Ground Idle - 7%, 26.0 minutes].

Lean Blow Out
The fuel-air ratio of a combustion chamber at 'flame out.'

Lean Pre-Mixed Pre-Vaporized
Description of principal combustor features.

Life-Cycle Cost
The cost of a good or service over its entire lifetime.

Log Normal
Function of the form $y(x) = (C1/x)*exp(-(lnx-lnx0)**2/C2)$, where C1, C2, and x0 are constants.

Long-Wave Range
The terrestrial spectral radiation range at wavelengths larger about 4 μm.

Low Emissivity
A property of materials that hinders or blocks the transmission of a particular band of radiation (e.g., that in the infrared).

Mach Number
Speed divided by the local speed of sound.

Mitigation
An anthropogenic intervention to reduce the effects of emissions or enhance the sinks of greenhouse gases.

NO_x
Oxides of nitrogen, defined as the sum of the amounts of nitric oxide (NO) and nitrogen dioxide (NO_2) with mass calculated as if the NO were in the form of NO_2.

Nucleation
Phase change of a substance to a more condensed state initiated at a certain loci within a less condensed state.

Optical Depth or Optical Thickness
The parameter of a transparent layer of gases or particles defined as the logarithm of the ratio between incident and transmitted radiative flux.

Organic Carbon
The carbonaceous fraction of ambient particulate matter consisting of a variety of organic compounds.

Overall Efficiency (η)
The ratio between mechanical work delivered by an engine relative to the chemical energy provided from burning a fuel [η = (thrust x speed)/(specific combustion heat x fuel consumption rate)].

Ozone

A gas that is formed naturally in the stratosphere by the action of ultraviolet radiation on oxygen molecules. A molecule of ozone is made of up three atoms of oxygen.

Ozone Hole

A substantial reduction below the naturally occurring concentration of ozone, mainly over Antarctica.

Ozone Layer

A layer of ozone gas in the stratosphere that shields the Earth from most of the harmful ultraviolet radiation coming from the Sun.

Particulate Mass Emission Index

The number of grams of particulate matter generated in the exhaust per kg of fuel burned.

Particulate Number Emission Index

The number of particles generated in the exhaust per kg of fuel burned.

Plume

The region behind an aircraft containing the engine exhaust.

Polar Stratospheric Clouds

Large, diffuse, ice-particle clouds that form in the stratosphere usually over polar regions.

Polar Vortex

In the stratosphere, a strong belt of winds that encircles the South Pole at mean latitudes of approximately 60°S to 70°S. A weaker and considerably more variable belt of stratospheric winds also encircles the North Pole at high latitudes during the colder months of the year.

Pressure Ratio

The ratio of the mean total pressure exiting the compressor to the mean total pressure of the inlet when the engine is developing take-off thrust rating in ISA sea level static conditions.

Primary Energy

The energy that is embodied in resources as they exist in nature (e.g., coal, crude oil, natural gas, uranium, or sunlight); the energy that has not undergone any sort of conversion.

Radiative Forcing

A change in average net radiation (in W m^{-2}) at the top of the troposphere resulting from a change in either solar or infrared radiation due to a change in atmospheric greenhouse gases concentrations; perturbance in the balance between incoming solar radiation and outgoing infrared radiation.

Rated Output

The maximum thrust available for take-off under normal operating conditions, as approved by the certificating authority.

Relative Humidity

The ratio of the partial pressure of water vapor in an air parcel to the saturation pressure (usually over a liquid unless specified otherwise).

Reservoir Molecules

Molecules in the atmosphere that bind with atoms or other molecules and prevent them from participating in chemical reactions.

Scavenging

The process of removal of gases or small particles in the atmosphere by uptake (condensation, nucleation, impaction, or coagulation) into larger (cloud or precipitation) particles.

Short-Wave Range

The solar spectral range from about 0.3 to 4 μm.

Soot

Carbon-containing particles produced as a result of incomplete combustion processes.

Specific Fuel Consumption

The fuel flow rate (mass per time) per thrust (force) developed by an engine.

Stakeholders

Person or entity holding grants, concessions, or any other type of value which would be affected by a particular action or policy.

Stoichiometric Ratio

The fuel-air ratio at which all oxygen is consumed (approximately 0.068).

Stratosphere

The stably stratified atmosphere above the troposphere and below the mesosphere, at about 10- to 50-km altitude, containing the main ozone layer.

Surface Area Density

Surface area of aerosol per unit volume of atmosphere.

Susceptibility

Probability for an individual or population of being affected by an external factor.

Sustainable

A term used to characterize human action that can be undertaken in such a manner as to not adversely affect environmental conditions (e.g., soil, water quality, climate) that are necessary to support those same activities in the future.

Tropopause

The boundary between the troposphere and the stratosphere, usually characterized by an abrupt change in lapse rate (vertical temperature gradient).

Troposphere

The layer of the atmosphere between the Earth's surface and the tropopause below the stratosphere (i.e., the lowest 10 to 18 km of the atmosphere) where weather processes occur.

Ultraviolet Radiation

Energy waves with wavelengths ranging from about 0.005 to 0.4 μm on the electromagnetic spectrum. Most ultraviolet rays coming from the Sun have wavelengths between 0.2 and 0.4 μm. Much of this high-energy radiation is absorbed by the ozone layer in the stratosphere.

Volatiles

Particles that evaporate at temperatures less than about 100°C.

Vulnerability

The extent to which climate change may damage or harm a system; it depends not only on a system's sensitivity, but also on its ability to adapt to new climatic conditions.

Wake

The turbulent region behind a body or aircraft.

Windmilling

Inoperative engine with ram airflow through it.

C

Acronyms, Abbreviations, and Units

ACRONYMS AND ABBREVIATIONS

1-D	One-Dimensional	DLR	Deutsches Zentrum für Luft- und Raumfahrt
2-D	Two-Dimensional		
3-D	Three-Dimensional	DOT	Department of Transportation
ACAC	Arab Civil Aviation Commission	DTI	Department of Trade and Industry
ADS	Automatic Dependent Surveillance	DTR	Diurnal Surface Temperature Range
AEA	Association of European Airlines	DU	Dobson Unit
AEAP	Atmosphere Effects of Aviation Project	EASG	Economic Analysis Subgroup
AER	Atmospheric and Environmental Research, Inc.	EATMS	European ATM System
		ECMWF	European Centre for Medium-Range Weather Forecasts
AESA	Atmospheric Effects of Stratospheric Aircraft		
		ECON	Most Efficient Cruise Speed
AFCAC	African Civil Aviation Commission	ECS	Engine Control System
AMIP	Atmospheric Model Intercomparison Project	EDF	Environmental Defense Fund
		EEI	Effective Emissions Index
ANCAT	Abatement of Noises Caused by Air Transport	EI	Emissions Index
		EIDG	Emissions Inventory Database Group
ANDES	Aircraft Noise Design Effects Study	EISG	Emissions Inventory Sub-Group
API	American Petroleum Institute	ENSO	El Niño Southern Oscillation
APU	Auxiliary Power Unit	ERAA	European Regions Airline Association
ASK	Available Seat-Kilometers	ETOPs	Extended Twin Operations
ASM	Air Space Management	EUROCONTROL	European Organisation for Safety and Navigation
AST	Advanced Subsonic Technology		
ATC	Air Traffic Control	FADEC	Full Authority Digital Engine Control
ATFM	Air Traffic Flow Management	FANS	Future Air Navigation System
ATM	Air Traffic Management	FEMs	Finite Element Models
ATP	Advanced Turboprop	FESG	Forecast and Economics Sub-Group
ATR	Air Traffic Region	FIR	Flight Information Region
ATS	Air Traffic Services	FLEM	Flights and Emissions model
ATTAS	Advanced Technology Testing Aircraft System	FMS	Flight Management System
		FPC	Focal Point on Charges
AVHRR	Advanced Very High-Resolution Radiometer	FSU	Former Soviet Union
		FUA	Flexible Use of Airspace
BC	Black Carbon	GAMA	General Aviation Manufacturers Association
BWB	Blended Wing Body		
CAEP	Committee on Aviation Environmental Protection	GCM	General Circulation Model
		GDP	Gross Domestic Product
CCM3	Community Climate Model 3	GFDC	Geophysical Fluid Dynamics Laboratory
CE	Centre for Energy Conservation and Environmental Technology		
		GISS	Goddard Institute for Space Studies
CFC	Chlorofluorocarbon	GNBS	Global Navigation Satellite System
CFD	Computational Fluid Dynamics	GOES	Geostationary Operational Environmental Satellite
CGCM	Coupled General Circulation Model		
CI	Chemi-Ion	GPS	Global Positioning System
CN	Condensation Nucleus	GSFC	Goddard Space Flight Center
CNS/ATM	Communications, Navigation, and Surveillance	GWP	Global Warming Potential
		HALOE	Halogen Occultation Experiment
CSIRO	Commonwealth Scientific and Industrial Research Organization	HC	Hydrocarbon
		HCFC	Hydrochlorofluorocarbon
CTM	Chemical Transport Model	HF	High Frequency
DAC	Dual Annular Combustor	HFC	Hydrofluorocarbon
DEF STAN	Defence Standards	HIRS	High-Resolution Infrared Radiation Sounder
DERA	Defence Evaluation and Research Agency		
		HSCT	High Speed Civil Transport
DISORT	Discrete Ordinate Radiative Transfer	HYPR	Supersonic/Hypersonic Transport
DJF	December-January-February	IATA	International Air Transport Association

ICAO	International Civil Aviation Organization	OA	Objectively Analyzed
ICAS	International Council on Aeronautical Sciences	OAG	Official Airline Guide
		OECD	Organisation for Economic Cooperation and Development
ICCAIA	International Coordinating Council of Aerospace Industries Association	OEW	Operating Empty Weight
IFR	Instrument Flight Rule	OPMET	Operational Meteorological
IHPTET	Integrated High Performance Turbine Engine	OPR	Overall Pressure Ratio
		PAI	Propulsion/Airframe Integration
IMC	Instrument Meteorological Conditions	PAN	Peroxyacetylnitrate
IPCC	Intergovernmental Panel on Climate Change	PIANO	Project Interactive Analysis and Optimization
IR	Infrared	PMS	Performance Management System
IS92a	IPCC Scenarios 1992a	PNA	Pacific North America
ISA	International Standard Atmosphere	ppbv	Parts per Billion by Volume
ISCCP	International Satellite Cloud Climatology Project	ppmm	Parts per Million by Mass
		ppmv	Parts per Million by Volume
IWC	Ice-Water Content	PSC	Polar Stratospheric Cloud
IWP	Ice-Water Path	PSC1	Type I Polar Stratospheric Cloud
JGR	Journal of Geophysical Research	PSC2	Type II Polar Stratospheric Cloud
JJA	June-July-August	RBQQ	Rich Burn Quick Quench
LACAC	Latin American Civil Aviation Commission	RF	Radiative Forcing
		RFI	Radiative Forcing Index
LaRC	Langley Research Center	RH	Relative Humidity
LBO	Lean Blow Out	RNAV	Area Navigation
LES	Large Eddy Simulation	RPK	Revenue Passenger-Kilometer
LIDAR	Light Detection and Ranging	RQL	Rich Quench Lean
LLNL	Lawrence Livermore National Laboratory	RVSM	Reduced Vertical Separation
		SAD	Surface Area Density
LPP	Lean Pre-Mixed Pre-Vaporized	SAGE	Stratospheric Aerosol and Gas Experiment
LRC	Long-Range Cruise		
LS	Lower Stratosphere	SAM	Stratospheric Aerosol Measurement
LTO	Landing and Take-Off	SAO	Background Sulfate Surface Area Density
LW	Long-Wave		
MD	Mass Density	SARP	Standard and Recommended Practice
MIT	Massachusetts Institute of Technology	SA1	Sulfate Surface Area Density Scenario based upon 500 HSCT Fleet with 50% Conversion of Fuel Sulfur to Particles
MRC	Maximum Range Cruise		
MS	Middle Stratosphere		
NASA	National Aeronautics and Space Administration	SA2	Sulfate Surface Area Density Scenario based upon 1000 HSCT Fleet with 50% Conversion of Fuel Sulfur to Particles
NAT	Nitric Acid Trihydrate		
NCAR	National Center for Atmospheric Research	SA3	Sulfate Surface Area Density Scenario based upon 500 HSCT Fleet with 1000% Conversion of Fuel Sulfur to Particles
NCEP	National Centers for Environmental Prediction	SA4	Sulfate Surface Area Density Scenario based upon 1000 HSCT Fleet with 100% Conversion of Fuel Sulfur to Particles
NSA	Nitro Sylsulfuric Acid		
NH	Northern Hemisphere	SA5	Sulfate Surface Area Density Scenario based upon 500 HSCT Fleet with 10% Conversion of Fuel Sulfur to Particles
NIPER	National Institute for Petroleum and Energy Research		
NMC	National Meteorological Center	SA6	Sulfate Surface Area Density Scenario based upon 1000 HSCT Fleet with 10% Conversion of Fuel Sulfur to Particles
NMHC	Non-Methane Hydocarbons		
NOA	North Atlantic Oscillation		
NOAA	National Oceanic and Atmospheric Administration	SA7	Sulfate Surface Area Density Scenario based upon 500 HSCT Fleet with 0% Conversion of Fuel Sulfur to Particles
NOXAR	Nitrogen Oxides and Ozone Measurements along Air Routes	SBSTA	Subsidiary Body for Scientific and Technological Advice
NPRA	National Petroleum Refiners Association	SBUV	Solar Backscater Ultraviolet

SH	Southern Hemisphere		T/W	Thrust/Weight
SLS	Sea Level Static		UARS	Upper Atmosphere Research Satellite
SME	Solar Mesospheric Explorer		UDF	Unducted Fan
SN	Smoke Number		UiO	Universitet I Oslo
SNIF	Subsonic Assessment Nearfield Interactions Flight		UKMO	United Kingdom Meteorological Office
			UN	United Nations
SST	Supersonic Transport		UNIVAQ	Universita' degli Studi-l' Acquila
STE	Stratospheric-Tropospheric Exchange		UNEP	United Nations Environment Programme
STS	Sulfate Ternary Solution		UT	Upper Troposphere
SUCCESS	Subsonic Aircraft Contrail and Cloud Effects Special Study		UV	Ultraviolet
			UV-B	Ultraviolet-B
SW	Short-Wave		UV_{ery}	Erythemal Dose Rate
TCA	Technology Concept Aircraft		VOC	Volatile Organic Compound
TMA	Terminal Maneuvering Area		WAFS	World Area Forecast System
TOA	Top-of-Atmosphere		WCRP	World Climate Research Programme
TOMS	Total Ozone Mapping Spectrometer		WMO	World Meteorological Organization
TUV	Tropospheric Ultraviolet and Visible		WWF	World Wide Fund for Nature

UNITS

SI (Systéme Internationale) Units

Physical Quantity	Name of Unit	Symbol
length	meter	m
mass	kilogram	kg
time	second	s
thermodynamic temperature	kelvin	K
amount of substance	mole	mol

Special Names and Symbols for Certain SI-Derived Units

Physical Quantity	Name of SI Unit	Symbol for SI Unit	Definition of Unit
force	newton	N	$kg\ m\ s^{-2}$
pressure	pascal	Pa	$kg\ m^{-1}\ s^{-2}\ (= Nm^{-2})$
energy	joule	J	$kg\ m^2\ s^{-2}$
power	watt	W	$kg\ m^2\ s^{-3}\ (= Js^{-1})$
frequency	hertz	Hz	s^{-1} (cycle per second)

Decimal Fractions and Multiples of SI Units Having Special Names

Physical Quantity	Name of Unit	Symbol for Unit	Definition of Unit
length	ångstrom	Å	$10^{-10}\ m = 10^{-8} cm$
length	micrometer	μm	$10^{-6}m = \mu m$
area	hectare	ha	$10^4\ m^2$
force	dyne	dyn	$10^{-5}\ N$
pressure	bar	bar	$10^5\ N\ m^{-2}$
pressure	millibar	mb	$1hPa$
weight	ton	t	$10^3\ kg$

D

List of Major IPCC Reports

Climate Change—The IPCC Scientific Assessment
The 1990 Report of the IPCC Scientific Assessment Working Group (also in Chinese, French, Russian, and Spanish)

Climate Change—The IPCC Impacts Assessment
The 1990 Report of the IPCC Impacts Assessment Working Group (also in Chinese, French, Russian, and Spanish)

Climate Change—The IPCC Response Strategies
The 1990 Report of the IPCC Response Strategies Working Group (also in Chinese, French, Russian, and Spanish)

Emissions Scenarios
Prepared for the IPCC Response Strategies Working Group, 1990

Assessment of the Vulnerability of Coastal Areas to Sea Level Rise–A Common Methodology
1991 (also in Arabic and French)

Climate Change 1992—The Supplementary Report to the IPCC Scientific Assessment
The 1992 Report of the IPCC Scientific Assessment Working Group

Climate Change 1992—The Supplementary Report to the IPCC Impacts Assessment
The 1992 Report of the IPCC Impacts Assessment Working Group

Climate Change: The IPCC 1990 and 1992 Assessments
IPCC First Assessment Report Overview and Policymaker Summaries, and 1992 IPCC Supplement

Global Climate Change and the Rising Challenge of the Sea
Coastal Zone Management Subgroup of the IPCC Response Strategies Working Group, 1992

Report of the IPCC Country Studies Workshop
1992

Preliminary Guidelines for Assessing Impacts of Climate Change
1992

IPCC Guidelines for National Greenhouse Gas Inventories
Three volumes, 1994 (also in French, Russian, and Spanish)

IPCC Technical Guidelines for Assessing Climate Change Impacts and Adaptations
1995 (also in Arabic, Chinese, French, Russian, and Spanish)

Climate Change 1994—Radiative Forcing of Climate Change and an Evaluation of the IPCC IS92 Emission Scenarios
1995

**Climate Change 1995—The Science of Climate Change – Contribution of Working Group I
to the Second Assessment Report**
1996

**Climate Change 1995—Impacts, Adaptations, and Mitigation of Climate Change: Scientific-Technical Analyses –
Contribution of Working Group II to the Second Assessment Report**
1996

**Climate Change 1995—Economic and Social Dimensions of Climate Change – Contribution of Working Group III
to the Second Assessment Report**
1996

**Climate Change 1995—IPCC Second Assessment Synthesis of Scientific-Technical Information Relevant to Interpreting
Article 2 of the UN Framework Convention on Climate Change**
1996 (also in Arabic, Chinese, French, Russian, and Spanish)

Technologies, Policies, and Measures for Mitigating Climate Change – IPCC Technical Paper I
1996 (also in French and Spanish)

An Introduction to Simple Climate Models used in the IPCC Second Assessment Report – IPCC Technical Paper II
1997 (also in French and Spanish)

Stabilization of Atmospheric Greenhouse Gases: Physical, Biological and Socio-economic Implications – IPCC Technical Paper III
1997 (also in French and Spanish)

Implications of Proposed CO$_2$ Emissions Limitations – IPCC Technical Paper IV
1997 (also in French and Spanish)

The Regional Impacts of Climate Change: An Assessment of Vulnerability – IPCC Special Report
1998

ENQUIRIES: IPCC Secretariat, c/o World Meteorological Organization, 7 bis, Avenue de la Paix, Case Postale 2300, 1211 Geneva 2, Switzerland